INTRODUCTION TO MODERN ALGEBRA

MONOGRAPHS AND TEXTBOOKS IN PURE AND APPLIED MATHEMATICS

1. *K. Yano,* Integral Formulas in Riemannian Geometry (1970)
2. *S. Kobayashi,* Hyperbolic Manifolds and Holomorphic Mappings (1970)
3. *V. S. Vladimirov,* Equations of Mathematical Physics (A. Jeffrey, editor; A. Littlewood, translator) (1970)
4. *B. N. Pshenichnyi,* Necessary Conditions for an Extremum (L. Neustadt, translation editor; K. Makowski, translator) (1971)
5. *L. Narici, E. Beckenstein, and G. Bachman,* Functional Analysis and Valuation Theory (1971)
6. *D. S. Passman,* Infinite Group Rings (1971)
7. *L. Dornhoff,* Group Representation Theory (in two parts). Part A: Ordinary Representation Theory. Part B: Modular Representation Theory (1971, 1972)
8. *W. Boothby and G. L. Weiss (eds.),* Symmetric Spaces: Short Courses Presented at Washington University (1972)
9. *Y. Matsushima,* Differentiable Manifolds (E. T. Kobayashi, translator) (1972)
10. *L. E. Ward, Jr.,* Topology: An Outline for a First Course (1972) *out of print*
11. *A. Babakhanian,* Cohomological Methods in Group Theory (1972)
12. *R. Gilmer,* Multiplicative Ideal Theory (1972)
13. *J. Yeh,* Stochastic Processes and the Wiener Integral (1973) *out of print*
14. *J. Barros-Neto,* Introduction to the Theory of Distributions (1973) *out of print*
15. *R. Larsen,* Functional Analysis: An Introduction (1973)
16. *K. Yano and S. Ishihara,* Tangent and Cotangent Bundles: Differential Geometry (1973)
17. *C. Procesi,* Rings with Polynomial Identities (1973)
18. *R. Hermann,* Geometry, Physics, and Systems (1973)
19. *N. R. Wallach,* Harmonic Analysis on Homogeneous Spaces (1973)
20. *J. Dieudonné,* Introduction to the Theory of Formal Groups (1973)

21. *I. Vaisman,* Cohomology and Differential Forms (1973)
22. *B.-Y. Chen,* Geometry of Submanifolds (1973)
23. *M. Marcus,* Finite Dimensional Multilinear Algebra (in two parts) (1973, 1975)
24. *R. Larsen,* Banach Algebras: An Introduction (1973)
25. *R. O. Kujala and A. L. Vitter (eds.),* Value Distribution Theory: Part A; Part B. Deficit and Bezout Estimates by Wilhelm Stoll (1973)
26. *K. B. Stolarsky,* Algebraic Numbers and Diophantine Approximation (1974)
27. *A. R. Magid,* The Separable Galois Theory of Commutative Rings (1974)
28. *B. R. McDonald,* Finite Rings with Identity (1974)
29. *J. Satake,* Linear Algebra (S. Koh, T. Akiba, and S. Ihara, translators) (1975)
30. *J. S. Golan,* Localization of Noncommutative Rings (1975)
31. *G. Klambauer,* Mathematical Analysis (1975)
32. *M. K. Agoston,* Algebraic Topology: A First Course (1976)
33. *K. R. Goodearl,* Ring Theory: Nonsingular Rings and Modules (1976)
34. *L. E. Mansfield,* Linear Algebra with Geometric Applications (1976)
35. *N. J. Pullman,* Matrix Theory and its Applications: Selected Topics (1976)
36. *B. R. McDonald,* Geometric Algebra Over Local Rings (1976)
37. *C. W. Groetsch,* Generalized Inverses of Linear Operators: Representation and Approximation (1977)
38. *J. E. Kuczkowski and J. L. Gersting,* Abstract Algebra: A First Look (1977)
39. *C. O. Christenson and W. L. Voxman,* Aspects of Topology (1977)
40. *M. Nagata,* Field Theory (1977)
41. *R. L. Long,* Algebraic Number Theory (1977)
42. *W. F. Pfeffer,* Integrals and Measures (1977)
43. *R. L. Wheeden and A. Zygmund,* Measure and Integral: An Introduction to Real Analysis (1977)
44. *J. H. Curtiss,* Introduction to Functions of a Complex Variable (1978)
45. *K. Hrbacek and T. Jech,* Introduction to Set Theory (1978)
46. *W. S. Massey,* Homology and Cohomology Theory (1978)
47. *M. Marcus,* Introduction to Modern Algebra (1978)

INTRODUCTION TO MODERN ALGEBRA

Marvin Marcus
Institute for the Interdisciplinary
Applications of Algebra and Combinatorics
University of California, Santa Barbara
Santa Barbara, California

MARCEL DEKKER, INC. New York and Basel

Library of Congress Cataloging in Publication Data

Marcus, Marvin [Date]
Introduction to modern algebra.

(Monographs and textbooks in pure and applied mathematics ; v. 47)
1. Algebra, Abstract. I. Title.
QA162.M37 512'.02 78-2482
ISBN 0-8247-6479-X

MARCEL DEKKER, INC.
270 Madison Avenue, New York, New York 10016

Current printing (last digit):
10 9 8 7 6 5 4 3 2 1

PRINTED IN THE UNITED STATES OF AMERICA

To my wife

Rebecca Elizabeth Marcus

Contents

Preface

This book is written as a text for a basic one-year course in algebra at the advanced undergraduate or beginning graduate level. The presentation is oriented toward the applications of algebra to other branches of mathematics and to science in general. This point of view is reflected in the selection of constructive methods of proof, the choice of topics, and the space devoted to items related to current applications of algebra. Thus, modules over a principal ideal domain are studied via elementary operations on matrices. Considerable space is devoted to such topics as permutation groups and the Pòlya counting theory; polynomial theory; canonical forms for matrices; applications of linear algebra to differential equations; representations of groups.

Prerequisites for a course based on this book are minimal: standard one-quarter courses in the theory of equations and elementary matrix theory suffice. Altogether there are 390 exercises and these constitute an integral part of the book. Problems that require somewhat intricate arguments are accompanied by complete solutions. The exercises contain a number of important results and several definitions. Occasionally they are used to remove technical arguments from the mainstream of a proof within the section. Students should at least read the exercises. Frequently exercises appearing at the end of a section are mentioned within the section so that they can be easily assigned at the appropriate time.

We have not hesitated to reiterate definitions and results throughout the book. For example, conjugacy classes are discussed in the chapter on group theory and again in the chapter on group representation theory. Moreover, some arguments are repeated if they are separated from their last occurrence by a substantial amount of intervening material. Each section of the book is followed by a Glossary which contains the page numbers for important definitions, "name" theorems, and special notations.

What follows is a description of the contents of each of the chapters. A diagram illustrating the interdependency of the various sections appears after the Preface.

Chapter 1, *Basic Structures*, introduces many of the basic ideas that occur later in the book. Section 1.1, *Sets and Functions*, contains the usual material on sets, functions, the de Morgan formulas, function composition, Cartesian products, equivalence relations, quotient sets, systems of distinct representatives, universal properties, partial and linear orderings, etc. Towards the end of the section, the Axiom of Choice is discussed in a heuristic way. The equivalence of Zorn's lemma and the Axiom of Choice is mentioned without going into much detail. Section 1.2, *Algebraic Structures*, introduces in order of increasing complexity some of the basic items developed in the remainder of the book. Thus groupoids, semigroups, monoids, groups, modules, vector spaces, algebras, and matrices appear here. In this section we also define permutation groups, free monoids, groupoid rings, polynomial rings, free power series, etc. The section contains an extensive list of elementary examples illustrating the definitions. The student can obtain considerable practice in the manipulation of these basic ideas in the exercise sections. Categories and morphisms appear in the exercises, but only peripherally.

Section 1.3, *Permutation Groups*, examines the details of permutation groups and their structure. The basic properties of permutations (including cycle structure and the simplicity of the alternating group of degree n for $n \geq 5$) appear here. Many of the basic ideas of group theory are illustrated in Section 1.3 in the context of permutation groups.

Chapter 2, *Groups*, is a rather thorough study of most of the major elementary theorems in group theory. Section 2.1, *Isomorphism Theorems*, carries the reader through the Jordan-Hölder theorem, properties of solvable groups, and composition series. Section 2.2, *Group Actions and the Sylow Theorems*, is devoted to a systematic study of the three major Sylow theorems. Since this section is highly combinatorial in nature, it seemed appropriate to include the Pòlya counting theorem and some interesting combinatorial applications.

Section 2.3, *Some Additional Topics*, contains a number of more advanced items in group theory, beginning with the Zassenhaus isomorphism lemma for groups. We then develop the Schreier refinement theorem for subnormal series of a (not necessarily finite) group. This section also includes the notion of a group with operators, admissible subgroups, and linear maps on vector spaces.

Chapter 3, *Rings and Fields*, is the longest chapter in the book. Section

3.1, *Basic Facts*, covers ring characteristics, universal factorization properties of quotient rings, and the three ring isomorphism theorems.

Section 3.2, *Introduction to Polynomial Rings*, shows how an indeterminate can be constructed over an arbitrary ring. The ring extension theorem is proved and used here to imbed a ring in a ring with an indeterminate. This section also contains material on polynomials in several variables, including the basis theorem for symmetric polynomials. Ascending and descending chain conditions for ideals in a ring are discussed and the Hilbert basis theorem for Noetherian rings is proved. Quotient fields of integral domains, and more generally, rings of fractions with respect to ideals appear toward the end of the section.

Section 3.3, *Unique Factorization Domains*, starts with the usual material on polynomial division, the division algorithm and the remainder theorem. The division algorithm is proved over a noncommutative ring—it is required later in the study of elementary divisors over matrix rings. The basic fact that any principal ideal domain is a unique factorization domain is proved in Theorem 3.6. Example 6 shows how to calculate the greatest common divisor of two Gaussian integers using the Euclidean algorithm. Nilradicals, quotients of ideals, and the Jacobson radical all occur at the end of this section.

Section 3.4, *Polynomial Factorization*, begins in the standard way with Gauss' lemma and goes on to show that unique factorization is inherited by the polynomial ring over a unique factorization domain. Considerable space is devoted here to the practical problem of factoring polynomials. Theorem 4.7 shows how to construct a splitting field for a polynomial, and Theorem 4.13 exhibits the relationship between any two such splitting fields. The section concludes with a proof of the primitive element theorem for fields of characteristic zero. The exercises in this chapter contain a good deal of material, but detailed solutions are included for all but the most routine problems.

Section 3.5, *Polynomials and Resultants*, deals with the classical theory of polynomials. Sylvester's determinant, homogeneous polynomials, resultants, and discriminants appear here, and the fundamental question of when two polynomials have a common factor is investigated in some detail. This section concludes with a statement and proof of the Hilbert invariant theorem and a discussion of algebraic independence.

Section 3.6, *Applications to Geometric Constructions*, applies the preceding material on field theory to problems of ruler and compass construction of regular polygons and angle trisection.

Section 3.7, *Galois Theory*, is devoted to the proof of the fundamental theorem of Galois theory for fields of characteristic zero and its application to the classical problem of the solvability of a general polynomial of degree n by radicals.

Chapter 4, *Modules and Linear Operators*, begins in Section 4.1, *The Hermite Normal Form*, with the derivation of a normal form under left equivalence of matrices over a principal ideal domain. This theorem is then applied to finitely generated modules, yielding many of the standard results in module theory as easy consequences. The Hermite normal form is also useful in showing how to compute generators for ideals in a matrix ring. This section also contains the basic theory of finite dimensional vector spaces, the Steinitz exchange theorem, and the theory of linear equations.

Section 4.2, *The Smith Normal Form*, shows how to compute canonical forms for matrices under right and left equivalence over a principal ideal domain. The Smith form is then used to analyze the structure of finitely generated modules as direct sums of free submodules. The fundamental theorem of abelian groups appears in Corollary 7. We then determine all low-order abelian groups in some of the examples and exercises. The cyclic primary decomposition of a module is carried out in the exercises, together with an analysis of finitely generated abelian groups given in terms of certain defining relations, i.e., group presentations.

Section 4.3, *The Structure of Linear Transformations*, develops the standard elementary divisor theory and matrix canonical forms over a field. Our approach is computational, and the canonical forms under similarity are derived in terms of the reduction of the characteristic matrix via equivalence over a polynomial ring. Most of the important normal forms for matrices under similarity occur here, e.g., the Frobenius normal form and the Jordan normal form. A considerable part of the section deals with the problem of computing the elementary divisors of a function of a matrix. These important results are used in other parts of mathematics, e.g., the theory of ordinary differential equations.

In the last section of the chapter, *Introduction to Multilinear Algebra*, we introduce symmetry classes of tensors and briefly study the tensor, Grassmann, and completely symmetric spaces. As an example of the use of an inner product in a symmetry class, we show how the famous van der Waerden conjecture concerning doubly stochastic matrices can be partially resolved.

The fifth and final chapter of the book, *Representations of Groups*, is essentially self-contained and could be used for a short course on group representation theory. The major part of the chapter is concerned with matrix representations of finite groups. This permits us to achieve deep penetration of the subject rather rapidly.

The contents of a course in algebra vary considerably and seem to depend more on individual tastes and prejudices than do corresponding courses in analysis. The present book is no exception. However, a good deal of the material included can be justified in terms of its applications to other parts of mathematics and science. We anticipate that a student who gains reasonable mastery of the contents will be ready for more advanced courses in algebra and the applications of algebra to a wide range of fields, e.g., computer science, control theory, algebraic coding theory, system theory, numerical linear algebra, quantum mechanics, and crystallography.

References

Each chapter is divided into sections. Thus Section 4.2 is Section 2 of Chapter 4. Definitions, theorems, and examples are numbered serially within a section. Thus Theorem 1.4 is the fourth theorem in Section 1. Any reference to a definition, theorem, or example within the chapter in which it appears does not identify the chapter. Any reference to a definition, theorem, or example occurring in another chapter includes the chapter and section number. The symbol ▮ is used to denote the end of a proof.

Acknowledgment

The author is pleased to acknowledge the assistance of Dr. Ivan Filippenko in reading the original manuscript and providing invaluable help in proofreading the printed copy.

Marvin Marcus

DEPENDENCE OF SECTIONS

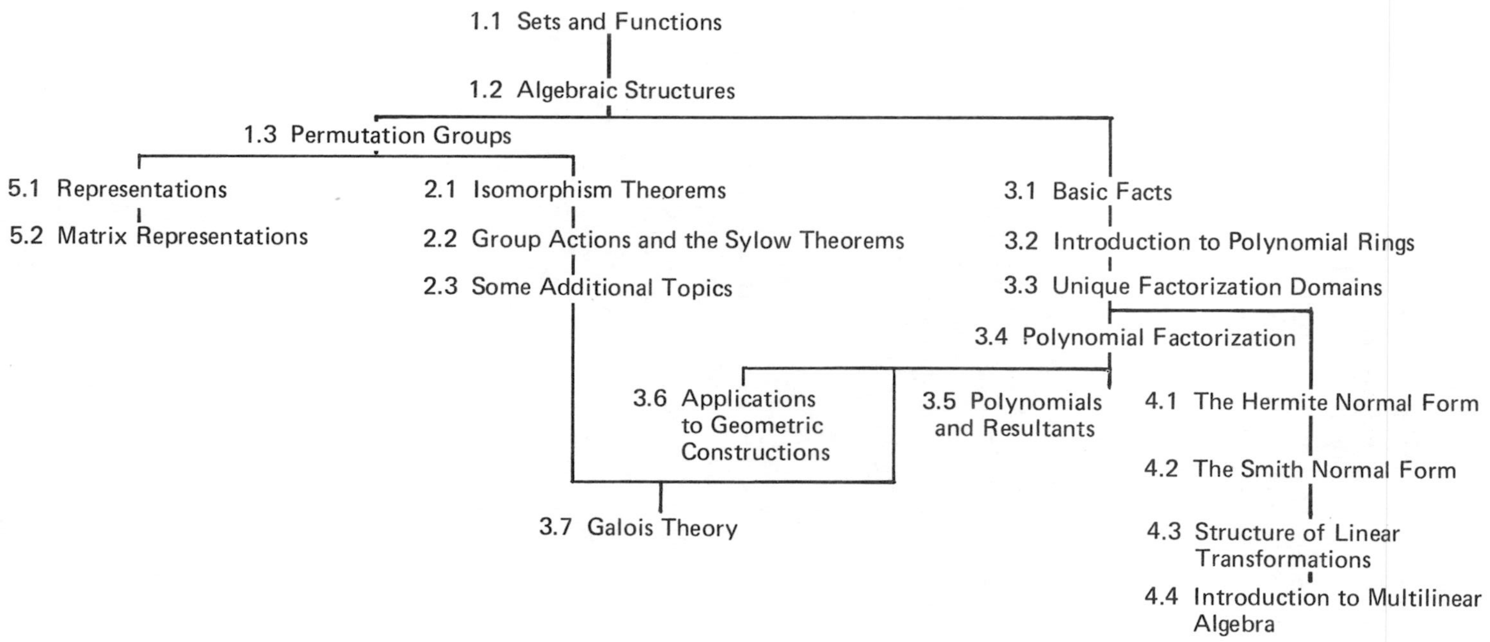

INTRODUCTION
TO
MODERN ALGEBRA

1

Basic Structures

1.1 Sets and Functions

We shall assume that the reader is familiar with the notion of a *set* or collection of objects. The purpose of this section is to set forth the notation and language used throughout this book.

If S is a set and x is a *member* or an *element* of S, we write

$$x \in S;$$

if x is not an element of S, we write

$$x \notin S.$$

If X is a set consisting of all elements x for which a certain proposition $p(x)$ is true, we write

$$X = \{x \mid p(x)\}.$$

Thus, for example,

$$\{x \mid x \text{ is an integer and } \tfrac{1}{2} \leq x < 5\}$$

is the set consisting of the integers 1, 2, 3, and 4. It is often feasible to explicitly write out the elements of a set, e.g.,

$$X = \{2,4,6\} \tag{1}$$

means that X consists of the numbers 2, 4, and 6. The curly bracket notation in (1) is usually reserved for finite sets, but sometimes infinite sets can be written this way by use of the ubiquitous "triple dot" notation, e.g.,

$$N = \{1, 2, 3, \ldots\}$$

is the set of positive integers.

If every element of the set X is in the set Y, we write

$$X \subset Y,$$

or

$$Y \supset X,$$

and call X a *subset* of Y. If $X \subset Y$ but $X \neq Y$, then X is a *proper* subset of Y. The *empty set* or *null set*, denoted by $\varnothing$, is the set with no elements; clearly,

$$\varnothing \subset X$$

for any X. The *power set* of a set X is the set of all subsets of X. It is denoted by $P(X)$:

$$P(X) = \{Y \mid Y \subset X\}.$$

If X contains only finitely many elements, we denote the number of elements in X by $|X|$. It is an easy exercise to verify that for a finite set X,

$$|P(X)| = 2^{|X|} \tag{2}$$

(See Exercise 1). For example, if X is the set (1), then the elements of $P(X)$ are the eight subsets

$$\varnothing, \quad \{2\}, \quad \{4\}, \quad \{6\}, \quad \{2,4\}, \quad \{2,6\}, \quad \{4,6\}, \quad X.$$

The *union* of two sets X and Y is the set of elements in either X or Y and is denoted by

$$X \cup Y.$$

The *intersection* of X and Y is the totality of elements in both X and Y and is denoted by

$$X \cap Y.$$

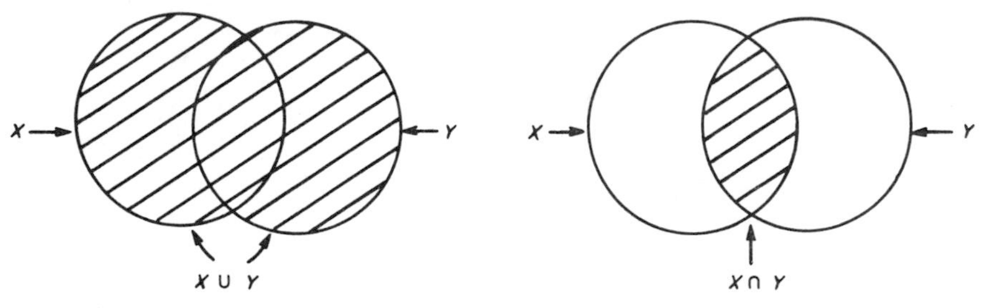

We say that I is an *indexing* set or *labeling* set for a family of sets if to each element of I there corresponds a well-defined set X_i in the family. The union and intersection of a family of sets indexed by I are written, respectively,

$$\bigcup_{i \in I} X_i$$

and

$$\bigcap_{i \in I} X_i.$$

Thus, $x \in \bigcup_{i \in I} X_i$ means that $x \in X_i$ for some $i \in I$, whereas $x \in \bigcap_{i \in I} X_i$ means that $x \in X_i$ for every $i \in I$. For example, if $I = N$ and X_i is the closed interval on the real line consisting of all x such that $1/i \leq x \leq 1$, then

$$\bigcup_{i \in N} X_i = \{x \mid 0 < x \leq 1\}$$

and

$$\bigcap_{i \in N} X_i = \{1\}.$$

If $\{X_i \mid i \in I\}$ is a family of sets and $X_i \cap X_j = \varnothing$ whenever $i \neq j$, we say that the sets in the family are *pairwise disjoint*. If

$$X = \bigcup_{i \in I} X_i$$

and the sets X_i are pairwise disjoint, then $\{X_i \mid i \in I\}$ is a *partition* of X.

If X and Y are sets, then the set of elements in Y but not in X is the *complement* of X *relative to* Y, denoted by

$$Y - X.$$

If $X \subset Y$, we write

$$X^c$$

instead of $Y - X$ when Y is understood. The *De Morgan formulas* connect the union, intersection, and complements of a family of subsets of Y:

$$\Big(\bigcup_{i \in I} X_i\Big)^c = \bigcap_{i \in I} X_i^c \tag{3}$$

and

$$\Big(\bigcap_{i \in I} X_i\Big)^c = \bigcup_{i \in I} X_i^c \tag{4}$$

(see Exercise 2).

If X and Y are sets, then a *function* (or *mapping* or *map*) from X to Y is a well-defined rule that associates with each element $x \in X$ an element $f(x) \in Y$. The set of all maps from X to Y is denoted by Y^X. We write

$$f: X \to Y$$

or in diagram form

$$X \xrightarrow{f} Y$$

to indicate that f is a function from X to Y.

The element $f(x) \in Y$ is the *value of f at x*, or the *image of x under f*;

the set X is called the *domain* of f, written dmn f; the set Y is called the *codomain* of f; the set

$$f(X) = \{y \in Y \mid y = f(x) \text{ for some } x \in X\}$$

is called the *image* or *range* of f, written im f.

If $Z \subset Y$, then $f^{-1}(Z)$, called the *inverse image* of Z, is the set

$$f^{-1}(Z) = \{x \in X \mid f(x) \in Z\}.$$

A map $f: X \to Y$ is *injective* (1–1) or an *injection* if $f(x_1) = f(x_2)$ implies $x_1 = x_2$; it is *surjective* (onto) or a *surjection* if $f(X) = Y$; it is *bijective* (1–1, onto) or a *bijection* if f is injective and surjective. Other words are *monomorphic* (injective); *epimorphic* (surjective); a *matching* (bijective) of X and Y.

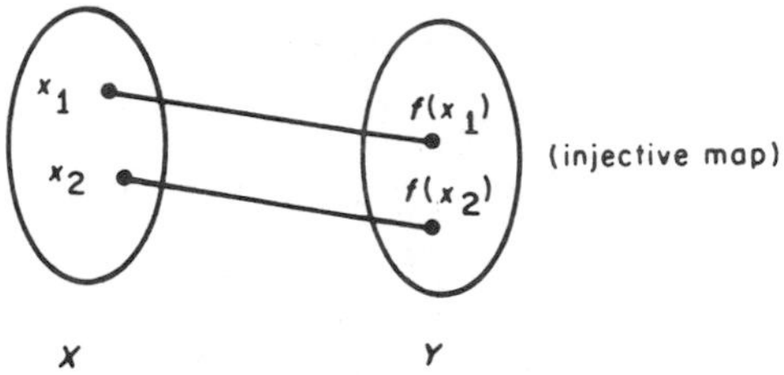

For example, the map $f: N \to N$ defined by the formula

$$f(n) = 2n$$

is an injection but certainly not a surjection. However, if by $2N$ we mean the set of positive even integers, then $f: N \to 2N$ is a bijection. If $g: N \to \{1, -1\}$ is defined by $g(n) = (-1)^n$, then g is an epimorphism or a surjection, and

$$g^{-1}(\{1\}) = 2N,$$
$$g^{-1}(\{-1\}) = N - 2N.$$

Two functions $f: X \to Y$ and $g: X \to Y$ are *equal* if and only if (hereafter abbreviated "iff")

$$f(x) = g(x)$$

for all $x \in X$.

The *composite* of two functions $f: X \to Y$ and $g: Y \to Z$ is the map $h: X \to Z$ whose value at any $x \in X$ is $h(x) = g(f(x))$. The composite of f and g is written gf or $g \cdot f$. The composition of maps is *associative*: If $X \xrightarrow{f} Y$, $Y \xrightarrow{g} Z$, and $Z \xrightarrow{h} W$, then $h \cdot (g \cdot f) = (h \cdot g) \cdot f$ (verify!). Thus the triple composite may simply be denoted by $h \cdot g \cdot f$.

If $\{Y_i \mid i \in I\}$ is a family of sets, then the *cartesian product* of the members of the family is the set of all functions $f: I \to \bigcup_{i \in I} Y_i$ such that $f(i) \in Y_i$ for each $i \in I$. The cartesian product is denoted by

$$\underset{i \in I}{\times} Y_i = \{f \mid f: I \to \bigcup_{i \in I} Y_i \text{ and } f(i) \in Y_i \text{ for each } i \in I\}.$$

If $\{Y_1, \ldots, Y_n\}$ is a family of n sets, their cartesian product is also written

$$Y_1 \times \cdots \times Y_n$$

and can be thought of as the totality of ordered n-tuples

$$f = (y_1, \ldots, y_n),$$

$y_i \in Y_i$, $i = 1, \ldots, n$; that is, $f(i) = y_i$, $i = 1, \ldots, n$. Two n-tuples $(y_1, \ldots, y_n)$ and $(z_1, \ldots, z_n)$ are equal iff $y_i = z_i$, $i = 1, \ldots, n$.

Suppose $I = [0,1]$, i.e., I is the closed interval on the real line consisting of all x for which $0 \leq x \leq 1$. For each $i \in I$ let $Y_i = I$. We assert that the cartesian product

$$\underset{i \in I}{\times} Y_i$$

is in fact the set of all maps from I to I, i.e.,

$$I^I = \underset{i \in I}{\times} Y_i.$$

For, any $f \in \times_{i \in I} Y_i$ is a function whose value at each $i \in I$ is an element of $Y_i = I$.

The special map $\iota_X: X \to X$, called the *identity map*, is defined by

$$\iota_X(x) = x$$

for each $x \in X$.

If $Z \subset X$ and $f: X \to Y$, then $f|Z$ is the function whose domain is Z and whose value for each $z \in Z$ is $f(z)$; $f|Z$ is called the *restriction* of f to Z, and f is called an *extension* of $f|Z$. If $Z \subset X$, then the map $\iota_X \mid Z$ is called the *canonical injection* of Z into X.

Compositions of maps are often depicted by *mapping diagrams*; for example,

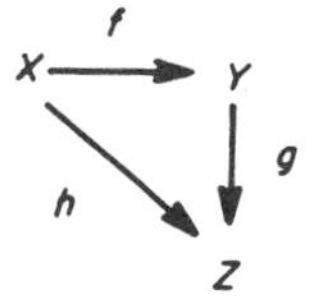

(5)

means that $f: X \to Y$, $g: Y \to Z$, $h: X \to Z$, and $h = g \cdot f$. Another example,

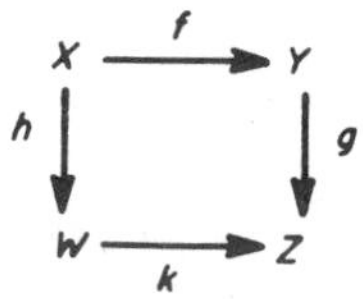

(6)

indicates that $g \cdot f = k \cdot h$. Diagrams showing the equality of compositions of sequences of maps, such as (5) and (6), are called *commutative diagrams*.

If $f\colon X \to Y$, $g\colon Y \to X$, and $gf = \iota_X$, then g is a *left inverse* of f; if $fg = \iota_Y$, then g is a *right inverse* of f. If g is a left and right inverse of f, then it is an *inverse* of f. (See Exercise 3.)

Theorem 1.1 *Assume $f\colon X \to Y$. Then*

(i) *f is injective iff it has a left inverse.*
(ii) *f is surjective iff it has a right inverse.*
(iii) *If f has a left inverse g and a right inverse h, then $g = h$.*
(iv) *f is bijective iff it has an inverse.*
(v) *If f has an inverse it is unique and is denoted by f^{-1}.*
(vi) *If f has an inverse, then $(f^{-1})^{-1} = f$.*

Proof: (i) If f has a left inverse $g\colon Y \to X$, then $f(x_1) = f(x_2)$ implies that $g(f(x_1)) = g(f(x_2))$ and hence that $x_1 = \iota_X(x_1) = (g \cdot f)(x_1) = g(f(x_1)) = g(f(x_2)) = (g \cdot f)(x_2) = \iota_X(x_2) = x_2$. Hence f is injective. Conversely, if f is injective, then for each $y \in f(X)$ there is exactly one element in X, call it $x_y \in X$, such that $f(x_y) = y$: define $g|f(X)$ by $g(y) = x_y$. For any other $z \in Y$, let $g(z) = x_0$, some fixed element in X. Obviously $(g \cdot f)(x) = g(f(x)) = x = \iota_X(x)$ for all $x \in X$, so g is a left inverse of f.

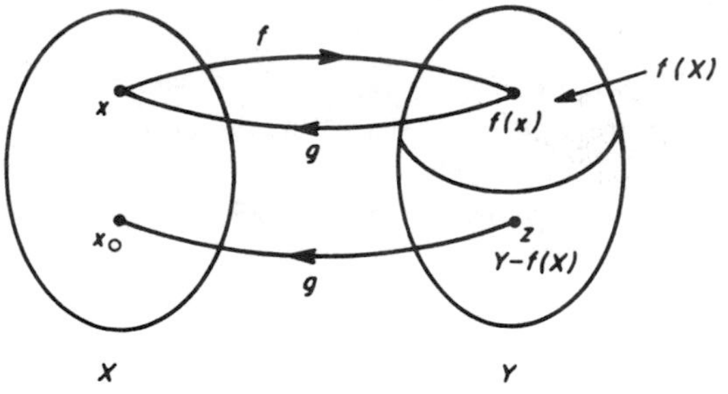

(ii) If $f\colon X \to Y$ is surjective, then $f(X) = Y$. Let $g\colon Y \to X$ be defined as follows: For each $y \in Y$ choose an $x_y \in f^{-1}(\{y\})$ and let $g(y) = x_y$. Then $(f \cdot g)(y) = f(g(y)) = f(x_y) = y = \iota_Y(y)$, i.e., $f \cdot g = \iota_Y$. Hence f has a right inverse. Conversely, if $g\colon Y \to X$ is a right inverse of f and $y \in Y$, then $y = \iota_Y(y) = (f \cdot g)(y) = f(g(y))$ and hence $y \in \operatorname{im} f$. Thus f is surjective.

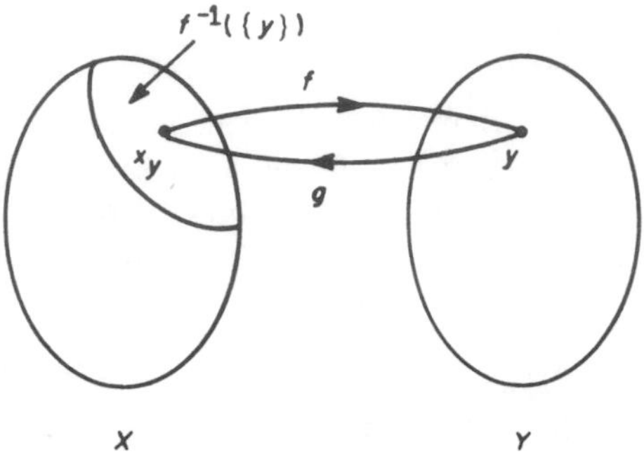

(iii) If $y \in Y$, then $g(y) = g(\iota_Y(y)) = g((f \cdot h)(y)) = (g \cdot f)(h(y)) = \iota_X(h(y)) = h(y)$. Hence $g = h$.
(iv) This follows immediately from (i), (ii), and (iii).
(v) If f has an inverse, it is a right and left inverse and by (iii) it is uniquely determined.
(vi) Let $g = f^{-1}$. Then $g \cdot f = \iota_X$ and $f \cdot g = \iota_Y$ so that f is a left and right inverse for g. Thus by (v), $g^{-1} = f$, that is, $(f^{-1})^{-1} = f$. ∎

In the proof of Theorem 1.1(ii) we were required to make (in general) an infinite number of choices of an element $x_y \in f^{-1}(\{y\})$. At the end of this section we shall discuss the *Axiom of Choice* which deals with the justification for this and similar arguments.

The set of all mappings from $\{1,2\}$ to a (nonempty) set X is called the *cartesian square* of X and denoted by X^2 (or $X \times X$), *i.e.*,

$$X^2 = X^{\{1,2\}}.$$

We can, of course, represent any $f \in X^2$ as $f = (x_1, x_2)$, where $x_1 = f(1)$ and $x_2 = f(2)$; (x_1,x_2) is called an *ordered pair*. If $D = \{(x,x) \mid x \in X\} \subset X^2$, then D is called the *diagonal* of X^2. Any subset $R \subset X^2$ is called a *relation on X*. If R is a relation on X, then R is called an *equivalence relation* on X if the following three conditions are satisfied:

$$D \subset R \tag{7}$$

$$(x_1,x_2) \in R \text{ implies } (x_2,x_1) \in R \tag{8}$$

$$(x_1,x_2) \in R \text{ and } (x_2,x_3) \in R \text{ implies } (x_1,x_3) \in R. \tag{9}$$

Properties (7), (8), and (9) are known as the *reflexive, symmetric,* and *transitive* properties of R, respectively. If $x \in X$, then $R(x)$ is the set

$$R(x) = \{y \mid y \in X \text{ and } (x,y) \in R\}$$

and is called the *R-equivalence class* containing x. The set of all R-equivalence classes is denoted by X/R and is called the *factor set* or *quotient set* of X *modulo R*. It is easy to see that if R is an equivalence relation on X, then the family of all R-equivalence classes forms a partition of X. We leave as an exercise the easy verification that either $R(x) \cap R(y) = \varnothing$ or $R(x) = R(y)$ (see Exercise 10). On the other hand, it is quite simple to see that if $X = \bigcup_{i \in I} X_i$ is a partition of X and

$$R = \{(x,y) \mid x \text{ and } y \text{ belong to the same } X_i\},$$

then R is an equivalence relation on X (see Exercise 11).

A *system of distinct representatives* (abbreviated S.D.R.), or a *transversal for R*, is a set T that contains precisely one element from each of the R-equivalence classes. The concept of an S.D.R. can be extended to a more general situation. Thus let $\mathfrak{A} = \{X_i \mid i \in I\}$ be a family of subsets of X (not necessarily pairwise disjoint or even distinct) indexed by an indexing set I. Let $T: I \to \bigcup_{i \in I} X_i$ be an injection satisfying

$$T(i) \in X_i, \qquad i \in I.$$

Then T is called an S.D.R. or a *transversal* for $\mathfrak{A}$. For example, suppose

$$X = \{1,2,3,4\}$$

and the family $\mathfrak{A} \subset P(X)$ consists of the subsets

$$X_1 = \{1,2,4\}, \quad X_2 = \{1,3\}, \quad X_3 = \{1,3,4\}, \quad X_4 = \{3,4\}.$$

Question: How many transversals are there for $\mathfrak{A}$? We can answer this rather easily by constructing an *incidence matrix* as follows. Write a 4×4 array

$$M = \begin{array}{c} \\ X_1 \\ X_2 \\ X_3 \\ X_4 \end{array} \begin{array}{c} \begin{matrix} 1 & 2 & 3 & 4 \end{matrix} \\ \begin{bmatrix} 1 & 1 & 0 & 1 \\ 1 & 0 & 1 & 0 \\ 1 & 0 & 1 & 1 \\ 0 & 0 & 1 & 1 \end{bmatrix} \end{array} \tag{10}$$

in which a 1 or 0 is entered as the (i, j) entry according as $x_j \in X_i$ or $x_j \notin X_i$, respectively. A transversal is then defined by a set of four nonzero entries in M, precisely one from each row and each column. Since the entries in M are either 1 or 0, it is clear that there is a one-to-one correspondence between the nonzero terms in the determinant expansion of M and the set of transversals. The reader can easily verify that the total number of such nonzero terms in the determinant is 3 (see Exercise 15). Hence there are three transversals for the family $\mathfrak{A}$.

If R is an equivalence relation on X, then the mapping $\nu\colon X \to X/R$ defined by

$$\nu(x) = R(x) \qquad \text{for } x \in X$$

is called the *natural map induced by* R. For example, suppose $X = Z$, the set of all integers, and let p be a fixed positive integer. Define

$$R = \{(m,n) \mid m - n \text{ is divisible by } p\}. \tag{11}$$

Then R is easily seen to be an equivalence relation on Z (see Exercise 12). The value $\nu(n) = R(n)$ is the set of all integers which differ from n by a multiple of p.

There is an elementary but important result which shows that any function defines an equivalence relation in a very simple way.

Theorem 1.2 *Let X and Y be nonempty sets and suppose $f\colon X \to Y$. Define*

$$R_f = \{(x_1,x_2) \mid f(x_1) = f(x_2)\}.$$

Then
(i) *R_f is an equivalence relation on X.*
(ii) *If $g\colon X \to Z$ and $R_f \subset R_g$, then there exists a unique map $\bar{g}\colon X/R_f \to Z$ such that the diagram*

$$
\begin{array}{ccc}
X & \xrightarrow{\;\nu\;} & X/R_f \\
 & \searrow{\scriptstyle g} & \downarrow{\scriptstyle \bar{g}} \\
 & & Z
\end{array}
\tag{12}
$$

commutes. In (12) ν is the natural map induced by R_f. [In particular, there exists a unique map $\bar{f}\colon X/R_f \to Y$ such that the diagram

$$
\begin{array}{ccc}
X & \xrightarrow{\;\nu\;} & X/R_f \\
 & \searrow{\scriptstyle f} & \downarrow{\scriptstyle \bar{f}} \\
 & & Y
\end{array}
$$

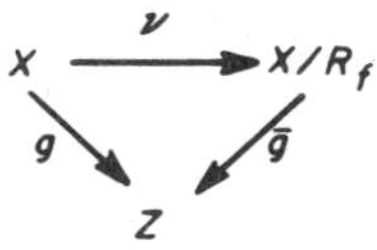

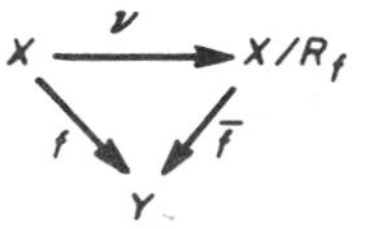

commutes.]

Proof: (i) See Exercise 13.

(ii) Observe first that if $R_f(x_1) = R_f(x_2)$, then $(x_1,x_2) \in R_f \subset R_g$ so that $g(x_1) = g(x_2)$. We can thus define a function $\bar{g}\colon X/R_f \to Z$ by $\bar{g}(R_f(x)) =$

$g(x)$. Of course $R_f(x) = \nu(x)$ so that $\bar{g} \cdot \nu = g$. Obviously the values of $\bar{g}$ are completely determined by g. ▮

The result in Theorem 1.2 is called the *factor theorem* for maps.

Let X be a nonempty set. A relation $R \subset X^2$ is *antisymmetric* if $(x_1,x_2) \in R$ and $(x_2,x_1) \in R$ together imply $x_1 = x_2$. If R is transitive, reflexive, and antisymmetric, then R is called a *partial ordering* in X. If, for any two elements x_1 and x_2 in X, $(x_1,x_2) \in R$ or $(x_2,x_1) \in R$, then the partial ordering R is called a *complete* or a *linear ordering*. If $C \subset X$ and C is linearly ordered with respect to the partial ordering R in X, then C is called a *chain* in X. If R is a partial ordering in X, then an element $a \in Y \subset X$ is a *least element* of Y if for every $y \in Y$, $(a,y) \in R$; similarly an element $b \in Y \subset X$ is a *greatest element* of Y if for every $y \in Y$, $(y,b) \in R$. It is customary to denote a partial ordering by the symbol $\leq$ and to write $(x_1,x_2) \in \leq$ as $x_1 \leq x_2$. Also, if $x_1 \leq x_2$ and $x_1 \neq x_2$, we write $x_1 < x_2$. Thus the definitions just given for a partial ordering read as follows:

(i) $x \leq x$ for every $x \in X$ (*reflexive*),
(ii) $x \leq y$ and $y \leq z$ imply $x \leq z$ (*transitive*),
(iii) $x \leq y$ and $y \leq x$ imply $x = y$ (*antisymmetric*).
(iv) If, in addition, $x_1 \in X$ and $x_2 \in X$ imply $x_1 \leq x_2$ or $x_2 \leq x_1$, then $\leq$ is a *linear ordering*.
(v) If $a \in Y \subset X$ and $a \leq y$ for all $y \in Y$, then a is a *least element* of Y. More generally, if $a \in X$ and $a \leq y$ for all $y \in Y$, then a is a *lower bound* of Y. An element $a \in X$ is *minimal* if there is no $x \in X$ such that $x < a$.
(vi) If $b \in Y \subset X$ and $y \leq b$ for all $y \in Y$, then b is a *greatest element* of Y. More generally, if $b \in X$ and $y \leq b$ for all $y \in Y$, then b is an *upper bound* of Y. An element $a \in X$ is *maximal* if there is no $x \in X$ such that $x > a$.

It is obvious from condition (iii) that there can be at most one least (greatest) element of a subset Y of a partially ordered set X (see Exercise 14).

The set of integers Z (i.e., positive, negative, and 0) is linearly ordered by the usual ordering: $x_1 \leq x_2$ iff $x_2 - x_1$ is nonnegative. Let X be a set and $P(X)$ its power set. Then the relation $\subset$ is a partial ordering in $P(X)$ but not a linear ordering; i.e., given two subsets of X, it is not generally true that one must be a subset of the other.

As an example let

$$X = \{0,1,2,3,4\},$$
$$R = \{(0,0), (1,1), (2,2), (3,3), (4,4), (0,2), (1,2), (3,4)\}.$$

We can construct a diagram of R:

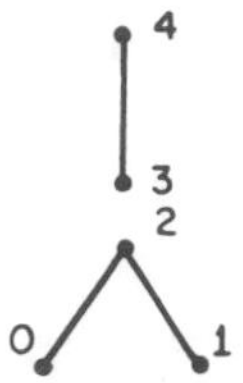

Thus $0 \leq 2$ simply means that 0 is below 2 and they are connected by a line segment. Note that since neither $2 \leq 3$ nor $3 \leq 2$ holds, R is not a linear ordering. Also, 0, 1, and 3 are all minimal elements because, e.g., there is no $x \in X$ for which $x < 3$. Similarly, 2 and 4 are maximal elements. The set X does not have a least or greatest element.

As another example let

$$X = P(\{0,1,2\}),$$

and let (ascending) inclusion be the ordering. The elements of X are

$$\varnothing, \quad \alpha = \{0\}, \quad \beta = \{1\}, \quad \gamma = \{2\},$$
$$\delta = \{0,1\}, \quad \mu = \{0,2\}, \quad \omega = \{1,2\}, \quad \tau = \{0,1,2\}.$$

The diagram that goes with this ordering is

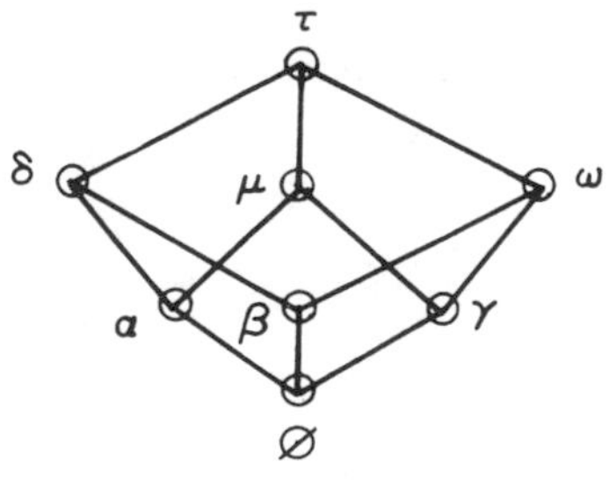

The circle around α indicates that $\alpha \subset \alpha$, etc. The line going upward from α to μ means that $\alpha \subset \mu$, etc. Finally, since the inclusion relation is transitive, we need not connect α to τ by a separate line segment. Note that

$$\{\varnothing, \beta, \omega, \tau\}$$

is a chain in X (identify some others), and that $\varnothing$ and τ are the least and greatest elements in X, respectively.

There is an important axiom in set theory known as *Zorn's lemma* which states:

If X is a partially ordered set and every chain C in X has an upper bound, then X contains at least one maximal element.

Zorn's lemma can be justified by the following heuristic argument. With the set X assumed to be nonempty, let $\alpha_0 \in X$. If α_0 is not maximal in X, there exists $\alpha_1 \in X$ such that

$$\alpha_0 < \alpha_1.$$

If α_1 is not maximal in X, then there exists an $\alpha_2 \in X$ such that

$$\alpha_0 < \alpha_1 < \alpha_2.$$

Either we obtain a maximal element or we construct an infinite chain

$$\alpha_0 < \alpha_1 < \alpha_2 < \alpha_3 < \cdots$$

which by assumption has an upper bound $\beta_0 \in X$. If β_0 is not maximal, proceed with β_0 exactly as with α_0, etc. Of course, the "etc." in this argument is really where all the action is and we are ultimately led to depend on another axiom of set theory which is perhaps superficially more plausible than Zorn's lemma. It is called the *Axiom of Choice.* The Axiom of Choice states the following:

Let $\{X_i \mid i \in I\}$ be a family of sets indexed by some set I, and assume that $X_i \neq \varnothing$ for all $i \in I$. Then

$$\mathop{\times}_{i \in I} X_i \neq \varnothing.$$

In other words, a cartesian product of nonempty sets must be nonempty. There is another way of saying this that makes the name of the axiom a little clearer. The elements of

$$\mathop{\times}_{i \in I} X_i$$

are functions f for which

$$f(i) \in X_i, \qquad i \in I.$$

Thus the Axiom of Choice asserts that there is a function f that "chooses" an element $f(i)$ out of each X_i.

By a rather lengthy and highly nontrivial argument, which we shall omit, it can be shown that Zorn's Lemma and the Axiom of Choice are equivalent, i.e., each implies the other. These connections (and several others) are admirably exhibited in the excellent book "Naive Set Theory" by P. R. Halmos, and the reader is referred to this source for further edification.

In any reasonable theory of sets no subset of a universe of elements U is allowed to be an element of itself, i.e.,

$$X \in X \tag{13}$$

is ruled out. The reasons for this lie in the famous Russell paradox. For suppose the relationship (13) is allowed and we let

$$A = \{X \mid X \subset U \text{ and } X \notin X\}.$$

If $A \in A$, then A is one of those subsets which is not an element of itself, i.e., $A \in A$ implies $A \notin A$. On the other hand, if $A \notin A$, then A is one of those sets which is not an element of itself and thus is a member of A, i.e., $A \notin A$ implies $A \in A$. This kind of unpleasantness is resolved by forbidding the relationship (13). In fact, the so-called *regularity axiom* asserts that no finite chain of memberships such as

$$X_1 \in X_2 \in X_3 \in \cdots \in X_p \in X_1$$

is permissible. The purpose of the preceding discussion is to enable us to state and prove the following result.

Theorem 1.3 *Let A be a set. Then there exist a set B and a bijection $\nu: A \to B$ such that $A \cap B = \varnothing$.*

Proof: Let $B = A \times \{A\}$ and define $\nu: A \to B$ by $\nu(x) = (x,A)$. Suppose that $x \in A \cap B$. Then $x = (y,A)$ for some $y \in A$. It should be observed (see Exercise 17) that an ordered pair (y,A) can be defined as the set

$$\{y, \{y,A\}\}.$$

Then

$$\begin{aligned} A &\in \{y,A\} \\ &\in \{y, \{y,A\}\} \\ &= x \\ &\in A, \end{aligned}$$

contradicting the regularity axiom. Thus $A \cap B = \varnothing$. Suppose that $\nu(x) = \nu(z)$. Then $(x,A) = (z,A)$ and, by Exercise 17, $x = z$. Hence ν is a bijection. ∎

As an immediate consequence of this result we have

Corollary 1 *Let A and R be sets. Then there exists a set B and a bijection $\nu: A \to B$ such that $B \cap R = \varnothing$.*

Proof: Let $A \cup R$ play the role of A in the preceding theorem. Obtain a set Γ such that $\Gamma \cap (A \cup R) = \varnothing$ and a bijection $\mu: A \cup R \to \Gamma$. Let $\mu(A) = B \subset \Gamma$ and set $\nu = \mu | A$. Then $B \cap R = \varnothing$ and ν is obviously a bijection. ∎

Exercises 1.1

1. Prove formula (2).
2. Prove formulas (3) and (4).
3. Let Z be the set of integers and define $f\colon Z \to Z$ and $g\colon Z \to Z$ by $f(n) = 2n$ and $g(n) = \begin{cases} n/2 & \text{if } n \text{ is even} \\ 0 & \text{if } n \text{ is odd} \end{cases}$
Then show that: (i) f is injective but not surjective; (ii) g is surjective but not injective; (iii) $gf = \iota_Z$, i.e., g is a left inverse of f; (iv) $fg \neq \iota_Z$, i.e., g is not a right inverse of f.
4. Prove that if $f\colon X \to Y$ and $g\colon Y \to Z$ and f and g are surjective (injective, bijective), then so is gf.
5. Complete the details of the proof of Theorem 1.1 (iv).
6. Show that if X is a set and $G = \{f \mid f\colon X \to X, f \text{ a bijection}\}$, then: (i) if f and g are in G, then $f \cdot g \in G$; (ii) if f, g, h are in G, then $(f \cdot g) \cdot h = f \cdot (g \cdot h)$; (iii) $\iota_X \in G$; (iv) if $f \in G$, then $f^{-1} \in G$.
7. Show that if $\{X_i \mid i \in I\}$ is a family of subsets of X and $f\colon X \to X$, then
$$f\Big(\bigcup_{i\in I} X_i\Big) = \bigcup_{i\in I} f(X_i),$$
$$f^{-1}\Big(\bigcup_{i\in I} X_i\Big) = \bigcup_{i\in I} f^{-1}(X_i),$$
and
$$f^{-1}\Big(\bigcap_{i\in I} X_i\Big) = \bigcap_{i\in I} f^{-1}(X_i).$$
8. Let $\{X_i \mid i \in I\}$ be a family of subsets of U and for a fixed $k \in I$ let $p_k\colon \times_{i\in I} X_i \to X_k$ be defined by $p_k(f) = f(k)$. Then p_k is called the *projection* on X_k. Take $X_i = X$, $i \in I$, and let $d\colon X \to \times_{i\in I} X$ be defined by $d(x) = f$, where $f(i) = x$ for all $i \in I$. Then d is called the *diagonal injection.* Prove that $p_k \cdot d = \iota_X$ for all $k \in I$.
9. Let $X_1, \ldots, X_n$ be subsets of U. Show that if $1 \leq k \leq n$, then there is a bijection
$$\iota\colon \Big(\mathop{\times}_{1}^{k} X_i\Big) \times \Big(\mathop{\times}_{k+1}^{n} X_i\Big) \to \mathop{\times}_{1}^{n} X_i.$$
10. Prove that if R is an equivalence relation, then the R-equivalence classes form a partition of X.
11. Prove that if $X = \bigcup_{i\in I} X_i$ is a partition of X and $R = \{(x,y) \mid x \text{ and } y \text{ belong to the same } X_i\}$, then R is an equivalence relation on X.
12. Prove that R as defined in (11) is an equivalence relation on Z.
13. Prove Theorem 1.2(i).
14. Prove that a least (greatest) element in a partially ordered set is unique, if it exists. *Hint*: Use antisymmetry.
15. Verify that the number of nonzero terms in the determinant expansion of M in (10) is 3.
16. Give an example of a partially ordered set which contains minimal elements but

no least element. *Hint*: Let $X = P(\{0,1\}) - \{\varnothing\}$ with inclusion as the ordering. Then $\{0\}$ and $\{1\}$ are minimal elements but neither is least.

17. Show that if (x,y) is defined to be the set $\{x, \{x,y\}\}$, then $(x,y) = (u,v)$ iff $x = u$ and $y = v$. *Hint*: First note that $x \neq \{x,y\}$; otherwise $x \in x$, contradicting the regularity axiom. Then from $\{x, \{x,y\}\} = \{u, \{u,v\}\}$, either (a) $x = u$ and $\{x,y\} = \{u,v\}$ or (b) $x = \{u,v\}$ and $\{x,y\} = u$. Clearly (b) is impossible; otherwise $x \in u \in x$. Hence (a) holds; so $x = u$ and $\{x,y\} = \{u,v\}$ becomes $\{u,y\} = \{u,v\}$. If $u = v$, then $\{u,v\} = \{u\}$ and $y \in \{u\}$, i.e., $y = u = v$. If $u \neq v$, then $y \neq u$ because $|\{u,v\}| = 2 = |\{u,y\}|$. Thus, $y \in \{u,v\}$ and $y \neq u$ imply $y = v$.

Glossary 1.1

antisymmetric, 10
Axiom of Choice, 12
bijection, 4
canonical injection, 5
Cartesian product, 4
$\times_{i \in I} Y_i$, 5
Cartesian square, 7
X^2, 7
chain, 10
codomain, 4
commutative diagrams, 6
complement, 3
$Y - X$, 3
X^c, 3
composite, 4
De Morgan formulas, 3
diagonal, 7
diagonal injection, 14
domain, 4
dmn f, 4
element, 1
empty set, 2
$\varnothing$, 2
epimorphic, 4
equivalence relation, 7
factor set, 8
X/R, 8
factor theorem for maps, 10
function, 3
$f: X \to Y$, 3
$X \xrightarrow{f} Y$, 3
greatest element, 10
identity, 5
identity map, 5
ι_X, 5
image, 3
$f(X)$, 4
im f, 4
incidence matrix, 8
inclusion, 2
$X \subset Y$, 2
indexing set, 2
injection, 4
intersection, 2
$X \cap Y$, 2
inverse, 6
inverse image, 4
$f^{-1}(Z)$, 4
least element, 10
left inverse, 6
linear ordering, 10
lower bound, 10
map, 3
mapping, 3
mapping diagrams, 5
matching, 4
maximal element, 10
member, 1
minimal element, 10
monomorphic, 4
N, 1
natural map induced by R, 9
number of elements, 2
$|X|$, 2

1.2 Algebraic Structures

Let S be a nonempty set. If $n \in N$, let $S^n = \times_1^n S$ be the *cartesian* nth *power* of S, and let $S^0 = \varnothing$. An *n-ary operation* on S is simply a function

$$\omega\colon S^n \to S.$$

We agree that a 0-ary operation ω is just some fixed element of S. A 1-ary operation is called *unary*, and a 2-ary operation is called *binary*. Let Ω be a collection of n-ary operations on S (n may vary). Then the set S together with the operations Ω is called an *Ω-algebra*. The set S is called the *carrier* of Ω and the Ω-algebra is often written (S,Ω).

If $\Omega = \{\omega\}$ and ω is binary, then (S,Ω) is called a *groupoid*. If we write $\omega(x,y) = x \cdot y$ (it is convenient to use juxtaposition for the values of ω, sometimes with a dot between x and y), it is natural to call xy the *product* of x and y and simply refer to "the groupoid S." Thus a groupoid is a nonempty set equipped with a binary operation whose values are in S; i.e., S is *closed* under the operation.

If S is a groupoid and for all x, y, and z in S the *associative law*

$$x(yz) = (xy)z \tag{1}$$

holds, then S is called a *semigroup*.

If a groupoid S possesses an element e such that

$$ex = xe = x \tag{2}$$

for all $x \in S$, then e is called a *neutral* or *identity* element.

A semigroup with an identity is called a *monoid*. Obviously a monoid can have only one identity, for if e and e' are identities, then $e = ee' = e'$ (why?).

If S is a monoid and $x \in S$, then y is an *inverse* of x if

$$\begin{aligned} xy &= yx \\ &= e. \end{aligned} \tag{3}$$

Obviously, if y' is also an inverse of x, then

$$y' = y'e = y'(xy) = (y'x)y = ey = y.$$

Hence an element of a monoid can have at most one inverse. A monoid in which every element has an inverse is called a *group*.

Summarizing: A groupoid S is simply a set with a binary operation taking on values in S; a semigroup is a groupoid in which the operation is associative; a monoid is a semigroup with an identity; a group is a monoid in which every element has an inverse.

$$\text{GROUP} \subset \text{MONOID} \subset \text{SEMIGROUP} \subset \text{GROUPOID}$$

$$\begin{aligned} \text{Groupoid} + \text{associativity} &= \text{semigroup} \\ \text{Semigroup} + \text{identity} &= \text{monoid} \\ \text{Monoid} + \text{inverses} &= \text{group.} \end{aligned}$$

For example, if Z denotes the set of integers

$$Z = \{0, \pm 1, \pm 2, \pm 3, \pm 4, \ . \ . \ . \}$$

and the operation is subtraction, then $(Z, \{-\})$ is a groupoid, but it is certainly not a semigroup:

$$(a - b) - c = a - (b - c)$$

can only hold for $c = 0$.

In general, a binary operation ω on S is *abelian* or *commutative* if

$$\omega(x,y) = \omega(y,x)$$

for all x and y in S.

Let $\Omega = \{\alpha,\beta\}$ with α and β binary operations on the carrier S. If

(i) $(S, \{\alpha\})$ is a group and α is abelian,

(ii) $(S, \{\beta\})$ is a groupoid,

(iii) $$\beta(\alpha(x,y),z) = \alpha(\beta(x,z),\beta(y,z)) \tag{4}$$

and $$\beta(z,\alpha(x,y)) = \alpha(\beta(z,x),\beta(z,y)) \tag{5}$$

for all x, y, and z in S, then the Ω-algebra (S,Ω) is called a *ring*. Usually $\alpha(x,y)$ is written $x + y$ and $\beta(x,y)$ is written $x \cdot y$ or simply, xy, with the obvious names of *addition* and *multiplication*, respectively. The formulas (4) and (5) then read

$$(x + y)z = xz + yz \tag{6}$$

and $$z(x + y) = zx + zy, \tag{7}$$

respectively, and are called the *distributive laws*. The identity of $(S, \{\alpha\})$ is usually written 0 and is called the *zero* of the ring. If $(S, \{\beta\})$ possesses an identity, then we denote it by 1 and the ring is said to have a *multiplicative identity*. If β is commutative, then (S,Ω) is called a *commutative ring*. If $(S, \{\beta\})$ is a semigroup, then (S,Ω) is called an *associative ring*.

If (S,Ω) is a ring and $xy = 0$, $x \neq 0$, $y \neq 0$, then x and y are called *divisors of* 0 (or *zero divisors*). A commutative associative ring (S,Ω) with multiplicative identity $1 \neq 0$ and which possesses no divisors of 0 is called an *integral domain*.

An associative ring (S,Ω) in which $(S - \{0\}, \{\beta\})$ is a group is called a *division ring* or a *skew-field*.

A division ring in which the multiplication is commutative is called a *field*.

Summarizing: A ring is a set S with two binary operations $+, \cdot$ such that $(S, \{+\})$ is an abelian group, $(S, \{\cdot\})$ is a groupoid, and the distributive laws (6) and (7) are satisfied; an integral domain is an associative commutative ring with at least two elements, having a multiplicative identity and no divisors of 0; a division ring is an associative ring in which the nonzero elements form a group with respect to the multiplication operation; a field is a division ring in which the multiplication is abelian.

If $S = 2Z$ denotes the set of even integers with the usual multiplication and addition, then S is clearly a commutative, associative ring without a multiplicative identity. The ring of integers Z with the usual operations is an integral domain. The set Q of all rational numbers p/q ($p,q \in Z$, $q \neq 0$) is a field with respect to the usual operations. Similarly **R** and **C**, the real and complex numbers, respectively, are fields with respect to the usual operations of addition and multiplication.

As a further example, let $X = \{0, 1\}$ and $S = P(X)$. Define $\omega(U,V) = U \cdot V = U - V = U \cap V^c$. Then $(S, \{\omega\})$ is a finite *groupoid*, but it is not a *semigroup*. For, take $U = V = W = X$. Then

$$(UV)W = (U \cap V^c) \cap W^c = U \cap V^c \cap W^c = \varnothing$$

whereas

$$\begin{aligned} U(VW) &= U \cap (VW)^c = U \cap (V \cap W^c)^c \\ &= U \cap (V^c \cup W) = X \cap (\varnothing \cup X) = X \\ &\neq \varnothing. \end{aligned}$$

Thus associativity fails.

Let $(M, \{+\})$ be an abelian group and let $(R, \{+, \cdot\})$ be an associative ring. Let $\rho: R \times M \to M$ satisfy

$$\rho(r, m_1 + m_2) = \rho(r,m_1) + \rho(r,m_2), \tag{8}$$

$$\rho(r_1 + r_2,m) = \rho(r_1,m) + \rho(r_2,m), \tag{9}$$

$$\rho(r_1 r_2,m) = \rho(r_1,\rho(r_2,m)), \tag{10}$$

for all r, r_1, r_2 in R and m, m_1, m_2 in M. Then $(M, \{+\})$ taken together with $(R, \{+\}, \cdot\})$ is called a *left R-module*. Usually we say simply that M is a left R-module. If R has a multiplicative identity 1 and

$$\rho(1,m) = m \tag{11}$$

for all $m \in M$, then M is called a *unital* R-module. Some standard notational confusions: the big $+$ in M and the small $+$ in R are both written (small) $+$. The value of $\rho(r,m)$ is simply written rm, and we say that M is an R-module (the word "left" being understood). Thus (8) to (11) can be written

$$\begin{aligned} r(m_1 + m_2) &= rm_1 + rm_2, \\ (r_1 + r_2)m &= r_1 m + r_2 m, \\ (r_1 r_2)m &= r_1(r_2 m), \\ 1m &= m. \end{aligned}$$

If M is a unital R-module and R is a field, then M is called a *vector space over R*, the elements of M are called *vectors*, and the elements of R are called *scalars*. The value rm is called the *scalar product* of m by r.

A *submodule* N of a (left) module M is a subset of M which is a (left) module over R in which the addition in N is the same as that in M and the scalar multiplication by elements of the ring R is also precisely the same as it is in M.

The following are some interesting elementary examples of the preceding structures.

Example 1 (*Multiples of Z*) If Z is the ring of integers and $0 \neq n \in Z$, then the subset $nZ = \{m \in Z \text{ and } n|m\}$ is a commutative ring. ($n|m$ means n divides m, i.e., there exists $q \in Z$ such that $m = nq$.)

Example 2 (*The Integers Modulo p*) Let $R \subset Z^2$ be the equivalence relation defined by $(m,n) \in R$ iff $p|m - n$, where p is some fixed positive integer. Then the quotient set Z/R can be made into a commutative ring, denoted by Z_p. The equivalence class $R(n)$ is denoted by $[n]$; addition and multiplication are defined by

$$[m] + [n] = [m + n], \tag{12}$$
$$[m][n] = [mn] \tag{13}$$

(the same notations are used for the operations in Z_p and Z). It is easily checked that these operations are well-defined. If p is a prime then Z_p is a field, and if p is composite then Z possesses divisors of 0. The ring Z_p is called the *ring of integers modulo p*, and the equivalence classes $[n]$ are called *residue classes modulo p*. The confirmations of these assertions concerning Z_p are easy. For example, if p is a prime and $[n] \neq [0]$, then n is not divisible by p, i.e., $p \nmid (n - 0)$, and thus the greatest common divisor of n and p is 1. The Euclidean algorithm then yields integers x and y such that $px + ny = 1$. Hence $[1] = [p][x] + [n][y] = [n][y]$, i.e., $[n]$ possesses the multiplicative inverse $[y]$.

The addition and multiplication in Z_3 are given in the following tables:

+	[0]	[1]	[2]
[0]	[0]	[1]	[2]
[1]	[1]	[2]	[0]
[2]	[2]	[0]	[1]

•	[0]	[1]	[2]
[0]	[0]	[0]	[0]
[1]	[0]	[1]	[2]
[2]	[0]	[2]	[1]

Note that each nonzero element in Z_3 has a multiplicative inverse.

The tables for Z_4 are

+	[0]	[1]	[2]	[3]
[0]	[0]	[1]	[2]	[3]
[1]	[1]	[2]	[3]	[0]
[2]	[2]	[3]	[0]	[1]
[3]	[3]	[0]	[1]	[2]

•	[0]	[1]	[2]	[3]
[0]	[0]	[0]	[0]	[0]
[1]	[0]	[1]	[2]	[3]
[2]	[0]	[2]	[0]	[2]
[3]	[0]	[3]	[2]	[1]

Note here that [2] has no multiplicative inverse in Z_4. It is relatively easy to see that $[n]$ has a multiplicative inverse in Z_p iff n and p are relatively prime (verify).

Example 3 (*n-Tuples*) Let R be an associative ring, and let $[1,n]$ be the segment of the first n positive integers. Let $M = R^{[1,n]}$, and define an addition in M by

$$(u + v)(j) = u(j) + v(j), \qquad j = 1, \ldots, n \tag{14}$$

and a scalar multiplication by

$$(ru)(j) = ru(j), \qquad j = 1, \ldots, n. \tag{15}$$

Then M is a left R-module; the elements of M are usually written out

$$(u(1), u(2), \ldots, u(n))$$

or

$$(u_1, u_2, \ldots, u_n)$$

and are called *n-tuples*. Usually M is written R^n, and if R is a field, then R^n is called the *vector space of n-tuples over R*.

If scalar multiplication is defined by

$$\rho(r,u)(j) = u(j)r, \qquad j = 1, \ldots, n,$$

rather than by (15), then M is a right R-module. A right R-module satisfies precisely the same conditions as a left R-module with the exception that (10) is replaced by

$$\rho(r_1r_2,m) = \rho(r_2,\rho(r_1,m)),$$

or

$$m(r_1r_2) = (mr_1)r_2.$$

If M is a vector space over a field R and $(M, \{+,\times\})$ is a ring which satisfies

$$r(m_1 \times m_2) = (rm_1) \times m_2 = m_1 \times (rm_2)$$

for $r \in R$ and m_1, m_2 in M, then M is called a *linear algebra* over R. If $(M, \{+,\times\})$

is associative with respect to the $\times$ operation, then M is called a *linear associative algebra* over R.

Example 4 (*Matrices*) Let R be a ring and define

$$M = R^{[1,m]\times[1,n]},$$

where m and n are positive integers. Define an addition in M by

$$(A+B)((i,j)) = A((i,j)) + B((i,j)), \qquad 1 \le i \le m, 1 \le j \le n,$$

and a left scalar multiplication by

$$(rA)((i,j)) = rA((i,j)).$$

Then M is a left R-module; the elements of M are called *$m \times n$ matrices* and im A is usually written out as a rectangular array:

$$A = \begin{bmatrix} a_{11} & a_{12} & \cdots & a_{1n} \\ a_{21} & a_{22} & \cdots & a_{2n} \\ \cdots & \cdots & \cdots & \cdots \\ a_{m1} & a_{m2} & \cdots & a_{mn} \end{bmatrix} \tag{16}$$

in which a_{ij} is the customary way of designating $A((i,j))$. The values a_{ij} are called *elements* or *entries* of A; $A|(\{i\} \times [1,n])$ is called the ith *row* of A and is designated by

$$A_{(i)} = [a_{i1} \cdots a_{in}];$$

$A|([1,m] \times \{j\})$ is called the jth *column* of A and is denoted by

$$A^{(j)} = \begin{bmatrix} a_{1j} \\ \vdots \\ a_{mj} \end{bmatrix}.$$

The R-module M is usually denoted by $M_{m,n}(R)$ or $R^{m\times n}$ and is called the *module of $m \times n$ matrices over R*. If $m = n$, then $M_{m,n}(R)$ is abbreviated to $M_n(R)$. There is a binary operation denoted by juxtaposition that can be defined from $M_{m,n}(R) \times M_{n,q}(R) \to M_{m,q}(R)$ as follows:

$$(AB)((i,j)) = \sum_{t=1}^{n} a_{it}b_{tj}, \qquad i = 1, \ldots, m, \qquad j = 1, \ldots, q. \tag{17}$$

The matrix $C = AB$ is called the *product* of A and B. Observe that the number of columns (n) of A must be equal to the number of rows of B before the product is defined. If the various products are defined and R is an associative ring, then

$$(AB)D = A(BD), \tag{18}$$

i.e., *matrix multiplication is associative*. The distributive laws hold:

$$A(B + D) = AB + AD$$

and

$$(B + D)A = BA + DA,$$

when the indicated products and sums are defined. If R is a field, then $M_n(R)$ together with matrix addition, scalar multiplication, and matrix multiplication, is a linear associative algebra over R called the *total matrix algebra over R*. Note that the multiplicative identity in $M_n(R)$ is the *identity matrix* I_n, i.e., $I_n(i,j) = \delta_{ij}$, $i,j = 1, \ldots, n$.

Example 5 (*Classical Groups*) Let R be a field and in $M_n(R)$, let $\mathrm{GL}(n,R)$ denote the set of all $n \times n$ matrices A which possess a two-sided multiplicative inverse A^{-1}. Then $\mathrm{GL}(n,R)$ with matrix multiplication is a group called the *full* or *general linear group*. It possesses a number of interesting subsets which are also groups with respect to matrix multiplication. Such a subset is obviously called a *subgroup*. For example, the *special linear group* is defined by

$$\mathrm{SL}_n(R) = \mathrm{GL}(n,R) \cap \{A \mid \det A = 1\}.$$

If $R = \mathbf{C}$ then the *unitary group* $U_n(\mathbf{C})$ is the totality of matrices $A \in \mathrm{GL}(n,\mathbf{C})$ for which

$$AA^* = I_n. \tag{19}$$

In (19), A^* is the *conjugate transpose* of A, i.e.,

$$A^*(i,j) = \overline{A(j,i)}, \qquad i,j = 1, \ldots, n.$$

If $R = \mathbf{R}$, the *real orthogonal group* $O_n(\mathbf{R})$ consists of those matrices $A \in \mathrm{GL}(n,\mathbf{R})$ which satisfy

$$AA^T = I_n. \tag{20}$$

In (20), A^T is the *transpose* of A, i.e.,

$$A^T(i,j) = A(j,i), \qquad i,j = 1, \ldots, n.$$

If $n = 2m$, let J denote the matrix in $\mathrm{GL}(n,\mathbf{C})$ defined by

$$J = \begin{bmatrix} 0 & I_m \\ -I_m & 0 \end{bmatrix},$$

where the zeros denote $m \times m$ matrices with every entry 0. Then the totality of matrices $A \in \mathrm{GL}(n,\mathbf{C})$ which satisfy

$$A^T J A = J$$

is called the *complex symplectic group* and is denoted by $\mathrm{Sp}_m(\mathbf{C})$. The reader will confirm (see Exercise 16) that $\mathrm{Sp}_m(\mathbf{C})$ is indeed a group with respect to matrix multiplication.

The preceding list of matrix groups are of crucial importance in the applications of group theory to geometry and physics.

Example 6 (*Real Quaternions*) In $M_2(\mathbf{C})$ let

$$1 = \begin{bmatrix} 1 & 0 \\ 0 & 1 \end{bmatrix}, \quad i = \begin{bmatrix} i & 0 \\ 0 & -i \end{bmatrix}, \quad j = \begin{bmatrix} 0 & 1 \\ -1 & 0 \end{bmatrix}, \quad k = \begin{bmatrix} 0 & i \\ i & 0 \end{bmatrix}.$$

It is easy to verify that

$$\begin{aligned} i^2 = j^2 = k^2 &= -1, & ij &= -ji = k, \\ jk = -kj &= i, & ki &= -ik = j. \end{aligned} \tag{21}$$

Let $Q(\mathbf{R})$ be the set of all matrices in $M_2(\mathbf{C})$ of the form

$$x = a1 + bi + cj + dk, \qquad a,b,c,d \in \mathbf{R};$$

then $$x = \begin{bmatrix} a+bi & c+di \\ -c+di & a-bi \end{bmatrix} \qquad (i \in \mathbf{C}, \quad i^2 = -1).$$

The operations of addition and multiplication in $Q(\mathbf{R})$ are the same as those in $M_2(\mathbf{C})$. Then $Q(\mathbf{R})$ is a *division ring* over $\mathbf{R}$ but not a field (i.e., multiplication is not commutative). It is called the division ring of *real quaternions*. Define the conjugate of x by

$$\bar{x} = a1 - bi - cj - dk.$$

Then it is easy to prove that $x = 0$ (the zero 2×2 matrix) iff $x\bar{x} = 0$. If $x \neq 0$, then

$$x^{-1} = \frac{1}{a^2 + b^2 + c^2 + d^2}\bar{x},$$

i.e.,
$$x\bar{x} = (a^2 + b^2 + c^2 + d^2)1.$$

It is worthwhile to point out here the distinction between right and left R-modules of n-tuples. Thus let $R = Q(\mathbf{R})$ and $M = R^2$ as in Example 3, and consider the two vectors (i,j) and $(k,-1)$ in M. Observe that from (21),

$$(i,j)1 + (k,-1)j = (i,j) + (-i,-j) = (0,0).$$

On the other hand suppose a and b in R were to satisfy

$$a(i,j) + b(k,-1) = (0,0).$$

Then
$$ai = -bk, \qquad aj = b,$$

and hence from (21),

$$ai = -ajk = -ai.$$

Thus $a = 0$ and $b = 0$. In other words, there exists a nontrivial sum of right scalar multiples of (i,j) and $(k, -1)$ that equals $(0,0)$, but no such nontrivial sum of left scalar multiples exist.

A good part of the remainder of this chapter will be devoted to the study of the class of groups described in the next example.

Example 7 (*Permutation Groups*) Let X be a nonempty set, and let S_X denote the set of all bijections on X. Then S_X is a group with respect to function composition. If $|X| = n$, then S_X is called the *symmetric group of degree n* on X, and the elements of S_X are called *permutations* on X. If $X = [1,n]$, then S_X is written S_n. It is easy to see that if $|X| = n$, then

$$|S_X| = n!.$$

The remaining examples in this section may not be as familiar to the reader.

Example 8 (*Free Monoid on X*) Let $\varnothing \neq X$, and let $S = \cup_{n=0}^{\infty} X^n$. If $f = (x_1, \ldots, x_p) \in X^p$ and $g = (y_1, \ldots, y_q) \in X^q$, define

$$\omega(f,g) = f \cdot g = (x_1, \ldots, x_p, y_1, \ldots, y_q) \in X^{p+q}.$$

The empty sequence is denoted by e. Then $(S, \{\omega\})$ is a *monoid*. It is called the *free monoid* on X. It is customary to drop the parentheses and commas in writing the elements of S, e.g., $f = x_1x_2 \cdots x_p$.

Example 9 (*Groupoid Ring of S over R*) Let S be a groupoid (with the operation denoted by juxtaposition). Let R be a ring with addition denoted by $+$ and multiplication by juxtaposition. Define $R[S]$ to be the subset of R^S consisting of those functions which are nonzero for at most a finite number of elements of S. Define two operations in $R[S]$:

$$(f+g)(s) = f(s) + g(s), \qquad s \in S, \tag{22}$$

and

$$(f \cdot g)(s) = \sum_{rt=s} f(r)g(t), \qquad s \in S. \tag{23}$$

The addition defined in (22) is ordinary function addition, and obviously $R[S]$ is *closed* with respect to this operation, i.e., $+\colon R[S]^2 \to R[S]$. The multiplication (23) needs some explanation: For $f,g \in R[S]$ we let $f \cdot g$ be the function whose value at $s \in S$ is the sum on the right in (23) in which the summation is over all pairs (r,t) for which $rt = s$ [if no such pair exists, then the value of $(f \cdot g)(s)$ is 0 by definition]. The multiplication in (23) is called *convolution*. Now clearly since $f(r) \neq 0$ and $g(t) \neq 0$ for at most a finite number of elements in S, the summation in (23) is a finite sum of elements of R. Moreover, there are only finitely many s for which $(f \cdot g)(s) \neq 0$. For if $r_1, \ldots, r_p$ are the values for which f is not 0 and $t_1, \ldots, t_q$ are the values for which g is not 0, then $r_i t_j$, $i = 1, \ldots, p$, $j = 1, \ldots, q$, are the only elements of S for which $f \cdot g$ can be different from 0. The assertion here is that $R[S]$ is a *ring*: It is called the *groupoid ring of S over R*. The fact that $(R[S], \{+\})$ is a commutative group is easy, with the "zero function" (the function identically 0) as the additive identity and with $(-f)(s) = -f(s)$, $s \in S$ as the obvious definition of the additive inverse, $-f$, of f. If S is a semigroup and R is an associative ring, it turns out that $R[S]$ is an associative ring. The only slightly troublesome point in verifying this is the associativity of multiplication:

$$\begin{aligned}((f \cdot g) \cdot h)(s) &= \sum_{rt=s} (f \cdot g)(r)h(t) = \sum_{rt=s} \Big(\sum_{uv=r} f(u)g(v)\Big)h(t) \\ &= \sum_{(uv)t=s} (f(u)g(v))h(t) = \sum_{u(vt)=s} f(u)(g(v)h(t)) \\ &= \sum_{ux=s} f(u) \sum_{vt=x} g(v)h(t) = \sum_{ux=s} f(u)(g \cdot h)(x) \\ &= (f \cdot (g \cdot h))(s). \end{aligned} \tag{24}$$

The distributive laws are also easy to verify (see Exercise 5). Finally, if S is a monoid with neutral element e and R is an associative ring with identity 1, then $R[S]$ is an associative ring with identity. Indeed, if $g \in R[S]$ is the function given by

$$g(s) = \begin{cases} 1 & \text{if } s = e \\ 0 & \text{if } s \neq e \end{cases},$$

then it is trivial to check that $f \cdot g = f = g \cdot f$ for all $f \in R[S]$.

Example 10 (*Free Ring on X over R*) Let S be the free monoid on X of Example 8; i.e., S consists of finite (but arbitrarily long) sequences of elements of X in which the operation is just "pasting" two of them together. Then the semigroup ring of S over the ring R is simply called the *free ring on X over R*, and the notation is simplified to $R[X]$. Thus $R[X]$ consists of finite R-valued functions on sequences of elements in X. Suppose X consists of a single element: $X = \{x\}$. Then the elements of S are e (the empty sequence), x, xx, xxx, $xxxx$, ... An element $f \in R[S]$ is a function which assigns an element of R to each of these sequences:

$$f(e) = r_0, \quad f(x) = r, \quad f(x^2) = r_2, \quad f(x^3) = r_3, \quad f(x^4) = r_4, \quad \ldots$$

where we have written $\overbrace{x \cdot x \cdot x \cdot x \cdots x}^{p \text{ times}}$ as x^p ($x^0 = e$) and $r_p \in R$, $p = 0, 1 \ldots$. The elements of $R[S]$ are called (obviously) *polynomials in x with coefficients* (the r_p) *in R*. We shorten the notation for $R[S]$ to $R[x]$, and $R[x]$ is called the *polynomial ring in x over R*. It is customary to write the values of f in front of the x^p and not to write e:

$$f = r_0 + r_1x + r_2x^2 + \cdots + r_nx^n.$$

Example 11 (*Free Power Series Ring on X over R*) Let S be the free monoid on X of Example 8; again, S consists of finite sequences $x_1x_2 \cdots x_n$ of elements of X. Observe that if $s = x_1 \cdots x_n$, then clearly there are only $n + 1$ pairs r, t in S for which $rt = s$, namely $r = e$, $t = s$; $r = x_1$, $t = x_2 \cdots x_n$; $r = x_1x_2$, $t = x_3 \cdots x_n$; $\cdots$; $r = x_1 \cdots x_{n-1}$, $t = x_n$; $r = x_1 \cdots x_n = s$, $t = e$. Let R be an associative ring as in Example 9 and consider R^S. Then it is easy to check that the definitions (22) and (23) make R^S into an associative ring. It is called the *free power series ring on X over R* and is denoted by $R[[X]]$. If R possesses a multiplicative identity 1, we define a subset of $R[[X]]$ by

$$M[[X]] = \{f \in R[[X]] \mid f(e) = 1\}.$$

We assert (see Exercise 6) that $M[[X]]$ is a group (with respect to the multiplication in $R[[X]]$).

Example 12 (*Reduced Free Group on Y*) Let Y be a nonempty set, and let $X = Y \times \{1,-1\}$. Let S be the free monoid on X of Example 8. Again, the elements of S are sequences or "*words*" in elements of X: $f = x_1 \cdots x_p$. Such a sequence is said to be *reduced* if $(y,1)$ and $(y,-1)$ never occur next to one another in f. Let G denote the set of reduced words in S, with the empty word e in G by convention. Define a binary operation $*$ in G as follows: $e * v = v$, $u * e = u$; in the word uv (pasting u and v together in S) cancel out all adjacent pairs $(y,1)$ and $(y,-1)$ (in either order). If everything cancels out, define $u * v = e$. If something is left, it is a reduced word and hence in G. We assert $(G, *)$ is a group, called the *reduced free group on Y* (see Exercise 7).

The structures introduced in this section are by no means intended to be exhaustive. They can be refined and extended in many directions, some of which will appear in later sections of the book.

Exercises 1.2

1. In Q define a binary operation by $\omega(a,b) = a * b = a + b + ab$. Show that 0 is the neutral element and $(Q, \{\omega\})$ is a monoid.

2. Show that the free monoid on X defined in Example 8 is a monoid.
3. Explain the steps in the computation (24).
4. Confirm that $(R[S], \{+\})$ is an abelian group in Example 9.
5. Verify the distributive laws for $(R[S], \{+, \cdot\})$ in Example 9.
6. Prove that $M = M[[X]]$ as defined in Example 11 is a group.
Hint: If $f, g \in M$, then $(f \cdot g)(e) = \sum_{rt=e} f(r)g(t)$. But there is only one solution to $rt = e$, namely $r = t = e$. Thus $(f \cdot g)(e) = 1$. Also multiplication is associative because it is in $R[[X]]$. If ι: $S \to R$ is defined by $\iota(s) = \delta_{se}$, then ι is a multiplicative identity in M. If $f \in M$, define g as follows: $g(e) = 1$; if x is a sequence of length 1, define $g(x) = -f(x)$ and observe that
$$(f \cdot g)(x) = \textstyle\sum_{rt=x} f(r)g(t) = f(e)g(x) + f(x)g(e)$$
$$= g(x) + f(x) = 0 = \iota(x).$$
If x_1x_2 is a sequence of length 2, then
$$(f \cdot g)(x_1x_2) = \sum_{rt=x_1x_2} f(r)g(t)$$
$$= f(e)g(x_1x_2) + f(x_1)g(x_2) + f(x_1x_2)g(e)$$
$$= g(x_1x_2) - f(x_1)f(x_2) + f(x_1x_2).$$
Thus define $g(x_1x_2) = f(x_1)f(x_2) - f(x_1x_2)$ so that $(f \cdot g)(x_1x_2) = 0 = \iota(x_1x_2)$, etc.
7. Show that $(G, \{*\})$ defined in Example 12 is a group.
8. Show that nZ as defined in Example 1 is a commutative ring.
9. Prove the assertions concerning Z_p made in Example 2.
10. Prove the assertions concerning R^n made in Example 3.
11. Prove the assertions concerning $M_{m,n}(R)$ and $M_n(R)$ made in Example 4.
12. Show that GL(n,R) is not closed with respect to matrix addition.
13. Confirm the formulas for x^{-1} and $x\bar{x}$ in Example 6.
14. Construct a table for S_3, the symmetric group of degree 3.
15. Show that in general matrix multiplication is not commutative.
16. With reference to Example 5 show that the classical groups GL(n,R), SL$_n(R)$, $U_n(\mathbf{C})$, $O_n(\mathbf{R})$, Sp$_m(\mathbf{C})$ are groups with matrix multiplication as the operation.
17. By attempting to fill out the multiplication table

•	e	a	b	c
e	e	a	b	c
a	a	•	•	•
b	b	•	•	•
c	c	•	•	•

show that there are only two groups of order 4, i.e., having four elements. We regard two groups as being the same if one can be obtained from the other by simply relabeling the elements. (In the table, e is the identity element.)

18. Show that there is only one group of order 3.

19. Let S be a nonempty set equipped with a binary operation which satisfies

(i) $$x(yz) = (xy)z \qquad (associativity);$$

(ii) for any a and b in S there exist x and y in S such that

$$ax = b$$

and $$ya = b.$$

Prove that S is a group. (It is obvious that any group satisfies these conditions, and thus the above conditions are equivalent to the defining properties of a group.) *Hint*: By the usual argument ($e_1 = e_1e_2 = e_2$), there is at most one identity in S. If $a \in S$, then by (ii) there exist e_1 and e_2 in S such that $ae_1 = a = e_2a$. If $z \in S$, obtain x_1 and x_2 in S such that $ax_1 = z = x_2a$. Then

$$ze_1 = (x_2a)e_1 = x_2(ae_1) = x_2a = z \tag{25}$$

and $$e_2z = e_2(ax_1) = (e_2a)x_1 = ax_1 = z. \tag{26}$$

Set $z = e_2$ in (25) and $z = e_1$ in (26) to obtain $e_2e_1 = e_2$ and $e_2e_1 = e_1$. Thus $e_1 = e_2 = e$ is the unique identity for S, and (25) and (26) show that $ze = ez = z$ for any $z \in S$. Next, if $x \in S$, obtain $y \in S$ such that $xy = e$. We show $yx = e$ also. First note that $y(xy) = ye = y$ and thus $(yx)y = y$. Choose $w \in S$ such that $yw = e$, and compute that

$$((yx)y)w = yw = e,$$

and $$((yx)y)w = (yx)(yw) = (yx)e = yx.$$

Hence $$yx = e.$$

20. If S is a monoid and x and y have inverses, show that xy has an inverse and $(xy)^{-1} = y^{-1}x^{-1}$. Also show that $(x^{-1})^{-1} = x$.

21. Let X denote the set $\mathbf{R} - \{0,1\}$. Define six functions f_i: $X \to X$ by $f_1(x) = x$, $f_2(x) = x^{-1}$, $f_3(x) = 1-x$, $f_4(x) = (1-x)^{-1}$, $f_5(x) = (x-1)x^{-1}$, and $f_6(x) = x(x-1)^{-1}$. Show that these six functions form a group with function composition as the operation. *Hint*: Associativity is automatic.

22. Let S be a set, $R \subset S \times S$, and ω: $R \to S$. Assume that

(i) $(\omega(x,y),z) \in R$ iff $(x,\omega(y,z)) \in R$, and then $\omega(\omega(x,y),z) = \omega(x,\omega(y,z))$;

(ii) if $(x,y) \in R$ and $(y,z) \in R$, then $(\omega(x,y),z) \in R$.

Then the pair $(S, \{\omega\})$ is called a *semigroupoid*. If $\omega(x,y)$ is denoted by juxtaposition, then a semigroupoid is a set and a binary operation defined for some, but not necessarily all, pairs in $S \times S$ which satisfies

(i) $(xy)z$ is defined iff $x(yz)$ is defined, and then $(xy)z = x(yz)$.

(ii) if xy is defined and yz is defined, then $(xy)z$, and by (i), $x(yz)$, are also defined with $(xy)z = x(yz)$.

Construct a semigroupoid that is not a groupoid. *Hint*: Below is a table for the operation in which $S = \{a,b,c\}$ and $R = \{(a,b), (b,c), (c,c), (a,c)\}$:

	a	b	c
a	$*$	c	c
b	$*$	$*$	c
c	$*$	$*$	c

23. An element u in a semigroupoid is called an *identity* if $ux = x$ whenever $(u,x) \in R$ and $xu = x$ whenever $(x,u) \in R$. A semigroupoid S is said to be *regular* if for every x there exist identities u and v such that $(u,x) \in R$ and $(x,v) \in R$. Prove that if S is regular, then for any $x \in S$ there exists a unique identity u such that $(u,x) \in R$ and hence $ux = x$. *Hint*: The definition of regular asserts the existence of one identity u such that $ux = x$. If μ is another such identity, then since $ux = x$ is defined and $x = \mu x$, it follows that $u(\mu x)$ is defined. Thus $u\mu$ is defined and since both u and μ are identities, $u = u\mu = \mu$.

24. Prove that if S is a regular semigroupoid, then for any $x \in S$ there exists a unique identity v such that $(x,v) \in R$ and hence $xv = x$.

25. Let X be a set and let

$$S = \{f \mid f\colon A \to B \quad \text{is a function for some } A,B \in P(X)\}.$$

Let $R \subset S \times S$ be the set of pairs (g,f) for which the domain of g and the codomain of f are the same, and define gf to be the composite of g and f. Show that S is a regular semigroupoid. *Hint*: Clearly, the conditions (i) and (ii) in Exercise 22 are satisfied. Suppose $f\colon A \to B$. Define $u = \iota_B$ so that $uf = f$. Similarly if $v = \iota_A$, then $fv = f$.

26. According to Exercises 23 and 24, for each element x in a regular semigroupoid S there exists a unique identity u_x which is a left identity for x and a unique identity v_x which is a right identity for x. Let $I(S)$ denote the totality of identities in S. Also let $\mathscr{L}(x) = u_x$ and $\mathscr{R}(x) = v_x$ for each $x \in S$. Prove that

$$\mathscr{L} \mid I(S) = \iota_{I(S)}$$

and

$$\mathscr{R} \mid I(S) = \iota_{I(S)}.$$

Hint: If $w \in I(S)$, then by definition $wx = x$ when the product is defined, and similarly $xw = x$ when the product is defined. Now $\mathscr{L}(w)$ is the unique left identity for w so that

$$\mathscr{L}(w)w = w.$$

But since w is an identity and $\mathscr{L}(w)w$ is defined from the preceding equation, we conclude that

$$\mathscr{L}(w)w = \mathscr{L}(w).$$

Thus $\mathscr{L}(w) = w$. Similarly, $\mathscr{R}(w) = w$.

27. Show that if S is a regular semigroupoid, then xy is defined iff

$$\mathscr{R}(x) = \mathscr{L}(y).$$

Hint: Suppose xy is defined. Then $\mathscr{L}(y)y = y$ so that $x(\mathscr{L}(y)y)$ is defined. Hence

$(x\mathscr{L}(y))y$ is defined and thus $x\mathscr{L}(y)$ is defined. Remember that $\mathscr{L}(y)$ is an identity so that by definition, if $x\mathscr{L}(y)$ is defined, then $x\mathscr{L}(y) = x$. But $x\mathscr{R}(x) = x$ and the uniqueness of a right identity implies

$$\mathscr{R}(x) = \mathscr{L}(y).$$

Conversely suppose $\mathscr{R}(x) = \mathscr{L}(y)$. Then $x\mathscr{R}(x)$ and $\mathscr{R}(x)y$ are defined and thus from (ii) in Exercise 22, $(x\mathscr{R}(x))y$ is defined. But $x\mathscr{R}(x) = x$, and so xy is defined.

28. Interpret Exercise 27 for the example of a semigroupoid in Exercise 25. *Hint*: First note that if $f: A \to B$, then $\mathscr{L}(f) = \iota_B$ and $\mathscr{R}(f) = \iota_A$. If $g: C \to D$, then gf is defined iff the domain of g is the same as the codomain of f, i.e., $C = B$. But $\mathscr{R}(g) = \iota_C$ and $\mathscr{L}(f) = \iota_B$ so that $\iota_C = \iota_B$ iff $C = B$.

29. Let S be a regular semigroupoid, and let the product $xy \in S$ be defined. Show

that $$\mathscr{L}(xy) = \mathscr{L}(x)$$
and $$\mathscr{R}(xy) = \mathscr{R}(y).$$

Hint: $\mathscr{L}(x)x$ is defined and xy is defined; thus by Exercise 22(ii),

$$(\mathscr{L}(x)x)y = \mathscr{L}(x)(xy)$$

is defined. But $\mathscr{L}(x)x = x$ so that the above equation becomes

$$\mathscr{L}(x)(xy) = xy.$$

Since $\mathscr{L}(x)$ is an identity, we conclude from Exercise 26 (Exercise 23) that

$$\mathscr{L}(x) = \mathscr{L}(xy).$$

Similarly, $$\mathscr{R}(xy) = \mathscr{R}(y).$$

30. An element x in a regular semigroupoid S has an *inverse* y if $xy = \mathscr{L}(x)$ and $yx = \mathscr{R}(x)$. Prove that if x has an inverse, then it is unique. The unique inverse, if it exists, is denoted by x^{-1}. *Hint*: Suppose $xy' = \mathscr{L}(x)$ and $y'x = \mathscr{R}(x)$. Both yx and xy' are defined, and hence the triple product

$$(yx)y' = y(xy')$$

is defined. Then

$$y(xy') = y\mathscr{L}(x).$$

Now $\mathscr{L}(x)$ is an identity; thus since $y\mathscr{L}(x)$ is defined, it follows from the definition of an identity that $y\mathscr{L}(x) = y$. Also

$$(yx)y' = \mathscr{R}(x)y' = y'$$

for the same reason. Hence $y = y'$.

31. If x is an element in a regular semigroupoid and x^{-1} exists, prove that

$$\mathscr{L}(x^{-1}) = \mathscr{R}(x)$$

and $$\mathscr{R}(x^{-1}) = \mathscr{L}(x).$$

Hint: Clearly $\mathscr{L}(x^{-1})x^{-1} = x^{-1}$ and xx^{-1} is defined; thus $x(\mathscr{L}(x^{-1})x^{-1})$ is defined. But then $x\mathscr{L}(x^{-1})$ is defined and $x\mathscr{L}(x^{-1}) = x$ since $\mathscr{L}(x^{-1})$ is an identity. But $x\mathscr{R}(x) = x$, and since both $\mathscr{L}(x^{-1})$ and $\mathscr{R}(x)$ are identities and right identities for x, it follows from Exercise 24 that $\mathscr{L}(x^{-1}) = \mathscr{R}(x)$. Similarly, $\mathscr{R}(x^{-1}) = \mathscr{L}(x)$.

32. A *category* $\mathscr{C}$ is a triple of items

$$\mathscr{C} = (F,S,\nu)$$

where S is a regular semigroupoid and

$$\nu: F \to I(S)$$

is a bijection of F onto the set $I(S)$ of identities of S. The elements of F are called the *objects* of the category, and the elements of S are called the *morphisms* of the category. Show that if for some set $X \neq \varnothing$ we have

$$F = P(X),$$
$$S = \{f \mid f: A \to B \text{ for some } A \text{ and } B \text{ in } P(X)\}$$

with function composition as the operation, and

$$\nu(A) = \iota_A, \qquad A \in P(X),$$

then $\mathscr{C} = (F, S, \nu)$ is a category. *Hint*: Exercise 25 tells us that S is indeed a regular semigroupoid. Obviously, the mapping $\nu: F \to I(S)$ is a bijection with codomain the set of identities in S.

33. Let $(M,\cdot)$ be a monoid. Let $F = \{M\}$, $S = (M,\cdot)$. Then $I(S) = \{e\}$, where e is the identity in M. Define $\nu(M) = e$. Show that $\mathscr{C} = (F,S,\nu)$ is a category.

Glossary 1.2

abelian, 17
associative law, 16
associative ring, 18
binary, 16
carrier, 16
category, 30
classical groups, 21
closed, 16
commutative, 17
commutative ring, 18
complex numbers, 18
C, 18
complex symplectic group, 22
Sp_m (**C**), 22
conjugate transpose, 22
A^*, 22
distributive laws, 18
division ring, 18
divisors of 0, 18
field, 18
free monoid on X, 23
free power series ring on X over R, 25
$R[[X]]$, 25
free ring on X over R, 24
general linear group, 22
GL(n,R), 21
group, 17
groupoid, 16
groupoid ring of S over R, 24
$R[S]$, 24
identity element, 16
identity matrix, 21
I_n, 21
integers, 17
integers modulo p, 19
integral domain, 18
inverse, 16
$I(S)$, 28
$\mathscr{L}$, 28
linear algebra, 20
linear associative algebra, 21
matrix, 21
matrix product, 21
monoid, 16
morphisms of a category, 30
multiplicative identity, 18
n-ary operation, 16

1.3 Permutation Groups

In this section we shall introduce a few of the elementary concepts concerning groups and apply them to the study of groups of permutations on finite sets (see Section 1.2, Example 7). As we shall soon see, every group of finite *order*, i.e., having a finite number of elements, appears as a finite group of permutations. Permutation groups are important throughout both pure and applied mathematics in such fields as the Galois theory of equations, the monodromy groups in complex function theory, quantum mechanics, crystallographic groups, and combinatorial analysis.

Let $(S, \{\alpha\})$ and $(T, \{\beta\})$ be semigroups, and let f: $S \to T$ be a map which satisfies

$$f(\alpha(s_1,s_2)) = \beta(f(s_1),f(s_2)), \qquad s_1,s_2 \in S. \tag{1}$$

Then f is called a *homomorphism*. If f is bijective, it is called an *isomorphism* and $(S, \{\alpha\})$ and $(T, \{\beta\})$ are called *isomorphic semigroups*. If f is simply injective, it is called a *monomorphism*; if f is surjective, it is called an *epimorphism*. If

$$(S, \{\alpha\}) = (T, \{\beta\}),$$

the term homomorphism (isomorphism) becomes *endomorphism* (*automorphism*). If juxtaposition is used for both binary operations α and β, then (1) reads

$$f(s_1 s_2) = f(s_1)f(s_2). \tag{2}$$

Isomorphic semigroups $(S, \{\alpha\})$ and $(T, \{\beta\})$ are really the same object from a structural standpoint. To clarify this idea, think of the elements of the group $(S, \{\alpha\})$ written down in a roster $x, y, z, \ldots$. Then since f is bijective, $\bar{x} = f(x)$, $\bar{y} = f(y)$, $\bar{z} = f(z), \ldots$ is a roster of the elements of $(T, \{\beta\})$; i.e., every element of T appears precisely once as a value of f. Moreover, if $xy = z$, then the value of $\bar{x}\bar{y}$ must be $\bar{z}$, for

$$\bar{x}\bar{y} = f(x)f(y) = f(xy) = f(z) = \bar{z}.$$

Thus except for the symbols used to designate the elements ($\bar{x}$ instead of x, etc.), the two semigroups $(S, \{\alpha\})$ and $(T, \{\beta\})$ are indistinguishable.

As a familiar example consider $(\mathbf{R}^+, \{\cdot\})$, the group of positive real numbers with ordinary multiplication as the operation. Let $(\mathbf{R}, \{+\})$ denote the group of all real numbers with addition as the operation, and define $f\colon \mathbf{R} \to \mathbf{R}^+$ by $f(r) = 10^r$. Then f is a bijection and

$$\begin{aligned} f(s_1 + s_2) &= 10^{s_1+s_2} = 10^{s_1} \cdot 10^{s_2} \\ &= f(s_1)f(s_2). \end{aligned}$$

Hence f is an isomorphism. It is easy to check that, in general, if $f\colon S \to T$ is an isomorphism, then $f^{-1}\colon T \to S$ is an isomorphism. The inverse of the isomorphism in this example is $f^{-1} = \log_{10}$.

Theorem 3.1 (*Cayley's Theorem*) *If X is a monoid, then there exists a monomorphism $f\colon X \to X^X$ mapping the monoid X into a monoid of functions in X^X in which function composition is the operation.*

Proof: For each $x \in X$, define $f(x) \in X^X$ to be the function whose value at each $z \in X$ is xz; i.e.,

$$f(x)\,(z) = xz.$$

Let e be the identity element in X. If $f(x) = f(y)$, then $f(x)\,(e) = f(y)\,(e)$, and so $x = xe = ye = y$. Hence f is injective. Now,

$$f(x_1 x_2)\,(z) = (x_1 x_2)z.$$

On the other hand,

$$\begin{aligned} (f(x_1) \circ f(x_2))\,(z) &= f(x_1)\,(f(x_2)\,(z)) \\ &= f(x_1)\,(x_2 z) = x_1(x_2 z) \\ &= (x_1 x_2)z. \end{aligned}$$

Thus $f(x_1) \circ f(x_2) = f(x_1x_2)$, and f is a monomorphism. ▮

Corollary 1 *If X is a group, then the monomorphism f of Theorem* 3.1 *maps X into S_X, the group of all bijections on X.*

Proof: We need only show that $f(x) \in S_X$. But $f(x)(z_1) = f(x)(z_2)$ implies $xz_1 = xz_2$, and hence $z_1 = z_2$ since X is a group. Also, if $z \in X$ then $f(x)(x^{-1}z) = xx^{-1}z = z$; thus $f(x)$ is a bijection. ▮

A *subsemigroup* T of a semigroup $(S, \{\alpha\})$ is a subset, $\varnothing \neq T \subset S$, such that im $\alpha | T^2 \subset T$; i.e, if $t_1, t_2 \in T$, then $\alpha(t_1, t_2) \in T$. In other words, T is *closed* with respect to α. More accurately, the subsemigroup is $(T, \{\alpha | T^2\})$. If $(S, \{\alpha\})$ is a monoid and $(T, \{\alpha | T^2\})$ is a subsemigroup which contains the identity element of S, then $(T, \{\alpha | T^2\})$ is called a *submonoid* of $(S, \{\alpha\})$. If $(T, \{\alpha | T^2\})$ is a subsemigroup of the semigroup $(S, \{\alpha\})$ and $(T, \{\alpha | T^2\})$ is a group, then we say that $(T, \{\alpha | T^2\})$ is a *subgroup* of the semigroup $(S, \{\alpha\})$. We usually drop the cumbersome notation and simply say T is a subsemigroup (or submonoid, or subgroup) of S.

A number of elementary facts about homomorphisms appear in the next result.

Theorem 3.2 *Let $h: X \to Y$ be a homomorphism of semigroups. If S is a subsemigroup of X, then $h(S)$ is a subsemigroup of Y. If T is a subsemigroup of Y, then $h^{-1}(T)$ is a subsemigroup of X. If X and Y are groups and S is a subgroup of X, then $h(S)$ is a subgroup of Y. If T is a subgroup of Y, then $h^{-1}(T)$ is a subgroup of X.*

Proof: The first and second assertions are trivial and are left to the reader (see Exercise 2).

We prove the third assertion. By the first assertion $h(S)$ is a subsemigroup of Y. Since S is a subgroup of X, $e \in S$, and $x \in S$ implies $x^{-1} \in S$ (verify). That is, the identity of X must be in any subgroup of X. We claim that $h(e)$ is the identity element e' of Y. For if $x \in X$, $h(e)h(x) = h(ex) = h(x) = e'h(x)$ and since $h(x)^{-1}$ exists, we have $h(e) = e'$. Of course, $h(e) \in h(S)$, and so $h(S)$ contains the identity of Y. Now if $x \in X$, then $h(x)h(x^{-1}) = h(xx^{-1}) = h(e) = e'$. Thus $h(x^{-1}) = h(x)^{-1}$ for $x \in X$; in particular $h(S)$ contains the inverse of each element in $h(S)$. Hence it is a subgroup of Y.

To prove the final assertion, we note the above statement that any subgroup of a group must contain the identity of the group. Thus $e' \in T$. Now $h(e) = e'$, and so $e \in h^{-1}(T)$. Also $x \in h^{-1}(T)$ implies $h(x) \in T$, and since $h(x^{-1}) = h(x)^{-1} \in T$, we have $x^{-1} \in h^{-1}(T)$. In other words, $h^{-1}(T)$ is a subgroup of X. ▮

In view of Theorem 3.2, we can rephrase Corollary 1 as follows:

Any group X is isomorphic to a subgroup of the group S_X of bijections on X.

Specializing this to finite groups, we have

Corollary 2 *If X is a finite group of order n, i.e., $|X| = n$, then X is isomorphic to a subgroup of the symmetric group of degree n on X.*

Corollary 2 justifies the study of symmetric groups: Among their subgroups all finite groups must appear to within isomorphism.

Let $|X| = n$; to simplify notation, we may as well assume $X = \{1, \ldots, n\} = [1, n]$, so that S_X becomes S_n. There are several convenient ways of writing permutations: If $\sigma \in S_n$, then we use the two-rowed array

$$\sigma = \begin{pmatrix} 1 & 2 & 3 & \cdots & n \\ \sigma(1) & \sigma(2) & \sigma(3) & \cdots & \sigma(n) \end{pmatrix} \tag{3}$$

to depict σ; i.e., $\sigma(i)$ is written directly under i. Of course the order in which the columns appear in (3) is immaterial; so we can write

$$\sigma^{-1} = \begin{pmatrix} \sigma(1) & \sigma(2) & \cdots & \sigma(n) \\ 1 & 2 & \cdots & n \end{pmatrix}. \tag{4}$$

If $\sigma \in S_n$, then

$$\sigma^0 = e = \iota_X,$$

$$\sigma^p = \overbrace{\sigma \cdots \sigma}^{p},$$

and

$$\sigma^{-p} = (\sigma^{-1})^p, \qquad p \in N.$$

If there exists a subset $K = \{i_1, \ldots, i_k\} \subset [1, n]$ such that

$$\sigma(i_t) = i_{t+1}, \qquad t = 1, \ldots, k-1,$$

$$\sigma(i_k) = i_1,$$

and

$$\sigma \mid K^c = \iota_{K^c},$$

then σ is called a *cycle of length k* or a *k-cycle*, and σ is written

$$\sigma = (i_1 i_2 \cdots i_k). \tag{5}$$

If $\sigma = e$, then of course the notation in (5) requires some explanation, e.g., $e = (1) = (2) = \cdots = (n)$ in this notation. For notational convenience we shall let e be designated by any of these cycles of length 1. A cycle of length 2 is called a *transposition*. Any cycle of length at least 2 is called a *circular* permutation.

Let $G \subset S_n$ be any subgroup of S_n. Define an equivalence relation $R \subset X^2$ by

$$R = \{(a, b) \mid \sigma(a) = b \text{ for some } \sigma \in G\} \tag{6}$$

(see Exercise 4). The equivalence classes in X/R are called the *orbits of G in* X, or simply the *G-orbits.* If $R(x)$ is a G-orbit and $|R(x)| = 1$, then $R(x)$ is called a *trivial orbit;* obviously, $R(x)$ is trivial iff $\sigma(x) = x$ for every $\sigma \in G$. In general, an orbit $R(x)$ consists of all the distinct values $\sigma(x)$ as σ runs over G.

If G is any group (not necessarily a group of permutations), it is quite simple to see (verify) that for $\sigma \in G$,

$$[\sigma] = \{\varphi \mid \varphi = \sigma^t, t \in Z\} \tag{7}$$

is a subgroup of G. The meaning of σ^t for a negative integer t is clear, e.g., $\sigma^{-2} = (\sigma^{-1})^2$. The subgroup $[\sigma]$ defined in (7) is called the *cyclic group generated by* σ. We also observe that if for some $t \neq 0$, $\sigma^t = e$ (where e is the identity in G), then

$$|\,[\sigma]\,| = k \tag{8}$$

where k is the least positive integer for which $\sigma^k = e$, and in fact

$$[\sigma] = \{e = \sigma^0, \sigma, \sigma^2, \ldots, \sigma^{k-1}\}. \tag{9}$$

For suppose $\sigma^t = e$ for some $t \neq 0$; we can assume $t \in N$ (since $\sigma^t = e$ iff $\sigma^{-t} = e$). If $n \in Z$, divide n by k,

$$n = qk + r, \qquad 0 \leq r < k.$$

Then

$$\sigma^n = \sigma^{qk+r} = (\sigma^k)^q \sigma^r = \sigma^r.$$

Also, if $0 \leq r < k$, $0 \leq s < k$, and $\sigma^r = \sigma^s$, then

$$\sigma^{|r-s|} = e$$

and the minimality of k shows that $r = s$. This confirms the equality (9).

As an immediate consequence of (9) it follows that the cyclic group generated by the k-cycle (5) is precisely the right side of (9), since $\sigma^t \neq e$ for $0 < t < k$ and $\sigma^k = e$.

Theorem 3.3 *Let $\sigma \in S_n$ and let $[\sigma]$ be the cyclic group generated by σ, $|\,[\sigma]\,| = k > 1$. Then σ is a k-cycle iff $[\sigma]$ has precisely one nontrivial orbit.*

Proof: If $\sigma = (i_1 i_2 \ldots i_k)$ is a k-cycle, then by the above remark

$$[\sigma] = \{e, \sigma, \ldots, \sigma^{k-1}\}.$$

Moreover, $\sigma^t(i_j) = i_{j+t}$, where $j + t$ is computed modulo k; i.e., if $j + t \geq k$, then the subscript is the remainder obtained upon dividing $j + t$ by k. It follows that $\{i_1, \ldots, i_k\}$ is a $[\sigma]$-orbit. Also since $\sigma | \{i_1, \ldots, i_k\}^c$ is the identity, $\{i_1, \ldots, i_k\}$ is the only nontrivial $[\sigma]$-orbit.

Conversely, suppose that $[\sigma]$ has precisely one nontrivial orbit $\mathcal{O}$. Let $j \in \mathcal{O}$. We assert that

$$\sigma = (j\ \sigma(j)\ \sigma^2(j) \cdots \sigma^{k-1}(j)). \tag{10}$$

To begin with, the elements $j = \sigma^0(j), \sigma(j), \ldots, \sigma^{k-1}(j)$ are distinct. For if not, then for some $s \neq t$ we have $\sigma^s(j) = \sigma^t(j)$ and hence $\sigma^m(j) = j$ for some m, $1 \leq m < k - 1$. But then for any integer r, $\sigma^m(\sigma^r(j)) = \sigma^r(\sigma^m(j)) = \sigma^r(j)$ so that σ^m holds each element of $\mathcal{O}$ fixed, and moreover holds any other element of $[1, n]$ fixed because σ does. In other words, $\sigma^m = e$ and this implies $|[\sigma]| \leq m < k$, in contradiction to $|[\sigma]| = k$. Also from (7) and (9) we have

$$[\sigma] = \{e, \sigma, \ldots, \sigma^{k-1}\} = \{\varphi \mid \varphi = \sigma^t, t \in Z\}$$

and hence $\mathcal{O} = \{j, \sigma(j), \ldots, \sigma^{k-1}(j)\}$, for any element of $\mathcal{O}$ is of the form $\sigma^n(j)$ and σ^n is one of $e, \sigma, \ldots, \sigma^{k-1}$. If $i \notin \mathcal{O}$, then $\sigma(i) = i$ since $\mathcal{O}$ is the only nontrivial $[\sigma]$-orbit. This proves (10). ∎

Let σ and π be any two permutations in S_n. If each nontrivial orbit of $[\sigma]$ is disjoint from the union of the nontrivial orbits of $[\pi]$, then we say that σ and π *act on disjoint subsets of* $[1, n]$, or more simply, are *disjoint* permutations. Clearly, if σ and π are disjoint, then $\sigma\pi = \pi\sigma$. For if j is in a nontrivial orbit of π, then $\pi(j)$ is in this orbit and hence is not in a nontrivial orbit of $[\sigma]$, i.e., $\sigma(\pi(j)) = \pi(j)$ and $\sigma(j) = j$, so that $\pi\sigma(j) = \pi(j)$ also. If $\pi(j) = j$, then $\sigma\pi(j) = \sigma(j)$. Now there are two possibilities for j: If it is in a nontrivial orbit of $[\sigma]$, then so is $\sigma(j)$, and $\pi\sigma(j) = \sigma(j)$; if $\sigma(j) = j$, then $\pi\sigma(j) = \pi(j) = j = \sigma(j) = \sigma\pi(j)$. Thus σ and π commute.

The fundamental result concerning cycles is the following so-called *cycle decomposition* theorem.

Theorem 3.4 *Let $e \neq \sigma \in S_n$. Then σ is a product of pairwise disjoint cycles of length at least 2. This representation is unique except for the order in which the cycles occur.*

Proof: Let $\mathcal{O}_1, \mathcal{O}_2, \ldots, \mathcal{O}_p$ be all the nontrivial $[\sigma]$-orbits. Each orbit $\mathcal{O}_t$ is of the form

$$\mathcal{O}_t = \{j_t, \sigma(j_t), \ldots, \sigma^{k_t-1}(j_t)\}, \qquad k_t > 1. \tag{11}$$

We assert that if

$$\sigma_t = (j_t\ \sigma(j_t) \cdots \sigma^{k_t-1}(j_t)),$$

then

$$\sigma = \sigma_1 \cdots \sigma_p. \tag{12}$$

Let $i \in [1, n]$. If i is in a trivial $[\sigma]$-orbit, then $\sigma(i) = i$, and since $i \notin \bigcup_{t=1}^{p} \mathcal{O}_t$, $\sigma_t(i) = i$, $t = 1, \ldots, p$. Hence $\sigma(i) = (\sigma_1 \cdots \sigma_p)(i)$. If $i \in \mathcal{O}_t$, then $i = \sigma^s(j_t)$ for some $0 \leq s < k_t$, and $\sigma(i) = \sigma(\sigma^s(j_t)) = \sigma^{s+1}(j_t)$. Also if $r \neq t$, then $\sigma_r(i) = i$ and $\sigma_r(\sigma_t(i)) = \sigma_t(i)$ since the $\mathcal{O}_t$ are pairwise disjoint, and thus

$$\begin{aligned}(\sigma_1 \cdots \sigma_p)(i) &= \sigma_t(i) = \sigma_t(\sigma^s(j_t)) \\ &= \sigma^{s+1}(j_t) \\ &= \sigma(i).\end{aligned}$$

Thus the equality (12) is established, and the cycles σ_t, $t = 1, \ldots, p$, are disjoint (see Exercise 7).

To prove uniqueness, suppose σ is is some way represented as a product of disjoint cycles of length at least 2 and π_1 is one of these cycles: $\pi_1 = (i_1 \cdots i_r)$. Let τ be the product of the remaining cycles in this second representation of σ. Then $\sigma = \pi_1\tau$. None of the cycles whose product is τ involve any of $i_1, \ldots, i_r$ (see Exercise 7); thus $\tau(i_j) = i_j, j = 1, \ldots, r$, and hence $\sigma(i_j) = \pi_1(i_j), j = 1, \ldots, r$. In other words, $\{i_1, \sigma(i_1), \ldots, \sigma^{r-1}(i_1)\} = \{i_1, \pi_1(i_1), \ldots, \pi_1^{r-1}(i_1)\}$ is a nontrivial $[\sigma]$-orbit, so π_1 must be one of the σ_t defined above. Thus each cycle such as π_1 in this second representation of σ must be one of the σ_t. Suppose $\sigma = \pi_1 \cdots \pi_m = \sigma_1 \cdots \sigma_p$. Then we can cancel the π_i's with the corresponding equal σ_t's. Clearly no σ_t's can be left, for otherwise we would have a product of disjoint cycles of length at least 2 equal to e. ∎

To find the disjoint cycle decomposition of any particular $\sigma \in S_n$ is relatively straightforward. For example, suppose $\sigma \in S_{10}$ is given by

$$\sigma = \begin{pmatrix} 1 & 2 & 3 & 4 & 5 & 6 & 7 & 8 & 9 & 10 \\ 3 & 4 & 6 & 7 & 5 & 1 & 9 & 10 & 8 & 2 \end{pmatrix}.$$

According to Theorem 3.4 [see (11) and (12)] we can start with any integer (1 is sensible) and determine the $[\sigma]$-orbit in which it lies. Thus

$$\sigma(1) = 3,$$
$$\sigma(3) = 6,$$
$$\sigma(6) = 1$$

so that

$$\mathcal{O}_1 = \{1, 3, 6\}$$

and

$$\sigma_1 = (1\ 3\ 6).$$

Choose any integer in $[1, 10] - \mathcal{O}_1$, say 2, and compute that

$$\sigma(2) = 4, \quad \sigma(4) = 7, \quad \sigma(7) = 9,$$
$$\sigma(9) = 8, \quad \sigma(8) = 10, \quad \sigma(10) = 2.$$

Thus

$$\mathcal{O}_2 = \{2, 4, 7, 9, 8, 10\},$$

and

$$\sigma_2 = (2\ 4\ 7\ 9\ 8\ 10).$$

Finally $\sigma(5) = 5$, and so 5 determines a trivial orbit. We have

$$\sigma = (1\ 3\ 6)\,(2\ 4\ 7\ 9\ 8\ 10).$$

Any cycle can be written as a product of transpositions: If $\sigma = (i_1 \cdots i_r)$, then

$$\sigma = (i_1 i_r)\,(i_1 i_{r-1})\,(i_1 i_{r-2}) \cdots (i_1 i_2).$$

From Theorem 3.4, any permutation can be written as a product of transpositions, but not uniquely, e.g.,

$$(1\ 2) = (3\ 4)\ (1\ 2)\ (3\ 4).$$

If $e \neq \sigma \in S_n$ and $\sigma = \sigma_1 \cdots \sigma_p$ is its unique (except for order) cycle decomposition, then we define

$$\begin{aligned} I(\sigma) &= (k_1 - 1) + (k_2 - 1) + \cdots + (k_p - 1) \\ &= k_1 + \cdots + k_p - p, \end{aligned}$$

where k_t is the length of σ_t. The integer $I(\sigma)$ is called the *Cauchy index* of σ. We define $I(e) = 0$. Now, although the above example shows that the representation of a permutation as a product of transpositions is not unique, we have the following theorem.

Theorem 3.5 *If $\sigma = \tau_1 \cdots \tau_k \in S_n$, where τ_i is a transposition for $i = 1, \ldots, k$, then $I(\sigma)$ and k have the same parity, i.e., they are both even or both odd.*

Proof: We first observe that if $a, b, x_1, \ldots, x_r, y_1, \ldots, y_s$ are distinct, then

$$(a\ b)(a\ x_1 \cdots x_r\ b\ y_1 \cdots y_s) = (a\ x_1 \cdots x_r)\ (b\ y_1 \cdots y_s), \tag{13}$$

$$(a\ b)(a\ x_1 \cdots x_r) = (a\ x_1 \cdots x_r\ b), \tag{14}$$

and

$$\begin{aligned} &(a\ b)(a\ x_1 \cdots x_r)\ (b\ y_1 \cdots y_s) \\ &\qquad = (a\ x_1 \cdots x_r\ b\ y_1 \cdots y_s). \end{aligned} \tag{15}$$

Thus if $(a\ b)$ is any transposition and $\sigma = \sigma_1 \cdots \sigma_p$ is the unique cycle decomposition of σ, we can assume that if a or b appear among the cycles $\sigma_1, \ldots, \sigma_p$, then the product $(a\ b)\sigma$ begins with (13), (14), or (15) (recall that disjoint cycles commute). But then clearly (see Exercise 9),

$$I((a\ b)\sigma) = I(\sigma) \pm 1 \tag{16}$$

in all cases. In other words, multiplying a permutation by a transposition changes the index of the permutation by 1. It follows immediately that

$$\begin{aligned} I(e) &= I(\tau_k \tau_{k-1} \cdots \tau_1 \sigma) \\ &= \overbrace{1 + \cdots + 1}^{m} + \overbrace{-1 + \cdots + -1}^{q} + I(\sigma) \\ &= m - q + I(\sigma). \end{aligned}$$

Now $I(e) = 0$ and $m - q$ has the same parity as $m + q = k$. Thus $I(\sigma)$ has the same parity as k. ∎

The *sign* or *signum* of a permutation σ is defined to be

$$\varepsilon(\sigma) = (-1)^{I(\sigma)}.$$

If $\varepsilon(\sigma) = 1$, the permutation σ is *even;* if $\varepsilon(\sigma) = -1$, the permutation σ is *odd.*

Corollary 1 *If $\sigma \in S_n$ and $\sigma = \tau_1 \cdots \tau_k$ where each τ_i is a transposition, then*

$$\varepsilon(\sigma) = (-1)^k. \tag{17}$$

Moreover, if $\theta \in S_n$, then

$$\varepsilon(\sigma\theta) = \varepsilon(\sigma)\varepsilon(\theta). \tag{18}$$

Proof: From Theorem 3.5, $I(\sigma)$ has the same parity as k and (17) follows. If $\theta = \mu_1 \cdots \mu_q$ where each μ_j is a transposition, then $\sigma\theta$ is a product of $k + q$ transpositions and (18) follows. ∎

Let A_n denote the set of all even permutations in S_n. It follows (see Exercise 10) immediately from (18) that A_n is a subgroup of S_n; it is called the *alternating group of degree n.*

If $\sigma \in S_n$, then according to Theorem 3.4, σ is a unique product of disjoint cycles

$$\sigma = \sigma_1\sigma_2 \cdots \sigma_p$$

$$\text{where} \quad \sigma_t = (j_t\ \sigma(j_t) \cdots \sigma^{k_t-1}(j_t)), \qquad t = 1, \ldots, p, \tag{19}$$

$$\text{and} \quad \mathcal{O}_t = \{j_t, \sigma(j_t), \ldots, \sigma^{k_t-1}(j_t)\}, \qquad t = 1, \ldots, p$$

are the nontrivial $[\sigma]$-orbits. Let $\{f_1\}, \ldots, \{f_{\lambda_1}\}$ denote the trivial orbits, i.e., $f_1, \ldots, f_{\lambda_1}$ are the elements mapped into themselves by σ. If there are λ_s orbits each with s elements in it, $s = 1, \ldots, n$, then we say that σ has the *cycle structure*

$$[1^{\lambda_1}, 2^{\lambda_2}, \ldots, n^{\lambda_n}]. \tag{20}$$

Thus σ has cycle structure (20) iff σ is a product of λ_s disjoint cycles of length s, $s = 1, \ldots, n$. (Of course, the cycles of length 1 are all e.)

Let G be any group, and define $R \subset G^2$ by

$$R = \{(\sigma,\varphi) \mid \text{there exists } \theta \in G \text{ such that } \varphi = \theta\sigma\theta^{-1}\}. \tag{21}$$

It is routine to verify that R is an equivalence relation on G (see Exercise 15). This relation is called *conjugacy* and the equivalence classes $R(\sigma) \in G/R$ are called *conjugacy classes.* The following result provides us with an interesting connection between cycle structure and conjugacy.

Theorem 3.6 (i) *Two permutations σ and φ in S_n are conjugate iff they have the same cycle structure.*

(ii) (*Cauchy's formula*) *The number of permutations in the conjugacy class determined by the cycle structure $[1^{\lambda_1}, 2^{\lambda_2}, \ldots, n^{\lambda_n}]$ is*

$$h(\lambda_1, \ldots, \lambda_n) = \frac{n!}{1^{\lambda_1}\lambda_1!\, 2^{\lambda_2}\lambda_2! \cdots n^{\lambda_n}\lambda_n!}. \tag{22}$$

Proof: (i) Suppose $\sigma = \sigma_1 \cdots \sigma_p$ where σ_t is given by (19), and let $\varphi = \theta\sigma\theta^{-1}$ for some $\theta \in S_n$. Then $\varphi = \theta\sigma_1\theta^{-1}\theta\sigma_2\theta^{-1} \cdots \theta\sigma_p\theta^{-1}$. Now set $\theta(\sigma_t^m(j_t)) = i_{t,m}$, $m = 0, \ldots, k_t - 1$, $t = 1, \ldots, p$. We compute

$$\begin{aligned}\theta\sigma_t\theta^{-1}(i_{t,m}) &= \theta\sigma_t(\sigma_t^m(j_t)) = \theta(\sigma_t^{m+1}(j_t))\\ &= i_{t,m+1}, \qquad m = 0, \ldots, k_t - 2,\end{aligned}$$

and

$$\begin{aligned}\theta\sigma_t\theta^{-1}(i_{t,k_t-1}) &= \theta\sigma_t(\sigma_t^{k_t-1}(j_t)) = \theta\sigma_t^{k_t}(j_t)\\ &= i_{t,0}.\end{aligned}$$

Now let $r \in \{i_{t,0}, \ldots, i_{t,k_t-1}\}^c$. Then $\theta^{-1}(r) \in \mathcal{O}_t^c$ because $\theta(\mathcal{O}_t) = \{i_{t,0}, \ldots, i_{t,k_t-1}\}$, and hence $\sigma_t\theta^{-1}(r) = \theta^{-1}(r)$. Thus $\theta\sigma_t\theta^{-1}(r) = \theta\theta^{-1}(r) = r$. We have proved that $\theta\sigma_t\theta^{-1} = (i_{t,0} \cdots i_{t,k_t-1})$, and moreover the sets $\{i_{t,0}, \ldots, i_{t,k_t-1}\} = \theta(\mathcal{O}_t)$ are disjoint. Also $r \in [1, n]$ is a fixed point of σ iff $\theta(r)$ is a fixed point of $\varphi = \theta\sigma\theta^{-1}$,

$$\theta\sigma\theta^{-1}(\theta(r)) = \theta\sigma(r).$$

Thus φ is the product of the disjoint cycles $(i_{t,0} \cdots i_{t,k_t-1})$, $t = 1, \ldots, p$, and has the same number of fixed points as σ does. It follows that φ and σ have the same cycle structure.

To prove the converse, suppose σ and φ have the same cycle structure:

$$\begin{aligned}\sigma &= (f_1) \cdots (f_{\lambda_1}) \cdots (j_{t,0} \cdots j_{t,k_t-1}) \cdots\\ \varphi &= (g_1) \cdots (g_{\lambda_1}) \cdots (i_{t,0} \cdots i_{t,k_t-1}) \cdots\end{aligned}$$

where we have written the corresponding cycles of equal length in φ directly below those in σ. Now define

$$\theta = \begin{pmatrix} f_1 \cdots f_{\lambda_1} \cdots j_{t,0} \cdots j_{t,k_t-1} \cdots \\ g_1 \cdots g_{\lambda_1} \cdots i_{t,0} \cdots i_{t,k_t-1} \cdots \end{pmatrix} \in S_n.$$

Then

$$\begin{aligned}\theta\sigma\theta^{-1}(g_\nu) &= \theta\sigma(f_\nu) = \theta(f_\nu) = g_\nu\\ &= \varphi(g_\nu), \qquad \nu = 1, \ldots, \lambda_1,\end{aligned}$$

and

$$\begin{aligned}\theta\sigma\theta^{-1}(i_{t,\mu}) &= \theta\sigma(j_{t,\mu}) = \theta(j_{t,\mu+1}) = i_{t,\mu+1}\\ &= \varphi(i_{t,\mu}), \qquad \mu = 0, \ldots, k_t - 1, t = 1, \ldots, p.\end{aligned}$$

Thus $\theta\sigma\theta^{-1} = \varphi$.

(ii) Permutations with cycle structure (20) can be constructed as follows:

$$\overbrace{(*)(*)\cdots(*)}^{\lambda_1}\ \overbrace{(*\,*)(*\,*)\cdots(*\,*)}^{\lambda_2}\ \cdots\ \overbrace{(*\,*\,*\,*\,*\,*)}^{\lambda_n} \tag{23}$$

where there are λ_s cycles of length s, and all the cycles are disjoint. Now $1\lambda_1 + 2\lambda_2 + \cdots + n\lambda_n = n$ and the n elements $1, \ldots, n$ are to be placed where the asterisks $(*)$ are in the scheme (23). There are $n!$ ways of doing this, but some of the resulting permutations are the same: Each of the λ_s cycles of length s can be permuted among themselves, and there are $\lambda_1! \cdots \lambda_n!$ ways altogether of doing this; also the elements in a given s-cycle can be shifted ahead $1, 2, \ldots, s$ places without altering the permutation [e.g., $(1\ 2\ 3) = (3\ 1\ 2) = (2\ 3\ 1)$], and there are

$$\overbrace{s\cdots s}^{\lambda_s} = s^{\lambda_s}$$

ways of doing this for the cycles of length s. Thus with each arrangement of $1, \ldots, n$ in the scheme (23), there are $1^{\lambda_1}\lambda_1!\ 2^{\lambda_2}\lambda_2! \cdots n^{\lambda_n}\lambda_n!$ of the permutations that are the same. Hence there are

$$\frac{n!}{1^{\lambda_1}\lambda_1!\ 2^{\lambda_2}\lambda_2! \cdots n^{\lambda_n}\lambda_n!}$$

distinct permutations with the cycle structure (20). ∎

As an example, suppose $n = 3$ and we want to list all the distinct permutations in S_3 with cycle structure

$$[1^1, 2^1].$$

We begin, as in the proof of Theorem 3.6 (ii), by listing all six arrangements of 1, 2, 3 in the scheme

$$(*)\ (**)$$

and identifying the equal ones:

$$\left.\begin{array}{l}(1)\ (2\ \ 3)\\(1)\ (3\ \ 2)\end{array}\right\}$$
$$\left.\begin{array}{l}(2)\ (1\ \ 3)\\(2)\ (3\ \ 1)\end{array}\right\}$$
$$\left.\begin{array}{l}(3)\ (1\ \ 2)\\(3)\ (2\ \ 1)\end{array}\right\}.$$

The bracketed permutations are equal and each cycle of the bracketed pair is obtained by cyclically permuting the integers in the other cycle of the same length. Here $\lambda_1 = \lambda_2 = 1$, $\lambda_3 = 0$ and

$$h(\lambda_1,\lambda_2,\lambda_3) = \frac{3!}{1^1 \cdot 1!\, 2^1 \cdot 1!}$$
$$= 3.$$

Somewhat less trivial to compute is the case $n = 4$ with prescribed cycle structure

$$[2^2].$$

Then the scheme to be filled in is

$$(*\ *)\ (*\ *).$$

We have the list

$$\left.\begin{matrix}(1\ 2)\ (3\ 4)\\(1\ 2)\ (4\ 3)\\(2\ 1)\ (3\ 4)\\(2\ 1)\ (4\ 3)\\(3\ 4)\ (1\ 2)\\(4\ 3)\ (1\ 2)\\(3\ 4)\ (2\ 1)\\(4\ 3)\ (2\ 1)\end{matrix}\right\} \qquad \left.\begin{matrix}(1\ 3)\ (2\ 4)\\(1\ 3)\ (4\ 2)\\(3\ 1)\ (2\ 4)\\(3\ 1)\ (4\ 2)\\(2\ 4)\ (1\ 3)\\(4\ 2)\ (1\ 3)\\(2\ 4)\ (3\ 1)\\(4\ 2)\ (3\ 1)\end{matrix}\right\} \qquad \left.\begin{matrix}(1\ 4)\ (2\ 3)\\(1\ 4)\ (3\ 2)\\(4\ 1)\ (2\ 3)\\(4\ 1)\ (3\ 2)\\(2\ 3)\ (1\ 4)\\(3\ 2)\ (1\ 4)\\(2\ 3)\ (4\ 1)\\(3\ 2)\ (4\ 1)\end{matrix}\right\}$$

The first 4 permutations in each list are obtained by cyclically permuting the integers in each of the cycles. The second 4 are obtained by interchanging cycles and then cyclically permuting the integers within a cycle. There are 3 distinct permutations with the cycle structure $[2^2]$. Note that $\lambda_1 = 0$, $\lambda_2 = 2$, $\lambda_3 = \lambda_4 = 0$, and

$$h(\lambda_1,\lambda_2,\lambda_3,\lambda_4) = \frac{4!}{2^2 \cdot 2!}$$
$$= 3.$$

In order to further unravel the structure of S_n, we introduce a few more general notions about groups.

Let G be any group (with the operation denoted simply by juxtaposition), and let $\varnothing \neq A \in P(G)$. Then A is called a *complex*. The product and inverses of complexes are given by

$$AB = \{ab \mid a \in A,\ b \in B\}$$

and

$$A^{-1} = \{a^{-1} \mid a \in A\}.$$

If H is a subgroup of G, we write

$$H < G.$$

If $a \in G$ and $H < G$, then the set $\{a\}H$ $(H\{a\})$ is called a *left* (*right*) *H-coset* in G. We write these cosets as aH or Ha. The element a is called a *representa-*

tive of the coset aH or Ha. It should be noted that if A is any complex and $g \in G$, then

$$|A| = |gA| = |Ag|.$$

The reader will also confirm (see Exercise 16) that for arbitrary complexes we have

$$(AB)^{-1} = B^{-1}A^{-1}; \tag{24}$$

$$A(BC) = (AB)C; \tag{25}$$

$$H < G \text{ iff } HH = H = H^{-1}; \tag{26}$$

$$\text{If } H < G \text{ and } K < G, \text{ then } HK = KH \text{ iff } HK < G. \tag{27}$$

The following important result shows that the order of a subgroup of a finite group must divide the order of the group.

Theorem 3.7 *(Lagrange's theorem)* (i) *If* $H < G$, *then*

$$R = \{(a, b) \mid a \in G, b \in G \text{ and } a^{-1}b \in H\}$$

is an equivalence relation in G, *and the elements* $R(a)$ *of* G/R *are precisely the left cosets* $R(a) = aH$, $a \in G$.

(ii) *If* $|G| = n < \infty$, $H < G$, *and* $|H| = m$, *then*

$$n = mi \tag{28}$$

where $i \in N$.

Proof: (i) Clearly $a^{-1}b \in H$ and $b^{-1}c \in H$ imply that $a^{-1}bb^{-1}c = a^{-1}c \in H$. This is the least trivial part of showing that R is an equivalence relation. Now for $a \in G$ we have $b \in R(a)$ iff $(a, b) \in R$, i.e., iff $a^{-1}b \in H$ or equivalently $b \in aH$. Thus $R(a) = aH$.

(ii) Let $a_1H, \ldots, a_iH$ be a list of all the distinct left H-cosets in G. Now $|a_tH| = |H| = m$ for $t = 1, \ldots, i$, so

$$n = |G| = \left|\bigcup_{t=1}^{i} a_tH\right| = \sum_{t=1}^{i} |a_tH|$$
$$= mi.$$

The third equality in the preceding calculation follows from the fact that the elements of G/R, i.e., the distinct equivalence classes, partition G (see Exercise 10, Section 1.1). ∎

The integer i in Theorem 3.7 is called the *index of* H *in* G and is written

$$i = [G : H].$$

Thus (28) reads

$$|G| = |H|\,[G : H]. \tag{29}$$

Theorem 3.8 *In the notation of Theorem 3.7(i),* $(G/R, \{\omega\})$ *is a group in* *which* $\omega(A, B) = AB$ *is complex multiplication, iff* $cH = Hc$ *for all* $c \in G$.

Proof: Assume $(G/R, \{\omega\})$ is a group. Then from $H^2 = H$ it follows that $H = eH$ is the neutral element of $(G/R, \{\omega\})$ (see Exercise 18). Hence for $c \in G$,

$$\begin{aligned} cH &= H(cH) = (Hc)H \\ &\supset Hc; \end{aligned} \tag{30}$$

then

$$H^{-1}c^{-1} \supset c^{-1}H^{-1}$$

or

$$Hc^{-1} \supset c^{-1}H, \tag{31}$$

and since c is arbitrary, we can replace c^{-1} by c in (31) to produce the inclusion

$$Hc \supset cH.$$

This combined with (30) yields $cH = Hc$.

Conversely, if $cH = Hc$ for all $c \in G$, then for all $a,b \in G$ we have

$$\begin{aligned} (aH)(bH) &= abH^2 \\ &= abH, \\ (aH)^{-1} &= H^{-1}a^{-1} = Ha^{-1} \\ &= a^{-1}H, \end{aligned}$$

and

$$\begin{aligned} (aH)^{-1}(aH) &= a^{-1}HaH = a^{-1}aH^2 = eH \\ &= H. \end{aligned}$$

Thus $(G/R, \{\omega\})$ is a group with H as the identity and $a^{-1}H$ the inverse of the coset aH. ∎

If $H < G$ and $cH = Hc$ for every $c \in G$, then H is called a *normal* subgroup of G or a *normal divisor* of G. In this event we write

$$H \triangleleft G.$$

The group $(G/R, \{\omega\})$ described in Theorem 3.8 is called the *quotient group* or *factor group* of G modulo H and is written

$$G/H.$$

As a corollary of the proof of Lagrange's theorem we have, in the case $|G| < \infty$, that $|G/H| = [G: H]$ and hence $|G| = |G/H|\,|H|$.

We define the *canonical homomorphism*

$$\nu: G \to G/H$$

by

$$\nu(a) = aH, \qquad a \in G. \tag{32}$$

Obviously

$$\begin{aligned} \nu(a)\nu(b) &= aHbH = abH^2 = abH \\ &= \nu(ab) \end{aligned}$$

so that ν is indeed a homomorphism and in fact an epimorphism. Also note that

$$\nu^{-1}(H) = H. \tag{33}$$

Now in general, recall from Theorem 1.2 that if $f: G \to K$ is any function, then $R_f = \{(x_1, x_2) \mid f(x_1) = f(x_2)\}$ is an equivalence relation on G. We assert that if G and K are groups and f is a homomorphism, then

$$R_f = \{(a, b) \mid a^{-1}b \in f^{-1}(\{e\})\}. \tag{34}$$

(Here we denote by e the neutral element in either G or K.) For, $a^{-1}b \in f^{-1}(\{e\})$ iff $f(a^{-1}b) = e$ iff $f(a) = f(b)$.

In particular, let $\nu: G \to G/H$ be the canonical homomorphism and consider the natural map $\mu: G \to G/R_\nu$ induced by R_ν, i.e.,

$$\mu(a) = R_\nu(a), \qquad a \in G.$$

According to (33) we have

$$R_\nu = \{(a,b) \mid a^{-1}b \in H\}, \tag{35}$$

and from Theorem 3.7(i), $R_\nu(a) = aH$. Hence $\mu(a) = aH = \nu(a)$. In other words, ν itself is the natural map induced by R_ν.

Theorem 3.9 *(Universal Factorization Theorem for Quotient Groups) Assume $H \lhd G$, and let $\nu: G \to G/H$ be the canonical homomorphism. The pair $(G/H, \nu)$ is universal in the following sense. Given any group K and any homomorphism $\varphi: G \to K$ for which $H \subset \varphi^{-1}(\{e\})$, there exists a unique homomorphism $\bar{\varphi}: G/H \to K$ such that $\varphi = \bar{\varphi}\nu$, i.e., the diagram*

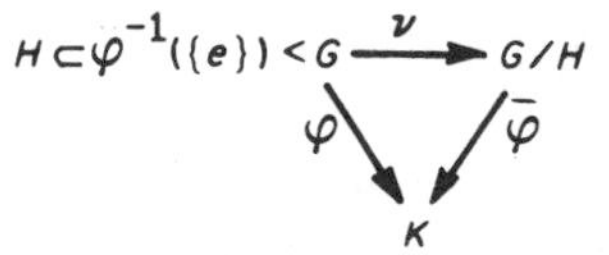

(36)

is commutative.

Proof: According to (34),

$$R_\varphi = \{(a,b) \mid a^{-1}b \in \varphi^{-1}(\{e\})\},$$

and from (35) and $H \subset \varphi^{-1}(\{e\})$ we have

$$R_\nu \subset R_\varphi.$$

But then by Theorem 1.2(ii) there exists a unique $\bar{\varphi}: G/R_\nu \to K$ such that

$$\varphi = \bar{\varphi}\nu.$$

Of course G/R_ν is just another way of writing G/H, and so $\bar{\varphi}$ completes the diagram (36), as advertised. Observe that

$$\bar{\varphi}(aHbH) = \bar{\varphi}(abH) = \bar{\varphi}\nu(ab) = \varphi(ab)$$
$$= \varphi(a)\varphi(b) = \bar{\varphi}(\nu(a))\bar{\varphi}(\nu(b))$$
$$= \bar{\varphi}(aH)\bar{\varphi}(bH)$$

so that $\bar{\varphi}$ is indeed a homomorphism. ▮

If $\varphi\colon G \to K$ is a homomorphism, then the inverse image $\varphi^{-1}(\{e\})$ of the neutral element e is called the *kernel* of φ and is denoted by

$$\ker \varphi.$$

It is very simple to verify that

$$\ker \varphi \lhd G. \tag{37}$$

For if $h \in \ker \varphi = H$, so that $\varphi(h) = e$, and if $c \in G$, then

$$\varphi(c^{-1}hc) = \varphi(c)^{-1}\varphi(h)\varphi(c)$$
$$= e,$$

and hence $c^{-1}hc \in H$. It follows that $cH = Hc$ (see Exercise 20).

Example 1 Let $G = \mathbf{R}^2$ and define the operation $+$ in G by

$$(x_1,x_2) + (y_1,y_2) = (x_1 + x_2,\ y_1 + y_2).$$

Then $(G,+)$ is clearly an abelian group, with $(0,0)$ as the identity. Let

$$H = \{(x_1,x_2) \mid x_2 = 2x_1\}.$$

Then it is simple to verify that $H \lhd G$. The subgroup H is automatically normal because G is abelian. The cosets in the group

$$G/H$$

will be denoted by

$$(x_1,x_2) + H.$$

Observe that geometrically G is the ordinary plane and H is the line $x_2 = 2x_1$ through the origin. A coset $(x_1,x_2) + H$ is a line through the point (x_1,x_2) parallel to H.

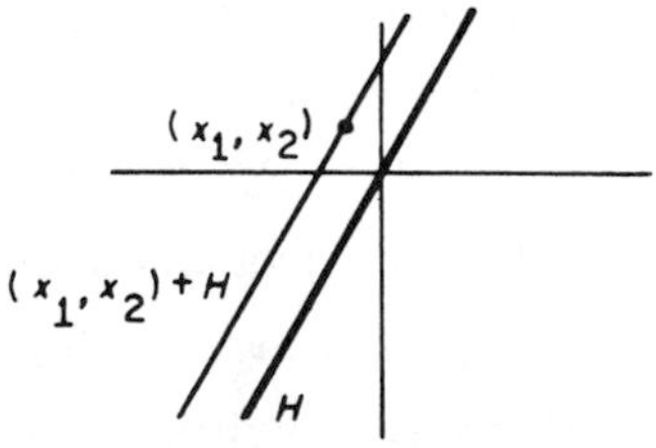

The canonical homomorphism

$$\nu: G \longrightarrow G/H$$

maps each point (x_1,x_2) to the line $(x_1,x_2) + H$:

$$\nu((x_1,x_2)) = (x_1,x_2) + H.$$

The proof of Theorem 3.8 tells us that in "adding" two cosets we can add any two representatives:

$$[(x_1,x_2) + H] + [(y_1,y_2) + H] = (x_1 + y_1, x_2 + y_2) + H.$$

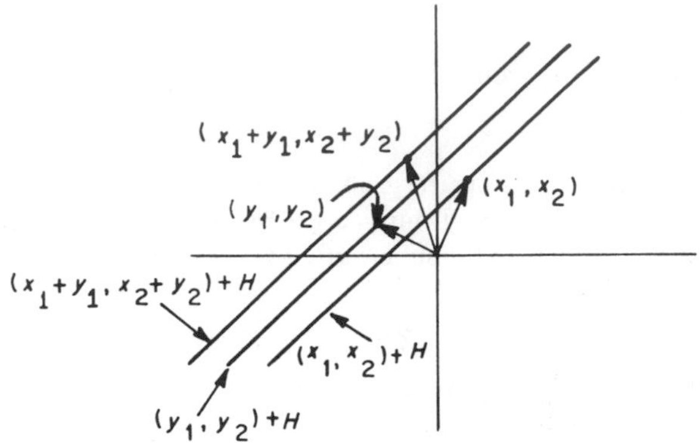

Thus the addition of two parallel lines in G/H is accomplished by choosing a point on each line and adding these points in a "vector" sense. Obviously, if K is any line through the origin other than H then we also have $K \lhd G$. Let

$$f: G/H \longrightarrow K$$

be defined by

$$f((x_1,x_2) + H) = (z_1,z_2),$$

where (z_1,z_2) is the point of intersection of the lines K and $(x_1,x_2) + H$. The reader can easily confirm that f is a group isomorphism.

Example 2 Let det: $GL(n,\mathbf{R}) \rightarrow \mathbf{R}$ denote the determinant function (see Section 1.2, Example 5). Observe that det is a group homomorphism:

$$\det(AB) = \det A \det B.$$

The kernel of this homomorphism is the group $SL_n(\mathbf{R})$. Thus (37) implies $SL_n(\mathbf{R}) \lhd GL(n,\mathbf{R})$. According to Theorem 3.8 the factor group

$$GL(n,\mathbf{R})/SL_n(\mathbf{R})$$

is well defined. The cosets are of the form

$$A\mathrm{SL}_n(\mathbf{R}), \qquad A \in \mathrm{GL}(n,\mathbf{R}).$$

Observe that

$$A\mathrm{SL}_n(\mathbf{R}) = B\mathrm{SL}_n(\mathbf{R})$$

iff

$$A^{-1}B \in \mathrm{SL}_n(\mathbf{R})$$

iff

$$\det(A^{-1}B) = 1$$

iff

$$\det A = \det B \neq 0.$$

Thus a coset in the factor group consists precisely of all nonsingular matrices with equal determinants.

Example 3 Let GL(2,Z) denote the group of all 2×2 matrices with integer entries and determinant ± 1. (Verify that this is a group with matrix multiplication as the operation.) Let $\mathrm{SL}_2(Z)$ denote the subgroup of GL(2,Z) consisting of those matrices with determinant 1. Then

$$\mathrm{SL}_2(Z) \lhd \mathrm{GL}(2,Z),$$

and the reader can easily check that the factor group

$$\mathrm{GL}(2,Z)/\mathrm{SL}_2(Z)$$

is isomorphic to the multiplicative group $\{1, -1\}$. The isomorphism f is defined by

$$f(A\mathrm{SL}_2(Z)) = \det A, \qquad A \in \mathrm{GL}(2,Z).$$

Before establishing some fundamental facts concerning A_n and S_n, we make the following definitions. A subgroup of a group G is *proper* if it is neither $\{e\}$ nor G itself. The group G is said to be *simple* if it contains no proper normal subgroup.

If A is any complex in a group G, then the totality of finite products of the form

$$a_1^{n_1}a_2^{n_2} \cdots a_r^{n_r}, \quad a_i \in A, n_i \in Z, \quad i = 1, \ldots, r,$$

together with the identity e ($= a^0$ for any $a \in A$) is obviously the smallest subgroup of G containing A. We call this the group *generated by A* and denote it by

$$[A],$$

in conformity with our earlier notation for cyclic groups.

Theorem 3.10 (i) $A_n < S_n$, $n \geq 1$.

(ii) Let $n \geq 3$. *If* $i_1, \ldots, i_n$ *are* $1, \ldots, n$ *in some order, then* A_n *is generated by the* $n - 2$ *3-cycles* $(i_1 i_2 i_k)$, $k = 3, \ldots, n$. *In particular,* A_n *is generated by the* $n - 2$ *3-cycles* $(1\ 2\ k)$, $k = 3, \ldots, n$.

(iii) *If* $n \geq 4$, $H \lhd A_n$, *and* H *contains a 3-cycle, then* $H = A_n$.

(iv) *If* $n \geq 5$, $H \lhd A_n$, *and the disjoint cycle factorization of some* $\sigma \in H$ *involves a cycle of length* $k > 3$, *then* $H = A_n$.

(v) *If* $n \geq 5$, *then* A_n *is simple*. (The groups A_1, A_2, A_3 are obviously simple.)

(vi) *Let* $n \geq 5$. *If* $K \lhd H < S_n$, $A_n < H$, *and* H/K *is abelian, then* $A_n < K$.

Proof: (i) Obviously $A_1 = S_1$. Let $n \geq 2$. Consider the sign function ε as a mapping of S_n onto the multiplicative group $\{1, -1\}$. Then ε is a group homomorphism and ker $\varepsilon = \varepsilon^{-1}(\{1\}) = A_n$. Apply (37).

(ii) Observe that $(i_1 i_k)(i_1 i_j)(i_1 i_k) = (i_j i_k)$ so that any permutation in S_n is generated by the special transpositions $(i_1 i_2), (i_1 i_3), \ldots, (i_1 i_n)$. Now each $\sigma \in A_n$ must be a product of an even number of such transpositions. But $(i_1 i_j)(i_1 i_k) = (i_1 i_k i_j) = (i_1 i_2 i_j)(i_1 i_2 i_j)(i_1 i_2 i_k)(i_1 i_2 i_j)$. Hence (ii) follows.

(iii) This is obvious for $n = 3$. Assume $n \geq 4$, and suppose $\sigma = (i_1 i_2 i_3) \in H$. Let $\theta = (i_3 i_2 i_k)$ where i_1, i_2, i_3, i_k are distinct. Then by Exercise 23 we have

$$\begin{aligned} \theta\sigma\theta^{-1} &= (\theta(i_1)\theta(i_2)\theta(i_3)) \\ &= (i_1 i_k i_2) \end{aligned}$$

and $\theta\sigma\theta^{-1} \in H$ since $H \lhd A_n$ [note that $\theta = (i_3 i_2 i_k) = (i_3 i_k)(i_3 i_2) \in A_n$]. Hence $(\theta\sigma\theta^{-1})^2 = (i_1 i_k i_2)^2 = (i_1 i_2 i_k) = H$. In other words, once $(i_1 i_2 i_3) \in H$ it follows that $(i_1 i_2 i_k) \in H$, $k = 3, \ldots, n$ (any order is alright). Now part (ii) implies $A_n = H$.

(iv) Write the disjoint cycle factorization of σ as $\sigma = \sigma_1 \cdots \sigma_p$ where $\sigma_1 = (i_1 i_2 \cdots i_k)$, $k > 3$. Then set $\theta = (i_1 i_2 i_3) \in A_n$. Since the cycles $\sigma_1, \sigma_2, \ldots, \sigma_p$ are disjoint, we compute that

$$\begin{aligned} \varphi = \theta\sigma\theta^{-1} &= \theta\sigma_1\theta^{-1}\sigma_2 \cdots \sigma_p \\ &= (\theta(i_1)\theta(i_2)\theta(i_3) \cdots \theta(i_k))\sigma_2 \cdots \sigma_p \\ &= (i_2 i_3 i_1 i_4 \cdots i_k)\,\sigma_2 \cdots \sigma_p. \end{aligned}$$

Then $\varphi \in H$ by the normality assumption, and hence $\sigma^{-1}\varphi \in H$. But

$$\begin{aligned} \sigma^{-1}\varphi &= \sigma_p^{-1}\sigma_{p-1}^{-1} \cdots \sigma_1^{-1}(i_2 i_3 i_1 i_4 \cdots i_k)\sigma_2 \cdots \sigma_p \\ &= \sigma_1^{-1}(i_2 i_3 i_1 i_4 \cdots i_k) \\ &= (i_k i_{k-1} i_{k-2} \cdots i_5 i_4 i_3 i_2 i_1)(i_2 i_3 i_1 i_4 \cdots i_k) \\ &= (i_1 i_3 i_k). \end{aligned}$$

Thus H contains a 3-cycle. By part (iii), $H = A_n$.

(v) Let $H \lhd A_n$ and assume H is a proper subgroup, i.e., it is neither

$\{e\}$ nor A_n itself. The argument is by contradiction. Let $e \neq \sigma \in H$. Then by part (iv), the disjoint cycle factorization of σ must involve only cycles of length 3 or less. There are several cases:

Case 1 σ involves a product of two (or more) disjoint cycles of length 3:

$$\sigma = (i_1 i_2 i_3)(i_4 i_5 i_6)\,\varphi \tag{38}$$

where φ is the product of the remaining disjoint cycles. Let $\theta = (i_2 i_3 i_4) \in A_n$, so that $\theta\sigma\theta^{-1} \in H$. We compute that

$$\begin{aligned}\theta\sigma\theta^{-1} &= \theta(i_1 i_2 i_3)\theta^{-1}\theta(i_4 i_5 i_6)\theta^{-1}\varphi \\ &= (\theta(i_1)\theta(i_2)\theta(i_3))(\theta(i_4)\theta(i_5)\theta(i_6))\varphi \\ &= (i_1 i_3 i_4)(i_2 i_5 i_6)\varphi\end{aligned}$$

and hence

$$\begin{aligned}\sigma^{-1}(\theta\sigma\theta^{-1}) &= \varphi^{-1}(i_6 i_5 i_4)(i_3 i_2 i_1)(i_1 i_3 i_4)(i_2 i_5 i_6)\varphi \\ &= (i_6 i_5 i_4)(i_3 i_2 i_1)(i_1 i_3 i_4)(i_2 i_5 i_6) \\ &= (i_1 i_2 i_4 i_3 i_6).\end{aligned} \tag{39}$$

But the 5-cycle (39) is a product of elements in H and hence belongs to H. Now part (iv) produces a contradiction.

Case 2 σ involves precisely one cycle of length 3 in its disjoint cycle decomposition:

$$\sigma = (i_1 i_2 i_3)\varphi,$$

and φ involves only 2-cycles so that $\varphi^2 = e$. But then

$$\begin{aligned}\sigma^2 &= (i_1 i_2 i_3)^2\varphi^2 = (i_1 i_3 i_2) \\ &\in H\end{aligned}$$

and H contains a 3-cycle. Part (iii) now produces a contradiction.

Case 3 No 3-cycles appear in the decomposition of σ, i.e., the disjoint cycle factorization of σ involves only 2-cycles. Then

$$\sigma = (i_1 i_2)(i_3 i_4)\varphi$$

since σ is not a single transposition (otherwise σ would be odd). Let $\theta = (i_2 i_3 i_4) \in A_n$ and compute easily that

$$\begin{aligned}\theta\sigma\theta^{-1} &= \theta(i_1 i_2)\theta^{-1}\theta(i_3 i_4)\theta^{-1}\varphi \\ &= (i_1 i_3)(i_4 i_2)\varphi,\end{aligned}$$

and hence

$$\begin{aligned}\sigma^{-1}(\theta\sigma\theta^{-1}) &= \varphi^{-1}(i_3 i_4)(i_1 i_2)(i_1 i_3)(i_4 i_2)\varphi \\ &= (i_1 i_4)(i_2 i_3).\end{aligned} \tag{40}$$

The permutation (40) is in H because it is a product of elements in H. Now let i_5 be distinct from i_1, i_2, i_3, i_4 and let $\gamma = (i_1 i_4 i_5) \in A_n$. Then

$$\gamma(i_1i_4)(i_2i_3)\gamma^{-1} = (i_4i_5)(i_2i_3). \tag{41}$$

Finally, multiplying (40) and (41) together, we have

$$(i_1i_4)(i_2i_3)(i_4i_5)(i_2i_3) = (i_1i_4i_5)$$

and again H contains a 3-cycle, contradicting (iii).

(vi) We will prove that K contains every 3-cycle, and hence $A_n < K$ by (ii). For let $\nu: H \to H/K$ be the canonical homomorphism. Let $\sigma = (i_1i_2i_3)$, $\varphi = (i_3i_4i_5)$, where i_1, i_2, i_3, i_4, i_5 are distinct but otherwise arbitrary integers in $[1,n]$; then $\sigma,\varphi \in A_n < H$. Since H/K is abelian, we have

$$\begin{aligned}\nu(\sigma^{-1}\varphi^{-1}\sigma\varphi) &= \nu(\sigma)^{-1}\nu(\sigma)\nu(\varphi)^{-1}\nu(\varphi)\\ &= K,\end{aligned}$$

and so

$$\sigma^{-1}\varphi^{-1}\sigma\varphi \in K.$$

But

$$\begin{aligned}\sigma^{-1}\varphi^{-1}\sigma\varphi &= (i_3i_2i_1)(\varphi^{-1}(i_1)\varphi^{-1}(i_2)\varphi^{-1}(i_3))\\ &= (i_3i_2i_1)(i_1i_2i_5)\\ &= (i_2i_5i_3).\end{aligned} \tag{42}$$

In other words, given any three distinct elements i_2, i_5, $i_3 \in [1, n]$, the 3-cycle (42) is in K. This completes the proof of the theorem. ∎

Example 4 In this example we determine the normal subgroups of A_n for $n \leq 4$. Obviously if $n \leq 2$, $A_n = \{e\}$. If $n = 3$, $|A_3| = 3$ and hence by Lagrange's Theorem [Theorem 3.7 (ii)], A_3 possesses no proper subgroups. There remains the case $n = 4$. It is routine to check that

$$K = \{e, (12)(34), (23)(14), (13)(24)\}$$

is a normal subgroup of A (see Exercise 25). Assume $H \lhd A_4$, and suppose $\sigma \in H$, $\sigma \neq e$. Since any cycle of even length is an odd permutation, it follows that σ must have either

$$[1^1, 2^0, 3^1, 4^0] \tag{43}$$

or

$$[1^0, 2^2, 3^0, 4^0] \tag{44}$$

as its cycle structure, i.e., σ is either a 3-cycle or a product of two 2-cycles. If σ is a 3-cycle, then Theorem 3.10 (iii) implies that $H = A_4$. Otherwise σ has cycle structure (44), i.e.,

$$\sigma = (i_1i_2)(i_3i_4),$$

and hence $\sigma \in K$. We conclude that if H is a proper normal subgroup of A_4, then $H < K$; in fact, by forming $\varphi\sigma\varphi^{-1} \in H$ for $\varphi = (i_1\, i_2\, i_3)$ and $\varphi = (i_1i_2i_4)$, we see that $H = K$. Thus the normal subgroups of A_4 are $\{e\}$, K, and A_4. If we let $a = (12)(34)$, $b = (23)(14)$, and $c = (13)(24)$, then the following table for K obtains:

	e	a	b	c
e	e	a	b	c
a	a	e	c	b
b	b	c	e	a
c	c	b	a	e

(45)

The abelian group K is called the *Klein four group.*

Exercises 1.3

1. Let $X = \{x_1, \ldots, x_n\}$. For each $f \in X^X$ define an $n \times n$ matrix $A(f)$ by
$$A(f)_{ij} = \delta_{f(x_j), x_i},$$
i.e., the (i,j)th entry of $A(f)$ is
$$\begin{cases} 1 & \text{if } f(x_j) = x_i \\ 0 & \text{if } f(x_j) \neq x_i \end{cases}.$$
Show that the mapping $A: X^X \to M_n(Z)$ is a monomorphism into the monoid of $n \times n$ matrices over Z (with matrix multiplication as the operation). *Hint*: $(A(f)A(g))_{ij} = \sum_{k=1}^n A(f)_{ik}A(g)_{kj} = \sum_{k=1}^n \delta_{f(x_k), x_i} \delta_{g(x_j), x_k}$. Now this last sum is 0 unless there is a k for which both $f(x_k) = x_j$ and $g(x_j) = x_k$, i.e., $f(g(x_j)) = x_i$. But there is at most one such k since $g(x_j)$ determines the x_k. Thus $(A(f)A(g))_{ij} = \delta_{(f \cdot g)(x_j), x_i} = A(fg)_{ij}$. To see that A is injective is easy, since $A(f)$ completely specifies f. The matrix $A(f)$ is called the *incidence matrix* for f.
2. Prove the first two assertions in Theorem 3.2.
3. Show that if $h: X \to Y$ is a homomorphism of monoids, it is not necessarily the case that $h(e)$ is the identity element of Y (where e is the identity element of X). *Hint*: Let $X = Y = (Z, \{\cdot\})$, the multiplicative monoid of integers. Define $h(x) = 0$ for all x.
4. Show that R, as defined in (6), is an equivalence relation in X.
5. In S_3 let $\sigma = (1\ \ 2\ \ 3)$ and let $G = \{e, \sigma, \sigma^2\}$. What are the orbits of G in $X = [1,3]$?
6. If $\sigma = (1\ \ 2)(3\ \ 4\ \ 5)$ in S_5, list the nontrivial orbits of $[\sigma]$.
7. Show that if $\omega = (i_1 \cdots i_r)$ and $\mu = (j_1 \cdots j_k)$ are in S_5 $(r, k > 1)$, then ω and μ are disjoint iff $\{i_1, \ldots, i_r\} \cap \{j_1, \ldots, j_k\} = \varnothing$. *Hint*: We need to show that the only nontrivial $[\omega]$-orbit is $\{i_1, \ldots, i_r\}$. But obviously, by definition, ω holds every $t \notin \{i_1, \ldots, i_r\}$ fixed, i.e., $\omega(t) = t$. Also $\omega(i_1) = i_2$, $\omega(i_2) = i_3, \ldots, \omega(i_{r-1}) = i_r$, $\omega(i_r) = i_1$, or equivalently $\omega(i_1) = i_2$, $\omega^2(i_1) = i_3, \ldots, \omega^{r-1}(i_1) = i_r$, $\omega^r(i_1) = i_1$. Hence $\{i_1, \ldots, i_r\}$ is the only nontrivial $[\omega]$-orbit.
8. Confirm formulas (13), (14), and (15).
9. Confirm formula (16).
10. Show that A_n is a subgroup of S_n.

11. List the elements of A_3 and of A_4.

12. Let $\sigma = \begin{pmatrix} 1 & 2 & 3 & 4 & \cdots & n \\ \sigma(1) & \sigma(2) & \sigma(3) & \sigma(4) & \cdots & \sigma(n) \end{pmatrix} \in S_n$. An *inversion* occurs in σ if some integer in the second row is preceded by a larger integer, i.e., $\sigma(i) > \sigma(j)$ for some $i < j$. Define $\iota(\sigma)$ to be the total number of such inversions, e.g.,

$$\iota\begin{pmatrix} 1 & 2 & 3 & 4 & 5 & 6 \\ 4 & 5 & 1 & 2 & 3 & 6 \end{pmatrix} = 2 + 2 + 2$$
$$= 6$$

because each of 1, 2, and 3 is preceded by two larger integers. Prove that

$$\varepsilon(\sigma) = (-1)^{\iota(\sigma)}.$$

Hint: Let a be the first integer in the second row of σ which is preceded by a larger integer (if no such a exists, then $\sigma = e$, $\iota(\sigma) = 0$, and we are done). We assert that the closest such larger integer b must immediately precede a; otherwise the second row of σ looks like

$$(\cdots b \cdots x \cdots a \cdots).$$

Now x is closer to a than b is, and so $x < a$. Also b cannot be less than x since $b > a$. Hence $b > x$. But then x precedes a and has a larger integer, namely b, preceding it. This contradicts the choice of a, and our assertion is established. We compute the product

$$(ab)\sigma = (ab)\begin{pmatrix} 1 & 2 & \cdots & i-1 & i & i+1 & \cdots & n \\ x_1 & x_2 & \cdots & x_{i-1} & b & a & y_{i+2} \cdots & y_n \end{pmatrix}$$
$$= \begin{pmatrix} 1 & 2 & \cdots & i-1 & i & i+1 & \cdots & n \\ x_1 & x_2 & \cdots & x_{i-1} & a & b & y_{i+2} \cdots & y_n \end{pmatrix}$$

Hence the number of inversions in $(ab)\sigma$ is one less than the number of inversions in σ, i.e., $\iota((ab)\sigma) = \iota(\sigma) - 1$. Repeating the argument with $(ab)\sigma$, etc., we see that if k inversions occur in σ, then by premultiplying σ by k transpositions, we arrive at a permutation with no inversions, i.e., the identity. We have $0 = \iota(e) = \iota(\tau_k \cdots \tau_1\sigma) = \iota(\sigma) - k$, and so $\iota(\sigma) = k$. Finally, $1 = \varepsilon(e) = \varepsilon(\tau_k \cdots \tau_1\sigma) = \varepsilon(\tau_k \cdots \tau_1)\varepsilon(\sigma) = (-1)^k\varepsilon(\sigma)$ implies $\varepsilon(\sigma) = (-1)^k = (-1)^{\iota(\sigma)}$.)

13. Show that S_n is generated by the $n - 1$ transpositions (1 2), (2 3), (3 4), . . ., $(n - 1\ n)$, i.e., any $\sigma \in S_n$ is a product of these transpositions. *Hint*: Using the notation of the previous hint, we compute that

$$\sigma(i\ i+1) = \begin{pmatrix} 1 & 2 & \cdots & i-1 & i & i+1 & i+2 & \cdots & n \\ x_1 & x_2 & \cdots & x_{i-1} & b & a & y_{i+1} & \cdots & y_n \end{pmatrix}(i\quad i+1)$$
$$= \begin{pmatrix} 1 & 2 & \cdots & i-1 & i & i+1 & i+2 & \cdots & n \\ x_1 & x_2 & \cdots & x_{i-1} & a & b & y_{i+1} & \cdots & y_n \end{pmatrix}.$$

Hence $\iota(\sigma(i\ i+1)) = \iota(\sigma) - 1$. In other words, if k inversions occur in σ, then postmultiplying σ by k of the required transpositions produces e.

14. Write $\sigma = \begin{pmatrix} 1 & 2 & 3 & 4 & 5 & 6 & 7 & 8 \\ 3 & 4 & 1 & 2 & 5 & 7 & 8 & 6 \end{pmatrix}$ as a product of transpositions of the form $(i \quad i+1)$. *Hint*: $\sigma = (6 \quad 7)(7 \quad 8)(2 \quad 3)(3 \quad 4)(1 \quad 2)(2 \quad 3)$.

15. Verify that the relation R in (21) is an equivalence relation on G.

16. Confirm (24), (25), (26) and (27). *Hint*: (24) and (25) are trivial. One direction of (26) is trivial. Conversely, if $H^2 = H$, H is closed; if $H = H^{-1}$, then H contains inverses, and $e \in H$ follows; (27) follows by application of (26) to HK.

17. Let $H < G$. Show that the mapping $\nu(aH) = Ha^{-1} = H^{-1}a^{-1}$ is bijective with domain the set of left H-cosets and codomain the set of right H-cosets.

18. Prove: If X is a group and $x^2 = x \in X$, then $x = e$, the neutral element in X.

19. Verify (33). *Hint*: If $\nu(a) = H$, then $aH = H$, and so $a \in H$.

20. Complete the proof of (37).

21. Show that if $G < S_n$, then either every $\sigma \in G$ is even or G contains precisely the same number of even and odd permutations. *Hint*: $s = \Sigma_{\sigma \in G}\varepsilon(\sigma) = \Sigma_{\sigma \in G}\, \varepsilon(\sigma\theta) = \varepsilon(\theta)\, \Sigma_{\sigma \in G}\varepsilon(\sigma) = \varepsilon(\theta)s$, where $\theta \in G$ is fixed. Thus $s = \varepsilon(\theta)s$, so that summing on θ yields $|G|s = s^2$. Hence $s = |G|$ or $s = 0$. This means that either $\varepsilon(\sigma) = 1$ for all $\sigma \in G$ or there are as many -1's as 1's in the sum s.

22. Show that $|A_n| = n!/2$ for $n \geq 2$. *Hint*: If $\tau \in S_n$ is a transposition, the mapping $\sigma \to \tau\sigma$ of A_n into S_n is an injection.

23. Show that if $\sigma = (i_1 \quad i_2 \cdots i_k) \in S_n$ and $\theta \in S_n$, then $\theta\sigma\theta^{-1} = (\theta(i_1)\theta(i_2) \cdots \theta(i_k))$.

24. Assume $H \lhd G$ and $\nu: G \to G/H$ is the canonical homomorphism. Prove that $\ker \nu = H$.

25. Verify that $K = \{e, (1\ 2)(3\ 4), (2\ 3)(1\ 4), (1\ 3)(2\ 4)\}$ is a normal subgroup of A_4.

26. Let G be the multiplicative 4-element group of complex numbers $\{1, -1, i, -i\}$. Show that G is not isomorphic to the Klein four group (45). *Hint*: In (45) the square of every element is the identity.

27. Let F be the class of all groups, and let S be the class of all group homomorphisms. Show that S is a regular semigroupoid with function composition as the operation. For each $A \in F$ define

$$\nu(A) = \iota_A.$$

Prove that $\mathscr{C} = (F,S,\nu)$ is a category, called the *category of groups* (see Section 1.2, Exercise 32). *Hint*: Suppose $f{:}A \to B$, $g{:}C \to D$, $h{:}E \to L$ are group homomorphisms. If $f(gh)$ is defined, then $L \subset C$ and $D \subset A$, so that obviously $(fg)h$ is defined and $f(gh) = (fg)h$. The composition of group homomorphisms is of course a group homomorphism when it is defined. Similarly, if hg and gf are defined, then $(hg)f$ is defined. The set $I(S)$ consists of the identity homomorphisms ι_A, $A \in F$.

Glossary 1.3

2

Groups

2.1 Isomorphism Theorems

The purpose of this chapter is to study in some detail the general structure of groups. As we mentioned in Section 1.3, groups play a fundamental role both within mathematics itself and in the applications of mathematics to other fields.

Some of the basic theorems in group theory have to do with classifying groups to within isomorphism. Isomorphic groups are, of course, identical except perhaps in the names given the elements. We begin with an elementary example in which we determine all additive subgroups of the additive group of integers.

Example 1 Let $\{0\} \neq H < (Z, \{+\})$. Then H must consist of all multiples of some unique positive integer d. Moreover, the function

$$f\colon Z \to H$$

defined by

$$f(n) = nd$$

is a group isomorphism. To see this, observe that since H contains some integer other than 0, it must contain a positive integer and hence a least positive integer d. Then of course

$$\text{im } f \subset H.$$

If $m \in H$, divide m by d:

$$m = qd + r, \qquad 0 \leq r < d,$$

or

$$r = m - qd.$$

Now $m \in H, qd \in H$ imply $r \in H$. But then the minimality of d implies that $r = 0$, and so $m = qd$. Thus H consists precisely of multiples of d and

$$\text{im } f = H.$$

The uniqueness of d and the fact that f is an isomorphism are easy verifications left to the reader.

As an instance of Example 1, let

$$H = \{z \mid z = 21x + 35y,\ x,y \in Z\}.$$

Clearly $\{0\} \neq H < (Z, \{+\})$. According to Example 1, H consists of all integral multiples of the least positive integer $d \in H$. Observe that $7 \in H$,

$$7 = 21 \cdot 2 - 35 \cdot 1,$$

and every integer in H is a multiple of 7. Thus H consists of all multiples of 7. We may use the notation for the cyclic group generated by 7 that was introduced in formula (7), Section 1.3:

$$H = [7]. \tag{1}$$

The important thing to remember in using (1) is that H is an *additive* group and [7] consists of all integral multiples of 7. The notation [7] could conceivably be confused with the notation for a residue class (see Example 2, Section 1.2). When both a cyclic group and a residue class modulo some positive integer p occur in the same argument, we shall use $[n]_p$ to denote the residue class.

The following very useful result is called the *group homomorphism theorem*. Part (iii) is known as the *first isomorphism theorem for groups.*

Theorem 1.1 *Let* $\varphi: G \to S$ *be a group homomorphism, and let* $H = \ker \varphi$. *Then:*

(i) $H \lhd G$ and $\varphi(G) < S$.

(ii) $\varphi(aH) = \{\varphi(a)\}$ *for all* $a \in G$.

(iii) If $\bar{\varphi}: G/H \to \varphi(G)$ *is defined by*

$$\bar{\varphi}(aH) = \varphi(a), \qquad a \in G,$$

then $\bar{\varphi}$ *is an isomorphism.*

(iv) *If* φ *is surjective and* $K \lhd G$, *then* $\varphi(K) \lhd S$.

Proof: Part (i) has already been proved [see Theorem 3.2 and formula (37) of Section 1.3]. Part (ii) follows trivially from the definition of H and the fact that φ is a homomorphism. By Theorem 3.9 of Section 1.3, the only part of (iii) requiring proof is that $\bar{\varphi}$ is injective. We have

$$\varphi(a) = \bar{\varphi}(aH) = \bar{\varphi}(bH) = \varphi(b)$$

iff

$$\varphi(b^{-1}a) = \varphi(e)$$

iff $$b^{-1}a \in \ker \varphi = H$$
iff $$a \in bH$$
iff $$aH = bH.$$

To prove (iv), observe that if φ is surjective, then any $s \in S$ is of the form $\varphi(g)$, $g \in G$. Hence if $\varphi(k) \in \varphi(K)$, we have

$$\begin{aligned} s^{-1}\varphi(k)s &= \varphi(g)^{-1}\varphi(k)\varphi(g) = \varphi(g^{-1}kg) \\ &\in \varphi(K), \end{aligned}$$

and so $\varphi(K)$ is normal. ▮

The homomorphism $\bar{\varphi}$ of Theorem 1.1(iii) is said to be *induced by* φ.

A *cyclic group* S (denoted multiplicatively here) just consists of all integral powers of some element a:

$$S = [a]. \tag{2}$$

If a is an element of an arbitrary group G, then the subgroup (2) is called the *cyclic subgroup generated by* a. If $|\,[a]\,| = k < \infty$, then the element a is said to be of *finite order* k; if $|\,[a]\,| = \infty$, then a is an element of *infinite order*.

As an easy application of Theorem 1.1 we can classify all cyclic groups.

Corollary 1 *If $S = [a]$, then either S is isomorphic to $(Z, \{+\})$ or S is isomorphic to $(Z_d, \{+\})$ for some positive integer d.*

Proof: Define

$$\varphi\colon (Z, \{+\}) \to S$$

by the formula

$$\varphi(n) = a^n, \qquad n \in Z.$$

Clearly φ is an epimorphism. There are two possibilities for φ. If φ is injective, then $(Z, \{+\})$ and S are isomorphic. Otherwise there exist $m, n \in Z, m \neq n$, such that

$$\varphi(n) = \varphi(m).$$

Then if $e = a^0$ is the identity in S, we have

$$a^{n-m} = e$$

so that $$\{0\} \neq \ker \varphi.$$

Then by Example 1, $H = \ker \varphi$ consists of all integral multiples of some positive integer d. By definition,

$$(Z, \{+\})/H = (Z_d, \{+\}).$$

But Theorem 1.1(iii) implies that the induced homomorphism

$$\bar{\varphi}\colon (Z_d, \{+\}) \to S$$

is an isomorphism. ▮

Corollary 1 is a simple example of the kind of result one hopes to obtain in investigating the classification of groups. It gives us complete information in the sense that we know to within isomorphism that any cyclic group is either the additive group of integers or the additive group of integers modulo some positive integer d.

Example 2 Let $\xi \in \mathbf{C}$ be a complex number of modulus 1. According to Corollary 1, the multiplicative group $[\xi]$ is isomorphic either to $(Z, \{+\})$ or to $(Z_d, \{+\})$ for some positive integer d. Write

$$\xi = e^{i\theta}, \qquad \theta \in \mathbf{R}.$$

Then $\xi^n = 1$ for some positive integer n iff $n\theta$ is an integral multiple of 2π. Thus ξ is an element of finite order and $[\xi]$ is isomorphic to $(Z_d, \{+\})$ for some d iff θ and π are commensurable, i.e., $\theta/\pi \in Q$.

It is convenient to have a notation indicating that two groups are isomorphic without actually exhibiting the isomorphism. Thus

$$S \cong T$$

will mean that S and T are isomorphic groups. It is simple to confirm that $\cong$ is an equivalence relation in the category of groups, i.e.,

$$S \cong S,$$
$$S \cong T \quad \text{iff} \quad T \cong S,$$
$$S \cong T \text{ and } T \cong G \text{ imply } S \cong G$$

for arbitrary groups S, T, and G.

Theorem 1.2 *Suppose $K < H < G$ and $H \lhd G$, $K \lhd G$. Then*

$$G/H \cong \frac{G/K}{H/K}. \tag{3}$$

Proof: Clearly H/K is defined, since $K \lhd G$ certainly implies $K \lhd H$. Consider the mapping diagram

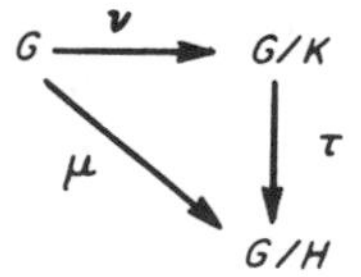

(4)

where ν and μ are the canonical epimorphisms. Then ker $\mu = H$ and $K < H$ so that by the fundamental mapping result, Theorem 3.9, Section 1.3, we obtain a homomorphism τ to complete the diagram (4), i.e.,

$$\tau\nu = \mu. \tag{5}$$

Now $gK = \nu(g) \in \ker \tau$ iff $\tau(\nu(g)) = \mu(g) = H$ iff $g \in \ker \mu$ iff $g \in H$. Thus

$$\ker \tau = \{gK \mid g \in H\}$$
$$= H/K.$$

Thus τ: $G/K \to G/H$ is an epimorphism with $\ker \tau = H/K$. It follows from Theorem 1.1(iii) that

$$\frac{G/K}{H/K} \cong G/H. \blacksquare$$

Corollary 2 *Suppose $K < H < G$ and $H \lhd G$, $K \lhd G$. If G/K is abelian, then G/H is abelian.*

Proof: By Theorem 1.2 we have

$$G/H \cong \frac{G/K}{H/K},$$

and a factor group of an abelian group is abelian. $\blacksquare$

The next two results are called the *third* and *second isomorphism theorems for groups,* respectively (we present them in reverse order). The first of these tells us about the structure of subgroups of a factor group.

Theorem 1.3 (*Third Isomorphism Theorem*) *Let $K \lhd G$, and let $\bar{H} < G/K$.*

(i) *There exists a unique subgroup H such that $K \lhd H < G$ and*

$$\bar{H} = H/K.$$

(ii) *If the group $\bar{H}$ satisfies $\bar{H} \lhd G/K$, then in (i) $H \lhd G$ and*

$$G/H \cong \frac{G/K}{H/K}.$$

Proof: (i) Let H be the union of all the K-cosets in $\bar{H}$, i.e.,

$$H = \bigcup_{gK \in \bar{H}} gK.$$

If h_1 and h_2 are in H, then h_1k and h_2k are in $\bar{H}$, so $(h_1K)(h_2K) = h_1h_2K \in \bar{H}$ and thus $h_1h_2 \in H$. Similarly, $h_1^{-1} \in H$, whence $H < G$. It is obvious that $K < H$. Since $K \lhd G$, K is automatically a normal subgroup of H as well. By definition, $\bar{H} = H/K$. To confirm uniqueness, suppose $H_1/K = H/K = \bar{H}$. Then each of the following statements implies its successor:

$$h_1 \in H_1,$$
$$h_1K \in H/K,$$
$$h_1K = hK \qquad \text{for some } h \in H,$$
$$h^{-1}h_1 \in K \subset H,$$
$$h_1 \in hH = H.$$

Thus $H_1 \subset H$. Similarly $H \subset H_1$, and so $H = H_1$.

(ii) Suppose $\bar{H} \lhd G/K$ and let $g \in G$. Then $(gK)\bar{H} = \bar{H}(gK)$, i.e.,

$$(gK)H/K = H/K(gK).$$

This says that

$$\{(gK)(hK) \mid h \in H\} = \{(hK)(gK) \mid h \in H\}$$

or

$$\{ghK \mid h \in H\} = \{hgK \mid h \in H\}.$$

Thus if $h \in H$ and g is an arbitrary element of G, then

$$ghK = h_1 gK$$

for some $h_1 \in H$. Since $K \lhd G$, we have

$$gh \in h_1 gK = h_1 Kg. \tag{6}$$

But $K < H$ and hence $h_1 K \subset H$. Now (6) implies that

$$gh \in Hg$$

or

$$ghg^{-1} \in H. \tag{7}$$

The element $g \in G$ was arbitrary; so it follows from (7) that $H \lhd G$. Thus $K \lhd H \lhd G$ and by Theorem 1.2 we have

$$G/H \cong \frac{G/K}{H/K}. \blacksquare$$

As a simple example of Theorem 1.3 we observe that in $(Z_6, \{+\})$, the additive group of integers modulo 6, the residue classes

$$[0]_6, \quad [2]_6, \quad [4]_6$$

constitute a subgroup $\bar{H}$. Take $G = (Z\ \{+\})$, $K = [6]$ (the cyclic subgroup of G generated by 6) and apply Theorem 1.3(i) to conclude that $\bar{H} = H/K$, where

$$H = [0]_6 \cup [2]_6 \cup [4]_6.$$

Thus the group H consists of all integers n of the form

$$n = \begin{cases} 6k \\ 6k + 2, & k \in Z. \\ 6k + 4 \end{cases} \tag{8}$$

Observe that $K \lhd H$ (indeed, H is abelian). Also note that the factor group $(Z, \{+\})/H$ is simple to compute. For by Theorem 1.3(ii),

$$G/H \cong \frac{G/K}{H/K} = (Z_6, \{+\})/\bar{H}.$$

Now $|(Z_6, \{+\})| = 6$ and $|\bar{H}| = 3$; so by Lagrange's theorem (Theorem 3.7, Section 1.3), $|G/H| = 2$. But there is only one group of order 2 to within isomorphism (verify); hence

$$G/H \cong (Z_2, \{+\}).$$

The next result is the so-called *second isomorphism theorem for groups.* We are given a group G and $H < G$, $K \lhd G$.

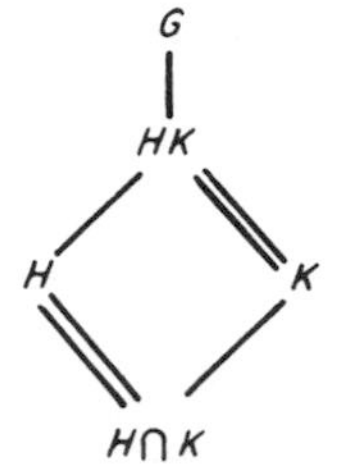

(9)

Reading from the bottom to the top of diagram (9), the single lines mean "is a subgroup of" and the double lines mean "is a normal subgroup of." The factor groups HK/K and $H/H \cap K$ are isomorphic:

$$HK/K \cong H/H \cap K.$$

More precisely we have the following theorem.

Theorem 1.4 (*Second Isomorphism Theorem*) *Assume* $H < G$, $K \lhd G$, *and let* $\nu\colon G \to G/K$ *be the canonical epimorphism. Then*

(i) $HK < G$, $K \lhd HK$, *and* $\nu(H) = HK/K$.

(ii) *If* $\mu = \nu | H$, *then* $\ker \mu = H \cap K$ *so that* $H \cap K \lhd H$.

(iii) *If* $\bar{\mu}\colon H/H \cap K \to G/K$ *is the homomorphism induced by* μ, *then* $\bar{\mu}$ *is injective and*

$$\bar{\mu}(H/H \cap K) = HK/K,$$

so that

$$H/H \cap K \cong HK/K.$$

Proof: (i) Since $K \lhd G$ it follows that $HK = KH$, and we can apply the assertion (27) in Section 1.3 to conclude that $HK < G$. Observe also that since $K \lhd G$, K automatically satisfies $K \lhd HK$. Moreover, if $h \in H$, then $\nu(h) = hK \in HK/K$; so

$$\nu(H) \subset HK/K.$$

On the other hand, $(hk)K = hK = \nu(h)$ for all $h \in H, k \in K$. Thus

$$\nu(H) = HK/K.$$

(ii) We have $\ker \mu \subset \ker \nu = K$. Since the domain of μ is H, it is clear that $\ker \mu = H \cap K$.

(iii) By Theorem 1.1(iii), $\ker \mu = H \cap K$ implies that

$$\bar{\mu}\colon H/H \cap K \to \bar{\mu}(H/H \cap K)$$

is an isomorphism. But $\bar{\mu}$ completes the diagram

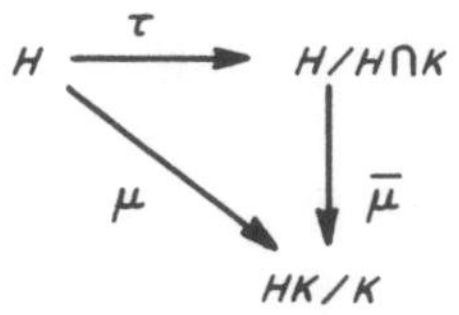

in which τ is the canonical epimorphism. Since $\mu(H) = \nu(H) = HK/K$ by (i) we have $\bar{\mu}\,(H/H \cap K) = HK/K$. Note that

$$\bar{\mu}(h(H \cap K)) = \mu(h) = hK \qquad \text{for } h \in H; \tag{10}$$

so the isomorphism connecting $H/H \cap K$ and HK/K is given by the correspondence

$$h(H \cap K) \leftrightarrow hK. \blacksquare$$

Let G be a group, written multiplicatively, and let $H_i < G$ for $i = 1, \ldots, r$. We say that G is the *internal direct product* of $H_1, \ldots, H_r$ if the following two conditions obtain:

(i) If $x_i \in H_i$ and $x_j \in H_j$, $i \neq j$, then $x_i x_j = x_j x_i$, i.e., elements from two different subgroups pairwise commute.

(ii) Each $x \in G$ has a unique expression $x = x_1 \cdots x_r$, where $x_i \in H_i$ for $i = 1, \ldots, r$.

Several notations are commonly used for the internal direct product:

$$G = H_1 H_2 \cdots H_r \qquad \text{(direct)}$$

or

$$G = H_1 \times H_2 \times \cdots \times H_r$$

or sometimes

$$G = \prod_{i=1}^{r}{}^{\cdot} H_i.$$

If the group G is written in additive notation, it is customary to call G the *direct sum* of $H_1, \ldots, H_r$ and write

$$G = H_1 + \cdots + H_r$$

or

$$G = H_1 \oplus \cdots \oplus H_r$$

or sometimes

$$G = \sum_{i=1}^{r}{}^{\cdot} H_i.$$

Example 3 In A_4 consider the permutations

$$\begin{aligned} &e, \\ a &= (1\quad 2)(3\quad 4), \\ b &= (2\quad 3)(1\quad 4), \\ c &= (1\quad 3)(2\quad 4). \end{aligned}$$

In Example 4 of Section 1.3 we computed that these four elements comprise a subgroup of A_4 called the Klein four group, denoted here by G.

	e	a	b	c
e	e	a	b	c
a	a	e	c	b
b	b	c	e	a
c	c	b	a	e

We have the subgroups $H_1 = \{e,a\}$ and $H_2 = \{e,b\}$ of G. The group G is abelian, so the elements in H_1 and H_2 pairwise commute. Also

$$\begin{aligned} H_1H_2 &= \{ee,eb,ae,ab\} = \{e,b,a,c\} \\ &= G, \end{aligned}$$

and it is obvious that each of e, a, b, c has a unique representation as a product of elements x_1x_2, where $x_1 \in H_1$ and $x_2 \in H_2$. Thus

$$G = H_1H_2 \qquad \text{(direct)}.$$

It is very simple to verify that if

$$G = H_1 \cdots H_r \qquad \text{(direct)},$$

then $H_i \triangleleft G$ for $i = 1, \ldots, r$. For suppose $x \in G$,

$$\begin{aligned} x &= x_1x_2 \cdots x_r, \\ x_i &\in H_i, \qquad i = 1, \ldots, r. \end{aligned}$$

Remembering that elements from different H_i pairwise commute and that $x_iH_i = H_i = H_ix_i$, we can immediately write

$$\begin{aligned} xH_i &= (x_1 \cdots x_{i-1})x_iH_i(x_{i+1} \cdots x_r) \\ &= (x_1 \cdots x_{i-1})H_ix_i(x_{i+1} \cdots x_r) \\ &= H_ix_1 \cdots x_r \\ &= H_ix, \qquad i = 1, \ldots, r. \end{aligned}$$

We also note that if $H_1 \triangleleft G$ and $H_2 \triangleleft G$ are such that $H_1 \cap H_2 = \{e\}$, then the elements of H_1 and H_2 pairwise commute. For if $h_i \in H_i$, $i = 1,2$, then

$$h_1^{-2}h_2^{-1}h_1h_2 = h_1^{-1}(h_2^{-1}h_1h_2) \tag{11}$$

$$= (h_1^{-1}h_2^{-1}h_1)h_2. \tag{12}$$

In (11), $h_2^{-1}h_1h_2 \in H_1$, and so the right side of (11) is in H_1. Similarly the right side of (12) is in H_2, and hence

$$h_1^{-1}h_2^{-1}h_1h_2 \in H_1 \cap H_2 = \{e\}$$

or

$$h_1^{-1}h_2^{-1}h_1h_2 = e$$

or
$$h_1h_2 = h_2h_1.$$

Let G_t, $t = 1, \ldots, m$, be groups (each with juxtaposition denoting the group operation). Define the group

$$G = G_1 \times \cdots \times G_m \qquad \text{(Cartesian product)}$$

with the operation

$$(h_1, \ldots, h_m)(k_1, \ldots, k_m) = (h_1k_1, \ldots, h_mk_m) \tag{13}$$

where $h_t, k_t \in G_t$, $t = 1, \ldots, m$. The group G is called the *external direct product* of the groups $G_1, \ldots, G_m$.

Theorem 1.5 (i) *Let G_t, $t = 1, \ldots, m$, be groups, and let $G = G_1 \times \cdots \times G_m$. Then there exist subgroups H_t of G, $t = 1, \ldots, m$, such that*

$$G = H_1 \cdots H_m \qquad (direct)$$

and

$$G_t \cong H_t, \qquad t = 1, \ldots, m.$$

(ii) *Let G be a group with subgroups H_t, $t = 1, \ldots, m$, such that*

$$G = H_1 \cdots H_m \qquad (direct).$$

Then $G \cong H_1 \times \cdots \times H_m$.

Proof: (i) For each $t = 1, \ldots, m$, let H_t be the totality of elements

$$(e, \ldots, e, h_t, e, \ldots, e), \qquad h_t \in G_t, \tag{14}$$

where h_t appears in the tth position in (14). We leave to the reader the verification that G is indeed the direct product of the subgroups H_t, $t = 1, \ldots, m$. Moreover, the obvious mapping

$$\pi_t \colon H_t \to G_t$$

defined by

$$\pi_t(e, \ldots, e, h_t, e, \ldots, e) = h_t, \qquad h_t \in G_t \tag{15}$$

is a group isomorphism.

(ii) Define $\varphi\colon G \to H_1 \times \cdots \times H_m$ by $\varphi(h_1 \cdots h_m) = (h_1, \ldots, h_m)$. Then φ is easily verified to be an isomorphism. ∎

Recall that a simple group is a group with no proper normal subgroups. A group is called *fully reducible* if it is a direct product of nontrivial simple subgroups. For example, consider the additive group $(Z_2 \times Z_2, \{+\}) = G$. Let

$$H = \{(0,0),(1,0)\},$$
$$K = \{(0,0),(0,1)\}.$$

Then it is clear that H and K are nontrivial simple subgroups of G such that

$$G = H \dotplus K.$$

Thus G is fully reducible.

We can easily establish a criterion for a product of groups to be direct. Recall from Section 1.3 that if A is a complex in a group G, then the subgroup of G generated by A is denoted by $[A]$. If $A_1, \ldots, A_m$ are arbitrary complexes in G, we write

$$[A_1, \ldots, A_m] = [\bigcup_{t=1}^{m} A_t]. \tag{16}$$

Theorem 1.6 *Suppose that G is a group, $H_i \lhd G$ for $i = 1, \ldots, m$, and*

$$G = [H_1, \ldots, H_m]. \tag{17}$$

If
$$H_t \cap [H_1, \ldots, H_{t-1}] = \{e\}, \qquad t = 2, \ldots, m, \tag{18}$$
then
$$G = H_1 \cdots H_m \qquad (\textit{direct}). \tag{19}$$

[Note that the converse of this theorem is also true: Let G be a group, $H_t < G$, $i = 1, \ldots, m$, and suppose that $G = H_1 \cdots H_m$ (direct). Then $G = [H_1, \ldots, H_m]$ and

$$H_t \cap [H_1, \ldots, H_{t-1}] = \{e\}, \qquad t = 2, \ldots, m.]$$

Proof: By (18), we have $H_i \cap H_j = \{e\}$ for $i \neq j$, so the remarks preceding Theorem 1.5 [see formulas (11) and (12)] imply that the elements of H_i and H_j pairwise commute. By (16), any element of G is a product of elements from the H_i, and since these pairwise commute, any $x \in G$ has the form

$$x = x_1 \cdots x_m, \qquad x_i \in H_i, i = 1, \ldots, m.$$

Suppose a second such representation of x is available:

$$x = y_1 \cdots y_m, \qquad y_i \in H_i, i = 1, \ldots, m.$$

Then unless $x_i = y_i$, $i = 1, \ldots, m$, there is a largest integer k such that $x_k \neq y_k$ but $x_j = y_j$, $j = k+1, \ldots, m$ (possibly $k = m$). Using the commutativity of elements from distinct H_i, we have

$$\begin{aligned} e = x^{-1}x &= x_1^{-1} \cdots x_m^{-1} y_1 \cdots y_m \\ &= (x_1^{-1}y_1) \cdots (x_k^{-1}y_k), \end{aligned}$$

or

$$y_k^{-1}x_k = (x_1^{-1}y_1) \cdots (x_{k-1}^{-1}y_{k-1}). \tag{20}$$

The left side of (20) is in H_k and the right side is in $[H_1, \ldots, H_{k-1}]$. By (18), if $k \geq 2$, we have

$$y_k^{-1}x_k = e$$

or

$$x_k = y_k,$$

a contradiction. Thus $k = 1$, i.e., $x_j = y_j$, $j = 2, \ldots, m$. But then again

$$x_1 \cdots x_m = y_1 \cdots y_m$$

implies $x_1 = y_1$. ∎

Theorem 1.7 *Let G be a fully reducible group, say*

$$G = H_1 \cdots H_m \qquad (direct)$$

where each H_i is a simple proper normal subgroup of G. Let H be a proper normal subgroup of G. Then there exist $H_{k_1}, \ldots, H_{k_r}$, an appropriate selection of $H_1, \ldots, H_m$, such that

$$G = HH_{k_1} \cdots H_{k_r} \qquad (direct).$$

Proof: Consider the product

$$G = HG = HH_1H_2 \cdots H_m. \tag{21}$$

Since H is normal in G, it follows that

$$H_1 \cap H \lhd H_1$$

and the simplicity of H_1 implies that either

$$H_1 \cap H = \{e\} \tag{22}$$

or

$$H_1 \cap H = H_1. \tag{23}$$

If (23) holds, then $H_1 \subset H$ so that $HH_1 = H$ and $G = HH_2 \cdots H_m$. In this case we simply start the argument over again.

Suppose that (22) holds and consider the intersection

$$H_2 = HH_1. \tag{24}$$

Clearly $HH_1 \lhd G$; the product of normal subgroups is always a subgroup (why?) and in fact is normal in G:

$$g^{-1}HH_1g = g^{-1}Hgg^{-1}H_1g$$

and $g^{-1}Hg = H$, $g^{-1}H_1g = H_1$. Then the intersection $H_2 \cap HH_1$ in (24) is normal in H_2. It follows from the simplicity of H_2 that either

$$H_2 \cap HH_1 = \{e\} \tag{25}$$

or

$$H_2 \cap HH_1 = H_2. \tag{26}$$

If (26) were to hold, then $H_2 \subset HH_1$, $HH_1H_2 = HH_1$, and

$$G = HH_1H_2H_3 \cdots H_m = HH_1H_3 \cdots H_m. \tag{27}$$

We again can start the argument with the decomposition in (27), but with the additional information that (22) holds. So suppose (25) holds. Again consider an intersection

$$H_3 \cap HH_1H_2. \tag{28}$$

The subgroup HH_1H_2 is a product of normal subgroups in G, so $HH_1H_2 \lhd G$ as before, and $H_3 \cap HH_1H_2$ is normal in H_3. The simplicity of H_3 then implies that either

$$H_3 \cap HH_1H_2 = \{e\} \tag{29}$$

or

$$H_3 \cap HH_1H_2 = H_3. \tag{30}$$

If (30) were to hold, then $H_3 \subset HH_1H_2$, $HH_1H_2H_3 = HH_1H_2$, and

$$\begin{aligned} G &= HH_1H_2H_3 \cdots H_m \\ &= HH_1H_2H_4 \cdots H_m. \end{aligned} \tag{31}$$

We once again simply start the argument with the decomposition (31) with the additional information that (22) and (25) hold.

Clearly, a repetition of this argument at most m times produces a selection of some r of the factors $H_1, \ldots, H_m$, say $H_{k_1}, \ldots, H_{k_r}$, such that

$$G = HH_{k_1} \cdots H_{k_r}$$

and

$$\begin{aligned} H_{k_1} \cap H &= \{e\}, \\ H_{k_2} \cap HH_{k_1} &= \{e\}, \\ &\cdots\cdots \\ H_{k_r} \cap HH_{k_1} \cdots H_{k_{r-1}} &= \{e\}. \end{aligned} \tag{32}$$

But the product of normal subgroups is always a (normal) subgroup as was remarked above, and thus

$$\begin{aligned} HH_{k_1} &= [H, H_{k_1}], \\ &\cdots\cdots \\ HH_{k_1} \cdots H_{k_{r-1}} &= [H, H_{k_1}, \ldots, H_{k_{r-1}}]. \end{aligned} \tag{33}$$

Combining the equations (32) and (33) permits us to apply Theorem 1.6 to conclude the desired result:

$$G = HH_{k_1} \cdots H_{k_r} \qquad \text{(direct).} \ \blacksquare$$

Example 4 Let G be the additive group $(Z_2 \times Z_2 \times Z_2, \{+\})$ and write

$$G = H_1 + H_2 + H_3$$

where

$$H_i = \{x = (x_1, x_2, x_3) \in Z_2 \times Z_2 \times Z_2 \mid x_j = 0, j \neq i\}, \qquad i = 1,2,3.$$

Define

$$H = \{x = (x_1, x_2, x_3) \in Z_2 \times Z_2 \times Z_2 \mid x_1 + x_2 + x_3 = 0\}.$$

Each of the subgroups of H, H_1, H_2, and H_3 is normal because G is abelian. We mimic the proof of Theorem 1.7.

Form the sum

$$G = H + H_1 + H_2 + H_3 \tag{34}$$

and consider the intersection

$$H \cap H_1. \tag{35}$$

It is clear from the definition of H and H_1 that the intersection (35) is the identity element in G, namely, $0 = (0,0,0)$. Next consider the intersection

$$H_2 \cap (H + H_1). \tag{36}$$

It is simple to confirm that $H + H_1 = G$, so (36) is in fact equal to H_2 and thus H_2 can be discarded in (34). We have

$$G = H + H_1 + H_3 \tag{37}$$

and

$$H \cap H_1 = \{0\}.$$

Finally, consider

$$H_3 \cap (H + H_1). \tag{38}$$

Again, since $H + H_1 = G$, (38) is equal to H_3 and we may discard H_3 in (37) to obtain

$$G = H \dot{+} H_1.$$

Observe that

$$G = H \dot{+} H_2$$

and

$$G = H \dot{+} H_3$$

as well, so there is nothing unique about the selection $H_{k_1}, \ldots, H_{k_r}$ in Theorem 1.7.

There is an important and interesting relationship between direct products and monotone chains of subgroups.

Theorem 1.8 *Assume that*

$$G = H_1 \cdots H_r \qquad (direct)$$

and set

$$G_i = H_{i+1} \cdots H_r, \qquad i = 0, \ldots, r-1.$$

Then

(i) $G_i < G$ *and*

$$G_i = H_{i+1} \cdots H_r \qquad (direct). \tag{39}$$

(ii) $G_i \lhd G$ *and hence setting* $G_r = \{e\}$, *we have*

$$\{e\} = G_r \lhd G_{r-1} \lhd G_{r-2} \lhd \cdots \lhd G_1 \lhd G_0 = G. \tag{40}$$

(iii) $G_{i-1}/G_i \cong H_i, \qquad i = 1, \ldots, r.$ (41)

Proof: (i) Using the commutativity of the factors in any direct product, we compute that

$$\begin{aligned} G_i^2 &= H_{i+1} \cdots H_r H_{i+1} \cdots H_r \\ &= H_{i+1}^2 \cdots H_r^2 = H_{i+1} \cdots H_r \\ &= G_i, \end{aligned}$$

and
$$\begin{aligned} G_i^{-1} &= (H_{i+1} \cdots H_r)^{-1} = H_r^{-1} \cdots H_{i+1}^{-1} \\ &= H_r \cdots H_{i+1} = H_{i+1} \cdots H_r \\ &= G_i. \end{aligned}$$

Hence $G_i < G$. Also, any element $x_{i+1} \cdots x_r$ in G_i is uniquely expressible as such a product because the same is true of the element $x_1 \cdots x_i x_{i+1} \cdots x_r$ in G (take $x_1 = \cdots = x_i = e$). This proves (39).

(ii) If $x \in G$, $x = x_1 \cdots x_r$, then

$$\begin{aligned} xG_i &= x_1 \cdots x_r H_{i+1} \cdots H_r \\ &= H_{i+1} \cdots H_r x_1 \cdots x_r \\ &= G_i x. \end{aligned}$$

Thus $G_i \lhd G$, and of course (40) follows.

(iii) Define $\varphi\colon G_{i-1} \to H_i$ by $\varphi(x) = x_i$, where x_i is the unique element in H_i for which $x = x_i x_{i+1} \cdots x_r$. If $y = y_i y_{i+1} \cdots y_r \in G_{i-1}$, we have

$$\begin{aligned} \varphi(xy) &= \varphi(x_i x_{i+1} \cdots x_r y_i y_{i+1} \cdots y_r) \\ &= \varphi(x_{i+1} \cdots x_r (x_i y_i) y_{i+1} \cdots y_r) \\ &= \varphi(x_{i+2} \cdots x_r (x_i y_i)(x_{i+1} y_{i+1}) \cdots y_r) \\ &= \cdots \\ &= \varphi((x_i y_i)(x_{i+1} y_{i+1}) \cdots (x_r y_r)) \\ &= x_i y_i \\ &= \varphi(x)\varphi(y). \end{aligned}$$

Thus φ is a homomorphism which is obviously surjective. Finally, $\varphi(x) = e$ iff $x_i = e$ iff $x = x_{i+1} \cdots x_r \in G_i$, so that ker $\varphi = G_i$. Then the induced homomorphism

$$\bar{\varphi}\colon G_{i-1}/G_i \to H_i$$

is an isomorphism (Theorem 1.1 (iii)); the formula for $\bar{\varphi}$ is

$$\bar{\varphi}((x_i x_{i+1} \cdots x_r)G_i) = x_i. \blacksquare$$

If $K \lhd G$ and $K \neq G$, we say that K is a *maximal normal subgroup* of G if $K \subset H \lhd G$ implies $H = K$ or $H = G$. In other words, K is not a proper subgroup of any proper normal subgroup of G.

Example 5 Let G be the cyclic group of order 6, $G = [a]$. Then $K = [a^2] = \{e, a^2, a^4\}$ is a subgroup of G (automatically normal since G is abelian). Observe that $|G/K| = 2$. If $K \lhd H \lhd G$, then

$$2 = |\,G/K\,| = \left|\frac{G/K}{H/K}\right| \,|\,H/K\,|.$$

Thus $|H/K| = 1$ or $|H/K| = 2$, so that $H = K$ or $|H| = 2|K| = 6$ and hence $H = G$. It follows that K is maximal.

The preceding example suggests the plausibility of the following result.

Theorem 1.9 *Let $K \lhd G$. Then K is maximal iff G/K is simple and $|G/K| > 1$.*

Proof: Assume K is maximal. Since $G \neq K$ by definition, it follows that $|G/K| > 1$. Suppose $\bar{H} \lhd G/K$. Then by Theorem 1.3, there is a unique $H \lhd G$ such that $K \lhd H$ and $\bar{H} = H/K$. The maximality of K implies $H = K$ or $H = G$. But then $H/K = \{K\}$ or $H/K = G/K$. Thus G/K contains no proper normal subgroup, i.e., G/K is simple.

Conversely, suppose that G/K is simple and $|G/K| > 1$, and assume that $K < H \lhd G$. Then of course $G \neq K$. By Theorem 1.2, $H/K \lhd G/K$ and

$$G/H \cong \frac{G/K}{H/K}.$$

But G/K is simple, and so $|H/K| = 1$ or $H/K = G/K$. These alternatives obviously imply that $H = K$ or $H = G$, and thus K is maximal. ∎

We remark that if $|G| < \infty$ and $S \lhd G$, $S \neq G$, then it is clear that there is a maximal subgroup containing S. For among the finite number of proper normal subgroups of G which contain S, there is at least one with a maximal number of elements.

Theorem 1.10 (*Refinement Theorem*) *If $|G| < \infty$ and*

$$G = G_0 \rhd G_1 \rhd G_2 \rhd \cdots \rhd G_r \rhd e, \tag{42}$$

then there is a series of subgroups of G

$$G = K_0 \rhd K_1 \rhd K_2 \cdots \rhd K_p = e \tag{43}$$

in which every G_i appears as one of the K_j and moreover K_i is maximal in K_{i-1}.

Proof: (Note: The symbol $G \rhd K$ means $K \lhd G$ and we write e instead of $\{e\}$.) The proof is by induction on $|G|$. The case $|G| = 1$ is trivial. Assume then that $|G| > 1$ and that the theorem holds for groups of order less than $|G|$. Note that we can assume the groups in (42) are distinct. The case $r = 1$ in (42) requires a special argument. For then (42) becomes

$$G \rhd e.$$

According to the remark immediately preceding the theorem, we can find K_1 such that K_1 is maximal in G:

$$G \rhd K_1 \rhd e. \tag{44}$$

Then since $|K_1| < |G|$, induction applied to $|K_1|$ will produce the required sequence. Now suppose $r > 1$ and consider the sequence

$$G_1 \rhd G_2 \rhd \cdots \rhd G_r = e. \tag{45}$$

We have assumed that $G_1 \neq G$, and so $|G_1| < |G|$. Applying the induction hypothesis to G_1, we secure a sequence

$$G_1 = M_0 \rhd M_1 \rhd \cdots \rhd M_q = e \tag{46}$$

in which all of $G_1, \ldots, G_r$ appear and M_i is maximal in M_{i-1}, $i = 1, \ldots, q$. Now consider the factor group

$$G/G_1 = \bar{H}_0.$$

Since $r > 1$, $G_1 \neq e$ and $G_1 \neq G$ so that by Theorem 3.7(ii), Chapter 1, $|G/G_1| = |\bar{H}_0| < |G|$ and $|\bar{H}_0| > 1$. Applying the induction hypothesis to the sequence

$$\bar{H}_0 \rhd \text{identity},$$

we obtain a sequence

$$G/G_1 = \bar{H}_0 \rhd \bar{H}_1 \rhd \bar{H}_2 \rhd \cdots \rhd \bar{H}_s = \text{identity}, \tag{47}$$

where $\bar{H}_1$ is maximal in $\bar{H}_0$, $\bar{H}_2$ is maximal in $\bar{H}_1$, etc. By Theorem 1.3(i), there exist subgroups $H_1, H_2, H_3, \ldots$ such that

$$\begin{aligned} \bar{H}_1 &= H_1/G_1, & H_1 &\lhd G \\ \bar{H}_2 &= H_2/G_1, & H_2 &\lhd H_1 \\ \bar{H}_3 &= H_3/G_1, & H_3 &\lhd H_2 \end{aligned}$$

etc. Observe that $H_s = G_1$ since $\bar{H}_s = H_s/G_1$ is the identity. Thus we have a sequence of groups

$$G = H_0 \rhd H_1 \rhd H_2 \rhd \cdots \rhd H_s = G_1. \tag{48}$$

If we can show that H_k is maximal in H_{k-1}, we can simply attach the sequence (48) on the left end of the sequence (46) to obtain the required sequence (43). For this purpose suppose that H_k is not maximal in H_{k-1}. Then by definition there exists a proper normal subgroup L of H_{k-1} which contains H_k, i.e.,

$$H_k \lhd L \lhd H_{k-1} \tag{49}$$

and the inclusions in (49) are proper. Now

$$H_k/G_1 < L/G_1 < H_{k-1}/G_1 \tag{50}$$

makes sense because $G_1 \lhd G$ (hence G_1 is normal in any subgroup of G). In fact the inclusions in (50) are proper since those in (49) are. Moreover, since the inclusions (49) are normal and $G_1 \lhd G$, it follows (see Exercise 3) that the inclusions in (50) are normal as well, i.e.,

$$\bar{H}_k \lhd L/G_1 \lhd \bar{H}_{k-1}.$$

This contradicts the maximality of $\bar{H}_k$ in $\bar{H}_{k-1}$ and the proof is complete. ▮

Theorem 1.11 *If the factor groups G_{i-1}/G_i in (42) are all abelian, then so are the factor groups K_{i-1}/K_i in (43).*

Proof: The proof is conducted precisely as the proof of Theorem 1.10, with the additional proviso in the induction hypothesis that the factors are abelian in the "refinement" sequence whenever they are abelian in the original sequence. The proof proceeds without alteration to obtain the sequence (46) for G_1 which now has abelian factors. We continue as before to obtain the sequence (47), in which (since $G/G_1 = \bar{H}_0$ is abelian) the induction hypothesis implies that each of

$$\bar{H}_{k-1}/\bar{H}_k$$

is also abelian. But by Theorem 1.2,

$$\bar{H}_{k-1}/\bar{H}_k = \frac{H_{k-1}/G_1}{H_k/G_1} \cong H_{k-1}/H_k.$$

Then all of the factors in (43) are indeed abelian. ▮

If

$$G = G_0 \triangleright G_1 \triangleright G_2 \triangleright \cdots \triangleright G_r = e, \tag{51}$$

then these groups comprise what is called a *subnormal series* for G. If in addition each G_i is maximal in G_{i-1}, (51) is called a *composition series* for G. For any subnormal series the factor groups G_{i-1}/G_i are called the *factors* of the series. If G is a group that has a subnormal series in which the factor groups are abelian, then G is said to be *solvable* or *metacyclic*. Any sequence of groups (43) in which all the subgroups (51) appear is called a *refinement* of (51).

We can reformulate Theorem 1.10 as follows: *A subnormal series for a finite group can always be refined to a composition series.*

Theorem 1.11 can also be restated: *If G is solvable, then there is a composition series for G in which the factors are abelian.*

Example 6 Let $G = [a]$ be a infinite cyclic group, i.e., G consists of all powers a^k of a, $k \in Z$. Then G does not have a composition series. For suppose $e \neq G_1 < G$. Then there is a least positive integer k_1 such that $a^{k_1} \in G_1$. Clearly, $[a^{k_1}] \subset G_1$. If $a^p \in G_1$, write

$$p = qk_1 + r, \qquad 0 \leq r < k$$

so that

$$a^p = a^{qk_1+r} = (a^{k_1})^q a^r$$

and

$$a^r = a^p(a^{k_1})^{-q} \in G_1.$$

Hence by the minimality of k_1, $r = 0$ and $a^p = (a^{k_1})^q$. Thus in fact, $G_1 = [a^{k_1}]$.

Observe, incidentally, that this argument shows that any subgroup of an infinite cyclic group is also cyclic. Applying the preceding argument to G_1, we conclude that if $e \neq G_2 < G_1$, then $G_2 = [a^{k_1 k_2}]$. Thus if

$$G = G_0 \rhd G_1 \rhd G_2 \rhd \cdots \rhd G_{r-1} \rhd G_r = e$$

were a composition series for G, e would be maximal in G_{r-1} and $G_{r-1} = [a^{k_1 \cdots k_{r-1}}] = [a^m]$. However, if $d > 1$, then $[a^{dm}]$ is a proper normal subgroup of $[a^m]$; so e is not maximal in G_{r-1}.

Example 7 (a) We construct a composition series for $G = S_4$. To begin with, $A_4 \lhd S_4$; therefore set $G_1 = A_4$. Next define

$$G_2 = \{e, (1\ \ 2)(3\ \ 4), (2\ \ 3)(1\ \ 4), (1\ \ 3)(2\ \ 4)\};$$

G_2 is the Klein four group whose table appears in Example 3. It is relatively easy to check that if τ is any transposition in S_4, then $\tau G_2 = G_2 \tau$. Since any permutation is a product of transpositions, we conclude that $G_2 \lhd S_4$ (see Exercise 15). Let $G_3 = \{e, (1\ 2)(3\ 4)\}$. The group G_3 has index 2 in G_2, so $G_3 \lhd G_2$ (see Exercise 16). Then

$$S_4 = G_0 \rhd G_1 \rhd G_2 \rhd G_3 \rhd e \tag{51}$$

is a composition series for S_4; each factor is of order 2 or 3, hence simple, and we can apply Theorem 1.9 to conclude that (51) is a composition series for S_4.

(b) Let G be the additive group $(Z_6, \{+\})$. Define $G_1 = \{[0], [2], [4]\}$ so that $|G_1| = 3$ and $|G/G_1| = 2$. Then G/G_1 is simple, and thus G_1 is maximal in G. Also, $|G_1| = 3$ implies G_1 is simple. Therefore

$$G = G_0 \rhd G_1 \rhd e$$

is a composition series whose factors are of orders 2 and 3, respectively, and hence are cyclic.

(c) Let $G = S_3$, $G_1 = A_3$. Then clearly $|A_3| = 3$ and $|G/G_1| = 2$. Hence

$$G = G_0 \rhd G_1 \rhd e$$

is a composition series for S_3 with factors which are cyclic of orders 2 and 3, respectively.

We see from Example 7(b) and (c) that the factors of a composition series do not suffice for a reconstruction of the entire group. For S_3 and the additive group $(Z_6, \{+\})$ are obviously not isomorphic, although they have composition series whose factors are isomorphic. However, we shall shortly prove a famous theorem (the Jordan-Hölder theorem) that shows the extent to which a finite group determines its composition series. We need a preliminary result.

Theorem 1.12 *Let G_1 and G_1' be distinct maximal subgroups of G, and set $K = G_1 \cap G_1'$. Then*

(i) *K is maximal in both G_1 and G_1'.*

(ii) $G/G_1' \cong G_1/K$ *and* $G/G_1 \cong G_1'/K$.

Proof: Since $G_1 \triangleleft G$ and $G_1' \triangleleft G$, we have $G_1G_1' \triangleleft G$ (see the proof of Theorem 1.7). Moreover, $G_1 < G_1G_1'$, and $G_1 \neq G_1'$ implies that $G_1G_1' \neq G_1$ (see Exercise 17). Thus G_1G_1' is a normal subgroup of G properly containing G_1. The maximality of G_1 in G forces $G = G_1G_1'$. Now we apply the first isomorphism theorem [see diagram (9)].

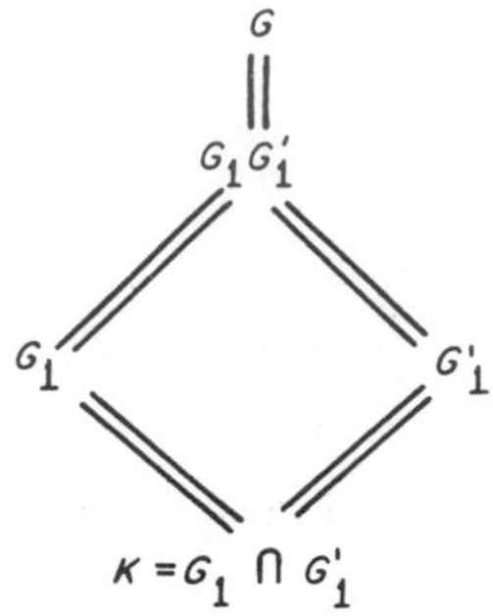

(52)

We have double bars in (52) because both G_1 and G_1' are normal in G [see the conditions on the groups in diagram (9)]. Then $K = G_1 \cap G_1'$ is normal in G, and hence in both G_1 and G_1'.
Also,

$$G_1/G_1 \cap G_1' \cong G_1G_1'/G_1',$$

or

$$G_1/K \cong G/G_1'.$$

Now G/G_1' is simple because G_1' is maximal in G (see Theorem 1.9). Thus G_1/K is simple and K is maximal in G_1. Similarly, $G_1'/K \cong G/G_1$, and K is maximal in G_1'. ∎

Our next result shows that although a finite group can have many composition series, the "length" and the factors are determined by the group. If $G = G_0 \triangleright G_1 \triangleright \cdots \triangleright G_r = e$ is a subnormal series for a group G, then the positive integer r is called the *length* of the series.

Theorem 1.13 (*Jordan-Hölder Theorem*) *Assume that* $1 < |G| < \infty$ *and let*

$$G = G_0 \triangleright G_1 \triangleright \cdots \triangleright G_r = e, \tag{53}$$

$$G = G_0' \triangleright G_1' \triangleright \cdots \triangleright G_s' = e \tag{54}$$

be two composition series for G. Then r = s, and the factors in the two series can be paired off in some order into isomorphic pairs.

Proof: The proof is by induction on $n = |G|$. If $n = 2$, the problem is trivial; thus assume that $n > 2$ and that the theorem holds for groups of order $n - 1$ or less. Let

$$K = G_1 \cap G_1',$$

and suppose first that $K = e$. Unless one of G_1 or G_1' is e itself, i.e., unless $r = 1$ or $s = 1$, it follows that G_1 and G_1' must be distinct. Suppose for a moment that $r = 1$. Then e is maximal in G, and so G is simple. Hence $s = 1$ and $G_1' = e$ as well; i.e., (53) and (54) both collapse to

$$G \rhd e$$

and there is nothing to prove. We may thus assume that in case $K = e$, we have $r > 1$ and $s > 1$. By Theorem 1.12, $K = e$ is maximal in both G_1 and G_1', and hence $r = s = 2$. Thus

$$G \rhd G_1 \rhd e \tag{55}$$

and

$$G \rhd G_1' \rhd e \tag{56}$$

are the composition series (53) and (54), respectively. Theorem 1.12 also tells us that

$$G/G_1' \cong G_1/K = G_1/e \cong G_1 \tag{57}$$

and

$$G/G_1 \cong G_1'/K = G_1'/e \cong G_1'. \tag{58}$$

Now the factors in (55) are

$$G/G_1 \qquad \text{and} \qquad G_1/e \cong G_1 \tag{59}$$

whereas the factors in (56) are

$$G/G_1' \qquad \text{and} \qquad G_1'/e \cong G_1'. \tag{60}$$

Then according to (57) and (58),

$$\{G/G_1',G_1\} \qquad \text{and} \qquad \{G/G_1,G_1'\} \tag{61}$$

are isomorphic pairings of the factors. This proves the result in the event $K = e$.

Now assume $K \neq e$. We can also assume that $G_1 \neq G_1'$; otherwise we simply apply the induction hypothesis to $G_1 = G_1'$ to obtain the result. Again, by Theorem 1.12 we know that $K = G_1 \cap G_1'$ is maximal in both G_1 and G_1'. Let

$$K \rhd K_1 \rhd K_2 \rhd \cdots \rhd K_p = e \tag{62}$$

be a composition series for K. Then the maximality of K in both G_1 and G_1' implies that

$$G_1 \rhd K \rhd K_1 \rhd \cdots \rhd K_p = e \tag{63}$$

and

$$G_1' \rhd K \rhd K_1 \rhd \cdots \rhd K_p = e \tag{64}$$

are composition series for G_1 and G_1', respectively, both of length $p + 1$. On the other hand,

$$G_1 \rhd \cdots \rhd G_r = e \tag{65}$$

and

$$G_1' > \cdots > G_s' = e \tag{66}$$

are also composition series for G_1 and G_1', respectively. Applying the induction hypothesis to G_1 and G_1', we obtain $p + 1 = r - 1$ and $p + 1 = s - 1$ so that $r = s$. Moreover, the factors

$$\{G_1/K,\ K/K_1,\ K_1/K_2, \ldots, K_{p-1}/K_p\} \tag{67}$$

and

$$\{G_1/G_2,\ G_2/G_3,\ G_3/G_4, \ldots, G_{r-1}/G_r\} \tag{68}$$

can be paired off into isomorphic pairs in some order, as can the factors

$$\{G_1'/K,\ K/K_1,\ K_1/K_2, \ldots, K_{p-1}/K_p\} \tag{69}$$

and

$$\{G_1'/G_2',\ G_2'/G_3', \ldots, G_{s-1}'/G_s'\} \tag{70}$$

($s = r$). Again applying Theorem 1.12, we have

$$G/G_1 \cong G_1'/K \tag{71}$$

and

$$G/G_1' \cong G_1/K. \tag{72}$$

Include G/G_1 in both the sets (67), (68) and include G/G_1' in both the sets (69), (70).

We then have

$$\begin{array}{lcl} \{G/G_1,\ G_1/G_2, \ldots, G_{r-1}/G_r\} & \cong & \{G/G_1,\ G_1/K,\ K/K_1, \ldots, K_{p-1}/K_p\} \quad (73) \\ & & \qquad \quad \nwarrow\!\nearrow \qquad \ \ \uparrow \qquad\qquad \uparrow \\ & & \qquad \quad \swarrow\!\searrow \qquad \ \ \downarrow \qquad\qquad \downarrow \\ \{G/G_1',\ G_1'/G_2', \ldots, G_{r-1}'/G_r'\} & \cong & \{G/G_1',\ G_1'/K,\ K/K_1, \ldots, K_{p-1}/K_p\} \quad (74) \end{array}$$

where "$\cong$" means isomorphic in pairs in some order. The arrows in (73) and (74) indicate isomorphic pairings in view of (71) and (72). Thus the two sets on the left in (73) and (74) can be paired into isomorphic pairs, and the induction is complete. ∎

In view of the Jordan-Hölder theorem we can refer to *the* factors in a composition series for a finite group, for these are determined, except for order, to within isomorphism.

Observe that the factors in the composition series for S_4 given in Example 4(a) are abelian, so that S_4 *is solvable*. In fact we have the following result for any finite group G.

Theorem 1.14 *If* $|G| < \infty$, *then* G *is solvable iff each factor in a composition series for* G *is of prime order.*

Proof: Assume first that G is solvable, i.e., G has a subnormal series with abelian factors. According to Theorem 1.11, this series can be refined to a composition series with abelian factors. By Theorem 1.9, these factors are simple groups. But we assert (see Exercise 1) that the only simple finite abelian groups are cyclic groups of prime order. By Theorem 1.13 the composition factors are determined by G to within isomorphism; so in any composition series for G the order of each factor must be prime.

On the other hand, if the order of each factor in a composition series for G is prime, each factor must be a cyclic group and hence abelian. Thus G itself is solvable. ▮

Example 8 (a) S_n is not solvable for $n \geq 5$. For consider the series

$$S_n \vartriangleright A_n \vartriangleright e. \tag{75}$$

Since $|S_n/A_n| = 2$, S_n/A_n is simple, and hence A_n is obviously maximal in S_n. According to Theorem 3.10(v), Chapter 1, A_n is a simple group for $n \geq 5$. Thus (75) is a composition series for S_n. But $A_n/e \cong A_n$ is not abelian (and A_n is simple of non-prime order; see Exercise 1). Now, of course, S_n cannot have any subnormal series with abelian factors; for Theorem 1.11 tells us that any such series can be refined to a composition series with abelian factors, and these factors are unique to within isomorphism by Theorem 1.13.

(b) We saw in Example 7(a) that S_4 has a composition series in which each factor has order 2 or 3. Therefore S_4 is solvable by Theorem 1.14. Next, consider S_3. We have $|A_3| = 3$; so A_3 is simple and hence

$$S_3 \vartriangleright A_3 \vartriangleright e$$

is a composition series with abelian quotients. Again, S_3 is solvable. (Of course, S_2 and S_1 are trivially solvable.)

There are several interesting and important results about finite solvable groups which we shall need.

Theorem 1.15 (i) *If G is solvable and $H < G$, then H is solvable.*

(i) *If G is solvable and $f\colon G \to G'$ is an epimorphism, then G' is solvable.*

(iii) *If $H \vartriangleleft G$, H is solvable, and G/H is solvable, then G is solvable.*

(iv) *If*

$$G = G_0 \vartriangleright G_1 \vartriangleright G_2 \vartriangleright \cdots \vartriangleright G_r = e \tag{76}$$

is a subnormal series with solvable factors, then G is solvable.

Proof: (i) Let

$$G = G_0 \vartriangleright G_1 \vartriangleright \cdots \vartriangleright G_r = e \tag{77}$$

be a subnormal series with abelian factors, and define

$$H_i = H \cap G_i, \qquad i = 0, \ldots, r.$$

Consider the diagram

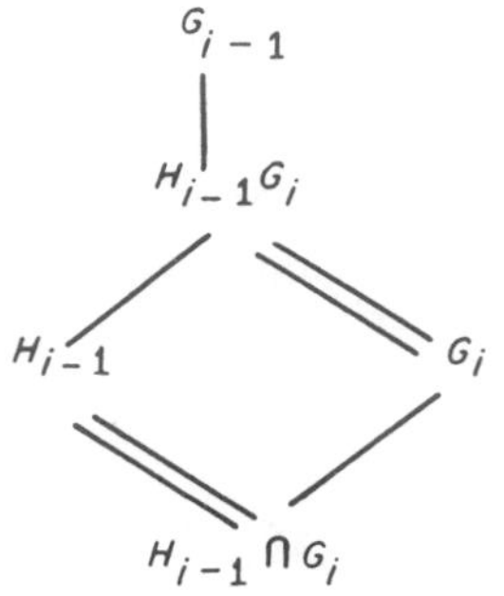

That is, in diagram (9) replace G by G_{i-1}, K by G_i, and H by H_{i-1}. This is permissible since $H_{i-1} < G_{i-1}$ and $G_i \lhd G_{i-1}$. Then the first isomorphism theorem (Theorem 1.4) tells us that $H_{i-1}G_i < G_{i-1}$ and

$$H_{i-1}/H_{i-1} \cap G_i \cong H_{i-1}G_i/G_i. \tag{78}$$

But

$$\begin{aligned} H_{i-1} \cap G_i &= H \cap G_{i-1} \cap G_i \\ &= H \cap G_i \\ &= H_i, \end{aligned}$$

and since $H_{i-1}G_i < G_{i-1}$, we have

$$H_{i-1}G_i/G_i < G_{i-1}/G_i. \tag{79}$$

The group on the right in (79) is abelian because it is one of the factors in the subnormal series (77). Thus by (78),

$$H_{i-1}/H_{i-1} \cap G_i < H_{i-1}/H_i$$

is isomorphic to a subgroup of an abelian group and hence is abelian. It follows that

$$H = H_0 \rhd H_1 \rhd \cdots \rhd H_r = e$$

is a subnormal series for H with abelian factors, so that H is solvable.

(ii) Let

$$G = G_0 \rhd G_1 \rhd \cdots \rhd G_r = e$$

be a subnormal series with abelian factors. Define $G_i' = f(G_i)$ and check (see Exercise 4) that G_{i-1}'/G_i' is a homomorphic image of G_{i-1}/G_i. It follows that G_{i-1}'/G_i' is abelian, so

$$f(G) = G' = G_0' \rhd G_1' \rhd \cdots \rhd G_r' = e$$

is a subnormal series for G' with abelian factors. Thus G' is solvable.

(iii) Since H is solvable, there is a subnormal series

$$H = K_0 \triangleright K_1 \triangleright \cdots \triangleright K_p = e \tag{80}$$

with abelian factors. Also G/H is solvable, so there exists a subnormal series for G/H with abelian factors, say

$$G/H = \bar{H}_0 \triangleright \bar{H}_1 \triangleright \bar{H}_2 \triangleright \cdots \triangleright \bar{H}_r = \text{identity}. \tag{81}$$

By Theorem 1.3(ii) there are unique subgroups H_i ($H_0 = G$) such that $H \triangleleft H_i \triangleleft H_{i-1}$, and $\bar{H}_i = H_i/H$, $i = 1, \ldots, r$; moreover

$$\begin{aligned} H_{i-1}/H_i &\cong \frac{H_{i-1}/H}{H_i/H} \\ &= \bar{H}_{i-1}/\bar{H}_i, \qquad i = 1, \ldots, r. \end{aligned} \tag{82}$$

The groups on the right in (82) are abelian, so each factor H_{i-1}/H_i is abelian. Observe that $H_r/H = \bar{H}_r =$ identity and hence $H_r = H$. Thus

$$G = H_0 \triangleright H_1 \triangleright \cdots \triangleright H_r = H \tag{83}$$

is a series with abelian factors that can be attached to the left end of (80) to yield a subnormal series with abelian factors for G itself. Therefore G is solvable.

(iv) The proof is by induction on r, the length of the series in (76). If $r = 1$, then $G = G/e$ is solvable and there is nothing to prove. Let $r > 1$, and assume the result holds for series of length $r - 1$ or less. Now

$$G_1 \triangleright G_2 \triangleright \cdots \triangleright G_r = e$$

is a subnormal series for G_1 with solvable factors. By the induction hypothesis G_1 is solvable; since G/G_1 is solvable by hypothesis, (iii) implies that G is solvable. ∎

We conclude this section by introducing three interesting ideas in group theory.

Let G be a group and $a,b \in G$. The *commutator of a and b* is the element

$$aba^{-1}b^{-1}.$$

The *commutator subgroup* of G is the group $G^{(1)}$ generated by the set of all commutators in G:

$$G^{(1)} = [\{z \mid z = aba^{-1}b^{-1}, a,b \in G\}].$$

Note that $(aba^{-1}b^{-1})^{-1} = bab^{-1}a^{-1}$, i.e., the inverse of a commutator is again a commutator. Thus $G^{(1)}$ consists precisely of all finite products of commutators in G. The higher-order commutator groups are defined inductively: Letting

$$G^{(0)} = G,$$

we set $$G^{(s)} = (G^{(s-1)})^{(1)}, \qquad s = 1, 2, \ldots.$$

The *normalizer* $N(X)$ of a subset $X \subset G$, $(X \neq \varnothing)$, is defined by

$$N(X) = \{g \in G \mid gX = Xg\}.$$

The *centralizer* $Z(X)$ of a subset $X \subset G$, $(X \neq \varnothing)$, is defined by

$$Z(X) = \{g \in G \mid gx = xg \text{ for all } x \in X\}.$$

In particular, $Z(G)$ is called the *center* of G:

$$Z(G) = \{g \in G \mid gx = xg \text{ for all } x \in G\}.$$

It is quite simple (see Exercise 5) to prove that for $\varnothing \neq X \subset G$,

$$Z(X) < N(X) < G. \tag{84}$$

In particular we have

$$Z(G) < G. \tag{85}$$

Theorem 1.16 *Let G be a group. Then*

(i) $G^{(1)} \lhd G$.

(ii) $G/G^{(1)}$ *is abelian.*

(iii) *If* $f\colon G \to S$ *is a homomorphism and S is abelian, then* $G^{(1)} < \ker f$.

(iv) *If* $K \lhd G$ *and* G/K *is abelian, then* $G^{(1)} < K$.

Proof: (i) If $g \in G$, then

$$g^{-1}\left(\prod_{t=1}^{n} a_t b_t a_t^{-1} b_t^{-1}\right) g = \prod_{t=1}^{n} \left((g^{-1}a_t g)(g^{-1}b_t g)(g^{-1}a_t g)^{-1}(g^{-1}b_t g)^{-1}\right)$$
$$\in G^{(1)}.$$

(ii) Let $a,b \in G$. Since $aba^{-1}b^{-1} \in G^{(1)}$, we have

$$ab(ba)^{-1} \in G^{(1)}$$

so that
$$(ab)G^{(1)} = G^{(1)}(ba),$$
$$(aG^{(1)})(bG^{(1)}) = (bG^{(1)})(aG^{(1)}).$$

(Note that in this computation we made use of the normality of $G^{(1)}$.) Thus $G/G^{(1)}$ is commutative.

(iii) Observe that since S is commutative,

$$f(a)f(b)f(a^{-1})f(b^{-1}) = e$$

and hence
$$f\left(\prod_{t=1}^{n} a_t b_t a_t^{-1} b_t^{-1}\right) = e.$$

Therefore
$$G^{(1)} < \ker f.$$

(iv) Let $\nu\colon G \to G/K$ be the quotient map. Since G/K is abelian and $K = \ker \nu$, we can apply (iii) to conclude that

$$G^{(1)} < K. \blacksquare$$

If $G^{(s)} = e$ for some s, then

$$G = G^{(0)} \rhd G^{(1)} \rhd \cdots \rhd G^{(s)} = e$$

is a subnormal series [by Theorem 1.16(i)] with abelian factors [by Theorem 1.16(ii)]. Thus G is solvable. The converse of this result is also true.

Theorem 1.17 *A necessary and sufficient condition that a group G be solvable is that*

$$G^{(s)} = e$$

for some $s \in N$.

Proof: We have proved the sufficiency of the condition. Suppose then that

$$G = G_0 \rhd G_1 \rhd \cdots \rhd G_r = e$$

is a subnormal series with abelian quotients. Now

$$G_1 \lhd G$$

and

$$G/G_1$$

is abelian, so by Theorem 1.16(iv) we have

$$G^{(1)} < G_1. \tag{86}$$

Similarly,

$$G_2 \lhd G_1$$

and

$$G_1/G_2$$

is abelian; hence

$$G_1{}^{(1)} < G_2. \tag{87}$$

But (86) implies that

$$G^{(2)} < G_1{}^{(1)},$$

and combined with (87) this yields

$$G^{(2)} < G_2.$$

Continuing in this fashion, we obtain

$$G^{(t)} < G_t, \qquad t = 1, 2, \ldots .$$

Since $G_r = e$, it follows that $G^{(s)} = e$ for some $s \leq r$. ▮

The importance of solvable groups will become clear when we study the problem of expressing the roots of a polynomial in terms of radicals and rational expressions in the coefficients of the polynomial. It turns out that a certain group of permutations of the roots, called the Galois group of the polynomial, is solvable iff the roots of the polynomial can be expressed in this way in terms of the coefficients. For polynomials of degree $n \geq 5$, the Galois group of the polynomial is in most cases S_n. Thus the following theorem is of considerable interest because of its applications in Galois theory. Although we proved this result in Example 8(a), we formally state it here and provide an independent proof which makes use of Theorem 1.17.

Theorem 1.18 *For $n \geq 5$, S_n is not solvable.*

Proof: First observe that if i, j, k, r, s are five distinct integers and

$$\sigma = (i \quad j \quad k),$$
$$\theta = (k \quad r \quad s),$$

then

$$\sigma^{-1}\theta^{-1}\sigma\theta = (i \quad k \quad j)(k \quad s \quad r)(i \quad j \quad k)(k \quad r \quad s)$$
$$= (j \quad s \quad k).$$

In other words, if a subgroup H of S_n contains every 3-cycle, then so does $H^{(1)}$. Thus

$$S_n^{(1)} \lhd S_n$$

must contain every 3-cycle, and similarly

$$S_n^{(2)} \lhd S_n^{(1)}$$

must contain every 3-cycle, etc. Clearly there can be no positive integer s such that

$$S_n^{(s)} = e,$$

and we may apply Theorem 1.17. ∎

In 1963, W. Feit and J. G. Thompson proved in a monumental paper (Pacific Journal of Mathematics, Vol. 13 (1963), pp. 775–1029) the following result, thereby resolving one of the most intransigent conjectures in the history of finite group theory.

Theorem 1.19 (*Feit-Thompson Theorem*) *Any group of odd order is solvable.*

The proof of this result depends on the theory of group characters, among other things, and is considerably beyond the scope of this book. However, we shall return to the study of group characters in a later section.

Exercises 2.1

1. Prove that if G is a finite abelian group, then G is simple iff $|G|$ is a prime. *Hint*: Suppose G is simple. If $e \neq a \in G$, then $[a] \lhd G$. Hence $G = [a]$ and by Corollary 1, $G \cong (Z_d, \{+\})$. Unless d is a prime, $(Z_d, \{+\})$ obviously contains a proper subgroup, e.g., $\{[0]_6, [2]_6, [4]_6\} < (Z_6, \{+\})$. Thus d must be prime. The converse is easily proved by using Theorem 3.7(ii), Chapter 1.
2. Prove that if

$$G = H_1 \cdots H_r \qquad \text{(direct)},$$

then
$$|G| = \prod_{k=1}^{r} |H_k|.$$

Hint: If G is finite, consider $G_1 = H_2 \cdots H_r$. Then by (41), $G/G_1 \cong H_1$ so that $|H_1| = |G/G_1| = |G|/|G_1| = |G|/|H_2| \cdots |H_r|$, by induction on r. Note that G is infinite iff one of the H_i is infinite.

3. Prove that if $H < G$, $L < G$, $K \lhd G$, and $K \lhd H \lhd L$, then $H/K \lhd L/K$. *Hint*: The group inclusion is obvious. To check normality, we have for $l \in L$ and $h \in H$ that

$$(lK)(hK)(lK)^{-1} = (lK)(hK)(l^{-1}K)$$
$$= (lhl^{-1})K.$$

But $lhl^{-1} \in H$ because $H \lhd L$. It follows that $H/K \lhd L/K$.

4. Let $f\colon G \to G'$ be an epimorphism. Let $K \lhd G$, $K' = f(K)$. Prove that there is an epimorphism τ mapping G/K onto G'/K'. *Hint*: We know by Theorem 1.1(iv) that $K' \lhd G'$. Define $\varphi\colon G \to G'/K'$ by $\varphi(g) = f(g)K'$. It is simple to verify that φ is an epimorphism and that $K < \ker \varphi$. Hence there exists a homomorphism τ which completes the diagram

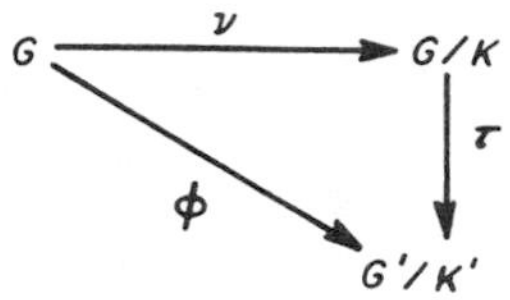

i.e., $\varphi = \tau\nu$, or $\tau(gK) = \varphi(g) = f(g)K'$; τ is obviously an epimorphism.

5. Confirm (84) and (85).

6. In $(\mathbf{R}^4, \{+\})$ let H be the subspace

$$H = \{x \mid x_1 + x_2 + x_4 = 0 \text{ and } x_1 - x_3 - x_4 = 0\}.$$

Define subspaces

$$H_i = \{x \in \mathbf{R}^4 \mid x_j = 0, j \neq i\}, \qquad i = 1, \ldots, 4.$$

Find $k_1, k_2, \ldots, k_r$ such that

$$\mathbf{R}^4 = H \dotplus H_{k_1} \dotplus H_{k_2} \dotplus \cdots \dotplus H_{k_r}.$$

In Exercises 7 through 14, G is a fixed finite group of order n, $Z(G)$ is its center, $C[a]$ is the conjugacy class to which an element $a \in G$ belongs, and $N(a)$ is the normalizer of the set $\{a\}$.

7. Show that $C[a] = \{a\}$ iff $a \in Z(G)$. Conjugacy classes consisting of a single element are called *trivial conjugacy classes.*

8. Let $C[a_1], \ldots, C[a_r]$ be the (distinct) nontrivial conjugacy classes in G, and let $c = |Z(G)|$. Prove that

$$n = c + \sum_{t=1}^{r} | C[a_t] |. \qquad (88)$$

Equation (88) is called the *class equation of G*. *Hint*: The conjugacy classes are disjoint.

9. Prove that $c|n$. *Hint*: $Z(G) < G$.

10. Let x and y be arbitrary elements of G. Show that

$$xN(a) = yN(a)$$

iff

$$xax^{-1} = yay^{-1}.$$

Hint: $xN(a) = yN(a)$ iff $x^{-1}y \in N(a)$ iff $(x^{-1}y)a(x^{-1}y)^{-1} = a$ iff $yay^{-1} = xax^{-1}$.

11. Let $G = \bigcup_{t=1}^{p} x_t N(a)$ be the left $N(a)$-coset decomposition of G. Prove that $p = |C[a]|$. *Hint*: Exercise 10 implies that $x_t a x_t^{-1} \neq x_s a x_s^{-1}$ when $s \neq t$. Thus the p elements $x_t a x_t^{-1}$, $t = 1, \ldots, p$, are distinct and by definition belong to $C[a]$. Hence $|C[a]| \geq p$. On the other hand, if $xax^{-1} \neq yay^{-1}$, then by Exercise 10 again, x and y are in different left $N(a)$-cosets. Thus the elements of $C[a]$ lie in distinct left $N(a)$-cosets, and so $p \geq |C[a]|$. We conclude that $p = |C[a]|$.

12. Prove that $|C[a]|\,|n$. *Hint*: By Exercise 11, $[G:N(a)] = p = |C[a]|$.

13. Prove that if G is abelian and p is a prime divisor of n, then G contains a cyclic subgroup of order p. Do this by completing the following steps. The argument is by induction on n, with nothing to prove for $n \leq 2$. Assume $n \geq 3$.

(i) The result follows immediately if G is cyclic. *Hint*: If $G = [g]$, let $h = g^{n/p}$. Then $|[h]| = p$.

(ii) The result is trivial if n is a prime.

(iii) If G is not cyclic, then G contains a cyclic subgroup H of order m, where $2 \leq m < n$. *Hint*: Let $h \in G$, $h \neq e$, and set $H = [h]$.

(iv) If $p|m$, the proof is complete. *Hint*: Induction.

(v) If $p \nmid m$, then $p\,|\,|G/H|$. *Hint*: $n = |G/H|\,|H| = |G/H|m$, $p|n$, and $p \nmid m$.)

(vi) G/H contains a cyclic subgroup $[\nu(g)]$ of order p, where $\nu: G \to G/H$ is the canonical homomorphism. *Hint*: Induction.

(vii) $g^p \in H$. *Hint*: $\nu(g)^p = \nu(e)$ since $[\nu(g)]$ is cyclic of order p.

(viii) $g^{mp} = e$. *Hint*: $g^p \in H$ and $|H| = m$, so $|[g^p]|\;|m$.

(ix) Let $k = g^m$; then $k = e$ or k has order p. *Hint*: $k^p = g^{mp} = e$. Therefore the order of k divides the prime p.

(x) $k \neq e$. *Hint*: If $k = e$, then $g^m = e$ so that $\nu(g)^m = \nu(g^m) = H$. Since $\nu(g)$ has order p, this implies $p|m$. But we are assuming [step (v)] that $p \nmid m$.

(xi) $[k]$ is the required cyclic subgroup of order p.

14. The hypothesis in Exercise 13 that G is abelian can be dropped; show this by completing the following steps. Assume that G is not abelian. The argument is by induction on n.

(i) If $p|c$, the proof is complete. *Hint*: Since G is not abelian, $Z(G) \neq G$. Use induction.

(ii) If $p \nmid c$, then $p \nmid |C[a_t]|$ for some t, say $t = 1$. *Hint*: Use (88).

(iii) $p\,|\,|N[a_1]|$. *Hint*: $|N(a_1)|\,|C[a_1]| = n$ (see Exercise 11), $p|n$, and $p \nmid |C[a_1]|$.

(iv) $N(a_1)$ contains a cyclic subgroup of order p; hence so does G. *Hint*: Since

$a_1 \notin Z(G)$, we have $|C[a_1]| > 1$; so $|N[a_1]| < n$ and the induction hypothesis applies.

15. Prove the assertion in Example 7(a) that $G_2 \lhd S_4$.

16. Prove that if $H < G$ and $[G:H] = 2$, then $H \lhd G$. *Hint*: Write $G = H \cup aH$, where a is any element not in H. This union is disjoint. Since $a \notin H$, $G = H \cup Ha$ is a right coset decomposition of H as well. Thus $aH = Ha$ and H is normal.

17. Prove the assertion in the proof of Theorem 1.12 that $G_1 \neq G_1'$ implies $G_1G_1' \neq G_1$. *Hint*: If $G_1G_1' = G_1$, then since $G_1' \subset G_1G_1'$ we have $G_1' \subset G_1$. Now G_1' is maximal (as is G_1), so this inclusion implies $G_1' = G_1$. But G_1' and G_1 are assumed to be distinct.

Glossary 2.1

center, 81
centralizer, 81
$Z(G)$, 81
$Z(X)$, 81
class equation of G, 85
commutator of two elements, 80
commutator subgroup, 80
$G^{(s)}$, 81
composition series, 73
cyclic group, 58
$[a]$, 58
cyclic subgroup generated by a, 58
direct sum, 63
$H_1 \dot{+} \cdots \dot{+} H_r$, 63
$H_1 \oplus \cdots \oplus H_r$, 63
$\sum_{i=1}^{r}{}^{\cdot} H_i$ 63
element of finite order, 58
element of infinite order, 58
external direct product, 65
factors of a series, 73
Feit-Thompson theorem, 83
fully reducible, 65
group generated by A, 66
$[A]$, 66
$[A_1, \ldots, A_m]$, 66
group homomorphism theorem, 57
group isomorphism, 59
$S \cong T$, 59
induced homomorphism, 58
$\bar{\varphi}$, 58
internal direct product, 63
$H_1H_2 \cdots H_r$ (direct), 63
$H_1 \times H_2 \times \cdots \times H_r$, 63
$\prod_{i=1}^{r}{}^{\cdot} H_i$, 63
isomorphism theorem for groups (first), 57
isomorphism theorem for groups (second), 62
isomorphism theorem for groups (third), 60
Jordan-Hölder theorem, 75
maximal normal subgroup, 70
metacyclic group, 73
normalizer, 81
$N(X)$, 81
order of an element, 58
refinement of a series, 73
refinement theorem, 71
S_4 is solvable, 77
S_n is not solvable for $n \geq 5$, 83
solvable group, 73
subnormal series, 73

2.2 Group Actions and the Sylow Theorems

Let H be an arbitrary group and X a set. If there exists a homomorphism

$$f: H \to S_X,$$

we say that H *acts on the set* X; this is indicated by the notation

$$H : X.$$

Since $f(H)$ is a group of bijections of the set X, we can consider the *$f(H)$-orbits in* X. The set X is the union of disjoint orbits of the form

$$\mathcal{O}_x = \{f(h)x \mid h \in H\}.$$

These are called the *orbits in X under the action of H*, or simply the *$H:X$ orbits*. If $H : X$ and $x \in X$, then the *stabilizer* of x is the subgroup $H_x < H$ defined by

$$H_x = \{h \mid f(h)x = x\}. \tag{1}$$

Henceforth in this section we assume that both $|H| < \infty$ and $|X| < \infty$.

Theorem 2.1 *If $H : X$, then for any $x \in X$ we have*

$$|H_x| \, |\mathcal{O}_x| = |H|. \tag{2}$$

Proof: Let $x \in X$. Suppose $|\mathcal{O}_x| = r$ and $h_1, \ldots, h_r$ are (distinct) elements of H such that

$$\mathcal{O}_x = \{f(h_1)x, \ldots, f(h_r)x\}. \tag{3}$$

We assert that

$$h_1H_x, \ldots, h_rH_x \tag{4}$$

is the complete list of left H_x-cosets in H. First, if $h_ih = h_jk$, $h, k \in H_x$, then

$$f(h_ih)x = f(h_i)f(h)x = f(h_i)x$$

and similarly $$f(h_jk)x = f(h_j)x.$$

But the elements on the right side of (3) are distinct (i.e., $|\mathcal{O}_x| = r$), so that $i = j$. In other words, the left cosets (4) are disjoint. Also, if $h \in H$ then $f(h)x = f(h_i)x$ for some i and hence

$$f(h_i^{-1}h)x = x.$$

This means that

$$h_i^{-1}h \in H_x,$$

or $$h \in h_iH_x.$$

Thus the left cosets (4) are precisely all the left H_x-cosets. Therefore

$$[H: H_x] = r = |\mathcal{O}_x|.$$

But by Lagrange's theorem [Theorem 3.7(ii), Chapter 1] we know that

$$|H| = |H_x| \, [H: H_x],$$

and (2) follows. ∎

The next result, known in some versions as *Burnside's lemma*, is an important tool in combinatorial analysis.

Theorem 2.2 *Let $H : X$ and let $\chi : H \to \mathbf{C}$ be a homomorphism of H into the multiplicative group of $\mathbf{C}$ (χ is called a character of H of degree 1). Let $W: X \to \mathbf{C}$ be a function which is constant on each $H : X$ orbit. Let $\Delta \subset X$ be an S.D.R.* (i.e., *system of distinct representatives*) *for the family of $H : X$ orbits. Define*

$$\bar{\Delta} = \{x \mid x \in \Delta \text{ and } H_x \subset \ker \chi\}.$$

Then

$$\sum_{x \in \bar{\Delta}} W(x) = \frac{1}{|H|} \sum_{h \in H} \chi(h) \sum_{f(h)x=x} W(x). \tag{5}$$

(*The summation on the right in* (5) *is over all $x \in X$ for which $f(h)x = x$.*)

Proof: Define a function δ: $H \times X \to \{0,1\}$ by

$$\delta(h,x) = \begin{cases} 1 & \text{if } f(h)x = x \\ 0 & \text{otherwise} \end{cases}.$$

Then for each $h \in H$,

$$\sum_{f(h)x=x} W(x) = \sum_{x \in X} \delta(h,x) W(x) \tag{6}$$

so that the double summation on the right side of (5) becomes

$$\sum_{h \in H} \chi(h) \sum_{x \in X} \delta(h,x) W(x) = \sum_{x \in X} \sum_{h \in H} \chi(h) W(x) \delta(h,x). \tag{7}$$

Now for a fixed $x \in X$, $\delta(h,x) = 0$ unless $f(h)x = x$, i.e., unless $h \in H_x$ in which case $\delta(h,x) = 1$. Therefore (7) becomes

$$\sum_{x \in X} [\sum_{h \in H_x} \chi(h) W(x)] \tag{8}$$

or

$$\sum_{x \in \Delta} \sum_{y \in \mathcal{O}_x} \sum_{h \in H_y} \chi(h) W(y). \tag{9}$$

The reason (9) follows from (8) is purely formal: (8) is of the form $\sum_{x \in X} q(x)$, and X is the disjoint union of the $\mathcal{O}_x$. Thus if one x is chosen from each $\mathcal{O}_x$, i.e., if x is chosen from Δ, then we can sum $q(y)$ over all $y \in \mathcal{O}_x$ and then sum over all $x \in \Delta$. Now, W is constant on $\mathcal{O}_x$ by hypothesis; so $W(y) = W(x)$ for all $y \in \mathcal{O}_x$. Thus (9) becomes

$$\sum_{x \in \Delta} W(x) \sum_{y \in \mathcal{O}_x} \sum_{h \in H_y} \chi(h). \tag{10}$$

We assert that if $y \in \mathcal{O}_x$, then

$$\sum_{h \in H_y} \chi(h) = \sum_{h \in H_x} \chi(h). \tag{11}$$

To confirm (11), first observe that if $y \in \mathcal{O}_x$, so that $y = f(g)x$ for some $g \in H$, then the following statements are equivalent:

$$\begin{aligned} h &\in H_y, \\ f(h)y &= y, \\ f(h)f(g)x &= f(g)x, \\ f(g)^{-1}f(h)f(g)x &= x, \\ f(g^{-1}hg)x &= x, \\ g^{-1}hg &\in H_x, \\ h &\in gH_xg^{-1}. \end{aligned}$$

In other words, $y \in \mathcal{O}_x$ implies

$$H_y = gH_xg^{-1}, \tag{12}$$

and hence

$$\begin{aligned} \sum_{h\in H_y} \chi(h) &= \sum_{h\in gH_xg^{-1}} \chi(h) = \sum_{h\in H_x} \chi(ghg^{-1}) \\ &= \sum_{h\in H_x} \chi(h). \end{aligned}$$

This establishes (11). Now the value of (11) is $|H_x|$ if $x \in \bar{\Delta}$ and 0 if $x \notin \bar{\Delta}$. For in general (see Exercise 1), if $\chi: G \to \mathbf{C}$ is a group homomorphism, we have

$$\sum_{g\in G} \chi(g) = \begin{cases} |G| & \text{if } \chi \equiv 1 \\ 0 & \text{otherwise} \end{cases}. \tag{13}$$

Thus (10) becomes

$$\begin{aligned} \sum_{x\in \Delta} \Big[W(x) \sum_{h\in H_x} \chi(h)\Big] \sum_{y\in \mathcal{O}_x} 1 &= \sum_{x\in \bar{\Delta}} W(x)\, |H_x|\, |\mathcal{O}_x| \\ &= |H| \sum_{x\in \bar{\Delta}} W(x) \qquad \text{(by(2))}. \end{aligned}$$

This completes the proof. ∎

If we specialize the formula (5) by taking $W(x) \equiv 1$, we obtain the following.

Corollary 1 *With the same notation as in Theorem 2.2,*

$$|\bar{\Delta}| = \frac{1}{|H|} \sum_{h\in H} \chi(h)\psi(h) \tag{14}$$

where $\psi(h)$ is the number of $x \in X$ such that $f(h)x = x$:

$$\psi(h) = |\{x \in X \mid f(h)x = x\}|.$$

If, further, $\chi \equiv 1$ *(so that* $\Delta = \bar{\Delta}$*), then*

$$|\Delta| = \frac{1}{|H|} \sum_{h\in H} \psi(h). \tag{15}$$

Theorem 2.3 *Let $H : X$, say $f\colon H \to S_X$ is a homomorphism, and let R be any nonempty finite set. Then H acts on R^X as follows: The function*

$$F\colon H \to S_{R^X}$$

defined by

$$F(h)(\varphi) = \varphi \cdot f(h^{-1}), \qquad \varphi \in R^X, \tag{16}$$

is a homomorphism of the indicated groups.

Proof: By definition, $f(h^{-1}) = f(h)^{-1} \in S_X$; so

$$F(h)(\varphi_1) = F(h)(\varphi_2)$$

implies that $\varphi_1 \cdot f(h)^{-1} = \varphi_2 \cdot f(h)^{-1}$ and hence $\varphi_1 = \varphi_2$. Thus $F(h)$ is an injection on R^X. Both X and R are finite so that $|R^X| < \infty$, and any injection of a finite set into itself is a bijection; therefore $F(h)$ is a bijection on R^X, i.e., $F(h) \in S_{R^X}$. To prove that F is a homomorphism, we must show

$$F(h_1h_2) = F(h_1) \cdot F(h_2), \qquad h_1,h_2 \in H.$$

But

$$\begin{aligned} F(h_1h_2)(\varphi) &= \varphi \cdot f((h_1h_2)^{-1}) \\ &= \varphi \cdot f(h_2^{-1}h_1^{-1}) \\ &= \varphi \cdot f(h_2^{-1}) \cdot f(h_1^{-1}) \\ &= F(h_1)(\varphi \cdot f(h_2^{-1})) \\ &= F(h_1)(F(h_2)(\varphi)) \\ &= (F(h_1) \cdot F(h_2))(\varphi). \ \blacksquare \end{aligned}$$

We now make the following specializations: Let

$$X = [1,n]$$

so that

$$S_X = S_n;$$

let H be a subgroup of S_n,

$$H < S_n;$$

let the homomorphism f be the canonical injection $\iota_H\colon H \to S_n$; finally, let

$$R = [1,m].$$

Thus

$$R^X = [1,m]^{[1,n]}$$

and hence

$$|R^X| = m^n.$$

According to Theorem 2.3, the group H acts on $[1,m]^{[1,n]}$ as follows:

$$F\colon H \to S_{[1,m]^{[1,n]}}$$

is given by

$$F(h)(\varphi) = \varphi f(h)^{-1}, \qquad \varphi \in [1,m]^{[1,n]}. \tag{17}$$

But $f = \iota_H$, so the formula (17) specializes to

$$F(h)(\varphi) = \varphi \cdot h^{-1}.$$

Let the character $\chi: H \to \mathbf{C}$ be $\chi \equiv 1$. We define the so-called *cycle index polynomial of* a subgroup H of S_n:

$$P(H: x_1, \ldots, x_n) = \frac{1}{|H|} \sum_{h \in H} x_1^{\lambda_1(h)} x_2^{\lambda_2(h)} \ldots x_n^{\lambda_n(h)},$$

where $\lambda_1(h)$ is the number of fixed points of h and $\lambda_t(h)$, $t \geq 2$, is the number of cycles of length t in the disjoint cycle decomposition of h. The following extraordinary result is due to G. Pòlya.

Theorem 2.4 (*Pòlya Counting Theorem*) *The number of* $H : [1,m]^{[1,n]}$ *orbits is*

$$P(H: m, \ldots, m) = \frac{1}{|H|} \sum_{h \in H} m^{\sum_{t=1}^{n} \lambda_t(h)}.$$

Proof: We apply (15) with the role of X being taken by $[1,m]^{[1,n]}$. Recall that Δ is a system of distinct representatives for the H: $[1,m]^{[1,n]}$ orbits, i.e., each H-orbit in $[1,m]^{[1,n]}$ contributes precisely one element to Δ. Formula (15) says that the number of such H-orbits is

$$|\Delta| = \frac{1}{|H|} \sum_{h \in H} \psi(h), \tag{18}$$

where in this case

$$\psi(h) = |\{\varphi \in [1,m]^{[1,n]} \mid \varphi \cdot h^{-1} = \varphi\}|.$$

In order to count the number of $\varphi \in [1,m]^{[1,n]}$ for which $\varphi \cdot \sigma = \varphi$, where $\sigma \in S_n$, let $\mathcal{O}$ be any $[\sigma]$-orbit in $[1,n]$:

$$\mathcal{O} = \{j, \sigma(j), \ldots, \sigma^{k-1}(j)\}, \qquad k = |[\sigma]|.$$

Then if $\varphi(\sigma(t)) = \varphi(t)$ for all $t \in [1,n]$, we have

$$\begin{aligned} \varphi(j) &= \varphi(\sigma(j)) = \varphi(\sigma(\sigma(j))) \\ &= \varphi(\sigma^2(j)) = \cdots \\ &= \varphi(\sigma^{k-1}(j)) = \varphi(\sigma^k(j)) \\ &= \varphi(j). \end{aligned}$$

In other words, $\varphi \cdot \sigma = \varphi$ implies φ is constant on every $[\sigma]$-orbit $\mathcal{O}$. Conversely, suppose φ is constant on every $[\sigma]$-orbit $\mathcal{O}$. If $j \in [1,n]$, then j belongs to some $[\sigma]$-orbit $\mathcal{O}$, so that $(\varphi \cdot \sigma)(j) = \varphi(\sigma(j)) = \varphi(j)$. Thus $\varphi \cdot \sigma = \varphi$. We have proved (since $[h] = [h^{-1}]$) that

$$\psi(h) = |\{\varphi \in [1,m]^{[1,n]} \mid \varphi \text{ is constant on every } [h]\text{-orbit } \mathcal{O}\}|. \tag{19}$$

Now suppose $h \in H$ has cycle structure

$$[1^{\lambda_1(h)}, 2^{\lambda_2(h)}, \ldots, n^{\lambda_n(h)}],$$

i.e., $[h]$ has $\lambda_t(h)$ orbits each containing t elements, $t = 1, \ldots, n$. We write out the cycle decomposition of h:

$$\overbrace{(*)(*)\cdots(*)}^{\lambda_1(h)} \quad \overbrace{(**)\cdots(**)}^{\lambda_2(h)} \cdots \overbrace{\underbrace{(*\cdots*)}_{t}\cdots\underbrace{(*\cdots*)}_{t}}^{\lambda_t(h)} \cdots. \tag{20}$$

The function values of an element φ in the set on the right side of (19) are the same on each of the integers appearing in each particular t-cycle. Thus there are as many such φ as there are ways of assigning an integer in $[1,m]$ to each of the cycles appearing in (20). There are m ways of doing this for any particular cycle, and there are a total of $\lambda_1(h)+\lambda_2(h)+\cdots+\lambda_t(h)+\cdots+\lambda_n(h)$ cycles. Therefore we have

$$m^{\sum_{t=1}^{n}\lambda_t(h)}$$

ways of defining such a φ, i.e.,

$$\psi(h) = m^{\sum_{t=1}^{n}\lambda_t(h)} \tag{21}$$

Combining (18) and (21) produces the result. ∎

Example 1 Five jewels chosen from emeralds [designated by (1)], diamonds (2), and sapphires (3) are to be set equally spaced around a crown. Clearly, two such settings are indistinguishable iff one can be obtained from the other by a rotation of the crown. The question is: How many distinct such settings are there? We proceed as follows: Label by 1, 2, 3, 4, 5 the five equally spaced positions on the crown into which the jewels are to be set. We can think of a setting as an element $\varphi \in [1,3]^{[1,5]}$, e.g., if $\varphi(1) = 2$, $\varphi(2) = 3$, $\varphi(3) = 1$, $\varphi(4) = 2$, $\varphi(5) = 3$, then an emerald (1) is placed in the position labeled 3, diamonds (2) are placed in positions 1 and 4, and sapphires (3) are placed in positions 2 and 5. Let $\sigma = (1\ 2\ 3\ 4\ 5) \in S_5$ and let $H = [\sigma]$. Then a rotation can be identified with an element of H. Two settings φ and θ are indistinguishable iff they are in the same H-orbit, i.e., $\varphi \cdot h = \theta$ for some $h \in H$. Thus the problem is to count the number of H-orbits in $[1,3]^{[1,5]}$. The cycle structure of each element $h \in H$ other than e is $[1^0,2^0,3^0,4^0,5^1]$ and of course the cycle structure of e is $[1^5,2^0,3^0,4^0,5^0]$. We have $|H| = 5$. Therefore the cycle index polynomial of H is

$$P(H\colon x_1,x_2,x_3,x_4,x_5) = \tfrac{1}{5}(4 \cdot x_5 + x_1^{5}).$$

By Theorem 2.4, the required number is

$$P(H\colon 3,3,3,3,3) = \tfrac{1}{5}(4\cdot 3 + 3^5) = 51.$$

Example 2 Determine the cycle index polynomial for the group of all rotations of a cube. By a rotation of a cube we mean a rotation about some axis which returns the cube to its original orientation in space. For an application we wish to make momentarily, we shall regard a rotation as a permutation of the eight vertices of the cube, labeled 1, . . ., 8. Two such rotations are distinct iff the positions of the vertices after performing the rotations are different. The rotations can be divided into five distinct types.

Type I. The identity permutation e with $\lambda_1(e) = 8$, $\lambda_t(e) = 0$ for $t \neq 1$. Thus x_1^{8} is contributed to the cycle index polynomial.

Type II. Rotations through $\pm\, 90°$ about an axis connecting the midpoints of two opposite faces. Clearly the permutations are of type $h = (1\ 2\ 3\ 4)\,(5\ 6\ 7\ 8)$ and hence $\lambda_4(h) = 2$, $\lambda_t(h) = 0$ for $t \neq 4$. For each of the three possible axes there are two such permutations, so there are six such permutations in the group altogether. These produce a total contribution of $6x_4^{2}$ to the cycle index polynomial.

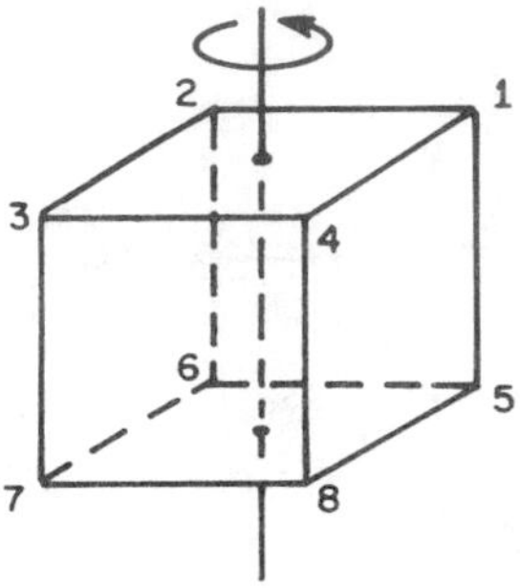

Type III. Rotations through $\pm 180°$ about one of the axes described above. The permutations are of the type $h = (1\ 3)(2\ 4)(5\ 7)(6\ 8)$, and hence $\lambda_2(h) = 4$, $\lambda_t(h) = 0$ for $t \neq 4$. Observe that rotations through $180°$ or $-180°$ are the same permutation, so that there is only one such permutation for each of the three axes. This gives a contribution of $3x_2^{4}$ to the cycle index polynomial.

Type IV. Rotation through $\pm 120°$ about an axis through a pair of opposite vertices.

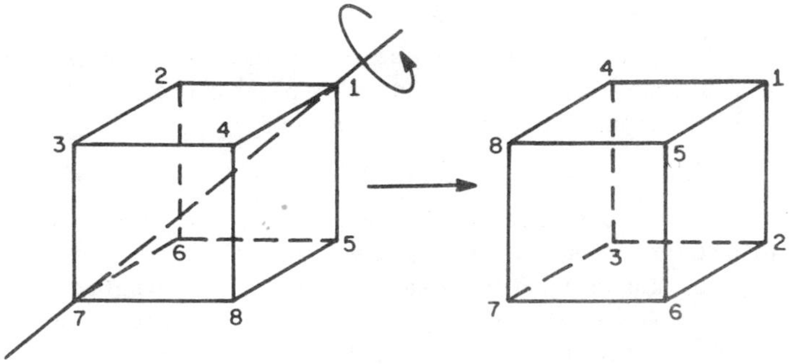

Such a permutation is of the type $h = (1)(7)(2\ 5\ 4)(3\ 6\ 8)$ and hence $\lambda_1(h) = 2$, $\lambda_3(h) = 2$, $\lambda_t(h) = 0$ for $t \neq 1, 3$. There are two rotations for each of the four axes of this kind, so the total contribution to the cycle index polynomial is $8x_1{}^2x_3{}^2$.

Type V. Rotations through $\pm 180°$ about an axis through the midpoints of a pair of opposite edges.

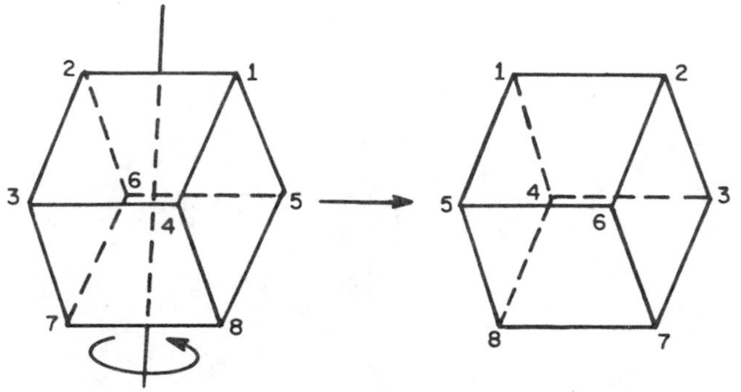

These permutations are of the type $h = (1\ 2)(3\ 5)(4\ 6)(7\ 8)$ and hence $\lambda_2(h) = 4$, $\lambda_t(h) = 0$ for $t \neq 2$. There is one rotation for each of the six axes of this kind, so the total contribution to the cycle index polynomial is $6x_2{}^4$.

The number of elements in H is $1 + 6 + 3 + 8 + 6 = 24$. Thus

$$\begin{aligned} P(H\colon x_1, x_2, \ldots, x_8) &= \tfrac{1}{24}(x_1{}^8 + 6x_4{}^2 + 3x_2{}^4 + 8_1{}^2x_3{}^2 + 6x_2{}^4) \\ &= \tfrac{1}{24}(x_1{}^8 + 9x_2{}^4 + 8x_1{}^2x_3{}^2 + 6x_4{}^2). \qquad (22) \end{aligned}$$

Example 3. The eight vertices of a cube are to be painted with four different colors. In how many different ways can this be done? We saw in (22) that the cycle index polynomial for the group H of rotations of the cube is

$$P(H\colon x_1, \ldots, x_8) = \tfrac{1}{24}(x_1{}^8 + 9x_2{}^4 + 8x_1{}^2x_3{}^2 + 6x_4{}^2).$$

A painting of the cube can be regarded as a function $\varphi \in [1,4]^{[1,8]}$, where $\varphi(i) = j$ indicates that vertex i is painted color j. Two paint jobs φ and θ on the vertices of the cube are indistinguishable iff one can be obtained from the other by a rotation $h \in H$, i.e., $\varphi \cdot h = \theta$. Thus the number of distinct paint jobs is the number of H-orbits in $[1,4]^{[1,8]}$. According to Theorem 2.4, this is precisely

$$\begin{aligned} P(H\colon 4, 4, 4, \ldots, 4) &= \tfrac{1}{24}(4^8 + 9 \cdot 4^4 + 8 \cdot 4^2 \cdot 4^2 + 6 \cdot 4^2) \\ &= 2916. \end{aligned}$$

Theorem 2.4 produces a count of the number of H-orbits when H acts on the set $R^X = [1,m]^{[1,n]}$ by $F(h) = \varphi \cdot h^{-1}$. The count followed from the formula (18)

$$|\Delta| = \frac{1}{|H|} \sum_{h \in H} \psi(h), \tag{23}$$

where $$\psi(h) = |\{\varphi \in [1,m]^{[1,n]} \mid \varphi \cdot h^{-1} = \varphi\}|. \tag{24}$$

We wish to extend the use of the result by regarding H as acting on some subset $M \subset R^X$ in precisely the same way. Clearly, the count of the number of distinct H-orbits in M will be given by (23) if

$$\psi(h) = |\{\varphi \mid \varphi \in M \quad \text{and} \quad \varphi \cdot h^{-1} = \varphi\}|.$$

Let $s_\lambda = [1^{\lambda_1}, \ldots, n^{\lambda_n}]$ be a cycle structure, so that $1\lambda_1 + 2\lambda_2 + \cdots + n\lambda_n = n$. An *$s_\lambda$-partition* of a set X with n elements is a partition of X into λ_i subsets of i elements each, $i = 1, \ldots, n$.

Let $M \subset [1,m]^{[1,n]}$ (i.e., $R = [1,m]$, $X = [1,n]$) be the subset

$$M = \{\varphi \mid |\varphi^{-1}(t)| = n_t, \quad t = 1, \ldots, m\}$$

where the n_t are prescribed integers satisfying $0 \leq n_t \leq n$, $t = 1, \ldots, m$. It is obvious that H acts on M. For if σ is any permutation in S_n and $\varphi \in M$, then a moment's reflection shows that $\varphi \cdot \sigma^{-1} \in M$ as well. Now let $h \in H$ have cycle structure s_λ. If $\varphi \in [1,m]^{[1,n]}$, we know that $\varphi \cdot h^{-1} = \varphi$ iff φ is constant on each of the sets of the s_λ-partition of $[1,n]$ induced by h [see the discussion centering around (19) and (20)]. Consider the set

$$C(h) = M \cap \{\varphi \mid \varphi \cdot h^{-1} = \varphi\};$$

in other words, $\varphi \in C(h)$ iff φ takes on the value t precisely n_t times, $t = 1, \ldots, m$, and φ is constant on each of the sets of the s_λ-partition of $[1,n]$ induced by h [$C(h)$ may very well be empty]. Let $h_1 \in H$, and suppose h_1 has the same cycle structure s_λ as h does, so that $h_1 = \tau^{-1} h \tau$ for some fixed permutation $\tau \in S_n$ [see Theorem 3.6(i), Chapter 1]. In general, of course, $C(h) \neq C(h_1)$, but we assert that $|C(h)| = |C(h_1)|$. For consider the function η defined on $C(h)$ by $\eta(\varphi) = \varphi \cdot \tau$. Obviously η is injective, and $\varphi \in M$ iff $\eta(\varphi) \in M$. Also

$$\varphi \in C(h)$$

iff $$\varphi \in M \text{ and } \varphi \cdot h = \varphi$$

iff $$\varphi\tau \in M \text{ and } \varphi\tau h_1 \tau^{-1} = \varphi$$

iff $$\varphi\tau \in M \text{ and } (\varphi\tau)h_1 = \varphi\tau$$

iff $$\eta(\varphi) = \varphi\tau \in C(h_1).$$

It follows that η: $C(h) \to C(h_1)$ is a bijection.

For all $h \in H$ with cycle structure s_λ we define

$$\mu(s_\lambda) = |\, C(h) \,|,$$

which according to the preceding discussion is independent of h and depends only on s_λ. We are now in a position to state the following result, also due to G. Pòlya.

Theorem 2.5 *Let $H < S_n$, and for each cycle structure*

$$s_\lambda = [1^{\lambda_1}, 2^{\lambda_2}, \ldots, n^{\lambda_n}]$$

let $\nu(s_\lambda)$ be the number of permutations $h \in H$ that have cycle structure s_λ. Then the number of $H : M$ orbits is

$$|\, \Delta \,| = \frac{1}{|\, H \,|} \sum_{s_\lambda} \nu(s_\lambda)\mu(s_\lambda). \tag{25}$$

Proof: From (23) and the remarks following this formula we have

$$|\, \Delta \,| = \frac{1}{|\, H \,|} \sum_{h \in H} \psi(h),$$

where
$$\begin{aligned}\psi(h) &= |\, \{\varphi \mid \varphi \in M \text{ and } \varphi \cdot h^{-1} = \varphi\} \,| \\ &= |\, \{\varphi \mid |\varphi^{-1}(t)| = n_t,\ t = 1, \ldots, m, \text{ and } \varphi \cdot h = \varphi\} \,| \\ &= |\, C(h) \,|.\end{aligned}$$

Thus $\psi(h)$ is a count of the number of $\varphi \in M$ which are constant on each of the sets of the partition of $[1, n]$ obtained from the cycle decomposition of h. This number, which we have called $\mu(s_\lambda)$, is the same for all h with a given cycle structure s_λ. Denote the conjugacy class in H corresponding to each s_λ by H_{s_λ}. Then for any $h \in H_{s_\lambda}$,

$$\psi(h) = |C(h)| = \mu(s_\lambda).$$

Since the conjugacy classes partition H, we have

$$\begin{aligned}\sum_{h \in H} \psi(h) &= \sum_{s_\lambda} \sum_{h \in H s_\lambda} \psi(h) = \sum_{s_\lambda} \sum_{h \in H s_\lambda} \mu(s_\lambda) \\ &= \sum_{s_\lambda} |\, H_{s_\lambda} \,|\, \mu(s_\lambda).\end{aligned} \tag{26}$$

But $|H_{s_\lambda}|$ is precisely $\nu(s_\lambda)$. The formula (25) now follows from (26). ∎

Example 4 The eight vertices of a cube are to be painted either red (1) or blue (2), four of them red and four of them blue. In how many different ways can this be done? Let H be the group of rotations of the cube described in Example 2. According to (25), if there are no elements in H with cycle structure s_λ, then $\nu(s_\lambda) = 0$ and no contribution to the sum results. Thus we must compute $\mu(s_\lambda)$ and $\nu(s_\lambda)$ for each s_λ such that there exists an $h \in H$ with cycle structure s_λ. We refer to Example 2.

Type I. The identity e: $s_\lambda = [1^8, 2^0, \ldots, 8^0]$, $\nu(s_\lambda) = 1$. To compute

$\mu(s_\lambda)$, we must count the number of $\varphi \in [1,2]^{[1,8]}$ such that

$$| \varphi^{-1}(1) | = 4, \qquad | \varphi^{-1}(2) | = 4, \tag{27}$$

and φ is constant on each subset in a partition of [1,8] into one element subsets. Of course, this last condition is vacuous and thus

$$\mu(s_\lambda) = \binom{8}{4}.$$

Type II. Permutations of the type (1 2 3 4)(5 6 7 8): $s_\lambda = [1^0, 2^0, 3^0, 4^2, 5^0, \ldots, 8^0]$, $\nu(s_\lambda) = 6$. To compute $\mu(s_\lambda)$, we must count the number of φ satisfying (27) which are constant on each subset in a partition of [1,8] into two four-element subsets. Since it is required that $|\varphi^{-1}(1)| = 4$, we have (for example) that either $\varphi^{-1}(1) = \{1,2,3,4\}$ or $\varphi^{-1}(1) = \{5,6,7,8\}$. There are two such φ; so $\mu(s_\lambda) = 2$. Hence

$$\nu(s_\lambda)\mu(s_\lambda) = 2 \cdot 6 = 12.$$

Type III and *Type V.* These permutations are of the type (1 3)(2 4)(5 7)(6 8): $s_\lambda = [1^0, 2^4, 3^0, \ldots, 8^0]$, $\nu(s_\lambda) = 3 + 6 = 9$. We must count the number of φ satisfying (27) and which are constant on each of the subsets {1,3}, {2,4}, {5,7}, {6,8}. Since $|\varphi^{-1}(1)| = 4$, we must choose two of these subsets so that φ takes the value 1 on their union (and of course φ takes on the value 2 on the union of the remaining two subsets). Thus $\mu(s_\lambda)$ is just a count of the number of ways of choosing 2 of the cycles out of the 4, so $\mu(s_\lambda) = \binom{4}{2} = 6$. Then

$$\nu(s_\lambda)\mu(s_\lambda) = 9 \cdot 6 = 54.$$

Type IV. Permutations of the type (1)(7)(2 5 4)(3 6 8): $s_\lambda = [1^2, 2^0, 3^2, 4^0, \ldots, 8^0]$, $\nu(s_\lambda) = 8$. We must count the number of φ satisfying (27) which are constant on each of the subsets {1}, {7}, {2,5,4}, {3,6,8}. Since $|\varphi^{-1}(1)| = 4$, it follows that $\varphi^{-1}(1)$ must consist of the union of exactly one of the two one-element subsets and exactly one of the two three-element subsets. Of course, $\varphi^{-1}(2)$ will consist of the union of the remaining subsets. There are $2 \cdot 2 = 4$ such φ, so $\mu(s_\lambda) = 4$. Hence

$$\nu(s_\lambda)\mu(s_\lambda) = 8 \cdot 4 = 32.$$

Recall that $|H| = 24$. We now have from (25) that the total number of distinct paint jobs of the desired type is

$$| \Delta | = \tfrac{1}{24}[\binom{8}{4} + 12 + 54 + 32] = 7.$$

The Lagrange theorem [Theorem 3.7(ii), Chapter 1] states that the order of a subgroup of a finite group must always divide the order of the group. We now turn our attention to the converse of this result. Namely, for a given integral divisor of the order of a finite group, under what circumstances does there exist a subgroup of this order? Results of this kind will tell us, for example, that a group of order 231 contains normal subgroups of orders 7 and 11; or if p and q are primes, $p > q$, and $[p]_q \neq [\pm 1]_q$, then any group of order p^2q must be abelian. The central ideas involved in approaching these questions are precisely those we have used in proving Burnside's Lemma (Theorem 2.2) and the Pòlya Counting Theorem (Theorem 2.4).

Let G be a finite group acting on a set X; say, $f\colon G \to S_X$ is a homomorphism. In the course of the proof of Theorem 2.2 [see formula (12)] we showed that for $x, y \in X$, if

$$y = f(g)x$$

for some $g \in G$ (i.e., if x and y are in the same G-orbit in X), then the stabilizer subgroups G_x and G_y are related by

$$G_y = gG_xg^{-1}. \tag{28}$$

It is perhaps worthwhile to recapitulate the proof of (28):

$$h \in G_y$$

iff $$f(h)y = y$$

iff $$f(h)f(g)x = f(g)x$$

iff $$f(g^{-1}hg)x = x$$

iff $$g^{-1}hg \in G_x$$

iff $$h \in gG_xg^{-1}.$$

Recall that two subgroups H and K of a group G are said to be *conjugate subgroups* if there exists $g \in G$ such that

$$K = gHg^{-1}.$$

Thus (28) states that the stabilizers of any two elements x and y in the same G-orbit are conjugate subgroups.

We have the following more specific information concerning these conjugate stabilizer subgroups.

Theorem 2.6 *Let the finite group G act on a set X, say $f: G \to S_X$ gives the group action, and let $\mathcal{O}_x$ be a G-orbit in X. Let r be the number of distinct subgroups among the stabilizers G_y of elements $y \in \mathcal{O}_x$. Then each G_y is the stabilizer of the same number of elements in $\mathcal{O}_x$, call this number s, and*

$$| \mathcal{O}_x | = rs. \tag{29}$$

Proof: By (28) the various stabilizers G_y, $y \in \mathcal{O}_x$, are conjugate subgroups. Let

$$g_tG_xg_t^{-1}, \qquad t = 1, \ldots, r$$

be all the distinct ones. Every stabilizer of an element in $\mathcal{O}_x$ appears among these r groups and since these groups are distinct, no two of them are the stabilizer of the same element in $\mathcal{O}_x$. Thus we can write

$$\mathcal{O}_x = U_1 \cup U_2 \cup \cdots \cup U_r \qquad \text{(disjoint)}, \tag{30}$$

where U_t is the set of elements in $\mathcal{O}_x$ for which $g_tG_xg_t^{-1}$ is the stabilizer. Consider the action of the element $g_ig_j^{-1}$ on an element $z \in U_j$. We have

$$z \in U_j$$

iff $\qquad f(g_j g g_j^{-1})z = z \qquad$ for all $g \in G_x$

iff $\qquad f(g_j g_i^{-1})f(g_i g g_i^{-1})f(g_i g_j^{-1})z = z \qquad$ for all $g \in G_x$

iff $\qquad f(g_i g g_i^{-1})f(g_i g_j^{-1})z = f(g_i g_j^{-1})z \qquad$ for all $g \in G_x$

iff $\qquad f(g_i g_j^{-1})z \in U_i.$

Thus $\qquad z \in U_j \qquad$ iff $f(g_i g_j^{-1})z \in U_i$;

since the group G acts on X by permuting the elements of X [i.e., $f(g_i g_j^{-1}) \in S_X$], it follows that $|U_i| = |U_j|$. Let s be this common number of elements. Then (30) implies that

$$|\mathcal{O}_x| = \sum_{t=1}^{r} |U_t| = \sum_{t=1}^{r} s = rs. \quad \blacksquare$$

The most important result in this development is the following theorem.

Theorem 2.7 *(The Sylow Theorems) Let G be a finite group, say*

$$|G| = n = p^\alpha q,$$

where $\alpha \geq 1$ and p is a prime. Then

(i) *G contains a subgroup of order p^α.*

(ii) *If $p \nmid q$ (i.e., p is not a divisor of q), then all subgroups of G of order p^α are conjugate.*

(iii) *In* (ii), *the number of these conjugate subgroups is congruent to* 1 *modulo p and divides q.*

Proof: (i) We first remark (see Exercise 12) that if p^β is the highest power of p that divides n, then $p^{\beta-\alpha}$ is the highest power of p that divides the binomial coefficient $\binom{n}{p^\alpha}$. Let

$$X = P(G) \cap \{S \mid |S| = p^\alpha\},$$

i.e., X is the family of all subsets of G which contain precisely p^α elements. Let $g \in G$ and define $f(g) \in S_X$ by

$$f(g)(S) = gS, \qquad S \in X. \tag{31}$$

The formula (31) makes sense because S is a subset of G. It is trivial to confirm that G operates on X according to the definition (31). Let

$$S = \{g_1, \ldots, g_{p^\alpha}\}$$

be a set in X and let $g \in G_S$, the stabilizer of S. Then

$$gS = S$$

so that for some k, $1 \leq k \leq p^\alpha$,

$$gg_1 = g_k,$$

or
$$g = g_k g_1^{-1}. \tag{32}$$
Thus any $g \in G_S$ must be one of the p^α elements (32). It follows that
$$|G_S| \leq p^\alpha. \tag{33}$$

Next, let $\mathcal{O}^1, \mathcal{O}^2, \ldots, \mathcal{O}^k$ be all the distinct G-orbits in X, so that
$$|X| = \sum_{t=1}^{k} |\mathcal{O}^t|. \tag{34}$$

We know that $|X| = \binom{n}{p^\alpha}$, i.e., there are $\binom{n}{p^\alpha}$ subsets of G which contain precisely p^α elements. Since $\binom{n}{p^\alpha}$ is not divisible by $p^{\beta-\alpha+1}$, (34) shows that $p^{\beta-\alpha+1}$ cannot divide every one of the numbers $|\mathcal{O}^t|$, $t = 1, \ldots, k$. Let $\mathcal{O}$ (dropping the superscript) be one of these orbits for which
$$p^{\beta-\alpha+1} \nmid |\mathcal{O}|. \tag{35}$$

If $S \in \mathcal{O}$, then
$$\mathcal{O} = \mathcal{O}_S = \{gS \mid g \in G\}$$
and by Theorem 2.1 we have
$$|G_S|\,|\mathcal{O}_S| = |G|,$$
or
$$|G_S|\,|\mathcal{O}| = p^\beta d \tag{36}$$
where again p^β is the highest power of p dividing n. Then
$$|\mathcal{O}| = \frac{p^\beta d}{|G_S|}$$
and hence the integer $|G_S|$ must contain p^α as a factor, for otherwise $|\mathcal{O}|$ would be divisible by $p^{\beta-\alpha+1}$, in contradiction to (35). This fact together with (33) implies that
$$|G_S| = p^\alpha. \tag{37}$$
Thus we have exhibited a subgroup of G, namely G_S, which contains precisely p^α elements. This proves (i).

(ii) To prove (ii), suppose that $\alpha = \beta$ in the proof of (i), i.e., p^α is the highest power of p dividing n. Then (35) says that
$$p \nmid |\mathcal{O}| \tag{38}$$
for some G-orbit $\mathcal{O}$ in X. Let $S \in \mathcal{O}$. We are going to show that if $H < G$ and $|H| = p^\alpha$, then $H = G_T$ where $T \in \mathcal{O}_S = \mathcal{O}$. This will prove (ii), for then $H = G_T$ and G_S are stabilizers of elements in the same G-orbit $\mathcal{O}_S$ and hence are conjugate.

Since G acts on X, it is clear that G acts on $\mathcal{O}$ and hence H acts on $\mathcal{O}$. Let $Q^1, \ldots, Q^m$ be all the H-orbits in $\mathcal{O}$, so that
$$|\mathcal{O}| = \sum_{t=1}^{m} |Q^t|. \tag{39}$$

Now (38) and (39) imply that

$$p \nmid |Q^t| \tag{40}$$

for at least one t. For this t, write $Q = Q^t$. If $T \in Q$, then

$$Q = Q_t = \{hT \mid h \in H\}$$

and Theorem 2.1 shows that

$$|H_T|\,|Q| = |H| = p^\alpha,$$

or

$$|Q| = \frac{p^\alpha}{|H_T|}. \tag{41}$$

But $p \nmid |Q|$ by (40), and so (41) immediately implies that

$$|H_T| = p^\alpha.$$

Since $H_T < H$ and $|H_T| = p^\alpha = |H|$, we conclude that

$$H = H_T. \tag{42}$$

Now $T \in \mathcal{O}$ and as we proved in part (i), $|G_T| = p^\alpha$. Since $H = H_T < G_T$, it follows that

$$H = G_T,$$

completing the proof of (ii).

Observe that we have proved that *if p^α is the highest power of p dividing n, then every subgroup of order p^α is the G-stabilizer of an element in the G-orbit $\mathcal{O}$, where $p \nmid |\mathcal{O}|$*. These various conjugate subgroups of order p^α are called the *p-Sylow subgroups* of G.

(iii) To prove (iii), suppose $H_1, \ldots, H_r$ are all the distinct (and conjugate) p-Sylow subgroups of G of order p^α (we are assuming that p^α is the highest power of p dividing n). According to what we have proved, these are precisely the distinct G- stabilizers of elements in the G-orbit $\mathcal{O}$, i.e., $H_1, \ldots, H_r$ are all the distinct groups of the form

$$G_S$$

where S varies over $\mathcal{O}$. Next, let P_1 be the subset of $\mathcal{O}$ consisting of those $S \in \mathcal{O}$ for which $H_1 = G_S$. That is, P_1 is the set of elements in $\mathcal{O}$ for which H_1 is the G-stabilizer. Of course, each element of P_1 itself comprises an H_1-orbit in $\mathcal{O}$. If $T \in \mathcal{O}$ and $T \notin P_1$ then the H_1-orbit U in $\mathcal{O}$ containing T must contain more than one element, for otherwise T is held fixed by the elements of H_1 and hence $T \in P_1$. Again using Theorem 2.1, we have

$$|H_{1T}|\,|U| = |H_1| \tag{43}$$

where H_{1T} is the H_1-stabilizer of T. But

$$|H_1| = p^\alpha, \qquad |U| > 1;$$

so (43) implies that $|U|$ must be a positive power of p. We have proved that except for the s one-element H_1-orbits which are the elements of P_1, any other H_1-orbit in $\mathcal{O}$ has a positive power of p elements in it. Since the H_1-orbits in $\mathcal{O}$ partition $\mathcal{O}$, it follows that

$$|\mathcal{O}| = s + mp$$

for some positive integer m, and hence

$$|\mathcal{O}| \equiv s \pmod{p}. \tag{44}$$

Recall that $p \nmid |\mathcal{O}|$, and so by (44), $s \not\equiv 0 \pmod{p}$. From Theorem 2.6 we know that $|\mathcal{O}|$ is the product of the number of distinct G-stabilizers of elements in $\mathcal{O}$, namely r (these are the groups $H_1, \ldots, H_r$), and the number of points held fixed by any one of these stabilizers, namely s (each of $H_1, \ldots,$ H_r holds the same number of elements in $\mathcal{O}$ fixed according to Theorem 2.6). Hence

$$|\mathcal{O}| = rs. \tag{45}$$

Now (44) implies that

$$rs \equiv s \pmod{p},$$
$$(r - 1)s \equiv 0 \pmod{p},$$

and since $s \not\equiv 0 \pmod{p}$ [see the remark following (44)], we have $r - 1 \equiv 0 \pmod{p}$, or

$$r \equiv 1 \pmod{p}.$$

Finally, for $S \in \mathcal{O}$ we have

$$
\begin{aligned}
|\mathcal{O}| &= \frac{|G|}{|G_S|} && \text{[recall (36) with } \beta = \alpha\text{]}\\
&= \frac{p^\alpha q}{p^\alpha} && \text{[see (37)]}\\
&= q;
\end{aligned}
$$

so (45) yields

$$rs = q$$

and hence $$r \mid q. \quad \blacksquare$$

Example 5 We show that if $|G| = 231$, then G contains normal subgroups of orders 7 and 11. To see this, first note that

$$231 = 3 \cdot 7 \cdot 11.$$

According to Theorem 2.7(i), (ii) with $p = 7$, G contains a subgroup H of order 7, and all subgroups of G of order 7 are conjugate. By Theorem 2.7(iii), the number r of such conjugate subgroups satisfies

$$r \equiv 1 \pmod{7} \tag{46}$$

and
$$r \mid 33. \tag{47}$$
Then (47) limits the possibilities for r to 1, 3, 11, 33 and 3, 11, 33 are excluded by (46). Hence $r = 1$. This means that the number of seven-element conjugate subgroups in G is 1; so
$$gHg^{-1} = H$$
for all $g \in G$, and we conclude that
$$H \lhd G.$$
A similar argument shows that G has a normal subgroup of order 11.

Example 6 We show in this example that although
$$|A_5| = 60,$$
A_5 has no subgroup of order 30. Assume that $H < A_5$ and $|H| = 30$. Then by Theorem 3.7(ii), Chapter 1,
$$[A_5 : H] = 2. \tag{48}$$
It follows that $H \lhd A_5$ (see Exercise 16, Section 2.1). This is a contradiction, since according to Theorem 3.10(v), Chapter 1, A_5 is simple.

There are a number of interesting and highly nontrivial consequences of Theorem 2.7 available.

Theorem 2.8 *Assume that p is a prime and G is a group of order p^α, $\alpha \geq 1$. Then G has a subgroup of order $p^{\alpha-1}$, and every subgroup of G of order $p^{\alpha-1}$ is normal in G.*

Proof: By Theorem 2.7(i) (with $\alpha - 1$ replacing α), we know that G has a subgroup H of order $p^{\alpha-1}$. Let
$$G = \bigcup_{t=1}^{p} a_t H \qquad \text{(disjoint)}$$
be the disjoint union of all the distinct left H-cosets in G; there are precisely p of these since
$$\begin{aligned}[G : H] &= |G| / |H| = p^\alpha / p^{\alpha-1} \\ &= p.\end{aligned}$$
Let
$$L = \{a_1 H, \ldots, a_p H\} \tag{49}$$
and observe that H acts on L by left multiplication as defined in (31). Decompose L into its disjoint H-orbits:
$$L = \bigcup_{t=1}^{r} \mathcal{O}^t \tag{50}$$
with the notation so chosen that $H \in \mathcal{O}^1$. Then it is clear that H is the only element in $\mathcal{O}^1$, since $h \in H$ implies $hH = H$. Thus $|\mathcal{O}^1| = 1$, and from (49) and (50) we obtain
$$\begin{aligned} p = |L| &= \sum_{t=1}^{r} |\mathcal{O}^t| \\ &= 1 + \sum_{t=2}^{r} |\mathcal{O}^t|. \end{aligned} \tag{51}$$

It is clear from (51) that

$$|\mathcal{O}^t| < p, \qquad t = 2, \ldots, r. \tag{52}$$

We next prove that for any $t \geq 2$, the stabilizer of an element $a_kH \in \mathcal{O}^t$ is given by

$$H_{a_kH} = H \cap a_kHa_k^{-1}. \tag{53}$$

For if $h \in H$, then

$$h \in H_{a_kH}$$

iff $$ha_kH = a_kH$$

iff $$a_k^{-1}ha_kH = H$$

iff $$a_k^{-1}ha_k \in H$$

iff $$h \in a_kHa_k^{-1}.$$

Note that since $H_{a_kH} < H$ and $|H| = p^{\alpha-1}$, we have

$$|H_{a_kH}| = p^\gamma \tag{54}$$

for some nonnegative integer γ. By Theorem 2.1 it follows that

$$\begin{aligned} |\mathcal{O}^t| &= |H| / |H_{a_kH}| = p^{\alpha-1}/p^\gamma \\ &= p^\nu, \qquad \nu \geq 0. \end{aligned} \tag{55}$$

Combining (52) and (55), we obtain $\nu = 0$,

$$|\mathcal{O}^t| = 1,$$

and $$|H| = |H_{a_kH}|.$$

Hence $$H_{a_kH} = H,$$

and so (53) implies that

$$a_kHa_k^{-1} = H.$$

Since the representatives a_k of thc cosets listed in (49) can be chosen as arbitrary elements of G, we conclude that

$$H \lhd G. \quad \blacksquare$$

Example 7 If G is a group of order 27, then G must contain a normal subgroup of order 9, and every subgroup of order 9 must be normal.

Example 8 There are no simple groups of order 148. Note that $148 = 37 \cdot 4$, so by Theorem 2.7(i), a group G of order 148 contains a Sylow subgroup H of order 37. If r is the number of conjugates of H, then Theorem 2.7(iii) implies that $r \mid 4$ and $r \equiv 1 \pmod{37}$. Hence $r = 1$ so that H is normal in G, and G is not simple.

Theorem 2.9 *Assume that p and q are primes, $p \geq q$, and*

$$p \not\equiv 1 \pmod{q}.$$

Let G be a group of order pq.

(i) *If $p > q$, then*

$$G = HK \qquad (direct)$$

where $|H| = p$, $|K| = q$; also G is cyclic (and hence abelian).

(ii) *If $p = q$ (so that $|G| = p^2$), then either G is cyclic or G is the direct product of two cyclic subgroups of order p; in either case G is abelian.*

Proof: (i) By Theorem 2.7(i), G contains Sylow subgroups H and K with

$$|H| = p, \qquad |K| = q$$

(H and K are Sylow subgroups since $p \nmid q$ and $q \nmid p$). Since $p \neq q$ the Lagrange theorem [Theorem 3.7(ii), Chapter 1] implies that

$$H \cap K = \{e\}. \tag{56}$$

By Theorem 2.7(iii), if r is the number of conjugate subgroups of H, then

$$r \equiv 1 \pmod{p} \tag{57}$$

and

$$r \mid q. \tag{58}$$

Since q is a prime, we have $r = 1$ or $r = q$. But $r = q$ is ruled out by (57), for then q would exceed p. Hence $r = 1$ and H is normal in G. Similarly if r_1 is the number of conjugates of K, then

$$r_1 \equiv 1 \pmod{q} \tag{59}$$

and

$$r_1 \mid p.$$

Again $r_1 = 1$ or $r_1 = p$. The latter alternative is ruled out by (59) and the initial assumption that $p \not\equiv 1 \pmod{q}$. Thus $r_1 = 1$ and K is normal in G.

Write

$$H = \{h_1, \ldots, h_p\}$$

and note that

$$h_1K, \ldots, h_pK$$

is the complete list of distinct left K-cosets in G. For since $|G| = pq$ and $|K| = q$, there are p distinct left K-cosets in G, and

$$h_sK = h_tK$$

implies

$$h_t^{-1}h_s \in K$$

so that $h_t = h_s$ by (56). It follows that any element of G must be of the form hk, $h \in H$, $k \in K$, i.e.,

$$G = HK.$$

Theorem 1.6 now shows that

$$G = HK \qquad \text{(direct).} \tag{60}$$

Since both H and K are groups of prime order, they are in fact cyclic. It is not difficult to verify that a direct product of cyclic groups of relatively prime orders is itself cyclic. Therefore G is cyclic.

(ii) Since $|G| = p^2$, we apply Theorem 2.8 to secure $H \lhd G$ with $|H| = p$. Let $c \in G - H$; then either the cyclic subgroup $[c]$ is G or

$$|[c]| = p. \tag{61}$$

If $[c] = G$, we are done. Otherwise (61) holds and we let

$$K = [c].$$

Observe that $H \cap K = \{e\}$; for

$$|H \cap K| \mid p$$

and $|H \cap K| = p$ would imply that $H = K$, contrary to the choice of c. Also observe that K is normal in G by Theorem 2.8. A repetition of the argument leading to (60) again yields the conclusion that $G = HK$ (direct) and G is abelian. ∎

Example 9 Theorem 2.9(i) shows that there is only one group of order 15 to within isomorphism, namely, the cyclic group of order 15.

Example 10 Let G be any group of order $245 = 7^2 \cdot 5$. We show that G is abelian. In fact, we claim that either G is the direct product of two cyclic groups of order 7 and a cyclic group of order 5, or it is the direct product of a cyclic group of order 49 and a cyclic group of order 5. For $245 = 7^2 \cdot 5$, so by Theorem 2.7(ii), G contains a Sylow subgroup H of order 7^2, and by Theorem 2.7(iii), the number r of conjugates of H satisfies $r \mid 5$, $r \equiv 1 \pmod 7$. Hence $r = 1$ and $H \lhd G$. Similarly G contains a Sylow subgroup K of order 5, and the number r_1 of conjugates of K satisfies $r_1 \mid 49$, $r_1 \equiv 1 \pmod 5$. Thus $r_1 = 1$ and $K < G$. Note that $|H \cap K| \mid 49$ and $|H \cap K| \mid 5$ so that $H \cap K = e$. Write $H = \{h_1, h_2, h_3, \ldots, h_{49}\}$, and consider the left cosets

$$h_1k, \ldots, h_{49}K. \tag{62}$$

If $h_sK = h_tK$, then $h_t^{-1}h_s \in K$ and of course $h_t^{-1}h_s \in H$. But $H \cap K = e$ implies $h_s = h_t$. Thus (62) is a complete list of left K-cosets in G (since $[G:K] = 49$), so that any element of G is of the form hk, $h \in H$, $k \in K$. We conclude by Theorem 1.6 that

$$G = HK \qquad \text{(direct).}$$

Finally, the group H of order 49 is either cyclic or the direct product of two cyclic groups of order 7 [Theorem 2.9(ii)].

Exercises 2.2

1. Show that if $\chi: G \to \mathbf{C}$ is a group homomorphism of the finite group G into the multiplicative group of nonzero complex numbers, then

$$\sum_{g\in G}\chi(g)=\begin{cases}|G| & \text{if } \chi\equiv 1\\ 0 & \text{otherwise.}\end{cases}$$

Hint: Note that $s=\sum_{g\in G}\chi(g)=\sum_{g\in G}\chi(hg)=\chi(h)\sum_{g\in G}\chi(g)=\chi(h)s$ for each $h\in G$. Sum on h to obtain $|G|s=s^2$.

2. Prove that if $|X|<\infty$, then any injection $f\colon X\to X$ is a bijection. *Hint*: Argue by contradiction.
3. Show that the cycle index polynomial $P(S_n\colon x_1,\ldots,x_n)$ is

$$\sum_{s_\lambda}\frac{1}{1^{\lambda_1}\lambda_1!\;2^{\lambda_2}\lambda_2!\cdots n^{\lambda_n}\lambda_n!}x_1^{\lambda_1}x_2^{\lambda_2}\cdots x_n^{\lambda_n},$$

where the summation is taken over all cycle structures $s_\lambda=[1^{\lambda_1},2^{\lambda_2},\ldots,n^{\lambda_n}]$. *Hint*: Recall Theorem 3.6, Chapter 1.
4. Let $Q_{m,n}$ denote the set of strictly increasing sequences of length m of integers chosen from $1,\ldots,n$, i.e., $\omega\in Q_{m,n}$ iff $\omega\in[1,n]^{[1,m]}$ and $\omega(i)<\omega(i+1)$, $i=1,\ldots,m-1$. Prove that $|Q_{m,n}|=\binom{n}{m}$.
5. Let $G_{m,n}$ denote the set of nondecreasing sequences of length m of integers chosen from $1,\ldots,n$, i.e., $G_{m,n}$ is the set of $f\in[1,n]^{[1,m]}$ which satisfy $f(1)\leq f(2)\leq\cdots\leq f(m)$. Show that $|G_{m,n}|=\binom{n+m-1}{m}$.
6. Let S_m act on $[1,n]^{[1,m]}$ by $F(\sigma)(\varphi)=\varphi\cdot\sigma^{-1}$. Prove that $\Delta=G_{m,n}$ is an S.D.R. for the S_m-orbits in $[1,n]^{[1,m]}$.
7. Using the results in Exercises 3, 5, and 6, together with Theorem 2.4, show that

$$\binom{n+m-1}{m}=\sum_{\substack{\lambda_1+2\lambda_2+\cdots+m\lambda_m=m\\ \lambda_i\geq 0,\,i=1,\ldots,m}}\frac{n^{\lambda_1+\lambda_2+\cdots+\lambda_m}}{1^{\lambda_1}\lambda_1!\,2^{\lambda_2}\lambda_2!\cdots m^{\lambda_m}\lambda_m!}.$$

8. If $\Delta=G_{m,n}$ as in Exercise 6 and $\chi\equiv\varepsilon$ [where $\varepsilon(\sigma)=(-1)^{I(\sigma)}$ for $\sigma\in S_m$], show that if

$$\bar{\Delta}=\{f\mid f\in\Delta \text{ and } (S_m)_f\subset\ker\varepsilon\},$$

then in fact

$$\bar{\Delta}=Q_{m,n}.$$

9. Label the faces of a cube with the integers $1,\ldots,6$ and regard the various rotations of the cube described in Example 2 as permutations in S_6. Write down the 24 elements of a group H of all these permutations, and compute the cycle index polynomial $P(H\colon x_1,\ldots,x_6)$. *Hint*: $P(H\colon x_1,\ldots,x_6)$ is

$$\frac{1}{24}(x_1^{\,6}+3x_1^{\,2}x_2^{\,2}+6x_1^{\,2}x_4+6x_2^{\,3}+8x_3^{\,2}).$$

10. Each face of the cube in the preceding exercise is to be painted either white or black. Two paint jobs are indistinguishable iff one can be obtained from the other by a rotation of the cube. Find the total number of such paint jobs. *Hint*: The answer is 10.
11. Find the number of paint jobs in Exercise 10 in which precisely three faces are painted white and three faces are painted black.
12. Let $n=p^\alpha q$ where $\alpha\geq 1$ and p is a prime, and let p^β be the highest power of p

that divides n. Prove that $p^{\beta-\alpha}$ is the highest power of p that divides $\binom{n}{p^\alpha}$. *Hint*: We have

$$\binom{n}{p^\alpha} = \frac{n(n-1)\cdots(n-p^\alpha+1)}{p^\alpha(p^\alpha-1)\cdots 1}$$

$$= \frac{n}{p^\alpha}\left[\frac{(n-1)(n-2)\cdots(n-p^\alpha+1)}{(p^\alpha-1)(p^\alpha-2)\cdots 1}\right]$$

$$= q\prod_{k=1}^{p^\alpha-1}\frac{n-k}{p^\alpha-k}. \tag{63}$$

Consider a pair $n-k = p^\alpha q - k$ and $p^\alpha - k$, and let p^s, $s < \alpha$, be the highest power of p that divides $p^\alpha - k$. Then $p^\alpha - k = Np^s$, so $k = p^\alpha - Np^s = p^s M$. If $p|M$, we could write $k = p^{s+1}M_1$ so that $p^\alpha - k = p^\alpha - p^{s+1}M_1$, and since $s+1 \le \alpha$, $p^\alpha - k$ would be divisible by p^{s+1}. Thus $k = p^s M$ and $p \nmid M$. Now $p^s|(p^\alpha q - k)$. Suppose $p^{s+1}|(p^\alpha q - k)$, i.e., $p^{s+1}|(p^\alpha q - p^s M)$. Then $p|(p^{\alpha-s}q - M)$. But since $\alpha > s$, it follows that $p|M$, a contradiction. In other words, both $n-k = p^\alpha q - k$ and $p^\alpha - k$ have the common highest factor p^s. Thus the right side of (63) has the form $q\dfrac{P}{Q} = B$, where neither P nor Q contains p as a factor; in each term $\dfrac{n-k}{p^\alpha - k}$, the common highest power of p in numerator and denominator has been cancelled. Since $qP = BQ$ and $p^{\beta-\alpha}|q$, we have $p^{\beta-\alpha}|BQ$ and hence $p^{\beta-\alpha}|B$. If B were divisible by $p^{\beta-\alpha+1}|$, say $B = B_1 p^{\beta-\alpha+1}$, then $qP = QB_1p^{\beta-\alpha+1}$ so that $p^{\beta-\alpha+1}|qP$. Since $p \nmid P$, this implies $p^{\beta-\alpha+1}|q$. But then $p^{\beta+1}$ divides $n = p^\alpha q$, which is impossible. Hence $p^{\beta-\alpha}$ is the highest power of p dividing $B = \binom{n}{p^\alpha}$.

13. Show that if $|G| = 4$ and G is not the Klein four group, then G is a cyclic group. *Hint*: By Theorem 2.9(ii), if G is not cyclic, then G is the direct product of two cyclic subgroups H and K of order 2. Take $H = K = \{-1,1\}$ and compute the external direct product $H \times K$. The elements of $H \times K$ are

$$e = (1,1)$$
$$a = (1,-1)$$
$$b = (-1,1)$$
$$c = (-1,-1)$$

and the multiplication table is

	e	a	b	c
e	e	a	b	c
a	a	e	c	b
b	b	c	e	a
c	c	b	a	e

This is precisely the table for the Klein four group in (46), Section 1.3.

14. Let p and q be primes, $p > q$, $p \not\equiv \pm 1 \pmod{q}$. Prove that if G is a group of order

p^2q, then either G is the direct product of cyclic subgroups of order p^2 and q, or G is the direct product of two cyclic subgroups of order p and a cyclic subgroup of order q. In either case G is abelian. *Hint*: Follow the method used in Example 10 with the role of 5 played by p and the role of 3 played by q. The argument is identical.

15. Classify all groups of order 147. *Hint*: Use Exercise 14.

16. Prove that any group of order 36 is abelian. *Hint*: The same argument as used in Example 10 will work.

Glossary 2.2

Burnside's lemma, 88
character of H of degree 1, 88
conjugate subgroups, 98
cycle index polynomial of H, 91
$P(H: x_1, \ldots, x_n)$, 91
Δ, 88
$\bar{\Delta}$, 88
group of rotations of the cube, 92
H acts on the set X, 87
$H: X$, 87
orbits, 87
$\mathcal{O}_x$, 87
Pôlya counting theorem, 91
$\psi(h)$, 89
stabilizer, 87
H_x, 87
s_λ-partition, 95
Sylow theorems, 99
p-Sylow subgroups, 101

2.3 Some Additional Topics

In this section we consider several more advanced topics in group theory.

We begin with a rather technical isomorphism theorem known as the *Zassenhaus lemma*.

Theorem 3.1 *Assume that* $H_0 \lhd H < G$, $K_0 \lhd K < G$. *Then*

$$H_0(H \cap K)/H_0(H \cap K_0) \cong K_0(K \cap H)/K_0(K \cap H_0). \tag{1}$$

Proof: First consider the product of the two groups

$$S = H_0(H \cap K_0)$$

and

$$R = H \cap K;$$

$$SR = H_0(H \cap K_0)(H \cap K) = H_0(H \cap K). \tag{2}$$

Clearly $H_0(H \cap K)$ is a subgroup of G, since H_0 is normal in H [see (27) of Section 1.3]. Set $D = S \cap R$, and observe that

$$D = (K \cap H_0)(H \cap K_0). \tag{3}$$

For if $x \in D$, then $x \in H_0(H \cap K_0)$ so that

$$x = h_0k_0, \quad h_0 \in H_0, \quad k_0 \in H \cap K_0. \tag{4}$$

Now, $D \subset R \subset K$ so that

$$x = h_0 k_0 \in K$$

and hence $$h_0 = x k_0^{-1} \in K.$$

Thus $$h_0 \in K \cap H_0, \tag{5}$$

and (4) implies

$$x \in (K \cap H_0)(H \cap K_0). \tag{6}$$

This shows that

$$D \subset (K \cap H_0)(H \cap K_0). \tag{7}$$

To prove the inclusion in the other direction, suppose that

$$x \in (K \cap H_0)(H \cap K_0)$$

so that $$x = h_0 k_0, \quad h_0 \in K \cap H_0, \quad k_0 \in H \cap K_0. \tag{8}$$

Then

$$x = h_0 k_0 \in H_0(H \cap K_0) = S; \tag{9}$$

since $h_0 k_0$ is a product of elements in both H and K by (8), we also have

$$x \in H \cap K = R. \tag{10}$$

Combining (9) and (10), we obtain

$$x \in S \cap R = D. \tag{11}$$

This shows that

$$(K \cap H_0)(H \cap K_0) \subset D. \tag{12}$$

The formula (3) now follows from (7) and (12).

We also assert that

$$S \lhd SR. \tag{13}$$

Since

$$S = H_0(H \cap K_0)$$

and [from (2)] $$SR = H_0(H \cap K),$$

we must show that

$$H_0(H \cap K_0) \lhd H_0(H \cap K). \tag{14}$$

Let

$$x \in H \cap K,$$
$$y \in H \cap K_0,$$
$$z, w \in H_0.$$

Then

$$x^{-1} y x \in H \tag{15}$$

because each factor belongs to H; also

$$x^{-1}yx \in K_0 \tag{16}$$

because $y \in K_0$, $x \in K$, and $K_0 \lhd K$. Combining (15) and (16), we obtain

$$x^{-1}yx \in H \cap K_0. \tag{17}$$

Since $x \in H$, $z \in H_0$, and $H_0 \lhd H$, we have

$$x^{-1}zx \in H_0. \tag{18}$$

A typical element in the left side of (14) has the form

$$zy; \tag{19}$$

conjugating (19) by x, we have, using (17) and (18), that

$$x^{-1}zyx = x^{-1}zxx^{-1}yx \in H_0(H \cap K_0).$$

Thus for $x \in H \cap K$,

$$x^{-1}H_0(H \cap K_0)x \subset H_0(H \cap K_0). \tag{20}$$

Next,

$$w^{-1}yw = w^{-1}(ywy^{-1})y \tag{21}$$

and since $y \in H \cap K_0$ and $w \in H_0 \lhd H$, we have

$$ywy^{-1} \in H_0. \tag{22}$$

Then (21) and (22) imply

$$w^{-1}yw \in H_0(H \cap K_0) \tag{23}$$

(the first two factors on the right in (21) are in H_0 and the third factor is in $H \cap K_0$). Thus for $w \in H_0$,

$$\begin{aligned} w^{-1}H_0(H \cap K_0)w &= (w^{-1}H_0w)w^{-1}(H \cap K_0)w \\ &\subset H_0H_0(H \cap K_0) \\ &= H_0(H \cap K_0). \end{aligned} \tag{24}$$

Now a typical element in the right side of (14) has the form

$$wx,$$

so that

$$\begin{aligned} (wx)^{-1}H_0(H \cap K_0)wx &= x^{-1}w^{-1}H_0(H \cap K_0)wx \\ &\subset x^{-1}H_0(H \cap K_0)x \qquad \text{[by (24)]} \\ &\subset H_0(H \cap K_0) \qquad \text{[by (20)].} \end{aligned}$$

This establishes (14) and hence (13).

Consider the diagram

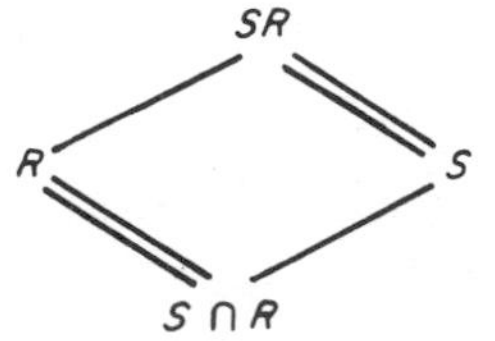

The hypotheses of Theorem 1.4 (the first isomorphism theorem) hold with the role of G in that theorem being played by SR itself, the role of H being played by R, and the role of K being played by S. Then

$$SR/S \cong R/S \cap R,$$

or in view of (2) and (3),

$$\frac{H_0(H \cap K)}{H_0(H \cap K_0)} \cong \frac{H \cap K}{(K \cap H_0)(H \cap K_0)}. \tag{25}$$

Interchanging H and K, H_0 and K_0 in (25) and observing that the right side of (25) is thereby unaltered, we have

$$\frac{K_0(K \cap H)}{K_0(K \cap H_0)} \cong \frac{H \cap K}{(K \cap H_0)(H \cap K_0)}. \tag{26}$$

Thus the left sides of (25) and (26) are isomorphic, proving the statement (1). ∎

Theorem 3.1 can be used to extend a modified version of Theorem 1.10 to groups which are not necessarily finite. First, recall that if

$$G = G_0 \rhd G_1 \rhd G_2 \rhd \cdots \rhd G_r = e \tag{27}$$

is a subnormal series of length r, then the subnormal series

$$K = K_0 \rhd K_1 \rhd K_2 \rhd \cdots \rhd K_s = e \tag{28}$$

is a *refinement* of (27) if every G_i occurs among the K_j. As in Theorem 1.13, two subnormal series for a group G are said to be *isomorphic* if they have the same length and there is a 1–1 isomorphism correspondence between the factor groups. We have the following theorem.

Theorem 3.2 (*The Schreier Refinement Theorem*) *Any two subnormal series for an arbitrary group G have isomorphic refinements.*

Proof: Let

$$G = H_0 \rhd H_1 \rhd \cdots \rhd H_r = e \tag{29}$$

and

$$G = K_0 \rhd K_1 \rhd \cdots \rhd K_s = e \tag{30}$$

be two subnormal series for G. For each $i = 1, \ldots, r$ define

$$H_{ij} = H_i(H_{i-1} \cap K_j), \qquad j = 0, \ldots, s.$$

Since $H_i \lhd H_{i-1}$ and $K_j \lhd K_{j-1}$, it follows from Theorem 3.1 that

$$\begin{aligned} H_{ij} &= H_i(H_{i-1} \cap K_j) \\ &\lhd H_i(H_{i-1} \cap K_{j-1}) \\ &= H_{ij-1}, \qquad j = 1, \ldots, s. \end{aligned} \tag{31}$$

Also note that

$$H_{i0} = H_i(H_{i-1} \cap K_0) = H_iH_{i-1} = H_{i-1}, \tag{32}$$

$$H_{is} = H_i(H_{i-1} \cap K_s) = H_i(H_{i-1} \cap e) = H_i. \tag{33}$$

Thus $$H_{i-1} = H_{i0} \triangleright H_{i1} \triangleright \cdots \triangleright H_{is} = H_i$$

so that the groups H_{ij}, $j = 1, \ldots, s-1$, are inserted between H_{i-1} and H_i. We obtain a refinement of (29) of length rs.

Similarly for each $j = 1, \ldots, s$, define

$$K_{ij} = K_j(K_{j-1} \cap H_i), \qquad i = 0, \ldots, r.$$

Theorem 3.1 again implies that

$$K_{ij} = K_j(K_{j-1} \cap H_i) \triangleleft K_j(K_{j-1} \cap H_{i-1}) = K_{i-1j}. \tag{34}$$

Also, $$K_{0j} = K_j(K_{j-1} \cap H_0) = K_jK_{j-1} = K_{j-1},$$

$$K_{rj} = K_j(K_{j-1} \cap H_r) = K_j(K_{j-1} \cap e) = K_j.$$

Thus, $$K_{j-1} = K_{0j} \triangleright K_{1j} \triangleright \cdots \triangleright K_{rj} = K_j$$

so that the groups K_{ij}, $i = 1, \ldots, r-1$, are inserted between K_{j-1} and K_j. We obtain a refinement of (30) of length rs. Finally Theorem 3.1 states that

$$\frac{H_{ij-1}}{H_{ij}} = \frac{H_i(H_{i-1} \cap K_{j-1})}{H_i(H_{i-1} \cap K_j)} \cong \frac{K_j(K_{j-1} \cap H_{i-1})}{K_j(K_{j-1} \cap H_i)}$$

$$= \frac{K_{i-1j}}{K_{ij}}, \qquad i = 1, \ldots, r, \quad j = 1, \ldots, s.$$

Thus the two refined series are isomorphic. ▮

We can now easily extend Theorem 1.13. Recall that an infinite group need not have a composition series (see Example 6, Section 2.1). However, the following theorem shows that if a composition series exists, then the factors are uniquely determined.

Theorem 3.3 (*Jordan-Hölder-Schreier Theorem*) *Let*

$$G = G_0 \triangleright G_1 \triangleright \cdots \triangleright G_r = e \tag{35}$$

and $$G = H_0 \triangleright H_1 \triangleright \cdots \triangleright H_s = e \tag{36}$$

be two composition series for the group G. Then the two series (35) and (36) are isomorphic, i.e., $r = s$ and the factors in the two series can be paired off in some order into isomorphic pairs.

Proof: Since the two series (35) and (36) are composition series, neither one of them can be (properly) refined, i.e., in either series each group is maximal in the group that contains it. But by Theorem 3.2 the two series have isomorphic refinements. Thus they must be initially isomorphic. ▮

Recall (Section 2.2) that a group M is said to *act on a set X* if there exists a homomorphism

$$f: M \to S_X.$$

An important modification of this idea is useful in the study of modules and vector spaces and in group representation theory.

Let M be an arbitrary set and G a group. Denote by

$$\text{end}\,(G) \tag{37}$$

the totality of endomorphisms of G (homomorphisms $\varphi: G \to G$). Let

$$f: M \to \text{end}\,(G) \tag{38}$$

be a function. Then the triple of objects

$$(M,G,f)$$

is called a *group with operators*; the elements of M are called the *operators*. Note that since $f(m) \in \text{end}\,(G)$ $(m \in M)$ we have for $g_1,g_2 \in G$ that

$$f(m)(g_1g_2) = f(m)(g_1)\,f(m)(g_2). \tag{39}$$

It is sometimes convenient to call G an *M-group* and abbreviate $f(m)(g)$ by

$$mg. \tag{40}$$

In the notation (40), the formula (39) becomes

$$m(g_1g_2) = (mg_1)(mg_2), \qquad m \in M,\ g_1, g_2 \in G.$$

Example 1 (a) Let G be an arbitrary group, take $M = G$, and define

$$f(m)g = m^{-1}gm, \qquad m \in M, g \in G;$$

$f(m)$ is called an *inner automorphism* or more precisely $f(m)$ is *conjugation by m*. Observe that

$$\begin{aligned} f(m)(g_1g_2) &= m^{-1}g_1g_2m = (m^{-1}g_1m)(m^{-1}g_2m) \\ &= f(m)(g_1)f(m)(g_2). \end{aligned}$$

(b) Let M be the set of all automorphisms of a group G and define $f = \iota_M$.

(c) Let $(R, \{+, \cdot\})$ be a ring, take $M = R$, and define

$$f(m)r = m \cdot r, \qquad m \in M, r \in R;$$

$f(m)$ is "left multiplication by m." Clearly

$$\begin{aligned} f(m)(r_1 + r_2) &= m(r_1 + r_2) = mr_1 + mr_2 \\ &= f(m)(r_1) + f(m)(r_2). \end{aligned}$$

A similar example can be constructed for right multiplication.

(d) Let G be a left R-module (see Section 1.2), and let $M = Z$ be the ring of integers. Define

$$f(m)g = \begin{cases} \sum_{i=1}^{m} g, & \text{if } m > 0 \\ -\sum_{i=1}^{m} g, & \text{if } m < 0 \\ 0 \in G, & \text{if } m = 0. \end{cases}$$

(e) Let K be a field and denote by $K[x]$ the set of all polynomials in x with coefficients in K (see Example 10, Section 1.2). Let V be a vector space over K (see Section 1.2). Let A be a fixed endomorphism of the additive group of V. Then take

$$M = K[x],$$
$$G = V,$$

and for any polynomial

$$m(x) = m_0 + m_1x + \cdots + m_px^p \in M$$

and any $v \in V$, define

$$f(m(x))v = m_0v + m_1Av + m_2A^2v + \cdots + m_pA^pv. \tag{41}$$

The reader can easily confirm that V is an M-group (see Exercise 1). Actually more is true. If $m_1(x)$ and $m_2(x)$ are any two polynomials in M, then

$$f(m_1(x) + m_2(x))v = f(m_1(x))v + f(m_2(x))v$$

and

$$f(m_1(x)m_2(x))v = f(m_1(x))f(m_2(x))v$$

(see Exercise 2). In fact, V is made into a left $K[x]$-module if the left multiplication of a vector $v \in V$ by a polynomial $m(x) \in K[x]$ is defined by (41) (see Exercise 3).

If (M,G,f) is an M-group and $H < G$, then H is said to be *admissible* if (M,H,f) is an M-group. This means that

$$f(m)H < H \qquad \text{for all } m \in M.$$

It is an easy exercise (see Exercise 4) to show that the intersection of an arbitrary family of admissible subgroups of an M-group G is again an admissible subgroup of G.

If (M,G,f_1) and (M,K,f_2) are M-groups and

$$\varphi: G \to K$$

is a group homomorphism, then φ is called an *operator homomorphism* or an *M-homomorphism* if

$$\varphi(f_1(m)(g)) = f_2(m)\varphi(g) \qquad \text{for all } m \in M,\ g \in G. \tag{42}$$

Example 2 Let G and K be vector spaces over a common field R, regarded as abelian groups. The scalar products ru and rv ($r \in R$, $u \in G$, $v \in K$) define R as a set of operators on G and K. To be precise, (R,G,f) is an M-group with $M = R$ and $f: R \to \text{end}\,(G)$ given by

$$f(r)u = ru, \qquad r \in R,\ u \in G.$$

If $\varphi: G \to K$ is an operator homomorphism, then (42) reads

$$\varphi(ru) = r\varphi(u) \qquad \text{for all } r \in R,\ u \in G; \tag{43}$$

φ is called a *linear map* or *linear transformation* from G to K. Observe that if $H < G$, then $u_1 + u_2 \in H$ for all $u_1,u_2 \in H$; for H to be admissible simply means that $ru \in H$ for all $r \in R$, $u \in H$. In this case H is called a *subspace* of G (see Section 1.2).

Theorem 3.4 *If G is an M-group and $H \lhd G$ is an admissible subgroup, then G/H can be made into an M-group in such a way that the canonical epimorphism*

$$\nu\colon G \to G/H$$

is an operator homomorphism.

Proof: Define

$$m(gH) = (mg)H, \qquad m \in M, g \in G. \tag{44}$$

To confirm that $m(gH)$ is well defined by (44), we must show that the value of $m(gH)$ is independent of the representative g of the left coset gH. Suppose that

$$g_1H = g_2H,$$

so that

$$g_2^{-1}g_1 \in H.$$

Since H is admissible, we have

$$m(g_2^{-1}g_1) \in H. \tag{45}$$

But

$$m(g_2^{-1}g_1) = (mg_2^{-1})(mg_1)$$

and

$$mg_2^{-1} = (mg_2)^{-1}$$

(see Exercise 5). Thus

$$m(g_2^{-1}g_1) = (mg_2)^{-1}(mg_1).$$

Now (45) implies

$$(mg_2)^{-1}(mg_1) \in H,$$

so that

$$(mg_2)H = (mg_1)H$$

and hence

$$m(g_2H) = m(g_1H).$$

Next, we have for $g,g' \in H$ that

$$\begin{aligned} m(gHg'H) &= m(gg'H) = m(gg')H \\ &= (mg)(mg')H = (mg)H(mg')H \\ &= m(gH)m(g'H). \end{aligned}$$

This shows that (44) makes G/H into an M-group.
Finally, the formula (44) can be rewritten

$$m\nu(g) = \nu(mg),$$

showing that $\nu\colon G \to G/H$ is an M-homomorphism. ▮

All the basic theorems (Theorems 1.1, 1.2, 1.3, 1.4, 3.1, 3.2, 3.3) carry over to M-groups with obvious modifications in their statements and proofs. We will restate and prove in detail Theorem 1.1 to exhibit the kinds of modifications that must be made. The reader may review the other theorems listed above by making similar modifications.

Theorem 3.5 *Let G and S be M-groups. Let $\varphi: G \to S$ be an M-homomorphism and set $H = \ker \varphi$. Then*

(i) *$H \lhd G$ and H is admissible; $\varphi(G) < S$ and $\varphi(G)$ is admissible.*

(ii) *$\varphi(aH) = \{\varphi(a)\}$ for all $a \in G$.*

(iii) *If $\bar{\varphi}: G/H \to \varphi(G)$ is defined by $\bar{\varphi}(aH) = \varphi(a)$, $a \in G$, then $\bar{\varphi}$ is an M-isomorphism.*

(iv) *If φ is surjective and $K \lhd G$ is admissible, then $\varphi(K) \lhd S$ and $\varphi(K)$ is admissible.*

Proof: (i) If $h \in H$ and $g \in G$, then

$$\varphi(g^{-1}hg) = \varphi(g)^{-1}\varphi(h)\varphi(g) = \varphi(g)^{-1}\varphi(g) = e$$

so that $g^{-1}hg \in H$ (e denotes the identity in both G and S). Thus $H \lhd G$. If $m \in M$ and $h \in H$, then

$$\varphi(mh) = m\varphi(h) = me = e \qquad \text{(see Exercise 5)},$$

i.e.,

$$mh \in H.$$

Hence H is admissible. It is clear that $\varphi(G) < S$. If $m \in M$ and $\varphi(g) \in \varphi(G)$, then $mg \in G$ so that

$$m\varphi(g) = \varphi(mg) \in \varphi(G).$$

Thus $\varphi(G)$ is admissible.

(ii) Clearly, $\varphi(ah) = \varphi(a)\varphi(h) = \varphi(a)$ for all $a \in G$, $h \in H$.

(iii) We have

$$\begin{aligned}\bar{\varphi}(aHbH) &= \bar{\varphi}(abH) = \varphi(ab) = \varphi(a)\varphi(b)\\ &= \bar{\varphi}(aH)\bar{\varphi}(bH)\end{aligned}$$

so that $\bar{\varphi}$ is a homomorphism. To see that $\bar{\varphi}$ is injective, note that $\bar{\varphi}(aH) = \bar{\varphi}(bH)$ iff $\varphi(a) = \varphi(b)$ iff $\varphi(b^{-1}a) = e$ iff $b^{-1}a \in H$ iff $aH = bH$. It is obvious that $\bar{\varphi}: G/H \to \varphi(G)$ is surjective. By Theorem 3.4, the definition

$$m(gH) = (mg)H$$

makes G/H into an M-group, and

$$\begin{aligned}\bar{\varphi}(m(gH)) &= \bar{\varphi}((mg)H) = \varphi(mg) = m\varphi(g)\\ &= m\bar{\varphi}(gH).\end{aligned}$$

Thus $\bar{\varphi}$ is an M-isomorphism.

(iv) We have

$$\varphi(K) = \varphi(g^{-1}Kg) = \varphi(g)^{-1}\varphi(K)\varphi(g),$$

and since φ is surjective, $\varphi(g)$ represents any element of S. Hence $\varphi(K) \lhd S$. Again, if $m \in M$, then $mK \subset K$ since K is admissible, so that

$$m\varphi(K) = \varphi(mK) \subset \varphi(K)$$

and $\varphi(K)$ is admissible. ∎

In the definition of an M-group

$$(M, V, f),$$

no restrictions are placed on the set M, and the only condition on the map f is that

$$\operatorname{im} f \subset \operatorname{end}(V).$$

(We use the letter V here instead of G for reasons that will soon be clear.) That is, $f(m)\colon V \to V$ is a group homomorphism for each $m \in M$. This latitude permits us to incorporate a definition that is useful in group representation theory. Let F be a field, R a ring with multiplicative identity 1_R, and V an additive abelian group. Let the set of operators M be defined by

$$M = F \cup R$$

and assume that V is an M-group. For $a,b \in F$, $r_1,r_2 \in R$, and $v \in V$ assume furthermore that the action of M on V satisfies:

$$(a + b)v = av + bv,$$
$$(ab)v = a(bv),$$
$$(r_1 + r_2)v = r_1v + r_2v,$$
$$(r_1r_2)v = r_1(r_2v),$$
$$1_Rv = v,$$

and

$$1_Fv = v, \tag{46}$$

where 1_F is the multiplicative identity in the field F. Then (for obvious reasons) V is called an *F-R module*. If we look back at Section 1.2, we see that V is a vector space over the field F and R is a set of operators on V. The third, fourth, and fifth conditions in (46) simply show how the additional structures of the ring R behave in their action on V. In fact, if R is an algebra over F, we will require an additional condition:

$$a(r_1v) = r_1(av). \tag{47}$$

It is instructive to have a nontrivial example of an F-R module available.

Example 3 Let the group V be the set of n-tuples over the field F (see Section 1.2):

$$V = F^n.$$

The elements of F act on V by scalar multiplication, i.e., $a(v(1), \ldots, v(n)) = (av(1), \ldots, av(n))$, so the first, second, and sixth of the conditions (46) are automatically satisfied. Let $R = F[S_n]$, the groupoid ring of S_n over F (see Example 9, Section 1.2). Denote the function in $F[S_n]$ whose value at $\sigma \in S_n$ is $c_\sigma \in F$ by $\sum_{\sigma\in S_n} c_\sigma\sigma$. The formulas (22) and (23) in Section 1.2 read

$$\sum_{\sigma\in S_n} c_\sigma\sigma + \sum_{\sigma\in S_n} d_\sigma\sigma = \sum_{\sigma\in S_n} (c_\sigma + d_\sigma)\sigma \tag{48}$$

and

$$\Big(\sum_{\sigma\in S_n} c_\sigma\sigma\Big)\Big(\sum_{\sigma\in S_n} d_\sigma\sigma\Big) = \sum_{\sigma\in S_n}\Big(\sum_{\theta\in S_n} c_\theta d_{\theta^{-1}\sigma}\Big)\sigma, \tag{49}$$

respectively. Observe that if we define

$$a \sum_{\sigma\in S_n} c_\sigma \sigma = \sum_{\sigma\in S_n} (ac_\sigma)\sigma, \qquad a \in F,$$

then in fact $F[S_n]$ becomes a linear associative algebra over F (see Exercise 12). Next, let

$$A(\sigma)$$

denote the incidence matrix for the permutation σ (see Exercise 1, Section 1.3). We now define the action of an element

$$r = \sum_{\sigma\in S_n} c_\sigma \sigma \in F[S_n]$$

on $V = F^n$ by

$$rv = \sum_{\sigma\in S_n} c_\sigma A(\sigma)v, \qquad v \in V. \tag{50}$$

The product $A(\sigma)v$ is the usual matrix product. That is, if $v = (v_1, \ldots, v_n)$ [in conformity with the usual notation for n-tuples, we write the function value $v(i)$ as v_i], then

$$A(\sigma)v = w$$

where
$$w_i = \sum_{j=1}^{n} A(\sigma)_{ij} v_j, \qquad i = 1, \ldots, m. \tag{51}$$

The definition (50) clearly makes R into a set of operators on V, and the third and fifth of the conditions (46) are readily seen to be satisfied (see Exercise 13); note that if

$$1_R = \sum_{\sigma\in S_n} c_\sigma \sigma,$$

then
$$c_\sigma = \begin{cases} 1_F, & \sigma = \text{identity} \\ 0, & \sigma \neq \text{identity} \end{cases},$$

and A (identity) $= I_n$ (the n-square identity matrix).It is also obvious that condition (47) obtains (recall that R is an algebra over F).

To complete the verification that V is an F-R module, it remains to check that the fourth of the conditions (46) holds. Thus, let

$$r_1 = \sum_{\sigma\in S_n} c_\sigma \sigma, \qquad r_2 = \sum_{\sigma\in S_n} d_\sigma \sigma,$$

and let $v \in V$. Using (49) and (50) and the fact that $A(\sigma_1\sigma_2) = A(\sigma_1)A(\sigma_2)$ (see Exercise 1, Section 1.3), we compute that

$$\begin{aligned} r_1(r_2v) &= r_1\Big(\sum_{\sigma\in S_n} d_\sigma A(\sigma)v\Big) = \sum_{\sigma\in S_n} r_1(d_\sigma A(\sigma)v) \\ &= \sum_{\sigma\in S_n} d_\sigma r_1(A(\sigma)v) = \sum_{\sigma\in S_n} d_\sigma \sum_{\theta\in S_n} c_\theta A(\theta)A(\sigma)v \\ &= \sum_{\sigma\in S_n} d_\sigma \sum_{\theta\in S_n} c_\theta A(\theta\sigma)v \\ &= \Big(\sum_{\sigma,\theta\in S_n} d_\sigma c_\theta A(\theta\sigma)\Big)v \\ &= \sum_{\tau\in S_n}\Big(\sum_{\theta\in S_n} c_\theta d_{\sigma^{-1}\tau}\Big)A(\tau)v. \end{aligned} \tag{52}$$

The second equality in the above computation is due to the fact that r_1 is an operator on V; the third comes from (47); the last is by the substitution $\theta\sigma = \tau$. On the other hand, observe that

$$r_1 r_2 = \sum_{\tau \in S_n} \Big(\sum_{\theta \in S_n} c_\theta d_{\theta^{-1}\tau} \Big) \tau;$$

so, by the definition (50),

$$(r_1 r_2)v = \sum_{\tau \in S_n} \Big(\sum_{\theta \in S_n} c_\theta d_{\theta^{-1}\tau} \Big) A(\tau)v.$$

Comparison with (52) now establishes the fourth of the conditions (46). Thus V is an F-R module.

The question arises in this example: What are the admissible subgroups of V? As we saw in Example 2, $W < V$ is an admissible subgroup with respect to F iff W is simply a subspace of F^n. Clearly W is admissible with respect to the operators in $F[S_n]$ as defined in (50) iff

$$A(\sigma)W \subset W \qquad \text{for all } \sigma \in S_n, \tag{53}$$

i.e.,

$$A(\sigma)w \in W$$

for all $w \in W$ and all $\sigma \in S_n$. It is instructive to work out the precise structure of an admissible subgroup W. We shall assume in the following discussion that $F = \mathbf{C}$. Moreover, we shall use a number of results from elementary linear algebra which are definitely not prerequisite to an understanding of the remainder of this book. To proceed, first observe that if $\sigma \in S_n$ and

$$(x_1, \ldots, x_n) \in \mathbf{C}^n,$$

then

$$A(\sigma)(x_1, \ldots, x_n) = (x_{\sigma^{-1}(1)}, \ldots, x_{\sigma^{-1}(n)}). \tag{54}$$

Now let $x_1, \ldots, x_n$ be any n complex numbers not all of which are the same. Let S be the subspace of $\mathbf{C}^n$ spanned by all vectors of the form (54); this is denoted by

$$S = \langle A(\sigma)(x_1, \ldots, x_n), \sigma \in S_n \rangle. \tag{55}$$

We assert that

(a) If $x_1 + \cdots + x_n = 0$, then $\dim S = n - 1$.
(b) If $x_1 + \cdots + x_n \neq 0$, then $\dim S = n$.

Since $x_1, \ldots, x_n$ are not all the same and $(x_{\sigma(1)}, \ldots, x_{\sigma(n)}) \in S$ for all $\sigma \in S_n$, we may assume that $(x_1, \ldots, x_n)$ has the form

$$(x_1, \ldots, x_n) = (x, \ldots, x, z_1, \ldots, z_q)$$

where $x \neq z_i$, $i = 1, \ldots, q$. Thus $x = x_1 = \cdots = x_{n-q}$. Set $v_1 = (x, \ldots, x, z_1, \ldots, z_q)$ and let v_i, $i = 2, \ldots, q + 1$, be the n-tuple obtained by exchanging z_{i-1} with the x in the first position:

$$v_1 = (x, \ldots, x, z_1, z_2, \ldots, z_{q-1}, z_q),$$
$$v_2 = (z_1, \ldots, x, x, z_2, \ldots, z_{q-1}, z_q),$$
$$\cdots\cdots\cdots\cdots\cdots\cdots\cdots\cdots\cdots\cdots$$
$$v_{q+1} = (z_q, \ldots, x, z_1, z_2, \ldots, z_{q-1}, x).$$

Then let v_{q+i}, $i = 2, \ldots, n - q$, be the n-tuple obtained by exchanging z_1 with the x occurring in the ith position:

$$\begin{aligned} v_{q+2} &= (x, z_1, x, x, \ldots, x, x, z_2, \ldots, z_q), \\ v_{q+3} &= (x, x, z_1, x, \ldots, x, x, z_2, \ldots, z_q), \\ &\cdots\cdots\cdots\cdots\cdots\cdots\cdots\cdots\cdots \\ v_n &= (x, x, x, x, \ldots, z_1, x, z_2, \ldots, z_q). \end{aligned}$$

Clearly $v_1, \ldots, v_n$ all belong to S. Let A be the n-square matrix whose rows are $v_1, \ldots, v_n$:

$$A = \begin{bmatrix} x & x & \cdots & x & z_1 & z_n & \cdots & z_q \\ z_1 & x & \cdots & x & x & z_2 & \cdots & z_q \\ \vdots & \vdots & & \vdots & \vdots & \vdots & & \vdots \\ z_q & x & \cdots & x & z_1 & z_2 & \cdots & x \\ x & z_1 & \cdots & x & x & z_2 & \cdots & z_q \\ \vdots & \vdots & & \vdots & \vdots & \vdots & & \vdots \\ x & x & \cdots & z_1 & x & z_2 & \cdots & z_q \end{bmatrix}.$$

We compute the rank of A as follows: Subtract, in turn, row 1 from rows $2, \ldots, n$ to obtain a matrix A_1; these are all elementary row operations and hence do not alter the rank of A. We then have the following form for A_1:

$$A_1 = \begin{bmatrix} x & x & x & \cdots & x & z_1 & z_2 & \cdots & z_{q-1} & z_q \\ z_1 - x & 0 & 0 & \cdots & 0 & x - z_1 & 0 & \cdots & 0 & 0 \\ \vdots & \vdots & \vdots & & \vdots & \vdots & \vdots & & \vdots & \vdots \\ z_q - x & 0 & 0 & \cdots & 0 & 0 & 0 & \cdots & 0 & x - z_q \\ 0 & z_1 - x & 0 & \cdots & 0 & x - z_1 & 0 & \cdots & 0 & 0 \\ \vdots & \vdots & \vdots & & \vdots & \vdots & \vdots & & \vdots & \vdots \\ 0 & 0 & 0 & \cdots & z_1 - x & x - z_1 & 0 & \cdots & 0 & 0 \end{bmatrix}. \tag{56}$$

Recall that $z_i \neq x$, $i = 1, \ldots, q$, so we can divide rows $2, \ldots, n$ in A_1 by suitable constants (another elementary operation) to obtain the matrix

$$A_2 = \begin{bmatrix} x & x & x & \cdots & x & x & z_1 & z_2 & \cdots & z_{q-1} & z_q \\ 1 & 0 & 0 & \cdots & 0 & 0 & -1 & 0 & \cdots & 0 & 0 \\ \vdots & \vdots & \vdots & & \vdots & \vdots & \vdots & \vdots & & \vdots & \vdots \\ 1 & 0 & 0 & \cdots & 0 & 0 & 0 & 0 & \cdots & 0 & -1 \\ 0 & 1 & 0 & \cdots & 0 & 0 & -1 & 0 & \cdots & 0 & 0 \\ \vdots & \vdots & \vdots & & \vdots & \vdots & \vdots & \vdots & & \vdots & \vdots \\ 0 & 0 & 0 & \cdots & 0 & 1 & -1 & 0 & \cdots & 0 & 0 \end{bmatrix}. \tag{57}$$

Finally, we add columns $2, \ldots, n$ to column 1 of A_2 to obtain

$$A_3 = \left[\begin{array}{c|ccc|cccc} (n-q)x + z_1 + \cdots + z_q & x & \cdots & x & z_1 & & \cdots & z_q \\ 0 & & & & & & & \\ 0 & & & & & & -I_q & \\ \vdots & & & & & & & \\ 0 & & & & & & & \\ \hline 0 & & & & -1 & 0 & \cdots & 0 \\ \vdots & & I_{n-q-1} & & \vdots & \vdots & & \vdots \\ 0 & & & & -1 & 0 & \cdots & 0 \end{array}\right].$$

Clearly the $(n-1)$-square submatrix of A_3 lying in rows and columns $2, \ldots, n$ has a nonzero determinant, and hence the rank of A_3 (which is equal to the rank of A) is

$$\begin{array}{ccc} n-1 & \text{if} & x_1 + \cdots + x_n = 0, \\ n & \text{if} & x_1 + \cdots + x_n \neq 0 \end{array}$$

[note that $(n-q)x + z_1 + \cdots + z_2 = x_1 + \cdots + x_n$]. Now (b) follows immediately, and (a) follows once we observe that $S \neq \mathbf{C}^n$ if $x_1 + \cdots + x_n = 0$. Note that in case (a), A is in fact the subspace of $\mathbf{C}^n$ spanned by all vectors $(y_1, \ldots, y_n)$ such that $y_1 + \cdots + y_n = 0$. For this latter space is a proper subspace of $\mathbf{C}^n$ containing the $(n-1)$-dimensional subspace S.

Now let $n \geq 3$, and let W be a proper subspace of $\mathbf{C}^n$ which is invariant under each $A(\sigma)$, $\sigma \in S_n$ (i.e., W is a proper admissible subgroup of $V = \mathbf{C}^n$). Suppose first that $\dim W > 1$. Then W contains a vector $(x_1, \ldots, x_n)$ such that not all the x_i are the same. Define

$$S = \langle A(\sigma)(x_1, \ldots, x_n), \sigma \in S_n \rangle$$

as in (55). Since $S \subset W \neq \mathbf{C}^n$ and $\dim S \geq n-1$, we must have $\dim S = n-1$ and hence $W = S$.

On the other hand, assume that $\dim W = 1$. Then W is spanned by a single (nonzero) vector $v = (v_1, \ldots, v_n)$ which is an eigenvector of each $A(\sigma)$. Thus for each $\sigma \in S_n$, there exists $\lambda(\sigma) \in \mathbf{C}$ such that

$$A(\sigma)v = \lambda(\sigma)v. \tag{58}$$

Now
$$A(\sigma)A(\tau)v = A(\sigma)\lambda(\tau)v = \lambda(\sigma)\lambda(\tau)v$$
and
$$A(\sigma)A(\tau)v = A(\sigma\tau)v = \lambda(\sigma\tau)v.$$

Thus $\lambda(\sigma\tau) = \lambda(\sigma)\lambda(\tau)$, $\sigma,\tau \in S_n$. Since

$$A(\sigma)v = (v_{\sigma^{-1}(1)}, \ldots, v_{\sigma^{-1}(n)}) = 0, \tag{59}$$

$\lambda(\sigma)$ cannot be zero for any $\sigma \in S_n$. Therefore λ is a homomorphism of S_n into the multiplicative group of nonzero complex numbers. We have $\lambda = \varepsilon$ or $\lambda \equiv 1$ (see Exercise 14). Suppose $\lambda = \varepsilon$, and assume without loss of generality that $v_1 \neq 0$. Then $(v_1,v_3,v_3) = -(v_1,v_2,v_3)$ and hence $v_1 = 0$, a contradiction. Thus $\lambda \equiv 1$, and (59) implies that

$$(v_{\sigma(1)}, \ldots, v_{\sigma(n)}) = (v_1, \ldots, v_n)$$

for all $\sigma \in S_n$. It follows that $v_1 = \cdots = v_n$ and hence v is a multiple of $e = (1, 1, \ldots, 1)$. We have shown that for $n \geq 3$, the only proper invariant subspaces of $\mathbf{C}^n$ are S and $\langle e \rangle$, where S is the $(n-1)$-dimensional subspace of $\mathbf{C}^n$ spanned by all vectors $(y_1, \ldots, y_n)$ such that $y_1 + \cdots + y_n = 0$.

It remains to consider the case $n = 2$. If W is a proper invariant subspace of $\mathbf{C}^2$, then W is a one-dimensional subspace of $\mathbf{C}^2$ spanned by a vector $v = (v_1,v_2)$ which is an eigenvector of each $A(\sigma)$:

$$A(\sigma)v = \lambda(\sigma)v, \qquad \lambda(\sigma) \in \mathbf{C}, \qquad \sigma \in S_n.$$

As before, $\lambda: S_n \to \mathbf{C}$ is a homomorphism of S_n into the multiplicative group of nonzero complex numbers, and so $\lambda = \varepsilon$ or $\lambda \equiv 1$. If $\lambda = \varepsilon$, then $(v_2, v_1) = -(v_1, v_2)$ and hence $v_1 + v_2 = 0$. In this case W is the one-dimensional subspace S defined above. If $\lambda \equiv 1$, then $(v_2, v_1) = (v_1, v_2)$ so that $v_1 = v_2$, and v is a multiple of $e = (1,1)$. Thus either $W = S$ or $W = \langle e \rangle$, the same result as when $n \geq 3$.

Exercises 2.3

1. Show that the Example 1(e) is an M-group.
2. Verify the two exhibited formulas in Example 1(e).
3. Show that V is a left $K[x]$-module in Example 1(e).
4. Prove that the intersection of admissible subgroups of an M-group is admissible.
5. Prove that if (M,G,f) is an M-group, then
$$me = e$$
and
$$(mg)^{-1} = mg^{-1}, \qquad m \in M, g \in G.$$
Hint: $me = f(m)(e)$ and $f(m) \in \text{end}\ (G)$. Thus $f(m)e = e$. Similarly $(mg)^{-1} = (f(m)(g))^{-1} = f(m)(g^{-1})$, again because $f(m)$ is an endomorphism of G.
6. State and prove Theorem 1.2 for M-groups.
7. State and prove Theorem 1.3 for M-groups.
8. State and prove Theorem 1.4 for M-groups.
9. State and prove Theorem 3.1 for M-groups.
10. State and prove Theorem 3.2 for M-groups.
11. State and prove Theorem 3.3 for M-groups.
12. Prove that $F[S_n]$ in Example 3 is a linear associative algebra over F.
13. Show that the third and fifth conditions in (46) are satisfied by the action of $r \in F[S_n]$ on V as defined in (50).
14. Show that if λ is a homomorphism of S_n into the multiplicative group of non-zero complex numbers, then $\lambda = \varepsilon$ or $\lambda \equiv 1$. *Hint*: Observe that if $\tau \in S_n$ is any transposition, then $\lambda(\tau) = \pm 1$. Since all transpositions are conjugate, it follows that either $\lambda \equiv 1$ or $\lambda(\tau) = -1$ for every transposition $\tau \in S_n$. But in the latter case we obtain $\lambda = \varepsilon$.

Glossary 2.3

admissible M-group, 115
F-R module, 118
group with operators, 114
inner automorphism, 114
isomorphic series, 112
Jordan-Hölder-Schreier theorem, 113
linear map, 115
linear transformation, 115
M-group, 114
M-homomorphism, 115
operator homomorphism, 115
refinement, 112
Schreier refinement theorem, 112
Zassenhaus lemma, 109

3

Rings and Fields

3.1 Basic Facts

Recall from Section 1.2 that $(R, \{+, \cdot\})$ is an *associative ring* if $(R, \{+\})$ is an abelian group, $(R, \{\cdot\})$ is a semigroup, and the two *distributive laws*

$$a(b + c) = ab + ac, \tag{1}$$

$$(b + c)a = ba + ca \tag{2}$$

hold. The neutral element for addition is denoted by 0. If a neutral element for multiplication exists, it is generally denoted by 1; $(R, \{+, \cdot\})$ is then said to be a ring with *identity*. If the multiplication is commutative, then the ring is said to be *commutative*. Ordinarily we will refer to "the ring R" without explicitly writing out the operations. Unless otherwise stated, the multiplication in a ring will always be associative.

The following facts (see Exercise 1) are easy consequences of the definition of a ring (here $r,s,t,a,b \in R$):

$$r \cdot 0 = 0 \cdot r = 0; \tag{3}$$

$$(-r)s = r(-s) = -(rs); \tag{4}$$

$$(-r)(-s) = rs; \tag{5}$$

$$r(s - t) = rs - rt; \tag{6}$$

$$(s - t)r = sr - tr; \tag{7}$$

If $r \neq 0$ is not a zero divisor, then $ra = rb$ implies $a = b$. (8)

If $\varnothing \neq S \in P(R)$, then S is a *subring* of R if $(S, \{+, \cdot\})$ is a ring. This means that $(S, \{+\}) < (R, \{+\})$ and $(S, \{\cdot\})$ is a semigroup. The distributive laws hold automatically in S because they hold in R. It is easy to check that if $\varnothing \neq S \in P(R)$, then S is a subring of R iff

$$r - s \in S \tag{9}$$

and

$$rs \in S \tag{10}$$

whenever r and s are in S. For if S satisfies (9), then since $S \neq \varnothing$, there exists $r \in S$ and hence $0 = r - r \in S$. Also $s \in S$ implies $0 - s = -s \in S$, and hence $r + s = r - (-s) \in S$. Thus $(S, \{+\}) < (R, \{+\})$. The condition (10) shows that $(S, \{\cdot\})$ is a semigroup (associativity holds automatically in S because it holds in R).

Example 1 (a) The set of even integers is a subring of Z. Thus a ring with identity can have a subring without identity.

(b) The *center* of a ring R is the set $Z(R) = \{c \in R \mid rc = cr \text{ for all } r \in R\}$. Obviously $Z(R)$ is a subring of R. (See Section 2.1 for the corresponding concept for groups.)

(c) In Z^2 define $(a,b) + (c,d) = (a + c, b + d)$ and $(a,b) \cdot (c,d) = (ac,bd)$. Then $(Z^2, \{+, \cdot\})$ is a ring and $Z \times \{0\}$ is a subring with (1,0) as identity element. But the identity element of Z^2 is (1,1). Thus a ring and a subring can have different identity elements. Observe that $(1,0)(0,1) = (0,0)$; so the identity in $Z \times \{0\}$ is a zero divisor in Z^2 (see Exercise 5).

(d) Let p be a positive integer. In Z_p (see Example 2, Section 1.2) we see that

$$\overbrace{[a] + \cdots + [a]}^{p} = [pa] = [0].$$

In other words, adding any element of Z_p to itself p times results in the zero of the ring.

In view of Example 1(d) we define an *integral multiple* of any ring element as follows: If $a \in R$, write

$$0 \cdot a = 0 \qquad \text{(on the left, } 0 \in Z\text{; on the right, } 0 \in R\text{)},$$

$$1 \cdot a = a \qquad \text{(here 1 denotes the integer)},$$

$$n \cdot a = \overbrace{a + \cdots + a}^{n} \qquad \text{if } n \in N,$$

$$n \cdot a = -(-n \cdot a) \qquad \text{if } -n \in N.$$

(The large dot in $n \cdot a$ is usually omitted.) The following formulas are simple to verify for all $n,m \in Z$ and all $a,b \in R$:

$$n(a + b) = na + nb, \tag{11}$$

$$(n + m)a = na + ma, \tag{12}$$

$$(na)b = n(ab) = a(nb), \tag{13}$$

$$(nm)a = n(ma), \tag{14}$$

$$(na)(mb) = (nm)(ab). \tag{15}$$

Let $$M = \{n \in N \mid na = 0 \text{ for all } a \in R\}.$$

If $M \neq \varnothing$, then the smallest integer $p \in M$ is called the *characteristic* of the ring R. If $M = \varnothing$, then R is said to have *characteristic* 0. The characteristic of a ring is denoted by

$$\text{char } R.$$

It is obvious that char $R = 1$ if and only if $R = \{0\}$.

Theorem 1.1 *If $R \neq \{0\}$ (i.e., $|R| \neq 1$) and R has no zero divisors, then either*

$$\text{char } R = 0$$

or

$$\text{char } R = p$$

where p is a prime number.

Proof: Suppose char $R \neq 0$. Then char $R = n \in N$; we have $na = 0$ for all $a \in R$. If $n = rs$ is a proper factorization of n, then by (13) and (15) it follows that for any $a \in R$,

$$\begin{aligned}(ra)(sa) &= (rs)a^2 = ((rs)a)a \\ &= (na)a = 0 \cdot a \\ &= 0.\end{aligned}$$

Since R has no zero divisors, we must have either $ra = 0$ or $sa = 0$. Suppose that $ra = 0$ for some $0 \neq a \in R$. Then for any $b \in R$,

$$\begin{aligned}a(rb) &= r(ab) = (ra)b = 0 \cdot b \\ &= 0.\end{aligned}$$

Hence $rb = 0$. In other words, if $ra = 0$ for some $a \neq 0$, then $rx = 0$ for all $x \in R$. We conclude that either

$$rx = 0, \qquad x \in R,$$

or

$$sx = 0, \qquad x \in R,$$

so that $r \in M$ or $s \in M$. But r and s are both less than n, and n is the smallest element of M. This contradiction shows that char $R = n$ is a prime number. ∎

Note that if char $R = 0$ and R has no zero divisors, the argument in the preceding proof shows that

$$na \neq 0, \quad 0 \neq a \in R, \quad n \in N. \tag{16}$$

Of course, (16) then holds for all $0 \neq n \in Z$. On the other hand, suppose char $R = p$ and $n \in Z$ is such that $na = 0$ for all $a \in R$. Write $n = pq + r$, where $0 \leq r < p$. Then

$$0 = na = (pq + r)a = ra$$

for all $a \in R$. The definition of p now implies that $r = 0$, so that $n = pq$. Thus $pZ = \{n \in Z \mid na = 0 \text{ for all } a \in R\}$.

We shall show that if R is any *integral domain* [i.e., R is a commutative ring with $1 \neq 0$ having no zero divisors (see Section 1.2)], then R contains an isomorphic copy of Z or an isomorphic copy of Z_p for some prime p. In order to formulate this statement precisely, we introduce the notion of ring homomorphism.

Let $(R, \{+, \cdot\})$ and $(S, \{\oplus, *\})$ be two rings, and let $\varphi: R \to S$ be a mapping which satisfies

$$\varphi(a + b) = \varphi(a) \oplus \varphi(b) \tag{17}$$

and

$$\varphi(a \cdot b) = \varphi(a) * \varphi(b) \tag{18}$$

for all $a,b \in R$. Then φ is a *ring homomorphism*. Thus φ is a homomorphism connecting both the additive groups and the multiplicative semigroups in R and S. If φ is injective, it is called a *ring monomorphism;* if φ is surjective, it is called a *ring epimorphism;* if φ is bijective, it is a *ring isomorphism* and R and S are called *isomorphic rings*. If $R = S$, then a homomorphism (isomorphism) $\varphi: R \to S$ is called an *endomorphism* (*automorphism*).

Example 2 (a) Let p be a fixed positive integer. Define $\varphi: Z \to Z_p$ by $\varphi(n) = [n]$. In other words, φ associates with each integer n the residue class in Z_p to which n belongs. Then (see Example 2, Section 1.2)

$$\begin{aligned}\varphi(n + m) &= [n + m] = [n] + [m]\\ &= \varphi(n) + \varphi(m),\end{aligned}$$

and

$$\begin{aligned}\varphi(n \cdot m) &= [nm] = [n][m]\\ &= \varphi(n)\varphi(m).\end{aligned}$$

Clearly φ is a ring epimorphism.

(b) Let $2Z$ denote the ring of even integers. Define $\varphi: Z \to 2Z$ by

$$\varphi(n) = 2n, \qquad n \in Z.$$

Then

$$\begin{aligned}\varphi(n + m) &= 2(n + m) = 2n + 2m\\ &= \varphi(n) + \varphi(m)\end{aligned}$$

and φ is obviously a group isomorphism connecting the additive groups $(Z, \{+\})$ and $(2Z, \{+\})$. But $\varphi(nm) = 2nm \neq (2n)(2m) = \varphi(n)\varphi(m)$ for most values of m and n. Thus φ is not a ring isomorphism.

(c) Let $(R, \{+, *\})$ be the ring consisting of all even integers with ordinary integer addition $+$ and multiplication $*$ defined by

$$m * n = \frac{m \cdot n}{2}, \qquad m,n \in R.$$

Define $\varphi : Z \to R$ by

$$\varphi(n) = 2n, \qquad n \in Z.$$

Then

$$\varphi(n + m) = \varphi(n) + \varphi(m).$$

Also
$$\varphi(n) * \varphi(m) = 2n * 2m = \frac{(2n)(2m)}{2} = 2nm$$
$$= \varphi(nm).$$

Thus φ is a ring homomorphism. Moreover, φ is clearly a bijection, so that Z and R are isomorphic rings.

(d) Let $Q[\sqrt{5}]$ denote the ring of all real numbers of the form $a+b\sqrt{5}$, $a,b \in Q$ with ordinary addition and multiplication as the operations (it is easy to see that every element of $Q[\sqrt{5}]$ has a unique representation in the form $a + b\sqrt{5}$). Define $\sigma\colon Q[\sqrt{5}] \to Q[\sqrt{5}]$ by $\sigma(a + b\sqrt{5}) = a - b\sqrt{5}$. Then

$$\begin{aligned}\sigma((a + b\sqrt{5}) + (c + d\sqrt{5})) &= \sigma(a + c + (b + d)\sqrt{5})\\ &= a + c - (b + d)\sqrt{5}\\ &= (a - b\sqrt{5}) + (c - d\sqrt{5})\\ &= \sigma(a + b\sqrt{5}) + \sigma(c + d\sqrt{5})\end{aligned}$$

and
$$\begin{aligned}\sigma((a + b\sqrt{5})(c + d\sqrt{5})) &= \sigma(ac + 5bd + (ad + bc)\sqrt{5})\\ &= ac + 5bd - (ad + bc)\sqrt{5}\\ &= (a - b\sqrt{5})(c - d\sqrt{5})\\ &= \sigma(a + b\sqrt{5})\sigma(c + d\sqrt{5}).\end{aligned}$$

Thus σ is a ring endomorphism. Clearly σ is a bijection and hence a ring automorphism.

(e) Let G be an additive abelian group, and let S be the set of all group endomorphisms of G. For $s,t \in S$ define $(s + t)(g) = s(g)+t(g)$, and $(s \cdot t)(g) = s(t(g))$, $g \in G$. It is routine to verify (see Exercise 8) that these operations make S into a ring. Then S is called the *endomorphism ring* of G and is denoted by end (G) [see (37) of Section 2.3]. If R is an arbitrary ring, define $\varphi\colon R \to \text{end}\,((R, \{+\}))$ by left multiplication:

$$\varphi(r)x = rx, \qquad r \in R, \quad x \in R.$$

Obviously, $\varphi(r) \in \text{end}\,((R, \{+\}))$ since

$$\begin{aligned}\varphi(r)(x + y) &= r(x + y) = rx + ry\\ &= \varphi(r)(x) + \varphi(r)(y).\end{aligned}$$

Now we assert that φ is a ring homomorphism. Indeed,

$$\begin{aligned}\varphi(r + s)(x) &= (r + s)x = rx + sx\\ &= \varphi(r)(x) + \varphi(s)(x)\\ &= (\varphi(r) + \varphi(s))(x)\end{aligned}$$

and
$$\begin{aligned}\varphi(rs)(x) &= (rs)x = r(sx)\\ &= \varphi(r)(\varphi(s)(x))\\ &= (\varphi(r) \cdot \varphi(s))(x).\end{aligned}$$

Thus $\varphi(r + s) = \varphi(r) + \varphi(s)$ and $\varphi(rs) = \varphi(r) \cdot \varphi(s)$, so φ is a ring homomorphism. Note that φ is a monomorphism if R has an identity 1; for then $\varphi(r)(1) = \varphi(s)(1)$ implies $r = s$. The set $\varphi(R)$ is easily seen to be a subring of end $((R, \{+\}))$. We conclude that *any ring with identity is isomorphic to a ring of endomorphisms of an additive abelian group.*

Let $\varphi\colon R \to S$ be a ring homomorphism and consider the *kernel* of φ:

$$\ker \varphi = \{x \in R \mid \varphi(x) = 0\}.$$

Then if $r \in R$ and $x \in \ker \varphi$, $\varphi(rx) = \varphi(r)\varphi(x) = 0 = \varphi(x)\varphi(r) = \varphi(xr)$. In other words,

$$R \ker \varphi \subset \ker \varphi \tag{19}$$

and

$$\ker \varphi \, R \subset \ker \varphi. \tag{20}$$

(Of course, if $X \subset R$, then $RX = \{rx \mid r \in R, x \in X\}$ and $XR = \{xr \mid x \in X, r \in R\}$.)

We make the following important definition. Let A be a subring of a ring R. If

$$RA \subset A, \tag{21}$$

then A is called a *left ideal* in R. If

$$AR \subset A, \tag{22}$$

then A is called a *right ideal* in R. If A is both a left and a right ideal in R, then A is called a *two-sided ideal* or simply an *ideal.* Thus (19) and (20) assert that $\ker \varphi$ is an ideal in R whenever $\varphi\colon R \to S$ is a ring homomorphism.

If A is an ideal in R, then the quotient group R/A is defined because A is automatically a normal subgroup of $(R, \{+\})$ (addition is always commutative in a ring). The A-cosets are denoted by $r + A$; from (the proof of) Theorem 3.8, Section 1.3, addition of A-cosets is given by

$$(r + A) + (s + A) = (r + s) + A. \tag{23}$$

Now consider the complex product $(r + A)(s + A)$ of $r + A$ and $s + A$; a typical element in the product is

$$\begin{aligned}(r + a)(s + b) &= rs + as + rb + ab \\ &\in rs + A.\end{aligned}$$

The A-cosets in R/A are disjoint; so the preceding calculation shows that the product of $r + A$ and $s + A$ as complexes is contained in the unique complex $rs + A$. We can now *define* a multiplication in R/A by

$$(r + A)(s + A) = rs + A. \tag{24}$$

We do not assert that the product as complexes of $r + A$ and $s + A$ is equal to the complex $rs + A$—only that it is a subset of $rs + A$ (see Exercise 9). But since the A-cosets are disjoint, there is only one A-coset, namely $rs + A$, containing the complex product. Formula (24) is to be interpreted as a *ring product* in R/A.

The following important result obtains.

Theorem 1.2 *If A is an ideal in R, then R/A is a ring with multiplication defined by* (24). *The canonical group homomorphism ν: $(R, \{+\}) \to R/A$ is also a ring epimorphism.*

Proof: The fact that R/A is a ring follows immediately from (23) and (24). That ν is a ring epimorphism is again a restatement of (23) and (24). ∎

The ring R/A in Theorem 1.2 is called the *factor ring* or *quotient ring of R modulo A*.

We shall use the notation $A \lhd R$ to indicate that A is an ideal in the ring R. If there is some possibility of confusion with the notation for a normal subgroup we shall specify what is meant. Of course, an ideal A is a normal subgroup of the additive group of R.

Suppose now that $A \lhd R$, $B \lhd S$, and φ: $R \to S$ is a ring homomorphism such that $\varphi(A) \subset B$. Consider the diagram

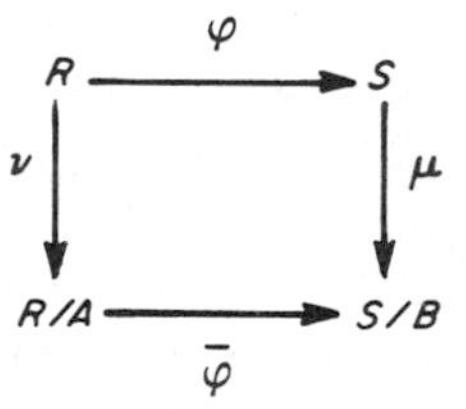

(25)

where ν: $R \to R/A$ and μ: $S \to S/B$ are the canonical epimorphisms. We assert that there is a unique ring homomorphism $\bar{\varphi}$: $R/A \to S/B$ which makes the diagram (25) commutative. Indeed, define

$$\bar{\varphi}(r + A) = \varphi(r) + B, \qquad r \in R. \tag{26}$$

To see that $\bar{\varphi}$ is well-defined by (26), suppose $r + A = s + A$ so that $r - s \in A$. Then $\varphi(r - s) \in \varphi(A) \subset B$, and hence $\varphi(r) - \varphi(s) \in B$. Thus $\varphi(r) \in \varphi(s) + B$, i.e., $\bar{\varphi}(r + A) = \varphi(r) + B = \varphi(s) + B = \bar{\varphi}(s + A)$. Moreover, it is trivial to verify that $\bar{\varphi}$ is a ring homomorphism (see Exercise 11). Obviously (26) is just another way of writing

$$\bar{\varphi}\nu = \mu\varphi \tag{27}$$

and $\bar{\varphi}$ is completely determined by φ.

We now state the fundamental universal property possessed by quotient rings. This result is clearly analogous to Theorem 3.9, Section 1.3.

Theorem 1.3 (*Universal Factorization Theorem for Quotient Rings*) *As-*

sume $A \lhd R$, and let $\nu\colon R \to R/A$ be the canonical ring epimorphism. The pair $(R/A, \nu)$ is universal in the following sense. Given any ring S and any ring homomorphism $\varphi\colon R \to S$ for which $A \subset \ker \varphi$, there exists a unique homomorphism $\bar{\varphi}\colon R/A \to S$ such that $\varphi = \bar{\varphi}\nu$, i.e., the diagram

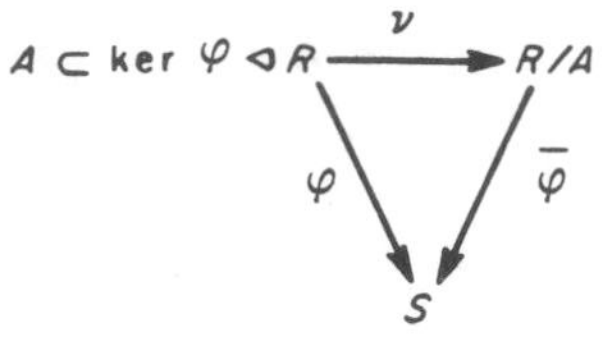

(28)

is commutative.

Proof: In diagram (25) let $B = \{0\}$ be the *zero ideal* in S; B is the trivial ideal in S whose only member is the 0 in S. Since $A \subset \ker \varphi$ we have $\varphi(A) = \{0\} = B$, so diagram (25) is indeed applicable. It is easy to see that in this case the map $\mu\colon S \to S/\{0\}$ is an isomorphism (see Exercise 15). To avoid a conflict in notation, we reproduce diagram (25) with φ^* denoting the map on the lower horizontal line:

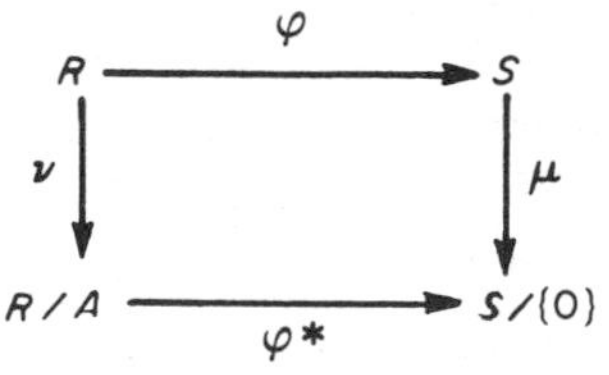

Then $\varphi^*\nu = \mu\varphi$, or

$$\varphi = \mu^{-1}\varphi^*\nu. \tag{29}$$

Now $\bar{\varphi} = \mu^{-1}\varphi^*\colon R/A \to S$ is a ring homomorphism and satisfies $\varphi = \bar{\varphi}\nu$. Obviously the values of $\bar{\varphi}$ are completely determined by the values of φ. ∎

The next result, which follows immediately from Theorem 1.3, is called the *first ring isomorphism theorem.*

Theorem 1.4 *If $\varphi\colon R \to S$ is a ring homomorphism, then $\varphi(R)$ is a subring of S and $R/\ker \varphi$ is isomorphic to $\varphi(R)$.*

Proof: It is simple to verify that $\varphi(R)$ is a subring of S (see Exercise 13). In diagram (28) replace S by $\varphi(R)$ and A by $\ker \varphi$. Then the diagram becomes

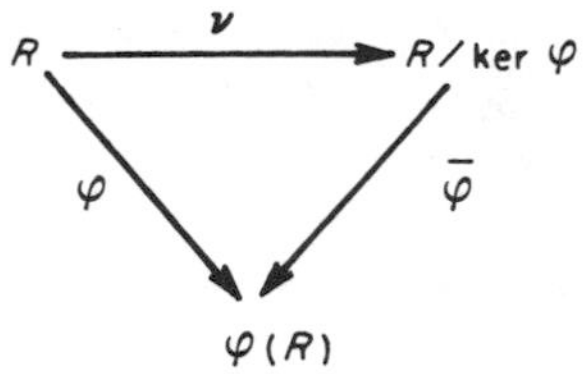

$\varphi = \bar{\varphi}\nu$. It is obvious that $\bar{\varphi}$ is surjective. We claim $\bar{\varphi}$ is injective. It suffices to prove that ker $\bar{\varphi}$ contains only the zero in $R/\ker \varphi$ (see Exercise 14). Now $\bar{\varphi}(\nu(r)) = 0$ implies (since $\varphi = \bar{\varphi}\nu$) that $\varphi(r) = 0$, i.e., $r \in \ker \varphi$. But then $\nu(r) = \ker \varphi$, the zero in $R/\ker \varphi$. ∎

If two rings R and S are isomorphic, this fact is denoted by

$$R \cong S.$$

Thus, for example, Theorem 1.4 states that

$$\varphi(R) \cong R/\ker \varphi.$$

The following interesting results for integral domains are applications of Theorem 1.4 (compare with Corollary 1, Section 2.1).

Theorem 1.5 *Let R be an integral domain. If* char $R = 0$, *then R contains a subring isomorphic to Z. If* char $R = p$, *then R contains a subring isomorphic to Z_p, the ring of integers modulo p.*

Proof: Define $\varphi: Z \to R$ by $\varphi(n) = n \cdot 1$. Clearly φ is a ring homomorphism. By Theorem 1.4, $\varphi(Z)$ is a subring of R and the mapping $\bar{\varphi}: Z/\ker \varphi \to \varphi(Z)$ given by

$$\bar{\varphi}(m + \ker \varphi) = \varphi(m), \qquad m \in Z$$

is an isomorphism. Now suppose char $R = 0$. Then (16) implies that $\ker \varphi = \{0\}$, and hence the canonical map $\nu: Z \to Z/\{0\}$ is an isomorphism. But then $\bar{\varphi}\nu: Z \to \varphi(Z)$ is an isomorphism.

On the other hand, suppose char $R = p$. By the discussion following Theorem 1.1, we have $\ker \varphi = \{n \in Z \mid n \cdot 1 = 0\} = \{n \in Z \mid na = 0 \text{ for all } a \in R\} = pZ$. Since $Z/pZ = Z_p$ (see Exercise 16), the proof is complete. ∎

Theorem 1.6 *If S is a homomorphic image of Z, then either S is isomorphic to Z or S is isomorphic to Z_n for some $n \in N$.*

Proof: Let $\varphi: Z \to S$ be an epimorphism; ker φ is a subring of Z (in fact, it is an ideal in Z). But the only subrings of Z other than $\{0\}$ are of the form

nZ for some $n \in N$ (see Exercise 17). Hence by Theorem 1.4, S is isomorphic to either $Z/\{0\} \cong Z$ or $Z/nZ = Z_n$. ∎

If A and B are arbitrary nonempty subsets of a ring R, define

$$A + B = \{a + b \mid a \in A, b \in B\}.$$

It is simple to verify that if S is a subring of R and $A \lhd R$, then $S + A$ is a subring of R and $S \cap A \lhd S$. We have the following result, called the *second ring isomorphism theorem.*

Theorem 1.7 *Let S be a subring of R and $A \lhd R$. Then*

$$(S + A)/A$$

and

$$S/S \cap A$$

are isomorphic rings, i.e.,

$$S/S \cap A \cong (S + A)/A.$$

Proof: Consider the canonical homomorphism

$$\nu: R \to R/A$$

and let $\mu = \nu | S$. If $s \in S$, then obviously $s \in S + A$ so that $\mu(s) = \nu(s) = s + A \in (S + A)/A$. Observe that $\mu: S \to (S + A)/A$ is an epimorphism. For if $s \in S$ and $a \in A$, then $(s + a) + A = s + A = \mu(s)$. To determine $\ker \mu$, we simply note that for $s \in S$, $\mu(s) = s + A$ is the zero of $(S + A)/A$ iff $s + A = A$ iff $s \in A$. Thus $\ker \mu = S \cap A$. From Theorem 1.4, it follows that $(S + A)/A$ and $S/S \cap A$ are isomorphic; in fact, the isomorphism is the map

$$\bar{\mu}: S/S \cap A \to (S + A)/A$$

induced by μ, satisfying

$$\bar{\mu}(s + S \cap A) = \mu(s) = s + A. \quad ∎$$

There is a standard diagram depicting the contents of Theorem 1.7 [compare diagram (9), Section 2.1]:

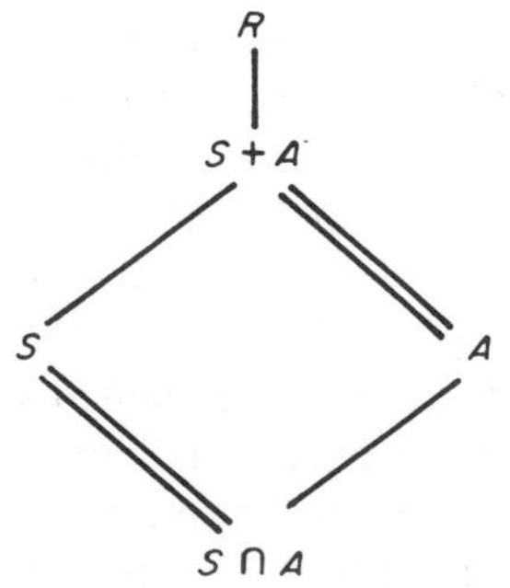

(30)

Reading up from the bottom of diagram (30), the double lines mean that the lower set is an ideal in the upper set. The factor rings corresponding to the opposite sides with double lines are isomorphic. Theorem 1.7 is also called the *parallelogram law* for factor rings.

Theorems 1.4 and 1.7 are the first two of a trilogy of results that are known in ring theory as the *three isomorphism theorems*. The *third ring isomorphism theorem* follows.

Theorem 1.8 *If R is a ring, $A \lhd R$, $B \lhd R$, and $A \subset B$, then $B/A \lhd R/A$ and*

$$\frac{R/A}{B/A} \cong R/B. \tag{31}$$

Proof: In the following diagram let μ and ν be the canonical epimorphisms:

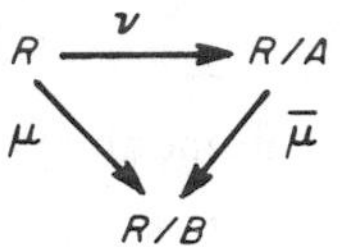

We have $\ker \mu = B \supset A$; so Theorem 1.3 provides a ring homomorphism $\bar{\mu}: R/A \to R/B$ such that

$$\mu = \bar{\mu}\nu.$$

Since μ is surjective, $\bar{\mu}$ is also surjective, i.e., $\bar{\mu}(R/A) = R/B$. Moreover, $r \in B$ iff $\bar{\mu}(\nu(r)) = B$ iff $\nu(r) \in \ker \bar{\mu}$. In other words, $r + A = \nu(r)$ is in the kernel of $\bar{\mu}$ iff $r \in B$; hence

$$\ker \bar{\mu} = \{r + A \mid r \in B\} = B/A.$$

In particular, $B/A \lhd R/A$. Now Theorem 1.4 tells us that $(R/A)/(B/A) \cong R/B$. ∎

Any ring R possesses at least two ideals, namely $\{0\}$ and R itself. If R is a field and $A \lhd R$ contains a nonzero element a, then $aa^{-1} \in A$ so that $1 \in A$. But then $A = R$. Hence a field (or more generally, a division ring) has only these two ideals. In general, if R is a ring, $|R| > 1$, and the only two ideals in R are $\{0\}$ and R, then R is called a *simple* ring. For example, if $p \in N$ is a prime, then Z_p is a simple ring (it is a field).

As we noted in Example 1(c) of Section 2.3, any ring R can be considered as a group with operators. The elements of R operate on R by left multiplication, or by right multiplication. Thus a left ideal in R is simply an admissible subgroup with respect to left multiplication, while a right ideal in R is an admissible subgroup with respect to right multiplication. An ideal, then, is admissible in both senses. A simple ring is an extension of the notion of a simple group to a group with operators. We can thus apply Theorem 1.3 (i), Section 1.3, to groups with operators to conclude that *if*

$$A \triangleleft R \tag{32}$$

and

$$\bar{B} \triangleleft R/A, \tag{33}$$

then there exists a unique ideal B such that $A \triangleleft B \triangleleft R$ *and*

$$\bar{B} = B/A. \tag{34}$$

However, it is instructive to verify this result directly. We have $\bar{B} \triangleleft R/A$; define

$$B = \bigcup_{b+A \in \bar{B}} b + A. \tag{35}$$

Then precisely as in the proof of the result noted above one can easily confirm that B is a subgroup of $(R, \{+\})$. Since $\bar{B} \triangleleft R/A$, it follows that for any $r \in R$ and $b + A \in \bar{B}$,

$$(r + A)(b + A) \in \bar{B}$$

and hence

$$rb + A \in \bar{B}.$$

Now for $r \in R$, $b + a \in B$ we have

$$r(b + a) = rb + ra,$$

and

$$ra \in A,$$

so that

$$r(b + a) \in rb + A \in \bar{B}.$$

In other words, $r(b + a) \in B$. The same argument works for right multiplication, showing that $B \triangleleft R$. By the definition of B,

$$\bar{B} = B/A.$$

Uniqueness is also proved precisely as in Theorem 1.3(i), Section 1.3.

An ideal M in a ring R is said to be *maximal* if $M \neq R$ and there is no ideal $C \triangleleft R$ such that

$$M \subset C \subset R \tag{36}$$

where the inclusions in (36) are proper. The following result is a direct application of Theorem 1.9, Section 2.1 to groups with operators. However, we shall prove it directly in order to increase the reader's familiarity with maximal ideals.

Theorem 1.9 *Let* $M \triangleleft R$. *Then* M *is maximal iff* R/M *is a simple ring.*

Proof: Assume M is maximal. Since $M \neq R$ by definition, it follows that $|R/M| > 1$ (this is one of the requirements for a simple ring). Suppose $\bar{B} \triangleleft R/M$. Then according to (34) there exists a unique ideal B in R such that

$$\bar{B} = B/M.$$

The maximality of M implies that $B = M$ or $B = R$. Hence $\bar{B}$ is the zero ring or $\bar{B} = R/M$. This shows that R/M contains no proper ideals and thus is simple.

Conversely, assume R/M is simple. Then $M \neq R$ since $|R/M| > 1$. Suppose that $C \lhd R$ and

$$M \subset C \subset R. \tag{37}$$

By Theorem 1.8 we have

$$C/M \lhd R/M$$

and

$$R/C \cong \frac{R/M}{C/M}.$$

But R/M is simple; so $|C/M| = 1$ or $C/M = R/M$. These alternatives obviously imply that $C = M$ or $C = R$. Thus M is maximal. ∎

In Section 3.2 it will be shown that if R is a commutative ring with identity, then $M \lhd R$ is maximal iff R/M is a field. We conclude this section with an application of Zorn's lemma.

Theorem 1.10 *Let R be a ring with identity* 1. *If $A \lhd R$ and $A \neq R$, then there exists a maximal ideal $M \lhd R$ such that $A \lhd M \lhd R$.*

Proof: We recall Zorn's lemma: If X is a partially ordered set and every chain C in X has an upper bound, then X contains at least one maximal element.

Define
$$X = \{B \mid B \lhd R, A \subset B, 1 \notin B\}.$$

Since $A \neq R$, it follows that $A \in X$; so $X \neq \varnothing$. It is also obvious that X is partially ordered by (ascending) inclusion. Let $\mathscr{C}$ be a chain in X, so that if B and D are in $\mathscr{C}$, then either $B \subset D$ or $D \subset B$. Set

$$K = \bigcup_{B \in \mathscr{C}} B.$$

If $a,b \in K$, then since $\mathscr{C}$ is a chain, there exists an ideal $B \in \mathscr{C}$ such that $a - b \in B$ and $ra \in B$ for all $r \in R$. Hence K is an ideal. Also $A \subset K$ and $1 \notin K$; so $K \in X$. It is obvious that K is an upper bound for $\mathscr{C}$. We conclude by Zorn's lemma that X contains a maximal element M. To confirm that M is a maximal ideal, first observe that $1 \notin M$ and hence $M \neq R$. Next, suppose that $M \lhd J \lhd R$ and $M \neq J$. Since M is a maximal element of X, we must have $J \notin X$. But $A \subset M \subset J$; so $J \notin X$ implies $1 \in J$ and hence $J = R$. Therefore M is a maximal ideal in R, and $A \lhd M$. ∎

Exercises 3.1

1. Prove assertions (3) to (8). *Hint*: To prove (3), observe that $r \cdot 0 = r \cdot (0 + 0) = r \cdot 0 + r \cdot 0$.

2. Let R be an integral domain in which $r^2 = r$ for every $r \in R$. Show that $|R| = 2$. *Hint*: $|R| \geq 2$ by the definition of an integral domain. Now $r^2 - r = 0$ implies $r(r - 1) = 0$, and R has no zero divisors.
3. Prove: A ring R has no zero divisors iff $ab = ac$ and $ba = ca$ both imply that $b = c$ for all $a \neq 0$. *Hint*: "Only if" is easy. For the "if" part, suppose $ab = 0$. If $a \neq 0$, then $ab = 0 = a \cdot 0$ implies $b = 0$. If $b \neq 0$, then $ab = 0 = 0 \cdot b$ implies $a = 0$.
4. Show that if S is a subring of R, then the additive identity in S is the same as the additive identity in R, and the additive inverse of an element in S is the same as it is in R. *Hint*: We have $(S, \{+\}) < (R, \{+\})$. If $0' \in S$ is the additive identity in S, then $0' = 0' + 0'$. Add the additive inverse in R of $0'$ to both sides. The left side becomes 0, the right side $0'$. The second conclusion is proved similarly.
5. Let S be a subring of R. Show that if S and R have different identities $1'$ and 1 respectively, then $1'$ is a divisor of 0 in R. *Hint*: Since $1'$ is not an identity in R, there is an $a \in R$, $a \neq 0$ such that $a \cdot 1' \neq a$ or $1' \cdot a \neq a$. Suppose $a \cdot 1' \neq a$. Then $a \cdot 1' - a \neq 0$, and $(a \cdot 1' - a)1' = a \cdot 1' - a \cdot 1' = 0$. Similarly if $1' \cdot a \neq a$.
6. Confirm (16). *Hint*: Suppose $n \in N$, $0 \neq a \in R$, and $na = 0$. If $b \in R$, then $a(nb) = n(ab) = (na)b = 0 \cdot b = 0$. Since R has no zero divisors, we must have $nb = 0$. It follows that $M \neq \varnothing$; so R has characteristic different from 0.
7. Show that if R is an integral domain and $|R| < \infty$, then char $R = p \in N$. *Hint*: If char $R = 0$, then according to (16), $na \neq 0$ for all $n \in N$, $0 \neq a \in R$. Fix $a \neq 0$ and consider the sequence $a, 2a, 3a, \ldots$. If $ka = ma$ for some $k > m$, then $(k - m)a = 0$. Hence the elements of the sequence are all distinct, contradicting $|R| < \infty$.
8. Verify that the endomorphism ring of an additive abelian group is a ring [see Example 2(*e*)].
9. Let $A \subset Z$ be the ideal $A = 4Z$, i.e., A consists of all integral multiples of 4. Show that the product as complexes of $2 + A$ and $2 + A$ is strictly contained in $4 + A = A$. *Hint*: $8 \in A$ but $(2 + 4m)(2 + 4n) \in 4(1 + 2m)(1 + 2n) = 8$ implies that $(1 + 2m)(1 + 2n) = 2$, and the product of odd integers is always odd.
10. Show that not every subring of a ring is necessarily an ideal. *Hint*: Consider the subring Z of the ring Q.
11. Verify that $\bar{\varphi}$ in diagram (25) is a ring homomorphism.
12. Let $\varphi: R \to S$ be a ring epimorphism. Show that if 1 is the identity in R, then $\varphi(1)$ is the identity in S. *Hint*: If $s \in S$, then $s = \varphi(r)$ for some $r \in R$, and $\varphi(1)s = \varphi(1)\varphi(r) = \varphi(1 \cdot r) = \varphi(r) = s$.
13. Show that if $\varphi: R \to S$ is a ring homomorphism, then $\varphi(R)$ is a subring of S.
14. Show that if $\varphi: R \to S$ is a ring homomorphism, then φ is injective (i.e., φ is a monomorphism) iff ker $\varphi = \{0\}$.
15. Prove that if $\nu: S \to S/\{0\}$ is the canonical homomorphism ($\{0\}$ is the zero ideal in S), then ν is an isomorphism.
16. Show that if $p \in N$, then $Z/pZ = Z_p$. *Hint*: By definition, the elements of Z/pZ

are cosets of the form $n + pZ = \{a \mid a = n + pm,\ m \in Z\} = [n]$. Moreover, addition and multiplication in Z/pZ are the same as previously defined for Z_p in Example 2, Section 1.2.

17. Find all subrings of Z. *Hint*: Let S be a subring of Z, $S \neq \{0\}$. Show that $S \cap N \neq \varnothing$, and let n be the least positive integer in S. Then $n + \cdots + n \in S$, and so $nZ \subset S$. If $m \in S$, then $m = nq + r$ where $0 \leq r < n$. The minimality of n implies $r = 0$. Hence $S \subset nZ \subset S$.

18. Let φ: $R \rightarrow S$ be a ring epimorphism. Let

$$\mathfrak{A} = \{X \mid X \text{ is a subring of } R \text{ and } \ker \varphi \subset X\},$$
$$\mathfrak{B} = \{Y \mid Y \text{ is a subring of } S\}.$$

Define Φ: $\mathfrak{A} \rightarrow \mathfrak{B}$ by $\Phi(X) = \varphi(X)$. Show that Φ is a bijection and $\Phi^{-1}(Y) = \varphi^{-1}(Y)$ for any $Y \in \mathfrak{B}$. *Hint*: We already know that $\Phi(X) = \varphi(X)$ is a subring of S (see Exercise 13). Moreover, if Y is any subring of S, set $X = \varphi^{-1}(Y)$. Since $0 \in Y$, it follows that $\ker \varphi \subset \varphi^{-1}(Y)$. Also $\varphi^{-1}(Y)$ is easily checked to be a subring of R. Thus $\varphi^{-1}(Y) \in \mathfrak{A}$. Since φ is an epimorphism, we have $\Phi(X) = \varphi(\varphi^{-1}(Y)) = Y$; so Φ is surjective. It is obvious that $X \subset \varphi^{-1}(\varphi(X))$ for any X. Now suppose $z \in \varphi^{-1}(\varphi(X))$. Then $\varphi(z) \in \varphi(X)$, i.e., $\varphi(z) = \varphi(x)$ for some $x \in X$. But then $z - x \in \ker \varphi \subset X$. Hence $z - x \in X$ and thus $z \in X$. In other words, $X = \varphi^{-1}(\varphi(X))$ for $X \in \mathfrak{A}$. Now if $\Phi(X) = \Phi(Z)$, then $\varphi(X) = \varphi(Z)$ and $X = \varphi^{-1}(\varphi(X)) = \varphi^{-1}(\varphi(Z)) = Z$. We conclude that Φ is a bijection.

19. Show that for the bijection Φ of the preceding exercise, $X \lhd R$ iff $\Phi(X) \lhd S$. *Hint*: Assume $X \lhd R$ and let $s \in S$, $x \in X$. Since $S = \varphi(R)$, $s = \varphi(r)$ for some $r \in R$. Then $s\varphi(x) = \varphi(r)\varphi(x) = \varphi(rx) \in \varphi(X)$. Thus $S\varphi(X) \subset \varphi(X)$. Similarly, $\varphi(X)S \subset \varphi(X)$. Also $\Phi(X) = \varphi(X)$ is a subring of S and hence an ideal in S. Conversely, assume $\Phi(X) = \varphi(X) \lhd S$. Let $r \in R$ and $x \in X$. Then $\varphi(rx) = \varphi(r)\varphi(x) \in \varphi(X)$; so $rx \in \varphi^{-1}(\varphi(X))$. Since $\varphi^{-1}(\varphi(X)) = X$ by Exercise 18, we have $rx \in X$. Thus $RX \subset X$, and similarly $XR \subset X$. Therefore $X \lhd R$.

20. Let $A \lhd R$. Let $\mathfrak{A} = \{X \mid X \text{ is a subring of } R \text{ and } A \subset X\}$ and $\mathfrak{B} = \{Y \mid Y \text{ is a subring of } R/A\}$. Show that there is a bijection Φ: $\mathfrak{A} \rightarrow \mathfrak{B}$ such that $X \lhd R$ iff $\Phi(X) \lhd R/A$. *Hint*: Apply the preceding two exercises to the canonical epimorphism ν: $R \rightarrow R/A$.

Glossary 3.1

characteristic of a ring, 126
characteristic 0, 126
char R, 126
commutative ring, 124
distributive laws, 124
endomorphism ring, 128
end G, 128
factor ring, 130
ideal, 129
isomorphic rings, 127
isomorphism theorem for rings (first), 131
isomorphism theorem for rings (second), 133
isomorphism theorem for rings (third), 134
kernel, 129
ker φ, 129
left ideal, 129
mapping diagram for quotient rings, 130
maximal ideal, 135
parallelogram law for factor rings, 134

3.2 Introduction to Polynomial Rings

In order to enrich our supply of rings, we will introduce the notion of an indeterminate over a ring and study the structure of polynomial rings.

In elementary algebra a polynomial in x is "defined" as an expression of the form

$$a_nx^n + a_{n-1}x^{n-1} + \cdots + a_0. \tag{1}$$

Operations of addition and multiplication are dfined, and the polynomial (1) is called the zero polynomial iff $a_n = a_{n-1} = \cdots = a_0 = 0$. There are several reasons why this is not altogether satisfactory. First, of course, what is x? One may attempt to answer this question by saying that x is just a dummy variable and that (1) is the formula for a function

$$p(x) = a_nx^n + \cdots + a_0.$$

But suppose the coefficients $a_n, \ldots, a_0$ belong to a ring such as Z_2, say

$$p(x) = [1]x^2 + [1]x. \tag{2}$$

Observe that

$$p([1]) = [0] = p([0]).$$

Thus, the value of the function $p: Z_2 \to z_2$ is equal to the zero of Z_2 for every specialization of x to an element of Z_2. But certainly neither of the coefficients of $p(x)$ in (2) is the zero in Z_2. In other words, polynomial functions can vanish identically without the coefficients being 0. For this reason a definition of a polynomial is required which is independent of the notion of a polynomial function. There are other serious difficulties: We would like polynomials to be defined for coefficients $a_0, \ldots, a_n$ belonging to a noncommutative ring, say $M_n(\mathbf{C})$.

We say that x is an *indeterminate with respect to a ring R* if there exists a ring with identity, S, containing x and a monomorphism $\iota: R \to S$ satisfying the following conditions:

(i) $x\iota(r) = \iota(r)x$ for each $r \in R$.

(ii) Every element of S is of the form $\sum_{k=0}^{n} a_k x^k$, where $a_k \in \iota(R)$, $n \in N \cup \{0\}$.

(iii) If $\sum_{k=0}^{n} a_k x^k = 0$, $a_k = \iota(r_k)$, $r_k \in R$, then $r_0 = \cdots = r_n = 0$.

Theorem 2.1 *Let R be a ring with identity* 1. *Then there exists an indeterminate x with respect to R.*

Proof: Let $(Z^+, \{+\})$ be the monoid of nonnegative integers with ordinary integer addition as the operation. We refer to Example 9 in Section 1.2 and denote by $R[Z^+]$ the groupoid ring of Z^+ over R. Recall that $R[Z^+]$ consists of those functions on Z^+ to R which are nonzero for at most a finite number of elements of Z^+. As was verified in the indicated example, $R[Z^+]$ is a ring with identity; the operations of addition and multiplication are defined respectively by

$$(f + g)(n) = f(n) + g(n), \qquad n \in Z^+$$

and

$$(fg)(n) = \sum_{k=0}^{n} f(k)g(n-k), \qquad n \in Z^+. \tag{3}$$

Now for $r \in R$, let $\iota(r) \in R[Z^+]$ be the function whose value at 0 is r and whose value at all other $n \in Z^+$ is 0, i.e.,

$$\iota(r)(n) = \delta_{0,n} r, \qquad n \in Z^+$$

($\delta_{s,t}$ is the Kronecker delta: $\delta_{s,t} = 0$ if $s \neq t$, $\delta_{s,s} = 1$). This defines a map $\iota\colon R \to R[Z^+]$ which is clearly injective. Also

$$\begin{aligned}\iota(r_1 + r_2)(n) &= \delta_{0,n}(r_1 + r_2)\\ &= \delta_{0,n} r_1 + \delta_{0,n} r_2\\ &= \iota(r_1)(n) + \iota(r_2)(n)\\ &= (\iota(r_1) + \iota(r_2))(n);\end{aligned}$$

and

$$\begin{aligned}(\iota(r_1)\iota(r_2))(n) &= \sum_{k=0}^{n} (\iota(r_1)(k))(\iota(r_2)(n-k))\\ &= \sum_{k=0}^{n} (\delta_{0,k} r_1)(\delta_{0,n-k} r_2).\end{aligned} \tag{4}$$

The sum on the right in (4) is 0 unless $k = 0$ and $n - k = 0$ for some k. But this requires that $n = k = 0$, and we have

$$(\iota(r_1)\iota(r_2))(n) = \delta_{0,n} r_1 r_2 = \iota(r_1 r_2)(n).$$

Thus ι is a ring monomorphism.

Next, let x be the element of $R[Z^+]$ defined by

$$x(n) = \delta_{1,n}, \qquad n \in Z^+.$$

We assert that if $f \in R[Z^+]$, then there exists $n \in Z^+$ such that

$$f = \sum_{k=0}^{n} \iota(f(k)) x^k, \tag{5}$$

where x^0 is the identity in $R[Z^+]$ (x^0 is the function whose value at 0 is 1 and whose other values are all 0). To verify (5), let n be the largest integer for which $f(n) \neq 0$ [if f is the zero function then obviously $f = \imath(f(0))x^0$; see Exercise 1]. Then for $q \in Z^+$ we compute that

$$\begin{aligned} x^k(q) &= (x^{k-1}x)(q) \\ &= \sum_{t=0}^{q} x^{k-1}(t)x(q-t) \\ &= \sum_{t=0}^{q} x^{k-1}(t)\delta_{1,q-t} \\ &= x^{k-1}(q-1) \\ &= x^{k-2}(q-2) \\ &= \cdots\cdots\cdots\cdots \\ &= x^0(q-k) \\ &= \delta_{k,q}. \end{aligned} \tag{6}$$

From (6) it follows that for $m \leq n$,

$$\begin{aligned} \sum_{k=0}^{n} (\imath(f(k))x^k)(m) &= \sum_{k=0}^{n} \left[\sum_{q=0}^{m} \imath(f(k))(q)x^k(m-q)\right] \\ &= \sum_{k=0}^{n}\sum_{q=0}^{m} \imath(f(k))(q)\delta_{k,m-q} \\ &= \sum_{k=0}^{n}\sum_{q=0}^{m} f(k)\delta_{0,q}\delta_{k,m-q} \\ &= \sum_{k=0}^{n} f(k)\delta_{k,m} \\ &= f(m). \end{aligned} \tag{7}$$

Of course, for $m > n$, the computation (7) shows that the value of the right side of (5) at m is $0 = f(m)$. Hence (5) is established; in particular, any element of $R[Z^+]$ is of the form

$$\sum_{k=0}^{n} a_k x^k, \qquad a_k \in \imath(R),\ n \in N \cup \{0\}. \tag{8}$$

Note that if $a_k = \imath(r_k)$, $r_k \in R$, $k = 0, \ldots, n$, then the same calculation as (7) shows that

$$\sum_{k=0}^{n} (a_k x^k)(m) = r_m, \qquad m = 0, \ldots, n.$$

Thus if $\sum_{k=0}^{n} a_k x^k = 0$, we have $r_0 = \cdots = r_n = 0$.

Finally, $(x\imath(r))(0) = 0 = (\imath(r)x)(0)$, and for $m \in N$,

$$\begin{aligned} (x\imath(r))(m) &= \sum_{k=0}^{m} x(k)\imath(r)(m-k) = \sum_{k=0}^{m} \delta_{1,k}\delta_{0,m-k}r \\ &= \delta_{0,m-1}r \end{aligned}$$

and $$(\iota(r)x)(m) = \sum_{k=0}^{m} \iota(r)(k)x(m-k) = \sum_{k=0}^{m} r\delta_{0,k}\delta_{1,m-k}$$
$$= r\delta_{1,m}.$$

Since $\delta_{1,m} = \delta_{0,m-1}$, x commutes with every element of $\iota(R)$. ∎

We denote the ring $R[Z^+]$ in Theorem 2.1 by $R[x]$. The element x is also called a *variable* or sometimes a *transcendental* over R. The ring $R[x]$ is called a *ring of polynomials in* x. The elements of $R[x]$ (*polynomials in* x) are typically denoted by symbols such as $f(x)$, $g(x)$, etc. If

$$f(x) = \sum_{k=0}^{n} a_k x^k$$

and $a_n \neq 0$ [it follows by condition (iii) in the definition of x that such a representation of $f(x)$ is unique], then n is called the *degree of* $f(x)$, and we write

$$\deg f(x) = n. \tag{9}$$

The degree is not defined for the zero polynomial. If $a_n = 1$, the polynomial $f(x)$ is called *monic*. The term $a_n x^n$ is the *leading term* in $f(x)$, a_n is the *leading coefficient*, and $a_0(= a_0 x^0)$ is the *constant term*. A polynomial of the form $a_n x^n$ is called a *monomial*.

Theorem 2.1 states that any ring R with identity can be injected into a ring $R[x] = R[Z^+]$ containing a transcendental x over R. It is convenient to actually have R appear as a subring of the polynomial ring $R[x]$. For example, we would prefer to regard the nonzero elements of R as zero degree polynomials in $R[x]$ instead of considering an isomorphic copy of R. This difficulty is easily resolved by applying some of the elementary set theoretic results in Section 1.1.

Theorem 2.2 *Let $(A, \{+, \cdot\})$ be a ring, and let R be an arbitrary set. Then there exists a ring $(B, \{\oplus, *\})$ such that $A \cong B$ and $B \cap R = \varnothing$.*

Proof: By Corollary 1 in Section 1.1, obtain a set B such that $B \cap R = \varnothing$ and a bijection $\nu: A \to B$. Define two binary operations $\oplus$ and $*$ in B by

$$x \oplus y = \nu(\nu^{-1}(x) + \nu^{-1}(y)), \tag{10}$$
$$x * y = \nu(\nu^{-1}(x) \cdot \nu^{-1}(y)). \tag{11}$$

We assert that $(B, \{\oplus, *\})$ is a ring. Once this is verified, then of course equations (10) and (11) show that $\nu^{-1}: B \to A$ is an isomorphism and hence $\nu: A \to B$ is an isomorphism.

For example, let $0' = \nu(0)$. Then

$$\begin{aligned} 0' \oplus x &= \nu(\nu^{-1}(0') + \nu^{-1}(x)) \\ &= \nu(0 + \nu^{-1}(x)) = \nu(\nu^{-1}(x)) \\ &= x \end{aligned}$$

for all $x \in B$. The verifications of the remaining ring axioms for $(B, \{\oplus, *\})$ are left to the reader (see Exercise 2). ∎

We are now in a position to state and prove a result of considerable utility in ring theory.

Theorem 2.3 (*The Ring Extension Theorem*) *Let R_1, R_2 and R_3 be rings satisfying the following two conditions:*

(i) *R_1 is a subring of R_2;*

(ii) *there exists a ring isomorphism $\varphi\colon R_1 \to R_3$.*

Then there exist a ring R_4 and an isomorphism $\theta\colon R_2 \to R_4$ such that

(iii) *R_3 is a subring of R_4;*

(iv) *$\theta | R_1 = \varphi$.*

Proof: By Theorem 2.2 there exists a ring B such that $R_2 \cong B$ and $B \cap R_3 = \varnothing$. Let $\nu\colon R_2 \to B$ be a fixed isomorphism.

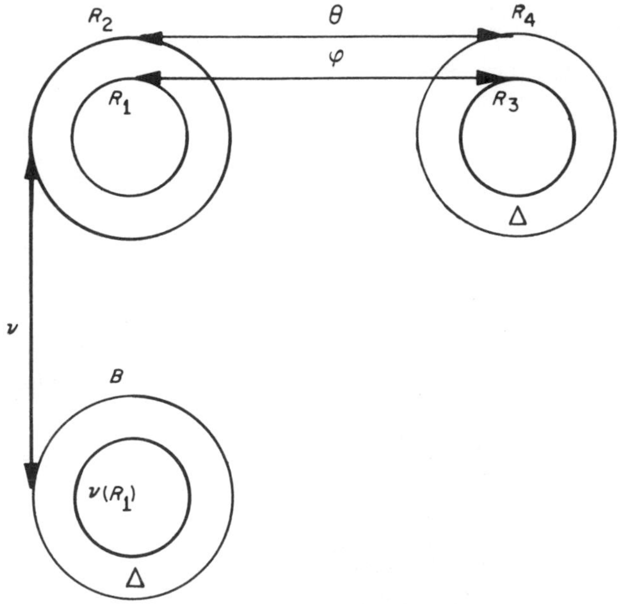

(12)

Write $\Delta = B - \nu(R_1)$, and set $R_4 = R_3 \cup \Delta$. Define $\theta: R_2 \to R_4$ by

$$\theta \mid R_1 = \varphi, \tag{13}$$

$$\theta \mid (R_2 - R_1) = \nu. \tag{14}$$

Since φ and ν are bijections, $\varphi(R_1) = R_3$, $\nu(R_2 - R_1) = \Delta$, and $\Delta \cap R_3 = \varnothing$, we conclude that θ is a bijection. Define binary operations $\oplus$ and $*$ in R_4 by

$$x \oplus y = \theta(\theta^{-1}(x) + \theta^{-1}(y)), \tag{15}$$

$$x * y = \theta(\theta^{-1}(x)\theta^{-1}(y)). \tag{16}$$

Then precisely as in Theorem 2.2, $(R_4, \{\oplus, *\})$ is a ring and $\theta: R_2 \to R_4$ is a ring isomorphism. The conclusion (iv) of the present theorem holds automatically by (13). It remains to check that R_3 is a subring of R_4. If $x,y \in R_3$, we have by (13) and (15) that

$$\begin{aligned} x \oplus y &= \theta(\theta^{-1}(x) + \theta^{-1}(y)) \\ &= \theta(\varphi^{-1}(x) + \varphi^{-1}(y)) \\ &= \varphi(\varphi^{-1}(x + y)) \\ &= x + y. \end{aligned}$$

Similarly, $x * y = xy$. In other words, the operations $\oplus$ and $*$ when specialized to the ring R_3 coincide with the ring operations in R_3. Hence R_3 is a subring of R_4. ∎

We will use the ring extension theorem to prove three important results: Any ring with identity R can be embedded in a ring of polynomials with coefficients from R; any integral domain can be embedded in a field; any ring can be embedded in a ring containing a multiplicative identity. The meaning of these results will be made precise as we go on.

First, we say that t is *an indeterminate over a ring R* if there exists a ring with identity S containing t and containing R as a subring such that the canonical injection $\iota = \iota_R: R \to S$ makes t into an indeterminate with respect to R (recall the definition immediately preceding Theorem 2.1). This means that

(i) $tr = rt$ for each $r \in R$;
(ii) every element of S is of the form $\sum_{k=0}^{n} r_k t^k$, where $r_k \in R$, $n \in N \cup \{0\}$;
(iii) if $\sum_{k=0}^{n} r_k t^k = 0$, $r_k \in R$, then $r_0 = \cdots = r_n = 0$.

The ring S is called *the ring of polynomials in the indeterminate t over R.*

Theorem 2.4 *Let R be a ring with identity* 1. *Then there exists an indeterminate t over R.*

Proof: By Theorem 2.1 there exists a ring with identity S containing an indeterminate x with respect to R. Thus we have a monomorphism

$$\iota: R \to S$$

such that

(i) x commutes with every element of the ring $\iota(R)$;

(ii) every element of S is of the form

$$\sum_{k=0}^{n} a_k x^k, \qquad a_k \in \iota(R), n \in N \cup \{0\};$$

(iii) if $\sum_{k=0}^{n} a_k x^k = 0$, $a_k = \iota(r_k)$, $r_k \in R$, then $r_0 = \cdots = r_n = 0$.

Now by Theorem 2.3, there exists a ring $\mathscr{R}$ and an isomorphism $\theta: S \to \mathscr{R}$ such that R is a subring of $\mathscr{R}$ and $\theta \mid \iota(R) = \iota^{-1}$ (or equivalently, $\theta^{-1} | R = \iota$).

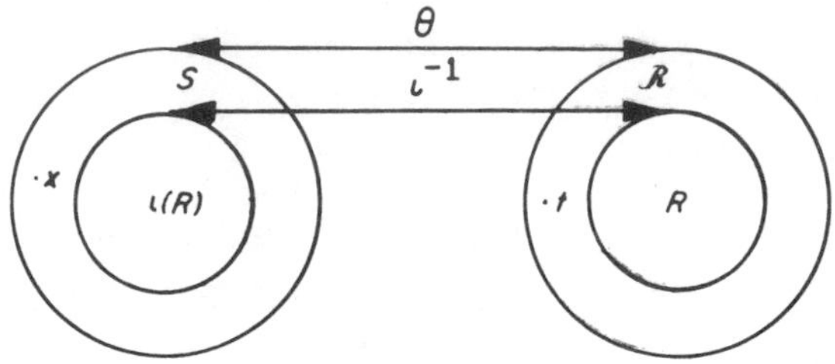

Set $t = \theta(x)$. Then if $r \in R$,

$$\begin{aligned} tr &= \theta(\theta^{-1}(tr)) = \theta(\theta^{-1}(t)\theta^{-1}(r)) \\ &= \theta(x \cdot \iota(r)) = \theta(\iota(r)x) \qquad \text{(by (i))} \\ &= \theta(\iota(r))\theta(x) = \iota^{-1}(\iota(r))t \\ &= rt. \end{aligned}$$

Thus t commutes with every element of R. Every element of S is of the form

$$f(x) = \sum_{k=0}^{n} \iota(r_k)x^k$$

and hence, since θ is surjective, every element of $\mathscr{R}$ is of the form

$$\begin{aligned} \theta(f(x)) &= \sum_{k=0}^{n} \theta(\iota(r_k))\theta(x)^k = \sum_{k=0}^{n} (\iota^{-1}\iota)(r_k)t^k \\ &= \sum_{k=0}^{n} r_k t^k. \end{aligned}$$

Finally,

$$\sum_{k=0}^{n} r_k t^k = 0$$

implies that

$$\sum_{k=0}^{n} \theta^{-1}(r_k)\theta^{-1}(t)^k = 0,$$

i.e.,
$$\sum_{k=0}^{n} \theta^{-1}(r_k)x^k = 0.$$
Since $\theta^{-1}(r_k) = \imath(r_k)$ we have
$$\sum_{k=0}^{n} \imath(r_k)x^k = 0,$$
and hence $r_0 = \cdots = r_n = 0$ by (iii). Thus t is an indeterminate over R. ▮

Theorem 2.4 tells us then that given any ring R with identity, we can construct a ring of polynomials in an indeterminate over R. We shall usually use x (instead of t) to denote an indeterminate over R, provided there is no conflict in the notation. The ring of polynomials in the indeterminate x over R will be denoted, as before, by $R[x]$. In conformity with standard usage the notation $R[x]$ will also be applied in a somewhat more general context. If R is a subring of a ring S and s is any element of S which commutes with every element of R, then $R[s]$ will denote the totality of elements of the form $\sum_{k=0}^{n} r_k s^k$, $r_k \in R$, $n \in N \cup \{0\}$. It is clear that $R[s]$ is a subring of S; the elements of $R[s]$ will be called *polynomials in s*. We say that s is *adjoined* to R. In particular the ring R is a subring of $R[x]$, the ring of polynomials in the indeterminate x over R:
$$R \subset R[x]. \tag{17}$$
It is obvious that the identity of the ring R is the identity of the ring $R[x]$. Also observe that if R is commutative, then $R[x]$ is commutative. In general, repeated use of the fact that x commutes with elements of R shows that
$$\sum_{p=0}^{r} a_p x^p \sum_{q=0}^{s} b_q x^q = \sum_{m=0}^{r+s} c_m x^m,$$
where
$$c_m = \sum_{p+q=m} a_p b_q. \tag{18}$$

Corollary 1 *If R is an integral domain, then so is* $R[x]$.

Proof: We have already seen that $R[x]$ is a commutative ring with identity when R is. Observe that if $f(x)g(x) = 0$, then (18) implies that the product of the leading coefficients is 0. Since R has no zero divisors, it follows that either $f(x)$ or $g(x)$ is 0. Finally, $1 \neq 0$ in the integral domain $R \subset R[x]$. Thus $R[x]$ is an integral domain. ▮

Suppose that R is a ring with identity and we construct a ring $R[x_1]$ of polynomials in an indeterminate x_1 over R. Then we can construct a ring $R[x_1][x_2]$ of polynomials in an indeterminate x_2 over $R[x_1]$. Continuing, we construct a ring $R[x_1][x_2] \cdots [x_n]$, where x_k is an indeterminate over $R[x_1] \cdots [x_{k-1}]$, $2 \leq k \leq n$. It is easy to see that x_k is an indeterminate over R

for $1 \leq k \leq n$ (see Exercise 3). It is also obvious that any element of $R[x_1][x_2] \cdots [x_n]$ can be written as

$$f(x_1, \ldots, x_n) = \sum a_{i_1 \cdots i_n} x_1^{i_1} x_2^{i_2} \ldots x_n^{i_n}, \qquad a_{i_1 \cdots i_n} \in R, \tag{19}$$

the summation being taken over $i_1 = 0, \ldots, r_1, i_2 = 0, \ldots, r_2, \ldots, i_n = 0, \ldots, r_n$. The element (19) is called a *polynomial in the indeterminates* $x_1, \ldots, x_n$ *over* R. A single term such as $ax_1^{i_1} \cdots x_n^{i_n}$ $(a \in R)$ is called a *monomial.* The ring $R[x_1] \ldots [x_n]$ is called the *ring of polynomials in several indeterminates* $x_1, \ldots, x_n$ *over* R, and we simplify the notation by writing

$$R[x_1, \ldots, x_n] = R[x_1] \cdots [x_n].$$

It is clear that if $\sigma \in S_n$ then $R[x_{\sigma(1)}, \ldots, x_{\sigma(n)}] = R[x_1, \ldots, x_n]$, since both sets consist of all polynomials (19) in $x_1, \ldots, x_n$ over R.

Corollary 2 *Any x_t is an indeterminate over $R[x_1, \ldots, x_{t-1}, x_{t+1}, \ldots, x_n]$.*

Proof: Fix t, $1 \leq t \leq n$, and let $R_t = R[x_1, \ldots, x_{t-1}, x_{t+1}, \ldots, x_n]$, $R_0 = R$. Obviously $R_t[x_t] = R[x_1, \ldots, x_n]$ because any polynomial $f(x_1, \ldots, x_n) \in R[x_1, \ldots, x_n]$ can be written as a polynomial in x_t with coefficients from R_t. Thus we need only show that if $g_k \in R_t$, $k = 0, \ldots, m$, and

$$\sum_{k=0}^{m} g_k x_t^k = 0 \tag{20}$$

then $g_0 = \cdots = g_m = 0$. Write each g_k as a polynomial in $x_1, \ldots, x_{t-1}, x_{t+1}, \ldots, x_n$,

$$g_k = \sum_{\gamma} a_{k\gamma} x_1^{\gamma_1} \ldots x_{t-1}^{\gamma_{t-1}} x_{t+1}^{\gamma_{t+1}} \ldots x_n^{\gamma_n}, \qquad a_{k\gamma} \in R, \tag{21}$$

substitute g_k in (20), $k = 0, \ldots, m$, and rewrite the resulting polynomial, call it $f(x_1, \ldots, x_n)$, as a polynomial in x_n with coefficients in $R_{n-1} = R[x_1, \ldots, x_{n-1}]$. Note that if $k \neq l$, no monomial appearing in $g_k x_t^k$ can appear in $g_l x_k^l$ because they involve different powers of x_t. Hence if we write
t

$$f(x_1, \ldots, x_n) = \sum_{j=0}^{p} f_j x_n^j,$$

where $f_j = \sum_{\alpha} b_{j\alpha} x_1^{\alpha_1} \ldots x_{n-1}^{\alpha_{n-1}} \in R_{n-1}, \qquad b_{j\alpha} \in R, j = 0, \ldots, p,$

then the $b_{j\alpha}$ are just the $a_{k\gamma}$ in some other order. Now $f(x_1, \ldots, x_n) = 0$ and since x_n is an indeterminate over $R[x_1] \ldots [x_{n-1}] = R[x_1, \ldots, x_{n-1}] = R_{n-1}$, it follows that each $f_j = 0$. But then an obvious induction on n implies that all the $b_{j\alpha}$, i.e., all the $a_{k\gamma}$, must be 0. In other words, $g_k = 0$, $k = 0, \ldots, m$. ∎

In view of Corollary 2 we say that $x_1, \ldots, x_n$ are *independent indeterminates over R.*

Corollary 3 *If R is an integral domain then so is $R[x_1, \ldots, x_n]$.*

Proof: This is obvious from Corollary 1. ∎

An extensive theory is available for polynomials in several indeterminates. At this point we will consider a few of the more elementary results. In order to do this, we introduce some convenient notation for sequence sets. If $n_1, \ldots, n_p$ are positive integers, let

$$\Gamma(n_1, \ldots, n_p) \tag{22}$$

denote the set of integer sequences γ which satisfy

$$1 \leq \gamma(i) \leq n_i, \qquad i = 1, \ldots, p.$$

If p and n are positive integers, let

$$\Gamma_p(n) = \Gamma(\overbrace{n, \ldots, n}^{p}); \tag{23}$$

$\Gamma_p(n)$ is the set of integer sequences γ which satisfy

$$1 \leq \gamma(i) \leq n, \qquad i = 1, \ldots, p.$$

If $1 \leq m \leq n$, define

$$G_{m,n} \tag{24}$$

to be the set of integer sequences γ which satisfy

$$1 \leq \gamma(1) \leq \gamma(2) \leq \cdots \leq \gamma(m) \leq n. \tag{25}$$

Also define the sets of integer sequences

$$D_{m,n} \tag{26}$$

and

$$Q_{m,n} \tag{27}$$

by the respective conditions:

$$\gamma(i) \neq \gamma(j) \qquad \text{when } i \neq j,\ i, j = 1, \ldots, m \tag{28}$$

and

$$1 \leq \gamma(1) < \gamma(2) < \cdots < \gamma(m) \leq n. \tag{29}$$

Thus $D_{m,n}$ is the set of sequences of m distinct integers chosen from $1, \ldots, n$, and $Q_{m,n}$ is the set of strictly increasing sequences of m integers chosen from $1, \ldots, n$. We have

$$Q_{m,n} \subset D_{m,n} \subset \Gamma_m(n)$$

and

$$Q_{m,n} \subset G_{m,n} \subset \Gamma_m(n).$$

Observe that

$$|\Gamma(n_1, \ldots, n_p)| = n_1 \cdots n_p, \tag{30}$$

$$|G_{m,n}| = \binom{n+m-1}{m}, \tag{31}$$

$$|D_{m,n}| = \frac{n!}{(n-m)!}, \tag{32}$$

and
$$|Q_{m,n}| = \binom{n}{m} \tag{33}$$

(see Exercise 4).

We may modify (22) slightly by defining
$$\Gamma^0(n_1, \ldots, n_p) \tag{34}$$
to be the set of all integer sequences γ which satisfy
$$0 \leq \gamma(i) \leq n_i, \qquad i = 1, \ldots, p.$$
Then a rather unpleasant expression such as (19) can be shortened to
$$f(x_1, \ldots, x_n) = \sum_{\gamma \in \Gamma^0(r_1, \ldots, r_n)} a_\gamma \prod_{t=1}^{n} x_t^{\gamma(t)}. \tag{35}$$
In the same way, (23) can be modified by writing
$$\Gamma_p{}^0(n) = \Gamma^0(\overbrace{n, \ldots, n}^{p}). \tag{36}$$
Then an arbitrary polynomial in which no x_t appears to any higher power than r can be written as
$$f(x_1, \ldots, x_n) = \sum_{\gamma \in \Gamma_n{}^0(r)} a_\gamma \prod_{t=1}^{n} x_t^{\gamma(t)}. \tag{37}$$

A polynomial $f(x_1, \ldots, x_n) \in R[x_1, \ldots, x_n]$ is said to be *symmetric* if
$$f(x_{\sigma(1)}, \ldots, x_{\sigma(n)}) = f(x_1, \ldots, x_n)$$
for all $\sigma \in S_n$. The most familiar and useful symmetric polynomials are the so called *elementary symmetric functions* (abbreviated e.s.f.) given by
$$E_m(x_1, \ldots, x_n) = \sum_{\gamma \in Q_{m,n}} \prod_{i=1}^{m} x_{\gamma(i)}, \qquad 1 \leq m \leq n, \tag{38}$$
$$E_0(x_1, \ldots, x_n) = 1.$$
The *completely symmetric functions* or *Wronski polynomials* are another class of symmetric polynomials; they are defined by
$$h_m(x_1, \ldots, x_n) = \sum_{\gamma \in G_{m,n}} \prod_{i=1}^{m} x_{\gamma(i)}, \qquad 1 \leq m \leq n, \tag{39}$$
$$h_0(x_1, \ldots, x_n) = 1.$$

The *power sums*
$$s_m(x_1, \ldots, x_n) = \sum_{t=1}^{n} x_t^m, \qquad m = 1, 2, \ldots, \tag{40}$$
$$s_0(x_1, \ldots, x_n) = n$$
are also symmetric polynomials.

For a sequence γ in $\Gamma(n_1, \ldots, n_p)$ or $\Gamma^0(n_1, \ldots, n_p)$, set

$$m_t(\gamma) = |\,\gamma^{-1}(t)\,|, \qquad 0 \le t \le \max_{1\le i\le p} n_i; \tag{41}$$

$m_t(\gamma)$ is the multiplicity of occurrence of t in im γ. Thus we can rewrite E_m and h_m respectively as

$$E_m(x_1, \ldots, x_n) = \sum_{\gamma\in Q_{m,n}} \prod_{t=1}^{n} x_t^{m_t(\gamma)}, \qquad 1 \le m \le n \tag{42}$$

and
$$h_m(x_1, \ldots, x_n) = \sum_{\gamma\in G_{m,n}} \prod_{t=1}^{n} x_t^{m_t(\gamma)}, \qquad 1 \le m \le n \tag{43}$$

(see Exercise 5).

Note that a sequence γ in $G_{m,n}$ is uniquely determined by the sequence of multiplicities $(m_1(\gamma), \ldots, m_n(\gamma))$. For example, if $m = 5$, $n = 7$, $m_1(\gamma) = 1$, $m_2(\gamma) = 3$, $m_5(\gamma) = 1$, then

$$\gamma = (1,2,2,2,5).$$

A *lexicographic ordering* can be introduced into any of the preceding sequence sets as follows: For sequences α and β, write $\alpha < \beta$ (α precedes β) if $\alpha \ne \beta$ and the first nonzero difference $\alpha(i) - \beta(i)$ is negative. We also write $\beta > \alpha$ to mean $\alpha < \beta$. If

$$a_\beta \prod_{t=1}^{n} x_t^{\beta(t)}, \qquad a_\alpha \prod_{t=1}^{n} x_t^{\alpha(t)}$$

are two nonzero terms in the polynomial (35), then the first of these is *higher* than the second (in lexicographic order) if $\beta > \alpha$. For example, the terms of the polynomial

$$3x_1^2x_3 - 2x_1x_2^2 - x_1 + 2x_2^4 - x_2^3x_3 + x_2x_3 + 5 \tag{44}$$

are arranged in decreasing order. The term which is highest in lexicographic order is called the *leading term* of the polynomial, and the coefficient of that term is called the *leading coefficient*. Thus $3x_1^2x_3$ is the leading term in the polynomial (44). As another example, the leading term in

$$E_m(x_1, \ldots, x_n)$$

is obviously

$$x_1 \ldots x_m.$$

When no confusion can result, we will frequently omit the indeterminates and abbreviate expressions such as $f(x_1, \ldots, x_n)$ simply by f. The *degree* of a nonzero polynomial

$$f(x_1, \ldots, x_n) = \sum_{\gamma\in\Gamma^0(r_1,\ldots,r_n)} a_\gamma \prod_{t=1}^{n} x_t^{\gamma(t)}$$

is the largest of the integers

$$\sum_{t=1}^{n} \gamma(t)$$

for which $a_\gamma \ne 0$, and is denoted by

$$\deg f.$$

The degree is not defined for the zero polynomial. For example, if

$$f(x_1,x_2,x_3) = x_1{}^2x_2 + 3x_2{}^2 + 5x_1x_2x_3{}^2$$

then $\deg f = 4$. It is clear that if R is an integral domain, then

$$\deg fg = \deg f + \deg g \tag{45}$$

(see Exercise 7). The single most important result concerning the e.s.f. follows.

Theorem 2.5 (*Basis Theorem for Symmetric Polynomials*) *Let R be an integral domain, and let $f(x_1, \ldots, x_n)$ be a symmetric polynomial in $R[x_1, \ldots, x_n]$. Then there exists a unique polynomial*

$$\psi(x_1, \ldots, x_n) = \sum_{\alpha \in \Gamma^0(r_1,\ldots,r_n)} c_\alpha \prod_{t=1}^{n} x_t{}^{\alpha(t)}$$

in $R[x_1, \ldots, x_n]$ such that

$$\begin{aligned} f(x_1, \ldots, x_n) &= \psi(E_1(x_1, \ldots, x_n), \ldots, E_n(x_1, \ldots, x_n)) \\ &= \sum_{\alpha \in \Gamma^0(r_1,\ldots,r_n)} c_\alpha \prod_{t=1}^{n} E_t(x_1, \ldots, x_n)^{\alpha(t)}. \end{aligned} \tag{46}$$

Proof: Let

$$a_\gamma \prod_{t=1}^{n} x_t{}^{\gamma(t)} \tag{47}$$

be the leading term of $f(x_1, \ldots, x_n)$. The leading term of a product of polynomials is clearly the product of the leading terms (see Exercise 6). Thus if $s_1, \ldots, s_n$ are nonnegative integers, the leading term of

$$E_1^{s_1} \cdots E_n^{s_n}$$

is

$$x_1{}^{s_1}(x_1x_2)^{s_2}(x_1x_2x_3)^{s_3} \cdots (x_1 \cdots x_n)^{s_n} = x_1{}^{\sigma_n}x_2{}^{\sigma_{n-1}} \cdots x_n{}^{\sigma_1}$$

where $\qquad \sigma_{n-k} = s_{k+1} + \cdots + s_n, \qquad k = 0, \ldots, n-1.$

We want to choose nonnegative integers $s_1, \ldots, s_n$ such that

$$\begin{aligned} \sigma_n &= s_1 + \cdots + s_n = \gamma(1), \\ \sigma_{n-1} &= s_2 + \cdots + s_n = \gamma(2), \\ &\cdots\cdots\cdots\cdots \\ \sigma_2 &= s_{n-1} + s_n = \gamma(n-1), \\ \sigma_1 &= s_n = \gamma(n). \end{aligned}$$

This forces the choice

$$\begin{aligned} s_n &= \gamma(n), \\ s_{n-1} &= \gamma(n-1) - \gamma(n), \\ &\cdots\cdots\cdots\cdots \\ s_2 &= \gamma(2) - \gamma(3), \\ s_1 &= \gamma(1) - \gamma(2). \end{aligned} \tag{48}$$

The question is whether the s_i in (48) are nonnegative integers; here we make use of the symmetry of f. Suppose $s_k < 0$ for some k, so that

$$\gamma(k) < \gamma(k+1),$$

and let $\sigma = (k \ \ k+1)$. Then since $f(x_{\sigma(1)}, \ldots, x_{\sigma(n)}) = f(x_1, \ldots, x_n)$, it follows that

$$\begin{aligned} &a_\gamma x_{\sigma(1)}^{\gamma(1)} \cdots x_{\sigma(k)}^{\gamma(k)} x_{\sigma(k+1)}^{\gamma(k+1)} \cdots x_{\sigma(n)}^{\gamma(n)} \\ &= a_\gamma x_1^{\gamma(1)} \cdots x_{k+1}^{\gamma(k)} x_k^{\gamma(k+1)} \cdots x_n^{\gamma(n)} \\ &= a_\gamma x_1^{\gamma(1)} \cdots x_k^{\gamma(k+1)} x_{k+1}^{\gamma(k)} \cdots x_n^{\gamma(n)} \end{aligned} \tag{49}$$

is a term of $f(x_1, \ldots, x_n)$. But the term (49) is higher than the leading term (47) of f; the first nonzero difference in the exponents is

$$\gamma(k+1) - \gamma(k) > 0.$$

This contradiction shows that $\gamma(1) \geq \gamma(2) \geq \cdots \geq \gamma(n)$; so the s_i in (48) are nonnegative integers. Now form the difference

$$f_1 = f - a_\gamma E_1^{s_1} \cdots E_n^{s_n}. \tag{50}$$

Obviously f_1 is symmetric, and its leading term (if $f_1 \neq 0$) is lower in lexicographic order than the leading term of f. We repeat the above procedure with f_1 and continue. Since there are only a finite number of sequences which precede γ in lexicographic order, the process must terminate. This establishes the existence of the desired polynomial $\psi(x_1, \ldots, x_n)$. Observe that the coefficients that arise in the procedure are obtained from the coefficients of f by addition and subtraction only.

It remains to prove uniqueness. Suppose $\varphi(x_1, \ldots, x_n)$ is a polynomial in $R[x_1, \ldots, x_n]$ such that

$$\varphi(E_1, \ldots, E_n) = f = \psi(E_1, \ldots, E_n). \tag{51}$$

Let

$$c_\alpha \prod_{t=1}^{n} x_t^{\alpha(t)} \qquad \text{and} \qquad d_\beta \prod_{t=1}^{n} x_t^{\beta(t)}$$

be the leading terms of ψ and φ, respectively. Clearly the leading term of $\psi(E_1, \ldots, E_n)$ is

$$c_\alpha x_1^{\alpha(1)}(x_1x_2)^{\alpha(2)} \cdots (x_1 \cdots x_n)^{\alpha(n)}$$

and the leading term of $\varphi(E_1, \ldots, E_n)$ is

$$d_\beta x_1^{\beta(1)}(x_1x_2)^{\beta(2)} \cdots (x_1 \cdots x_n)^{\beta(n)}.$$

These are equal by (51), so that $c_\alpha = d_\beta$ and

$$\begin{aligned}\alpha(1) + \cdots + \alpha(n) &= \beta(1) + \cdots + \beta(n),\\ \alpha(2) + \cdots + \alpha(n) &= \beta(2) + \cdots + \beta(n),\\ &\cdots\cdots\\ \alpha(n-1) + \alpha(n) &= \beta(n-1) + \beta(n),\\ \alpha(n) &= \beta(n).\end{aligned} \tag{52}$$

Since (52) obviously implies that $\alpha = \beta$, the leading terms of ψ and φ are identical. Subtracting this common term from both ψ and φ, applying the above argument to the resulting polynomials, and continuing the process, we see that $\psi = \varphi$. This completes the proof. ∎

Example 1 In $Z[x_1,x_2,x_3]$, we express

$$f(x_1,x_2,x_3) = x_1^2x_2 + x_1x_2^2 + x_1^2x_3 + x_1x_3^2 + x_2^2x_3 + x_2x_3^2$$

as a polynomial in $E_1(x_1,x_2,x_3)$, $E_2(x_1,x_2,x_3)$, $E_3(x_1,x_2,x_3)$. We follow the procedure given in the proof of Theorem 2.5. The leading term of f is $x_1^2x_2$; so we choose

$$\begin{aligned}s_3 &= \gamma(3) = 0,\\ s_2 &= \gamma(2) - \gamma(3) = 1,\\ s_1 &= \gamma(1) - \gamma(2) = 1\end{aligned}$$

and form the difference

$$\begin{aligned}f_1 &= f - E_1^{s_1}E_2^{s_2}E_3^{s_3}\\ &= f - (x_1 + x_2 + x_3)(x_1x_2 + x_1x_3 + x_2x_3)\\ &= -3x_1x_2x_3 = -3E_3.\end{aligned} \tag{53}$$

Thus $f - E_1^{s_1}E_2^{s_2}E_3^{s_3} = -3E_3$ and hence

$$f = E_1E_2 - 3E_3. \tag{54}$$

The following classical result exhibits the connection between arbitrary polynomials in a single indeterminate and the e.s.f.

Theorem 2.6 *Let x be an indeterminate over the commutative ring R. If $c_1, \ldots, c_n$ are elements of R, then*

$$\prod_{j=1}^{n} (x - c_j) = \sum_{k=0}^{n} (-1)^k E_k(c_1, \ldots, c_n)x^{n-k}. \tag{55}$$

Proof: The meaning of $E_k(c_1, \ldots, c_n)$ is obvious:

$$\begin{aligned}E_k(c_1, \ldots, c_n) &= \sum_{\omega\in Q_{k,n}} \prod_{t=1}^{k} c_{\omega(t)}, \qquad 1 \le k \le n,\\ E_0(c_1, \ldots, c_n) &= 1.\end{aligned}$$

We argue by induction on n. The result is immediate for $n = 1$. Assume that $n > 1$ and that (55) holds for $n - 1$, i.e.,

$$\prod_{j=1}^{n-1} (x - c_j) = \sum_{k=0}^{n-1} (-1)^k E_k(c_1, \ldots, c_{n-1})x^{n-1-k}. \tag{56}$$

Multiplying both sides by $x - c_n$ yields

$$\prod_{j=1}^{n} (x - c_j) = \sum_{k=0}^{n-1} (-1)^k E_k(c_1, \ldots, c_{n-1}) x^{n-k} + \sum_{k=0}^{n-1} (-1)^{k+1} E_k(c_1, \ldots, c_{n-1}) c_n x^{n-1-k}. \tag{57}$$

If in the second summation in (57) we replace the dummy index k by $k - 1$, this summation becomes

$$\sum_{k=1}^{n} (-1)^k E_{k-1}(c_1, \ldots, c_{n-1}) c_n x^{n-k}. \tag{58}$$

Separating the first ($k = 0$) term from the first summation in (57) and the last ($k = n$) term from (58), we obtain

$$\prod_{j=1}^{n} (x - c_j) = x^n + (-1)^n E_{n-1}(c_1, \ldots, c_{n-1}) c_n + \sum_{k=1}^{n-1} (-1)^k [E_k(c_1, \ldots, c_{n-1}) + E_{k-1}(c_1, \ldots, c_{n-1}) c_n] x^{n-k}. \tag{59}$$

But
$$E_{n-1}(c_1, \ldots, c_{n-1}) c_n = E_n(c_1, \ldots, c_n)$$
and
$$E_k(c_1, \ldots, c_{n-1}) + E_{k-1}(c_1, \ldots, c_{n-1}) c_n = E_k(c_1, \ldots, c_n) \tag{60}$$
(see Exercise 8). Thus (55) follows from (59). ∎

Example 2 Suppose that

$$\prod_{j=1}^{9} (x - c_j) = x^9 + x^7 + x^3 + x^2 + x - 1;$$

then

$$\sum_{j=1}^{9} c_j^{\,2} = -2.$$

Indeed, it is easy to see that

$$\sum_{j=1}^{9} c_j^{\,2} = E_1(c_1, \ldots, c_9)^2 - 2E_2(c_1, \ldots, c_9).$$

Since $E_1(c_1, \ldots, c_9) = 0$ and $E_2(c_1, \ldots, c_9) = 1$ by Theorem 2.6, we obtain the desired result.

Example 3 (*Newton Identities*) These classical identities relate the e.s.f. and the power sums (40) in n indeterminates:

If $k \leq n$, then

$$s_k - E_1 s_{k-1} + E_2 s_{k-2} - \cdots + (-1)^{k-1} E_{k-1} s_1 + (-1)^k k E_k = 0; \tag{61}$$

If $k > n$, then

$$s_k - E_1 s_{k-1} + E_2 s_{k-2} - \cdots + (-1)^n E_n s_{k-n} = 0. \tag{62}$$

To see these identities, we first remark that if $x, x_1, \ldots, x_n$ are all indeterminates over an integral domain R, then the same argument used in establishing (55) shows that

$$\prod_{j=1}^{n} (x - x_j) = \sum_{t=0}^{n} (-1)^t E_t(x_1, \ldots, x_n)x^{n-t}. \tag{63}$$

Successively substituting each x_k for x in (63), we have

$$0 = \sum_{t=0}^{n} (-1)^t E_t x_k^{\,n-t}, \qquad k = 1, \ldots, n,$$

and adding these equations yields

$$\sum_{t=0}^{n} (-1)^t E_t \sum_{k=1}^{n} x_k^{\,n-t} = 0,$$

i.e.,

$$\sum_{t=0}^{n} (-1)^t E_t s_{n-t} = 0$$

or

$$s_n - E_1 s_{n-1} + E_2 s_{n-2} - \cdots + (-1)^{n-1} E_{n-1} s_1 + (-1)^n n E_n = 0,$$

precisely (61) for $k = n$. To obtain (62), simply multiply both sides of (63) by x^{k-n}, replace x by $x_1, \ldots, x_n$ successively, and add as above. To confirm (61) for $k < n$, we argue by induction on the number of indeterminates $x_1, \ldots, x_n$. We know that (61) holds if $n = k$; so we assume it holds with $k \leq n$ for n indeterminates, and we prove it holds with $k \leq n$ for $n + 1$ indeterminates. Consider then the left side of (61) for $n + 1$ indeterminates $x_1, \ldots, x_{n+1}$:

$$\sum_{t=0}^{k-1} (-1)^t E_t s_{k-t} + (-1)^k k E_k = f(x_1, \ldots, x_n, x_{n+1}). \tag{64}$$

(For the purposes of this example we shall freely set some of the indeterminates equal to 0 in a polynomial equality. This process will be dealt with in detail at the beginning of Section 3.3.) If we set $x_{n+1} = 0$ in (64), we are reduced to the case of n indeterminates and (64) is 0 by induction. Now write

$$f = \sum_{t=0}^{k} f_t(x_1, \ldots, x_n)x_{n+1}^t \tag{65}$$

so that $f(x_1, \ldots, x_n, 0) = 0$ implies that $f_0(x_1, \ldots, x_n) = 0$, i.e.,

$$f = p(x_1, \ldots, x_n, x_{n+1})x_{n+1} \tag{66}$$

where $p \in R[x_1, \ldots, x_{n+1}]$. Since f is symmetric in $x_1, \ldots, x_{n+1}$, it must also vanish when we set $x_n = 0$. Thus if we write

$$p = \sum h_t(x_1, \ldots, x_{n-1}, x_{n+1})x_n^{\,t},$$

then

$$\begin{aligned} f(x_1, \ldots, x_{n-1}, 0, x_{n+1}) &= p(x_1, \ldots, x_{n-1}, 0, x_{n+1})x_{n+1} \\ &= h_0(x_1, \ldots, x_{n-1}, x_{n+1})x_{n+1} \\ &= 0 \end{aligned}$$

implies $h_0 = 0$. Hence $p = q(x_1, \ldots, x_n, x_{n+1})x_n$ for some $q \in R[x_1, \ldots, x_n, x_{n+1}]$, and

$$f = q x_n x_{n+1}.$$

Continuing in this way, we obtain

$$f(x_1, \ldots, x_{n+1}) = r(x_1, \ldots, x_{n+1})x_1 \cdots x_n x_{n+1} \tag{67}$$

where $r(x_1, \ldots, x_{n+1}) \in R[x_1, \ldots, x_{n+1}]$. But from (64),

$$\deg f = k < n + 1$$

which contradicts (67) unless $r = 0$, i.e., unless $f = 0$.

Our next goal in studying polynomials is the Hilbert Basis Theorem. This result was proved and subsequently developed by David Hilbert over a period of several years beginning in 1888. The ideas introduced by Hilbert show the power of abstract methods and the theorem can in some sense be regarded as the beginning of modern algebra. Hilbert's results stemmed from earlier investigations on the structure of algebraic invariants, a topic to which we shall return.

We begin the development leading to the Hilbert Basis Theorem by briefly discussing various operations on ideals in a ring.

Let $\{A_i \mid i \in I\}$ be a family of ideals in a ring R, indexed by some indexing set I. Then the *intersection*

$$\bigcap_{i \in I} A_i$$

is easily seen to be an ideal in R (see Exericise 9). The *sum*

$$\sum_{i \in I} A_i$$

is defined to be the totality of finite sums of the form

$$\sum a_i, \qquad a_i \in A_i;$$

it is also clearly an ideal in R (see Exercise 10).

In general, the product as complexes of two ideals in a ring is not necessarily an ideal. For example, in $Z[x]$ let

$$A = (2,x) = 2Z[x] + xZ[x]$$

and

$$B = (3,x) = 3Z[x] + xZ[x].$$

The complex product of A and B, call it C, consists of all polynomials in $Z[x]$ of the form

$$[2f(x) + xg(x)][3k(x) + xm(x)] = 6f(x)k(x) + x[3g(x)k(x) + 2f(x)m(x)] + x^2 g(x)m(x). \tag{68}$$

In particular, the choice $f(x) = m(x) = 0$ and $g(x) = k(x) = 1$ yields $3x \in C$, whereas the choice $f(x) = m(x) = 1$ and $g(x) = k(x) = 0$ yields $2x \in C$. If C is an ideal, it follows that $x = 3x - 2x \in C$, and hence x is of the form (68). Writing $f(x) = f_0 + f_1 x +$ (higher degree terms in x) and using similar notation for $g(x)$, $k(x)$, and $m(x)$ in the expression for x, we conclude that

$$f_0 k_0 = 0,$$

$$6(f_0 k_1 + f_1 k_0) + 3g_0 k_0 + 2f_0 m_0 = 1.$$

Since Z is an integral domain, either $f_0 = 0$ or $k_0 = 0$. Suppose $f_0 = 0$. Then $6f_1k_0 + 3g_0k_0 = 1$ which is impossible in Z. On the other hand, $k_0 = 0$ implies $6f_0k_1 + 2f_0m_0 = 1$, which is also impossible. This contradiction shows that C is not an ideal in $Z[x]$.

We are thus led to define the *product* of (a finite number of) ideals $A_1, \ldots, A_n$ in a ring R. The product is the set of all finite sums of the form

$$\sum a_1a_2 \ldots a_n, \qquad a_i \in A_i, i = 1, \ldots, n. \tag{69}$$

It is simple to verify that the product is an ideal in R. Despite possible conflict with the notation for the product as complexes, we will (in conformity with standard usage) designate the product of ideals $A_1, \ldots, A_n$ simply by juxtaposition:

$$A_1A_2 \cdots A_n.$$

If A is an ideal in R, we define $A^n = \overbrace{A \cdots A}^{n}$ for $n \in N$, and we agree to write $A^0 = R$. For ideals A and B the inclusion $AB \subset A \cap B$ holds; it follows that

$$A^n \supset A^{n+1}, \qquad n = 0, 1, 2, \ldots.$$

Let $X \subset R$, and let $\mathfrak{A}$ be the family of all ideals in R containing X (there is at least one such, namely R itself). Then

$$(X) = \bigcap_{A \in \mathfrak{A}} A$$

is called the *ideal generated* by X. In case X is a finite set, the ideal (X) is said to be *finitely generated.* If $X = \{a_1, \ldots, a_n\}$, the notation for (X) is

$$(a_1, \ldots, a_n).$$

The ideal (a) generated by a single element $a \in R$ is called a *principal ideal.* It is obvious (see Exercise 11) that

$$(a) = aR + Ra + RaR + Z \cdot a \tag{70}$$

where $Z \cdot a$ denotes the set of all integral multiples of a and RaR is the set of all finite sums of the form

$$\sum r_ias_i.$$

If R is a ring with identity 1, then since $na = (n \cdot 1)a \in Ra$, the term $Z \cdot a$ in (70) can be omitted. If R is a commutative ring, then the terms aR (or Ra) and RaR in (70) can be omitted. In particular, for an integral domain R we have

$$(a) = Ra = aR$$

(this is the case, of course, if R is just a commutative ring with identity).

If every ideal in a ring R is generated by a single element, i.e., if every ideal in R is a principal ideal, then R is called a *principal ideal ring.*

Let $\{A_i \mid i = 1, 2, \ldots\}$ be a family of ideals in a ring R. Suppose that

$$A_1 \subset A_2 \subset A_3 \subset \cdots. \tag{71}$$

Then the sequence (71) is called an *ascending chain of ideals* in R. The ring R is said to satisfy the *ascending chain condition* if every ascending chain of ideals in R involves only a finite number of distinct ideals, i.e., if for every ascending chain

$$A_1 \subset A_2 \subset A_3 \subset \cdots$$

there exists a positive integer n such that $A_i = A_n$ for all $i \geq n$. It is customary to abbreviate by saying that "R is a.c.c."

Similarly, if $\{A_i \mid i = 1, 2, \ldots\}$ is a family of ideals in a ring R such that

$$A_1 \supset A_2 \supset A_3 \supset \cdots, \tag{72}$$

then the sequence (72) is called a *descending chain of ideals* in R. The ring is said to satisfy the *descending chain condition* if every descending chain of ideals in R involves only a finite number of distinct ideals, i.e., if for every descending chain

$$A_1 \supset A_2 \supset A_3 \supset \cdots$$

there exists a positive integer n such that $A_i = A_n$ for all $i \geq n$. Again, the abbreviated terminology "R is d.c.c." is customary.

Recall that an ideal A in a ring R is said to be finitely generated if there exist elements $a_1, \ldots, a_n$ in R such that

$$A = (a_1, \ldots, a_n).$$

For a commutative ring R with identity, the ideal $A = (a_1, \ldots, a_n)$ is precisely the set

$$\left\{a \mid a = \sum_{i=1}^{n} r_i a_i, r_i \in R, \ i = 1, \ldots, n\right\}.$$

Indeed, any ideal in R containing $a_1, \ldots, a_n$ must contain this set, which is itself such an ideal.

Theorem 2.7 *A ring R is a.c.c. iff every ideal in R is finitely generated.*

Proof: Assume that every ideal in R is finitely generated and let

$$A_1 \subset A_2 \subset A_3 \subset \cdots \tag{73}$$

be an ascending chain of ideals in R. Set

$$A = \bigcup_{k \in N} A_k;$$

it is simple to confirm that $A \lhd R$. Then A is finitely generated by hypothesis, so there exist elements $a_1, \ldots, a_n \in R$ such that

$$A = (a_1, \ldots, a_n).$$

Suppose $a_i \in A_{k_i}$, $i = 1, \ldots, n$, and let $k = \max_{1 \le i \le n} k_i$. Since $A_{k_i} \subset A_k$ for each i, we conclude that $A \subset A_k$. Thus

$$A \subset A_k \subset A_{k+1} \subset \cdots \subset A$$

and hence there are only a finite number of distinct ideals in the chain (73). This shows that R is a.c.c.

On the other hand, assume that R is a.c.c., and suppose A is an ideal in R which is not finitely generated. Choose $a_1 \in A$. Then $(a_1) \neq A$, so we may choose $a_2 \in A - (a_1)$ [i.e., $a_2 \in A$ but $a_2 \notin (a_1)$]. Since $(a_1,a_2) \neq A$, we may choose $a_3 \in A - (a_1,a_2)$. Continuing in this manner, we obtain an ascending chain

$$(a_1) \subset (a_1,a_2) \subset (a_1, a_2, a_3) \subset \cdots$$

which involves infinitely many distinct ideals. This contradicts the hypothesis that R is a.c.c. Thus every ideal in R is finitely generated. ∎

A commutative ring with identity which is a.c.c. is called a *Noetherian ring* (after Emmy Noether). With this terminology, Theorem 2.7 admits the following immediate corollary.

Corollary 4 *A commutative ring with identity R is Noetherian iff every ideal in R is finitely generated.*

Example 4 (a) If R is a commutative principal ideal ring with identity, then every ideal in R is obviously finitely generated (by a single element of R) and hence R is Noetherian.

(b) Any field is Noetherian (why?).

(c) Let $R = (P(N), \{+, \cdot\})$, where for $a,b \in P(N)$ we define

$$a + b = a \cup b - a \cap b$$

and

$$a \cdot b = a \cap b.$$

Clearly R is a commutative ring with identity $1_R = N$. Let

$$A = \{a \in R \mid \ |a| < \infty\}$$

and observe that $A \triangleleft R$. We assert that A is not finitely generated so that R is not Noetherian. For suppose there exist elements $a_1, \ldots, a_n \in R$ such that

$$A = (a_1, \ldots, a_n).$$

Then any $a \in A$ is of the form

$$a = \sum_{i=1}^{n} r_i a_i, \qquad r_i \in R, i = 1, \ldots, n.$$

Since $r_i a_i \subset a_i$ for each i, it follows that $a \subset a_1 \cup a_2 \cup \cdots \cup a_n$. In other words, any finite subset a of N is contained in the fixed finite subset $a_1 \cup \cdots \cup a_n$; this is obviously an absurdity.

We can now state and prove the celebrated Hilbert Basis Theorem.

Theorem 2.8 (*Hilbert Basis Theorem*) *If R is a Noetherian ring and x is an indeterminate over R, then $R[x]$ is Noetherian. That is, if every ideal in the commutative ring with identity R is finitely generated, then the same is true of the polynomial ring $R[x]$.*

Proof: Let A be an ideal in $R[x]$. For each nonnegative integer k, define

$$B_k = \{b \in R \mid b \text{ is the leading coefficient of some polynomial } f(x) \in A \text{ with } \deg f(x) = k, \text{ or } b = 0\}.$$

Clearly, $B_k \lhd R$. If $f(x) = b_k x^k + \cdots + b_0 \in A$, then $xf(x) = b_k x^{k+1} + \cdots + b_0 x$ and $xf(x) \in A$, so that $b_k \in B_{k+1}$. Thus

$$B_k \subset B_{k+1}, \qquad k = 0, 1, 2, \ldots. \tag{74}$$

Since R is Noetherian, the chain (74) involves only a finite number of distinct ideals, i.e., there exists a nonnegative integer m such that

$$B_0 \subset B_1 \subset \cdots \subset B_m = B_{m+1} = \cdots.$$

Now each $B_k \lhd R$ is finitely generated; so for each $k = 0, \ldots, m$ there exist a positive integer p_k and nonzero elements $\beta_{k1}, \ldots, \beta_{kp_k} \in R$ such that

$$B_k = (\beta_{k1}, \ldots, \beta_{kp_k}).$$

By the definition of the B_k, there exist polynomials

$$f_{kt}(x) = \beta_{kt} x^k + \cdots + b_{kt} \in A, \qquad t = 1, \ldots, p_k,\ k = 0, \ldots, m.$$

Let

$$A' = (f_{kt}(x) \colon t = 1, \ldots, p_k, k = 0, \ldots, m)$$

be the ideal in $R[x]$ generated by the $f_{kt}(x)$. Obviously

$$A' \subset A.$$

We assert that $A \subset A'$ as well, and hence $A = A'$.

For let $f(x) \in A$, say $\deg f(x) = n$. If $n = 0$, then $f(x)$ is a polynomial in A of degree 0 and so belongs to B_0; since $B_0 \subset A'$ as is easily seen, we obtain $f(x) \in A'$. Now assume inductively that $n > 0$ and that any polynomial in A of degree $n - 1$ or less belongs to A'. Write

$$f(x) = \beta x^n + \cdots + b.$$

Thus $\beta \in B_n$. If $n \leq m$, we have

$$\beta = \sum_{t=1}^{p_n} r_t \beta_{nt}, \qquad r_t \in R, t = 1, \ldots, p_n.$$

But then the coefficient of x^n in

$$g(x) = f(x) - \sum_{t=1}^{p_n} r_t f_{nt}(x)$$

is 0; so $g(x)$ is a polynomial in A of degree at most $n - 1$ [or $g(x) = 0$]. Hence $g(x) \in A'$ by the induction hypothesis. Since $\sum_{t=1}^{p_n} r_t f_{nt}(x) \in A'$, we conclude that $f(x) \in A'$. If $n > m$, then $\beta \in B_m$ because $B_m = B_{m+1} = \cdots = B_n$; again we have

$$\beta = \sum_{t=1}^{p_m} r_t \beta_{mt}, \qquad r_t \in R, t = 1, \ldots, p_m.$$

The polynomial

$$g(x) = f(x) - \sum_{t=1}^{p_m} r_t x^{n-m} f_{mt}(x)$$

has 0 as the coefficient of x^n. Thus either $g(x) = 0$ or $g(x)$ is a polynomial in A of degree at most $n - 1$; by the induction hypothesis, $g(x) \in A'$. Since $\sum_{t=1}^{p_m} r_t x^{n-m} f_{mt}(x) \in A'$, we again conclude that $f(x) \in A'$.

This completes the induction showing that $A \subset A'$ (and hence $A = A'$). In particular, A is finitely generated. Thus $R[x]$ is a Noetherian ring. ∎

Actually, the following corollary to Theorem 2.8 is also called the Hilbert Basis Theorem.

Corollary 5 *If R is a Noetherian ring and $x_1, \ldots, x_n$ are independent indeterminates over R, then $R[x_1, \ldots, x_n]$ is Noetherian.*

Proof: Since $R[x_1, \ldots, x_n] = R[x_1] \cdots [x_n]$, simply apply Theorem 2.8 n times. ∎

We will return to the study of polynomial rings after the necessary factorization theory is developed in Section 3.3. At present we wish to apply the ring extension theorem to show how an integral domain (e.g., $R[x_1, \ldots, x_n]$) can be embedded in a field.

Theorem 2.9 *Let D be an integral domain.*

(i) *There exists a field K such that that D is a subring of K and every element of K is of the form ab^{-1}, $a, b \in D$.*

(ii) *If K_1 is any other field satisfying (i), then there exists precisely one isomorphism $\nu: K \to K_1$ such that $\nu \mid D = \iota_D$.*

Proof: (i) Let $S = D \times (D - \{0\})$ and define $\rho \subset S^2$ by

$$\rho = \{((a,b), (c,d)) \in S^2 \mid ad = bc\}. \qquad (75)$$

It is routine to verify that ρ is an equivalence relation on S (see Exercise 12).

Let a/b denote the equivalence class in S/ρ to which $(a,b) \in S$ belongs. We define two binary operations $\oplus$ and $\otimes$ on S/ρ by

$$a/b \oplus c/d = (ad + bc)/bd \tag{76}$$

and

$$a/b \otimes c/d = ac/bd. \tag{77}$$

It must be checked that the operations given by (76) and (77) are well defined, i.e., that they do not depend upon the choice of representatives of equivalence classes. So, suppose that $a/b = \alpha/\beta$ and $c/d = \gamma/\delta$. Then $a\beta = b\alpha$, $c\delta = d\gamma$, and hence

$$(a\beta)(d\delta) + (c\delta)(b\beta) = (b\alpha)(d\delta) + (d\gamma)(b\beta),$$

or

$$(ad + bc)\beta\delta = (\alpha\delta + \beta\gamma)bd. \tag{78}$$

But (78) implies that

$$(ad + bc)/bd = (\alpha\delta + \beta\gamma)/\beta\delta.$$

This shows that the operation $\oplus$ is well defined, and the argument for $\otimes$ is similar.

It is very easy (see Exercise 13) to verify that $(S/\rho, \{\oplus,\otimes\})$ is a field; in particular, we have

(a) 0/1 and 1/1 are additive and multiplicative identities in S/ρ, respectively.

(b) If $(a,b) \in S$, then $a/b \oplus (-a)/b = 0/1$.

(c) If $a/b \neq 0/1$, then $a/b \otimes b/a = 1/1$.

(d) $\{a/b \mid (a,b) \in S\}$ is an abelian group with respect to $\oplus$.

(e) $\{a/b \mid (a,b) \in S, a \neq 0\}$ is an abelian group with respect to $\otimes$.

(f) The distributive laws hold in $(S/\rho, \{\oplus, \otimes\})$.

Now let $S_1 = \{a/1 \mid a \in D\}$, and observe that S_1 is a subring of S/ρ. Define $\varphi: S_1 \to D$ by

$$\varphi(a/1) = a, \qquad a \in D;$$

φ is obviously a well defined ring isomorphism. Hence by Theorem 2.3, there exist a ring K and an isomorphism $\theta: S/\rho \to K$ such that D is a subring of K and $\theta/S_1 = \varphi$. Since S/ρ is a field and $K \cong S/\rho$, it follows easily (see Exercise 14) that K is a field. We will denote the operations in K in the same way as those in D (e.g., $a + b$ and ab).

Next observe that if $a,b \in D$, $b \neq 0$, then

$$a/b = a/1 \otimes 1/\mathrm{b} = a/1 \otimes (b/1)^{-1},$$

so that

$$\begin{aligned}\theta(a/b) &= \theta(a/1 \otimes (b/1)^{-1}) = \theta(a/1)\theta((b/1)^{-1}) \\ &= \varphi(a/1)(\theta(b/1))^{-1} = a(\varphi(b/1))^{-1} \\ &= ab^{-1}.\end{aligned}$$

Since θ is surjective, we conclude that every element of K is of the form ab^{-1}, $a,b \in D$. This completes the proof of (i).

(ii) Suppose $(K_1, [+, \times])$ is another field containing D as a subring such that every element of K_1 is of the form $a \times b^{-1}$, $a,b \in D$. Define $\nu: K \to K_1$ by

$$\nu(ab^{-1}) = a \times b^{-1}.$$

To see that ν is well defined, we must check that $uv^{-1} = ab^{-1}$ implies $u \times v^{-1} = a \times b^{-1}$. Now $uv^{-1} = ab^{-1}$ iff

$$ub = av, \tag{79}$$

and $u \times v^{-1} = a \times b^{-1}$ iff

$$u \times b = a \times v. \tag{80}$$

But the multiplications in (79) and (80) both take place in D, so (79) trivially implies (80). Thus ν is well defined. Moreover, it is simple to check that ν is an isomorphism satisfying $\nu | D = \iota_D$. If $\nu_1: K \to K_1$ is an isomorphism with $\nu_1 | D = \iota_D$, then

$$\nu_1(ab^{-1}) = \nu_1(a) \times \nu_1(b)^{-1} = a \times b^{-1} = \nu(ab^{-1}).$$

Thus $\nu: K \to K_1$ is the unique isomorphism such that $\nu | D = \iota_D$, and (ii) is established. ∎

Any one of the fields K described in Theorem 2.9 (they are all isomorphic) is called a *quotient field* or *field of fractions* of D. We will continue to denote the elements of K by $a/b = ab^{-1}$. The mapping $\iota: D \to K$ defined by $\iota(a) = a/1 = a$ is called the *canonical injection* of D into K.

Example 5 (a) The quotient field of Z is Q.

(b) The quotient field of the integral domain $R[x_1, \ldots, x_n]$ of polynomials over an integral domain R is denoted by $R(x_1, \ldots, x_n)$. The elements of $R(x_1, \ldots, x_n)$ are written

$$\frac{f(x_1, \ldots, x_n)}{g(x_1, \ldots, x_n)}$$

and are called *rational functions*. The field $R(x_1, \ldots, x_n)$ is called the *rational function field in* $x_1, \ldots, x_n$ *over* R.

(c) If D is a field, then the quotient field of D is trivially D itself. For obviously D is a subring of itself, and every element of D is of the form ab^{-1}, $a,b, \in D$.

The quotient field of an integral domain satisfies a certain universal mapping property. Before stating this result, we recall the notion of a *unit* in a ring. Let R be a ring with identity 1. If $a \in R$ and there exists $b \in R$ such that

$$ab = ba = 1, \tag{81}$$

then a is called a *unit* in R. Observe that if $ac = ca = 1$ also, then

$$\begin{aligned} b &= b \cdot 1 = b(ac) \\ &= (ba)c = 1 \cdot c \\ &= c. \end{aligned}$$

Thus $a \in R$ is a unit iff a possesses a unique multiplicative inverse, generally denoted by a^{-1}, in R.

Example 6 (a) The units in Z are 1 and -1.

(b) The units in $M_2(Z)$ are the matrices with determinant 1 or -1 (see Exercise 15).

(c) If R is an integral domain and x is an indeterminate over R, then the units in $R[x]$ are precisely the units in R. It is obvious that the units in R are units in $R[x]$. On the other hand, if $f(x)$ is a unit in $R[x]$, there exists an element $g(x) \in R[x]$ such that

$$f(x)g(x) = 1 \qquad [= g(x)f(x)]. \tag{82}$$

But
$$\deg f(x)g(x) = \deg f(x) + \deg g(x)$$

(see Exercise 7). Hence $\deg f(x) = \deg g(x) = 0$ so that $f(x) = f_0 \in R$ and $g(x) = g_0 \in R$. By (82), we have $f_0 g_0 = g_0 f_0 = 1$, i.e., $f(x) = f_0$ is a unit in R.

(d) Every nonzero element of a field is a unit.

(e) The set U of all units in a ring with identity R is a subgroup of the multiplicative semigroup in R. Indeed, $1 \in U$. If $a \in U$, then obviously $a^{-1} \in U$. Finally, $a,b \in U$ implies $(ab)(b^{-1}a^{-1}) = (b^{-1}a^{-1})(ab) = 1$ so that $ab \in U$ as well.

Theorem 2.10 *Let D be an integral domain, and K a quotient field of D. Let $\varphi: D \to R$ be a homomorphism of D into a commutative ring R with identity such that $\varphi(a)$ is a unit in R for all $a \in D$, $a \neq 0$. Then there exists a unique ring homomorphism $\bar{\varphi}: K \to R$ such that $\varphi = \bar{\varphi}\iota$, where $\iota: D \to K$ is the canonical injection.*

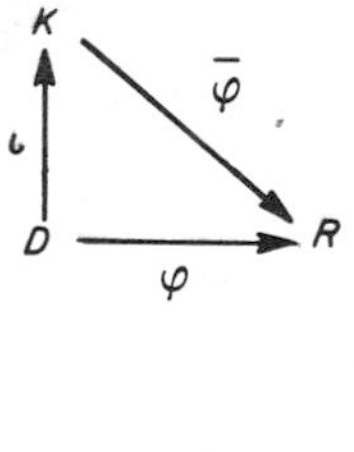

Proof: Define $\bar{\varphi}: K \to R$ by

$$\bar{\varphi}(a/b) = \varphi(a)\varphi(b)^{-1}, \qquad a,b \in D,\ b \neq 0.$$

Observe that if $\alpha/\beta = a/b$, then $\alpha b = a\beta \in D$ so that $\varphi(\alpha)\varphi(b) = \varphi(a)\varphi(\beta)$. Since $b \neq 0$ and $\beta \neq 0$, $\varphi(b)$ and $\varphi(\beta)$ are units in R and $\varphi(\alpha)\varphi(\beta)^{-1} = \varphi(a)\varphi(b)^{-1}$(recall that R is commutative). Thus $\bar{\varphi}(a/b)$ does not depend on the particular representative for a/b, i.e., $\bar{\varphi}$ is well defined. We compute that

$$\begin{aligned}\bar{\varphi}(a/b + c/d) &= \bar{\varphi}((ad + bc)/bd) = \varphi(ad + bc)\varphi(bd)^{-1} \\ &= \varphi(ad)\varphi(bd)^{-1} + \varphi(bc)\varphi(bd)^{-1} \\ &= \bar{\varphi}(ad/bd) + \bar{\varphi}(bc/bd) \\ &= \bar{\varphi}(a/b) + \bar{\varphi}(c/d).\end{aligned}$$

Similarly, $\bar{\varphi}((a/b)(c/d)) = \bar{\varphi}(a/b)\bar{\varphi}(c/d)$; so $\bar{\varphi}$ is a ring homomorphism. Now $\varphi(1)$ is a unit in R by hypothesis, and $\varphi(1) = \varphi(1 \cdot 1) = \varphi(1)\varphi(1)$ implies that $\varphi(1)$ is the identity in R. Then for $a \in D$ we have $\bar{\varphi}(a) = \bar{\varphi}(a/1) = \varphi(a)\varphi(1) = \varphi(a)$; in other words, $\varphi = \bar{\varphi}\iota$. It is clear that $\bar{\varphi}: K \to R$ is completely determined by the values of φ. ▮

The construction of the quotient field can be generalized to a larger class of rings than just integral domains. Let R be a commutative ring with identity 1, and let A be a submonoid of the multiplicative monoid in R. Let $S = R \times A$ and define $\rho \subset S^2$ by

$$\rho = \{((a,b), (c,d)) \in S^2 \mid (ad - bc)u = 0 \text{ for some } u \in A\}. \tag{83}$$

It is not difficult to show that ρ is an equivalence relation on S (see Exercise 16). As before, we let a/b denote the equivalence class in S/ρ to which $(a,b) \in S$ belongs. We define two binary operations $\oplus$ and $\otimes$ on S/ρ by

$$a/b \oplus c/d = (ad + bc)/bd \tag{84}$$

and

$$a/b \otimes c/d = ac/bd. \tag{85}$$

Once again it is relatively simple to confirm that (84) and (85) are coherent definitions in the sense that they do not depend on the particular representatives (see Exercise 17). It follows that $(S/\rho, \{\oplus, \otimes\})$ is a commutative ring with identity (see Exercise 18). If we define $\iota: R \to S/\rho$ by

$$\iota(a) = a/1, \quad a \in R, \tag{86}$$

then ι is a ring homomorphism which is not necessarily injective. Of course, if no element of A is a zero divisor and $0 \notin A$, then ι is an injection [$\iota(a) = \iota(c)$ iff $a/1 = c/1$ iff $(a - c)u = 0$ for some $u \in A$ iff $a - c = 0$]. The ring S/ρ is called the *ring of fractions over R with respect to A*.

Example 7 (a) Let $R = Z$ and $A = Z - 3Z$; A consists of all integers which are not multiples of 3. Clearly $1 \in A$ and if $a,b \in A$, then $ab \in A$ (otherwise 3 divides ab and hence divides either a or b). Now $((a,b),(c,d)) \in \rho$ iff $(ad - bc)u = 0$ for some $u \in A$. Since $0 \notin A$, we have $((a,b),(c,d)) \in \rho$ iff $ad - bc = 0$. Thus a/b just consists of all pairs of integers (c,d) such that $ad = bc$, and d is not divisible by 3. We can identify S/ρ with the set of all rational numbers a/b such that b is not a multiple of 3. In this example the mapping $\iota: R \to S/\rho$ is an injection.

(b) Let $R = Z_6$ be the ring of integers modulo 6, and let $A = \{[1],[2],[4]\}$. Then A is a submonoid of the multiplicative monoid in Z_6. But $[5]/[1] = [2]/[1]$ in S/ρ because $([5] - [2])[2] = [3][2] = [0]$. Thus the mapping ι given by (86) is not an injection.

In constructing the ring of fractions over R with respect to A, it was stipulated that A be a submonoid of the multiplicative monoid in R. There is a routine way of constructing such A that leads us to consider two important kinds of ideals in a ring.

Recall that an ideal M in a ring R is *maximal* if $M \neq R$ and there is no ideal $C \lhd R$ such that $C \neq M$, $C \neq R$, and

$$M \subset C \subset R$$

[see formula (36) in Section 3.1].

Let R be an arbitrary commutative ring (not necessarily possessing an identity) and let P be an ideal in R, $P \neq R$. Then P is called a *prime ideal* if $R - P$ is a subsemigroup of the multiplicative semigroup in R.

Example 8 (a) Let $R = Z$ and $P = pZ$, where p is a prime integer. If $a,b \in Z - pZ$, then neither a nor b is divisible by p, and hence ab is not divisible by p, i.e., $ab \in Z - pZ$. Thus pZ is a prime ideal in Z.

(b) Let P be an ideal in a commutative ring R, $P \neq R$. Then P is prime iff $ab \in P$ implies $a \in P$ or $b \in P$. For if P is prime and $a \notin P$, $b \notin P$, then $a,b \in R - P$, and hence $ab \in R - P$, i.e., $ab \notin P$. Conversely, assume the given condition; then if $a,b \in R - P$, we obviously must have $ab \in R - P$. Thus $R - P$ is a subsemigroup of the multiplicative semigroup in R, i.e., P is a prime ideal.

(c) Let x be an indeterminate over Z and let $R = Z[x]$. Let (x) be the ideal in R generated by x: $(x) = xR = \{xh(x) \mid h(x) \in R\}$. Then (x) is a prime ideal in R. For, if $f(x)g(x) \in (x)$, then $f(x)g(x) = xh(x)$ for some $h(x) \in R$. For the moment we shall assume that the reader accepts the fact that if x divides a product of polynomials with integer coefficients, then x divides at least one of the factors (this will be proved in the next section). Thus x divides $f(x)$ or x divides $g(x)$, i.e., $f(x) \in (x)$ or $g(x) \in (x)$.

(d) As in part (c), let x be an indeterminate over Z, and let $R = Z[x]$. Let $(2, x)$ be the ideal in R generated by 2 and x:

$$(2,x) = \{2h(x) + xk(x) \mid h(x),k(x) \in R\}.$$

We assert that $(2,x)$ is a prime ideal. Indeed, a moment's reflection shows that $(2,x)$ is precisely the set of all polynomials in $Z[x]$ with constant term a multiple of 2. It is obvious that if $f(x)g(x)$ has an even constant term, then so does either $f(x)$ or $g(x)$. Thus $(2,x)$ is a prime ideal.

Notice that

$$(x) \subset (2,x) \subset Z[x],$$

and the inclusions are strict. Hence a prime ideal [e.g., (x)] need not be maximal. However, we shall see later that any maximal ideal is necessarily prime.

(e) If p is a prime integer, then the prime ideal $P = pZ$ in Z [see part (a)] is maximal. For suppose C is an ideal properly containing P, and let $n \in C - P$. Since n is not divisible by the prime p, n and p are relatively prime. This means that there exist integers r and s such that $rn + sp = 1$. But $rn \in C$ and $sp \in P \subset C$. Hence $1 \in C$ and it follows that $C = Z$.

(f) Consider the ring S/ρ in Example 7(a): S/ρ is the ring of all rational numbers a/b such that b is not a multiple of 3. Let $M = \{a/b \in S/\rho \mid a \text{ is a multiple of } 3\}$. We assert that M is a maximal ideal in S/ρ and that S/ρ has no other maximal ideals. The fact that M is an ideal is obvious (also note that $M \neq S/\rho$). Suppose $c/d \notin M$. Then $c \notin 3Z$, and hence $d/c \in S/\rho$. Thus if C is an ideal in S/ρ containing an element c/d not in M, then the multiplicative inverse d/c of c/d is in S/ρ. It follows that $1 \in C$, and hence $C = S/\rho$. We have shown that if C is an ideal in S/ρ and C is not a subset of M, then $C = S/\rho$. Consequently M is the unique maximal ideal in S/ρ. Rings which possess precisely one maximal ideal are called *local rings*.

There is an interesting and important connection between prime and maximal ideals and quotient rings (see also Theorem 1.9, Section 3.1).

Theorem 2.11 *Let R be a commutative ring with identity* 1, *and let A be an ideal in R. Then*
(i) *A is prime iff R/A is an integral domain.*
(ii) *A is maximal iff R/A is a field.*

Proof: (i) Suppose that A is prime. We know that R/A is a commutative ring; it remains to show that R/A has an identity not equal to A and contains no zero divisors. Since $A \neq R$, it follows that $1 + A \neq A$, and $1 + A$ is obviously an identity in R/A. Now assume that $(r + A)(s + A) = A$. Then $rs \in A$ and so $r \in A$ or $s \in A$ (since A is prime). Therefore $r + A = A$ or $s + A = A$, showing that R/A has no zero divisors.

Suppose conversely that R/A is an integral domain; thus $A \neq R$. If $rs \in A$, then $(r + A)(s + A) = A$. But R/A contains no zero divisors, so we have $r + A = A$ or $s + A = A$, i.e., $r \in A$ or $s \in A$. Hence A is a prime ideal.

(ii) Suppose that A is maximal. Again, R/A is a commutative ring with identity $1 + A \neq A$. Let $\nu\colon R \to R/A$ be the canonical epimorphism given by $\nu(r) = r + A$. To show that R/A is a field, it suffices to show that if $\nu(r) \neq A$, then $\nu(r)$ possesses an inverse in R/A. So assume $\nu(r) \neq A$. Then $r \notin A$, and $C = A + rR$ is clearly an ideal in R properly containing A. Hence $C = R$ because A is maximal. There exist elements $a \in A$ and $s \in R$ such that $1 = a + rs$. Then $\nu(1) = \nu(a) + \nu(rs) = \nu(r)\nu(s)$, i.e., $\nu(s)$ is the multiplicative inverse of $\nu(r)$ in R/A.

Conversely, suppose that R/A is a field; in particular, $A \neq R$. Let C be an ideal in R properly containing A, and let $r \in C - A$. Then $\nu(r) \neq A$, so $\nu(r)$ possesses an inverse in $R(A)$. Let $s \in R$, and consider the element $\nu(r)^{-1}\nu(s) \in R/A$. Choose $u \in R$ such that $\nu(u) \in \nu(r)^{-1}\nu(s)$. We have

$$\nu(ur) = \nu(u)\nu(r) = \nu(r)^{-1}\nu(s)\nu(r) = \nu(s).$$

In other words, $s - ur \in A \subset C$. But $ur \in C$ implies $s \in C$. Since s was chosen to be an arbitrary element in R, we conclude that $C = R$. Thus A is a maximal ideal. ∎

Corollary 6 *Let R be a commutative ring with identity. Then any maximal ideal in R is a prime ideal.*

Proof: If A is a maximal ideal in R, then R/A is a field by Theorem 2.11 (ii). In particular, R/A is an integral domain, and hence A is a prime ideal by Theorem 2.11(i). ∎

Example 9 (a) The ring Z_p of integers modulo p is a field iff p is a prime integer. We have $Z_p = Z/pZ$; if p is prime, then pZ is a maximal ideal in Z by Example 8(e), and Theorem 2.11(ii) implies that Z_p is a field. Conversely, suppose p is composite, say $p = mn$ for positive integers $m,n \neq 1$. Then obviously $[m][n] = [0]$, so Z_p possesses zero divisors and hence is not a field.

(b) Let p be a prime integer. Since Z_p is a field [see part (a)], $Z_p - \{[0]\}$ is a multiplicative group of order $p - 1$. In any (multiplicative) group of order m with identity e, $g^m = e$ for all elements g in the group (see Exercise 21). Hence if $[a] \neq [0]$, then $[a]^{p-1} = [1]$. This means that if the integer a is not a multiple of p, then $a^{p-1} - 1$ is divisible by p. For example, if $p = 3$ and $a = 5$, we obtain that $5^2 - 1$ is divisible by 3. This result is known as *Fermat's theorem*. An obvious corollary is that if a is any integer, then $a^p - a$ is divisible by p.

(c) Let x be an indeterminate over the real field $\mathbf{R}$ and consider the polynomial $p(x) = x^2 + 1 \in \mathbf{R}[x]$. For the purposes of this important example, we shall assume that $p(x)$ is an irreducible polynomial in $\mathbf{R}[x]$, meaning that it cannot be factored. Hence if $f(x) \in \mathbf{R}[x]$ and $f(x) \notin p(x)\mathbf{R}[x]$, then $f(x)$ and $p(x)$ have a greatest common divisor of 1 [we shall thoroughly investigate greatest common divisors (g.c.d.'s) in polynomial rings in the next section, but we assume here that the reader is familiar with the process of finding the g.c.d. of two polynomials with real coefficients]. It follows that there exist polynomials $s(x)$ and $t(x)$ in $\mathbf{R}[x]$ such that $s(x)f(x) + t(x)p(x) = 1$. Thus if C is an ideal in $\mathbf{R}[x]$ containing $p(x)\mathbf{R}[x]$ and if $f(x) \in C$, then $1 = s(x)f(x) + t(x)p(x) \in C$ so that $C = \mathbf{R}[x]$. This shows that $p(x)\mathbf{R}[x]$ is a maximal ideal. We conclude by Theorem 2.11(ii) that $\mathbf{R}[x]/p(x)\mathbf{R}[x]$ is a field.

Let

$$\nu\colon \mathbf{R}[x] \to \mathbf{R}[x]/p(x)\mathbf{R}[x]$$

be the canonical epimorphism and let $\mu = \nu|\mathbf{R}$ be the restriction of ν to $\mathbf{R}$. Suppose

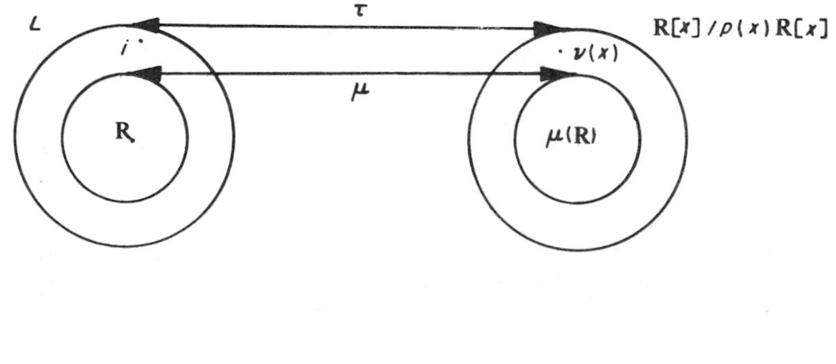

$\mu(r_1) = \mu(r_2)$; then $r_1 - r_2 = (x^2 + 1)f(x)$ for some $f(x) \in \mathbf{R}[x]$, and hence $r_1 = r_2$. Thus

$$\mu\colon \mathbf{R} \to \mathbf{R}[x]/p(x)\mathbf{R}[x]$$

is a monomorphism and $\mu(\mathbf{R})$ is a field contained in the field $\mathbf{R}[x]/p(x)\mathbf{R}[x]$ (see Exercises 14 and 22). At this point we use Theorem 2.3 to secure a ring L containing $\mathbf{R}$ as a subring and an isomorphism $\tau\colon L \to \mathbf{R}[x]/p(x)\mathbf{R}[x]$ such that $\tau|\mathbf{R} = \mu$. Of course L is a field because $L \cong \mathbf{R}[x]/p(x)\mathbf{R}[x]$. Now set $i = \tau^{-1}(\nu(x)) \in L$ and compute that

$$\begin{aligned}\tau(i^2 + 1) &= \tau(i)^2 + \tau(1) = \nu(x)^2 + \nu(1)\\ &= \nu(x^2 + 1) = \nu(p(x))\\ &= p(x)\mathbf{R}[x].\end{aligned}$$

In other words, $\tau(i^2 + 1)$ is the zero element of $\mathbf{R}[x]/p(x)\mathbf{R}[x]$. Since τ is injective, it follows that $i^2 + 1 = 0$. Let $f(x) \in \mathbf{R}[x]$, say $f(x) = \sum_{t=0}^{n} r_t x^t$; then

$$\begin{aligned}\tau^{-1}(\nu(f(x))) &= \tau^{-1}\left(\sum_{t=0}^{n} \nu(r_t)\nu(x)^t\right) = \sum_{t=0}^{n} \tau^{-1}(\nu(r_t))(\tau^{-1}(\nu(x)))^t\\ &= \sum_{t=0}^{n} \tau^{-1}(\nu(r_t))i^t = \sum_{t=0}^{n} \tau^{-1}(\mu(r_t))i^t = \sum_{t=0}^{n} (\tau^{-1}\tau)(r_t)i^t\\ &= \sum_{t=0}^{n} r_t i^t.\end{aligned}$$

But $\tau^{-1}\nu\colon \mathbf{R}[x] \to L$ is surjective, and so we conclude that every element of L is a polynomial in i. Thus $L = \mathbf{R}[i]$. Furthermore, since $i^2 + 1 = 0$, we have $i^2 = -1$, and hence every element of $\mathbf{R}[i]$ is of the form $a + bi$, $a,b \in \mathbf{R}$.

We have constructed a field $\mathbf{R}[i]$ containing $\mathbf{R}$, containing an element i such that $i^2 + 1 = 0$, and consisting of all elements of the form $a + bi$, $a,b \in \mathbf{R}$. Of course, $\mathbf{R}[i]$ is what we ordinarily call the field of complex numbers, $\mathbf{C}$. As we shall see, the construction given here can be carried out for any irreducible polynomial $p(x)$, but we must first systematically investigate the problem of factorization in a general polynomial ring $R[x]$.

As a final application of the ring extension theorem in this section, we indicate a method of embedding an arbitrary ring in a ring with a multiplicative identity.

Theorem 2.12 *Let R be a ring. Then there exists a ring S with a multiplicative identity such that R is a subring of S.*

Proof: Let $T = Z \times R$ and define two binary operations $\oplus$ and $\otimes$ on T by

$$(m,r_1) \oplus (n,r_2) = (m + n,\ r_1 + r_2) \tag{87}$$

and

$$(m,r_1) \otimes (n,r_2) = (mn,\ n \cdot r_1 + m \cdot r_2 + r_1 r_2). \tag{88}$$

It is easy to verify that $(T, \{\oplus,\otimes\})$ is a ring and that $(1,0)$ is the multiplicative identity in T (see Exercise 24). Moreover, $L = \{0\} \times R$ is a subring of T and the mapping $\nu\colon R \to L$ given by $\nu(r) = (0,r)$ is a ring isomorphism (see Exercise 25). By the ring extension theorem there exist a ring S containing R as a subring and an isomorphism $\tau: S \to T$ such that $\tau | R = \nu$. Since $S \cong T$ it follows that S contains a multiplicative identity. ∎

Corollary 7 *Let R be a ring. Then there exist a ring T containing R, a ring with identity S containing R, and an indeterminate x over S such that T is the subring of $S[x]$ consisting precisely of those polynomials with coefficients in R.*

Proof: By Theorem 2.12, there exists a ring with identity S containing R. By Theorem 2.4, there exists an indeterminate x over S. Let T be the totality of polynomials in $S[x]$ whose coefficients belong to R. Then T is obviously a subring of $S[x]$. ∎

The point of this corollary is that even if R is a ring without identity, one can still formulate a notion of a polynomial ring in an indeterminate x over R. We denote the ring T of the corollary simply by $R[x]$.

Example 10 Let $R = 2Z$ be the ring of even integers. Then $R[x]$ is the ring of polynomials with even coefficients in an indeterminate x over Z.

Exercises 3.2

1. Show that if $f \in R[Z^+]$ is the zero function, then $f = \iota(f(0))x^0$ [see formula (5) for notation].
2. Complete the verification that $(B, \{\oplus, *\})$ in Theorem 2.2 is a ring.
3. Let R be a ring with identity. Let x_1 be an indeterminate over R and x_k an indeterminate over $R[x_1] \cdots [x_{k-1}]$, $2 \leq k \leq n$. Show that each x_k is an indeterminate over R.
4. Prove formulas (30) to (33).
5. Confirm formulas (42) and (43).
6. Show that if R is an integral domain, the leading term of a product of polynomials in $R[x_1, \ldots, x_n]$ is the product of the leading terms. *Hint*: Prove this for the product of two polynomials and proceed by induction.
7. Let x be an indeterminate over R, and let $f(x) \in R[x]$ be a nonzero polynomial whose leading coefficient is not a zero divisor in R. Show that if $g(x) \in R[x]$ is any other nonzero polynomial, then

$$\deg f(x)g(x) = \deg f(x) + \deg g(x).$$

Show that in general, if $f(x), g(x) \in R[x]$ and $f(x)g(x) \neq 0$, then

$$\deg f(x)g(x) \leq \deg f(x) + \deg g(x).$$

Finally, confirm formula (45).

8. Prove formula (60).
9. Show that if $\{A_i \mid i \in I\}$ is a family of right ideals in a ring R, then $\bigcap_{i \in I} A_i$ is a right ideal in R. Show that the same result is true for left and two-sided ideals.

10. Show that if $\{A_i \mid i \in I\}$ is a family of right ideals in a ring R, then $\sum_{i \in I} A_i$ is a right ideal in R. Show that the same result is true for left and two-sided ideals.

11. Confirm formula (70). *Hint*: Clearly $aR + Ra + RaR + Z \cdot a$ is an ideal in R containing a; so $(a) \subset aR + Ra + RaR + Z \cdot a$. But if A is any ideal in R containing a, then $aR \subset A$, $Ra \subset A$, $RaR \subset A$, and $Z \cdot a \subset A$.

12. Show that the relation ρ defined by (75) is an equivalence relation on S.

13. Verify assertions (a) to (f) in the proof of Theorem 2.9.

14. Show that if $\theta: K \to L$ is a ring isomorphism and K is a field, then L is a field.

15. Show that the units in $M_2(Z)$ are the matrices with determinant ± 1.

16. Show that the relation ρ defined by (83) is an equivalence relation on S. *Hint*: It is obvious that ρ is reflexive and symmetric. To show that ρ is transitive, suppose $((a,b),(c,d)) \in \rho$ and $((c,d),(e,f)) \in \rho$ so that there exist $u,v \in A$ such that $(ad - bc)u = 0$ and $(cf - de)v = 0$. Multiply the first of the equalities by fv and the second by bu: $adfvu - bcfvu = 0$, $cfbuv - debuv = 0$. Now add these equations to obtain $adfvu - debuv = 0$, $(af - eb)duv = 0$. Since $d, u, v \in A$ we have $duv \in A$. Hence $((a,b),(e,f)) \in \rho$.

17. Show that the definitions (84) and (85) of addition $\oplus$ and multiplication $\otimes$ in the ring of fractions over R with respect to A do not depend on the particular representatives. *Hint*: Suppose $a/b = \alpha/\beta$, $c/d = \gamma/\delta$ so that $(a\beta - b\alpha)u = 0$, $(c\delta - d\gamma)v = 0$ for some $u,v \in A$. We must show that

(i) $(ad + bc)/bd = (\alpha\delta + \beta\gamma)/\beta\delta$

and

(ii) $ac/bd = \alpha\gamma/\beta\delta$.

Now

$$\begin{aligned} ((ad + bc)\beta\delta - (\alpha\delta + \beta\gamma)bd)uv &= ((a\beta - b\alpha)d\delta + (c\delta - d\gamma)b\beta)uv \\ &= (a\beta - b\alpha)ud\delta v + (c\delta - d\gamma)vb\beta u \\ &= 0, \end{aligned}$$

and $uv \in A$. This proves (i), and (ii) is proved similarly.

18. Show that the ring of fractions over R with respect to A is a commutative ring with identity. *Hint*: The identity is $1/1$.

19. Let R be a commutative ring with identity 1, and let P be a prime ideal in R. Set $A = R - P$, and consider the ring of fractions S/ρ over R with respect to A. Prove that S/ρ is a local ring. *Hint*: Following Example 8(f), let $M = \{a/b \in S/\rho \mid a \in P\}$. Then obviously M is an ideal in S/ρ and $M \neq S/\rho$. Suppose C is an ideal in S/ρ which is not a subset of M. Choose $c/d \in C$ such that $c/d \notin M$. Then $c \in R - P = A$ so that $d/c \in S/\rho$. Hence $1/1 = d/c \cdot c/d \in C$, and it follows that $C = R$. Thus M is the unique maximal ideal in S/ρ.

20. Show that the ideal $(2,x)$ in Example 8(d) is a maximal ideal in $Z[x]$. *Hint*: $(2,x)$ consists of the polynomials in $Z[x]$ with even constant terms. Suppose C is an ideal in $Z[x]$ containing $(2,x)$ and a polynomial $f(x)$ with an odd constant term, say $f(x) = 2m - 1 + xg(x)$, where $m \in Z$ and $g(x) \in Z[x]$. Consider the ideal $M = (2,x,f(x)) = 2Z[x] + xZ[x] + f(x)Z[x]$. We have $2m \in M$, $xg(x) \in$

M, and $f(x) \in M$; hence $1 \in M$. It follows that $M = Z[x]$, and since $M \subset C$, we obtain $C = \mathbb{Z}[x]$. Thus $(2,x)$ is a maximal ideal.

21. Let a group G with identity element e be written multiplicatively, and suppose $|G| = m$. Show that $g^m = e$ for all $g \in G$. *Hint*: If $g \in G$, the elements $g^0 = e$, $g^1, \ldots, g^m$ cannot all be distinct. It follows that $g^k = e$ for some $k \geq 1$. Then $[g] = \{g^0, \ldots, g^{k-1}\}$ is a subgroup of G, and Lagrange's theorem implies that $k \mid m$. Hence $g^m = e$.

22. Let K be a field, L a ring, and $\mu: K \to L$ a ring homomorphism. Show that if $\mu(K) \neq \{0\}$, then $\mu(K)$ is a field. *Hint*: Since $\ker \mu$ is an ideal in the field K, we have $\ker \mu = K$ or $\ker \mu = \{0\}$. But $\mu(K) \neq \{0\}$ so that $\ker \mu = \{0\}$. Thus $\mu(K)$ is the isomorphic image of the field K and hence is itself a field (see Exercise 14).

23. Following Example 9(*c*), show how to construct a field containing Q and an element r such that $r^2 - 2 = 0$.

24. In the proof of Theorem 2.12, show that $(T, \{\oplus, \otimes\})$ is a ring with $(1,0)$ as multiplicative identity.

25. Show that the mapping $\nu: R \to L$ in the proof of Theorem 2.12 is a ring isomorphism.

Glossary 3.2

ascending chain condition, 158
a.c.c., 158
chain of ideals (ascending, descending), 158
completely symmetric functions, 149
constant term, 142
degree of $f(x)$, 142
$\deg f(x)$, 142
descending chain condition, 158
d.c.c., 158
elementary symmetric functions, 149
e.s.f., 149
Fermat's theorem, 168
field of fractions, 163
finitely generated ideal, 157
$(a_1, \ldots, a_n)$, 157
Hilbert Basis Theorem, 160
ideal generated by X, 157
(X), 157
independent indeterminates, 147
indeterminate over a ring R, 144
indeterminate with respect to a ring R, 139
leading coefficient, 142,150
leading term, 142,150
local ring, 167
maximal ideal, 166
monic, 142
monomial, 142
Newton identities, 154
Noetherian ring, 159
quotient field, 163
prime ideal, 166
principal ideal, 157
principal ideal ring, 157
rational functions, 163
rational function field, 163
$R(x_1, \ldots, x_n)$, 163
ring extension theorem, 143
ring of fractions over R with respect to A, 165
ring of polynomials in the indeterminates $x_1, \ldots, x_n$ over R, 147
$R[x_1, \ldots, x_n]$, 147
ring of polynomials in x, 142
$R[x]$, 142
symmetric polynomial, 149
transcendental over R, 142
unit in a ring, 163
variable over R, 142

3.3 Unique Factorization Domains

A *principal ideal domain*, abbreviated P.I.D., is a principal ideal ring which is also an integral domain. Thus a ring R is a P.I.D. iff R is a commutative ring with identity containing no zero divisors and such that any ideal in R is of the form

$$(a) = Ra.$$

We know from elementary arithmetic that any integer is a product of prime integers. Moreover, this factorization is unique except possibly for order and factors of ± 1, e.g.,

$$6 = 2 \cdot 3 = (-2)\,(-3) = 3 \cdot 2 = (-3)\,(-2).$$

In this section we will show that if R is a P.I.D., then a unique factorization into primes is possible in R as it is in Z.

Example 1 Z is a P.I.D., for Z is certainly an integral domain, and if $A < Z$, then $A = nZ = (n)$ for some $n \in Z$ (see Exercise 17, Section 3.1).

If R is a field and x is an indeterminate over R, then $R[x]$ is a principal ideal domain. In order to establish this important fact, we need to study divisibility in polynomial rings. We will conduct this study in somewhat greater generality so that our results may be used later in discussing canonical forms for matrices.

Let R be a subring of an arbitrary ring S, and let x be an indeterminate over R. Let t be a fixed element of S, and define two mappings, φ_r: $R[x] \to S$ and φ_l: $R[x] \to S$, as follows: for $f(x) = a_0 + a_1x + \cdots + a_nx^n \in R[x]$, put

$$\varphi_r(f(x)) = a_0 + a_1t + \cdots + a_nt^n$$

and

$$\varphi_l(f(x)) = a_0 + ta_1 + \cdots + t^na_n;$$

we also denote $\varphi_r(f(x))$ by $f_r(t)$ and $\varphi_l(f(x))$ by $f_l(t)$. The function φ_r is called *right substitution* or *right specialization at t*, and $f_r(t)$ is called the *right-hand value of f(x) at t*. Similarly, the function φ_l is called *left substitution* or *left specialization at t* and $f_l(t)$ is the *left-hand value of f(x) at t*. If t commutes with the elements of R, then obviously $f_r(t) = \varphi_r(f(x)) = \varphi_l(f(x)) = f_l(t)$; in this case we designate φ_r and φ_l simply by φ and call $f_r(t) = f_l(t)$ the *value of f(x) at t*, denoting it by $f(t)$. If R is a subring of a commutative ring S and $c \in S$ is such that $f(c) = 0$, then c is called a *root* or a *zero of f(x) in S*. The following result is evident from the definitions, and its proof is left to the reader (see Exercise 1).

Theorem 3.1 *Let R be a subring of a ring S, x an indeterminate over R, and t a fixed element of S.*

(i) *The mappings φ_r and φ_l are group homomorphisms of $(R[x], \{+\})$ into $(S, \{+\})$.*

(ii) *If t commutes with every element of R, then $\varphi: R[x] \to S$ is a ring homomorphism.*

(iii) *If t is an indeterminate over R, then $\varphi: R[x] \to R[t]$ is a ring isomorphism.*

The key elementary result concerning polynomials is

Theorem 3.2 (*Division Algorithm*) *Let R be a ring with identity and x an indeterminate over R. Let*

$$a(x) = a_0 + a_1x + \cdots + a_nx^n \in R[x]$$

and

$$b(x) = b_0 + b_1x + \cdots + b_mx^m \in R[x],$$

where $a_n \neq 0$ and b_m is a unit in R. Then there exist unique polynomials $q(x)$ and $r(x)$ in $R[x]$ such that

$$a(x) = q(x)b(x) + r(x) \tag{1}$$

and $r(x) = 0$ or $\deg r(x) < \deg b(x)$. *Similarly, there exist unique polynomials $s(x)$ and $u(x)$ in $R[x]$ such that*

$$a(x) = b(x)s(x) + u(x) \tag{2}$$

and $u(x) = 0$ or $\deg u(x) < \deg b(x)$.

Proof: We first establish the existence of the polynomials $q(x)$ and $r(x)$. The proof is by induction on n. Suppose $n = 0$, so that $a(x) = a_0$. If $m > 0$, set $q(x) = 0$ and $r(x) = a_0$. If $m = 0$, set $q(x) = a_0b_0^{-1}$ and $r(x) = 0$. Now assume that $n > 0$ and that (1) holds for all polynomials of degree $n - 1$ or less. If $m > n$, set $q(x) = 0$ and $r(x) = a(x)$. If $m \leq n$, consider the polynomial

$$\alpha(x) = a(x) - a_nb_m^{-1}x^{n-m}b(x). \tag{3}$$

It is obvious that $\alpha(x) = 0$ or $\deg \alpha(x) \leq n - 1$. If $\alpha(x) = 0$, set $q(x) = a_nb_m^{-1}x^{n-m}$ and $r(x) = 0$. Otherwise by the induction hypothesis there exist polynomials $q_1(x)$ and $r_1(x)$ such that

$$\alpha(x) = q_1(x)b(x) + r_1(x) \tag{4}$$

and $r_1(x) = 0$ or $\deg r_1(x) < m$. Substituting (4) in (3) yields

$$a(x) = [a_nb_m^{-1}x^{n-m} + q_1(x)]b(x) + r_1(x);$$

now set $q(x) = a_nb_m^{-1}x^{n-m} + q_1(x)$ and $r(x) = r_1(x)$ to obtain (1).

To show that $q(x)$ and $r(x)$ are uniquely determined, suppose we also have

$$a(x) = Q(x)b(x) + \rho(x), \tag{5}$$

where $\rho(x) = 0$ or $\deg \rho(x) < \deg b(x) = m$. Combining (1) and (5) yields

$$[Q(x) - q(x)]\, b(x) = r(x) - \rho(x). \tag{6}$$

If $Q(x) \neq q(x)$, then since b_m is not a zero divisor in R (being a unit), it follows that $\deg [Q(x) - q(x)]\, b(x) \geq m$ whereas $\deg[r(x) - \rho(x)] < m$. Thus $Q(x) = q(x)$ and [by (6)] $r(x) = \rho(x)$. This establishes the uniqueness of the polynomials $q(x)$ and $r(x)$. The proof of the second conclusion [involving (2)] is entirely analogous. ▮

Example 2 Let x be an indeterminate over $M_2(Z)$. Suppose

$$a(x) = \begin{bmatrix} 2 & 1 \\ 0 & 0 \end{bmatrix} x^2 + \begin{bmatrix} 0 & 0 \\ 1 & 0 \end{bmatrix} x + \begin{bmatrix} 1 & 0 \\ 0 & 1 \end{bmatrix},$$

$$b(x) = \begin{bmatrix} 0 & 1 \\ -1 & 0 \end{bmatrix} x + \begin{bmatrix} 1 & 1 \\ 1 & 1 \end{bmatrix}.$$

Then $b_1 = \begin{bmatrix} 0 & 1 \\ -1 & 0 \end{bmatrix}$ is a unit in $M_2(Z)\colon b_1^{-1} = \begin{bmatrix} 0 & -1 \\ 1 & 0 \end{bmatrix}$.

We find the polynomials $q(x), r(x) \in M_2(Z)[x]$ such that $a(x) = q(x)b(x) + r(x)$ and $r(x) = 0$ or $\deg r(x) < \deg b(x)$. Mimicking the proof of Theorem 3.2, we set

$$\alpha(x) = a(x) - \begin{bmatrix} 2 & 1 \\ 0 & 0 \end{bmatrix} \begin{bmatrix} 0 & -1 \\ 1 & 0 \end{bmatrix} x b(x)$$

$$= \begin{bmatrix} 1 & 1 \\ 1 & 0 \end{bmatrix} x + \begin{bmatrix} 1 & 0 \\ 0 & 1 \end{bmatrix}. \tag{7}$$

Since $\deg \alpha(x) = 1 = \deg b(x)$, we repeat the process with $\alpha(x)$ and compute

$$\alpha(x) - \begin{bmatrix} 1 & 1 \\ 0 & 1 \end{bmatrix} \begin{bmatrix} 0 & -1 \\ 1 & 0 \end{bmatrix} b(x) = \begin{bmatrix} 1 & 0 \\ 1 & 2 \end{bmatrix},$$

so that

$$\alpha(x) = \begin{bmatrix} 1 & -1 \\ 0 & -1 \end{bmatrix} b(x) + \begin{bmatrix} 1 & 0 \\ 1 & 2 \end{bmatrix}. \tag{8}$$

Substituting (8) into the first equality in (7) yields

$$a(x) = \begin{bmatrix} 1 & -2 \\ 0 & 0 \end{bmatrix} x b(x) + \begin{bmatrix} 1 & -1 \\ 0 & -1 \end{bmatrix} b(x) + \begin{bmatrix} 1 & 0 \\ 1 & 2 \end{bmatrix}.$$

Therefore

$$q(x) = \begin{bmatrix} 1 & -2 \\ 0 & 0 \end{bmatrix} x + \begin{bmatrix} 1 & -1 \\ 0 & -1 \end{bmatrix},$$

$$r(x) = \begin{bmatrix} 1 & 0 \\ 1 & 2 \end{bmatrix}.$$

Observe that

$$b(x)q(x) + r(x) = \begin{bmatrix} 0 & 0 \\ -1 & 2 \end{bmatrix} x^2 + \begin{bmatrix} 1 & -3 \\ 0 & -1 \end{bmatrix} x + \begin{bmatrix} 2 & -2 \\ 2 & 0 \end{bmatrix} \neq a(x);$$

thus in general $q(x) \neq s(x)$ and $r(x) \neq u(x)$.

The polynomials $a(x)$, $b(x)$, $q(x)$, $r(x)$, $s(x)$, and $u(x)$ in Theorem 3.2 are called the *dividend, divisor, right quotient, right remainder, left quotient* and *left remainder*, respectively. Of course, if R is a commutative ring, there is no distinction between "right" and "left"; $q(x) = s(x)$ is called the *quotient* and $r(x) = u(x)$ is called the *remainder*. If R is commutative and $r(x) = 0$ so that

$$a(x) = q(x)b(x) = b(x)q(x),$$

then we say $b(x)$ *divides* $a(x)$ and write

$$b(x) \mid a(x).$$

The reader will easily confirm the following facts (see Exercise 2):

If $f(x) \mid g(x)$ and $g(x) \mid h(x)$, then $f(x) \mid h(x)$. (9)

If $f(x) \mid g(x), f(x) \mid h(x)$, and $m(x), n(x) \in R[x]$, then

$$f(x) \mid [m(x)g(x) + n(x)h(x)]. \tag{10}$$

If the leading coefficient of $f(x)$ is not a zero divisor, then

$$f(x)g(x) = f(x)h(x) \text{ implies } g(x) = h(x). \tag{11}$$

If $f(x) \mid g(x)$, $g(x) \mid f(x)$, and the leading coefficients of $f(x)$ and $g(x)$ are not zero divisors, then

$$f(x) = rg(x) \text{ for some unit } r \in R. \tag{12}$$

Theorem 3.3 (*Remainder theorem*) *Let R be a ring with identity and x an indeterminate over R. If $a(x) \in R[x]$, $a(x) \neq 0$, and $c \in R$, then there exist unique polynomials $q(x)$ and $s(x)$ in $R[x]$ such that*

$$a(x) = q(x)(x - c) + a_r(c), \tag{13}$$

$$a(x) = (x - c)s(x) + a_l(c). \tag{14}$$

Proof: By Theorem 3.2 there exist unique polynomials $q(x)$, $r(x) \in R[x]$ such that

$$a(x) = q(x)(x - c) + r(x) \tag{15}$$

and $r(x) = 0$ or $\deg r(x) < \deg (x - c) = 1$. In any event we have $r(x) = d \in R$, so that

$$a(x) = q(x)(x - c) + d. \tag{16}$$

Let φ_r: $R[x] \to R$ be right substitution at c; recall that φ_r is a homomorphism of the additive group $(R[x], \{+\})$. Applying φ_r to both sides of (16) [say $q(x) = \sum_{k=0}^{n} q_k x^k$], we compute that

$$\begin{aligned} a_r(c) &= \varphi_r(a(x)) = \varphi_r[q(x)(x - c)] + \varphi_r(d) \\ &= \varphi_r\left[\left(\sum_{k=0}^{n} q_k x^k\right)(x - c)\right] + d = \varphi_r\left(\sum_{k=0}^{n} q_k(x^{k+1} - cx^k)\right) + d \end{aligned}$$

$$= \varphi_r\left(\sum_{k=0}^{n} q_k x^{k+1}\right) - \varphi_r\left(\sum_{k=0}^{n} q_k c x^k\right) + d$$
$$= \sum_{k=0}^{n} q_k c^{k+1} - \sum_{k=0}^{n} q_k c c^k + d = \sum_{k=0}^{n} q_k c^{k+1} - \sum_{k=0}^{n} q_k c^{k+1} + d$$
$$= d.$$

Thus $a(x) = q(x)(x - c) + a_r(c)$, proving (13). The proof of (14) is similar. ∎

Corollary 1 *If R is a commutative ring with identity and $a(x) \in R[x]$, $a(x) \neq 0$, then $c \in R$ is a root of $a(x)$ iff $(x - c)|a(x)$.*

Proof: If c is a root of $a(x)$, then (13) reads $a(x) = q(x)(x - c)$ so that $(x - c)|\ a(x)$. On the other hand, if $(x - c)|a(x)$, the uniqueness of the remainder in (13) implies that $a(c) = a_r(c) = 0$, i.e., c is a root of $a(x)$. ∎

Corollary 2 *If R is an integral domain, $a(x) \in R[x]$, and* $\deg a(x) = n \geq 0$, *then $a(x)$ has at most n roots in R.*

Proof: The proof is by induction on n. The result is trivial if $n = 0$. Suppose $n = 1$ and $c,d \in R$ are roots of $a(x) = a_1x + a_0$. Then $a_1c + a_0 = 0 = a_1d + a_0$ implies $a_1(c - d) = 0$. Since R is an integral domain, we obtain $c = d$. Now assume that $n > 1$ and the result holds for polynomials of degree less than n. If $a(x)$ has a root $c \in R$, then by Corollary 1

$$a(x) = q(x)(x - c)$$

for some $q(x) \in R[x]$. Note that $d \neq c$ is a root of $a(x)$ iff $0 = a(d) = q(d)(d - c)$ iff $q(d) = 0$, since R is an integral domain. Thus the only roots of $a(x)$ other than c are those that $q(x)$ might possess. But $\deg q(x) = n - 1$ (see Exercise 7, Section 3.2), so $q(x)$ has at most $n - 1$ roots in R by the induction hypothesis. We conclude that $a(x)$ has at most n roots in R. ∎

Corollary 3 *Let R be an integral domain, and let $f(x), g(x) \in R[x]$ be polynomials of degree at most $n \geq 0$. Suppose $c_1, \ldots, c_{n+1}$ are $n + 1$ distinct elements of R. Then $f(c_k) = g(c_k)$ for $k = 1, \ldots, n + 1$ iff $f(x) = g(x)$.*

Proof: Apply Corollary 2 to the polynomial $f(x) - g(x)$. ∎

Corollary 4 *Let R be an integral domain, let $x_1, \ldots, x_n$ be independent indeterminates over R, and let $f(x_1, \ldots, x_n) \in R[x_1, \ldots, x_n]$. Suppose R has an infinite subset S such that $f(s_1, \ldots, s_n) = 0$ for arbitrary specializations of the x_k to $s_k \in S$, $k = 1, \ldots, n$. Then $f(x_1, \ldots, x_n) = 0$.*

Proof: We first remark that if $f(x_1, \ldots, x_n) = \sum_\gamma a_\gamma x_1^{\gamma(1)} \cdots x_n^{\gamma(n)}$, then $f(s_1, \ldots, s_n)$ has the obvious meaning

$$f(s_1, \ldots, s_n) = \sum a_\gamma s_1^{\gamma(1)} \cdots s_n^{\gamma(n)}. \tag{17}$$

The proof is by induction on n. If $n = 1$, the result follows trivially from Corollary 2. Assume that $n > 1$ and that the result holds for polynomials in fewer than n indeterminates over R. Write $f(x_1, \ldots, x_n)$ as a polynomial in x_n with coefficients in $R[x_1, \ldots, x_{n-1}]$:

$$f(x_1, \ldots, x_n) = \sum_{j=0}^{r} g_j(x_1, \ldots, x_{n-1})x_n{}^j. \tag{18}$$

Let $s_k{}^0$, $k = 1, \ldots, n-1$, be arbitrary but fixed elements of S. From (18) we obtain

$$f(s_1{}^0, \ldots, s_{n-1}^0, x_n) = \sum_{j=0}^{r} g_j(s_1{}^0, \ldots, s_{n-1}^0)x_n{}^j. \tag{19}$$

Now the polynomial (19) belongs to $R[x_n]$ and vanishes for more than r specializations of x_n to elements of R. We conclude by Corollary 2 that $f(s_1{}^0, \ldots, s_{n-1}^0, x_n) = 0$, i.e., $g_j(s_1{}^0, \ldots, s_{n-1}^0) = 0$ for $j = 0, \ldots, r$. It follows by the induction hypothesis that $g_j(x_1, \ldots, x_{n-1}) = 0$ for $j = 0, \ldots, r$. Finally, (18) implies that $f(x_1, \ldots, x_n) = 0$. ∎

Corollary 5 *Let R be an integral domain, let $x_1, \ldots, x_n$ be independent indeterminates over R, and let $f(x_1, \ldots, x_n), g(x_1, \ldots, x_n) \in R[x_1, \ldots, x_n]$. Suppose R has an infinite subset S such that $f(s_1, \ldots, s_n) = g(s_1, \ldots, s_n)$ for arbitrary specializations of the x_k to $s_k \in S$, $k = 1, \ldots, n$. Then $f(x_1, \ldots, x_n) = g(x_1, \ldots, x_n)$.*

Proof: Apply Corollary 4 to the polynomial $f(x_1, \ldots, x_n) - g(x_1, \ldots, x_n)$. ∎

The most important consequence of Theorem 3.2 is the following result.

Theorem 3.4 *If R is a field and x is an indeterminate over R, then $R[x]$ is a P.I.D.*

Proof: We know that $R[x]$ is an integral domain (see Corollary 1, Section 3.2). Suppose $A \lhd R[x]$, $A \neq \{0\}$, and let $b(x) \in A$ be a polynomial in A of least degree. If $0 \neq a(x) \in A$, then by Theorem 3.2 there exist unique polynomials $q(x), r(x) \in R[x]$ such that

$$a(x) = q(x)b(x) + r(x)$$

and $r(x) = 0$ or $\deg r(x) < \deg b(x)$ [note that this application of Theorem 3.2 depends on the fact that the leading coefficient of $b(x)$ is a unit in the field R]. If $r(x) \neq 0$, then

$$r(x) = a(x) - q(x)b(x) \in A,$$

and so $r(x)$ is a polynomial in A of degree less than $\deg b(x)$. This contradiction implies that $r(x) = 0$. Hence $a(x) = q(x)b(x)$, i.e., $a(x) \in b(x)R[x] = (b(x))$. Since $a(x)$ was chosen to be an arbitrary nonzero element of A, we have $A = (b(x))$. Thus every ideal in the integral domain $R[x]$ is principal; so $R[x]$ is a P.I.D. ∎

Example 3 If x is an indeterminate over the real field $\mathbf{R}$, then $\mathbf{R}[x]$ is a P.I.D. by Theorem 3.4. In this example we will show that on the other hand, if x_1 and x_2 are independent indeterminates over $\mathbf{R}$, it is not true that $\mathbf{R}[x_1,x_2]$ is a P.I.D. Let $A \subset \mathbf{R}[x_1,x_2]$ be the set consisting of all polynomials in $\mathbf{R}[x_1,x_2]$ with 0 constant term: $a_{00} + a_{10}x_1 + a_{01}x_2 + \cdots \in A$ iff $a_{00} = 0$. It is trivial to see that $A \lhd \mathbf{R}[x_1,x_2]$. Suppose A is a principal ideal in $\mathbf{R}[x_1,x_2]$, i.e., suppose $A = (b(x_1,x_2))$ for some $b(x_1,x_2) \in \mathbf{R}[x_1,x_2]$. Then since $x_1 \in A$, there exists $q(x_1,x_2) \in \mathbf{R}[x_1,x_2]$ such that

$$x_1 = q(x_1,x_2)b(x_1,x_2). \tag{20}$$

Regarding each side of (20) first as a polynomial in $\mathbf{R}[x_1][x_2]$ and then as a polynomial in $\mathbf{R}[x_2][x_1]$, we have

$$0 = \deg_{x_2} q(x_1,x_2) + \deg_{x_2} b(x_1,x_2)$$

and

$$1 = \deg_{x_1} q(x_1,x_2) + \deg_{x_1} b(x_1,x_2) \tag{21}$$

where the notations $\deg_{x_2}$ and $\deg_{x_1}$ are self-explanatory. It follows that

$$\deg_{x_2} q(x_1,x_2) = 0 = \deg_{x_2} b(x_1,x_2)$$

and either

$$\deg_{x_1} q(x_1,x_2) = 1, \qquad \deg_{x_1} b(x_1,x_2) = 0$$

or

$$\deg_{x_1} q(x_1,x_2) = 0, \qquad \deg_{x_1} b(x_1,x_2) = 1.$$

Since $b(x_1,x_2)$ has a 0 constant term, $\deg_{x_1} b(x_1,x_2) = 0$ would imply that $b(x_1,x_2) = 0$ and hence that $A = \{0\}$, which is obviously nonsense. Thus $\deg_{x_1} q(x_1,x_2) = 0$ and $\deg_{x_1} b(x_1,x_2) = 1$; so

$$q(x_1,x_2) = q_0 \text{ for some } q_0 \in \mathbf{R}$$

and

$$b(x_1,x_2) = b_1x_1 + b_0 \qquad \text{for some } b_1,b_0 \in \mathbf{R},\ b_1 \neq 0.$$

Now (20) reads

$$x_1 = b_1q_0x_1 + b_0q_0;$$

then $q_0 \neq 0$, $b_0q_0 = 0$, and hence $b_0 = 0$. We have shown that $b(x_1,x_2) = b_1x_1$, $b_1 \neq 0$; reversing the roles of x_1 and x_2 in the above argument yields $b(x_1, x_2) = \beta_1x_2$, $\beta_1 \neq 0$. But

$$b_1x_1 = \beta_1x_2, \qquad b_1 \neq 0, \beta_1 \neq 0$$

is impossible since x_1 and x_2 are independent indeterminates over $\mathbf{R}$. Thus A cannot be a principal ideal in $\mathbf{R}[x_1,x_2]$, and so $\mathbf{R}[x_1,x_2]$ is not a P.I.D.

Our next major goal is to show that in a P.I.D. (e.g., $R[x]$), any element can be "factored" into a product of "primes" in the same sense as in Z.

Let R be a commutative ring with identity, and let $a,b \in R$. If $a \neq 0$, we say that *a divides b* and write

$$a \mid b$$

if there exists $c \in R$ such that $b = ac$. If $a = ub$ for some unit $u \in R$, then a is said to be an *associate* of b. If $a|b$ and a is neither an associate of b nor a unit in R, then a is called a *proper divisor* of b. It is easy to check (see Exercise 3) that the relation $\rho \subset R^2$ defined by

$$\rho = \{(a,b) \in R^2 \mid a \text{ is an associate of } b\} \tag{22}$$

is an equivalence relation on R. Thus R/ρ consists of classes of associated elements; a system of distinct representatives for R/ρ is called a *complete system of nonassociates in R*.

Example 4 (a) The units in Z are 1 and -1, so $N \cup \{0\}$ is a complete system of nonassociates in Z.

(b) In $\mathbf{R}[x]$ the units are the nonzero elements of $\mathbf{R}$. Thus $a(x),b(x) \in \mathbf{R}[x]$ are associates iff $a(x) = rb(x)$ for some $0 \neq r \in \mathbf{R}$.

(c) Consider the subring $Z[i]$ of $\mathbf{C}$ defined by

$$Z[i] = \{a + bi \in \mathbf{C} \mid a,b \in Z\}.$$

It is clear that $1, -1, i$, and $-i$ are units in $Z[i]$; we claim that these are all the units in $Z[i]$. For suppose $a + bi$ is a unit in $Z[i]$. Then $(a + bi)(c + di) = 1$ for some $c,d \in Z$, and hence $ac - bd = 1$, $ad + bc = 0$. If $d = 0$, then $ac = 1$, $bc = 0$ so that $a = \pm 1$, $c = \pm 1$, and $b = 0$; in this case we obtain $a + bi = \pm 1$. If $d \neq 0$, then $a = -bc/d$ implies $-(bc/d)c - bd = 1$ and hence $-b(c^2 + d^2) = d$. Thus $(c^2 + d^2) \mid d$, and since $d \neq 0$, we must have $c = 0$. Then $a = 0$, $-bd^2 = d$, $-bd = 1$, and $b = \pm 1$; in this case we obtain $a + bi = \pm i$. This establishes our claim. It follows that the associates of an element $x + yi \in Z[i]$ are $x + yi$, $-x - yi$, $-y + xi$, and $y - xi$. The ring $Z[i]$ is called the *ring of Gaussian integers*. Of course, $Z[i]$ is an integral domain since $\mathbf{C}$ is.

We leave to the reader to verify a number of simple facts concerning divisibility in a commutative ring R with identity 1 (see Exercise 4); let $a,b,c \in R$, $a \neq 0$:

$$a \mid 0; \tag{23}$$

$$1 \mid a; \tag{24}$$

$$a \mid a; \tag{25}$$

$$a \mid 1 \text{ iff } a \text{ is a unit.} \tag{26}$$

$$\text{If } a \mid b \text{ and } ac \neq 0, \text{ then } \quad ac \mid bc. \tag{27}$$

$$\text{If } a \mid b \text{ and } b \mid c, \text{ then } a \mid c. \tag{28}$$

$$\text{If } b_i,c_i \in R \text{ for } i = 1, \ldots, n \text{ and } a \mid b_i \text{ for each } i, \text{ then } a \mid \sum_{i=1}^{n} b_ic_i; \tag{29}$$

$$a \mid b \text{ iff } (b) \subset (a). \tag{30}$$

If a is not a zero divisor, then

$$(a) = (b) \tag{31}$$

iff a and b are associates.

If a is not a zero divisor, then a is a proper divisor of b iff $(b) \subset (a) \subset R$ and the inclusions are strict. (32)

Let R be a commutative ring with identity, and let $0 \neq m \in R$. Define the relation μ on R by

$$\mu = \{(a,b) \in R^2 \mid m \mid (a - b)\}. \tag{33}$$

It is evident that μ is an equivalence relation on R and hence partitions R into a set of equivalence classes R/μ. These equivalence classes are called *residue classes modulo m*, and a system of distinct representatives for R/μ is called a *complete system of residues modulo m*. For example, in Z a complete system of residues modulo m is the set $\{0,1,2, \ldots, m - 1\}$. As another example, take $m = 1 + i$ in the Gaussian integers $Z[i]$. It is easy to see that $1 + i \mid a + bi$ in $Z[i]$ iff $2 \mid a + b$ and $2 \mid a - b$ in Z, i.e., iff a and b have the same parity. Thus $1 + i$ divides $(a + bi) - (c + di)$ iff $a - c$ and $b - d$ have the same parity. It follows that $2 + 3i$ and $4 + 2i$ do not belong to the same residue class modulo $1 + i$.

Henceforth in the present section D will designate an integral domain, unless otherwise stated. Suppose $0 \neq q \in D$ is not a unit. If q has no proper divisors in D (i.e., the only divisors of q are units and associates of q), then q is called an *irreducible element* of D; otherwise q is a *reducible element* of D. If $0 \neq p \in D$ is not a unit and

$$p \mid ab \text{ implies } p \mid a \text{ or } p \mid b \tag{34}$$

for any $a,b \in D$, then p is called a *prime* in D.

Example 5 (a) The primes and irreducible elements in Z coincide.

(b) An element $p \in D$ is a prime iff (p) is a proper prime ideal in D. Indeed, $p \neq 0$ iff $(p) \neq \{0\}$, p is not a unit iff $(p) \neq D$, and $p \mid a$ if $a \in (p)$ [so the statement "$p \mid ab$ implies $p \mid a$ or $p \mid b$" is equivalent to the statement "$ab \in (p)$ implies $a \in (p)$ or $b \in (p)$"].

(c) Any prime element in D is irreducible. For suppose $p \in D$ is prime and $r \mid p$. Then $p = rs$ for some $s \in D$, and since p is a prime, we have $p \mid r$ or $p \mid s$. Suppose $p \mid s$ so that $s = tp$ for some $t \in D$. Then $p = rs = rtp$ and hence $rt = 1$, i.e., r is a unit. On the other hand, if $p \mid r$, then since also $r \mid p$ it follows that p and r are associates. Thus the only divisors of p are units and associates of p, and so p is irreducible.

(d) In this example we show that not every irreducible element in an integral domain is necessarily a prime. Consider the subdomain

$$D = Z[\sqrt{-5}\,] = \{a + b\sqrt{-5} \mid a,b \in Z\}$$

of **C**. We assert that $\alpha = 2 + \sqrt{-5} \in D$ is irreducible but is not a prime. To verify that α is not a unit, suppose $(a + b\sqrt{-5})(c + d\sqrt{-5}) = 1$. Multiplying both sides by $a - b\sqrt{-5}$ yields $(a^2 + 5b^2)(c + d\sqrt{-5}) = a - b\sqrt{-5}$, so that

$$(a^2 + 5b^2)c = a \tag{35}$$

and

$$(a^2 + 5b^2)d = -b. \tag{36}$$

An elementary argument (see Exercise 5) shows that (35) and (36) can hold iff $a = \pm 1$, $b = 0$. Thus the only units in D are ± 1, and $\alpha = 2 + \sqrt{-5}$ is therefore not a unit. Observe next that

$$(2 + \sqrt{-5})(2 - \sqrt{-5}) = 9 = 3 \cdot 3$$

and hence

$$2 + \sqrt{-5} \mid 3 \cdot 3.$$

On the other hand, it is simple to check that $2 + \sqrt{-5}$ is not a divisor of 3 (see Exercise 6). Thus $\alpha = 2 + \sqrt{-5}$ is not a prime.

It remains to show that α is irreducible. Suppose $a + b\sqrt{-5}$ is a proper divisor of α; we have

$$(a + b\sqrt{-5})(c + d\sqrt{-5}) = 2 + \sqrt{-5} \tag{37}$$

for some $c,d \in Z$. Take the complex conjugate of both sides of (37) to obtain

$$(a - b\sqrt{-5})(c - d\sqrt{-5}) = 2 - \sqrt{-5}, \tag{38}$$

and then multiply (37) and (38) together:

$$(a^2 + 5b^2)(c^2 + 5d^2) = 9. \tag{39}$$

Thus $a^2 + 5b^2$ divides 9 in Z, and this can happen iff $a^2 + 5b^2 = 9$ or $a = \pm 1$, $b = 0$. The latter eventuality is ruled out because $a + b\sqrt{-5}$ is supposed to be a proper divisor of α. But then $c^2 + 5d^2 = 1$, so that $c = \pm 1$, $d = 0$. In other words, (37) becomes $(a + b\sqrt{-5}) = \pm(2 + \sqrt{-5})$; this means that $a + b\sqrt{-5}$ is an associate of α and contradicts the hypothesis that $a + b\sqrt{-5}$ is a proper divisor of α. We conclude that the nonunit $\alpha = 2 + \sqrt{-5}$ has no proper divisors in $Z[\sqrt{-5}]$, i.e., α is irreducible.

Let $a_1, \ldots, a_n$ be nonzero elements of D. Suppose $d \in D$ satifies the following two conditions:

(i) $d \mid a_i$, $i = 1, \ldots, n$.
(ii) If $d_1 \in D$ and $d_1 \mid a_i$, $i = 1, \ldots, n$, then $d_1 \mid d$.

Then d is called a *greatest common divisor* of $a_1, \ldots, a_n$ in D. For example, a greatest common divisor (hereafter abbreviated g.c.d.) of the integers 1820 and 1287 is 13; note that

$$13 = 1820 \cdot 29 + 1287 \cdot (-41).$$

As another example, in $Q[x]$ a g.c.d. of $2x^5 + 2x^4 + 4x^3 - x^2 - x - 2$ and $2x^4 + 2x^3 + 5x^2 + x + 2$ is $3x^2 + 3x + 6$; again, note that

$$3x^2 + 3x + 6 = (2x - 2)(2x^5 + 2x^4 + 4x^3 - x^2 - x - 2) + (-2x^2 + 2x + 1)(2x^4 + 2x^3 + 5x^2 + x + 2).$$

If $a_1, \ldots, a_n$ are nonzero elements of D and the only elements $d \in D$ satisfying (i) above are units, then $a_1, \ldots, a_n$ are said to be *relatively prime* in D. For example, 4 and 15 are relatively prime in Z.

We have the following interesting facts related to the notion of divisibility.

Theorem 3.5 *Assume that D is a P.I.D. and $a_1, \ldots, a_n$ are nonzero elements of D.*

(i) *There exists a g.c.d. d of $a_1, \ldots, a_n$, d is uniquely determined except for associates, and there exist $r_1, \ldots, r_n \in D$ such that*

$$d = \sum_{i=1}^{n} r_i a_i. \tag{40}$$

(ii) *$a_1, \ldots, a_n$ are relatively prime iff there exist $s_1, \ldots, s_n \in D$* such that

$$1 = \sum_{i=1}^{n} s_i a_i. \tag{41}$$

(iii) *$p \in D$ is irreducible iff (p) is a proper maximal ideal in D.*

(iv) *$p \in D$ is prime iff p is irreducible; thus in a P.I.D. there is no distinction between prime elements and irreducible elements.*

(v) *If $0 \neq a \in D$ is not a unit, then there exists a prime $p \in D$ such that $p \mid a$.*

(vi) *If $a_0 \mid a_1 a_2 \cdots a_k$ and a_0 and a_1 are relatively prime, then $a_0 \mid a_2 \cdots a_k$.*

Proof: (i) Consider the ideal $A = (a_1, \ldots, a_n)$ in D generated by $a_1, \ldots, a_n$. Since D is a commutative ring with identity, A consists of all finite sums of the form

$$\sum_{i=1}^{n} r_i a_i, \qquad r_i \in D,\ i = 1, \ldots, n. \tag{42}$$

Since D is a P.I.D., we have $A = (d)$ for some $d \in D$, and d is of the form (42). Now $a_i \in (d)$ implies $d \mid a_i$, $i = 1, \ldots, n$. If $d_1 \mid a_i$ for $i = 1, \ldots, n$, then clearly $d_1 \mid \sum_{i=1}^{n} r_i a_i$, i.e., $d_1 \mid d$. Thus d is a g.c.d. of $a_1, \ldots, a_n$. Suppose δ is a g.c.d. of $a_1, \ldots, a_n$. Then $\delta \mid a_i$, $i = 1, \ldots, n$, and hence $\delta \mid d$, say $d = u\delta$ for $u \in D$. Reversing the roles of d and δ, we see that $\delta = vd$ for some $v \in D$. It follows that $d = u\delta = uvd$, $uv = 1$, and u is a unit. Thus d and δ are associates.

(ii) If $a_1, \ldots, a_n$ are relatively prime and d is a g.c.d. of $a_1, \ldots, a_n$, then d is a unit. Thus (41) follows from (40). Conversely, if (41) holds and $d \mid a_i$, $i = 1, \ldots, n$, then $d \mid 1$ so that d is a unit; hence $a_1, \ldots, a_n$ are relatively prime.

(iii) Suppose $p \in D$ is irreducible. Then $(p) \neq \{0\}$ since $p \neq 0$, and $(p) \neq D$ since p is not a unit. Let $q \notin (p)$ and choose d such that $(d) = (p,q)$. Then $d \mid p$, and the irreducibility of p implies d is a unit or an associate of p. But if

d were an associate of p, we could conclude from $d|q$ that $p|q$, contradicting $q \notin (p)$. Thus d is a unit and $(p,q) = D$. In other words, the ideal generated by p and any element not in (p) is D, and hence (p) is maximal. Conversely, suppose (p) is a proper maximal ideal in D. Then $(p) \neq \{0\}$ and $(p) \neq D$; so $p \neq 0$ and p is not a unit. If $q|p$, then $(p) \subset (q)$. The maximality of (p) implies that $(q) = D$ or $(q) = (p)$. If $(q) = D$, then q is a unit. If $(q) = (p)$, then q is an associate of p. Thus the only divisors of p are units and associates of p. Hence p is irreducible.

(iv) Example 5(c) shows that any prime in D is irreducible. Conversely, if $p \in D$ is irreducible, then by (iii), (p) is a proper maximal ideal. Hence (p) is a proper prime ideal by Corollary 6, Section 3.2. Now Example 5(b) shows that p is a prime.

(v) If a is irreducible, we may may choose $p = a$ by (iv). Otherwise there exists $a_1 \in D$ such that $a_1|a$ and a_1 is neither a unit nor an associate of a. Hence $(a) \subset (a_1)$, $(a_1) \neq D$, and $(a) \neq (a_1)$. Now if a_1 is irreducible, choose $p = a_1$. Otherwise there exists $a_2 \in D$ such that $a_2|a_1$ and a_2 is neither a unit nor an associate of a_1. Then $(a) \subset (a_1) \subset (a_2) \subset D$ and the inclusions are strict. Continuing, either we obtain an irreducible (i.e., prime) element a_k at some stage, or we obtain an infinite sequence

$$(a) \subset (a_1) \subset (a_2) \subset (a_3) \subset \cdots \subset (a_n) \subset (a_{n+1}) \subset \cdots \subset D \quad (43)$$

of ideals in D in which the inclusions are strict. Suppose the latter eventuality holds; set $a_0 = a$ and consider

$$A = \bigcup_{n=0}^{\infty} (a_n).$$

It is obvious that A is an ideal, and hence $A = (d)$ for some $d \in D$. Since $d \in A$ there exists a nonnegative integer n_0 such that $d \in (a_{n_0})$. But then $(a_{n_0+1}) \subset A = (d) \subset (a_{n_0})$, contradicting the strictness of the inclusions in (43). We conclude that some element a_k is a prime. Of course, $a_k|a$ and we let $p = a_k$.

(vi) Since a_0 and a_1 are relatively prime, by (ii) there exist elements $s,t \in D$ such that

$$1 = sa_0 + ta_1. \quad (44)$$

Multiply both sides of (44) by $a_2 \cdots a_k$ to obtain

$$a_2 \cdots a_k = sa_0a_2 \cdots a_k + ta_1a_2 \cdots a_k. \quad (45)$$

By the hypothesis that $a_0|a_1a_2 \ldots a_k$, a_0 divides the right side of (45) and hence divides $a_2 \cdots a_k$. ∎

In view of Theorem 3.5(i), we denote any g.c.d. of the nonzero elements $a_1, \ldots, a_n$ by

$$\text{g.c.d.}\ (a_1, \ldots, a_n);$$

these g.c.d.'s are all associates.

We now introduce the central idea of the present section. An integral domain D is called a *unique factorization domain*, abbreviated U.F.D., provided that any nonzero, nonunit element of D is a product of a finite number of primes and this factorization is unique except for order and associates. Thus if

$$\prod_{i=1}^{n} p_i = \prod_{j=1}^{m} q_j$$

and the p_i $(i = 1, \ldots, n)$ and q_j $(j = 1, \ldots, m)$ are primes in D, then $m = n$ and there exists $\sigma \in S_n$ such that p_i and $q_{\sigma(i)}$ are associates, $i = 1, \ldots, n$. It is easy to show that in a U.F.D., the set of prime elements and the set of irreducible elements coincide (see Exercise 7).

The following result is fundamental.

Theorem 3.6 *If D is a P.I.D., then D is a U.F.D.*

Proof: Let $0 \neq a \in D$ be a nonunit. By Theorem 3.5(v), there exists a prime $p_1 \in D$ such that $p_1 | a$, say $a = p_1 b_1$ for $b_1 \in D$. If b_1 is a unit, then a is itself a prime. Otherwise there exists a prime $p_2 \in D$ such that $p_2 | b_1$; we have $b_1 = p_2 b_2$ for some $b_2 \in D$. If b_2 is a unit, then b_1 is a prime and $a = p_1 b_1$ is the product of two primes. Otherwise there exists a prime $p_3 \in D$ such that $p_3 | b_2$. Continuing this process, either at some stage we express a as a product of a finite number of primes, or for each $n \in N$ we have

$$a = p_1 \cdots p_n b_n \tag{46}$$

where $p_1, \ldots, p_n$ are primes and b_n is not a unit.

Suppose the latter eventuality obtains, and consider the principal ideals

$$A_0 = (a),$$
$$A_n = (b_n), \qquad n = 1, 2, \ldots$$

in D. Since $b_1 | a$, $b_2 | b_1$, $b_3 | b_2, \ldots, b_{n+1} | b_n, \ldots$ it follows that

$$A_0 \subset A_1 \subset A_2 \subset \cdots \subset A_n \subset A_{n+1} \subset \cdots \subset D. \tag{47}$$

The inclusions in (47) are easily seen to be strict; for example, if $n \geq 1$ and $A_{n+1} = A_n$, then b_{n+1} and b_n are associates, implying that p_{n+1} is a unit and contradicting the fact that p_{n+1} is a prime. Let

$$A = \bigcup_{n=0}^{\infty} A_n.$$

The set A is obviously an ideal in D, and hence $A = (d)$ for some $d \in D$. Since $d \in A$, there exists a nonnegative integer n_0 such that $d \in A_{n0}$. But then

$$A_{n0} \subset A = (d) \subset A_{n0},$$

contradicting the strictness of the inclusions in (47). We conclude that a factors into a finite number of primes in D.

Next, suppose

$$p_1 \cdots p_n = q_1 \cdots q_m \tag{48}$$

for primes p_i ($i = 1, \ldots, n$) and q_j ($j = 1, \ldots, m$) in D. Then $p_1 | q_1 \cdots q_m$. If p_1 is not an associate of q_1, then p_1 and q_1 are relatively prime (see Exercise 8), and hence $p_1 | q_2 \cdots q_m$ [by Theorem 3.5(vi)]. After at most m repetitions of this argument we see that p_1 and some q_j must be associates. A suitable reordering of the q_j allows us to assume that p_1 and q_1 are associates, say $p_1 = uq_1$ for some unit $u \in D$. We have from (48) that

$$uq_1p_2 \cdots p_n = q_1q_2 \cdots q_m$$

and hence

$$up_2 \cdots p_n = q_2 \cdots q_m. \tag{49}$$

Now it is easy to see that if $n = 1$ in (48), i.e., if

$$p_1 = q_1 \cdots q_m,$$

then $m = 1$ and $p_1 = q_1$ (in particular, p_1 and q_1 are associates). Since up_2 in (49) is a prime, it follows by induction on n that $m - 1 = n - 1$ and the primes

$$up_2, \ldots, p_n$$

and

$$q_2, \ldots, q_n$$

can be paired off as associates in some order. Hence $m = n$ and the primes

$$p_1, \ldots, p_n$$

and

$$q_1, \ldots, q_n$$

can be paired off as associates in some order [one such pair being (p_1, q_1)]. This completes the proof. ∎

There are several important consequences of the preceding result.

Corollary 6 *If R is a field and x is an indeterminate over R, then R[x] is a U.F.D.*

Proof: By Theorem 3.4, $R[x]$ is a P.I.D. Apply Theorem 3.6. ∎

Corollary 7 *Z is a U.F.D.*

Proof: By Example 1, Z is a P.I.D. Apply Theorem 3.6. ∎

Corollary 8 *The ring of Gaussian integers Z[i] is a P.I.D., and hence a U.F.D.*

Proof: Let $B \lhd Z[i]$, $B \neq \{0\}$, and choose a nonzero element $\beta \in B$ of least absolute value. We will show that $B = (\beta)$.

Let $\alpha \in B$; clearly $\alpha/\beta \in Q[i] = \{a + bi \in \mathbf{C} \mid a,b \in Q\}$.

Write $\alpha/\beta = r + si$, $r,s \in Q$, and choose integers p and q closest to r and s, respectively:

$$0 \leq | r - p | \leq \tfrac{1}{2}, \qquad 0 \leq | s - q | \leq \tfrac{1}{2}. \tag{50}$$

Let $\gamma = p + qi \in Z[i]$, and compute from (50) that

$$\begin{aligned}\left|\frac{\alpha}{\beta} - \gamma\right| &= \sqrt{(r-p)^2 + (s-q)^2} \\ &\leq \sqrt{\frac{1}{4} + \frac{1}{4}} = \frac{1}{\sqrt{2}} \\ &< 1. \end{aligned} \tag{51}$$

Now let $\theta = \alpha - \beta\gamma$. Since $\alpha \in B$, $\beta \in B$, and $\gamma \in Z[i]$, it follows that $\theta \in B$. Moreover, from (51),

$$\begin{aligned}| \theta | = | \alpha - \beta\gamma | &= \left|\beta\left(\frac{\alpha}{\beta} - \gamma\right)\right| \\ &= | \beta | \left|\frac{\alpha}{\beta} - \gamma\right| < | \beta |.\end{aligned}$$

Unless $\theta = 0$, this contradicts the choice of $\beta \in B$. Thus $\theta = 0$, and we have $\alpha = \beta\gamma \in (\beta)$. This shows that $B = (\beta)$ and completes the proof. ∎

The following corollary of Theorem 3.5 concerning matrices over P.I.D.'s is of importance in studying modules, finitely generated abelian groups, and the reduction of matrices to Hermite normal form. These topics will be covered in subsequent chapters.

Corollary 9 [*Hermite (1848)*] *Let $a_1, \ldots, a_n$ $(n \geq 2)$ be nonzero elements in a P.I.D. D, and let $d_n = \text{g.c.d.}(a_1, \ldots, a_n)$. Then there exists a matrix $A_n \in M_n(D)$ such that the first row of A_n is $[a_1 \ \ldots \ a_n]$ and* $\det(A_n) = d_n$.

Proof: The proof is by induction on n. If $n = 2$, use Theorem 3.5(i) to obtain elements $r,s \in D$ such that $d_2 = ra_1 - sa_2$. Then set

$$A_2 = \begin{bmatrix} a_1 & a_2 \\ s & r \end{bmatrix}.$$

Assume now that $n > 2$ and the result holds for any $n - 1$ nonzero elements of D. Let d_{n-1} be a g.c.d. of $a_1, \ldots, a_{n-1}$. Then d_n is a g.c.d. of d_{n-1} and a_n (see Exercise 9). By the induction hypothesis, there exists $A_{n-1} \in M_{n-1}(D)$ with $[a_1 \ \ldots \ a_{n-1}]$ as first row and $\det(A_{n-1}) = d_{n-1} = \text{g.c.d.}(a_1, \ldots, a_{n-1})$. Since $d_n = \text{g.c.d.}(d_{n-1}, a_n)$, we again use Theorem 3.5(i) to obtain elements $r,s \in D$ such that $d_n = rd_{n-1} - sa_n$. Set

$$A_n = \left|\begin{array}{ccc|c} & & & a_n \\ & A_{n-1} & & 0 \\ & & & \vdots \\ & & & 0 \\ \hline \dfrac{a_1 s}{d_{n-1}} & \cdots & \dfrac{a_{n-1}s}{d_{n-1}} & r \end{array}\right| . \tag{52}$$

Obviously the matrix $A_n \in M_n(D)$ has first row $[a_1 \ . \ . \ . \ a_n]$. Let $B_{n-1} \in M_{n-1}(D)$ be the matrix whose row t is the same as row $t + 1$ of A_{n-1}, $t = 1, \ldots, n - 2$, and whose row $n - 1$ is $[a_1 s/d_{n-1} \ . \ . \ . \ a_{n-1}s/d_{n-1}]$. Expand $\det(A_n)$ by the last column to obtain

$$\begin{aligned} \det(A_n) &= (-1)^{n+1} a_n \det(B_{n-1}) + r \det(A_{n-1}) \\ &= (-1)^{n+1} a_n \det(B_{n-1}) + r d_{n-1}. \end{aligned} \tag{53}$$

Now observe that the same matrix results if we

(i) multiply row $n - 1$ of B_{n-1} by d_{n-1}.
(ii) multiply row 1 of A_{n-1} by s, and then interchange successively the first and second rows, the second and third rows, and so on, finally interchanging the $(n - 2)$nd and $(n - 1)$st rows for a total of $n - 2$ interchanges.

It follows that

$$d_{n-1} \det(B_{n-1}) = (-1)^{n-2} s \det(A_{n-1}) = (-1)^{n-2} s d_{n-1}. \tag{54}$$

Multiplying (53) on both sides by d_{n-1} and substituting (54), we have

$$\begin{aligned} d_{n-1} \det(A_n) &= (-1)^{n+1} a_n d_{n-1} \det(B_{n-1}) + r d_{n-1}^2 \\ &= (-1)^{n+1} a_n (-1)^{n-2} s d_{n-1} + r d_{n-1}^2 \\ &= d_{n-1}(r d_{n-1} - s a_n). \end{aligned}$$

Hence $\det(A_n) = r d_{n-1} - s a_n = d_n$, completing the proof. ∎

Example 6 (a) According to Corollary 8, $Z[i]$ is a P.I.D. Hence by Theorem 3.5(i), a g.c.d. of any finite set of nonzero elements of $Z[i]$ exists. Let us find g.c.d.$(11 + 7i, 3 + 7i)$. The method is to consider the ideal $B = (11 + 7i, 3 + 7i)$ and find a nonzero element $\beta \in B$ of least absolute value, for then $B = (\beta)$ as we saw in the proof of Corollary 8. Thus $\beta \mid 11 + 7i$ and $\beta \mid 3 + 7i$; also if $\beta_1 \in Z[i]$ is any divisor of $11 + 7i$ and $3 + 7i$, then $\beta_1 \mid \beta$ since $\beta = \rho(11 + 7i) + \sigma(3 + 7i)$ for some $\rho, \sigma \in Z[i]$. Thus such a β is a g.c.d. of $11 + 7i$ and $3 + 7i$.

We obtain β by a constructive procedure reminiscent of the proof of Corollary 8. First compute in $Q[i]$ that

$$\frac{11 + 7i}{3 + 7i} = \frac{(11 + 7i)(3 - 7i)}{58}$$

$$= \frac{82}{58} - \frac{56}{58}i = \frac{41}{29} - \frac{28}{29}i.$$

The element $a + bi \in Z[i]$ such that a and b are the closest integers to $\frac{41}{29}$ and $-\frac{28}{29}$, respectively, is clearly $1 - i$: $\left|\frac{41}{29} - 1\right| = \frac{12}{29} < \frac{1}{2}$ and $\left|-\frac{28}{29} + 1\right| = \frac{1}{29} < \frac{1}{2}$. We have

$$\frac{11+7i}{3+7i} = \frac{41}{29} - \frac{28}{29}i = \left(1 + \frac{12}{29}\right) - \left(1 - \frac{1}{29}\right)i$$
$$= (1 - i) + \left(\frac{12}{29} + \frac{1}{29}i\right),$$

and multiplication of both sides by $3 + 7i$ yields

$$11 + 7i = (1 - i)(3 + 7i) + \left(\frac{12}{29} + \frac{1}{29}i\right)(3 + 7i)$$
$$= (1 - i)(3 + 7i) + (1 + 3i). \tag{55}$$

It follows that $1 + 3i = (11 + 7i) - (1 - i)(3 + 7i) \in B$. Next, divide $3 + 7i$ by $1 + 3i$ in $Q[i]$ to obtain

$$\frac{3+7i}{1+3i} = \frac{(3+7i)(1-3i)}{10}$$
$$= \frac{12}{5} - \frac{1}{5}i = 2 + \left(\frac{2}{5} - \frac{1}{5}i\right).$$

Thus
$$3 + 7i = 2(1 + 3i) + \left(\frac{2}{5} - \frac{1}{5}i\right)(1 + 3i)$$
$$= 2(1 + 3i) + (1 + i)$$

so that
$$1 + i = (3 + 7i) - 2(1 + 3i) \in B. \tag{56}$$

Finally, dividing $1 + 3i$ by $1 + i$ in $Q[i]$, we have

$$\frac{1+3i}{1+i} = \frac{(1+3i)(1-i)}{2} = 2 + i,$$

i.e., $1 + i$ divides $1 + 3i$ in $Z[i]$. It now follows from (56) and (55) that $1 + i$ divides $3 + 7i$ and $11 + 7i$. Since $1 + i \in B$, we conclude that $1 + i = \text{g.c.d.}(11 + 7i, 3 + 7i)$. Note that $\beta = 1 + i$ is indeed a nonzero element of least absolute value in B. For if λ is a nonzero element of least absolute value in B, then $\lambda \mid \beta$ since $B = (\lambda)$, whereas $\beta \mid \lambda$ since λ is of the form $\rho(11 + 7i) + \sigma(3 + 7i)$ for some $\rho, \sigma \in Z[i]$. Hence β and λ are associates in $Z[i]$. But the units in $Z[i]$ are ± 1 and $\pm i$ [see Example 4(*c*)], and so $|\beta| = |\lambda|$.

(b) We construct a matrix $A_3 \in M_3(Z)$ such that the first row of A_3 is [6 10 15] and $\det(A_3) = 1 = \text{g.c.d.}(6,10,15)$. Following the proof of Corollary 9, note that $d_2 = \text{g.c.d.}(6,10) = 2 = 2 \cdot 6 - 1 \cdot 10$ and set

$$A_2 = \begin{bmatrix} 6 & 10 \\ 1 & 2 \end{bmatrix}.$$

Now $1 = d_3 = \text{g.c.d.}(6,10,15) = \text{g.c.d.}(d_2,15) = \text{g.c.d.}(2,15) = 8\cdot 2 - 1\cdot 15$. As in (52), set

$$A_3 = \begin{bmatrix} 6 & 10 & 15 \\ 1 & 2 & 0 \\ \dfrac{6\cdot 1}{2} & \dfrac{10\cdot 1}{2} & 8 \end{bmatrix} = \begin{bmatrix} 6 & 10 & 15 \\ 1 & 2 & 0 \\ 3 & 5 & 8 \end{bmatrix}.$$

We have

$$\det(A_3) = 15\cdot(-1) + 8\cdot 2 = 1.$$

Exercises 3.3

1. Prove Theorem 3.1.
2. Verify assertions (9) to (12).
3. Show that the relation ρ defined by (22) is an equivalence relation on R.
4. Verify assertions (23) to (32).
5. Show that (35) and (36) imply $a = \pm 1$, $b = 0$. *Hint*: If $c = 0$, then $a = 0$ and $5b^2d = -b$. Since $a + b\sqrt{-5}$ is assumed to be a unit, we have $b \neq 0$. Hence $5bd = -1$ which is impossible. Thus $c \neq 0$ and (35) implies $a \neq 0$. Then $b = 0$, for otherwise $|(a^2 + 5b^2)c| > |a|$ in contradiction to (35). We conclude that $a^2c = a$, $ac = 1$, and finally $a = \pm 1$.
6. Show that $2 + \sqrt{-5}$ does not divide 3 in $Z[\sqrt{-5}]$. *Hint*: Suppose $3 = (2 + \sqrt{-5})\ (a + b\sqrt{-5})$. Multiplying both sides by $2 - \sqrt{-5}$ yields $6 - 3\sqrt{-5} = 9a + 9b\sqrt{-5}$. But $9a = 6$ is impossible.
7. Let D be a U.F.D. Show that the set of prime elements in D and the set of irreducible elements in D coincide. *Hint*: We already know that any prime in an integral domain is irreducible [see Example 5(c)]. Suppose $a \in D$ is irreducible. Since D is a U.F.D., there exist primes $p_1, \ldots, p_n \in D$ such that $a = p_1 \cdots p_n$. Now a is irreducible, $p_1 | a$, and p_1 is not a unit; so p_1 must be an associate of a. Then a is an associate of the prime p_1 and hence is itself a prime.
8. Show that if p and q are two primes in a P.I.D. which are not associates, then p and q are relatively prime. *Hint*: If $d | p$ and $d | q$, then d must be a unit because the divisors of a prime (i.e., irreducible) element are units and associates.
9. Let D be a P.I.D., and suppose $a_1, \ldots, a_n$ are nonzero elements of D. Let $\alpha = \text{g.c.d.}(a_1, \ldots, a_k)$, $\beta = \text{g.c.d.}(a_{k+1}, \ldots, a_n)$, and $d = \text{g.c.d.}(a_1, \ldots, a_n)$, all unique to within unit multiples. Show that to within unit multiples, $d = \text{g.c.d.}(\alpha,\beta)$. *Hint*: $d | a_i$, $i = 1, \ldots, n$, so $d | \alpha$ and $d | \beta$. Suppose $d_1 | \alpha$ and $d_1 | \beta$. Since $d_1 | \alpha | a_i$, $i = 1, \ldots, k$, and $d_1 | \beta | a_i$, $i = k + 1, \ldots, n$, we have $d_1 | a_i$, $i = 1, \ldots, n$. It follows that $d_1 | d$. Thus $d = \text{g.c.d.}(\alpha,\beta)$.
10. Show that $1 - 2i$ is a prime in $Z[i]$. *Hint*: $|1 - 2i|^2 = 5$. Thus if $\gamma = a + bi \in Z[i]$ and $\gamma | (1 - 2i)$, then $|\gamma|^2 = |a + bi|^2 = a^2 + b^2$ is either 1 or 5. The possibilities are $a = \pm 1$, $b = 0$; $a = 0$, $b = \pm 1$; $a = \pm 1$, $b = \pm 2$; $a = \pm 2$, $b = \pm 1$. In the first two cases γ is a unit.

11. Let D be a P.I.D., and let $0 \neq a \in D$. Show that there are only a finite number of ideals in D containing a. *Hint*: Consider the factorization $a = p_1 \cdots p_n$ of a into a product of primes, unique except for order and associates. Then $b|a$ iff $b = up_{i_1} \cdots p_{i_k}$ where u is a unit in D and $\{i_1, \ldots, i_k\}$ is a k-element subset of $[1,n]$ ($k = 0$ is not excluded; in this case $b = u$). Since any ideal containing a is principal, it must be of the form (b) for some $b \in D$ dividing a. Hence there are at most 2^n such ideals.

12. Show that $Z[\sqrt{-5}]$ is not a P.I.D. . *Hint*: Example 5(d) shows that $2 + \sqrt{-5} \in Z[\sqrt{-5}]$ is irreducible but not prime, so $Z[\sqrt{-5}]$ cannot be a P.I.D. [see Theorem 3.5(iv)].

13. Imitate the proof of Corollary 8 to show that the subdomain $Z[\sqrt{-2}]$ of **C** is a P.I.D. and hence a U.F.D.

14. Show that the subdomain $Z[\sqrt{-3}]$ of **C** is not a P.I.D. *Hint*: $(1 + \sqrt{-3})(1 - \sqrt{-3}) = 4 = 2 \cdot 2$, but $1 + \sqrt{-3}$ does not divide 2.

15. Let R be a commutative ring with identity. An element $x \in R$ is said to be *nilpotent* if $x^n = 0$ for some $n \in N$. Show that if $\mathcal{N}$ is the set of all nilpotent elements in R, then $\mathcal{N}$ is an ideal in R. *Hint*: If $x,y \in \mathcal{N}$, then $x^n = 0 = y^m$ for $n,m \in N$. Now

$$(x + y)^{n+m-1} = \sum_{k=0}^{n+m-1} \binom{n+m-1}{k} x^k y^{n+m-1-k}.$$

For each k we clearly have either $k \geq n$ or $n + m - 1 - k \geq m$, since otherwise $n + m - 1 = k + n + m - 1 - k < n + m - 1$. Hence for each k either $x^k = 0$ or $y^{n+m-1-k} = 0$; thus $(x + y)^{n+m-1} = 0$. Also if $r \in R$, then $(rx)^n = r^n x^n = 0$.

16. Show that for the ideal $\mathcal{N}$ defined in Exercise 15, the quotient ring $R/\mathcal{N}$ has no nonzero nilpotent elements. *Hint*: Let $\nu: R \to R/\mathcal{N}$ be the canonical epimorphism. Suppose $\nu(r)^n = 0$ for some $n \in N$. Then $r^n \in \mathcal{N}$, so $(r^n)^m = 0$ for some $m \in N$. Hence $r \in \mathcal{N}$ and $\nu(r) = 0$.

17. The ideal $\mathcal{N}$ described in Exercises 15 and 16 is called the *nilradical of* R and is denoted by $\mathcal{N}(R)$. Prove that $\mathcal{N}(R)$ is the intersection of all prime ideals in R:

$$\mathcal{N}(R) = \bigcap_{A \lhd R, A \text{ prime}} A$$

(if R has no prime ideals, this intersection is taken to be R itself). *Hint*: Let $P = \bigcap_{A \lhd R, A \text{ prime}} A$ If $x^n = 0$, then $x^n = 0 \in A$ for any ideal A; if A is prime, it follows that $x \in A$. Thus $\mathcal{N}(R) \subset P$.

Now suppose that $x \notin \mathcal{N}(R)$ so that $x^n \neq 0$ for all $n \in N$. Then there are ideals A such that $x^n \notin A$ for all $n \in N$, e.g., $A = \{0\}$. Let $\mathfrak{A}$ be the collection of all such ideals A. Clearly, if $\{A_i \mid i \in I\}$ is a family of ideals in $\mathfrak{A}$ indexed by some set I, and if for any $i,j \in I$ either $A_i \subset A_j$ or $A_j \subset A_i$, then $\bigcup_{i \in I} A_i \in \mathfrak{A}$. For $\bigcup_{i \in I} A_i$ is obviously an ideal, and $x^n \notin \bigcup_{i \in I} A_i$ for all $n \in N$ since otherwise $x^n \in A_i \in \mathfrak{A}$ for some $i \in I$.

Apply Zorn's lemma to the collection of ideals $\mathfrak{A}$ to obtain an ideal $A \in \mathfrak{A}$ not properly contained in any other ideal in $\mathfrak{A}$. We assert that A is a prime ideal. Certainly $A \neq R$, by the definition of $\mathfrak{A}$. Let $y,z \in R - A$ and consider the ideals $A + (y)$ and $A + (z)$, both of which properly contain A. Since A is a maximal element of $\mathfrak{A}$, we have $A + (y) \notin \mathfrak{A}$ and $A + (z) \notin \mathfrak{A}$. Then by the definition of $\mathfrak{A}$, there exist $m,n \in N$ such that

$$x^m \in A + (y)$$

and

$$x^n \in A + (z).$$

Now

$$\begin{aligned} x^{m+n} = x^m x^n &\in [A + (y)][A + (z)] \\ &\subset A + (yz) \qquad \text{(why?),} \end{aligned}$$

so $A + (yz) \notin \mathfrak{A}$ (recall the definition of $\mathfrak{A}$). Hence $yz \notin A$. Thus $R - A$ is multiplicatively closed, and we conclude that A is a prime ideal in R. Since $x \notin A$, it follows that $x \notin \bigcap_{A \triangleleft R,\, A \text{ prime}} A = P$. We have proved that $x \notin \mathscr{N}(R)$ implies $x \notin P$, i.e., $P \subset \mathscr{N}(R)$. Therefore $P = \mathscr{N}(R)$.

18. Let R be a commutative ring with identity, and let U be the group of all units in R. Prove that

$$U = R - \bigcap_{A \triangleleft R,\, A \text{ maximal}} A .$$

Hint: A unit in R cannot belong to any maximal ideal, since such an ideal is distinct from R. Conversely, suppose $u \in R$ does not belong to any maximal ideal. Consider the principal ideal (u). If $(u) \neq R$, then by Theorem 1.10, Section 3.1, there exists a maximal ideal in R containing (u). Thus $(u) = R$, and hence u is a unit in R.

19. Let A and B be ideals in a commutative ring R with identity. Define $[A{:}B]$ by

$$[A{:}B] = \{x \in R \mid xB \subset A\};$$

$[A{:}B]$ is called the *quotient of A and B*. If $B = (b)$, we write $[A{:}b]$ instead of $[A{:}(b)]$.

Suppose now that $R = Z$, $0 \neq m,n \in Z$, $A = (m)$, and $B = (n)$; then $[A{:}B] = \{x \in Z \mid m \mid xn\}$. Clearly $[A{:}B]$ is an ideal in Z, and hence $[A{:}B] = (x_0)$, where x_0 is the least positive integer such that $m \mid x_0 n$. We can assume m and n are positive. Let $d = \text{g.c.d.}(m,n)$, and let $c = m/d$. Show that $x_0 = c$. *Hint*: Let $p_1, \ldots, p_k$ be all the distinct primes occurring in the prime factorizations of m or n. Then $m = p_1^{s_1} \cdots p_k^{s_k}$ and $n = p_1^{t_1} \cdots p_k^{t_k}$ for nonnegative integers s_i, t_i, $i = 1, \ldots, k$. Since $m \mid x_0 n$, $x_0 n$ contains each of the $p_i^{s_i}$ in its prime factorization. Since x_0 is the least positive integer such that $m \mid x_0 n$, it is obvious that x_0 contains no primes other than $p_1, \ldots, p_k$ in its prime factorization, say $x_0 = p_1^{r_1} \cdots p_k^{r_k}$ for nonnegative integers r_i, $i = 1, \ldots, k$. Then

$$x_0 n = p_1^{r_1} \cdots p_k^{r_k} p_1^{t_1} \cdots p_k^{t_k} = p_1^{r_1 + t_1} \cdots p_k^{r_k + t_k}.$$

Clearly the choice for each r_i is $r_i = s_i - t_i$ if $s_i > t_i$ and $r_i = 0$ if $s_i \leq t_i$, i.e., $r_i = \max\{0, s_i - t_i\} = s_i - \min\{s_i, t_i\}$. Thus

$$x_0 = p_1^{s_1} \cdots p_k^{s_k} / p_1^{\min\{s_1, t_1\}} \cdots p_k^{\min\{s_k, t_k\}} = \frac{m}{d} = c.$$

20. Prove that, in general, the quotient $[A{:}B]$ in Exercise 19 is an ideal in R.

21. Let R be a commutative ring with identity 1. The *Jacobson radical of R*, denoted by $J(R)$, is defined as the intersection of all maximal ideals in R: $J(R) = \bigcap_{A \triangleleft R,\, A \text{ maximal}} A$.
Prove that

$$J(R) = \{x \in R \mid 1 - xy \text{ is a unit for all } y \in R\}.$$

Hint: Suppose $x \in J(R)$ and $1 - xy$ is not a unit for some $y \in R$. Then $(1 - xy) \neq R$; so by Theorem 1.10, Section 3.1, there exists a maximal ideal A in R containing $1 - xy$. But then $1 - xy \in A$ and $x \in J(R) \subset A$ imply $1 \in A$, contradicting the fact that $A \neq R$. Thus $1 - xy$ is a unit for all $y \in R$. Conversely, suppose $x \notin A$ for some maximal ideal A in R. The maximality of A implies $A + (x) = R$, so that

$$1 = a + xy \qquad \text{for some } a \in A,\ y \in R.$$

Hence $1 - xy = a \in A$, and since $A \neq R$, we conclude that $1 - xy$ is not a unit.

Glossary 3.3

associate, 180
complete system of nonassociates, 180
complete system of residues modulo m, 181
dividend, 176
division algorithm, 174
divisor, 176
Gaussian integers, 180
greatest common divisor, 182
g.c.d., 182
Hermite's theorem, 187
irreducible element, 181
Jacobson radical, 193
$J(R)$, 193
left-hand value of $f(x)$ at t, 173
$f_l(t)$, 173
left quotient, 176
left remainder, 176
left specialization at t, 173
left substitution at t, 173
nilpotent element, 191
nilradical, 191
prime element, 181
principal ideal domain, 173
P.I.D., 173
proper divisor, 180
quotient of ideals, 192
$[A:B]$, 192
reducible element, 181
relatively prime, 183
remainder theorem, 176
residue classes modulo m, 181
right-hand value of $f(x)$ at t, 173
$f_r(t)$, 173
right quotient, 176
right remainder, 176
right specialization at t, 173
right substitution at t, 173
root of $f(x)$, 173
unique factorization domain, 185
U.F.D., 185
value of $f(x)$ at t, 173
$f(t)$, 173
zero of $f(x)$, 173
$Z[i]$, 180
φ_r, 173
φ_l, 173
$b(x) \mid a(x)$, 176
$a \mid b$, 179

3.4 Polynomial Factorization

We begin this section with a sequence of arguments showing that a polynomial ring $D[x]$ over a U.F.D. D is necessarily a U.F.D.. It should be noted that since $Z[x]$ is not a P.I.D. [e.g., $(2,x)$ is not a principal ideal], there is no hope of proving that $Z[x]$ is a U.F.D. by using Theorem 3.6., Section 3.3. However, we can prove the following analogue of Theorem 3.5(i).

Theorem 4.1 *If D is a U.F.D. and $a_1, \ldots, a_n$ are nonzero elements of D, then there exists a g.c.d. of $a_1, \ldots, a_n$ which is unique except for associates.*

Proof: Since D is a U.F.D., there exist (nonassociate) primes $p_1, \ldots, p_k \in D$ and units $\varepsilon_1, \ldots, \varepsilon_n \in D$ such that

$$a_r = \varepsilon_r \prod_{j=1}^{k} p_j^{e_{jr}}, \qquad r = 1, \ldots, n, \tag{1}$$

where the e_{jr} are nonnegative integers (of course, if a_r is a unit, we have $e_{jr} = 0, j = 1, \ldots, k$). Let

$$e_j = \min_{1 \leq r \leq n} e_{jr}, \qquad j = 1, \ldots, k$$

and set

$$d = p_1^{e_1} \cdots p_k^{e_k}.$$

Obviously $d | a_r$, $r = 1, \ldots, n$. Suppose $\delta \in D$ is such that $\delta | a_r$, $r = 1, \ldots, n$. Since D is a U.F.D., it follows that the factorization of δ into primes involves no primes other than associates of $p_1, \ldots, p_k$; moreover, after collecting the appropriate units, we see that each p_j can occur with exponent at most e_{jr}, $r = 1, \ldots, n$. In short, $\delta = \varepsilon p_1^{f_1} \cdots p_k^{f_k}$, where $f_j \leq \min_{1 \leq r \leq n} e_{jr} = e_j$, $j = 1, \ldots, k$, and ε is a unit. Thus $\delta | d$, showing that d is a g.c.d. of $a_1, \ldots, a_n$. The proof that d is unique except for associates is as in Theorem 3.5(i). ∎

Let D be a U.F.D., and let x be an indeterminate over D. If $0 \neq f(x) \in D[x]$, then the *content of* $f(x)$ is the g.c.d. of all the coefficients of $f(x)$; it is denoted by $c(f(x))$. Thus $c(f(x))$ is defined only to within a unit factor. If $c(f(x)) = 1$ [i.e., if $c(f(x))$ is a unit], then $f(x)$ is said to be a *primitive polynomial* in $D[x]$. If $d \in D$ and d divides every coefficient of $f(x)$, we of course write $d \mid f(x)$.

Theorem 4.2 (*Gauss' Lemma*) *Let D be a U.F.D. and x an indeterminate over D. Suppose $f(x),g(x) \in D[x]$ are such that $c(f(x)) = c(g(x)) = 1$. Then $c(f(x)g(x)) = 1$. In other words, the product of primitive polynomials is primitive.*

Proof: Let $f(x) = a_0 + a_1x + \cdots + a_nx^n$, $g(x) = b_0 + b_1x + \cdots$

$+ b_m x^m$, and assume that $c(f(x)g(x)) \neq 1$. Since D is a U.F.D., it follows that there exists a prime $p \in D$ such that $p \mid f(x)g(x)$. Since $f(x)$ and $g(x)$ are primitive, not all the coefficients of either $f(x)$ or $g(x)$ are divisible by p. Let r be the least integer for which $p \nmid a_r$, and similarly let s be the least integer for which $p \nmid b_s$. Then the coefficient of x^{r+s} in $f(x)g(x)$ is

$$\sum_{i=0}^{r+s} a_i b_{r+s-i} = \sum_{i=0}^{r-1} a_i b_{r+s-i} + \sum_{j=0}^{s-1} a_{r+s-j} b_j + a_r b_s. \tag{2}$$

By the choice of r and s, the first and second terms on the right in (2) are divisible by p, and since $p \mid f(x)g(x)$, we have that p divides the left side of (2) as well. Hence $p \mid a_r b_s$, implying that $p \mid a_r$ or $p \mid b_s$ since p is a prime; this is a contradiction. Thus $c(f(x)g(x)) = 1$. ∎

Theorem 4.3 *Let D be a U.F.D., and let F be the quotient field of D. Let x be an indeterminate over F, and suppose $0 \neq \varphi(x) \in F[x]$. Then $\varphi(x) = (a/b)\, g(x)$, where $a,b \in D$, $g(x) \in D[x]$, and $g(x)$ is primitive. Moreover, both $g(x)$ and a/b are uniquely determined by $\varphi(x)$ to within unit multiples in D.*

Proof: Write

$$\varphi(x) = \sum_{i=0}^{n} \frac{a_i}{b_i} x^i, \qquad a_i, b_i \in D,\ a_n \neq 0.$$

Let $b = b_0 \cdots b_n \in D$ and $c_i = a_i \prod_{j \neq i} b_j \in D$, $i = 0, \ldots, n$. Consider

$$f(x) = \sum_{i=0}^{n} c_i x^i \in D[x];$$

since $a_n \neq 0$, we have $c_n \neq 0$ so that $f(x) \neq 0$, and

$$\frac{1}{b} f(x) = \sum_{i=0}^{n} \frac{c_i}{b} x^i = \sum_{i=0}^{n} \frac{a_i}{b_i} x^i = \varphi(x).$$

If we write $c(f(x)) = a$, then $f(x) = ag(x)$, where $g(x) \in D[x]$ is primitive (see Exercise 1), and hence

$$\varphi(x) = \frac{a}{b} g(x).$$

To establish uniqueness, suppose $\varphi(x) = (e/d)h(x)$, where $e,d \in D$ and $h(x) \in D[x]$ is primitive. Then

$$adg(x) = beh(x).$$

Note that $ad \neq 0$ and $be \neq 0$ because $\varphi(x) \neq 0$. Since $g(x)$ and $h(x)$ are primitive, $c(adg(x)) = ad\varepsilon_1$ and $c(beh(x)) = be\varepsilon_2$ for some units $\varepsilon_1, \varepsilon_2 \in D$. It follows that $g(x) = \varepsilon h(x)$ and $a/b = \varepsilon'(e/d)$, where ε and ε' are units in D. ∎

With the same notation as in Theorem 4.3 we have the following:

Theorem 4.4 (i) *$\varphi(x)$ is irreducible in $F[x]$ iff $g(x)$ is irreducible in $D[x]$.*

(ii) *If a polynomial in $D[x]$ is of positive degree, then it has two factors of positive degree in $D[x]$ iff it has two factors of positive degree in $F[x]$.*

Proof: (i) Assume $\varphi(x)$ is not irreducible in $F[x]$. If deg $\varphi(x) = 0$, then deg $g(x) = 0$; since $g(x)$ is primitive, this implies that $g(x)$ is a unit in D, and hence $g(x)$ is not irreducible in $D[x]$. If deg $\varphi(x) > 0$, then $\varphi(x) = h(x)k(x)$ where $h(x),k(x) \in F[x]$ have positive degree. By Theorem 4.3 we may write $h(x) = (a_1/b_1)h_1(x)$ and $k(x) = (a_2/b_2)k_1(x)$, where $a_1,b_1,a_2,b_2 \in D$ and $h_1(x)$, $k_1(x)$ are primitive polynomials in $D[x]$. Then

$$\varphi(x) = \frac{a_1a_2}{b_1b_2}h_1(x)k_1(x)$$

and $h_1(x)k_1(x) \in D[x]$ is a primitive polynomial by Theorem 4.2. The uniqueness part of Theorem 4.3 implies that $g(x) = \varepsilon h_1(x)k_1(x)$ for some unit $\varepsilon \in D$; so again $g(x)$ is not irreducible in $D[x]$.

Conversely, assume $g(x)$ is not irreducible in $D[x]$. If deg $g(x) = 0$, then $0 \neq \varphi(x) \in F$ is a unit in $F[x]$ and hence is not irreducible in $F[x]$. If deg $g(x) > 0$, then since $g(x)$ is primitive, it follows that $g(x)$ is a product of two polynomials in $D[x]$ of positive degree. Again, this obviously implies that $\varphi(x)$ is not irreducible in $F[x]$.

(ii) Suppose $\varphi(x) \in D[x]$ is of positive degree. If $\varphi(x)$ has two factors of positive degree in $D[x]$, then obviously these factors also belong to $F[x]$. Conversely, assume $\varphi(x)$ has two factors of positive degree in $F[x]$, i.e., $\varphi(x)$ is not irreducible in $F[x]$. Write $\varphi(x) = ag(x)$, where $a \in D$ and $g(x) \in D[x]$ is primitive (see Exercise 1). Since deg $g(x) > 0$ and $g(x)$ is primitive, it follows from (i) that $g(x)$ has two factors of positive degree in $D[x]$. Hence the same is true of $\varphi(x)$. ∎

Corollary 1 *If D is a U.F.D. and x is an indeterminate over D, then the set of prime elements and the set of irreducible elements in $D[x]$ coincide.*

Proof: Recall that any prime in an integral domain is irreducible [see Example 5(c), Section 3.3]. On the other hand, suppose $\varphi(x) \in D[x]$ is irreducible in $D[x]$. If $\varphi(x)$ is not a primitive polynomial, then there exists a prime $p \in D$ which divides every coefficient of $\varphi(x)$. Since $\varphi(x)$ is irreducible in $D[x]$ and $p \mid \varphi(x)$, it follows that $\varphi(x)$ is an associate of p and hence is itself a prime in D. Now Exercise 2 implies that $\varphi(x)$ is a prime in $D[x]$.

So assume $\varphi(x)$ is a primitive irreducible polynomial in $D[x]$; then we must have deg $\varphi(x) > 0$. Suppose $\varphi(x) \mid \alpha(x)\beta(x)$ for some $0 \neq \alpha(x),\beta(x) \in D[x]$. Let F be the quotient field of D. Since $F[x]$ is a P.I.D. by Theorem 3.4, every irreducible element of $F[x]$ is a prime. We conclude from Theorem 4.4(ii) that $\varphi(x)$ is irreducible in $F[x]$ and hence is a prime in $F[x]$. Regarding the division $\varphi(x) \mid \alpha(x)\beta(x)$ as taking place in $F[x]$, it follows that $\varphi(x) \mid \alpha(x)$ or $\varphi(x) \mid \beta(x)$ in $F[x]$. Assume without loss of generality that

$$\alpha(x) = \gamma(x)\varphi(x) \qquad \text{for some } \gamma(x) \in F[x].$$

Write $\alpha(x) = d_1\alpha_1(x)$, $\gamma(x) = (d_2/d_3)\gamma_1(x)$, where $d_1,d_2,d_3 \in D$ and $\alpha_1(x)$, $\gamma_1(x)$ are primitive polynomials in $D[x]$. Then

$$d_1\alpha_1(x) = \frac{d_2}{d_3}\gamma_1(x)\varphi(x),$$

and $\gamma_1(x)\varphi(x)$ is a primitive polynomial in $D[x]$ by Theorem 4.2. The uniqueness part of Theorem 4.3 implies that $d_2/d_3 = \varepsilon d_1$ for some unit $\varepsilon \in D$. Hence

$$\gamma(x) = \frac{d_2}{d_3}\gamma_1(x) = \varepsilon d_1\gamma_1(x) \in D[x],$$

i.e., $\varphi(x)$ divides $\alpha(x)$ in $D[x]$. This shows that $\varphi(x)$ is a prime in $D[x]$. ∎

Theorem 4.4(i) implies the important fact that *an irreducible polynomial in $D[x]$ is either an irreducible element of D or a primitive polynomial of positive degree in $D[x]$ which is irreducible in $F[x]$*. Using Corollary 1, we obtain the equivalent statement that *a prime polynomial in $D[x]$ is either a prime in D or a primitive polynomial of positive degree in $D[x]$ which is prime in $F[x]$*.

The result of our work so far is the following important statement.

Theorem 4.5 *If D is a U.F.D. and x is an indeterminate over D, then $D[x]$ is a U.F.D..*

Proof: Let $0 \neq f(x) \in D[x]$ be a nonunit in $D[x]$. If $f(x) \in D$, then since D is a U.F.D., $f(x)$ can be written as a product of primes in D (and hence in $D[x]$). So assume $\deg f(x) \geq 1$ and write $f(x) = dg(x)$, where $d \in D$ and $g(x) \in D[x]$ is a primitive polynomial of positive degree. Let F be the quotient field of D. Since $F[x]$ is a U.F.D. by Theorems 3.4 and 3.6, regarding $g(x)$ as an element of $F[x]$ we can write

$$g(x) = \varphi_1(x) \cdots \varphi_r(x), \tag{3}$$

where $\varphi_i(x) \in F[x]$ is a prime polynomial of positive degree in $F[x]$, $i = 1, \ldots, r$. By Theorem 4.3 we have for each i that

$$\varphi(x) = \frac{a_i}{b_i} g_i(x) \tag{4}$$

where $a_i,b_i \in D$, $g_i(x) \in D[x]$, and $g_i(x)$ is primitive. For each i, $g_i(x)$ is prime in $F[x]$ since $\varphi_i(x)$ is, and hence $g_i(x)$ is a prime polynomial in $D[x]$ by the remark following Corollary 1. Combining (3) and (4), we have

$$g(x) = \frac{a_1 \cdots a_r}{b_1 \cdots b_r} g_1(x) \cdots g_r(x), \tag{5}$$

and $g_1(x) \cdots g_r(x)$ is a primitive polynomial in $D[x]$ by Theorem 4.2.

Since $g(x) \in D[x]$ is primitive, the uniqueness part of Theorem 4.3 implies that

$$\frac{a_1 \cdots a_r}{b_1 \cdots b_r} = \varepsilon \tag{6}$$

for some unit $\varepsilon \in D$. Using (5) and (6), we obtain

$$\begin{aligned} f(x) = dg(x) &= d\frac{a_1 \cdots a_r}{b_1 \cdots b_r} g_1(x) \cdots g_r(x) \\ &= d\varepsilon g_1(x) \cdots g_r(x). \end{aligned} \tag{7}$$

Now D is a U.F.D. so there exist a unit $\delta \in D$ and primes $d_1, \ldots, d_s \in D$ such that $d\varepsilon = \delta d_1 \cdots d_s$ (if d is a unit in D no primes are required). Then from (7),

$$f(x) = \delta d_1 \cdots d_s g_1(x) \cdots g_r(x) \tag{8}$$

is a factorization of $f(x)$ into a product of a unit $\delta \in D$, primes $d_1, \ldots, d_s$ in D (and hence in $D[x]$), and prime polynomials of positive degree $g_1(x), \ldots, g_r(x)$ in $D[x]$.

Suppose that

$$f(x) = \mu e_1 \cdots e_n h_1(x) \cdots h_m(x) \tag{9}$$

is another such factorization. The $g_i(x)$ ($i = 1, \ldots, r$) and $h_j(x)$ ($j = 1, \ldots, m$) are prime polynomials of positive degree in $D[x]$, and hence are primitive polynomials in $D[x]$ which are prime in $F[x]$ (see the remark following Corollary 1). Since $F[x]$ is a U.F.D., $\delta d_1 \cdots d_s g_1(x) \cdots g_r(x) = \mu e_1 \cdots e_n h_1(x)$, and $\delta d_1 \cdots d_s$ and $\mu e_1 \cdots e_n$ are units in $F[x]$, it follows that $r = m$ and after a suitable reordering, $h_i(x) = (k_i/l_i)g_i(x)$ for some $k_i, l_i \in D$, $i = 1, \ldots, r$. Hence there exists a unit $\mu' \in D$ such that $\delta d_1 \cdots d_s = \mu' e_1 \cdots e_n$, and the fact that D is a U.F.D. completes the proof. ∎

Thus we see that *if D is a* U.F.D., *then any polynomial $f(x) \in D[x]$ of positive degree has a factorization of the form*

$$f(x) = \delta d_1 \cdots d_s g_1(x) \cdots g_r(x),$$

where δ is a unit in D, $d_1, \ldots, d_s$ are primes in D, and $g_1(x), \ldots, g_r(x)$ are prime polynomials of positive degree in $D[x]$ (which are thereby primitive in $D[x]$ and prime in $F[x]$).

Corollary 2 *If D is a U.F.D. and $x_1, \ldots, x_n$ are independent indeterminates over D, then $D[x_1, \ldots, x_n]$ is a U.F.D..*

Proof: Since $D[x_1, \ldots, x_n] = D[x_1, \ldots, x_{n-1}][x_n]$, this result follows from Theorem 4.5 by induction. ∎

Example 1 (a) The polynomial $x^2 + x + 1 \in Z[x]$ is a primitive polynomial of positive degree in $Z[x]$ which is easily seen to be irreducible in $Q[x]$; thus $x^2 + x + 1$ is irreducible in $Z[x]$ by the remark following Corollary 1. On the other hand, $2x^2 + 2x + 2 = 2(x^2 + x + 1)$ is not an irreducible polynomial in $Z[x]$.

(b) Let D be an integral domain and x an indeterminate over D. Let r be a nonnegative integer, and let $c_k, d_k \in D$, $k = 0, \ldots, r$, where the d_k are $r + 1$ distinct elements of D. We assert that there is at most one polynomial $f(x) \in D[x]$ of degree at most r such that $f(d_k) = c_k$, $k = 0, \ldots, r$. First of all, if $r = 0$ we simply set $f(x) = c_0$ (uniqueness is obvious). Assume that $r \geq 1$, let F be the quotient field of D, and consider

$$f(x) = \sum_{k=0}^{r} c_k \frac{\Pi_{j \neq k}(x - d_j)}{\Pi_{j \neq k}(d_k - d_j)} \in F[x]. \tag{10}$$

Since $f(x)$ is clearly a polynomial of degree at most r satisfying $f(d_k) = c_k$, $k = 0, \ldots, r$, if $f(x) \in D[x]$ it is a polynomial of the desired sort, and in this case the uniqueness of $f(x)$ follows from Corollary 3, Section 3.3.

A somewhat different argument can be used to show that $f(x) \in D[x]$ is unique if it exists. Suppose

$$f(x) = \sum_{k=0}^{r} \alpha_k x^k, \qquad g(x) = \sum_{k=0}^{r} \beta_k x^k \in D[x]$$

are polynomials of degree at most r such that

$$f(d_j) = c_j = g(d_j), \qquad j = 0, \ldots, r.$$

Then we obtain

$$(\alpha_0 - \beta_0) + (\alpha_1 - \beta_1)d_j + \cdots + (\alpha_r - \beta_r)d_j^r = 0, \qquad j = 0, \ldots, r. \tag{11}$$

Regard (11) as a system of $r + 1$ equations in the $r + 1$ "unknowns" $\alpha_k - \beta_k$ with coefficient matrix

$$A = \begin{bmatrix} 1 & d_1 & d_1^2 & \cdots & d_1^r \\ 1 & d_2 & d_2^2 & \cdots & d_2^r \\ \cdots & \cdots & \cdots & \cdots & \cdots \\ 1 & d_r & d_r^2 & \cdots & d_r^r \end{bmatrix}. \tag{12}$$

The distinctness of the d_k implies that $\det A \neq 0$ (see Exercise 3). Thus by an elementary result in matrix algebra the system (11) has only the solution $\alpha_k - \beta_k = 0$, $k = 0, \ldots, r$.

(c) In this example we exhibit an explicit constructive procedure for finding the factors of a polynomial in $D[x]$, where D is a U.F.D. of characteristic 0 with a finite number of units. Let F be the quotient field of D. Suppose $f(x)$ and $g(x)$ are in $D[x]$ and $\deg g(x) = r < \deg f(x) = n$. Let $\xi_0, \ldots, \xi_r$ be $r + 1$ distinct elements of D such that $\eta_0 = f(\xi_0), \ldots, \eta_r = f(\xi_r)$ are all different from 0. Let $\delta_{k1}, \delta_{k2}, \ldots, \delta_{km_k}$ (not Kronecker deltas!) be the complete set of all divisors of η_k, $k = 0, \ldots, r$. (There are only finitely many of these in a U.F.D. with finitely many units; see Exercise 4.) We construct a table:

ξ	η	δ
ξ_0	η_0	$\delta_{01}\delta_{02} \cdots \delta_{0m_0}$
$\vdots$	$\vdots$	$\vdots$
ξ_r	η_r	$\delta_{r1}\delta_{r2} \cdots \delta_{rm_r}$

(13)

Clearly, if $g(x) \mid f(x)$ in $D[x]$, then $g(\xi_k) \mid f(\xi_k)$, $k = 0, \ldots, r$. Thus a necessary condition for $g(x) \mid f(x)$ is that $g(\xi_k)$ must be one of $\delta_{k0}, \ldots, \delta_{km_k}$, say

$$g(\xi_k) = \delta_{kt_k}, \qquad k = 0, \ldots, r. \tag{14}$$

We construct a unique polynomial $g(x) \in F[x]$ of degree r which satisfies (14). If $g(x) \notin D[x]$, we reject it. If $g(x) \in D[x]$, we test by dividing to see if $g(x)$ is actually a divisor of $f(x)$ in $D[x]$. Thus the procedure is to take all possible choices $\delta_{0t_0}, \ldots, \delta_{rt_r}$ of δ's, one out of each row of table (13), construct $g(x)$ satisfying (14), see if it is in $D[x]$, and finally test whether or not it divides $f(x)$. Surely any divisor $g(x) \in D[x]$ of $f(x)$ of degree r must appear in this process. For if $g(x) \mid f(x)$, then $g(\xi_k) = \delta_{kt_k}$, $k = 0, \ldots, r$, for some choice of $t_0, \ldots, t_r$ and $g(x)$ is uniquely determined by these $r + 1$ equalities. Of course, what one wants to do here is make as many of the η_k as possible into primes so that the number of δ's is diminished.

(d) We try to find a quadratic factor of $f(x) = x^5 + x - 1$ in $Z[x]$ by the preceding method. Here $r = 2$ and we can take $\xi_0 = -1$, $\xi_1 = 0$, $\xi_2 = 1$ so that $\eta_0 = f(\xi_0) = -3$, $\eta_1 = f(0) = -1$, $\eta_2 = 1$. Table (13) becomes

ξ	η	δ			
-1	-3	3	-3	1	-1
0	-1	-1	1		
1	1	-1	1		

For example, take 3, 1, 1 as the choice of δ's and construct an interpolating polynomial $g(x)$ using formula (10):

$$\begin{aligned} g(x) &= 3\frac{x(x-1)}{(-1)(-1-1)} + 1\frac{(x+1)(x-1)}{(1)(-1)} + 1\frac{(x+1)x}{(1)[1-(-1)]} \\ &= x^2 - x + 1. \end{aligned}$$

If we divide $g(x)$ into $f(x)$, we see that

$$x^5 + x - 1 = (x^2 - x + 1)(x^3 + x^2 - 1).$$

Although the method of Example 1(c) is finite, it can obviously become extraordinarily tedious even for polynomials of small degree. There is a very useful result for testing irreducibility in $D[x]$.

Theorem 4.6 (*Eisenstein Criterion*) *Let D be a U.F.D., let $p \in D$ be a* *prime, and let* $f(x) = \sum_{k=0}^{n} a_k x^k \in D[x]$, $a_n \neq 0$. *Assume that*

(i) $a_n \notin (p)$.
(ii) $a_0 \notin (p^2)$.
(iii) $a_k \in (p)$, $0 \leq k < n$.

Then $f(x)$ *has no divisors of positive degree in* $D[x]$, *and hence is irreducible in* $F[x]$, *where* F *is the quotient field of* D.

Proof: Suppose

$$f(x) = (b_r x^r + \cdots + b_0)(c_s x^s + \cdots + c_0),$$

$b_r c_s \neq 0$, $1 \leq r < n$. Now $a_0 = b_0 c_0$, and since a_0 is not divisible by p^2 [by (ii)], it follows that not both b_0 and c_0 are divisible by p, say $b_0 \notin (p)$. But $p | a_0$ by (iii), so $p | b_0 c_0$. Thus $p | b_0$ or $p | c_0$, i.e., $p | c_0$. Similarly, we conclude from (i) and $a_n = b_r c_s$ that p is not a divisor of either b_r or c_s. Thus p divides c_0, but it does not divide c_s. We pick the least integer $m \geq 1$ for which c_m is not divisible by p. Then

$$a_m = b_0 c_m + b_1 c_{m-1} + \cdots + b_m c_0.$$

Now the minimality of m implies that

$$p | c_k, \qquad 0 \leq k \leq m - 1. \tag{15}$$

Thus $a_m - b_0 c_m \in (p)$. But $b_0 \notin (p)$ and $c_m \notin (p)$, and since p is prime it follows that $b_0 c_m \notin (p)$. Thus $a_m \notin (p)$ [otherwise $b_0 c_m \in (p)$]. But $m \leq s < n$ in contradiction to (iii). ∎

Example 2 (a) The polynomial $f(x) = x^5 - 6x + 3$ is irreducible in $Z[x]$. To begin with, $f(x)$ is primitive, so if it were reducible, it would have factors of positive degree. But take $p = 3$ in Theorem 4.6. Then conditions (i) to (iii) are obviously satisfied, so $f(x)$ has no factors of positive degree in $Z[x]$ or $Q[x]$.

(b) $F(x_1,x_2) = x_1^5 + x_1^3(x_2 + 1) + x_1x_2 + x_2^2$ is irreducible in $Z[x_1,x_2]$. First, write $F(x_1,x_2)$ as a polynomial in x_2 with coefficients in the U.F.D. $Z[x_1]$:

$$x_2^2 + (x_1^2 + 1)x_1x_2 + (x_1^2 + 1)x_1^3.$$

Now let $p = x_1^2 + 1$, a prime in $Z[x_1]$. Then conditions (i) to (iii) are clearly satisfied, so $F(x_1,x_2)$ cannot be factored in $Z[x_1][x_2]$ into a product of polynomials of positive degree in x_2. Thus, if $F(x_1,x_2) = g(x_1,x_2)h(x_1,x_2)$ is a factorization in $Z[x_1,x_2]$ into nonconstant polynomials, then $\deg_{x_2} g(x_1,x_2) = 0$ or $\deg_{x_2} h(x_1,x_2) = 0$. Assume that $\deg_{x_2} g(x_1,x_2) = 0$ so that $g(x_1,x_2) = g(x_1) \in Z[x_1]$. Now $\deg_{x_1} g(x_1) \neq 0$; otherwise $g(x_1,x_2) = g(x_1)$ is a constant. Thus there exists $n_0 \in Z$ such that $g(n_0)$ is distinct from ± 1 or ± 2 (why?). Since $g(x_1) | F(x_1,x_2)$, we have $g(n_0) | F(n_0,m)$ for any specialization of x_2 to an integer m. But

$$\begin{aligned} F(n_0,m) &= n_0^5 + n_0^3(m + 1) + n_0 m + m^2 \\ &= m^2 + Am + B \end{aligned}$$

where A and B are in Z. We assert that $m^2 + Am + B$ is not always divisible by $g(n_0)$. For if so, then

$$g(n_0)\,|\,[F(n_0,m+1) - F(n_0,m)], \tag{16}$$

where

$$F(n_0,m+1) - F(n_0,m) = (m+1)^2 - m^2 + A$$
$$= 2m + 1 + A.$$

If A is odd, then choose m so that $2m + 1 + A = 2$. If A is even, choose m so that $2m + 1 + A = 1$. In either case it is not true that $g(n_0)$ satisfies (16). Thus $F(x_1,x_2)$ has no factors of positive degree in $Z[x_1,x_2]$. On the other hand, it is obvious that $F(x_1,x_2)$ has no nonunit integral factor either. Thus $F(x_1,x_2)$ is irreducible in $Z[x_1,x_2]$.

In Example 9(c), Section 3.2, we constructed a field containing the roots of the polynomial $p(x) = x^2 + 1 \in \mathbf{R}[x]$. We can extend this construction now to arbitrary polynomials over a field. We first make the following definitions for fields: If K and L are fields and K is a subfield (i.e., subring) of L, we say that L is an *extension* of K. The ordinary set inclusion symbol $K \subset L$ for fields K and L will denote that L is an extension of K. If $K \subset L$ and S is any subset of L, then $K(S)$ denotes the intersection of all subfields of L containing $K \cup S$; this is obviously a field (see Exercise 5). If $S = \{u_1, \ldots, u_n\}$, then $K(S)$ is written $K(u_1, \ldots, u_n)$. If $n = 1$, $K(u_1)$ is called a *simple field extension.* Let x be an indeterminate over K and $f(x)$ a polynomial in $K[x]$. Let L be an extension field of K, and assume that x is an indeterminate over L as well. If $f(x)$ factors into linear factors (i.e., first-degree factors) or *splits in* $L[x]$, and if moreover L is the smallest such field in the sense that there is no field F satisfying $K \subset F \subset L$, with the second inclusion strict, over which $f(x)$ splits, then L is called a *splitting field for* $f(x)$.

Theorem 4.7 *Let K be a field, x an indeterminate over K, and $p(x) \in K[x]$ a prime polynomial. Then there exists an extension field L of K such that:*

(i) *L contains a root θ of $p(x)$ and $L = K(\theta)$.*

(ii) *Every nonzero element of L has a unique representation of the form*

$$f(\theta) = a_0 + a_1\theta + a_2\theta^2 + \cdots + a_p\theta^p \tag{17}$$

where $0 \le p < n = \deg p(x)$ and $a_p \ne 0$.

(iii) *If $f(\theta) = 0$ in (17), then $a_0 = a_1 = \cdots = a_p = 0$.*

(iv) *x is an indeterminate over L.*

Proof: The ideal $(p(x))$ in $K[x]$ is maximal, for $K[x]$ is a P.I.D. (Theorem 3.4, Section 3.3) and $p(x)$ is prime and hence irreducible [Theorem 3.5(iv), Section 3.3] so that Theorem 3.5(iii), Section 3.3, applies. Thus from Theorem 2.11(ii), Section 3.2, $K[x]/(p(x))$ is a field. Let

$$\nu\colon K[x] \to K[x]/(p(x))$$

be the canonical epimorphism and set $\mu = \nu\,|\,K$. Observe that $\deg p(x) \ge 1$ [i.e., $p(x)$ is not a unit because it is prime] so that $\mu(k_1) = \mu(k_2)$ implies

$k_1 - k_2 \in (p(x))$ and hence $k_1 = k_2$. Hence $\mu: K \to \mu(K)$ is a monomorphism, and of course $\mu(K)$ is a subfield of $K[x]/(p(x))$. Now by Theorem 2.3, Section 3.2, secure a ring $\mathscr{L}$ such that K is a subring of $\mathscr{L}$ and an isomorphism $\tau: \mathscr{L} \to K[x]/(p(x))$ for which $\tau | K = \mu$. We know that $\mathscr{L}$ is an extension of K. Now set $\alpha = \tau^{-1}(\nu(x)) \in \mathscr{L}$ and suppose $p(x) = \sum_{i=0}^{n} k_i x^i$, $k_i \in K$, $i = 0, \ldots, n$. Then

$$\begin{aligned}\tau(p(\alpha)) &= \tau\left(\sum_{i=0}^{n} k_i \alpha^i\right)\\ &= \sum_{i=0}^{n} \tau(k_i)\tau(\alpha)^i = \sum_{i=0}^{n} \mu(k_i)\nu(x)^i\\ &= \sum_{i=0}^{n} \nu(k_i)\nu(x^i) = \nu\left(\sum_{i=0}^{n} k_i x^i\right)\\ &= \nu(p(x)).\end{aligned}$$

Thus $\tau(p(\alpha))$ is the zero in $K[x]/(p(x))$, and since τ is an isomorphism, it follows that $p(\alpha) = 0$. Hence $\alpha \in \mathscr{L}$ is a root of the polynomial $p(x)$. If $f(x) \in K[x]$, divide $f(x)$ by $p(x)$ to obtain a remainder $r(x) = \sum_{i=0}^{\rho} a_i x^i$, $a_i \in K$, $i = 0, \ldots, \rho$, $0 \le \rho \le n - 1$:

$$f(x) = q(x)p(x) + r(x).$$

Then

$$f(\alpha) = q(\alpha)p(\alpha) + r(\alpha) = r(\alpha) = \sum_{i=0}^{\rho} a_i \alpha^i. \tag{18}$$

Now suppose $f(x) = \sum_{i=0}^{m} b_i x^i$. Then

$$\begin{aligned}\tau^{-1}(\nu(f(x))) &= \tau^{-1}\left(\sum_{i=0}^{m} \mu(b_i)\nu(x)^i\right)\\ &= \sum_{i=0}^{m} b_i \tau^{-1}(\nu(x))^i = \sum_{i=0}^{m} b_i \alpha^i\\ &= f(\alpha).\end{aligned} \tag{19}$$

But τ^{-1} is surjective so that everything in $\mathscr{L}$ is of the form $\tau^{-1}(\nu(f(x)))$, and hence from (18) and (19) everything in $\mathscr{L}$ is of the form (17). To confirm the uniqueness of the representation (17), suppose $r(x)$ and $s(x)$ are polynomials in $K[x]$, neither of which has degree exceeding $n - 1$ and which satisfy $r(\alpha) = s(\alpha)$. Thus we have a polynomial $g(x) = r(x) - s(x)$ which satisfies $g(\alpha) = 0$, and if $g(x) \neq 0$, then $\deg g(x) < \deg p(x)$; hence $g(x) \notin (p(x))$. By the maximality of $(p(x))$, $(p(x), g(x)) = K[x]$, and hence there exist polynomials $u(x)$ and $v(x)$ in $K[x]$ such that

$$u(x)p(x) + v(x)g(x) = 1.$$

If we specialize x to α, we obtain the contradiction $0 = 1$. Thus $g(x) = 0$, and the uniqueness of the representation (17) follows. We have proved that

$\mathscr{L}$ is a field consisting of polynomials in α, and it is obvious that $\mathscr{L} = K(\alpha)$, i.e., it is the smallest field containing α and K. It is also a consequence of the uniqueness that if $a_0 + a_1\alpha + \cdots + a_p\alpha^p = 0$, $0 < p \leq n$, then $a_0 = \cdots = a_p = 0$.

We have proved the first three parts of the theorem for the field $\mathscr{L}$ and the root $\alpha \in \mathscr{L}$. The remaining part (iv) is something of a technicality that requires replacing $\mathscr{L}$ by an isomorphic copy as follows. Let y be an indeterminate over $\mathscr{L}$, and consider the polynomial ring $\mathscr{L}[y]$. Since $K \subset \mathscr{L}$, it is clear that y is an indeterminate over K as well and $K[y]$ is a subring of $\mathscr{L}[y]$. Moreover it is obvious that the mapping

$$\psi\colon K[y] \to K[x]$$

defined by $\psi(\sum_{i=0}^m b_i y^i) = \sum_{i=0}^m b_i x^i$ is an isomorphism. [We write $\psi(f(y)) = f(x)$.] Thus by Theorem 2.3, Section 3.2, there exists a ring $\mathscr{R}$ and an isomorphism $\varphi\colon \mathscr{L}[y] \to \mathscr{R}$ such that $K[x] \subset \mathscr{R}$ and $\varphi | K[y] = \psi$.

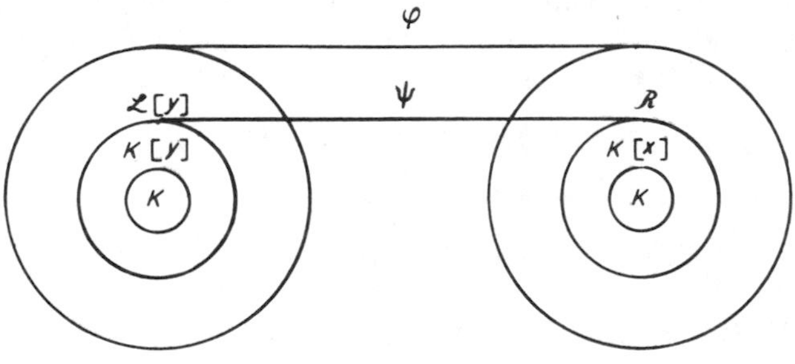

Now let $\varphi(\mathscr{L}) = L$, and note that since $\psi | K = \iota_K$ it follows that L is an extension field of K. Moreover, for $\theta = \varphi(\alpha) \in L$ we have

$$\begin{aligned}
p(\theta) &= \sum_{i=0}^n k_i\theta^i = \sum_{i=0}^n k_i\varphi(\alpha)^i \\
&= \sum_{i=0}^n \varphi(k_i)\varphi(\alpha)^i \qquad \text{(remember that } \varphi|K[y] = \psi \text{ and } \psi|K = \iota_K) \\
&= \varphi\left(\sum_{i=0}^n k_i\alpha^i\right) = \varphi(p(\alpha)) \\
&= \varphi(0) = 0.
\end{aligned}$$

The rest of the conclusions (i) to (iii) concerning L and θ are also obvious consequences of the same statements for $\mathscr{L}$ and α. Now y is an indeterminate over $\mathscr{L}$, so $\varphi(y) = \psi(y) = x$ must be an indeterminate over $\varphi(\mathscr{L}) = L$. Finally, anything in $\mathscr{R}$ is of the form $\varphi(h(y))$, $h(y) \in \mathscr{L}[y]$. But then since $\varphi(y) = x$, anything in $\mathscr{R}$ is of the form $h(x)$ where $h(x)$ has coefficients in L, i.e., $\mathscr{R} = L[x]$. ▌

In the notation of Theorem 4.7 we have the following:

Corollary 3 *If* $f(x) \in K[x]$ *and* $f(\theta) \neq 0$, *then there exists a polynomial* $u(x) \in K[x]$ *such that* $\deg u(x) < n = \deg p(x)$ and $u(\theta) = 1/f(\theta) \in K(\theta)$.

Proof: This follows immediately from Theorem 4.7 because $L = K(\theta)$ is a field. ▮

Example 3 (a) If a method for computing the g.c.d. of two polynomials in $K[x]$ is available, then the polynomial $u(x)$ in Corollary 3 can be computed. For if $f(\theta) \neq 0$, then $f(x) \notin (p(x))$ and hence $(f(x),p(x)) = K[x]$ [i.e., $(p(x))$ is maximal]. But then g.c.d.$(f(x),p(x)) = 1$, and we can find $u(x)$ and $g(x)$ in $K[x]$ such that $u(x)f(x) + g(x)p(x) = 1$. Specializing x to θ in L, we have

$$u(\theta)f(\theta) + g(\theta)p(\theta) = 1.$$

But $p(\theta) = 0$, so $u(\theta)f(\theta) = 1$.

(b) We express $1/[(1+i)^3 + 3(1+i) + 2]$ as a polynomial in $1 + i$ with rational number coefficients. Let $\theta = 1 + i$. Then $(\theta - 1)^2 = i^2 = -1$, and hence $\theta^2 - 2\theta + 2 = 0$. Let $p(x) = x^2 - 2x + 2 \in Q[x]$. Then $p(x)$ is irreducible (use Theorem 4.6) in $Q[x]$. Now we want to compute g.c.d.$(f(x),p(x))$ where $f(x) = x^3 + 3x + 2$. Divide $f(x)$ by $p(x)$ to obtain

$$f(x) = p(x)(x + 2) + 5x - 2,$$

and then divide $p(x)$ by $5x - 2$ to obtain

$$p(x) = (5x - 2)\left(\frac{x}{5} - \frac{8}{25}\right) + \frac{34}{25}.$$

Combining these equations, we obtain

$$p(x) = [f(x) - p(x)(x + 2)]\left(\frac{x}{5} - \frac{8}{25}\right) + \frac{34}{25}$$

or

$$\frac{34}{25} = p(x)\left[1 + (x + 2)\left(\frac{x}{5} - \frac{8}{25}\right)\right] - f(x)\left(\frac{x}{5} - \frac{8}{25}\right).$$

Evaluating at θ, we have

$$\frac{34}{25} = -f(\theta)\left(\frac{\theta}{5} - \frac{8}{25}\right)$$

or

$$\frac{1}{f(\theta)} = \frac{4}{17} - \frac{5}{34}\theta.$$

The field $L = K(\theta)$ of Theorem 4.7 is called a *simple algebraic extension of* K. The element $\theta \in L$ is said to be *algebraic over* K because it satisfies the equation $p(\theta) = 0$ in L: An element of an extension field L of a field K is *algebraic over* K if it is the root of a polynomial in $K[x]$. If every element of L is algebraic over K, then L is called an *algebraic extension* of K. The degree of the polynomial $p(x)$ is called the *degree of the extension* $L = K(\theta)$ *over* K; it is denoted by

$$[L:K].$$

More generally, if L is an extension of K and there exist elements $u_1, \ldots, u_n$ in L such that every element $u \in L$ has a representation of the form

$$u = \sum_{i=1}^{n} k_i u_i, \qquad k_i \in K, \tag{20}$$

then L is called a *finite extension* of K and $\{u_1, \ldots, u_n\}$ is called a *spanning set for L over K*. If no finite spanning set exists, then L is called an *infinite extension of K*. If the coefficients $k_1, \ldots, k_n$ in (20) are uniquely determined by u for each $u \in L$, we call $\{u_1, \ldots, u_n\}$ a *basis of L over K*, and the integer n, also denoted by $[L:K] = n$, is called the *degree of L over K*. It is quite easy to prove (see Exercise 14) that *if* $\{u_1, \ldots, u_n\}$ *is a spanning set for L over K, then some subset of* $\{u_1, \ldots, u_n\}$ *is a basis of L over K*. Thus in Theorem 4.7, $K(\theta)$ is a finite extension of K with $\{1, \theta, \ldots, \theta^{n-1}\}$ as a basis of $K(\theta)$ over K. We shall see shortly that any finite extension L of a field K is an algebraic extension, and that any two bases of L over K must have the same number of elements in them.

Theorem 4.8 *Let $p(x) \in K[x]$ be irreducible, and let θ be a root of $p(x)$ in some extension field F of K (e.g., the field $F = L$ provided by Theorem 4.7). Let ν: $K[x] \to K[x]/(p(x))$ be the canonical epimorphism. Then the mapping*

$$\sigma\colon \frac{K[x]}{(p(x))} \to K(\theta)$$

defined by

$$\sigma(\nu(f(x))) = f(\theta), \qquad f(x) \in K[x], \tag{21}$$

is an isomorphism.

Proof: First define μ: $K[x] \to K[\theta]$ by $\mu(f(x)) = f(\theta)$, an obvious homomorphism. If $f(x) \in \ker \mu$, then $f(\theta) = 0$. Now unless $f(x) \in (p(x))$, we know from the maximality of $(p(x))$ that g.c.d.$(f(x),p(x)) = 1$. Hence there exist $u(x)$ and $v(x)$ such that $u(x)p(x) + f(x)v(x) = 1$, which is clearly a contradiction since $0 = p(\theta) = f(\theta)$. Thus $p(x) \mid f(x)$, and it follows that $\ker \mu = (p(x))$. But then Theorem 1.4, Section 3.1, tells us that there exists a unique isomorphism σ: $K[x]/(p(x)) \to K[\theta]$ such that $\sigma\nu = \mu$. Finally, $K[x]/(p(x))$ is a field so that $K[\theta]$ is a field, and hence $K[\theta] = K(\theta)$ (see Exercise 7). ∎

Corollary 4 *If $p(x) \in K[x]$ is irreducible and θ_1 and θ_2 are roots of $p(x)$ in extension fields F_1 and F_2 of K, respectively, then there exists precisely one isomorphism ω: $K(\theta_1) \to K(\theta_2)$ which satisfies $\omega(\theta_1) = \theta_2$ and $\omega \mid K = \iota_K$.*

Proof: Let σ_i: $K[x]/(p(x)) \to K(\theta_i)$, $i = 1,2$, be the two isomorphisms (21). Set $\omega = \sigma_2\sigma_1^{-1}$ so that ω: $K(\theta_1) \to K(\theta_2)$ is an isomorphism. From (21), for $f(x) \in K[x]$ we have

$$\sigma_2\sigma_1^{-1}(f(\theta_1)) = \sigma_2(\nu(f(x))) = f(\theta_2),$$

and hence $\omega(\theta_1) = \theta_2$, $\omega|K = \iota_K$. Since everything in $K(\theta_1)$ has the form $f(\theta_1)$, we conclude that ω is uniquely determined by $\omega(\theta_1) = \theta_2$ and $\omega|K = \iota_K$. ∎

Two extensions L_1 and L_2 of K are *equivalent extensions of* K if there exists an isomorphism $\omega: L_1 \to L_2$ such that $\omega|K = \iota_K$. Thus $K(\theta_1)$ and $K(\theta_2)$ in Corollary 4 are equivalent extensions of K.

In order to investigate finite field extensions, we need the following elementary fact.

Theorem 4.9 *Let* L *be a finite extension of* K *and suppose* $\{u_1, \ldots, u_n\}$ *is a basis of* L *over* K. *If* $v_1, \ldots, v_m$ *are any* m *elements of* L, $m > n$, *then there exist* $k_j \in K$, $j = 1, \ldots, m$, *not all* 0, *such that*

$$\sum_{j=1}^{m} k_j v_j = 0. \tag{22}$$

Proof: Write

$$v_j = \sum_{i=1}^{n} c_{ij} u_i, \qquad j = 1, \ldots, m.$$

To satisfy (22), we must have

$$0 = \sum_{j=1}^{m} k_j v_j = \sum_{i=1}^{n} \left(\sum_{j=1}^{m} c_{ij} k_j \right) u_i$$

and thus we want $k_1, \ldots, k_m$ to satisfy

$$\sum_{j=1}^{m} c_{ij} k_j = 0, \qquad i = 1, \ldots, n. \tag{23}$$

We prove the existence of $k_1, \ldots, k_m$ satisfying (23) by induction on n. We can begin by assuming that no $v_j = 0$; otherwise if $v_{j_0} = 0$, simply take $k_{j_0} = 1$ and $k_j = 0$ for $j \neq j_0$ in (22). If $n = 1$, then $v_j = c_{1j}u_1$, $j = 1, \ldots, m$, and no v_j is 0. Take $k_1 = c_{12}$, $k_2 = -c_{11}$, $k_3 = \cdots = k_m = 0$, so that

$$\sum_{j=1}^{m} k_j v_j = k_1 v_1 + k_2 v_2 = c_{12}c_{11}u_1 - c_{11}c_{12}u_1$$
$$= 0.$$

Suppose then that $n > 1$. Since not all $c_{ij} = 0$ we can, without loss of generality, assume that $c_{1m} \neq 0$. (Otherwise simply renumber both the u_i's and v_j's.) Then a nonzero m-tuple $(k_1, \ldots, k_m)$ satisfying (23) exists iff such an m-tuple exists for the system of equations obtained from (23) by performing the following operations: From equation i substract $c_{im}c_{1m}^{-1}$ times equation 1, $i = 2, \ldots, n$. By induction the last $n - 1$ equations of the resulting system then have a solution $k_1^0, \ldots, k_{m-1}^0$ not all 0. Put these in the first equation

$$c_{11}k_1{}^0 + \cdots + c_{1\ m-1}k_{m-1}^0 + c_{1m}k_m = 0; \tag{24}$$

since $c_{1m} \neq 0$, $k_m = k_m^0$ can be determined from (24). ▮

Theorem 4.10 *Let L be a finite extension of K. Then*

(i) *Any two bases of L over K contain the same number of elements, say n.*

(ii) *Any element $\theta \in L$ is algebraic over K; in fact, θ satisfies an irreducible polynomial $p(x) \in K[x]$ with* $\deg p(x) \leq n$.

Proof: (i) Suppose $\{u_1, \ldots, u_n\}$ and $\{v_1, \ldots, v_m\}$ are two bases of L over K and $m > n$. Then by Theorem 4.9 there exist $k_1, \ldots, k_m \in K$ not all 0 such that $\sum_{j=1}^{m} k_j v_j = 0$. But this contradicts the uniqueness of the representation of 0 in terms of $v_1, \ldots, v_m$.

(ii) Consider the $n + 1$ elements $1, \theta, \ldots, \theta^n$ in L. Then by Theorem 4.9 again there exist $k_0, \ldots, k_n$ in K not all 0 such that

$$k_0 + k_1\theta + \cdots + k_n\theta^n = 0.$$

Thus θ is a root of a polynomial $f(x) = k_0 + \cdots + k_n x^n \in K[x]$, of degree at most n. Now $K[x]$ is a U.F.D., so we can factor $f(x)$ into a product of prime polynomials in $K[x]$. Then obviously $p(\theta) = 0$ for one of these primes $p(x)$. ▮

Corollary 5 *If L is a finite extension of K and $\theta \in L$, then there is precisely one monic polynomial $p(x) \in K[x]$ of least degree such that $p(\theta) = 0$, and $p(x)$ is irreducible in $K[x]$. Moreover, any polynomial $f(x) \in K[x]$ for which $f(\theta) = 0$ is divisible by $p(x)$ in $K[x]$.*

Proof: According to Theorem 4.10(ii), there are polynomials in $K[x]$ having θ as a root. Let $A \lhd K[x]$ be the ideal consisting of all such polynomials. Since $K[x]$ is a P.I.D., let $A = (p(x))$. Clearly if $f(\theta) = 0$, then $f(x) \in A = (p(x))$ and $p(x) | f(x)$ in $K[x]$. Thus $p(x)$ has least degree among all polynomials $f(x) \in K[x]$ for which $f(\theta) = 0$. The generator of A is unique to within unit multiples, i.e., multiples in K, so that if $p(x)$ is chosen to be monic it is uniquely determined. It is obviously irreducible, since otherwise θ would satisfy a polynomial of lower degree in $K[x]$. ▮

The unique monic polynomial $p(x) \in K[x]$ of least degree that an element θ in a finite extension L of K satisfies (which is irreducible according to Corollary 5) is called the *characteristic polynomial* of θ. Observe that if $f(x)$ is an irreducible polynomial in $K[x]$ for which $f(\theta) = 0$, then $f(x) = kp(x)$, $k \in K$.

Example 4 Let $L = Q(\sqrt{2}, \sqrt{3})$.

(a) We have $L = Q(\sqrt{2})(\sqrt{3})$. To see this, we note that anything in $Q(\sqrt{2}, \sqrt{3})$ is a quotient of polynomials in $\sqrt{2}$ and $\sqrt{3}$ with rational coefficients because the totality of such quotients is a field containing Q, $\sqrt{2}$, and $\sqrt{3}$, and any such field must contain these quotients. But any polynomial in $\sqrt{2}$ and $\sqrt{3}$ with coefficients in Q can obviously be expressed as a polynomial in $\sqrt{3}$ with coefficients in $Q(\sqrt{2})$, i.e., as an element in $Q(\sqrt{2})(\sqrt{3})$. Hence $Q(\sqrt{2}, \sqrt{3}) \subset Q(\sqrt{2})(\sqrt{3})$ and the other inclusion is also obvious. (See Exercise 8 for the general statement.)

(b) The polynomial $p(x) = x^2 - 3$ is irreducible in $Q(\sqrt{2})[x]$. First, any element of $Q(\sqrt{2})$ is of the form $a + b\sqrt{2}$, $a, b \in Q$, because $Q(\sqrt{2}) = Q[\sqrt{2}]$. (This was proved in general in the argument for Theorem 4.8.) Thus if $x^2 - 3$ is reducible, it must have a root of the form $a + b\sqrt{2}$. But $(a + b\sqrt{2})^2 - 3 = a^2 + 2b^2 + 2ab\sqrt{2} - 3 = 0$ would imply that $\sqrt{2} \in Q$ unless $a = 0$ or $b = 0$. If $b = 0$, then $a^2 = 3$, impossible since $\sqrt{3}$ is irrational. If $a = 0$, $2b^2 = 3$. But $2x^2 - 3$ is irreducible in $Q[x]$ by Theorem 4.6.

(c) We claim that $\{1, \sqrt{2}, \sqrt{3}, \sqrt{6}\}$ is a basis of $Q(\sqrt{2}, \sqrt{3})$ over Q. Observe first by (*b*) and Theorem 4.7(ii) [with $K = Q(\sqrt{2})$] that $\{1, \sqrt{3}\}$ is a basis of $Q(\sqrt{2}, \sqrt{3})$ over $Q(\sqrt{2})$. Now $\{1, \sqrt{2}\}$ is a basis of $Q(\sqrt{2})$ over Q, so obviously anything in $Q(\sqrt{2}, \sqrt{3})$ can be written in the form

$$a_1 + a_2\sqrt{2} + a_3\sqrt{3} + a_4\sqrt{6}, \qquad a_i \in Q.$$

Suppose

$$b_1 + b_2\sqrt{2} + b_3\sqrt{3} + b_4\sqrt{6} = a_1 + a_2\sqrt{2} + a_3\sqrt{3} + a_4\sqrt{6},$$

or

$$d_1 + d_2\sqrt{2} + d_3\sqrt{3} + d_4\sqrt{6} = 0, \qquad d_i = a_i - b_i, \qquad i = 1, \ldots, 4.$$

Then $(d_1 + d_2\sqrt{2}) + (d_3 + d_4\sqrt{2})\sqrt{3} = 0$, and since $\{1, \sqrt{3}\}$ is a basis of $Q(\sqrt{2})(\sqrt{3})$ over $Q(\sqrt{2})$, we conclude that $d_1 + d_2\sqrt{2} = 0$ and $d_3 + d_4\sqrt{2} = 0$. But then since $\{1, \sqrt{2}\}$ is a basis of $Q(\sqrt{2})$ over Q, we have $d_1 = d_2 = d_3 = d_4 = 0$.

(d) From Corollary 5 we know that there is a characteristic polynomial in $Q[x]$ for any element of $Q(\sqrt{2}, \sqrt{3})$. Let $\theta = \sqrt{2} + \sqrt{6}$. Then $(\theta - \sqrt{2})^2 = 6$, $\theta^2 - 2\theta\sqrt{2} + 2 = 6$, $2\theta\sqrt{2} = \theta^2 - 4$, $8\theta^2 = \theta^4 - 8\theta^2 + 16$, $\theta^4 - 16\theta^2 + 16 = 0$. We assert that $p(x) = x^4 - 16x^2 + 16$ is the characteristic polynomial of θ. Now $p(x)$ clearly has no linear factors in $Z[x]$ and hence in $Q[x]$ because none of ± 1, ± 2, ± 4, ± 8, ± 16 is a root of $p(x)$. If $p(x)$ has a quadratic factor in $Q[x]$, it would follow that $\sqrt{2} + \sqrt{6}$ is the root of a quadratic polynomial in $Q[x]$ and hence that $[Q(\sqrt{2} + \sqrt{6}): Q] = 2$. But $(\sqrt{2} + \sqrt{6})^2 = 8 + 4\sqrt{3}$, so $\sqrt{3} \in Q(\sqrt{2} + \sqrt{6})$. Also $1/(\sqrt{2} + \sqrt{6}) = -\frac{1}{4}(\sqrt{2} - \sqrt{6})$, so $\sqrt{2} - \sqrt{6} \in Q(\sqrt{2} + \sqrt{6})$ and hence $\sqrt{2} = \frac{1}{2}(\sqrt{2} + \sqrt{6} + \sqrt{2} - \sqrt{6}) \in Q(\sqrt{2} + \sqrt{6})$. We conclude that $Q(\sqrt{2} + \sqrt{6}) = Q(\sqrt{2}, \sqrt{3})$. But by (*c*), $[Q(\sqrt{2}, \sqrt{3}): Q] = 4$. Hence $p(x)$ is irreducible in $Q[x]$ and therefore it is the characteristic polynomial for $\sqrt{2} + \sqrt{6}$.

Using Theorem 4.7 we can easily show that any polynomial has a splitting field.

Theorem 4.11 *Let K be a field, x an indeterminate over K, and $f(x)$ a polynomial in $K[x]$. Then there exists a splitting field for $f(x)$.*

Proof: Obviously we can assume $f(x)$ is monic. Let $f(x) = p_1(x) \cdots p_m(x)$ be the prime factorization of $f(x)$ in the U.F.D. $K[x]$. By Theorem 4.7, let θ_1 be a root of $p_1(x)$ in some extension field $L_1 = K(\theta_1)$ over which x is also an indeterminate. From Corollary 1, Section 3.3, $(x - \theta_1) \mid f(x)$ in $L_1[x]$ so that

$$f(x) = (x - \theta_1)\, g(x)$$

where $g(x) \in L_1[x]$ and $g(x)$ is monic. We can now proceed by a routine induction on $n = \deg f(x)$. Indeed, $\deg g(x) = n - 1$, and hence there exists a splitting field L_n for $g(x)$. That is, there exists a field extension L_n of L_1 such that x is an indeterminate over L_n and $g(x)$ factors into linear factors: $g(x) = (x - \theta_2) \cdots (x - \theta_n)$ in $L_n[x]$ (so that $\theta_2, \ldots, \theta_n$ are in L_n). Then

$$f(x) = (x - \theta_1)\, g(x) = (x - \theta_1)(x - \theta_2) \cdots (x - \theta_n) \tag{25}$$

is a factorization of $f(x)$ into linear factors in $L_n[x]$. Now let $L = K(\theta_1, \ldots, \theta_n)$. Then by definition L is the intersection of all extension fields F of K for which $K \cup \{\theta_1, \ldots, \theta_n\} \subset F$. Since L_n is one of these fields, we know that $L \subset L_n$ and hence x is an indeterminate over L as well. Clearly the factorization (25) of $f(x)$ into linear factors is in $L[x]$. Now suppose F is a field, $K \subset F \subset L$, and $f(x)$ "splits" into linear factors in $F[x]$,

$$f(x) = (x - \alpha_1) \cdots (x - \alpha_n). \tag{26}$$

Both (25) and (26) are factorizations of $f(x)$ in $L[x]$ into prime factors, and since $L[x]$ is a U.F.D. it follows that (in some order) the $x - \alpha_i$ and $x - \theta_i$ are identical. Thus $\{\alpha_1, \ldots, \alpha_n\} = \{\theta_1, \ldots, \theta_n\}$, and hence $\{\theta_1, \ldots, \theta_n\} \subset F$. We conclude that $L = K(\theta_1, \ldots, \theta_n) \subset F \subset L$ and the inclusion $F \subset L$ cannot be strict. ∎

We can show that the splitting field of a polynomial is unique to within equivalence. First we prove an extension of Corollary 4 that involves a bit of diagram chasing.

Theorem 4.12 *Let $\varphi: F \to K$ be a field isomorphism, and x and y indeterminates over F and K, respectively. Let $p(x) = \sum_{k=0}^{n} f_k x^k$ be an irreducible polynomial in $F[x]$, and let $q(y) = \sum_{k=0}^{n} \varphi(f_k) y^k \in K[y]$. Let α be a root of $p(x)$ in some extension field of F, and β a root of $q(y)$ in some extension field of K. Then φ can be extended to an isomorphism ψ: $F(\alpha) \to K(\beta)$ (i.e., $\psi \mid F = \varphi$) such that $\psi(\alpha) = \beta$.*

Proof: It is obvious that $q(y)$ is irreducible in $K[y]$ (see Exercise 12). Let $\sigma_1: F[x]/(p(x)) \to F(\alpha), \sigma_2: K[y]/(q(y)) \to K(\beta)$ be the isomorphisms of Theorem

4.8, and denote by ν_1 and ν_2 the canonical epimorphisms $\nu_1\colon F[x] \to F[x]/$
$(p(x))$, $\nu_2\colon K[y] \to K[y]/(q(y))$. Consider the diagram

$$\tag{27}$$

in which $\delta\colon F[x] \to K[y]$ is the obvious ring isomorphism satisfying $\delta|F = \varphi$
and $\delta(x) = y$. The existence of a ring (in this case field) homomorphism μ
satisfying

$$\mu\nu_1 = \nu_2\delta \tag{28}$$

is guaranteed by the basic mapping diagram (25) in Section 3.1. Since δ
and ν_2 are surjective, μ is surjective. Also from (28), $\nu_1(f(x)) \in \ker \mu$ iff
$\nu_2(\delta(f(x)))$ is the zero in $K[y]/(q(y))$ iff $\delta(f(x)) \in (q(y))$ iff $\delta(f(x)) = a(y)q(y)$
iff $f(x) = \delta^{-1}(a(y))\delta^{-1}(q(y)) = \delta^{-1}(a(y))\,p(x) \in (p(x))$ iff $\nu_1(f(x))$ is the
zero in $F[x]/(p(x))$. Hence $\ker \mu$ consists of 0 only, and μ is an isomorphism.
Now consider the diagram

$$
\begin{array}{ccc}
F[x] & \xrightarrow{\;\delta\;} & K[y] \\
\nu_1 \downarrow & & \downarrow \nu_2 \\
F[x]/(p(x)) & \xrightarrow{\;\mu\;} & K[y]/(q(y)) \\
\sigma_1 \downarrow & & \downarrow \sigma_2 \\
F(\alpha) & \xrightarrow{\;\psi\;} & K(\beta)
\end{array}
\qquad (29)
$$

and define $\psi\colon F(\alpha) \to K(\beta)$ by $\psi = \sigma_2\mu\sigma_1^{-1}$, a composition of isomorphisms
and hence an isomorphism. Observe that

$$
\begin{aligned}
\psi(\alpha) &= \sigma_2\mu\sigma_1^{-1}(\alpha) \\
&= \sigma_2\mu\nu_1(x) && \text{[by (21)]} \\
&= \sigma_1\nu_2\delta(x) && \text{[by (29)]} \\
&= \sigma_2\nu_2(y) \\
&= \beta && \text{[by (21)]}.
\end{aligned}
$$

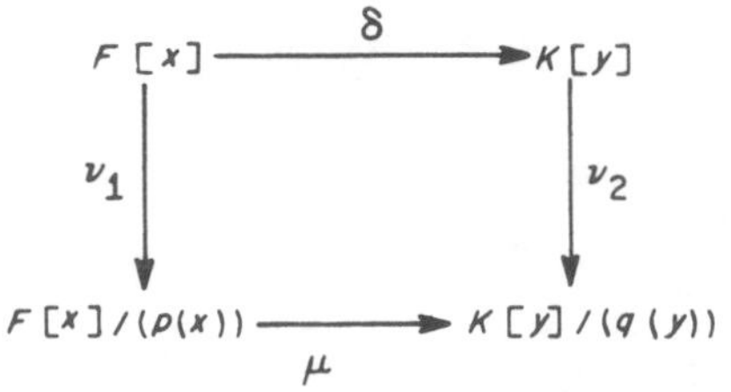

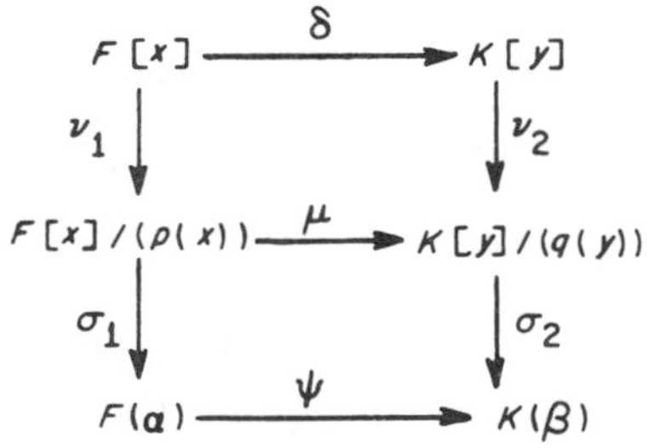

Also, if $f \in F$, then

$$\begin{aligned} \psi(f) &= \sigma_2 \mu \sigma_2^{-1}(f) \\ &= \sigma_2 \mu \nu_1(f) \qquad \text{[by (21)]} \\ &= \sigma_2 \nu_2 \delta(f) \qquad \text{[by (29)]} \\ &= \sigma_2 \nu_2(\varphi(f)) \\ &= \varphi(f) \qquad \text{[by (21)].} \end{aligned}$$

In other words, $\psi | F = \varphi$. ∎

Theorem 4.13 *Let φ: $F \to K$ be a field isomorphism, and x and y indeterminates over F and K, respectively. Let δ_φ: $F[x] \to K[y]$ be the ring isomorphism defined by $\delta_\varphi(h(x)) = \sum_{k=0}^{n} \varphi(h_k) y^k$, where $h(x) = \sum_{k=0}^{n} h_k x^k \in F[x]$. Let $f(x) \in F[x]$, $\deg f(x) = n$, and let $F(\alpha_1, \ldots, \alpha_n)$ be a splitting field for $f(x)$. Let $g(y) = \delta_\varphi(f(x))$, and let $K(\beta_1, \ldots, \beta_n)$ be a splitting field for $g(y)$. Then φ may be extended to an isomorphism θ: $F(\alpha_1, \ldots, \alpha_n) \to K(\beta_1, \ldots, \beta_n)$ in which $\theta(\alpha_i) = \beta_{\tau(i)}$ for some $\tau \in S_n$.*

Proof: Let $p(x)$ be a prime factor of $f(x)$ in $F[x]$, and let $q(y) = \delta_\varphi(p(x))$ be the corresponding prime factor of $g(y)$ in $K[y]$. Without loss of generality (it amounts to reordering the α's and β's), suppose α_1 is a root of $p(x)$ and β_1 is a root of $q(y)$. Then by Theorem 4.12, φ can be extended to an isomorphism ψ: $F(\alpha_1) \to K(\beta_1)$, $\psi(\alpha_1) = \beta_1$. Now let δ_ψ: $F(\alpha_1)[x] \to K(\beta_1)[y]$ be the obvious polynomial ring isomorphism defined by $\delta_\psi(m(x)) = \sum_{k=0}^{r} \psi(m_k) y^k$, where $m(x) = \sum_{k=0}^{r} m_k x^k \in F(\alpha_1)[x]$. We have the factorizations

$$\begin{aligned} f(x) &= (x - \alpha_1)h(x) \qquad (\text{in } F(\alpha_1)[x]), \\ g(y) &= (y - \beta_1)k(y) \qquad (\text{in } K(\beta_1)[x]). \end{aligned}$$

Then

$$\begin{aligned} (y - \beta_1)k(y) &= g(y) = \delta_\varphi(f(x)) \\ &= \delta_\psi(f(x)) \qquad (\text{since } \psi | F = \varphi) \\ &= \delta_\psi((x - \alpha_1)h(x)) = \delta_\psi(x - \alpha_1)\delta_\psi(h(x)) \\ &= (y - \beta_1)\delta_\psi(h(x)). \end{aligned} \tag{30}$$

Since $K(\beta_1)[y]$ is a U.F.D. (all we need really is an integral domain here), we have from (30) that

$$\delta_\psi(h(x)) = k(y).$$

Also observe that the roots of $h(x)$ are $\alpha_2, \ldots, \alpha_n$ (because $F(\alpha_1, \ldots, \alpha_n)[x]$ is a U.F.D. and $(x - \alpha_1)h(x) = f(x) = c(x - \alpha_1)(x - \alpha_2) \cdots (x - \alpha_n)$, $c \in F$), and similarly the roots of $k(y)$ are $\beta_2, \ldots, \beta_n$.

We are now in a position to use induction on $n = \deg f(x)$, for we now have the following setup: ψ: $F(\alpha_1) \to K(\beta_1)$ is an isomorphism; $\delta_\psi(h(x)) =$

$k(y)$; the splitting field of $h(x)$ is $F(\alpha_1)(\alpha_2, \ldots, \alpha_n)$ (see Exercise 13); the splitting field for $k(y)$ is $K(\beta_1)(\beta_2, \ldots, \beta_n)$. Thus by induction, ψ can be extended to an isomorphism θ: $F(\alpha_1)(\alpha_2, \ldots, \alpha_n) \to K(\beta_1)(\beta_2, \ldots, \beta_n)$ in which $\theta(\alpha_t) = \beta_{\tau'(t)}$, $t = 2, \ldots, n$, and τ' is a bijection on $[2,n]$. Since $\theta | F(\alpha_1) = \psi$ and $\psi | F = \varphi$, we have $\theta | F = \varphi$, and hence θ is an extension of φ. Moreover, since $F(\alpha_1)(\alpha_2, \ldots, \alpha_n) = F(\alpha_1, \ldots, \alpha_n)$ (see Exercise 8), the proof is complete. ∎

We are now in a position to prove that the splitting field of a polynomial is uniquely determined to within isomorphism.

Theorem 4.14 *Let K be a field, x an indeterminate over K, and $f(x) \in K[x]$.*

(i) *The splitting field for $f(x)$ is uniquely determined to within equivalent extensions.*

(ii) *If two splitting fields for $f(x)$ are subfields of a common field L over which x is an indeterminate, then in fact they must be equal.*

Proof: (i) Make the following specializations in Theorem 4.13: $F = K$, $x = y$, $\varphi = \iota_K$. Then of course $f(x)$ and $g(y)$ collapse to the same polynomial, and we conclude that ι_K may be extended to an isomorphism θ: $K(\alpha_1, \ldots, \alpha_n) \to K(\beta_1, \ldots, \beta_n)$ with $\theta(\alpha_i) = \beta_{\tau(i)}$, $\tau \in S_n$.

(ii) Suppose

$$f(x) = a(x - \alpha_1) \cdots (x - \alpha_n) \tag{31}$$

and

$$f(x) = a(x - \beta_1) \cdots (x - \beta_n) \tag{32}$$

are factorizations of $f(x)$ in $K(\alpha_1, \ldots, \alpha_n)[x]$ and $K(\beta_1, \ldots, \beta_n)[x]$ [in which $a \in K$ is the leading coefficient of $f(x)$]. Then (31) and (32) can be considered as factorizations in $L[x]$, a U.F.D. Thus the prime factors $x - \alpha_i$ and $x - \beta_i$ must match up to within order and unit multiples. Since all these are monic, it follows that $\alpha_i = \beta_{\tau(i)}$, $i = 1, \ldots, n$, $\tau \in S_n$, and hence $K(\alpha_1, \ldots, \alpha_n) = K(\beta_1, \ldots, \beta_n)$. ∎

Actually Theorem 4.14 shows more, namely that *if $K(\alpha_1, \ldots, \alpha_n)$ and $K(\beta_1, \ldots, \beta_n)$ are splitting fields for $f(x) \in K[x]$, then there exists an isomorphism*

$$\theta: K(\alpha_1, \ldots, \alpha_n) \to K(\beta_1, \ldots, \beta_n) \tag{33}$$

such that

$$\theta | K = \iota_K \tag{34}$$

and

$$\theta(\alpha_i) = \beta_{\tau(i)}, \qquad i = 1, \ldots, n, \ \tau \in S_n. \tag{35}$$

Also the proof of (ii) implies a bit more: *If $K(\alpha_1, \ldots, \alpha_n)$ and $K(\beta_1, \ldots, \beta_n)$ are splitting fields for $f(x) \in K[x]$ and are subfields of a common field L over which x is an indeterminate, then the α_i and β_i are the same to within order.*

As our final major result in this section we prove that successive finite algebraic extensions of a field of characteristic 0 can essentially be accomplished with a simple algebraic extension. We saw an instance of this in Example 4(d) in which we confirmed that $Q(\sqrt{2})(\sqrt{3}) = Q(\sqrt{2}, \sqrt{3}) = Q(\sqrt{2} + \sqrt{6})$ and $p(x) = x^4 - 16x^2 + 16$ is the characteristic polynomial of $\sqrt{2} + \sqrt{6}$.

Theorem 4.15 (*Primitive Element Theorem*) *Assume that K is a field of characteristic* 0. *Let $\theta_1, \ldots, \theta_n$ be elements in an extension field M of K, and assume that each θ_i is algebraic over K, $i = 1, \ldots, n$. Then there exists an element $\theta \in K(\theta_1, \ldots, \theta_n)$, algebraic over K, such that*

$$K(\theta) = K(\theta_1, \ldots, \theta_n).$$

Proof: The proof is by induction on n with nothing to prove for $n = 1$. Now consider $K(\theta_1, \ldots, \theta_n) = K(\theta_1, \ldots, \theta_{n-1})(\theta_n)$, and by induction obtain an element $\xi \in K(\theta_1, \ldots, \theta_{n-1}) = K(\xi)$. Thus the problem reduces to the case $n = 2$, i.e., we must prove the theorem for $K(\xi,\theta_n)$. Let x be an indeterminate over M and hence over $K(\xi,\theta_n)$, and let $f(x)$ and $g(x)$ be the monic characteristic polynomials in $K[x]$ for ξ and θ_n, respectively. Let L be the splitting field for $f(x)g(x)$ regarded as a polynomial in $K(\xi,\theta_n)[x]$ [so that $K(\xi,\theta_n) \subset L$]; in $L[x]$ we have the factorizations

$$f(x) = (x - \xi)(x - \alpha_1) \cdots (x - \alpha_r), \qquad \alpha_i \in L, \qquad i = 1, \ldots, r,$$
$$g(x) = (x - \theta_n)(x - \beta_1) \cdots (x - \beta_s), \qquad \beta_i \in L, \qquad i = 1, \ldots, s.$$

The result of Exercise 26 tells us that the elements $\xi, \alpha_1, \ldots, \alpha_r$ (and of course $\theta_n, \beta_1, \ldots, \beta_s$) are distinct, so we can consider the elements

$$\frac{\beta_j - \theta_n}{\xi - \alpha_i} \in L, \qquad i = 1, \ldots, r, \quad j = 1, \ldots, s. \tag{36}$$

Now K has characteristic 0, so there are more than a finite number of multiples of 1 in K. Let $0 \neq c \in K$ be distinct from the rs elements in (36), and set

$$\theta = \theta_n + c\xi.$$

Obviously $K(\theta) \subset K(\xi,\theta_n)$. Now

$$f(\xi) = 0 \tag{37}$$

and

$$g(\theta - c\xi) = g(\theta_n) = 0. \tag{38}$$

Also the polynomial $h(x) = g(\theta - cx) \in K(\theta)[x]$, and from (38),

$$h(\xi) = 0. \tag{39}$$

Thus in $L[x]$ the polynomials $f(x)$ and $h(x)$ must, in view of (37) and (39), have a common factor of $x - \xi$. Suppose that as polynomials in $L[x]$, $f(x)$

and $h(x)$ were both divisible by some $x - \alpha_t$. Then it would follow that

$$0 = h(\alpha_t) = g(\theta - c\alpha_t)$$

so that $\theta - c\alpha_t$ must be one of the roots of $g(x)$. Thus

$$\theta - c\alpha_t = \theta_n$$

or

$$\theta - c\alpha_t = \beta_j$$

for some j. But then

$$\theta = \theta_n + c\alpha_t$$

or

$$\theta = \beta_j + c\alpha_t.$$

Hence since $\theta = \theta_n + c\xi$ by definition, it follows by substitution in the preceding two equations that

$$c(\xi - \alpha_t) = 0$$

or

$$c = \frac{\beta_j - \theta_n}{\xi - \alpha_t}. \tag{40}$$

Since $\xi - \alpha_t \neq 0$ (i.e., $\xi, \alpha_1, \ldots, \alpha_r$ are distinct), $c = 0$ or c satisfies (40). But c was chosen to exclude both of these possibilities. Thus $h(x)$ is not divisible by any $x - \alpha_t$ in $L[x]$, and since these are the prime factors of $f(x)$ in $L[x]$, we conclude that in $L[x]$ the monic g.c.d. of $f(x)$ and $h(x)$ is precisely $x - \xi$. But as polynomials in $K(\theta)[x]$ they have precisely the same monic g.c.d. as they have as polynomials in $L[x]$ (see Exercise 20). Hence

$$x - \xi \in K(\theta)[x],$$

and it follows that $\xi \in K(\theta)$ and hence that $\theta_n = \theta - c\xi \in K(\theta)$. We have proved that $K(\theta_n,\xi) \subset K(\theta)$ and thus that

$$K(\theta_n,\xi) = K(\theta).$$

But then

$$K(\theta_1, \ldots, \theta_n) = K(\theta)$$

and θ is algebraic over K by Exercise 18. ∎

The element θ in Theorem 4.15 is called a *primitive element* for $K(\theta_1, \ldots, \theta_n)$ over K.

Example 5 (a) Find a primitive element for $Q(i,2^{1/2})$ over Q. We follow the proof of Theorem 4.15. Monic characteristic polynomials for i and $2^{1/2}$ are $f(x) = x^2 + 1$ and $g(x) = x^2 - 2$, respectively. Then

$$f(x) = (x - i)(x + i),$$
$$g(x) = (x - 2^{1/2})(x + 2^{1/2}).$$

The number

$$\frac{-2^{1/2} - 2^{1/2}}{i - (-i)} = -\frac{2^{1/2}}{i} = 2^{1/2}i$$

corresponds to (36). We can take $c = 1$ and define $\theta = i + c\cdot 2^{1/2} = i + 2^{1/2}$. Then the proof of Theorem 4.15 tells us that $Q(i, 2^{1/2}) = Q(\theta) = Q(i + 2^{1/2})$. Observe that

$$(\theta - i)^2 = 2$$

so that

$$\theta^2 - 2i\theta + i^2 = 2$$

or

$$2i\theta = \theta^2 - 3.$$

But then

$$-4\theta^2 = \theta^4 - 6\theta^2 + 9$$

or

$$\theta^4 - 2\theta^2 + 9 = 0.$$

Consider the polynomial

$$p(x) = x^4 - 2x^2 + 9 \in Q[x].$$

We claim that $p(x)$ is the characteristic polynomial for θ. It is obvious that $p(x)$ has no linear factors in $Z[x]$ because they must be of the form $x - a$, where $a|9$ and none of the divisors of 9 is a root of $p(x)$. On the other hand, if $p(x)$ has a quadratic factor in $Z[x]$, then it would follow that θ satisfies a polynomial of degree 2 over Q. But then $[Q(\theta):Q] = 2$. Since $Q(i, 2^{1/2}) = Q(\theta)$, we conclude that $[Q(2^{1/2})(i):Q] = 2$. Then by the degree theorem (Exercise 17), $2 = [Q(2^{1/2})(i):Q] = [Q(2^{1/2}):Q][Q(2^{1/2})(i):Q(2^{1/2})]$. We know that $[Q(2^{1/2}):Q] = 2$ so that $[Q(2^{1/2})(i):Q(2^{1/2})] = 1$. But then it follows that $i \in Q(2^{1/2})$, which is obvious nonsense.

(b) Find a primitive element for $Q(2^{1/3}, \omega)$ over Q, where $\omega^3 = 1$, $\omega \neq 1$. The monic characteristic polynomials for $2^{1/3}$ and ω are, respectively,

$$f(x) = x^3 - 2$$

and

$$g(x) = x^2 + x + 1.$$

The remaining roots of $f(x)$ are $\alpha_1 = \omega 2^{1/3}$ and $\alpha_2 = \omega^2 2^{1/3}$, and the remaining root of $g(x)$ is $\beta_1 = \omega^2$. Then the numbers (36) become

$$\frac{\omega^2 - \omega}{2^{1/3} - \omega 2^{1/3}} \quad \text{and} \quad \frac{\omega^2 - \omega}{2^{1/3} - \omega^2 2^{1/3}}$$

or

$$\frac{-\omega}{2^{1/3}}$$

and

$$\frac{\omega^2 - \omega}{2^{1/3}(1 - \omega^2)} = \frac{\omega(\omega - 1)}{2^{1/3}(1 - \omega)(1 + \omega)}$$
$$= \frac{-\omega}{2^{1/3}(1 + \omega)} = \frac{-\omega}{2^{1/3}(-\omega^2)}$$
$$= \frac{1}{2^{1/3}\omega} = \frac{\omega^2}{2^{1/3}}.$$

Thus the choice of $c \in Q$ need only be different from 0, $-\omega/2^{1/3}$, and $\omega^2/2^{1/3}$. Take $c = 1$ so that

$$\theta = 2^{1/3} + \omega.$$

Exercises 3.4

1. Let D be a U.F.D. and $a_1, \ldots, a_n$ nonzero elements of D. If $d =$ g.c.d. $(a_1, \ldots, a_n)$, show that $a_1/d, \ldots, a_n/d$ are relatively prime. *Hint*: Suppose $k \in$

D and $k\,|\,(a_i/d)$, $i = 1, \ldots, n$. If k is not a unit, it has a factorization into primes; call one of these primes q. Then $q\,|\,(a_i/d)$, $i = 1, \ldots, n$, so there exist $c_i \in D$ such that $a_i/d = qc_i$, $i = 1, \ldots, n$. Hence $a_i = dqc_i$ or $dq\,|\,a_i$, $i = 1, \ldots, n$. It follows that $dq\,|\,d$, and so $d = dql$ for some $l \in D$. But then $ql = 1$, contradicting the fact that q is a prime. Thus k must be a unit.

2. Show that if D is a U.F.D. and p is a prime in D, then p is a prime in the polynomial ring $D[x]$. *Hint*: Suppose $\alpha(x)$ and $\beta(x)$ are polynomials in $D[x]$ and p divides $\alpha(x)\beta(x)$. Write $\alpha(x)\beta(x) = p\cdot q(x)$, $q(x) \in D[x]$, and set $\alpha(x) = d_1\alpha_1(x)$, $\beta(x) = d_2\beta_1(x)$, $q(x) = d_3q_1(x)$, where $d_i \in D$, $i = 1, 2, 3$, and $\alpha_1(x)$, $\beta_1(x)$, and $q_1(x)$ are primitive polynomials in $D[x]$. Then $d_1d_2\alpha_1(x)\beta_1(x) = pd_3q_1(x)$, $\alpha_1(x)\,\beta_1(x)$ is primitive in $D[x]$, and the uniqueness part of Theorem 4.3 tells us that $d_1d_2 = pd_3\varepsilon$ for some unit $\varepsilon \in D$. But then since p is a prime in D, $p\,|\,d_1$ or $p\,|\,d_2$. It follows that p divides $\alpha(x)$ or p divides $\beta(x)$.
3. Show that $\det A \neq 0$ for the matrix in (12). It is called a *Vandermonde matrix*.
4. Let D be a U.F.D. with a finite number of units. Show that there are only a finite number of divisors of any element of D.
5. Show that if L is an extension field of K and $S \subset L$, then $K(S)$ is a field.
6. Show that $p(x) = x^2 + 1$ is irreducible in $Z[x]$. *Hint*: Consider $p(x + 1)$ and use Eisenstein's criterion.
7. Show that if $K[\theta]$ is a field, then $K[\theta] = K(\theta)$. *Hint*: $K(\theta)$ is the smallest field containing K and θ. Since it must contain all polynomials in θ with coefficients in K, it follows that $K \subset K[\theta] \subset K(\theta)$. But if $K[\theta]$ is a field, then $K[\theta] = K(\theta)$.
8. Let $K \subset L$ be fields. If S and T are subsets of L, show that $K(S \cup T) = K(S)(T)$. *Hint*: If F is a field containing $S \cup T$ and K, then clearly $K(S) \subset F$ and hence $K(S)(T) \subset F$. It follows that $K(S)(T) \subset K(S \cup T)$. Conversely, if F is a field containing $K(S)$ and T, it contains K and $S \cup T$ and hence contains $K(S \cup T)$. Thus $K(S \cup T) \subset K(S)(T)$.
9. Let L be a finite extension field of K, $[L{:}K] = n$. Show that if $\{u_1, \ldots, u_n\}$ is a basis of L over K and $\sum_{i=1}^{n} k_iu_i = 0$, $k_i \in K$, $i = 1, \ldots, n$, then in fact $k_1 = \cdots = k_n = 0$.
10. Show that $\cos 20°$ is not a rational number and find $[Q(\cos 20°){:}Q]$. *Hint*: By De Moivre's theorem, $(\cos\theta + i\sin\theta)^3 = \cos 3\theta + i\sin 3\theta$, and hence $\cos 3\theta = 4\cos^3\theta - 3\cos\theta$. Thus if $u = \cos 20°$, we have $\cos 60° = 4u^3 - 3u$ or $8u^3 - 6u - 1 = 0$. Set $v = 2u$ so $v^3 - 3v - 1 = 0$. Investigate the irreducibility of $p(x) = x^3 - 3x - 1$ in $Z[x]$.
11. Is $f(x) = x^3 + 2x^2 + 2x + 4$ irreducible in $Q[x]$?
12. Prove the assertion in the proof of Theorem 4.12 concerning the irreducibility of $q(y)$. *Hint*: Suppose $q(y) = a(y)b(y) = \sum_{k=0}^{n}(\sum_{t=0}^{k}a_{k-t}b_t)y^k = \sum_{k=0}^{n}\varphi(f_k)y^k$. Since φ is an isomorphism, $f_k = \sum_{t=0}^{k}\varphi^{-1}(a_{k-t})\varphi^{-1}(b_t)$. Define $c(x) = \sum_{k=0}^{n}\varphi^{-1}(a_k)x^k$ and $d(x) = \sum_{k=0}^{n}\varphi^{-1}(b_k)x^k$. Then $p(x) = c(x)d(x)$.
13. Let $F(\alpha_1, \ldots, \alpha_n)$ be a splitting field for $f(x) \in F[x]$, and let $f(x) = (x-\alpha_1)h(x)$, a factorization in $F(\alpha_1)[x]$. Show that $F(\alpha_1)(\alpha_2, \ldots, \alpha_n) = F(\alpha_1, \ldots, \alpha_n)$ is a splitting field for $h(x)$ as a polynomial over $F(\alpha_1)$. *Hint*: Clearly the roots

of $h(x)$ are $\alpha_2, \ldots, \alpha_n$. If $F(\alpha_1) \subset L \subset F(\alpha_1, \ldots, \alpha_n)$ and the second inclusion is strict, then surely $h(x)$ cannot factor into linear factors in $L[x]$, for otherwise so could $f(x)$ and this would contradict the fact that $F(\alpha_1, \ldots, \alpha_n)$ is a splitting field for $f(x)$.

14. Let $K \subset L$ be fields. Show that if $L = K(u_1, \ldots, u_n)$ and $\{u_1, \ldots, u_n\}$ is a spanning set for L over K, then a basis of L over K may be chosen from $\{u_1, \ldots, u_n\}$. *Hint*: A subset $\{u_{i_1}, \ldots, u_{i_k}\}$ ($i_1, \ldots, i_k$ distinct) of $\{u_1, \ldots, u_n\}$ will be called *linearly independent* (*l.i.*) if: $a_{i_1}u_{i_1} + \cdots + a_{i_r}u_{i_r} = 0$, $a_{i_t} \in K$, $t = 1, \ldots, r$, iff $a_{i_1} = \cdots = a_{i_r} = 0$. Let r be the largest integer k such that there exists a l.i. subset $\{u_{i_1}, \ldots, u_{i_k}\}$. Without loss of generality assume it is $u_1, \ldots, u_r$. Then if $s > r$, it follows that $\{u_1, \ldots, u_r, u_s\}$ is not l.i., and hence there exist $a_1, \ldots, a_r, b$ not all 0 such that $a_1u_1 + \cdots + a_ru_r + bu_s = 0$. Now $b \neq 0$ (why?), so $u_s = \sum_{i=1}^{r}(-a_ib^{-1})u_i$. Thus any u_s can be written in the form $\sum_{i=1}^{r} c_iu_i$, $c_i \in K$. Since $\{u_1, \ldots, u_n\}$ is a spanning set for L over K, it follows that any element of L can be written in the same way (why?).

15. Let L be a finite extension of K, and $\{u_1, \ldots, u_n\}$ a basis of L over K, i.e., $L = K(u_1, \ldots, u_n)$, $[L:K] = n$. Is it true that $[K(u_1, \ldots, u_{n-1}):K] = n - 1$? *Hint*: No. Consider $L = Q(\sqrt{2}, \sqrt{3}) = Q(1, \sqrt{2}, \sqrt{3}, \sqrt{6}) = Q(\sqrt{2}, \sqrt{3}, \sqrt{6})$.

16. Let L be an extension of K, and let $\theta \in L$ be algebraic over K. Show that $K(\theta)$ is a finite extension of K and thereby that every element of $K(\theta)$ is algebraic over K. *Hint*: By hypothesis, θ is the root of some polynomial $f(x) \in K[x]$ (x is an indeterminate over K). Thus $f(\theta) = 0$. Factor $f(x)$ into prime factors in $K[x]$ so that θ must be the root of some prime factor of $f(x)$ of degree p. But then $[K(\theta):K] = p$ by Theorem 4.7(ii). Hence $K(\theta)$ is a finite extension of K, and we can apply Theorem 4.10(ii).

17. (*The Degree Theorem*) Let $K \subset L \subset M$ be fields. Prove that if $[L:K] = m < \infty$ and $[M:L] = n < \infty$, then $[M:K] = mn$. Thus a finite extension of a finite extension is still a finite extension. *Hint*: Let $\{l_1, \ldots, l_m\}$ and $\{\mu_1, \ldots, \mu_n\}$ be bases for L over K and M over L, respectively. If $z \in M$, write

$$z = \sum_{t=1}^{n} c_t\mu_t, \qquad c_t \in L,$$

and

$$c_t = \sum_{j=1}^{m} d_{tj}l_j, \qquad d_{tj} \in K.$$

Then

$$z = \sum_{t=1}^{n}\left(\sum_{j=1}^{m} d_{tj}l_j\right)\mu_t = \sum_{t=1, j=1}^{n, m} d_{tj}l_j\mu_t.$$

If $z = 0$, then

$$\sum_{t=1}^{n}\left(\sum_{j=1}^{m} d_{tj}l_j\right)\mu_t = 0,$$

and since $\{\mu_1, \ldots, \mu_n\}$ is a basis of M over L, $\sum_{j=1}^{m} d_{tj}l_j = 0$, $t = 1, \ldots, n$. But then $d_{tj} = 0$, $j = 1, \ldots, m$, $t = 1, \ldots, n$, because $\{l_1, \ldots, l_m\}$ is a basis of L over K. Thus $\{l_j\mu_t \mid j = 1, \ldots, m,\ t = 1, \ldots, n\}$ is a basis of M over K.

18. Generalize the result of Exercise 16 as follows: Let L be an extension of K, and let $\theta_1, \ldots, \theta_n$ be elements of L each algebraic over K. Show that $K(\theta_1, \ldots, \theta_n)$ is a finite extension of K and hence [Theorem 4.10(ii)] every element of $K(\theta_1, \ldots, \theta_n)$ is algebraic over K. *Hint*: For $n = 1$, use Exercise 16. So assume that $K(\theta_1, \ldots, \theta_{n-1})$ is a finite extension of K. Then, since θ_n is algebraic over K, it is algebraic over $K(\theta_1, \ldots, \theta_{n-1})$. Hence $K(\theta_1, \ldots, \theta_n) = K(\theta_1, \ldots, \theta_{n-1})(\theta_n)$ is a finite extension of $K(\theta_1, \ldots, \theta_{n-1})$ by Exercise 16. Apply the degree theorem (Exercise 17) with $L = K(\theta_1, \ldots, \theta_{n-1})$ and $M = K(\theta_1, \ldots, \theta_{n-1}, \theta_n)$.

19. (*The Euclidean Algorithm*) Let K be a field and x an indeterminate over K. Then the g.c.d. of two polynomials $f(x)$ and $g(x)$ in $K[x]$ can be computed as follows: Write

$$\begin{aligned}
f(x) &= g(x)q_1(x) + r_1(x), && \deg r_1(x) < \deg g(x)\\
g(x) &= r_1(x)q_2(x) + r_2(x), && \deg r_2(x) < \deg r_1(x)\\
r_1(x) &= r_2(x)q_3(x) + r_3(x), && \deg r_3(x) < \deg r_2(x)\\
&\cdots\cdots\cdots\cdots\cdots\cdots\cdots\cdots\\
r_{k-1}(x) &= r_k(x)q_{k+1}(x) + r_{k+1}(x), && \deg r_{k+1}(x) < \deg r_k(x)\\
r_k(x) &= r_{k+1}(x)q_{k+2}(x).
\end{aligned}$$

Then $r_{k+1}(x) = \text{g.c.d.}(f(x), g(x))$. In other words, perform the indicated sequence of divisions and since

$$0 \le \deg r_{k+1}(x) < \cdots < \deg r_1(x) < \deg g(x),$$

the procedure must terminate with an exact division (i.e., 0 remainder) after at most $\deg g(x)$ steps. Observe that the procedure also produces $m(x)$ and $n(x)$ in $K[x]$ such that $m(x)f(x) + n(x)g(x) = r_{k+1}(x)$.

20. Let K be a field and L an extension field of K. Assume that x is an indeterminate over L and that $f(x)$ and $g(x)$ are polynomials in $K[x]$. Then the monic g.c.d. of $f(x)$ and $g(x)$, regarded as polynomials in $K[x]$, is the same as the monic g.c.d. of $f(x)$ and $g(x)$ regarded as polynomials in $L[x]$. *Hint*: Any two g.c.d.'s of $f(x)$ and $g(x)$ in $L[x]$ differ only by nonzero multiples of elements from L. Now in $L[x]$ apply the Euclidean algorithm of Exercise 19 to produce a g.c.d. $r_{k+1}(x)$. The entire process involves only rational operations in $L[x]$ and starts with the polynomials $f(x)$ and $g(x)$ in $K[x]$. Hence the process leads to a polynomial $r_{k+1}(x) \in K[x]$. To render it monic requires multiplication by a nonzero element of K.

21. In the same notation as in Exercise 20, show that $f(x)$ and $g(x)$ are relatively prime regarded as elements of $K[x]$ iff they are relatively prime regarded as elements of $L[x]$. *Hint*: $\text{g.c.d.}(f(x), g(x)) = 1$. Apply Exercise 20.

22. Let R be a commutative ring and x an indeterminate over R. If $f(x) = \sum_{k=0}^{n} a_k x^k \in R[x]$, then the *derivative* of $f(x)$ is the polynomial $f'(x) = \sum_{k=1}^{n} k a_k x^{k-1}$ [if $f(x) \in R, f'(x) = 0$]. Verify the following formulas:

(i) $(f(x) + g(x))' = f'(x) + g'(x)$.

(ii) $(f(x)g(x))' = f'(x)g(x) + f(x)g'(x)$.

(iii) $(\prod_{k=1}^{n} f_k(x))' = \sum_{k=1}^{n} f_k'(x) \prod_{j \ne k} f_j(x)$.

(iv) $f'(x) = 0$ iff $k \cdot a_k = \overbrace{a_k + \cdots + a_k}^{k} = 0,\ k = 1, \ldots, n$.

(v) If R is an integral domain of characteristic 0, then $f'(x) = 0$ iff $f(x) \in R$.
(vi) If R is an integral domain of characteristic p, p a prime, then $f'(x) = 0$ iff $f(x) \in R[x^p]$.
(vii) If R is an integral domain and x and h are independent indeterminates over R, then

$$f(x+h) - f(x) = h(f'(x) + hg(x,h))$$

where $g(x,h) \in R[x,h]$.

23. Let R be an integral domain, x an indeterminate over R, $f(x) \in R[x]$, and $\alpha \in R$. Then α is a *root of $f(x)$ of multiplicity* $k \geq 1$ if $(x-\alpha)^k \mid f(x)$ and $(x-\alpha)^{k+1} \nmid f(x)$ ($\nmid$ means "does not divide"). If $k = 1$, α is a *simple root* of $f(x)$. Prove that if α is a root of $f(x)$ of multiplicity k, then α is a root of $f'(x)$ of multiplicity at least $k-1$. *Hint*: Write $f(x) = (x-\alpha)^k g(x)$. Then by Exercise 22, $f'(x) = k(x-\alpha)^{k-1}g(x) + (x-\alpha)^k g'(x) = (x-\alpha)^{k-1}[kg(x) + (x-\alpha)g'(x)]$.

24. In the notation of Exercise 23, prove that if α is a simple root of $f(x)$, then α is not a root of $f'(x)$. *Hint*: $f'(x) = g(x) + (x-\alpha)g'(x)$. Then $f'(\alpha) = g(\alpha)$ and $g(\alpha) \neq 0$, since otherwise $(x-\alpha) \mid g(x)$ and hence $(x-\alpha)^2 \mid f(x)$.

25. Let K be a field, x an indeterminate over K, $f(x) \in K[x]$, $\deg f(x) = n \geq 1$, and let L be a splitting field for $f(x)$. Prove that $f(x)$ has fewer than n distinct roots in L iff either deg g.c.d.$(f(x), f'(x)) \geq 1$ or $f'(x) = 0$. Moreover, $f'(x) = 0$ only in case char $K \neq 0$. *Hint*: First note that in view of Theorem 4.14, the number of distinct roots $f(x)$ has in L is independent of L. Assume $f(x) = a\Pi_{i=1}^r(x-\xi_i)^{m_i}$, $m_i > 1$, $i = 1, \ldots, s$, $m_{s+1} = \cdots = m_r = 1$, and $\xi_1, \ldots, \xi_r$ are the distinct roots of $f(x)$ in L, $r < n$. Compute that $f'(x) = a\Sigma_{i=1}^r m_i(x-\xi_i)^{m_i-1}$ $\Pi_{j\neq i}(x-\xi_j)^{m_j} = (x-\xi_1)^{m_1-1} \cdots (x-\xi_s)^{m_s-1}h(x)$, $h(x) \in L[x]$. It is possible that $f'(x) = 0$ (e.g., if $f(x) = x^p \in Z_p[x]$). If $f'(x) \neq 0$, then clearly $(x-\xi_1)^{m_1-1}$ $\cdots (x-\xi_s)^{m_s-1}$ is a factor of $f(x)$ and $f'(x)$ in $L[x]$. But then g.c.d.$(f(x), f'(x))$ in $L[x]$ is of positive degree. In view of Exercise 20, the (monic) g.c.d.$(f(x), f'(x))$ in $K[x]$ is the same as it is in $L[x]$. To prove the converse, assume first that $f'(x) = 0$. Then in view of Exercise 22(v) and (vi), char $K = p$ and $f(x) = \varphi(x^p) \in K[x^p]$, where $\varphi(y) \in K[y]$ for y some indeterminate over K. Let F be a splitting field for $\varphi(y)$ so that $\varphi(y) = a\Pi_{i=1}^r(y-\eta_i)$, $\eta_1, \ldots, \eta_r \in F$. Note that since $\deg f(x) = n$ and $f(x) = \varphi(x^p) = a\Pi_{i=1}^r(x^p - \eta_i)$, we have $rp = n$. Extend F (call the new field F also) so that it contains elements $\nu_1, \ldots, \nu_r$ for which $\nu_i{}^p = \eta_i$, $i = 1, \ldots, r$. Then since $K \subset F$, F also has characteristic p because $p \cdot 1 = 0$ in K and F both. Hence (why?) $(x-\nu_i)^p = x^p - \nu_i{}^p = x^p - \eta_i$. Thus in $F[x]$, $f(x) = a\Pi_{i=1}^r(x^p - \eta_i) = a(\Pi_{i=1}^r(x-\nu_i))^p$. The roots of $f(x)$ are $\nu_1, \ldots, \nu_r$ and since $rp = n$, $p > 1$, they are fewer than n in number. As we noted above, this fact is independent of the field containing the roots of $f(x)$. Next, assume $f'(x) \neq 0$ and $d(x) =$ g.c.d. $(f(x), f'(x))$, $\deg d(x) \geq 1$. Then $f(x) = d(x)h(x)$, $f'(x) = d(x)h'(x) + d'(x)h(x)$, and $\deg h(x) \leq n-1$ because $h(x) \mid f(x)$ and $\deg d(x) \geq 1$. Also $d(x) \mid d'(x)h(x)$. Case (i): $d'(x) = 0$. Then as in the above argument, K is of finite characteristic p and $d(x) = (l(x))^p$ in some extension $F[x]$. Thus since $d(x) \mid f(x)$, $f(x)$ must have multiple roots in some extension. Case (ii): $d'(x) \neq 0$. Then since $d(x) \mid d'(x)h(x)$ and deg $d'(x) <$ deg $d(x)$, it follows that $d(x)$ and $h(x)$

must have a common factor $\mu(x)$ of positive degree. Thus since $f(x) = d(x)h(x)$, it follows that $\mu(x)^2 \mid f(x)$ and again $f(x)$ has multiple roots.

26. In the notation of Exercise 25, prove that if char $K = 0$ and $f(x)$ is irreducible in $K[x]$, $\deg f(x) \geq 1$, then the roots of $f(x)$ are distinct. *Hint*: If $f(x)$ is irreducible, then $f(x)$ and $f'(x)$ are relatively prime in $K[x]$ because $f'(x) \neq 0$ and $\deg f'(x) < \deg f(x)$. In other words, g.c.d.$(f(x), f'(x)) = 1$. Apply Exercise 25.

27. Let p be a prime integer, K a field, x an indeterminate over K, and $\alpha \in K$. Prove that $x^p - \alpha$ is irreducible in $K[x]$ iff $x^p - \alpha$ has no roots in K. *Hint*: If $x^p - \alpha$ has a root $\xi \in K$, then trivially $x - \xi \in K[x]$ and $x - \xi \mid x^p - \alpha$. To prove the converse, clearly $\alpha \neq 0$, and we take cases: (i) char $K = p$. In some extension field F of K let ξ be a root of $x^p - \alpha$ so that $\xi^p = \alpha$. Then $(x - \xi)^p = x^p - \xi^p = x^p - \alpha$ (why?) so that all roots of $x^p - \alpha$ are equal to ξ, i.e., ξ is a root of multiplicity p. Suppose $x^p - \alpha = g(x)h(x)$ is a proper factorization in $K[x]$, $1 \leq \deg g(x) = r < p$. Since $x^p - \alpha = (x - \xi)^p$ and $F[x]$ is a U.F.D., we conclude that $g(x) = f(x - \xi)^r$, $f \in F$. But the leading coefficient of $g(x) = f(x - \xi)^r$ is f and $g(x) \in K[x]$. Thus $f \in K$. The constant term is $(-1)^r f\xi^r \in K$, again because $g(x) \in K[x]$, and so $\xi^r \in K$. Now $r < p$ and p is prime, so g.c.d.$(p,r) = 1$ and there exist integers s and t such that $sp + tr = 1$. Then $\xi^{tr} = (\xi^r)^t \in K$ and $\xi^{tr} = \xi^{1-sp} = \xi \cdot (\xi^p)^{-s} = \xi\alpha^{-s}$. Hence $\xi\alpha^{-s} \in K$. But then $\xi = (\xi\alpha^{-s})\alpha^s \in K$. In other words the root ξ is in K. Case (ii): char $K \neq p$. If ξ and η are any two roots of $x^p - \alpha$ in an extension field F, then $(\xi\eta^{-1})^p = \xi^p\eta^{-p} = 1$. Thus $\xi\eta^{-1} = \omega$ where $\omega^p = 1$, i.e., $\xi = \eta\omega$. List the roots of $x^p - \alpha$ as follows: $\xi\omega_1$ $(\omega_1 = 1)$, $\xi\omega_2, \ldots, \xi\omega_p$, $\omega_i{}^p = 1$, $i = 1, \ldots, p$. Again, if $x^p - \alpha = g(x)h(x)$ is a proper factorization in $K[x]$, $\deg g(x) = r < p$, then $g(x) = f(x - \xi\omega_1)(x - \xi\omega_2) \cdots (x - \xi\omega_r)$, $f \in K$ (after a suitable reordering of the roots). The constant term of $g(x)$ is $f(-1)^r\xi^r\omega_1 \cdots \omega_r \in K$. Let $\omega = \omega_1 \cdots \omega_r$ so that $\beta = \xi^r\omega \in K$, $\omega^p = 1$. Again $sp + tr = 1$ and $\xi^{tr} = (\xi^r)^t = (\beta\omega^{-1})^t = \beta^t\omega^{-t}$. Also, $\xi^{tr} = \xi^{1-sp} = \xi \cdot (\xi^p)^{-s} = \xi\alpha^{-s}$. Hence, $\xi\alpha^{-s} = \beta^t\omega^{-t}$, or $\xi = \beta^t\alpha^s\omega^{-t}$. But then $\alpha = \xi^p = (\beta^t\alpha^s)^p\omega^{-tp} = (\beta^t\alpha^s)^p$. In other words, $\beta^t\alpha^s \in K$ is a root of $x^p - \alpha$.

28. Discuss the irreducibility of the following polynomials over the indicated fields:
   (i) $x^7 - 2$ in $Q[x]$. *Hint*: $x^7 - 2$ has no linear factor in $Z[x]$, and hence no linear factor in $Q[x]$. Apply Exercise 27.
   (ii) $x^3 - [2]$ in $Z_5[x]$.
   (iii) $x^5 - [3]$ in $Z_7[x]$.

29. Let K be a field, char $K = p$ (prime, of course!), x an indeterminate over K, and $\alpha \in K$. Then $x^p - x - \alpha$ is irreducible in $K[x]$ iff $x^p - x - \alpha$ has no root in K. *Hint*: If $\xi^p = \xi + \alpha$, then $(\xi + k)^p = \xi^p + k^p = \xi + \alpha + k^p$. Let $k = l \cdot 1$, $1 \leq l \leq p - 1$. Then since char $K = p$, it follows that $\{l \cdot 1 \mid 1 \leq l \leq p - 1\}$ is a multiplicative group of order $p - 1$, and hence (see Exercise 21, Section 3.2) if k is any element of this group we have $k^{p-1} = 1$, or $k^p = k$. Thus $(\xi + l \cdot 1)^p = \xi + \alpha + l \cdot 1$, $l = 0, \ldots, p - 1$, and we conclude that the p distinct roots of $x^p - x - \alpha$ are $\xi, \xi + 1, \xi + 2, \ldots, \xi + (p - 1)$. Suppose that $x^p - x - \alpha = g(x)h(x)$ is a proper factorization in $K[x]$, $1 \leq \deg g(x) = r < p$. The sum β of the roots of $g(x)$ is an element of K, and these roots are some r of the elements $\xi, \xi + 1, \ldots, \xi + (p - 1)$. Thus $\beta = r\xi + m \cdot 1$, $m \in N$. But $0 < r$

$< p$, so $r \cdot 1 \neq 0$ and hence has an inverse in K. We conclude $\xi = (r \cdot 1)^{-1}(\beta - m \cdot 1) \in K$.

30. Discuss the irreducibility of the following polynomials over the indicated fields.
(i) $x^3 - x - [2]$ in $Z_3[x]$.
(ii) $x^5 - x - [4]$ in $Z_5[x]$.
(iii) $x^7 - x - [5]$ in $Z_7[x]$.

31. Let D be an integral domain and let $\delta: D - \{0\} \to N \cup \{0\}$ be a function satisfying
(i) $\delta(cd) \geq \delta(c)$ if $cd \neq 0$;
(ii) If $b \neq 0$ and $a \in D$, then there exist q and r in D such that $a = bq + r$ and $r = 0$ or $\delta(r) < \delta(b)$.

Then the pair (D,δ) is called a *Euclidean domain*, δ is called a *valuation* or a *norm* or a *stathm*, and b, q ,and r are called the *divisor*, *quotient*, and *remainder*, respectively.
(i) Show that (Z,δ) is a Euclidean domain where $\delta(n) = |n|$, $n \neq 0$.
(ii) Show that if K is a field and x is an indeterminate over K, then $(K[x],\delta)$ is a Euclidean domain where $\delta(f(x)) = \deg f(x)$, $f(x) \neq 0$.
(iii) Show that $(Z[i],\delta)$ is a Euclidean domain where $\delta(a + ib) = a^2 + b^2$ $a,b \neq 0$. *Hint*: For (iii), see the proof of Corollary 8, Section 3.3.

32. Let (D,δ) be a Euclidean domain. Prove:
(i) If $d \neq 0$, then $\delta(d) \geq \delta(1)$.
(ii) If c and d are associates, then $\delta(c) = \delta(d)$.
(iii) e is a unit iff $\delta(e) = \delta(1)$.
(iv) The quotient q and remainder r in Exercise 31 are uniquely determined (given a and b) iff $\delta(a + b) \leq \max\{\delta(a),\delta(b)\}$.
(v) D is a P.I.D.
(vi) D is a U.F.D.

Hint: (i) $\delta(d) = \delta(d \cdot 1) \geq \delta(1)$; (ii) $c = ed$, e a unit, implies $\delta(c) = \delta(ed) \geq \delta(d) = \delta(e^{-1}c) \geq \delta(c)$; (iii) obvious by (i) and (ii); (iv) Suppose $bq + r = bQ + R$, $b(q - Q) = R - r$. If $R - r \neq 0$, then $\delta(b) \leq \delta(b(q - Q)) = \delta(R - r) \leq \max\{\delta(R),\delta(-r)\} = \max\{\delta(R),\delta(r)\} < \delta(b)$. Thus $R - r = 0$, $Q - q = 0$. Conversely, suppose $\delta(a + b) > \max\{\delta(a),\delta(b)\}$ for some a and b. Then $b = 0(a + b) + b = 1(a + b) - a$ and $\delta(-a) = \delta(a) < \delta(a + b)$, $\delta(b) < \delta(a + b)$. (v) If $A \lhd D$, choose an element $a \in A$ of least norm and use Exercise 31, part (ii) of the definition, to show that $A = (a)$. (vi) Apply Theorem 3.6, Section 3.3.

33. If $n \in Z$ is not divisible by the square of any integer (other than ± 1), then n is said to be *square free* and $Q(\sqrt{n})$ is called a *quadratic number field.*
(i) If m and n are square free, then $m = n$ iff $Q(\sqrt{m}) = Q(\sqrt{n})$.
(ii) If $\alpha = a + b\sqrt{n} \in Q(\sqrt{n})$, $a,b \in Q$, then $a - b\sqrt{n}$ is called the *conjugate* of α and is denoted by $\bar{\alpha}$. Show that f: $Q(\sqrt{n}) \to Q(\sqrt{n})$ defined by $f(\alpha) = \bar{\alpha}$ is an isomorphism.
(iii) Let $N(\alpha) = a^2 - b^2 n$ and $\delta(\alpha) = |N(\alpha)|$. Then $\delta(\alpha) = 0$ iff $\alpha = 0$; $\delta(\alpha\beta) = \delta(\alpha)\delta(\beta)$; $\delta(1) = 1$; if $\alpha \neq 0$, then $\bar{\alpha}/N(\alpha) = \alpha^{-1}$.
(iv) Set $Z[\sqrt{n}] = \{a + b\sqrt{n} \mid a,b \in Z\}$. Then $\delta(\alpha) = 1$ iff α is a unit in $Z[\sqrt{n}]$; if $\delta(\alpha)$ is a prime, then α is irreducible. *Hint*: (i) If $Q(\sqrt{m}) = Q(\sqrt{n})$,

then $\sqrt{m} = a + b\sqrt{n}$, $a,b \in Q$. Hence $m = a^2 + nb^2 + 2ab\sqrt{n}$. If $ab \neq 0$, then $\sqrt{n} \in Q$. Consider $x^2 - n$. If $n < 0$, then this is obviously irreducible over Q; if $n > 0$, it would be reducible iff $\sqrt{n} \in Z$. But n is square free. Thus $\sqrt{n}$ is irrational and we conclude $ab = 0$. By similar reasoning (since $\sqrt{m}$ is irrational) $b \neq 0$. Hence $a = 0$, $\sqrt{m} = b\sqrt{n}$, $m = b^2n$, $b^2 \mid m$, a contradiction unless $b^2 = 1$, i.e., $m = n$. (ii) obvious. (iii) $\delta(\alpha) = |N(\alpha)| = |\alpha\bar{\alpha}|$, so $\delta(\alpha\beta) = |\alpha\beta\overline{\alpha\beta}| = |\alpha\bar{\alpha}|\,|\beta\bar{\beta}| = \delta(\alpha)\delta(\beta)$. Also $(\bar{\alpha}/N(\alpha))\cdot\alpha = \bar{\alpha}\alpha/\bar{\alpha}\alpha = 1$. (iv) If $\alpha\beta = 1$, then $\delta(\alpha)\delta(\beta) = 1$ and $\delta(\alpha)$ and $\delta(\beta)$ are positive integers. Conversely if $\delta(\alpha) = 1$, then $N(\alpha) = \pm 1$ and $\bar{\alpha}/N(\alpha) \in Z[\sqrt{n}]$. The rest of (iv) is easy.

34. Let D be a U.F.D., x an indeterminate over D, and let α be algebraic over D. Let K be the quotient field of D. An element $\xi \in K(\alpha)$ is said to be an *algebraic integer* if $\varphi(\xi) = 0$ for some monic polynomial $\varphi(x) \in D[x]$. Prove that if $f(x) \in D[x]$ is a primitive polynomial of least degree in $D[x]$ for which $f(\xi) = 0$, then the leading coefficient of $f(x)$ is a unit in D. *Hint*: By the usual argument, any polynomial in $K[x]$ of which ξ is a root is divisible in $K[x]$ by $f(x)$. Thus suppose $\varphi(x) \in D[x]$ is monic and $\varphi(\xi) = 0$. Then $\varphi(x) = h(x)f(x)$, $h(x) \in K[x]$. Write $h(x) = (a/b)h_1(x)$ in which $h_1(x)$ is primitive and $a,b \in D$. Then $\varphi(x) = (a/b)\,h_1(x)f(x)$ or $b\varphi(x) = ah_1(x)f(x)$. Since $\varphi(x)$ is monic, it is obviously primitive, and $h_1(x)f(x)$ is primitive by Theorem 4.2. Thus $a = \varepsilon b$, where ε is a unit in D. Since $\varphi(x) = \varepsilon h_1(x)f(x)$, we see by comparing leading coefficients that $1 = \varepsilon cd$, where c is the leading coefficient of $h_1(x)$ and d is the leading coefficient of $f(x)$. But then d is a unit.

35. In the notation of Exercise 34, prove that if $\psi(x)$ is the characteristic polynomial of ξ, then $\psi(x) \in D[x]$. *Hint*: Obviously $f(x)$ must be a multiple of $\psi(x)$, i.e., $\psi(x) = kf(x)$ $k \in K$. But by Exercise 34 the leading coefficient of $f(x)$ is a unit in D. Hence since by definition $\psi(x)$ is monic, k must be a unit in D. But then $\psi(x) \in D[x]$.

36. Recall the notation: $a \equiv b \pmod{m}$ in Z means simply that $[a] = [b]$ in Z_m, or equivalently that $m \mid a - b$. Let $m \neq 1$, a square free integer (positive or negative). Prove that the algebraic integers in $Q(\sqrt{m})$ are:
(i) Numbers of the form $a + b\sqrt{m}$, a,b in Z, if $m \equiv 2 \pmod 4$ or $m \equiv 3 \pmod 4$;
(ii) Numbers of the form $a + b\tau$, $\tau = (\sqrt{m} - 1)/2$, a, b in Z if $m \equiv 1 \pmod 4$.
Hint: Let ξ be an algebraic integer in $Q(\sqrt{m})$. Then we can write $\xi = (a + b\sqrt{m})/c$, where $a, b, c \in Z$, $c > 0$, and g.c.d.$(a,b,c) = 1$ (why?). If $b = 0$ there is no problem, for the characteristic polynomial of a/c in $Q[x]$ is $x - a/c$, but this must be in $Z[x]$ by Exercise 35, i.e., $c \mid a$. If $b \neq 0$ then $(c\xi - a)^2 = b^2m$ or $c^2\xi^2 - 2ac\xi + a^2 - b^2m = 0$. Let $\psi(x) = x^2 - (2a/c)x + (a^2 - b^2m)/2 \in Q[x]$, clearly the characteristic polynomial for ξ. By Exercise 35, $\psi(x) \in Z[x]$ and hence $c \mid 2a$ and $c^2 \mid a^2 - b^2m$. Let $d =$ g.c.d.(a,c). Then $d^2 \mid a^2$, $d^2 \mid c^2$ and $d^2 \mid a^2 - b^2m$. Thus $d^2 \mid b^2m$ and since m is square free, $d \mid b$ (why? Factor d,b,m into prime factors to see it). But then $d \mid a$, $d \mid b$, $d \mid c$, and g.c.d.$(a,b,c) = 1$. Thus $d = 1$ and a and c are relatively prime. But $c \mid 2a$, $c > 0$; so $c = 1$ or $c = 2$. If $c = 2$, then since g.c.d. $(a,c) = 1$, it follows that a is odd and hence $a^2 \equiv 1 \pmod 4$. But $c^2 \mid a^2 - b^2m$, so $a^2 - b^2m \equiv 0 \pmod 4$. Thus $b^2m \equiv 1 \pmod 4$. Now b must be odd; otherwise $b^2 \equiv 0 \pmod 4$ and it would follow that $b^2m \equiv 0 \pmod 4$. Now since b is

odd, $b^2 \equiv 1 \pmod 4$, and hence $1 \equiv b^2m \equiv m \pmod 4$. We have proved: If $\xi = (a + b\sqrt{m})/c, c > 0$, g.c.d.$(a,b,c) = 1$, then $c = 1$ or $c = 2$, and if $c = 2$ then $m \equiv 1 \pmod 4$. Thus if $m \equiv 2 \pmod 4$ or $m \equiv 3 \pmod 4$, we can conclude that $c = 1$, $\xi = a + b\sqrt{m}$, i.e., case (i). Thus suppose $m \equiv 1 \pmod 4$. We can assume $c = 2$ since if $c = 1$, $\xi = a + b\sqrt{m} = a + b + 2b(\sqrt{m} - 1)/2 = a + b + 2b\tau$ which is of the form in (ii). If $c = 2$ then $\xi = (a + b\sqrt{m})/2 = (a + b)/2 + b(\sqrt{m} - 1)/2$. Both a and b are odd as we saw above, and so $(a + b)/2$ is an integer.

37. Prove that if $m = -11, -7, -3, -2, -1, 2, 3, 5$, then the set D of all algebraic integers in $Q(\sqrt{m})$ is a Euclidean domain in which the valuation δ is given by $\delta(\alpha) = |N(\alpha)|$ (see Exercise 33). *Hint*: Case (i): $m \not\equiv 1 \pmod 4$. Then $m = -2, -1, 2, 3$, and by Exercise 36, $D = \{a + b\sqrt{m} \mid a,b \in Z\}$ and it is obvious that D is an integral domain. Let α and β be in D, $\beta \neq 0$, and write $\alpha/\beta = x + y\sqrt{m}$, $x,y \in Q$. Choose integers r and s such that $|x - r| \leq \frac{1}{2}$, $|y - s| \leq \frac{1}{2}$. Let $\gamma = r + s\sqrt{m} \in D$, $\rho = \beta[(x - r) + (y - s)\sqrt{m}]$. Then $\alpha = \beta(x + y\sqrt{m}) = \beta\gamma + \rho$. Also $\rho = \alpha - \beta\gamma \in D$. Assume $\rho \neq 0$. Then compute that $\delta(\rho) = |N(\beta)|\,|N((x - r) + (y - s)\sqrt{m})| = \delta(\beta)|(x - r)^2 - m(y - s)^2| \leq \delta(\beta)\{|x - r|^2 + |m|\,|y - s|^2\} \leq \delta(\beta)\{(\frac{1}{2})^2 + |m|(\frac{1}{2})^2\} \leq \delta(\beta)\{\frac{1}{4} + \frac{3}{4}\} = \delta(\beta)$. Equality can occur only when $|x - r| = |y - s| = \frac{1}{2}$ and $m = 3$, but then $|(x - r)^2 - m(y - s)^2| = |\frac{1}{4} - \frac{3}{4}| = \frac{1}{2}$ and $\delta(\rho) = \frac{1}{2}\delta(\beta)$. Thus in fact $\delta(\rho) < \delta(\beta)$ and (D,δ) is Euclidean. Case (ii): $m \equiv 1 \pmod 4$. Then $m = -11, -7, -3, 5$. By Exercise 36, $D = \{a + b\tau \mid a,b \in Z, \tau = (\sqrt{m} - 1)/2\}$, obviously an integral domain (why?). Observe that $a + b(\sqrt{m} + 1)/2 \in D$ for any $a,b \in Z$ (why?). Let α and β be in D, $\beta \neq 0$, and again write $\alpha/\beta = x + y\sqrt{m}$, $x,y \in Q$. Obtain $s \in Z$ such that $|2y - s| \leq \frac{1}{2}$, $|y - s/2| \leq \frac{1}{4}$. Also $x - s/2 \in Q$ so choose $r \in Z$ such that $|(x - s/2) - r| \leq \frac{1}{2}$. Let $\gamma = r + s(1 + \sqrt{m})/2 \in D$, let $\rho = \beta[(x - r - s/2) + (y - s/2)\sqrt{m}]$, and check that $\alpha = \beta(x + y\sqrt{m}) = \beta\gamma + \rho$. Since $\alpha, \beta, \gamma \in D$ it follows that $\rho \in D$. Moreover, if $\rho \neq 0$, we have

$$\delta(\rho) = |N(\beta)|\left|N\left(\left(x - r - \frac{s}{2}\right) + \left(y - \frac{s}{2}\right)\sqrt{m}\right)\right| = \delta(\beta)\left|\left(x - r - \frac{s}{2}\right)^2 - m\left(y - \frac{s}{2}\right)^2\right| \leq \delta(\beta)\left(\frac{1}{4} + |m|\left(\frac{1}{4}\right)^2\right) \leq \delta(\beta)\left(\frac{1}{4} + \frac{11}{16}\right) < \delta(\beta).$$

The problem of determining the entire set of values of $m \in Z$ (square free) for which the integers in $Q(\sqrt{m})$ form a Euclidean domain is very difficult but known: These are $m = -11, -7, -3, -2, -1, 2, 3, 5, 6, 7, 11, 13, 17, 19, 21, 29, 33, 37, 41, 57, 73$. It is also known that $m = 2, 3, 5, 6, 7, 11, 13, 14, 17, 19, 21, 22, 23, 29, 31, 33, 37, 38, 41, 43, 46, 47, 53, 57, 59, 61, 62, 67, 69, 71, 73, 77, 83, 86, 89, 93, 94, 97$ are values of m for which the algebraic integers in $Q(\sqrt{m})$ form a U.F.D. It is conjectured that there exist infinitely many such $m > 0$, but this has not been proved.

Glossary 3.4

algebraic extension, 205
algebraic integer, 223
basis, 206
characteristic polynomial of an algebraic number, 208
content of a polynomial, 194

3.5 Polynomials and Resultants

As we saw in Corollary 2, Section 3.4, if D is a U.F.D., then the domain of polynomials in n independent indeterminates, $D[x_1, \ldots, x_n]$, is also a U.F. D. We can then apply Theorem 4.1, Section 3.4, to immediately conclude the following.

Theorem 5.1 *Let D be a* U.F.D.; *$x_1, \ldots, x_n$, n independent indeterminates over D; and $\varphi_i(x_1, \ldots, x_n)$, $i = 1, \ldots, n$, nonzero polynomials in $D[x_1, \ldots, x_n]$. Then there exists a g.c.d. of these polynomials in $D[x_1, \ldots, x_n]$ which is unique except for associates.*

We specialize a number of earlier results to $D[x_1, \ldots, x_n]$.

Corollary 1 *If f, g, h are in $D[x_1, \ldots, x_n]$, $f \mid gh$ and* g.c.d.$(f, g) = 1$, *then $f \mid h$.*

Proof: Let $p_1, \ldots, p_m$ be all the distinct prime factors that appear in the factorizations of f, g, h. Write

$$f = p_1^{a_1} \cdots p_m^{a_m},$$
$$g = p_1^{b_1} \cdots p_m^{b_m},$$
$$h = p_1^{c_1} \cdots p_m^{c_m},$$

where the a_i, b_i, and c_i are nonnegative integers. Now g.c.d.$(f, g) = 1$ means that $a_i b_i = 0$, $i = 1, \ldots, m$, and $f \mid gh$ is equivalent to

$$a_i \leq b_i + c_i, \qquad i = 1, \ldots, m. \tag{1}$$

Thus if $a_i > 0$, then $b_i = 0$ and (1) implies that $a_i \leq c_i$. If $a_i = 0$, then trivially $a_i \leq c_i$. Hence $f | h$. ∎

Corollary 2 *Let* $0 \neq \varphi_i \in D[x_1, \ldots, x_n]$, $i = 1, \ldots, m$, *and set*

$$d = \text{g.c.d.}(\varphi_1, \ldots, \varphi_m)$$

so that

$$\varphi_i = dq_i \tag{2}$$

where

$$q_i \in D[x_1, \ldots, x_n], \qquad i = 1, \ldots, m.$$

Then

$$\text{g.c.d.}(q_1, \ldots, q_m) = 1. \tag{3}$$

Proof: Let $p_1, \ldots, p_k$ be all the distinct primes that appear in the factorization of the φ_i, $i = 1, \ldots, m$, so that

$$\varphi_i = \prod_{t=1}^{k} p_t^{e_{it}}$$

$e_{it} \in N \cup \{0\}$, $t = 1, \ldots, k$, $i = 1, \ldots, m$. Let

$$e_t = \min_{1 \leq i \leq m} e_{it}, \qquad t = 1, \ldots, k \tag{4}$$

and as in Theorem 4.1, Section 3.4, we have

$$d = \prod_{t=1}^{k} p_t^{e_t}.$$

By (4), for each $t = 1, \ldots, k$ there is at least one i_t such that the power of p_t occurring in φ_{i_t} is the same as the power of p_t occurring in d. Thus p_t does not occur in the prime factorization of q_{i_t}. In other words, if

$$q_i = \prod_{t=1}^{k} p_t^{a_{it}}$$

then for each t there is an i_t such that

$$a_{i_t t} = 0.$$

Hence

$$\min_{1 \leq i \leq m} a_{it} = 0, \qquad t = 1, \ldots, k$$

and it follows that

$$\text{g.c.d.}(q_1, \ldots, q_m) = 1. \quad ∎$$

Corollary 3 *If* $p \in D[x_1, \ldots, x_n]$ *is prime and* $\text{g.c.d.}(p, f) \neq 1$, *then* $p | f$.

Proof: Let $d = \text{g.c.d.}(p, f)$. Since $d \neq 1$ and $d | p$, it follows that $d = \varepsilon p$, ε a unit in D. Hence $p | f$. ∎

Corollary 4 *Let* $f_i \in D[x_1, \ldots, x_n]$, $i = 1, 2$, *and assume* g.c.d. $(f_1, f_2) = 1$. *If* $f_i | \varphi$, $i = 1, 2$, *then*

$$f_1 f_2 \mid \varphi.$$

Proof: Write

$$\varphi = q_1 f_1. \tag{5}$$

Then

$$f_2 \mid \varphi$$

and g.c.d.$(f_1, f_2) = 1$ imply (by Corollary 1) that

$$f_2 \mid q_1.$$

Hence from (5), $f_1 f_2 \mid \varphi$. ∎

Example 1 (a) In $D[x_1,x_2]$ observe that g.c.d.$(x_1,x_2) = 1$. However, there do not exist polynomials $f,g \in D[x_1,x_2]$ such that

$$x_1 f(x_1,x_2) + x_2 g(x_1,x_2) = 1, \tag{6}$$

for the left side of (6) has degree at least 1 [see the definition of degree preceding formula (45) in Section 3.2] and thus formula (6) is not possible.

(b) $D[x_1,x_2]$ is not a P.I.D., for otherwise f and g satisfying (6) would exist.

(c) $D[x_1,x_2]$ is not a Euclidean domain, for otherwise it would be a P.I.D. [see Exercise 33(v), Section 3.4].

(d) Let f, g, φ, θ be nonzero polynomials in $D[x_1, \ldots, x_n]$, and assume that f/g is in *lowest terms* in the rational function field $D(x_1, \ldots, x_n)$, i.e., g.c.d.$(f,g) = 1$. Let $S \subset D$ be an infinite set and suppose

$$\frac{f(s_1, \ldots, s_n)}{g(s_1, \ldots, s_n)} = \frac{\varphi(s_1, \ldots, s_n)}{\theta(s_1, \ldots, s_n)}$$

for all specializations of x_i to s_i in S. Then there exists $h \in D[x_1, \ldots, x_n]$ such that

$$\varphi = hf \quad \text{and} \quad \theta = hg.$$

For let $\psi = f\theta - g\varphi \in D[x_1, \ldots, x_n]$ and use Corollary 4, Section 3.3, to conclude that $\psi = 0$, i.e.,

$$f\theta - g\varphi = 0$$

or

$$f\theta = g\psi. \tag{7}$$

Now $f \mid g\varphi$ from (7) and g.c.d.$(f,g) = 1$, so Corollary 1 implies that $f \mid \varphi$, say $\varphi = kf$. Similarly $\theta = hg$, (7) becomes

$$fhg = gkf,$$

and hence $h = k$.

Let m be a nonnegative integer. A polynomial

$$f = \sum_{\gamma \in \Gamma_n 0(q)} a_\gamma \prod_{t=1}^{n} x_t^{\gamma(t)} \tag{8}$$

in $D[x_1, \ldots, x_n]$ is *homogeneous of degree m*, or a *form of degree m*, if whenever $a_r \neq 0$ we have

$$\sum_{t=1}^{n} \gamma(t) = \deg \prod_{t=1}^{n} x_t^{\gamma(t)} = m.$$

(The zero polynomial is homogeneous of arbitrary degree.) Thus the elementary symmetric function (e.s.f.)

$$E_m = \sum_{r \in Q_{m,n}} \prod_{t=1}^{m} x_{\gamma(t)}, \qquad 1 \leq m \leq n,$$

and the completely symmetric function

$$h_m = \sum_{\gamma \in G_{m,n}} \prod_{t=1}^{m} x_{\gamma(t)}, \qquad 1 \leq m \leq n,$$

are both homogeneous of degree m.

A simple test for homogeneity is given in the next result.

Theorem 5.2 *Let $x_1, \ldots, x_n, t$ be independent indeterminates over the integral domain D. Let $f(x_1, \ldots, x_n) \in D[x_1, \ldots, x_n]$. Then f is homogeneous of degree m iff*

$$f(tx_1, tx_2, \ldots, tx_n) = t^m f(x_1, \ldots, x_n) \tag{9}$$

in $D[x_1, \ldots, x_n, t]$.

Proof: Let f be homogeneous of degree m and given by (8). Observe that whenever $a_\gamma \neq 0$,

$$\begin{aligned} \prod_{i=1}^{n} (tx_i)^{\gamma(i)} &= t^{\sum_{i=1}^{n} \gamma(i)} \prod_{i=1}^{n} x_i^{\gamma(i)} \\ &= t^m \prod_{i=1}^{n} x_i^{\gamma(i)}, \end{aligned}$$

and (9) follows. Conversely, let f be any nonzero polynomial in $D[x_1, \ldots, x_n]$ and group together the nonzero terms of equal degree in f,

$$f = \sum_{i=1}^{k} f_{n_i}, \tag{10}$$

in which $f_{n_i} \neq 0$ is the sum of all homogeneous terms of degree n_i appearing in f and $0 \leq n_1 < n_2 < \cdots < n_k$. Then from (10),

$$\begin{aligned} f(tx_1, \ldots, tx_n) &= \sum_{i=1}^{k} f_{n_i}(tx_1, \ldots, tx_n) \\ &= \sum_{i=1}^{k} t^{n_i} f_{n_i}(x_1, \ldots, x_n). \end{aligned} \tag{11}$$

Now if

$$f(tx_1, \ldots, tx_n) = t^m f(x_1, \ldots, x_n), \tag{12}$$

we combine (11) and (12) to obtain

$$t^m f(x_1, \ldots, x_n) - \sum_{i=1}^{k} t^{n_i} f_{n_i}(x_1, \ldots, x_n) = 0. \tag{13}$$

Since t is an indeterminate over $D[x_1, \ldots, x_n]$, k must be 1. For if $k \geq 2$, the left side of (13) is a nonzero polynomial in $D[x_1, \ldots, x_n][t]$. We also must have

$$m = n \quad \text{and} \quad f = f_{n_1};$$

thus f is homogeneous of degree m. ∎

Example 2 (a) The *discriminant polynomial* in $D[x_1, \ldots, x_n]$ is the polynomial

$$\Delta(x_1, \ldots, x_n) = \prod_{1 \leq i < j \leq n} (x_i - x_j)^2.$$

Observe that

$$\begin{aligned}\Delta(tx_1, \ldots, tx_n) &= \prod_{1 \leq i < j \leq n} (tx_i - tx_j)^2 \\ &= t^{n(n-1)} \Delta(x_1, \ldots, x_n).\end{aligned}$$

Hence by Theorem 5.2, Δ is homogeneous of degree $n(n-1)$.

(b) Observe that $\Delta(x_1, \ldots, x_n)$ is a symmetric polynomial and therefore by Theorem 2.5, Section 3.2, Δ is a (unique) polynomial in the e.s.f. In fact, let V be the matrix

$$V = \begin{bmatrix} 1 & 1 & 1 & \cdots & 1 \\ x_1 & x_2 & x_3 & \cdots & x_n \\ x_1^2 & x_2^2 & x_3^2 & \cdots & x_n^2 \\ \cdots & \cdots & \cdots & \cdots & \cdots \\ x_1^{n-1} & x_2^{n-1} & x_3^{n-1} & \cdots & x_n^{n-1} \end{bmatrix}.$$

It is easily checked that

$$\det V = \prod_{1 \leq i < j \leq n} (x_j - x_i).$$

Hence $$\Delta = \det (V)^2 = \det VV^{\mathsf{T}}.$$

But $$\begin{aligned}(VV^{\mathsf{T}})_{ij} &= \sum_{k=1}^{n} x_k^{i-1} x_k^{j-1} = \sum_{k=1}^{n} x_k^{i+j-2} \\ &= s_{i+j-2}(x_1, \ldots, x_n).\end{aligned}$$

In other words, Δ is the determinant of the matrix

$$VV^{\mathsf{T}} = \begin{bmatrix} n & s_1 & s_2 & \cdots & s_{n-1} \\ s_1 & s_2 & s_3 & \cdots & s_n \\ s_2 & s_3 & s_4 & \cdots & s_{n+1} \\ \cdots & \cdots & \cdots & \cdots & \cdots \\ s_{n-1} & s_n & s_{n+1} & \cdots & s_{2n-2} \end{bmatrix}.$$

From the Newton identities (see Example 3, Section 3.2), each s_j can be expressed (recursively) in terms of the e.s.f. $E_1, \ldots, E_{2n-2}$, and hence Δ may be so expressed.

Let $f(x_1, \ldots, x_n) \in D[x_1, \ldots, x_n]$ be a homogeneous polynomial. If D is a U.F.D., then $D[x_1, \ldots, x_n]$ is a U.F.D., and hence f can be factored uniquely into a product of primes. The question arises, are the factors also homogeneous? We have the following theorem.

Theorem 5.3 *If D is a U.F.D., then any factor of a homogeneous polynomial in $D[x_1, \ldots, x_n]$ must also be homogeneous.*

Proof: It suffices to prove that if $f \in D[x_1, \ldots, x_n]$ is not a homogeneous polynomial, then fg is not homogeneous for any $0 \neq g \in D[x_1, \ldots, x_n]$. Write

$$f = \sum_{i=1}^{k} f_{n_i}$$

in which $k \geq 2$ as in (10), $f_{n_i} \neq 0$ homogeneous of degree n_i, $0 \leq n_1 < \cdots < n_k$. Similarly, write

$$g = \sum_{i=1}^{p} g_{r_i},$$

$g_{r_i} \neq 0$ homogeneous of degree r_i, $0 \leq r_1 < \cdots < r_p$. Then

$$fg = (f_{n_1} g_{r_1}) + \cdots + (f_{n_k} g_{r_p}), \tag{14}$$

in which the terms homogeneous of a given degree are grouped together. Clearly

$$\begin{aligned} \deg f_{n_1} g_{r_1} &= n_1 + r_1 < n_k + r_p \\ &= \deg f_{n_k} g_{r_p} \end{aligned}$$

and the intermediate terms in (14) must have degrees strictly between $n_1 + r_1$ and $n_k + r_p$. But then (14) is certainly not a sum of monomials in $x_1, \ldots, x_n$ all of the same degree. ∎

Let $u_0, u_1, \ldots, u_m$ and $v_0, v_1, \ldots, v_n$ be $m + n + 2$ independent indeterminates over D, and define an $(m + n)$-square matrix

$$S = \begin{bmatrix} u_0 & u_1 & u_2 & \cdots & u_m & & \cdots & \\ \cdot & u_0 & u_1 & \cdots & u_{m-1} & u_m & \cdots & \\ \cdots & \cdots & \cdots & \cdots & \cdots & \cdots & \cdots & \cdots \\ \cdot & \cdot & \cdot & u_0 & u_1 & \cdots & & u_m \\ v_0 & v_1 & v_2 & \cdots & v_n & \cdots & & \\ \cdot & v_0 & v_1 & \cdots & v_{n-1} & v_n & \cdots & \\ \cdots & \cdots & \cdots & \cdots & \cdots & \cdots & \cdots & \cdots \\ \cdot & \cdot & \cdot & v_0 & v_1 & \cdot & \cdot & v_n \end{bmatrix} \begin{matrix} \left.\vphantom{\begin{matrix}a\\a\\a\\a\end{matrix}}\right\} n \text{ rows} \\ \left.\vphantom{\begin{matrix}a\\a\\a\\a\end{matrix}}\right\} m \text{ rows} \end{matrix} ; \tag{15}$$

the elements in the matrix (15) not explicitly designated are all 0. Then the polynomial

$$R = \det S \in D[u_0, u_1, \ldots, u_m, v_0, v_1, \ldots, v_n] \tag{16}$$

is called *Sylvester's determinant.*

In general, if $u_0, \ldots, u_m, v_0, \ldots, v_n, \ldots, w_0, \ldots, w_k$ are independent indeterminates over D, then a term

$$cu_0^{\alpha(0)}u_1^{\alpha(1)} \cdots u_m^{\alpha(m)}v_0^{\beta(0)} \cdots v_n^{\beta(n)} \cdots w_0^{\gamma(0)} \cdots w_k^{\gamma(k)}, \qquad c \neq 0$$

is said to have *weight*

$$\sum_{t=0}^{m} t\alpha(t) + \sum_{t=0}^{n} t\beta(t) + \cdots + \sum_{t=0}^{k} t\gamma(t). \tag{17}$$

A polynomial in these indeterminates is said to be *isobaric* if every term has the same weight, and the *weight* of an isobaric polynomial is the common weight of all its terms. Thus for example

$$\det \begin{bmatrix} u_0 & u_1 & u_2 & 0 \\ 0 & u_0 & u_1 & u_2 \\ v_0 & v_1 & v_2 & 0 \\ 0 & v_0 & v_1 & v_2 \end{bmatrix} = u_0^2v_2^2 - u_0u_1v_1v_2 + u_0u_2v_1^2 - u_0u_2v_0v_2 \\ + u_1^2v_0v_2 - u_0u_2v_0v_2 - u_1u_2v_0v_1 + u_2^2v_0^2$$

is an isobaric polynomial of weight 4.

Theorem 5.4 *Sylvester's determinant R in* (16) *is homogeneous of degree $m + n$ and isobaric of weight mn.*

Proof: Write $R = R(u_0, \ldots, u_m, v_0, \ldots, v_n)$ so that

$$R(tu_0, \ldots, tu_m, tv_0, \ldots, tv_n) \tag{18}$$

multiplies each row in the matrix (16) by t. By elementary properties of determinants we see that (18) is equal to

$$t^{m+n}R(u_0, \ldots, u_m, v_0, \ldots, v_n).$$

Apply Theorem 5.2 to conclude that R is homogeneous of degree $m + n$.

Observe next that the (i, j) entry of S is given by

$$S_{ij} = \begin{cases} u_{j-i}, & 1 \leq i \leq n \\ v_{n+j-i}, & n + 1 \leq i \leq m + n \end{cases}$$

where we define $u_k = 0$ if $k < 0$ or $k > m$ and $v_k = 0$ for $k < 0$ or $k > n$. A typical term in $R = \det S$ is

$$\varepsilon(\sigma) \prod_{i=1}^{m+n} s_{i\sigma(i)} = \pm \prod_{i=1}^{n} u_{\sigma(i)-i} \prod_{i=n+1}^{m+n} v_{n+\sigma(i)-i}. \tag{19}$$

If a term (19) is not 0, its weight [see (17)] is

$$\sum_{i=1}^{n} \sigma(i) - i + \sum_{i=n+1}^{m+n} n + \sigma(i) - i = mn + \sum_{i=1}^{m+n} \sigma(i) - \sum_{i=1}^{m+n} i \\ = mn. \quad \blacksquare$$

If

$$f(x) = a_m x^m + a_{m-1}x^{m-1} + \cdots + a_0 \tag{20}$$

and

$$g(x) = b_n x^n + b_{n-1}x^{n-1} + \cdots + b_0 \tag{21}$$

are in $D[x]$, then the *resultant* of these two polynomials, denoted by $R(f, g)$, is the specialization of Sylvester's determinant

$$R(u_0, \ldots, u_m, v_0, \ldots, v_n)$$

to

$$R(a_0, \ldots, a_m, b_0, \ldots, b_n).$$

We shall assume in what follows that D has characteristic 0 and $a_m b_n \neq 0$, $m \geq 1$, $n \geq 1$, so that $\deg f(x) = m$, $\deg g(x) = n$. The following result provides a useful test for deciding if (20) and (21) have a common factor of positive degree.

Theorem 5.5 *Assume that D is a U.F.D. Then the polynomials* (20) *and* (21) *have a common factor of positive degree in* $D[x]$ *iff* $R(f, g) = 0$.

Proof: We first show that f and g have a common factor of positive degree iff there exist nonzero polynomials φ and ψ in $D[x]$, $\deg \varphi < m$, $\deg \psi < n$ such that

$$\psi f = \varphi g. \tag{22}$$

Suppose first that h is a common factor of positive degree so that

$$f = h\varphi, \qquad g = h\psi.$$

Then obviously

$$\psi f = \psi h \varphi = \varphi g.$$

On the other hand, if (22) holds then every prime divisor of g of positive degree must be a divisor of ψf. They cannot all be divisors of ψ because $\deg \psi < n = \deg g$. Hence at least one of the prime divisors of g must divide f, and hence f and g have a common factor of positive degree.

Thus in proving the theorem it suffices to show that $R(f, g) = 0$ iff there exist polynomials φ and ψ,

$$0 \neq \varphi = \alpha_{m-1}x^{m-1} + \cdots + \alpha_0,$$
$$0 \neq \psi = \beta_{n-1}x^{n-1} + \cdots + \beta_0$$

such that (22) holds. The condition (22) is then equivalent to the system of linear equations

$$\begin{aligned} a_0\beta_0 &= b_0\alpha_0, \\ a_1\beta_0 + a_0\beta_1 &= b_0\alpha_1 + b_1\alpha_0, \\ a_2\beta_0 + a_1\beta_1 + a_0\beta_2 &= b_0\alpha_2 + b_1\alpha_1 + b_2\alpha_0, \end{aligned} \tag{23}$$

$$a_3\beta_0 + a_2\beta_1 + a_1\beta_2 + a_0\beta_3 = b_0\alpha_3 + b_1\alpha_2 + b_2\alpha_1 + b_3\alpha_0,$$
$$\vdots$$
$$a_m\beta_{n-1} = b_n\alpha_{m-1}.$$

Regard (23) as a system of linear equations over the quotient field of D for the determination of the $m + n$ unknowns $\alpha_0, \ldots, \alpha_{m-1}, \beta_0, \ldots, \beta_{n-1}$. The matrix of coefficients is

$$\begin{matrix} \beta_0 & \beta_1 & \cdots & \beta_{n-1} & \alpha_0 & \alpha_1 & \cdots & \alpha_{m-1} \end{matrix}$$
$$\begin{bmatrix} a_0 & 0 & \cdots & 0 & -b_0 & 0 & \cdots & 0 \\ a_1 & a_0 & 0 \cdots & 0 & -b_1 & -b_0 & 0 \cdots & 0 \\ a_2 & a_1 & a_0 \; 0 \cdots & 0 & -b_2 & -b_1 & -b_0 \; 0 \cdots & 0 \\ \vdots & & & \vdots & \vdots & & & \vdots \\ 0 & \cdots & 0 & a_m & 0 & \cdots & 0 & -b_n \end{bmatrix} \tag{24}$$

where the columns in (24) are ordered according to the indicated unknowns. By taking the transpose of (24) and multiplying the last m rows of the resulting matrix by -1, we obtain Sylvester's matrix specialized to the coefficients of $f(x)$ and $g(x)$. But then the homogeneous system (23) has a nonzero solution iff the determinant of the coefficient matrix is 0, i.e., iff $R(f, g) = 0$. Of course, any nonzero solution in the quotient field of D produces a nonzero solution in D to the homogeneous system (23) by simply multiplying through every equation by the same common denominator. ∎

Example 3 (a) We construct the system of equations (23) and the matrix (24) for $m = 3$, $n = 2$. The product $f\psi$ is

$$(a_3x^3 + a_2x^2 + a_1x + a_0)(\beta_1x + \beta_0) = a_0\beta_0 + (a_1\beta_0 + a_0\beta_1)x$$
$$+ (a_2\beta_0 + a_1\beta_1)x^2 + (a_3\beta_0 + a_2\beta_1)x^3 + a_3\beta_1x^4.$$

Similarly the product $g\varphi$ is

$$(b_2x^2 + b_1x + b_0)(\alpha_2x^2 + \alpha_1x + \alpha_0)$$
$$= b_0\alpha_0 + (b_1\alpha_0 + b_0\alpha_1)x + (b_2\alpha_0 + b_1\alpha_1 + b_0\alpha_2)x^2$$
$$+ (b_2\alpha_1 + b_1\alpha_2)x^3 + b_2\alpha_2x^4.$$

The condition $f\psi = g\varphi$ then becomes

$$a_0\beta_0 = b_0\alpha_0,$$
$$a_1\beta_0 + a_0\beta_1 = b_1\alpha_0 + b_0\alpha_1,$$
$$a_2\beta_0 + a_1\beta_1 = b_2\alpha_0 + b_1\alpha_1 + b_0\alpha_2,$$
$$a_3\beta_0 + a_2\beta_1 = b_2\alpha_1 + b_1\alpha_2,$$
$$a_3\beta_1 = b_2\alpha_2.$$

The matrix (24) is

$$\begin{array}{ccccc} \beta_0 & \beta_1 & \alpha_0 & \alpha_1 & \alpha_2 \end{array}$$
$$\begin{bmatrix} a_0 & 0 & -b_0 & 0 & 0 \\ a_1 & a_0 & -b_1 & -b_0 & 0 \\ a_2 & a_1 & -b_2 & -b_1 & -b_0 \\ a_3 & a_2 & 0 & -b_2 & -b_1 \\ 0 & a_3 & 0 & 0 & -b_2 \end{bmatrix}. \tag{25}$$

Taking the transpose of the matrix (25) and multiplying the last three rows of the resulting matrix by -1, we obtain the matrix

$$\begin{bmatrix} a_0 & a_1 & a_2 & a_3 & 0 \\ 0 & a_0 & a_1 & a_2 & a_3 \\ b_0 & b_1 & b_2 & 0 & 0 \\ 0 & b_0 & b_1 & b_2 & 0 \\ 0 & 0 & b_0 & b_1 & b_2 \end{bmatrix}.$$

Thus f and g have a common factor of positive degree in $D[x]$ iff the determinant $R(f,g)$ of this matrix is 0.

(b) Let $f(x) = a_3x^3 + a_2x^2 + a_1x + a_0$, $g(x) = x - b$. Then

$$R(f,g) = \det \begin{bmatrix} a_0 & a_1 & a_2 & a_3 \\ -b & 1 & 0 & 0 \\ 0 & -b & 1 & 0 \\ 0 & 0 & -b & 1 \end{bmatrix}. \tag{26}$$

Now add b times column 2 to column 1, b^2 times column 3 to column 1, and finally b^3 times column 4 to column 1 to obtain

$$R(f,g) = \det \begin{bmatrix} f(b) & a_1 & a_2 & a_3 \\ 0 & 1 & 0 & 0 \\ 0 & -b & 1 & 0 \\ 0 & 0 & -b & 1 \end{bmatrix} = f(b).$$

Thus f and g have a common nonconstant factor iff $f(b) = 0$. But this is precisely the content of Corollary 1, Section 3.3. This calculation can obviously be generalized to a polynomial $f(x)$ of arbitrary degree.

Corollary 5 *Let D be a* U.F.D. *and let* $f(x) \in D[x]$, $\deg f(x) \geq 2$.

(a) *If $g(x) \in D[x]$ is irreducible,* $\deg g(x) \geq 1$, *then $g^2 | f$ iff $g | f$ and $g | f'$.*

(b) *There exists a polynomial* $g(x) \in D[x]$, $\deg g(x) \geq 1$, *such that $g^2 | f$ iff* $R(f, f') = 0$.

Proof: (a) If $g | f$ then $f = gh$,

$$f' = g'h + gh', \tag{27}$$

so that $g \mid f'$ implies from (27) that $g \mid g'h$. Now deg $g' <$ deg g, and since g is irreducible it follows that $g \mid h$. Hence $g^2 \mid f$. Conversely, if $g^2 \mid f$ then $f = g^2 k$,

$$f' = 2gg'k + g^2 k'$$

and hence $g \mid f'$

(b) By Theorem 5.5, f and f' have a common nonconstant factor iff $R(f, f') = 0$. Apply (a). ∎

The resultant of f and f', $R(f, f')$, is called the *discriminant* of f.

Corollary 6 *Assume that D_1 and D_2 are U.F.D.'s and $D_1 \subset D_2$. Let x be an indeterminate over D_2. If the polynomials (20) and (21) in $D_1[x]$ have a common factor of positive degree in $D_2[x]$, then they have a common factor of positive degree in $D_1[x]$.*

Proof: By Theorem 5.5 applied to D_2, $R(f, g) = 0$. But $R(f, g) \in D_1$ so that we can apply Theorem 5.5 again to conclude that f and g have a common factor of positive degree in $D_1[x]$. ∎

Theorem 5.6 *Let D be a* U.F.D.. *Then for the polynomials $f(x)$ and $g(x)$ in (20) and (21) there exist polynomials $\alpha(x)$ and $\beta(x)$ in $D[x]$,* deg $\alpha \leq n - 1$, deg $\beta \leq m - 1$, *such that*

$$R(f,g) = \alpha f + \beta g. \tag{28}$$

Proof: Write

$$\begin{aligned}
f &= a_0 + a_1 x + a_2 x^2 + \cdots + a_m x^m, \\
xf &= \quad a_0 x + a_1 x^2 + \cdots + a_{m-1} x^m + a_m x^{m+1}, \\
&\vdots \\
x^{n-1} f &= \quad a_0 x^{n-1} + \cdots + a_m x^{m+n-1}, \\
g &= b_0 + b_1 x + b_2 x^2 + \cdots + b_n x^n, \\
xg &= \quad b_0 x + b_1 x^2 + \cdots + b_{n-1} x^n + b_n x^{n+1}, \\
&\vdots \\
x^{m-1} g &= \quad b_0 x^{m-1} + \cdots + b_n x^{m+n-1}.
\end{aligned} \tag{29}$$

Observe that the coefficient matrix $S^0 = S(a_0, \ldots, a_m, b_0, \ldots, b_n)$ of 1, $x, x^2, \ldots, x^{m+n-1}$ on the right in (29) is precisely Sylvester's matrix (15), specialized to the coefficients of f and g. Now let

$$c_t = (-1)^{1+t} \det S^0(t \mid 1), \qquad t = 1, \ldots, m+n$$

be the cofactors of the elements in column 1 of S^0. It follows from the Laplace expansion theorem that

$$\sum_{t=1}^{m+n} c_t S^0_{tk} = \delta_{1k} \det S^0 = \delta_{1k} R(f, g). \tag{30}$$

Multiplying the tth equation in (29) by c_t and adding all $m + n$ equations, we have

$$(c_1 + c_2x + \cdots + c_nx^{n-1})f + (c_{n+1} + c_{n+2}x + \cdots + c_{n+m}x^{m-1})g = R(f, g).$$

Simply let α and β be the coefficients of f and g, respectively, in the preceding equation to obtain (28). ∎

Example 4 (a) In Corollary 6, let $D_1 = \mathbf{R}$, $D_2 = \mathbf{C}$, $f(x) \in \mathbf{R}[x]$ and assume $f(\alpha) = 0$, $\alpha = a + ib \in \mathbf{C}$, $b \neq 0$. Form the real polynomial

$$g(x) = (x - \alpha)(x - \bar{\alpha})$$

which, of course, is irreducible in $\mathbf{R}[x]$. Now $R(f,g) = 0$ in $\mathbf{C}$ because f and g have the common factor $x - \alpha$. But then $R(f,g) = 0$ in $\mathbf{R}$ so that f and g must have a common factor of positive degree in $\mathbf{R}[x]$. Since $g(x)$ is irreducible, it follows that this common factor must be g, $g \mid f$, and hence $f(\bar{\alpha}) = 0$ also. In other words, we have as a consequence of Corollary 6 the well-known result that complex roots of a real polynomial occur in complex conjugate pairs.

(b) Let $f(x) = x^3 + px + q \in \mathbf{R}[x]$. We compute that the discriminant of f is

$$R(f,f') = \det \begin{bmatrix} q & p & 0 & 1 & 0 \\ 0 & q & p & 0 & 1 \\ p & 0 & 3 & 0 & 0 \\ 0 & p & 0 & 3 & 0 \\ 0 & 0 & p & 0 & 3 \end{bmatrix} = 27q^2 + 4p^3.$$

Thus if $27q^2 + 4p^3 = 0$, then $f(x)$ must be divisible by $g(x)^2$, $\deg g(x) \geq 1$. Obviously $\deg g = 1$, and hence $f(x)$ has a real root of multiplicity at least 2.

Let $x_1, \ldots, x_p$ be independent indeterminates over the U.F.D. D. If f and g are in $D[x_1, \ldots, x_p]$ and

$$\deg_{x_p} f = m,$$

$$\deg_{x_p} g = n,$$

write

$$f = a_0 + a_1x_p + \cdots + a_mx_p^{\,m}, \tag{31}$$

$$g = b_0 + b_1x_p + \cdots + b_nx_p^{\,n} \tag{32}$$

in which the polynomials $a_0, \ldots, a_m$, $b_0, \ldots, b_n$ in (31) and (32) are in $D[x_1, \ldots, x_{p-1}]$. Then the resultant

$$R(f,g)$$

is obviously a polynomial $R(x_1, \ldots, x_{p-1})$ in $D[x_1, \ldots, x_{p-1}]$. Of course if f is homogeneous of degree m in $x_1, \ldots, x_p$, then clearly $a_t(x_1, \ldots, x_{p-1})$ must be homogeneous of degree $m - t$, $t = 0, \ldots, m$. Similarly for g. Then $R(tx_1, \ldots, tx_{p-1})$ is given by

$$\det \begin{bmatrix} t^m a_0 & t^{m-1}a_1 & \cdots & ta_{m-1} & a_m & 0 & \cdots \\ & t^m a_0 & \cdots & \cdots & \cdots & a_m & \cdots \\ & & \ddots & & & & \\ & & t^m a_0 & \cdots & & & a_m \\ t^n b_0 & t^{n-1}b_1 & \cdots & b_n & 0 & \cdots & \\ & t^n b_0 & \cdots & b_n & \cdots & & \\ & & \ddots & & & & \\ & & t^n b_0 & t^{n-1}b_1 & \cdots & & b_n \end{bmatrix} \begin{matrix} \left.\right\} n \text{ rows} \\ \\ \left.\right\} m \text{ rows} \end{matrix} \tag{33}$$

Multiply the ith row by t^{n-i+1}, $i = 1, \ldots, n$ and multiply the $(n + i)$ th row by t^{m-i+1}, $i = 1, \ldots, m$. Then the jth column of the resulting matrix will have $t^{n+m-j+1}$ as the power of t occurring in all $m + n$ rows, $j = 1, \ldots, n + m$. The effect of the row multiplications is to multiply $R(tx_1, \ldots, tx_{p-1})$ by t^σ, where

$$\begin{aligned} \sigma &= \sum_{i=1}^{n} n - i + 1 + \sum_{i=1}^{m} m - i + 1 \\ &= \frac{n(n+1)}{2} + \frac{m(m+1)}{2}. \end{aligned}$$

Taking $t^{n+m-j+1}$ out of the jth column we have

$$t^\sigma R(tx_1, \ldots, tx_{p-1}) = t^\theta R(x_1, \ldots, x_{p-1})$$

where $\qquad \theta = \sum_{j=1}^{n+m} n + m - j + 1 = \dfrac{(n+m)(n+m+1)}{2}.$

Thus

$$\begin{aligned} R(tx_1, \ldots, tx_{p-1}) &= t^{\theta-\sigma} R(x_1, \ldots, x_{p-1}) \\ &= t^{nm} R(x_1, \ldots, x_{p-1}). \end{aligned} \tag{34}$$

In other words, *if* $f(x_1, \ldots, x_p)$ *and* $g(x_1, \ldots x_p)$ *are homogeneous of degrees m and n, respectively, then the resultant* $R(f,g)$ *with respect to* x_p *(or, of course, any other* x_j*) is either* 0 *or homogeneous of degree mn*. This observation permits a relatively easy proof of the following important result.

Theorem 5.7 *Let* $f(x)$ *and* $g(x)$ *be the polynomials in (20) and (21). If* $\xi_1, \ldots, \xi_m$ *and* $\theta_1, \ldots, \theta_n$ *are the roots of* $f(x)$ *and* $g(x)$*, respectively, in an appropriate extension field of the quotient field of D, then*

$$R(f, g) = (-1)^{mn} a_m^n \prod_{t=1}^{m} g(\xi_t) \tag{35}$$

$$= b_n^m \prod_{t=1}^{n} f(\theta_t). \tag{36}$$

Proof: First observe that if c, d are in D then

$$R(cf, dg) = c^n d^m R(f, g) \tag{37}$$

(see Exercise 1). Thus we may begin by assuming $f(x)$ and $g(x)$ are monic: $a_m = 1$, $b_n = 1$, and

$$f(x) = \prod_{t=1}^{m} (x - \xi_t),$$

$$g(x) = \prod_{t=1}^{n} (x - \theta_t).$$

Let $x_1, \ldots, x_m, y_1, \ldots, y_n$ be indeterminates so that $x, x_i, y_j, i = 1, \ldots, m, j = 1, \ldots, n$ are independent over the splitting field for f and g and define

$$F(x, x_1, \ldots, x_m, y_1, \ldots, y_n) = \prod_{t=1}^{m} (x - x_t), \tag{38}$$

$$G(x, x_1, \ldots, x_m, y_1, \ldots, y_n) = \prod_{t=1}^{n} (x - y_t), \tag{39}$$

$$H(x_1, \ldots, x_m, y_1, \ldots, y_n) = \prod_{i=1}^{m} \prod_{j=1}^{n} (x_i - y_j). \tag{40}$$

Now F is homogeneous of degree m, G is homogeneous of degree n, and H is homogeneous of degree mn. Let

$$R(x_1, \ldots, x_m, y_1, \ldots, y_n) = R(F, G) \tag{41}$$

denote the resultant of (38) and (39) with respect to x. By the observation immediately preceding the statement of the theorem,

$$R(x_1, \ldots, x_m, y_1, \ldots, y_n)$$

is homogeneous of degree mn or possibly 0. That is, (41) is the resultant of homogeneous polynomials. Clearly $R(x_1, \ldots, x_m, y_1, \ldots, y_n)$ is not the zero polynomial. For we can specialize the x_t and y_t to elements of D so that the resulting polynomials have no common roots (remember that D has characteristic 0), hence no common factors, and thus their resultant with respect to x must be different from 0 (Theorem 5.5). On the other hand, if we substitute y_j for x_i in (38), then the resulting polynomials will have a common factor and the resultant (41) must be 0, i.e.,

$$R(x_1, \ldots, x_{i-1}, y_j, x_{i+1}, \ldots, x_m, y_1, \ldots, y_n) = 0.$$

But then regarding (41) as a polynomial in x_i over $D[x_1, \ldots, x_{i-1}, x_{i+1}, \ldots, x_m, y_1, \ldots, y_n]$, it follows that $(x_i - y_j) \mid R(x_1, \ldots, x_m, y_1, \ldots, y_n)$. Since this is true for any binomial $x_i - y_j$, it follows that

$$H \mid R(x_1, \ldots, x_m, y_1, \ldots, y_n). \tag{42}$$

But both polynomials in (42) are homogeneous of degree mn so that (42) implies that

$$R(x_1, \ldots, x_m, y_1, \ldots, y_n) = aH(x_1, \ldots, x_n, y_1, \ldots, y_n) \tag{43}$$

where $a \in D$, $a \neq 0$, and a is completely independent of $f(x)$ and $g(x)$ and depends only on the polynomials (38) and (39). In other words, as polynomials in $x_1, \ldots, x_m, y_1, \ldots, y_n$ the following equality holds:

$$R(F(x, x_1, \ldots, x_m, y_1, \ldots, y_n), G(x, x_1, \ldots, x_m, y_1, \ldots, y_n)) = a \prod_{i=1}^{m} \prod_{j=1}^{n} (x_i - y_j). \tag{44}$$

If we specialize x_i to ξ_i and y_j to θ_j in (44), $i = 1, \ldots, m, j = 1, \ldots, n$, then since

$$F(x, \xi_1, \ldots, \xi_m, \theta_1, \ldots, \theta_n) = f(x) = \prod_{t=1}^{m} (x - \xi_t)$$

and
$$G(x, \xi_1, \ldots, \xi_m, \theta_1, \ldots, \theta_n) = g(x) = \prod_{t=1}^{n} (x - \theta_t)$$

we have

$$\begin{aligned} R(f(x), g(x)) &= a \prod_{i=1}^{m} \prod_{j=1}^{n} \xi_i - \theta_j \\ &= a \prod_{i=1}^{m} g(\xi_i) \qquad (45) \\ &= a \prod_{j=1}^{n} \left(\prod_{i=1}^{m} - (\theta_j - \xi_i) \right) \\ &= a \prod_{j=1}^{n} (-1)^m \prod_{i=1}^{m} (\theta_j - \xi_i) \\ &= a \prod_{j=1}^{n} (-1)^m f(\theta_j) \\ &= (-1)^{mn} a \prod_{j=1}^{n} f(\theta_j). \qquad (46) \end{aligned}$$

To determine a, we argue as follows. Regard the coefficients of $f(x)$ and $g(x)$ as additional independent indeterminates over D. Then the coefficient of

$$a_0{}^n b_n{}^m = a_0{}^n \text{ (since } b_n = 1)$$

in the determinant expansion of Sylvester's matrix is precisely 1. On the other hand, in (46) take every θ_i to be 1 [we can do this, since a depends only on the polynomials (38) and (39) which are defined completely independently of $f(x)$ and $g(x)$] to obtain

$$\begin{aligned} R(f(x),g(x)) &= (-1)^{mn} a \prod_{i=1}^{n} (a_0 + a_1 + \cdots + a_m) \\ &= (-1)^{mn} a (a_0{}^n + \cdots). \qquad (47) \end{aligned}$$

Thus matching coefficients, $a = (-1)^{mn}$, and the formulas (35) and (36) follow immediately from (45) and (46), respectively (remembering our assumption that $a_m = b_n = 1$). ∎

If we take $g(x) = f'(x)$ in (35), then since $n = \deg f'(x) = m - 1$, the discriminant becomes

$$R(f,f') = (-1)^{m(m-1)} a_m^{\ m-1} \prod_{t=1}^{m} f'(\xi_t) \tag{48}$$

where $\xi_1, \ldots, \xi_m$ are the roots of $f(x)$. Now write

$$f(x) = a_m \prod_{t=1}^{m} x - \xi_t$$

so that
$$f'(x) = a_m \sum_{t=1}^{m} \prod_{i \neq t} (x - \xi_i),$$

and hence
$$f'(\xi_t) = a_m \prod_{i \neq t} (\xi_t - \xi_i). \tag{49}$$

Substituting (49) into (48), we obtain

$$\begin{aligned} R(f, f') &= (-1)^{m(m-1)} a_m^{\ m-1} a_m^{\ m} \prod_{t=1}^{m} \prod_{i \neq t} (\xi_t - \xi_i) \\ &= (-1)^{m(m-1)} a_m^{\ 2m-1} \prod_{t=1}^{m} \prod_{i \neq t} (\xi_t - \xi_i). \end{aligned} \tag{50}$$

For a given pair of indices ν and μ, $1 \leq \nu < \mu \leq m$, the term $\xi_\nu - \xi_\mu$ occurs precisely twice in (50), once with coefficient 1, once with coefficient -1. There are a total of $m^2 - m$ terms in (50) altogether so that there are $(m^2 - m)/2$ such oppositely signed pairs, $\xi_\nu - \xi_\mu$, $1 \leq \nu < \mu \leq m$. Thus

$$R(f, f') = a_m^{\ 2m-1}(-1)^{m(m-1)/2} \prod_{1 \leq \nu < \mu \leq m} (\xi_\nu - \xi_\mu)^2. \tag{51}$$

We have the following theorem.

Theorem 5.8 *Let $x, x_1, \ldots, x_m$ be independent indeterminates over D and define*

$$\begin{aligned} f(x) &= a_m \prod_{t=1}^{m} (x - x_t) \\ &\in D[x_1, \ldots, x_m][x], \qquad a_m \neq 0. \end{aligned}$$

Then as polynomials in $D[x_1, \ldots, x_m]$,

$$R(f(x), f'(x)) = a_m^{\ 2m-1}(-1)^{m(m-1)/2} \prod_{1 \leq \nu < \mu \leq m} (x_\nu - x_\mu)^2. \tag{52}$$

Proof: As we saw in (51), the formula (52) holds for arbitrary specializations of the x_t to $\xi_t \in D$. But then since D has characteristic 0, (52) must be an equality of polynomials. ∎

From formula (35) we see immediately that a root ξ_t of $f(x)$ is a root of $g(x)$ iff $R(f,g) = 0$. Thus *a pair of polynomials can have a common root iff their resultant vanishes.* This elementary observation can be used to solve pairs of equations in two indeterminates.

Example 5 (a) Find all pairs of complex numbers ξ, η such that

$$\begin{aligned}\xi^2 + \eta^2 &= 2,\\ \xi^2 + \eta^2 &= 1.\end{aligned} \tag{53}$$

Of course we know that the system (53) has no common solutions. However, consider the polynomials

$$\begin{aligned}f(x_1,x_2) &= x_1^{\,2} + x_2^{\,2} - 2,\\ g(x_1,x_2) &= x_1^{\,2} + x_2^{\,2} - 1.\end{aligned}$$

Regarding these as polynomials in $\mathbf{R}[x_1][x_2]$, we compute the resultant with respect to x_2:

$$R = \det\begin{bmatrix} x_1^{\,2}-2 & 0 & 1 & 0\\ 0 & x_1^{\,2}-2 & 0 & 1\\ x_1^{\,2}-1 & 0 & 1 & 0\\ 0 & x_1^{\,2}-1 & 0 & 1\end{bmatrix} = -1.$$

Thus there is no specialization of x_1 to $\xi \in \mathbf{R}$ (or $\mathbf{C}$) for which the equations (53) have a solution.

(b) Find all pairs of numbers ξ and η in $\mathbf{C}$ for which

$$\begin{aligned}\xi^3 + 2\xi^2\eta + 2\eta(\eta - 2)\xi + \eta^2 - 4 &= 0,\\ \xi^2 + 2\xi\eta + 2\eta^2 - 5\eta \quad + 2 &= 0.\end{aligned} \tag{54}$$

Again, consider the polynomials

$$\begin{aligned}f(x_1,x_2) &= x_1^{\,3} + 2x_1^{\,2}x_2 + 2x_2(x_2 - 2)x_1 + x_2^{\,2} - 4,\\ g(x_1,x_2) &= x_1^{\,2} + 2x_1x_2 + 2x_2^{\,2} - 5x_2 + 2.\end{aligned}$$

Rearrange these polynomials in ascending powers of x_2:

$$\begin{aligned}f &= x_1^{\,3} - 4 + (2x_1^{\,2} - 4x_1)x_2 + (2x_1 + 1)x_2^{\,2},\\ g &= x_1^{\,2} + 2 + (2x_1 - 5)x_2 + 2x_2^{\,2}.\end{aligned}$$

Next, compute the resultant with respect to x_2:

$$R = \det\begin{bmatrix} x_1^{\,3}-4 & 2x_1^{\,2}-4x_1 & 2x_1+1 & 0\\ 0 & x_1^{\,3}-4 & 2x_1^{\,2}-4x_1 & 2x_1+1\\ x_1^{\,2}+2 & 2x_1-5 & 2 & 0\\ 0 & x_1^{\,2}+2 & 2x_1-5 & 2\end{bmatrix}.$$

A somewhat tedious calculation shows that

$$R = x_1(x_1 + 4)^2(x_1 + 5).$$

Thus $\xi = 0, -4, -5$ are the only numbers for which

$$\begin{aligned}f(\xi, \eta) &= 0,\\ g(\xi, \eta) &= 0\end{aligned}$$

have simultaneous solutions for η. Since the equations (54) are quadratic in η, it is a simple routine calculation to find these common solutions. Further tedious work then shows that (0,2), (−4,2), (−5,3) are all the pairs (ξ,η) which satisfy the system (54).

The next result shows that the resultant of two polynomials with indeterminate coefficients is in fact an irreducible polynomial in these coefficients.

Theorem 5.9 *Let $x, u_0, \ldots, u_m, v_0, \ldots, v_n$ be independent indeterminates over the U.F.D. D, and let*

$$f(x) = u_0 + u_1x + \cdots + u_mx^m$$

and

$$g(x) = v_0 + v_1x + \cdots + v_nx^n.$$

Then $R(f(x),g(x))$ is an irreducible polynomial in $D[u_0, \ldots, u_m,v_0, \ldots, v_n]$.

Proof: It is easy to check that $u_0{}^n v_n{}^m$ occurs with the coefficient 1 in Sylvester's determinant (16) so that $R(f(x),g(x))$ has no nonunit factors in D. The proof is by induction on $m + n$. If $m = n = 1$, then

$$R(f(x), g(x)) = \det \begin{bmatrix} u_0 & u_1 \\ v_0 & v_1 \end{bmatrix} = u_0v_1 - u_1v_0,$$

obviously irreducible (see Exercise 2). Thus assume $m + n > 2$, so that $m + n \geq 3$. Let

$$g_1(x) = v_0 + v_1x + \cdots + v_{n-1}x^{n-1}$$

and assume inductively that $R(f(x),g_1(x))$ is irreducible in $D[u_0, \ldots, u_m,v_0, \ldots, v_{n-1}]$. It is simple (see Exercise 3) to check from the form of the Sylvester determinant (16) that

$$R(f, g) \,|_{v_n=0} = \pm u_mR(f, g_1) \tag{55}$$

where for any polynomial P, $P\,|_{v_n=0}$ denotes the specialization of v_n to 0. Suppose then that P is an irreducible factor of positive degree of $R(f,g)$ in $D[u_0, \ldots, u_m, v_0, \ldots, v_n]$. By Theorem 5.4 $R(f,g) = R$ is homogeneous in $v_0, \ldots, v_n$ and hence by Theorem 5.3 P must be homogeneous in $v_0, \ldots, v_n$. Write

$$R = PQ \cdots M, \tag{56}$$

the factorization of R into irreducible factors in $D[u_0, \ldots, u_m,v_0, \ldots, v_n]$. Then

$$R\,|_{v_n=0} = P\,|_{v_n=0}Q\,|_{v_n=0} \cdots M\,|_{v_n=0}$$

so that by (55) (absorbing the ± 1)

$$u_mR(f_1,g) = P\,|_{v_n=0}Q\,|_{v_n=0} \cdots M\,|_{v_n=0}. \tag{57}$$

Now u_m and $R(f,g_1)$ (by the induction) are both irreducible so that by unique factorization each of the factors on the right in (57) must have one of the following forms:

$$\begin{aligned} &du_m, \\ &dR(f, g_1), \\ &du_m R(f, g_1), \\ &d. \end{aligned} \tag{58}$$

Moreover $d \neq 0$ since $R(f, g)|_{v_n=0} = \pm u_m R(f, g_1)$ and $R(f, g_1) \neq 0$. It follows immediately from (58) that each of the factors on the right in (56) must have one of the following forms:

$$du_m + v_n F, \tag{59}$$

$$dR(f,g_1) + v_n F, \tag{60}$$

$$du_m R(f,g_1) + v_n F, \tag{61}$$

$$d + v_n F \tag{62}$$

where $F \in D[u_0, \ldots, u_m, v_0, \ldots, v_n]$. Observe that since $d \neq 0$, neither (59) nor (62) is homogeneous in $v_0, \ldots, v_n$ and hence is excluded as a factor in (56). On the other hand, there cannot be two (or more) factors of the form (60) or (61) in (56). To see this, note that since v_n does not occur in $R(f,g_1)$, none of the terms in $dR(f,g_1)$ can be cancelled by the terms in $v_n F$ in (60), and similarly none of the terms in $du_m R(f,g_1)$ can be cancelled by the terms in $v_n F$ in (61). Thus

$$\deg [dR(f,g_1) + v_n F] \geq \deg R(f,g_1) = m + n - 1,$$

and

$$\deg [du_m R(f,g_1) + v_n F] \geq m + n.$$

But then the appearance of two terms such as (60) or (61) on the right in (56) would imply that

$$\deg R \geq 2(m + n - 1). \tag{63}$$

Since $m + n \geq 3$, the right side of (63) exceeds $m + n$. We have proved that every factor on the right in (56) must be either of the two forms (60) or (61) but that only one such factor can appear. Hence $R = R(f(x),g(x))$ is irreducible. ∎

Example 6 (a) Let $\varphi(x_1,x_2)$ and $f(x_1,x_2)$ be two polynomials in $D[x_1,x_2]$, and assume φ and f are relatively prime. Then there are only a finite number of pairs (ξ, η), ξ, η in D, such that

$$\varphi(\xi,\eta) = 0$$

and

$$f(\xi,\eta) = 0$$

hold simultaneously. For write

$$\varphi(x_1,x_2) = a_0(x_2) + a_1(x_2)x_1 + \cdots + a_m(x_2)x_1^{\,m}$$

and $$f(x_1,x_2) = b_0(x_2) + b_1(x_2)x_1 + \cdots + b_n(x_2)x_1^n.$$

Now let $R(x_2)$ denote the resultant of φ and f with respect to x_1 regarded as polynomials in x_1. If $R(x_2)$ were identically 0, this would imply by Theorem 5.5 that φ and f have a common factor of positive degree (in x_1) in $D[x_2][x_1] = D[x_1,x_2]$, contrary to assumption. Thus from Theorem 5.6,

$$R(x_2) = \alpha(x_1,x_2)f(x_1,x_2) + \beta(x_1,x_2)\varphi(x_1,x_2). \tag{64}$$

If

$$(\xi_i,\eta_i), \qquad i = 1,2, \ldots$$

is an infinite list of values for which φ and f simultaneously vanish, we can assume that there are an infinite number of distinct η_i in the list (otherwise reverse the roles of x_1 and x_2) so that from (64),

$$R(\eta_i) = 0, \qquad i = 1, 2, \ldots. \tag{65}$$

[Note also that φ and f must have positive degree in x_1; otherwise at least one of them would be a polynomial in x_2 vanishing for an infinite number of specializations, which is impossible. Thus $R(x_2)$ can indeed be defined, i.e., $m \geq 1$, $n \geq 1$.] But (65) implies $R \equiv 0$, a contradiction.

(b) Suppose $\varphi(x_1,x_2)$ and $f(x_1,x_2)$ are two irreducible polynomials in $\mathbf{R}[x_1,x_2]$ and the graphs of the curves

$$\varphi(x_1,x_2) = 0$$

and

$$f(x_1,x_2) = 0$$

coincide and consist of an infinite number of points. Then in fact $\varphi = cf$, $c \in \mathbf{R}$. For by (a), if φ and f were not associates, they would be relatively prime and hence could intersect in only a finite number of points.

(c) Let $f(x) = u_0 + u_1x + \cdots + u_{m-1}x^{m-1} + x^m$, $m \geq 1$, where u_0, u_1, $\ldots$, u_{m-1}, x are independent indeterminates over D. If $\xi_1, \ldots, \xi_m$ are roots of $f(x)$ in some splitting field, then Theorem 5.8 tells us that the discriminant satisfies

$$R(f(x),f'(x)) = (-1)^{m(m-1)/2} \prod_{1 \leq i < j \leq m} (\xi_i - \xi_j)^2. \tag{66}$$

Of course, the resultant is also a polynomial in $u_0, \ldots, u_{m-1}$, say $R(u_0, \ldots, u_{m-1})$, and thus $R \neq 0$ is equivalent to the statement that the roots of $f(x)$ are distinct. It is quite easy to prove that in fact

$$R(u_0, \ldots, u_{m-1})$$

is an irreducible polynomial in the indeterminates $u_0, \ldots, u_{m-1}$. For suppose

$$R(u_0, \ldots, u_{m-1}) = F_1(u_0, \ldots, u_{m-1})F_2(u_0, \ldots, u_{m-1}) \tag{67}$$

is a proper factorization, i.e., F_1 and F_2 have positive degree. Now

$$u_t = (-1)^{m-t}E_{m-t}(\xi_1, \ldots, \xi_n), \qquad t = 0, \ldots, m-1,$$

so that F_1 and F_2 can be expressed as symmetric polynomials $\varphi_1(\xi_1, \ldots, \xi_m)$ and $\varphi_2(\xi_1, \ldots, \xi_m)$, respectively. Then from (66) and (67),

$$\varphi_1(\xi_1, \ldots, \xi_m)\varphi_2(\xi_1, \ldots, \xi_m) = \pm \prod_{1 \leq i < j \leq m} (\xi_i - \xi_j)^2. \tag{68}$$

The factors $\xi_i - \xi_j$ on the right in (68) are irreducible in $\xi_1, \ldots, \xi_m$ so that $\varphi_1(\xi_1, \ldots, \xi_m)$ and $\varphi_2(\xi_1, \ldots, \xi_m)$ are each composed of some of these factors. But if $\xi_1 - \xi_2$ is a factor of φ_1, say, then the symmetry of φ_1 implies that all of the binomials $\xi_i - \xi_j$, $i < j$, must also be factors of φ_1. Moreover unless $(\xi_1 - \xi_2)^2$ is also a factor of φ_1, i.e., if $\xi_1 - \xi_2$ occurs as a factor of φ_1 to the first power only, then by the symmetry again all of the binomials $\xi_i - \xi_j$, $i < j$, can occur as factors of φ_1 to the first power only. Thus $\varphi_1 = c \prod_{1 \le i < j \le m} (\xi_i - \xi_j)$, $c \in D$. But the polynomial

$$\prod_{1 \le i < j \le m} \xi_i - \xi_j$$

is not symmetric (see Exercise 4). Thus for any $i < j$, $(\xi_i - \xi_j)^2$ must be a factor of φ_1, and it follows that

$$\begin{aligned} \varphi_1 &= d \prod_{1 \le i < j \le m} (\xi_i - \xi_j)^2 \\ &= dR(u_0, \ldots, u_{m-1}), \qquad d \in D. \end{aligned}$$

Hence
$$F_1(u_0, \ldots, u_{m-1}) = dR(u_0, \ldots, u_{m-1}), \qquad d \in D,$$

and (67) is not a proper factorization as assumed. If we refer to Example 4(b), we can conclude as a consequence of the irreducibility of the discriminant that the polynomial

$$R(p,q) = 27q^2 + 4p^3$$

is irreducible.

(d) Let

$$f(x) = u_0 + u_1 x + \cdots + u_{m-1} x^{m-1} + x^m$$

and define the $m \times m$ *companion matrix* of $f(x)$ by

$$C(f(x)) = \begin{bmatrix} 0 & \cdots & 0 & -u_0 \\ 1 & & \vdots & -u_1 \\ & \ddots & \vdots & \vdots \\ & & 0 & -u_{m-2} \\ \bigcirc & & 1 & -u_{m-1} \end{bmatrix}. \tag{69}$$

[If $m = 1$, so that $f(x) = x + u_0$, then $C(f(x)) = [-u_0]$ by definition.] For example, if $f(x) = u_0 + u_1 x + x^2$, then

$$C(f(x)) = \begin{bmatrix} 0 & -u_0 \\ 1 & -u_2 \end{bmatrix}. \tag{70}$$

Note that for the matrix (70),

$$\begin{aligned} \det (xI_2 - C(f(x))) &= \det \begin{bmatrix} x & u_0 \\ -1 & x + u_1 \end{bmatrix} \\ &= f(x). \end{aligned}$$

An elementary induction argument, together with Laplace expansion on the first row (which will appear in a later section), shows in general that

$$\det (xI_m - C(f(x)) = f(x).$$

According to part (c) above, if we let

$$R(u_0, \ldots, u_{m-1})$$

denote the discriminant of the polynomial $f(x)$ as a polynomial in the coefficients $u_0, \ldots, u_{m-1}$, then R is irreducible. Next, let $x_{ij}, i, j = 1, \ldots, m$ be m^2 independent indeterminates over D. Let $\xi_1, \ldots, \xi_m$ be the eigenvalues of the matrix

$$X = [x_{ij}]$$

in a splitting field over the rational function field $D(x_{11}, x_{12}, \ldots, x_{mm})$, and consider the discriminant

$$\Delta = \prod_{1 \leq i < j \leq m} (\xi_i - \xi_j)^2. \tag{71}$$

Since the right-hand side of (71) is symmetric in $\xi_1, \ldots, \xi_m$, it is an integral polynomial in the elementary symmetric functions

$$E_k(\xi_1, \ldots, \xi_m), \qquad k = 0, \ldots, m \tag{72}$$

(see Theorem 2.5, Section 3.2.) It is an elementary theorem in linear algebra that

$$E_k(\xi, \ldots, \xi_m) = \sum_{\omega \in Q_{k,m}} \det X[\omega \mid \omega], \tag{73}$$

that is, the kth elementary symmetric function of the eigenvalues of a matrix is a polynomial in the entries, namely the sum of all $k \times k$ principal subdeterminants of the matrix (see M. Marcus and H. Minc, A Survey of Matrix Theory and Matrix Inequalities, Allyn and Bacon, Boston, 1964, p. 22). Thus we can write

$$\Delta = \Delta(X), \tag{74}$$

a polynomial in $D[x_{11}, \ldots, x_{mm}]$. We assert that in fact $\Delta(X)$ *is an irreducible polynomial* (see Exercise 5). Note also that $\Delta(X)$ is homogeneous of degree $m(m-1)$; for the eigenvalues of tX are $t\xi_i$, $i = 1, \ldots, m$, and from (71) we have

$$\begin{aligned}\Delta(tX) &= \prod_{1 \leq i < j \leq m} (t\xi_i - t\xi_j)^2 = t^{m(m-1)} \prod_{1 \leq i < j \leq m} (\xi_i - \xi_j)^2 \\ &= t^{m(m-1)} \Delta(X).\end{aligned}$$

If we specialize X to $C(f(x))$, the matrix in (69), then $\Delta(X)$ specializes to $\Delta(C(f(x)))$, a polynomial in $u_0, \ldots, u_{m-1}$. The eigenvalues of $C(f(x))$ are the roots of $f(x)$, and thus from (66), $\Delta(C(f(x)))$ differs only in sign (possibly) from the discriminant:

$$\begin{aligned}R(f(x), f'(x)) &= (-1)^{m(m-1)/2} \prod_{1 \leq i < j \leq m} (\xi_i - \xi_j)^2 \\ &= (-1)^{m(m-1)/2} \Delta(C(f(x))).\end{aligned}$$

According to part (c) above, $R(f(x), f'(x)) = R(u_0, \ldots, u_{m-1})$ is irreducible so that $\Delta(C(f(x)))$ is irreducible as a polynomial in $u_0, \ldots, u_{m-1}$.

For the purposes of our next result involving the resultant we shall assume that D is an *algebraically closed field* of characteristic 0. That is, every polynomial in $D[x]$ has all its roots in D. The field of complex numbers, **C**, is an example of an algebraically closed field but this result, the so-called *fundamental theorem of algebra*, is not a theorem in algebra despite its name.

Theorem 5.10 (*Study's Lemma*) *Assume that $x_1, \ldots, x_n$ are independent indeterminates over an algebraically closed field D, and let $p \in D[x_1, \ldots, x_n]$ be an irreducible polynomial. Let $f \in D[x_1, \ldots, x_n]$, and assume that $f(a) = 0$ whenever $p(a) = 0$, $a = (a_1, \ldots, a_n)$, $a_i \in D$, $i = 1, \ldots, n$. Then $p \mid f$ in $D[x_1, \ldots, x_n]$.*

Proof: If $p \in D$, the result is obvious. Thus without loss of generality we may assume that $deg_{x_n} p = m \geq 1$ and write

$$p = p_0 + p_1 x_n + \cdots + p_m x_n{}^m, \qquad 0 \neq p_m \in D[x_1, \ldots, x_{n-1}]. \tag{75}$$

Also, it is obvious that there is nothing to prove unless $f \neq 0$. If $f \in D[x_1, \ldots, x_{n-1}]$, then since $fp_m \neq 0$, it follows (see Corollary 4, Section 3.3) that there exist $a_1, \ldots, a_{n-1}$ such that

$$p_m(a_1, \ldots, a_{n-1}) f(a_1, \ldots, a_{n-1}) \neq 0. \tag{76}$$

Let $a_n \in D$ be a root of the equation

$$p(a_1, \ldots, a_{n-1}, x_n) = 0,$$

and set $a = (a_1, \ldots, a_{n-1}, a_n)$. Then obviously $p(a) = 0$ but from (76), $f(a) \neq 0$, contradicting the assumptions of the theorem. Thus

$$deg_{x_n} f(x_1, \ldots, x_n) \geq 1$$

and we may form the resultant of $p(x_1, \ldots, x_n)$ and $f(x_1, \ldots, x_n)$ with respect to x_n, call it $R(x_1, \ldots, x_{n-1})$. From Theorem 5.6 there exist $\alpha \in D[x_1, \ldots, x_{n-1}][x_n]$ and $\beta \in D[x_1, \ldots, x_{n-1}][x_n]$ such that

$$\alpha f + \beta p = R(x_1, \ldots, x_{n-1}). \tag{77}$$

Observe that if $a = (a_1, \ldots, a_n)$ and $p(a) = 0$ then by assumption $f(a) = 0$ and hence from (77) $R(a) = R(a_1, \ldots, a_{n-1}) = 0$. In other words $R(x_1, \ldots, x_{n-1})$ is a polynomial in $D[x_1, \ldots, x_{n-1}]$ which vanishes whenever $p(a_1, \ldots, a_{n-1}, a_n) = 0$. By the argument we made in case f itself is in $D[x_1, \ldots, x_{n-1}]$ we can conclude that in fact $R = 0$. But then by Theorem 5.5 $f(x_1, \ldots, x_n)$ and $p(x_1, \ldots, x_n)$ have a common factor of positive degree in x_n. Since p is irreducible this common factor must be a constant multiple of p and we conclude the desired result that $p \mid f$. ▮

Example 7 (a) If $x_{11}, \ldots, x_{nn}$ are independent indeterminates over D, then $\det X = \det [x_{ij}]$ is a prime polynomial in $D[x_{11}, \ldots, x_{nn}]$. For suppose that

$$\det X = f(x_{11}, \ldots, x_{nn}) g(x_{11}, \ldots, x_{nn})$$

and assume that $\deg_{x_{pq}} f = 1$. If $\deg_{x_{pk}} g = 1$ then we could write

$$f = x_{pq} A_1 + A_0, \quad A_1 \neq 0,$$

$$g = x_{pk}B_1 + B_0, \quad B_1 \neq 0,$$

where A_i does not involve x_{pq} and B_i does not involve x_{pk}, $i = 0,1$. Hence

$$\det X = x_{pq}x_{pk}A_1B_1 + \cdot \cdot \cdot, \ A_1B_1 \neq 0.$$

But the product $x_{pq}x_{pk}$ does not occur in any term in det X. Thus if x_{pq} occurs in f, then x_{pk} cannot occur in g. Similarly, neither can x_{kq} occur in g. By the same argument, if x_{rs} occurs in g, then x_{rk} and x_{ks} cannot occur in f. So suppose x_{pq} occurs in f and x_{rs} occurs in g. We have proved that if the element labeled 1 in the following schematic does occur in one of the polynomials, then the elements labeled 2 and 3 do not occur in the other polynomial:

$$\begin{matrix} 1 & 2 \\ 3 & \end{matrix}$$

Now suppose we let the element 1 be x_{ps} and assume x_{ps} occurs in f. But then x_{rs} (the element labeled 3) does not occur in g, contrary to assumption. Thus x_{ps} cannot occur in f. Similarly if x_{ps} occurs in g, then x_{pq} (the element labeled 2) cannot occur in f, contrary to assumption. But this means that x_{ps} cannot occur in either f or g. We know, however, that x_{ps} occurs in det X. Thus one of f and g cannot involve any of the indeterminates; i.e., det X is irreducible.

(b) Let $f(X) = f([x_{ij}])$ be a polynomial of degree n such that $f(A) = 0$ for all specializations of the indeterminates $X = [x_{ij}]$ to a singular matrix $A \in M_n(D)$, and assume that D is an algebraically closed field of characteristic 0. Then

$$f(X) = c \det X, \qquad c \in D.$$

Indeed, simply apply Theorem 5.10 with det playing the role of p, and use part (a). Thus, the determinant is essentially characterized as a polynomial in n^2 indeterminates x_{ij} which vanishes for any specialization of the indeterminates to a singular matrix.

(c) Let $h(x_1, \ldots, x_{n+1})$ be a homogeneous polynomial in $D[x_1, \ldots, x_{n+1}]$ and assume that $h(x_1, \ldots, x_n, 0) \neq 0$. Set

$$\bar{h}(x_1, \ldots, x_n) = h(x_1, \ldots, x_n, 1).$$

Then $\bar{h}$ is called the *nonhomogeneous polynomial associated with the homogeneous polynomial* h. (Although $\bar{h}$ is obtained from h by specializing x_{n+1} to 1, the arguments that follow work equally well for any other x_j with obvious modifications in notation.) Assume that deg $h = m$, and write

$$h = x_{n+1}^m h_m(x_1, \ldots, x_n) + \cdots + h_0(x_1, \ldots, x_n), \tag{78}$$

where $h_0 \neq 0$, i.e., $h(x_1, \ldots, x_n, 0) \neq 0$. Then

$$\bar{h} = h_m(x_1, \ldots, x_n) + \cdots + h_t(x_1, \ldots, x_n) + \cdots + h_0(x_1, \ldots x_n). \tag{79}$$

In (78), h_t is a homogeneous polynomial of degree $m - t$ in $x_1, \ldots, x_n$. Observe that from (78) and (79) the correspondence between h and $\bar{h}$ is bijective. That is, to recapture h from $\bar{h}$, we first collect all the terms in $\bar{h}$ of degree $m - t$ together and call the resulting homogeneous polynomial h_t, $t = 0, \ldots, m$. Then we simply multiply h_t by x_{n+1}^t and add the products to form h. Note: since $h(x_1, \ldots, x_n, 0)$

$\neq 0$ and h is homogeneous of degree m, it follows that $h_0(x_1, \ldots, x_n)$ is homogeneous of degree m as well. Moreover, if h_t, $t > 0$, is not the 0 polynomial, then $\deg h_t < m$. Hence

$$\deg \bar{h} = \deg h. \tag{80}$$

Also note that if

$$h = \prod_{k=1}^{r} p_k, \qquad p_k \in D[x_1, \ldots, x_{n+1}],$$

then by Theorem 5.3 each p_k is also homogeneous. Moreover, it is obvious that

$$\bar{h} = \prod_{k=1}^{r} \bar{p}_k$$

and by (80), $\deg \bar{p}_k = \deg p_k$. Thus h is irreducible iff $\bar{h}$ is irreducible. For example, if $D = Z$ and

$$h(x_1,x_2) = x_1^2 + x_1x_2 + x_2^2,$$

then

$$\begin{aligned} \bar{h}(x_1) &= h(x_1,1) \\ &= x_1^2 + x_1 + 1. \end{aligned}$$

Obviously $\bar{h}(x_1)$ is irreducible in $Z[x_1]$; so h is irreducible in $Z[x_1,x_2]$.

(d) If

$$h(x_1,x_2) = a_0x_1^m + a_1x_1^{m-1}x_2 + a_2x_1^{m-2}x_2^2 + \cdots + a_mx_2^m$$

and

$$k(x_1,x_2) = b_0x_1^n + b_1x_1^{n-1}x_2 + b_2x_1^{n-2}x_2^2 + \cdots + b_nx_2^n$$

are two homogeneous polynomials, then we define the resultant $R(h,k)$ to be the specialization of Sylvester's determinant (16) to

$$R(a_0, \ldots, a_m, b_0, \ldots, b_n). \tag{81}$$

Observe that if $a_m = b_n = 0$, then h and k have the common factor x_1 and $R = 0$. If $a_mb_n \neq 0$, then by specializing x_1 to 1 in both h and k, we obtain $\bar{h}(x_2)$ and $\bar{k}(x_2)$, polynomials with nonzero leading coefficients. Then (81) is precisely the resultant $R(\bar{h},\bar{k})$ which is 0 iff $\bar{h}$ and $\bar{k}$ have a common nonconstant factor, or equivalently by part (c), iff h and k have a common nonconstant factor. Suppose next that $a_m = 0$ but $b_n \neq 0$. Then write $h(x_1,x_2) = x_1^r h_1(x_1,x_2)$, where $h_1(x_1,x_2)$ is not divisible by x_1, i.e.,

$$h_1(x_1,x_2) = a_0x_1^{m-r} + a_1x_1^{m-r-1}x_2 + a_2x_1^{m-r-2}x_2^2 + \cdots + a_rx_2^{m-r}, \qquad a_r \neq 0. \tag{82}$$

From Exercise 3 we have

$$R(h,k) = \pm b_n^r\, R(h_1, k) \tag{83}$$

so that $R(h,k) = 0$ iff $R(h_1,k) = 0$. Thus by the argument we just used in the case the leading coefficients of the polynomials in question are both different from 0, we conclude that h_1 and k, and hence h and k, have a common nonconstant factor iff $R(h,k) = 0$. Thus under all circumstances, with no assumptions on the coefficients, $R(h,k) = 0$ *iff h and k have a common nonconstant factor*. This should be compared to Theorem 5.5, in which the same result is proved for polynomials in a single variable with nonzero leading coefficients.

(e) Let $h(x_1,x_2)$ be the polynomial in part (d) and write

$$h(x_1,x_2) = x_1{}^r h_1(x_1,x_2)$$

where $h_1(x_1,x_2)$ is as in (82). Then obviously $\deg_{x_2} \bar{h}_1(1,x_2) = \deg \bar{h}(x_2) = m - r$. If D is an algebraically closed field or at least contains all the roots of $\bar{h}(x_2)$, then we can write

$$\bar{h}(x_2) = \alpha \prod_{j=1}^{m-r} (x_2 - d_j), \tag{84}$$

where $\alpha, d_j \in D$, $j = 1, \ldots, m - r$. Now note from (82) that in the rational function field $D(x_1,x_2)$,

$$\begin{aligned} h_1(x_1,x_2) &= x_1{}^{m-r}\left[a_0 + a_1 \frac{x_2}{x_1} + a_2\left(\frac{x_2}{x_1}\right)^2 + \cdots + a_r\left(\frac{x_2}{x_1}\right)^{m-r}\right] \\ &= x_1{}^{m-r}\bar{h}\left(\frac{x_2}{x_1}\right) \\ &= x_1{}^{m-r}\alpha \prod_{j=1}^{m-r}\left(\frac{x_2}{x_1} - d_j\right) \qquad \text{[from (84)]} \\ &= \alpha \prod_{=1}^{m-r} (x_2 - d_j x_1). \end{aligned}$$

Then
$$h(x_1,x_2) = \alpha x_1{}^r \prod_{j=1}^{m-r} (x_2 - d_j x_1). \tag{85}$$

We have proved (and in fact shown how to compute) that if $h(x_1,x_2)$ is homogeneous, then there exist constants α, c_i and d_i, $i = 1, \ldots, m$, such that

$$h(x_1,x_2) = \alpha \prod_{i=1}^{m} (c_i x_2 - d_i x_1). \tag{86}$$

For example, let

$$h(x_1,x_2) = 2x_1{}^3 - x_1{}^2 x_2 - 2x_1 x_2{}^2 + x_2{}^3.$$

Then
$$\begin{aligned} \bar{h}(x_2) = h(1,x_2) &= 2 - x_2 - 2x_2{}^2 + x_2{}^3 \\ &= (x_2 - 1)(x_2 + 1)(x_2 - 2). \end{aligned}$$

Then
$$h(x_1,x_2) = (x_2 - x_1)(x_2 + x_1)(x_2 - 2x_1).$$

As a final topic in this section we shall apply some of the ideas developed in the study of polynomials in several indeterminates to discuss the celebrated theorem of D. Hilbert on invariants. Invariant theory was the "modern algebra" of the last twenty years of the nineteenth century. The subject is enjoying a modern revival with the more recent work of J. A. Dieudonné, H. Weyl, D. Mumford, G. C. Rota and others.

Let K be a field of characteristic 0, and let G be a finite group of nonsingular $n \times n$ matrices over K, $|G| = r$. Let $x_1, \ldots, x_n$ be independent indeterminates over K and for each $f \in K[x_1, \ldots, x_n]$ and $A \in G$, define $f_A \in K[x_1, \ldots, x_n]$ by the formula

$$f_A(x) = f(A^{-1}x), \tag{87}$$

where in (87) $x = (x_1, \ldots, x_n)$ is the column n-tuple and $A^{-1}x$ is the ordinary matrix-vector product. The polynomial f is an *invariant of* G if

$$f_A(x) = f(x) \tag{88}$$

for all $A \in G$.

An example of this is easy to construct. Let G be the group of $n \times n$ permutation matrices, $|G| = n!$. Let $f(x_1, \ldots, x_n) = f(x)$ be any symmetric polynomial in the n indeterminates. Then obviously $f_A(x) = f(x)$ for all $A \in G$, and f is an invariant of G.

Next, let $I_G \subset K[x_1, \ldots, x_n]$ be the set of all invariants of G in $K[x_1, \ldots, x_n]$. Note that I_G is a subring of $K[x_1, \ldots, x_n]$:

$$\begin{aligned}(f+g)_A(x) &= (f+g)(A^{-1}x) = f(A^{-1}x) + g(A^{-1}x) \\ &= f_A(x) + g_A(x) = (f_A + g_A)(x),\end{aligned} \tag{89}$$

$$\begin{aligned}(fg)_A(x) &= (fg)(A^{-1}x) = f(A^{-1}x)g(A^{-1}x) \\ &= f_A(x)g_A(x) = (f_A g_A)(x).\end{aligned} \tag{90}$$

Thus if $f, g \in I_G$, so that $f_A = f$, $g_A = g$, then $(f+g)_A = f + g$, $(fg)_A = fg$ and also, of course, $(cf)_A = cf$.

For any $f \in K[x_1, \ldots, x_n]$ define

$$f^s(x) = \frac{1}{r} \sum_{A \in G} f_A(x). \tag{91}$$

For example, if G is the group of 3×3 permutation matrices and $f(x) = x_1 + 2x_2$, then

$$\begin{aligned}f^s(x) &= \tfrac{1}{6}\{(x_1 + 2x_2) + (x_2 + 2x_1) + (x_3 + 2x_2) + (x_1 + 2x_3) \\ &\qquad + (x_2 + 2x_3) + (x_3 + 2x_1)\} \\ &= x_1 + x_2 + x_3.\end{aligned}$$

Similarly, if $f(x) = x_1x_2$, then

$$f^s(x) = \tfrac{1}{3}\{x_1x_2 + x_1x_3 + x_2x_3\}.$$

Note that $f^s(x)$ is symmetric in both instances. From (89) we see immediately that

$$(f+g)^s = f^s + g^s. \tag{92}$$

Also observe that if $f \in I_G$, so that $f_A = f$ for all $A \in G$, then

$$f^s = f. \tag{93}$$

Conversely, if $f^s = f$ and $B \in G$, then

$$
\begin{aligned}
f_B(x) &= (f^s)_B(x) \\
&= \frac{1}{r} \sum_{A \in G} (f_A)_B(x) \qquad \text{[from (89)]} \\
&= \frac{1}{r} \sum_{A \in G} f(B^{-1}A^{-1}x) \\
&= \frac{1}{r} \sum_{A \in G} f_{AB}(x) \\
&= \frac{1}{r} \sum_{A \in G} f_A(x) \\
&= f^s(x) \quad = f(x).
\end{aligned}
$$

Thus $f_B(x) = f(x)$ for all $B \in G$; so $f \in I_G$.

If $f \in K[x_1, \ldots, x_n]$, write f (uniquely) as in (10):

$$f = \sum_{i=1}^{k} f_{n_i}$$

where $f_{n_i} \neq 0$ is homogeneous of degree n_i, $0 \leq n_1 < n_2 < \cdots < n_k$. Then

$$f_A(x) = \sum_{i=1}^{k} f_{n_i}(A^{-1}x)$$

and clearly $$f_{n_i}(A^{-1}tx) = t^{n_i} f_{n_i}(A^{-1}x)$$

so that $f_{n_i}(A^{-1}x)$ is also homogeneous of degree n_i. If $f \in I_G$, then $f_A = f$ so that

$$
\begin{aligned}
f(x) = \sum_{i=1}^{k} f_{n_i}(x) &= \sum_{i=1}^{k} f_{n_i}(A^{-1}x) \\
&= f_A(x)
\end{aligned}
$$

and the uniqueness of the homogeneous components in $f(x)$ implies that

$$f_{n_i}(A^{-1}x) = f_{n_i}(x), \qquad i = 1, \ldots, k.$$

In other words, $f \in I_G$ iff $f_{n_i} \in I_G$, $i = 1, \ldots, k$. Thus the study of general polynomial invariants of G is reduced to the study of homogeneous invariants.

Let

$$J_G$$

be the ideal in $K[x_1, \ldots, x_n]$ generated by those elements of I_G with a zero constant term. Clearly any polynomial in J_G also has a zero constant term. Since $K[x_1, \ldots, x_n]$ is Noetherian (by the Hilbert Basis theorem), it follows that J_G is finitely generated so that by the preceding remarks there exists a finite set of nonzero homogeneous polynomials

$$f_i(x_1, \ldots, x_n),$$

$\deg f_i(x) = q_i$, $i = 1, \ldots, p$, such that

$$J_G = (f_1, f_2, \ldots, f_p). \tag{94}$$

Note that $q_i \geq 1$ because $f_i \neq 0$ and f_i has a zero constant term. Now let $f \in I_G$, $\deg f = q \geq 1$, and f homogeneous. Write

$$f = \sum_{i=1}^{p} v_i f_i, \tag{95}$$

where $v_i \in K[x_1, \ldots, x_n]$. From (95), Exercise 6, and the fact that $f^s = f$ and $f_i^s = f_i$, $i = 1, \ldots, p$, we have

$$\begin{aligned} f = f^s &= \sum_{i=1}^{p} (v_i f_i)^s \\ &= \sum_{i=1}^{p} v_i^s f_i. \end{aligned}$$

But $h^s \in I_G$ for any $h \in K[x_1, \ldots, x_n]$, i.e., we have already seen that $(h^s)_B(x) = h^s(x)$ for any $B \in G$. Hence by the above remarks we can write v_i^s as

$$v_i^s = \sum_{t=0}^{N} w_{it}, \qquad i = 1, \ldots, p, \tag{96}$$

where each w_{it} is homogeneous of degree t (or is the zero polynomial), $w_{it} \in I_G$, and N is a positive integer at least as large as $\max_{1 \leq i \leq p} \deg v_i^s$. Substituting (96) in (95), we have

$$\begin{aligned} f &= \sum_{i=1}^{p} f_i v_i = \sum_{i=1}^{p} f_i \sum_{t=0}^{N} w_{it} \\ &= \sum_{i=1}^{p} f_i w_{i,q-q_i} + \sum_{i=1}^{p} \sum_{t \neq q-q_i} f_i w_{it}. \end{aligned} \tag{97}$$

Each nonzero summand in the first summation in (97) has degree

$$q_i + q - q_i = q,$$

whereas for $t \neq q - q_i$ any nonzero term in the second summation in (97) has degree

$$\begin{aligned} \deg f_i w_{it} &= q_i + t \\ &\neq q. \end{aligned}$$

Thus the second summation must be 0 because f is homogeneous of degree q. Hence we may write

$$f = \sum_{i=1}^{p} f_i u_i, \tag{98}$$

where $u_i \in I_G$ and if $u_i \neq 0$ then u_i is homogeneous of degree $q - q_i$. Moreover since $q_i \geq 1$ [see the remark following (94)], we see that any nonzero u_i has degree strictly less than q.

Theorem 5.10 (*The Hilbert Invariant Theorem*) *The ring of invariants* I_G

consists of all polynomials with coefficients in K in a fixed set of homogeneous polynomial invariants $f_1, \ldots, f_p$, *i.e.,*

$$I_G = K[f_1(x), \ldots, f_p(x)]. \tag{99}$$

Proof: Let $f \in I_G$ be homogeneous of degree q. If $q = 0$, then $f = k \in K$ and trivially $f \in K[f_1(x), \ldots, f_p(x)]$. Now assume $q \geq 1$. Then since f is a homogeneous invariant of positive degree, it has no constant term and hence is in J_G. By (98) and the subsequent remark,

$$f = \sum_{i=1}^{p} f_i u_i \tag{100}$$

and $u_i \in I_G$ is homogeneous of degree $q - q_i < q$. By induction on the degree of f we can immediately conclude that $u_i \in K[f_1(x), \ldots, f_p(x)]$, $i = 1, \ldots, p$, and hence $f \in K[f_1(x), \ldots, f_p(x)]$ follows immediately from (100). As we noted above, any invariant of G is a sum of homogeneous invariants, and thus $I_G \subset K[f_1(x), \ldots, f_p(x)]$. The inclusion in the other direction is trivial. ∎

Theorem 5.10 is of course of considerable interest, but it must be observed that it is nonconstructive. That is, the proof does not show one how to construct the basic homogeneous invariants of G, but tells us only that they exist. Moreover, the invariants $f_1, \ldots, f_p$ may or may not be *algebraically independent*. Polynomials $w_1(x_1, \ldots, x_n), \ldots, w_p(x_1, \ldots, x_n)$ in $K[x_1, \ldots, x_n]$ are *algebraically dependent* if there exists a nonzero polynomial $\Omega(x_1, \ldots, x_n) \in K[x_1, \ldots, x_n]$ such that

$$\Omega(w_1(x_1, \ldots, x_n), \ldots, w_p(x_1, \ldots, x_n)) = 0.$$

Otherwise, $w_1, \ldots, w_p$ are *algebraically independent*. A polynomial Ω for which $\Omega(f_1, \ldots, f_p) = 0$ is called a *syzygy*.

Example 8 Let G be the group of all $n \times n$ permutation matrices. Then I_G is the ring of all polynomial invariants of G, i.e., $f \in I_G$ iff

$$f(A^{-1}x) = f(x)$$

for all $A \in G$. Hence I_G is precisely the ring of all symmetric polynomials in the indeterminates $x_1, \ldots, x_n$. According to Theorem 2.5, Section 3.2, any $f \in I_G$ is a polynomial in $1 = E_0, E_1, \ldots, E_n$, the elementary symmetric functions. That is,

$$I_G = K[E_1, \ldots, E_n].$$

Thus the E_i play the roles of the f_i in the Hilbert Invariant Theorem. There is, of course, nothing unique about the E_i, for the Newton identities (Example 3, Section 3.2) show that

$$I_G = K[s_1, \ldots, s_n],$$

where $s_k = \sum_{j=1}^{n} x_j^k$ is the kth power sum.

The polynomials $E_1, \ldots, E_n$ are a minimal polynomial basis of I_G in the following sense: No E_k is a polynomial in the remaining E_i, $i \neq k$. More generally it is quite easy to see that $E_1, \ldots, E_n$ are algebraically independent. For, suppose that $\Omega(x_1, \ldots, x_n)$ is a nonzero polynomial in $K[x_1, \ldots, x_n]$. From Corollary 4, Section 3.3 (we assume char $K = 0$) there exist $k_i \in K$, $i = 1, \ldots, n$, such that

$$\Omega(k_1, \ldots, k_n) \neq 0. \tag{101}$$

Consider the polynomial

$$p(t) = t^n - k_1 t^{n-1} + k_2 t^{n-2} + \cdots + (-1)^n k_n \in K[t]$$

and let $\theta_1, \ldots, \theta_n$ be the roots of $p(t)$ in some splitting field for $p(t)$, $K(\theta_1, \ldots, \theta_n)$. Then from Theorem 2.6, Section 3.2,

$$E_i(\theta_1, \ldots, \theta_n) = k_i, \qquad i = 1, \ldots, n$$

and from (101),

$$\Omega(E_1(\theta_1, \ldots, \theta_n), \ldots, E_n(\theta_1, \ldots, \theta_n)) \neq 0. \tag{102}$$

Of course, if

$$\Omega(E_1(x_1, \ldots, x_n), \ldots, E_n(x_1, \ldots, x_n))$$

were the zero polynomial in $K[x_1, \ldots, x_n]$, then all its coefficients would be 0 and hence a specialization of x_i to θ_i in the extension field $K(\theta_1, \ldots, \theta_n)$ would yield the value 0, in contradiction to (102).

Exercises 3.5

1. Confirm the formula (37). *Hint*: The determinant is a linear function of the rows.
2. Show that $u_0v_1 - u_1v_0$ is irreducible in $D[u_0,u_1,v_0,v_1]$. *Hint*: Use Theorem 5.3.
3. Confirm formula (55). *Hint*: Use Laplace expansion for det S on the last column. For example, if $m = n = 2$ and $v_2 = 0$ in

$$\det\begin{bmatrix} u_0 & u_1 & u_2 & 0 \\ 0 & u_0 & u_1 & u_2 \\ v_0 & v_1 & v_2 & 0 \\ 0 & v_0 & v_1 & v_2 \end{bmatrix},$$

then

$$R(f(x),g(x))|_{v_2=0} = \det\begin{bmatrix} u_0 & u_1 & u_2 & 0 \\ 0 & u_0 & u_1 & u_2 \\ v_0 & v_1 & 0 & 0 \\ 0 & v_0 & v_1 & 0 \end{bmatrix} = \pm u_2 \det\begin{bmatrix} u_0 & u_1 & u_2 \\ v_0 & v_1 & 0 \\ 0 & v_0 & v_1 \end{bmatrix}$$

$$= \pm u_2 R(f,g_1).$$

4. Show that if $\sigma \in S_m$, $\prod_{1 \le i < j \le m} \xi_{\sigma(i)} - \xi_{\sigma(j)} = \varepsilon(\sigma) \prod_{1 \le i < j \le m} \xi_i - \xi_j$.
5. Prove that $\Delta(X)$ as defined in (74) is an irreducible polynomial in the m^2 in-

determinates x_{ij}. *Remark*: It would be interesting to have a simple proof of this fact.

6. Let $v, f \in K[x_1, \ldots, x_n]$ and assume $f^s = f$. Prove that $(vf)^s = v^s f$ [see the definition (91)]. *Hint*:

$$(vf)^s(x) = \frac{1}{r} \sum_{A \in G} (vf)(A^{-1}x) = \frac{1}{r} \sum_{A \in G} v(A^{-1}x)f(A^{-1}x).$$

But it is simple to check that if $f^s = f$, then $f(A^{-1}x) = f(x)$. Hence

$$(vf)^s(x) = \frac{1}{r} \sum_{A \in G} v(A^{-1}x)f(x) = v^s(x)f(x).$$

7. Calculate the two resultants with respect to x_1 and x_2 of

$$f(x_1,x_2) = x_1^2x_2 + x_1x_2^2 - x_1x_2 + 1,$$
$$g(x_1,x_2) = x_1x_2 + x_1 + x_2 - 1.$$

Hint: In obvious notation

$$R_2(f,g) = \det \begin{bmatrix} 1 & x_1^2 - x_1 & x_1 \\ -1 + x_1 & 1 + x_1 & 0 \\ 0 & -1 + x_1 & 1 + x_1 \end{bmatrix}$$
$$= (1 + x_1^2)(1 + 2x_1 - x_1^2);$$
$$R_1(f,g) = (1 + x_2^2)(1 + 2x_2 - x_2^2).$$

8. Find all pairs of complex numbers (ξ,η) such that $f(\xi,\eta) = 0$ and $g(\xi,\eta) = 0$, where f and g are the polynomials in Exercise 7. *Hint*: For a fixed ξ, η is a common root of $f(\xi,x_2)$ and $g(\xi,x_2)$ iff $R_2(f(\xi,x_2), g(\xi,x_2))$ is 0. But $R_2(f(\xi,\eta), g(\xi,\eta)) = (1 + \xi^2)(1 + 2\xi - \xi^2) = 0$ iff $\xi = \pm i, 1 \pm \sqrt{2}$. Similarly, $\eta = \pm i, 1 \pm \sqrt{2}$. Thus the possible points of intersection are among the sixteen possible pairs. These must be separately checked to see that $(i, -i)$, $(-i, i)$, $(1 + \sqrt{2}, 1 - \sqrt{2})$, $(1 - \sqrt{2}, 1 + \sqrt{2})$ are the required pairs. The symmetry of f and g can be used to diminish the labor here.

9. Prove that $f(x_1,x_2,x_3,x_4) = x_1^2 + x_2^2 - 2x_1x_2 + 4x_3x_4$ is irreducible over $Z[x_1,x_2,x_3,x_4]$. *Hint*: Use Theorem 5.3.

10. Let $f(x)$, $g(x)$, $h(x)$ be nonzero polynomials in $D[x]$. Prove that

$$R(f,gh) = R(f,g)R(f,h).$$

Hint: From Theorem 5.7,

$$R(f,gh) = (-1)^{m(p+q)} a_m^n \prod_{t=1}^{m} g(\xi_t)h(\xi_t),$$

where $\deg g = p$, $\deg h = q$, $n = p + q$, $f = a_0 + \cdots + a_m x^m$. Then

$$R(f,gh) = (-1)^{mp} a_m^p \prod_{t=1}^{m} g(\xi_t) \cdot (-1)^{mq} a_m^q \prod_{t=1}^{m} h(\xi_t)$$
$$= R(f,g)R(f,h).$$

Glossary 3.5

3.6 Applications to Geometric Constructions

In this section of the chapter we apply some of the elementary concepts of field extension theory to several classical geometric construction problems. The most famous result is Theorem 6.5, in which we exhibit an angle that cannot be trisected using a straightedge and compass alone.

We recapitulate the contents of Exercise 17, Section 3.4—the degree theorem for extension fields. Let M be an extension field of K. A set of elements $\{x_1, \ldots, x_k\}$ in M is *linearly independent* (l.i.) *over* K if

$$\sum_{j=1}^{k} c_j x_j = 0, \qquad c_j \in K, \quad j = 1, \ldots, k$$

can hold iff $c_j = 0, j = 1, \ldots, k$ (see Exercise 14, Section 3.4). Any set of elements not l.i. is said to be *linearly dependent* (l.d.) over K.

Theorem 6.1 *Let $M = K(u_1, \ldots, u_n)$ be a finite extension of K in which $\{u_1, \ldots, u_n\}$ is a basis of M over K. If L is a field and $K \subset L \subset M$, then L is a finite extension of K and $[L:K] \leq n$. Moreover, any basis of L over K can be extended to a basis of M over K.*

Proof: From Theorem 4.9, Section 3.4, any set of l.i. elements of L over K can contain at most n members. Let $\{v_1, \ldots, v_m\}$ be a maximal (i.e., m is maximal) l.i. set of elements of L over K; then $m \leq n$. If $x \in L$, the maximality of m implies that $v_1, \ldots, v_m, x$ are l.d. over K: we have

$$\sum_{j=1}^{m} c_j v_j + cx = 0, \tag{1}$$

and not all of $c_1, \ldots, c_m, c$ are 0. Obviously $c \neq 0$; otherwise $v_1, \ldots, v_m$ would be l.d. Thus from (1),

$$x = \sum_{j=1}^{m} (-c^{-1}c_j)v_j$$

and $\{v_1, \ldots, v_m\}$ is a spanning set for L over K which is l.i. over K. In other words, $\{v_1, \ldots, v_m\}$ is a basis of L over K, and hence $[L:K] \leq n$. Now consider all l.i. sets in M over K which contain $\{v_1, \ldots, v_m\}$. Again, by Theorem 4.9, Section 3.4, none of these sets can contain more than n elements. Let $\{v_1, \ldots, v_m, v_{m+1}, \ldots, v_p\}$ be a maximal such l.i. set and let $x \in M$. By an exact repetition of the preceding argument, using the maximality of p, we can conclude that $x = \sum_{j=1}^{p} d_j v_j$, $d_j \in K$, $j = 1, \ldots, p$. Thus $\{v_1, \ldots, v_p\}$ is a basis of M over K. By Theorem 4.10(i), Section 3.4, $p = n$. ∎

We can now restate in somewhat greater generality the contents of Exercise 17, Section 3.4.

Theorem 6.2 (*The Degree Theorem*) *Let $K \subset L \subset M$ be fields.*

(i) *If M is a finite extension of K, then L is a finite extension of K and M is a finite extension of L.*

(ii) *If $[L:K] = m < \infty$ and $[M:L] = n < \infty$, then $[M:K] = mn$.*

(iii) *If L is a finite extension of K and $c \in L$, then $[K(c):K] \mid [L:K]$.*

Proof: (i) The fact that L is a finite extension of K is contained in Theorem 6.1. If $\{u_1, \ldots, u_p\}$ is a basis of M over K, it is surely a spanning set of M over L because everything in M is of the form $\sum_{j=1}^{p} c_j u_j$, where $c_j \in K \subset L$, $j = 1, \ldots, p$. But then by Exercise 14, Section 3.4, a basis of M over L can be selected from $\{u_1, \ldots, u_p\}$. Thus M is a finite extension of L.

(ii) This is precisely the content of Exercise 17, Section 3.4.

(iii) Clearly $K \subset K(c) \subset L$, and since L is a finite extension of K, $K(c)$ must be a finite extension of K from (i). But then (ii) finishes the argument. ∎

The principal goal of this section is to determine precisely those geometric figures in the plane that can be constructed using straightedge and compass only. Our results will allow us to answer several classical questions concerning such constructions which arose in Greek mathematics:

(i) Can an arbitrary angle be trisected?

(ii) Can the edge of a cube having twice the volume of a given cube be constructed?

(iii) Can a regular polygon of seven sides be constructed?

There are of course many questions beyond these that can be asked: Precisely which angles can be trisected? Which regular polygons can be constructed? Etc. We shall set up the machinery to deal with these questions, but not all the answers will be provided.

We must first decide precisely which constructions are legal. Our definitions for *legal constructions* follow.

1. The points (0,0) and (1,0) in the Cartesian plane are given.
2. If P and Q are points already constructed, then the line $\overline{PQ}$ (or line segment) joining them is constructible.
3. If P is a point already constructed and r is a length already constructed, i.e., the length of a constructible line segment, then the circle with center P and radius r is constructible.
4. The intersections of lines already constructed are constructible points.
5. The intersections of already constructed circles are constructible points.
6. The intersection of an already constructed line and circle are constructible points.

A complex number $z = a + bi \in \mathbf{C}$ is called *constructible* if the point $(a,b) = P$ is a constructible point starting with (0,0) and (1,0) and using a finite sequence of legal constructions. Obviously (a,b) is constructible iff $|a|$ and $|b|$ are constructible lengths (see Exercise 3). Let $\mathscr{C}$ be the totality of constructible numbers in $\mathbf{C}$.

Theorem 6.3 *$\mathscr{C}$ is a subfield of* $\mathbf{C}$.

Proof: It is obvious that $\mathscr{C}$ is closed with respect to addition (see Exercise 4). Now consider the product $(a + ib)(c + id) = ac - bd + i(bc + ad)$. The product will be constructible if we can show that the product of two constructible lengths α and β is constructible. The reader will recall the following construction from high school geometry:

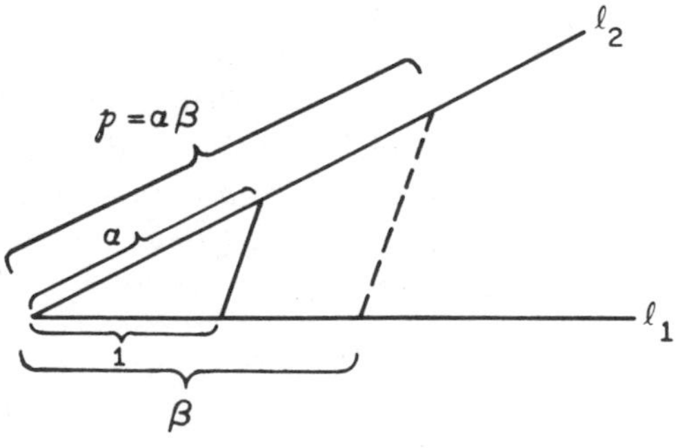

By similar triangles we have $\alpha/p = 1/\beta$, and hence $p = \alpha\beta$.

We remark that l_1 can be taken as the line joining (0,0) and (1,0) and l_2 as the line joining (0,0) and $(1,\frac{1}{2})$, both obviously legal constructions (why?).

If $a + ib \neq 0$, then $(a + ib)^{-1} = a/(a^2 + b^2) - i[b/(a^2 + b^2)]$. Thus if we show that the quotient α/β of two constructible lengths α and β is constructible, then it will follow that $(a + ib)^{-1}$ is constructible. The following construction shows this:

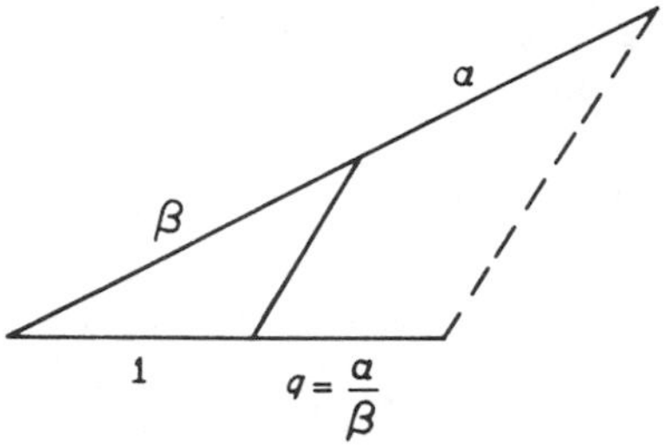

Thus $\mathscr{C}$ is closed with respect to addition and multiplication, and the inverse of every nonzero element in $\mathscr{C}$ is in $\mathscr{C}$. It follows that $\mathscr{C}$ is a subfield of **C**. ∎

If $F \subset \mathscr{C}$ is a field, then F is called a *constructible number field*. Note that since 0 and 1 are in F, it follows from the arguments in Theorem 6.3 that $Q \subset F$. Thus any constructible number field is an extension field of Q.

Theorem 6.4 *Let F be a constructible number field. If F is a finite extension of Q, then $[F{:}Q] = 2^m$ for some nonnegative integer m.*

Proof: By Theorems 4.10(ii) and 4.15, Section 3.4, any finite extension F of Q is of the form $F = Q(\theta)$, where θ is algebraic over Q. Now clearly $\bar{\theta}$ (the complex conjugate of θ) is also algebraic over Q and in fact satisfies the same polynomial with rational coefficients that θ does. It is obvious then that $Q(\theta,\bar{\theta})$ is a finite extension of Q. Let $\theta = a + ib$. Then a and b are constructible numbers, and we can consider the real extension field $Q(a,b)$ of Q. We assert first that $Q(a,b)$ is a finite extension of Q. For $a = (\theta + \bar{\theta})/2 \in Q(\theta,\bar{\theta})$, and so a is algebraic over Q. Also $ib = (\theta - \bar{\theta})/2 \in Q(\theta,\bar{\theta})$, and hence ib is algebraic over Q. Suppose $p(x) = \sum_{k=0}^{m} a_k x^k \in Q[x]$ and $p(ib) = 0$. Then

$$\sum_{k=0}^{m} a_k i^k b^k = 0. \tag{2}$$

If we separate off the real part of (2), we see that b is algebraic over Q. Thus a and b are both algebraic over Q, and it follows that $Q(a,b)$ is a real finite extension of Q. Moreover it consists of constructible numbers (since a and b are constructible). Suppose we can prove that $[Q(a,b){:}Q] = 2^r$. Then since $x^2 + 1$ is obviously irreducible over $Q(a,b)$, we have

$$[Q(a,b)(i)\colon Q(a,b)] = 2. \tag{3}$$

Hence by Theorem 6.2,

$$[Q(a,b)(i)\colon Q] = 2^{r+1}. \tag{4}$$

But $Q \subset Q(\theta) \subset Q(a,b)(i)$ (since $\theta = a + ib$) so that

$$[Q(a,b)(i){:}Q] = [Q(a,b)(i){:}Q(\theta)][Q(\theta){:}Q]. \tag{5}$$

Thus $[Q(\theta){:}Q] \mid 2^{r+1}$ and it follows that $[Q(\theta){:}Q] = 2^m$, or $[F{:}Q] = 2^m$.

Thus we need only prove the result for a real finite constructible extension of Q. It is obvious from elementary algebra that if θ is a real number obtained by a finite sequence of legal constructions starting with Q then $\theta \in Q(\alpha_1, \ldots, \alpha_n)$ and α_t is in at worst a real quadratic extension of $Q(\alpha_1, \ldots, \alpha_{t-1})$, $t = 1, \ldots, n$. For a number determined by a legal construction is the solution of the equations for two lines, a line and a circle, or two circles. In every case this leads to at most a real quadratic extension. But then $[Q(\alpha_1, \ldots, \alpha_t)\colon Q(\alpha_1, \ldots, \alpha_{t-1})] \leq 2$, and it follows from the degree theorem again that $[Q(\alpha_1, \ldots, \alpha_n)\colon Q] = 2^n$. Since $\theta \in Q(\alpha_1, \ldots, \alpha_n)$, we can apply Theorem 6.2(iii) to conclude that $[Q(\theta)\colon Q] \mid 2^n$. ▮

Theorem 6.5 *Starting with points in the plane whose coordinates are rational, it is not possible to trisect a* 60° *angle by legal constructions.*

Proof: If 20° were constructible, it would follow immediately that cos 20° is constructible (why?). However, we saw in Exercise 11, Section 3.4, that if $v = 2\cos 20°$, then $p(v) = 0$, where $p(x) = x^3 - 3x - 1 \in Z[x]$. Now we assert that $p(x)$ is irreducible over $Q[x]$. To see this, we need only show that $p(x)$ has no factor of the form $x - \alpha \in Z[x]$. But if it did, then clearly $\alpha = \pm 1$, and neither of these values is a root. Hence $[Q(\cos 20°){:}Q] = 3$ by Theorem 4.7, Section 3.4. But if cos 20° is constructible, then obviously $Q(\cos 20°)$ is a constructible number field and from Theorem 6.4, $[Q(\cos 20°)\colon Q] = 2^m$. ▮

Theorem 6.6 *Starting with points in the plane whose coordinates are rational, it is not possible to legally construct a regular polygon of seven sides.*

Proof: Let $\alpha = 360°/7$. Clearly, if a regular seven-sided polygon were constructible, then $u = \cos\alpha$ would be a constructible number. From trigonometry we have

$$\begin{aligned}\cos 4\alpha &= \cos(360 - 4\alpha) = \cos(7\alpha - 4\alpha)\\ &= \cos 3\alpha = 4\cos^3\alpha - 3\cos\alpha, \end{aligned}\tag{6}$$

$$\begin{aligned}\cos 4\alpha &= \cos(2\cdot 2\alpha) = 2\cos^2 2\alpha - 1\\ &= 2(2\cos^2\alpha - 1)^2 - 1. \end{aligned}\tag{7}$$

Equating (6) and (7), we have

$$2(2u^2 - 1)^2 - 1 = 4u^3 - 3u,$$

or multiplying both sides by 2,

$$4(2u^2 - 1)^2 - 2 = (2u)^3 - 3(2u),$$
$$[(2u)^2 - 2]^2 - 2 = (2u)^3 - 3(2u). \tag{8}$$

Set $v = 2u$ in (8):

$$(v^2 - 2)^2 - 2 = v^3 - 3v,$$
$$v^4 - 4v^2 + 2 = v^3 - 3v,$$
$$v^4 - 4v^2 - (v^3 - 3v - 2) = 0,$$
$$v^2(v^2 - 4) - (v - 2)(v^2 + 2v + 1) = 0,$$
$$(v - 2)[v^2(v + 2) - (v^2 + 2v + 1)] = 0,$$
$$(v - 2)(v^3 + v^2 - 2v - 1) = 0. \tag{9}$$

Now, if $v = 2\cos\alpha = 2$, then $\cos\alpha = 1$, i.e., α is a multiple of 360°, obvious nonsense. Consider the polynomial

$$p(x) = x^3 + x^2 - 2x - 1$$
$$\in Z[x].$$

If it were reducible in $Q[x]$, it would be reducible in $Z[x]$ and hence divisible by $x - c$, $c \in Z$. But then $c = \pm 1$, neither of which is a root of $p(x)$. Thus $p(x)$ is irreducible and $p(2\cos\alpha) = 0$. It follows that $[Q(u):Q] = 3$, and Theorem 6.4 again completes the argument. ∎

Theorem 6.7 *Given the length* 1 *it is impossible to legally construct the length of the edge of a cube of volume* 2. ∎

Proof: Clearly, if this were possible then the number $\sqrt[3]{2}$ would be constructible. But $x^3 - 2$ is irreducible over the rationals, and hence $[Q(\sqrt[3]{2}) : Q] = 3$. Apply Theorem 6.4. ∎

It is perhaps worthwhile to exhibit an example of an illegal construction that can be used to trisect any angle θ.

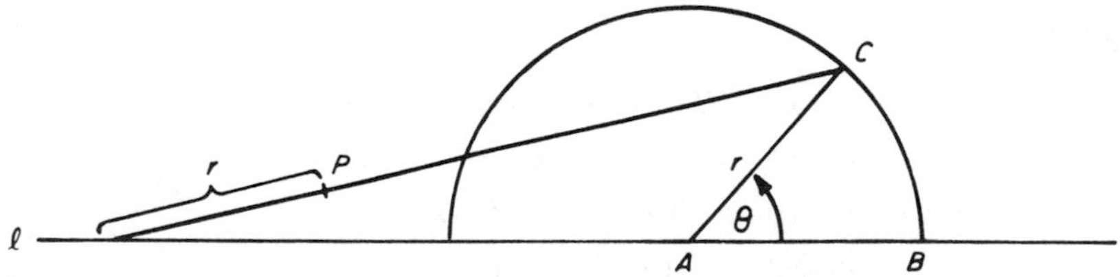

(i) Put the ruler with left end at A, and mark P on the ruler.

(ii) Put the left end of the ruler on the (left) extension of AB, and put the edge at C.

(iii) Slide the ruler along l keeping C on the edge until P hits the circle (illegal!).

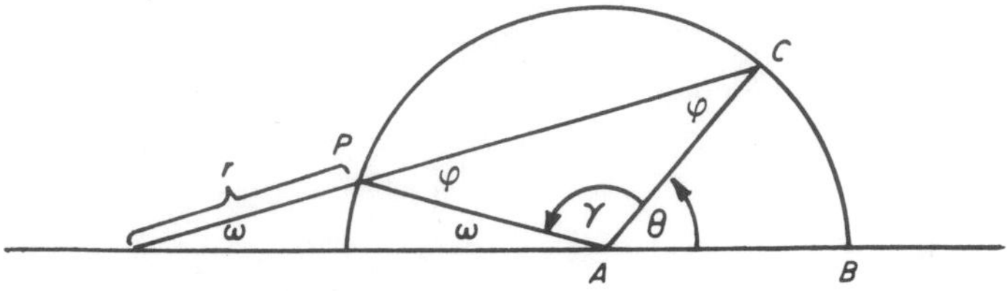

We assert $\omega = \theta/3$. For

$$\theta + \gamma + \omega = \pi,$$
$$\pi - (\pi - 2\omega) = \varphi,$$
$$\gamma = \pi - 2\varphi.$$

Thus

$$\gamma = \pi - 2\varphi$$
$$= \pi - 4\omega,$$
$$\theta + (\pi - 4\omega) + \omega = \pi,$$
$$3\omega = \theta,$$
$$\omega = \frac{\theta}{3}.$$

Exercises 3.6

1. Find $[Q(i, \sqrt{2}):Q]$.
2. Find $[Q(\sqrt{3}, \sqrt{5}):Q]$. *Hint*: The problem is to show that $x^2 - 5$ is irreducible over $Q(\sqrt{3})$. If not, then there exists a root $a + b\sqrt{3}$, $a,b \in Q$, such that $(a + b\sqrt{3})^2 = 5$.
3. Prove that $a + ib \in \mathbf{C}$ is a constructible number iff $|a|$ and $|b|$ are constructible lengths. *Hint*: We are given (0,0) and (1,0) and hence it is obvious that the line through (0,0) and (1,0), i.e., the x-axis, is constructible. Also the line perpendicular to the x-axis at (0,0) is constructible, i.e., the y-axis is constructible. Now if $|a|$ and $|b|$ are constructible lengths, then a and b are constructible numbers (if $a < 0$, then $|a| = -a$ is constructible). It follows that $a + ib$ is a constructible number. Conversely, if $a + ib$ is constructible, then both a and b are constructible by dropping perpendiculars. But then $|a|$ and $|b|$ are constructible lengths.

4. Prove that the sum of constructible numbers is constructible.
5. Exhibit an example of a constructible number field which is not a finite extension of Q.
6. Give an example of an algebraic number $\theta = a + ib$ over Q such that $Q(a,b,i)$ properly contains $Q(\theta,\bar{\theta})$. *Hint*: Let $\theta = i\sqrt{2}$, i.e., $\theta^2 + 2 = 0$. Then $\bar{\theta} = -i\sqrt{2}$ and $Q(\theta,\bar{\theta}) = Q(\theta)$. Observe that $[Q(\theta):Q] = 2$. Now obviously $Q(a,b,i) = Q(\sqrt{2})(i)$. But $x^2 + 1$ is irreducible over $Q(\sqrt{2})$, and hence $[Q(\sqrt{2},i):Q] = [Q(\sqrt{2}):Q][Q(\sqrt{2})(i):Q(\sqrt{2})] = 2 \cdot 2 = 4$.
7. Show that starting with points in the plane whose coordinates are rational it is impossible to legally construct a regular polygon of nine sides.

Glossary 3.6

$\mathscr{C}$, 259
constructible number, 259
constructible number field, 260
cube of volume 2, 262
degree theorem, 257
illegal trisection of an angle, 262,
legal constructions, 259
linearly dependent, 257
l.d., 257
linearly independent, 257
l.i., 257
regular polygon of seven sides, 261
trisection of an angle, 261

3.7 Galois Theory

In this section we shall investigate the intricate interplay between finite groups and finite field extensions that provides us with the necessary results to decide about the solvability of algebraic equations.

Let G be a group, K a field, and

$$\lambda\colon G \to K$$

a homomorphism of G into the multiplicative group of nonzero elements of K. Then λ is called a *character* (*of degree* 1) of G. It is easy to construct group characters from field automorphisms. For suppose

$$\sigma\colon K \to K$$

is an automorphism. Then $\sigma(1) = 1$, $\sigma(hk) = \sigma(h)\sigma(k)$ so that σ is a character of K (of degree 1).

Example 1 (a) $\sigma\colon \mathbf{C} \to \mathbf{C}$ defined by $\sigma(z) = \bar{z}$ is a field automorphism.

(b) If $\sigma\colon \mathbf{R} \to \mathbf{R}$ is an automorphism, then $\sigma(r) = r$ for each $r \in Q$; i.e., an automorphism of $\mathbf{R}$ must hold Q pointwise fixed. For $\sigma(1) = 1$ implies that for any positive integer p,

$$\sigma(p) = \sigma(\overbrace{1 + \cdots + 1}^{p}) = \overbrace{\sigma(1) + \cdots + \sigma(1)}^{p} = p.$$

Also $\sigma(0) = 0$; $\sigma(-p) = -\sigma(p) = -p$; $\sigma(q^{-1}) = \sigma(q)^{-1} = q^{-1}$. Hence $\sigma(p/q) = \sigma(pq^{-1}) = \sigma(p)\sigma(q)^{-1} = p/q$.

(c) If $\sigma: \mathbf{R} \to \mathbf{R}$ is an automorphism, then $a > 0$ implies $\sigma(a) > 0$, for if $a = b^2$, then $\sigma(a) = \sigma(b)^2 > 0$. Also note that $a > c$ implies $\sigma(a) > \sigma(c)$ since $\sigma(a - c) > 0$.

(d) The only automorphism of $\mathbf{R}$ is ι_R, the identity function. For let $a \in R$ and let $r < a < s$, where r and s are in Q. Then $\sigma(r) = r < \sigma(a) < \sigma(s) = s$, $r < \sigma(a) < s$. Thus given any rational numbers r and s such that $r < a < s$, both a and $\sigma(a)$ are in the interval between r and s. Clearly, $\sigma(a) = a$.

(e) If $\sigma: Q(\sqrt{2}) \to Q(\sqrt{2})$ is defined by

$$\sigma(a + b\sqrt{2}) = a - b\sqrt{2},$$

then σ is an automorphism but σ is not a continuous function. For let $\lim_{n\to\infty} r_n = \sqrt{2}$, $r_n \in Q$. Then $\sigma(r_n) = r_n$ so that $\lim_{n\to\infty} \sigma(r_n) = \sqrt{2} \neq -\sqrt{2} = \sigma(\sqrt{2}) = \sigma(\lim_{n\to\infty} r_n)$.

Let K and F be fields, $F \subset K$, and let G be a set of automorphisms of K. Define

$$A(K) = \{\sigma \mid \sigma \text{ is an automorphism of } K\}; \tag{1}$$

$$A(K|F) = \{\sigma \in A(K) \mid \sigma(f) = f \text{ for each } f \in F\}; \tag{2}$$

$$K(G) = \{f \in K \mid \sigma(f) = f \text{ for all } \sigma \in G\}. \tag{3}$$

The group $A(K)$ is called the *automorphism group of K*; $A(K|F)$ is called the *automorphism group* or *Galois group of K over F*; $K(G)$ is called the *fixed field of G*. It is an easy exercise to verify that $A(K) < S_K$, the symmetric group on K; $A(K|F) < A(K)$; $K(G)$ is a subfield of K (see Exercises 1, 2, 3).

Example 2 We compute $A(\mathbf{C}|\mathbf{R})$. Note that $\sigma(i)^2 = \sigma(i^2) = \sigma(-1) = -\sigma(1) = -1$. Hence $\sigma(i) = i$ or $\sigma(i) = -i$. Then $\sigma(a + bi) = \sigma(a) + \sigma(b)\sigma(i) = a + b\sigma(i)$. Thus $\sigma \in A(\mathbf{C}|\mathbf{R})$ implies that $\sigma(a + ib) = a + ib$ or $\sigma(a + ib) = a - ib$ depending on whether $\sigma(i) = i$ or $\sigma(i) = -i$. Thus $A(\mathbf{C}|\mathbf{R})$ is a group of order 2 consisting of e and σ, $\sigma(z) = \bar{z}$.

Example 3 Let $K = Q(2^{1/4}, i)$ and let $G = \{e, \varphi\}$, where $\varphi \in A(K)$ satisfies $\varphi(2^{1/4}) = -2^{1/4}$, $\varphi(i) = -i$, $\varphi(r) = r$, $r \in Q$. Note that

$$\varphi^2(2^{1/4}) = \varphi(-2^{1/4}) = -\varphi(2^{1/4}) = 2^{1/4},$$
$$\varphi^2(i) = \varphi(-i) = -\varphi(i) = i,$$

and hence $\varphi^2 = e$. We compute the fixed field $K(G)$ in a perfectly general way. We know from Theorem 6.2, Section 3.6, that the elements

$$1,\quad 2^{1/4},\quad 2^{1/2},\quad 2^{3/4},\quad i,\quad i2^{1/4},\quad i2^{1/2},\quad i2^{3/4}$$

comprise a basis of K over Q. Now let

$$f = a_1 + a_2 2^{1/4} + a_3 2^{1/2} + a_4 2^{3/4} + b_1 i + b_2 i2^{1/4} + b_3 i2^{1/2} + b_4 i2^{3/4}$$

be a typical element of K. Then $f \in K(G)$ iff

$$\varphi(f) = f,$$

i.e., $\quad f = a_1 - a_2 2^{1/4} + a_3 2^{1/2} - a_4 2^{3/4} - b_1 i + b_2 i 2^{1/4} - b_3 i 2^{1/2} + b_4 i 2^{3/4}$
so that $a_2 = a_4 = b_1 = b_3 = 0$. Hence $K(G)$ consists of all elements of the form

$$f = a_1 + a_3 2^{1/2} + b_2 i 2^{1/4} + b_4 i 2^{3/4}. \tag{4}$$

Observe that $[K(G):Q] = 4$ and $K(G) = Q(i2^{1/4})$.

Our first important result is the following theorem.

Theorem 7.1 *Let K be a field and $G \subset A(K)$, $|G| = n$. Then*

$$[K: K(G)] \geq n. \tag{5}$$

Proof: We remark that G is simply a set of automorphisms of K, not necessarily a group. We know in general that $K(G)$ is a subfield of K. Suppose that

$$[K: K(G)] = r < n,$$

and let $\{\eta_1, \ldots, \eta_r\}$ be a basis of K over $K(G)$. Consider the system of linear equations

$$\sum_{j=1}^{n} \sigma_j(\eta_i)\xi_j = 0, \qquad i = 1, \ldots, r, \tag{6}$$

where $\{\sigma_1, \ldots, \sigma_n\} = G$; since $r < n$, there is a nontrivial solution in K to (6). Observe that if $k = \sum_{t=1}^{r} c_t\eta_t$, $c_t \in K(G)$, and $\sigma = \sum_{j=1}^{n} \xi_j\sigma_j$, then

$$\begin{aligned}\sigma(k) &= \sum_{j=1}^{n} \xi_j\sigma_j\Big(\sum_{t=1}^{r} c_t\eta_t\Big) = \sum_{j=1}^{n}\sum_{t=1}^{r} \xi_j c_t \sigma_j(\eta_t)\\ &= \sum_{t=1}^{r} c_t \sum_{j=1}^{n} \xi_j\sigma_j(\eta_t) = 0.\end{aligned}$$

Thus

$$\sum_{j=1}^{n} \xi_j\sigma_j = 0, \tag{7}$$

and not all of $\xi_1, \ldots, \xi_n$ in K are 0. However, we can regard $\sigma_1, \ldots, \sigma_n$ as distinct group characters of the multiplicative group of nonzero elements in K, and by Exercise 4 they must be 1.i. over K. This contradicts (7). ∎

Our next result shows that in the event G is a group, we can substantially improve Theorem 7.1.

Theorem 7.2 *Let K be a field and $G = \{\sigma_1, \ldots, \sigma_n\}$ a group of automorphisms of K. Let x be an indeterminate over K. Then*

(i) $[K:K(G)] = n$.

(ii) *If $\alpha \in K$, define*

$$p(x) = \prod[x - \sigma_i(\alpha)], \tag{8}$$

where the product is taken over all those i, $1 \leq i \leq n$, for which the corresponding $\sigma_i(\alpha)$ are distinct. Then $p(x)$ is a monic irreducible polynomial over $K(G)$.

It is, in fact, the characteristic polynomial for α in $K(G)[x]$, i.e., *$p(x)$ is the monic polynomial of least degree in $K(G)[x]$ having α as a root.*

Proof: (i) To prove that $[K:K(G)] = n$, we are going to prove that any m elements of K, $m > n$, must be 1.d. over $K(G)$. Then $[K:K(G)] \leq n$, which combined with Theorem 7.1 produces the required equality.

Thus let $k_1, \ldots, k_m$ be m elements of K, $m > n$, and consider the system of homogeneous linear equations

$$\sum_{j=1}^{m} \sigma_i(k_j)\xi_j = 0, \qquad i = 1, \ldots, n. \tag{9}$$

Since $m > n$, the system (9) has a nontrivial solution $\xi = (\xi_1, \ldots, \xi_m)$. We can assume $\xi_1 \neq 0$ (by renumbering the k_j if necessary). Now according to Exercise 4, $\sigma_1, \ldots, \sigma_n$ are 1.i. over K so that $\sum_{i=1}^{n} \sigma_i \neq 0$. Thus there exists $\theta \in K$, $\theta \neq 0$, such that

$$\sum_{i=1}^{n} \sigma_i(\theta) \neq 0. \tag{10}$$

Choose $\lambda \in K$ such that $\lambda\xi_1 = \theta$, i.e., $\lambda = \xi_1^{-1}\theta$. Then $\lambda\xi$ is a solution to (9) and $(\lambda\xi)_1 = \theta$ satisfies (10). Thus we can assume that ξ is already chosen so that (9) is satisfied and

$$\sum_{i=1}^{n} \sigma_i(\xi_1) \neq 0. \tag{11}$$

Let t be fixed for the moment, and evaluate σ_t on both sides of (9) to obtain

$$\sum_{j=1}^{m} \sigma_t\sigma_i(k_j)\sigma_t(\xi_j) = 0, \qquad i = 1, \ldots, n. \tag{12}$$

Of course, $\sigma_t\sigma_i = \sigma_{i'}$ and as i runs over $\{1, \ldots, n\}$ so does i'. Thus (12) becomes

$$\sum_{j=1}^{m} \sigma_{i'}(k_j)\sigma_t(\xi_j) = 0, \qquad i' = 1, \ldots, n, \tag{13}$$

and we have a system such as (13) for each $t = 1, \ldots, n$. In other words, the n vectors

$$(\sigma_t(\xi_1), \ldots, \sigma_t(\xi_m)), \qquad t = 1, \ldots, n, \tag{14}$$

all satisfy the homogeneous system (9) and hence their sum

$$\eta = \left(\sum_{t=1}^{n} \sigma_t(\xi_1), \ldots, \sum_{t=1}^{n} \sigma_t(\xi_m)\right) \tag{15}$$

also satisfies (9). Note that (11) states that $\eta_1 \neq 0$, and hence η is a nontrivial solution to (9). Also observe that

$$\sigma_s(\eta_j) = \sum_{t=1}^{n} \sigma_s\sigma_t(\xi_j) = \sum_{t=1}^{n} \sigma_t(\xi_j)$$

$$= \eta_j, \qquad s = 1, \ldots, n, \; j = 1, \ldots, m,$$

and hence

$$\eta_j \in K(G), \qquad j = 1, \ldots, m.$$

Observe also that if for a fixed i' we sum (13) for $t = 1, \ldots, n$, then we obtain (dropping the prime on i)

$$\sum_{j=1}^{m} \sigma_i(k_j)\eta_j = 0, \qquad i = 1, \ldots, n. \tag{16}$$

In particular, there is an i such that $\sigma_i = e$, and for this i (16) becomes

$$\sum_{j=1}^{m} k_j\eta_j = 0,$$

a linear dependence relation for $k_1, \ldots, k_m$ over $K(G)$.

(ii) It is routine to verify that the polynomial (8) is equal to

$$p(x) = x^p - E_1(\sigma_1(\alpha), \ldots, \sigma_p(\alpha))x^{p-1} + \cdots + (-1)^p E_p(\sigma_1(\alpha), \ldots, \sigma_p(\alpha)) \tag{17}$$

where $E_r(t_1, \ldots, t_p)$ is the rth elementary symmetric function of $t_1, \ldots, t_p$, and the ordering of the σ_i is chosen so that $\sigma_1(\alpha), \ldots, \sigma_p(\alpha)$ are all the distinct values of $\sigma_j(\alpha)$, $j = 1, \ldots, n$ (see Exercises 5 to 7). But then for any $k = 1, \ldots, n$,

$$\sigma_k\sigma_1(\alpha), \ldots, \sigma_k\sigma_p(\alpha) \tag{18}$$

are distinct and hence must be $\sigma_1(\alpha), \ldots, \sigma_p(\alpha)$ in some order [since $\sigma_1(\alpha), \ldots, \sigma_p(\alpha)$ are all the distinct values of $\sigma_i(\alpha)$, $i = 1, \ldots, n$, and the p values (18) are distinct]. Now (see Exercise 8)

$$\begin{aligned}\sigma_k E_r(\sigma_1(\alpha), \ldots, \sigma_p(\alpha)) &= E_r(\sigma_k\sigma_1(\alpha), \ldots, \sigma_k\sigma_p(\alpha)) \\ &= E_r(\sigma_1(\alpha), \ldots, \sigma_p(\alpha)).\end{aligned}$$

In other words, the coefficients of the polynomial $p(x)$ are in $K(G)$ and, in fact, $\sigma_1(\alpha), \ldots, \sigma_p(\alpha)$ are the roots of $p(x)$. Of course, since G is a group one of these values is α itself. Suppose $f(x) \in K(G)[x]$ is a polynomial for which $f(\alpha) = 0$; then for any $i = 1, \ldots, n$,

$$0 = \sigma_i(f(\alpha)) = f(\sigma_i(\alpha))$$

so that $\sigma_i(\alpha)$ is a root of $f(x)$ as well. Since there are p distinct values among the $\sigma_i(\alpha)$, $i = 1, \ldots, n$, it follows that $\deg f(x) \geq p$. In other words, $p(x)$ is the monic polynomial of least degree in $K(G)[x]$ having α as a root, i.e., it is the characteristic polynomial for α and hence is irreducible. ∎

Example 4 Let $K = Q(2^{1/3})$. The question we wish to consider is this: Does there exist a group $G < A(K)$ such that $Q = K(G)$? That is, does there exist a group of automorphisms of $Q(2^{1/3})$ for which Q is the fixed field? Observe first that $\sigma(2^{1/3})^3 = \sigma(2) = 2$ for any $\sigma \in A(K)$ holding Q elementwise fixed. The elements of K are real numbers, and the only real cube root of 2 is $2^{1/3}$. Thus if $\sigma \in A(K)$ and σ holds Q elementwise fixed, then $\sigma(2^{1/3}) = 2^{1/3}$ and hence $\sigma = e$. But $K(e) = K \neq Q$, so there is no group $G < A(K)$ such that $Q = K(G)$.

Our next result formalizes a fact we have verified several times before.

Theorem 7.3 *Let F be a field, $f(x) \in F[x]$, and let K be a splitting field for $f(x)$. Let $E = \{\alpha_1, \ldots, \alpha_m\}$ be the set of distinct roots of $f(x)$ in K. Then $A(K|F)|E < S_E$, the symmetric group on E, where $A(K|F)|E = \{\sigma|E \mid \sigma \in A(K|F)\}$.*

Proof: First observe that if $\sigma \in A(K|F)$, then σ is completely determined by its values on the roots in E, i.e., by the values $\sigma(\alpha_i)$. Indeed, any element of K is a linear combination of products of powers of the α_i with coefficients from F [i.e., K is a splitting field for $f(x)$]. Now if $\sigma \in A(K|F)$, then for $t = 1, \ldots, m$ we have

$$\begin{aligned} f(\sigma(\alpha_t)) &= \sigma(f(\alpha_t)) = \sigma(0) \\ &= 0; \end{aligned}$$

so $\sigma(\alpha_t)$ is a root of $f(x)$ and hence must be one of $\alpha_1, \ldots, \alpha_m$. Since σ is injective, we conclude that $\sigma|E \in S_E$. ∎

We also have the following:

Theorem 7.4 *Let K be a field, $G < A(K)$, $|G| = n$, and let $f(x) \in K(G)[x]$ be an irreducible polynomial with some root $\alpha \in K$. Then, in fact, K contains all the roots of $f(x)$, i.e., K contains a splitting field for $f(x)$.*

Proof: From Theorem 7.2 we know that the polynomial $p(x)$ defined in (8) has the following properties: $p(x) \in K(G)[x]$; $p(x)$ is the characteristic polynomial for α in $K(G)[x]$; all the roots of $p(x)$ are in K and one of these roots is α. Since $p(x)$ is the characteristic polynomial for α and $f(\alpha) = 0$, we know that $p(x)|f(x)$ in $K(G)[x]$. But the irreducibility of $f(x)$ implies that, in fact, $f(x) = cp(x)$, $c \in K(G)$. Thus all the roots of $f(x)$ are in K. ∎

If $f(x) \in K[x]$ and every irreducible factor of $f(x)$ in $K[x]$ has distinct roots [in the splitting field for $f(x)$], then $f(x)$ is called a *separable polynomial.* If we look back at Exercises 25 and 26, Section 3.4, we see that if char $K = 0$, then any polynomial is separable. If char $K = p$, however, then there exist irreducible polynomials in $K[x]$ which do not have distinct roots. We shall exhibit an example of an irreducible polynomial which is not separable, i.e., which does not have distinct roots. It is a fact that no such example exists if K is a finite field, but we shall not investigate this problem further. We content ourselves with the following rather unpleasant example.

Example 5 Let x be an indeterminate over Z_2, and let $F = Z_2(x)$. Let y be an indeterminate over F and consider the polynomial ring $F[y]$. First observe that char $F = 2$. Now consider the polynomial $f(y) = y^2 - x$ in $F[y]$. The derivative (in y) of $f(y)$ is $[2]y = 0$, so by Exercise 25, Section 3.4, $f(y)$ has fewer than 2 distinct roots. On the other hand, if $f(y)$ were reducible in $F[y]$, there would exist an $\alpha \in F$ such that $\alpha^2 = x$. But $\alpha = p(x)/q(x)$, $p(x)$ and $q(x)$ in $Z_2[x]$, and hence $p(x)^2 = xq(x)^2$. But deg $p(x)^2 = 2$ deg $p(x) =$ deg $xq(x)^2 =$ deg $x + 2$ deg $q(x) = 1 + 2$ deg $q(x)$, which is clearly impossible. Actually, the reason $y^2 - x$ has only one root (aside from the argument given above) is that its splitting field is $F(\sqrt{x})$, $y^2 - x = (y - \sqrt{x})(y + \sqrt{x})$, and $F(\sqrt{x})$ also has characteristic 2. Thus $-\sqrt{x} = \sqrt{x}$ and there is only one root of this irreducible polynomial.

If K is an extension field of F, then K is a *normal extension* of F if there exists a finite group $G < A(K)$ such that $F = K(G)$. In other words, K is a normal extension of F if F is precisely the fixed field of some finite group of automorphisms of K. Observe that if $k \in K$ and $k \notin F$ (where K is a normal extension of F), then there must be some $\sigma \in G$ such that $\sigma(k) \neq k$; otherwise k would be in F.

The following result is of crucial importance in understanding the structure of normal extension fields.

Theorem 7.5 *If $F \subset K$ are fields, then K is a normal extension of F iff K is the splitting field of a separable polynomial in $F[x]$. In fact, if K is the splitting field of the separable polynomial $p(x) \in F[x]$, then*

$$F = K(A(K|F)), \tag{19}$$

i.e., F is precisely the fixed field of the Galois group of K over F.

Proof: We begin by proving that if $F = K(G)$, $|G| = n$, then K is the splitting field of a separable polynomial in $F[x]$. From Theorem 7.2(i) we know that $[K:F] = n$ so that

$$K = F(\alpha_1, \ldots, \alpha_n) \tag{20}$$

where $\alpha_t \in K$, $t = 1, \ldots, n$, comprise a basis of K over F. According to Theorem 7.2(ii) we can construct the characterrstic polynomial $p_t(x) \in F[x]$ of each α_t. Moreover, $p_t(x) \in F[x]$ has distinct roots [they are, in fact, all the distinct values $\sigma_i(\alpha_t)$ obtained as σ_i runs over G], and of course $p_t(x)$ splits in $K[x]$, $t = 1, \ldots, n$. If we let

$$p(x) = p_1(x) \cdots p_n(x) \in F[x],$$

then $p(x)$ is obviously a separable polynomial. Also K is the splitting field for $p(x)$ because K contains all the roots of $p(x)$, and in view of (20) it is the smallest such field since $\alpha_1, \ldots, \alpha_n$ are among the roots of $p(x)$.

In order to prove the sufficiency, we must show that if K is the splitting field of a separable polynomial $p(x) \in F[x]$, then

$$F = K(A(K|F)), \tag{21}$$

so that K is indeed a normal extension of F [notice that $A(K|F)$ is a finite group as a consequence of Theorem 7.3]. Thus suppose

$$p(x) = c(x - \alpha_1) \cdots (x - \alpha_r)p_1(x) \cdots p_m(x) \tag{22}$$

is the prime factorization over $F[x]$ of the separable polynomial $p(x)$, where each of the irreducible polynomials $p_i(x)$ (of degree greater than 1) has distinct roots. We argue by induction on $k = n - r$, where $n = \deg p(x)$. Namely, we propose to prove that (21) holds if K is the splitting field of (22). For $k = 0$, i.e., $r = n$, $p(x)$ splits in $F[x]$ so that $K = F$, $A(K|F) = \{e\}$ and (21) is trivially true. We thus assume that (21) holds whenever $p(x)$ has $r + 1$ linear factors, $0 \leq r \leq n - 1$, and try to prove that (21) holds when $p(x)$ has r linear factors [i.e., we go from $k = n - (r + 1)$ to $k + 1 = n - r$]. Let $\alpha_{r+1} \in K$ be a root of $p_1(x)$, and let $F(\alpha_{r+1}) \subset K$ be the corresponding extension field of F. Let $\alpha_{r+1}, \ldots, \alpha_{r+s}$, all in the splitting field K, be the complete set of roots of $p_1(x)$, $\deg p_1(x) = s$. These are distinct by the separability of $p(x)$. The extensions $F(\alpha_{r+1}), \ldots, F(\alpha_{r+s})$ are all equivalent extensions of F. In fact, by Theorem 4.12, Section 3.4, there exist field isomorphisms $\sigma_1, \ldots, \sigma_s$ such that $\sigma_t: F(\alpha_{r+1}) \to F(\alpha_{r+t})$, σ_t holds F pointwise fixed, and

$$\sigma_t(\alpha_{r+1}) = \alpha_{r+t}, \qquad 1 \leq t \leq s. \tag{23}$$

Now K is a splitting field for $p(x)$ regarded as a polynomial in $F(\alpha_{r+1})[x]$ and also regarded as a polynomial in $F(\alpha_{r+t})[x]$; σ_t is a field isomorphism of $F(\alpha_{r+1})$ onto $F(\alpha_{r+t})$ in which the coefficients of $p(x)$ are held pointwise fixed. If we apply Theorem 4.13, Section 3.4, we can extend σ_t to an isomorphism of K onto itself, i.e., to an automorphism of K. We denote this extended automorphism by σ_t also. Thus $\sigma_1, \ldots, \sigma_s$ are automorphisms of K for which (23) holds and which hold F pointwise fixed, so that

$$\sigma_t \in A(K|F), \qquad t = 1, \ldots, s. \tag{24}$$

Now rewrite (22) as

$$p(x) = c(x - \alpha_1) \cdots (x - \alpha_r)(x - \alpha_{r+1})q_1(x)p_2(x) \cdots p_m(x), \tag{25}$$

regarded as a factorization in $F(\alpha_{r+1})[x]$, also with splitting field K. The induction hypothesis applies to the factorization (25) in which the coefficient field is $F(\alpha_{r+1})$, i.e., $F(\alpha_{r+1})$ is playing the role of F. Thus by induction [we have at least one more linear factor in (25) than in (22)] we have

$$F(\alpha_{r+1}) = K(A(K|F(\alpha_{r+1}))), \tag{26}$$

i.e., $F(\alpha_{r+1})$ is precisely the fixed field of the automorphism group of K over $F(\alpha_{r+1})$. Now suppose

$$\theta \in K(A(K|F)). \tag{27}$$

Since obviously $A(K|F(\alpha_{r+1})) < A(K|F)$ it follows from (27) that if $\sigma \in A(K|F(\alpha_{r+1}))$, then

$$\sigma(\theta) = \theta,$$

i.e.,

$$\theta \in K(A(K|F(\alpha_{r+1})))$$

or, by (26),

$$\theta \in F(\alpha_{r+1}). \tag{28}$$

Recall that α_{r+1} is a root of the irreducible polynomial $p_1(x)$, $\deg p_1(x) = s$; so (28) means that

$$\theta = c_0 + c_1\alpha_{r+1} + \cdots + c_{s-1}\alpha_{r+1}^{s-1} \tag{29}$$

where $c_i \in F, i = 0, \ldots, s-1$. Now (27) means that θ is held fixed by every $\sigma \in A(K|F)$, so by (24) we have

$$\theta = \sigma_t(\theta) = c_0 + c_1\alpha_{r+t} + \cdots + c_{s-1}\alpha_{r+t}^{s-1} \qquad \text{[see (23)]}$$

or

$$c_{s-1}\alpha_{r+t}^{s-1} + \cdots + c_1\alpha_{r+t} + (c_0 - \theta) = 0, \qquad t = 1, \ldots, s. \tag{30}$$

Consider the polynomial

$$f(x) = c_{s-1}x^{s-1} + \cdots + c_1x + (c_0 - \theta) \in K[x].$$

The s elements $\alpha_{r+1}, \ldots, \alpha_{r+s}$ are distinct and according to (30) are roots of $f(x)$. But $\deg f(x) = s - 1$ so that in fact $f(x) = 0$. We conclude that

$$c_0 - \theta = 0$$

or

$$\theta = c_0 \in F.$$

We have proved that

$$K(A(K|F)) \subset F$$

and the reverse inclusion is trivial. ∎

Corollary 1 *Let K be the splitting field of a separable polynomial $p(x) \in F[x]$. Then*

(i) $[K : F] = |A(K|F)|$.

(ii) *If $f(x) \in F[x]$ is any irreducible polynomial with a root in K, then the splitting field of $f(x)$ is a subfield of K.*

Proof: (i) We have from Theorem 7.5 that

$$F = K(A(K|F)). \tag{31}$$

But then Theorem 7.2(i) implies that

$$[K : F] = [K : K(A(K|F))] = |A(K|F)|.$$

(ii) Again, (31) implies that

$$f(x) \in F[x] = K(A(K|F))[x].$$

Now apply Theorem 7.4 ∎

Example 6 Let K be the splitting field for $p(x) = x^5 - 3x^4 - 2x + 6 \in Q[x]$. Then $p(x) = (x-3)(x^4-2) = (x-3)(x^2-\sqrt{2})(x^2+\sqrt{2}) = (x-3)(x-2^{1/4})$ $(x+2^{1/4}) \cdot (x+i2^{1/4})(x-i2^{1/4})$. Thus K must contain $2^{1/4}$ and $i2^{1/4}$ and hence i. But then $K = Q(2^{1/4}, i)$, and it is trivial to see that x^2+1 is irreducible over $Q(2^{1/4})$ as is x^4-2 over Q. Thus by the degree theorem (Theorem 6.2, Section 3.6), $[K: F] = [Q(2^{1/4}): Q][Q(2^{1/4})(i): Q(2^{1/4})] = 4 \cdot 2 = 8$. Hence $|A(K|Q)| = A(Q(2^{1/4}, i): Q) = 8$ by Corollary 1(i).

In general, if $p(x) \in F[x]$ is separable and K is the splitting field for $p(x)$, then $A(K|F)$ is called the *Galois group of the polynomial* $p(x)$.

Our next result, which will take a bit of proving, is called the *Fundamental Theorem of Galois Theory* (hereafter abbreviated to F.T.G.T.). *We assume that the underlying field has characteristic* 0 *so that any polynomial is separable.*

Theorem 7.6 (F.T.G.T.) *Let F be a field of characteristic* 0, $p(x) \in F[x]$ *a (separable) polynomial. Let K be the splitting field for $p(x)$. (In other words, we let K be a normal extension of F.) Then*

(i) *If $F \subset B \subset K$, B a field, then*

$$B = K(A(K|B)). \tag{32}$$

(ii) *If Φ is the set of all subfields of K containing F and Ω is the set of all subgroups of $A(K|F)$, then the mapping $\mu: \Phi \to \Omega$ defined by*

$$\mu(B) = A(K|B) \tag{33}$$

is a bijection. Moreover, if A and B are in Φ, then $A \subset B$, $A \neq B$ implies $\mu(B) \subset \mu(A)$, $\mu(B) \neq \mu(A)$.

(iii) *If $B \in \Phi$, then*

$$[B : F] = \frac{|A(K|F)|}{|A(K|B)|}. \tag{34}$$

(iv) *If $B \in \Phi$, then B is a normal extension of F iff*

$$A(K|B) \lhd A(K|F). \tag{35}$$

Also, if B is a normal extension of F, then

$$A(B|F) \cong A(K|F)/A(K|B). \tag{36}$$

Proof: (i) Regard $p(x)$ as a polynomial in $B[x]$ and apply Theorem 7.5 to conclude (32).

(ii) We first prove that μ is a surjection. Let $H < A(K|F)$, and let $B = K(H)$. Obviously, $F \subset B \subset K$. According to (32),

$$B = K(A(K|B)). \tag{37}$$

From Theorem 7.2(i) we have

$$[K{:}B] = [K{:}K(H)]$$
$$= |H|$$

and also from (37)

$$[K{:}B] = [K{:}K(A(K|B))]$$
$$= |A(K|B)|$$

so that

$$|H| = |A(K|B)|. \tag{38}$$

But $H < A(K|B)$ because $A(K|B)$ consists of *all* automorphisms of K which hold B pointwise fixed. Thus (38) implies that

$$H = A(K|B),$$

i.e.,

$$\mu(B) = H$$

and $\mu\colon \Phi \to \Omega$ is surjective. Suppose that $\mu(A) = \mu(B)$, i.e.,

$$A(K|A) = A(K|B).$$

Then by (i),

$$A = K(A(K|A)) = K(A(K|B))$$
$$= B.$$

Thus μ is an injection. Obviously, if $A \subset B$ then $A(K|B) \subset A(K|A)$ and since μ is an injection, it preserves proper inclusion.

(iii) From Theorem 6.2(ii), Section 3.6 (the degree theorem), we have

$$[B{:}F] = \frac{[K{:}F]}{[K{:}B]}. \tag{39}$$

Now $F = K(A(K|F))$ and $B = K(A(K|B))$ by (i), and hence from Theorem 7.2(i),

$$[K{:}F] = |A(K|F)|, \tag{40}$$

$$[K{:}B] = |A(K|B)|. \tag{41}$$

Combining this with (39) produces (34).

(iv) Let $B \in \Phi$. Then by Exercise 9, B is a normal extension of F iff

$$A(K|F)B \subset B. \tag{42}$$

Assume then that B is a normal extension of F, $\sigma \in A(K|F)$, $\tau \in A(K|B)$, and $b \in B$. Then $\sigma(b) \in B$ by (42) so that $\tau(\sigma(b)) = \sigma(b)$ and hence

$$\sigma^{-1}\tau\sigma(b) = \sigma^{-1}\sigma(b)$$
$$= b,$$

i.e., $\sigma^{-1}\tau\sigma \in A(K|B)$. Thus (35) follows. Conversely, assume (35). Then $\sigma^{-1}\tau\sigma \in A(K|B)$ and hence for any $b \in B$

$$\sigma^{-1}\tau\sigma(b) = b,$$
$$\tau(\sigma(b)) = \sigma(b). \tag{43}$$

Now (43) asserts that $\sigma(b)$ is held fixed by any $\tau \in A(K|B)$. But by (i),

$$B = K(A(K|B))$$

so that any element held fixed by all $\tau \in A(K|B)$ must be in B, i.e., $\sigma(b) \in B$. Hence (42) holds and B is a normal extension of F.

To complete the proof we must confirm (36), and this is done by exhibiting a group homomorphism

$$f: A(K|F) \to A(B|F)$$

whose kernel is precisely $A(K|B)$. Thus assume B is a normal extension of F so that (42) holds. Then since $\sigma(B) \subset B$ for any $\sigma \in A(K|F)$, we can define

$$f(\sigma) = \sigma|B$$

which obviously belongs to $A(B|F)$. Also f is clearly a homomorphism. Now $\sigma \in \ker f$ iff $\sigma|B = \imath_B$ iff $\sigma \in A(K|B)$. Hence $A(K|B) = \ker f$. Thus

$$\operatorname{im} f \cong \frac{A(K|F)}{\ker f} = \frac{A(K|F)}{A(K|B)}. \tag{44}$$

But

$$\left|\frac{A(K|F)}{A(K|B)}\right| = \frac{|A(K|F)|}{|A(K|B)|}$$
$$= [B:F] \qquad \text{[by (34)]}.$$

However, $B = K(A(B|F))$ by (i), so by Theorem 7.2(i) we have

$$[B:F] = |A(B|F)|.$$

Thus we have proved that $f: A(K|F) \to A(B|F)$ satisfies $\operatorname{im} f < A(B|F)$, and

$$|\operatorname{im} f| = [B:F]$$
$$= |A(B|F)|.$$

Thus $\operatorname{im} f = A(B|F)$ and (44) produces the result (36). ∎

Example 7 Let $p(x) = x^4 - 2 \in Q[x]$; this is a separable polynomial in $Q[x]$. Obviously the splitting field of $p(x)$ is $K = Q(2^{1/4}, i)$ (why?) and $[K:Q] = 8$ (why?). We compute the Galois group $A(K|Q)$ which by Theorem 7.5 and Theorem 7.2(i) is of order 8. Any $\varphi \in A(K|Q)$ is completely determined by what it does to the roots of $x^4 - 2$, and these in fact are simply permuted by φ (Theorem 7.3). Note that $\varphi(i)^2 = -1$ so that $\varphi(i) = i$ or $\varphi(i) = -i$. Note also that the values of $\varphi(2^{1/4})$ and $\varphi(i)$ determine φ completely. We construct the following table in which we stipulate that $\varphi|Q = i_Q$.

φ	e	σ	σ^2	σ^3	τ	$\sigma\tau$	$\sigma^2\tau$	$\sigma^3\tau$
$\varphi(2^{1/4})$	$2^{1/4}$	$i2^{1/4}$	$-2^{1/4}$	$-i2^{1/4}$	$2^{1/4}$	$i2^{1/4}$	$-2^{1/4}$	$-i2^{1/4}$
$\varphi(i)$	i	i	i	i	$-i$	$-i$	$-i$	$-i$

It is not difficult to check that these eight functions satisfy

$$\sigma^4 = e, \quad \tau^2 = e, \quad \tau\sigma = \sigma^3\tau, \quad \tau\sigma^2 = \sigma^2\tau, \quad \tau\sigma^3 = \sigma\tau. \tag{45}$$

Moreover, we can verify directly that $\sigma(ab) = \sigma(a)\sigma(b)$, $\sigma(a + b) = \sigma(a) + \sigma(b)$, $\tau(ab) = \tau(a)\tau(b)$, $\tau(a + b) = \tau(a) + \tau(b)$ for all a, b in K and that both σ and τ are bijections. Thus σ and τ are in $A(K \mid Q)$. Moreover, since these eight automorphisms are distinct, we conclude that they comprise $A(K \mid Q)$. Consider, for example, the subgroup $G_{2,4} = \{e, \sigma^2\tau\}$. (The notation will become clear in a moment.) We wish to compute $K(G)$. We have from the table: $\sigma^2\tau(1) = 1$, $\sigma^2\tau(2^{1/4}) = -2^{1/4}$, $\sigma^2\tau(2^{1/2}) = \sigma^2\tau(2^{1/4}2^{1/4}) = 2^{1/2}$, $\sigma^2\tau(2^{3/4}) = \sigma^2\tau((2^{1/4})^3) = -2^{3/4}$, $\sigma^2\tau(i) = -i$, $\sigma^2\tau(i2^{1/4}) = i2^{1/4}$, $\sigma^2\tau(i2^{1/2}) = -i2^{1/2}$, $\sigma^2\tau(i2^{3/4}) = i2^{3/4}$. Thus, if $a_i \in Q$, $i = 1, \ldots, 8$, and

$$\alpha = a_1 + a_2 2^{1/4} + a_3 2^{1/2} + a_4 2^{3/4} + a_5 i + a_6 i2^{1/4} + a_7 i2^{1/2} + a_8 i2^{3/4} \in K,$$

then $\sigma^2\tau(\alpha) = \alpha$ iff

$$\begin{aligned}\sigma^2\tau(\alpha) &= a_1 - a_2 2^{1/4} + a_3 2^{1/2} - a_4 2^{3/4} - a_5 i + a_6 i2^{1/4} - a_7 i2^{1/2} + a_8 i2^{3/4} \\ &= \alpha,\end{aligned}$$

or
$$\alpha = a_1 + a_3 2^{1/2} + a_6 i2^{1/4} + a_8 i2^{3/4}.$$

Thus $K(G) = Q(i2^{1/4})$.

In general, we know from the F.T.G.T. that there is a distinct subfield of K containing Q corresponding to each of the subgroups of $A(K \mid Q)$. If we define

$$\begin{aligned}
G_8 &= A(K \mid Q), \\
G_{4,1} &= \{e,\sigma,\sigma^2,\sigma^3\}, \\
G_{4,2} &= \{e,\sigma^2,\tau,\sigma^2\tau\}, \\
G_{4,3} &= \{e,\sigma^2,\sigma\tau,\sigma^3\tau\}, \\
G_{2,1} &= \{e,\sigma^2\}, \\
G_{2,2} &= \{e,\tau\}, \\
G_{2,3} &= \{e,\sigma\tau\}, \\
G_{2,4} &= \{e,\sigma^2\tau\}, \\
G_{2,5} &= \{e,\sigma^3\tau\}, \\
G_1 &= \{e\},
\end{aligned} \tag{46}$$

then by using the relations (45) we can check that these are subgroups of $G_8 = A(K \mid Q)$ (see Exercise 10).

We can use the preceding material to investigate the classical problem of expressing the roots of a polynomial in terms of the coefficients. We say that a polynomial $p(x) \in F[x]$ is *solvable by radicals* if a splitting field K for the polynomial is contained in an extension of F of the form $F(\alpha_1, \alpha_2, \ldots, \alpha_r)$, where $\alpha_i^{n_i} \in F(\alpha_1, \ldots, \alpha_{i-1})$, n_i positive integers ($\alpha_1^{n_1} \in F$). The field K is then called an *extension by radicals* of F. Observe that if we set $a_{i-1} = \alpha_i^{n_i} \in F(\alpha_1, \ldots, \alpha_{i-1})$, then α_i is a root of the polynomial $x^{n_i} - a_i \in F(\alpha_1, \ldots, \alpha_{i-1})[x]$.

Theorem 7.7 *Let F be a field of characteristic* 0, $0 \neq a \in F$. *There exists a* *root* ξ *of* $x^n - 1 \in F[x]$ *in an extension* Γ *of* F *such that if* r *is any root of* $x^n - a \in F[x]$ *(in a further extension* Ω *of* Γ *if necessary), then*

(i) $K = F(r, \xi r, \ldots, \xi^{n-1}r)$ *is a splitting field for* $x^n - a$ *and* $r, \xi r, \ldots, \xi^{n-1}r$ *are the* n *distinct roots of* $x^n - a$.

(ii) $K = F(\xi, r)$.

(iii) $A(K|F(\xi))$ *is abelian.*

(iv) $A(F(\xi)|F)$ *is abelian.*

(v) $A(K|F)$ *is solvable.*

Proof: (i) Observe first that the derivative of $x^n - 1$ is $nx^{n-1} \neq 0$ (F has characteristic 0). Clearly g.c.d.$(x^n - 1, nx^{n-1}) = 1$ so that the roots of $x^n - 1$ in any extension of F are distinct (see Exercise 25, Section 3.4). The roots of $x^n - 1$ clearly form an abelian group (they are in a field); indeed, $\alpha^n = 1$, $\beta^n = 1$ implies $(\alpha\beta)^n = 1$, etc. We assert (see Exercise 11) that in fact they form a cyclic group, and since the roots are distinct, there is a generator ξ so that $1 = \xi^0, \xi, \ldots, \xi^{n-1}$ are the n distinct roots of $x^n - 1$ in Γ. Let r be any root of $x^n - a$. Then obviously $(\xi^k r)^n = (\xi^n)^k r^n = r^n = a$ so that $r, \xi r, \ldots, \xi^{n-1}r$ are also roots of $x^n - a$. Since $a \neq 0$, $r \neq 0$, and hence $r, \xi r, \ldots, \xi^{n-1} r$ are distinct and therefore account for all the roots of $x^n - a$. This proves (i).

(ii) Since $\xi r/r = \xi$ is in $F(r, \xi r, \ldots, \xi^{n-1}r) = K$, we have $F(\xi, r) \subset K$, and obviously $K \subset F(\xi, r)$.

(iii) Since $K = F(\xi, r) = F(\xi)(r)$, it follows that any $\sigma \in A(K|F(\xi))$ is completely determined by its value $\sigma(r)$. Also since σ holds the coefficients of $x^n - a$ fixed, we know that $\sigma(r)$ is a root of $x^n - a$ also. Hence for any $\sigma \in A(K|F(\xi))$ there is an integer m_σ, $0 \leq m_\sigma \leq n - 1$ such that

$$\sigma(r) = r\xi^{m_\sigma},$$

and the correspondence $\sigma \leftrightarrow m_\sigma$ is injective. We conclude that for any two elements σ and τ in $A(K|F(\xi))$ we have

$$\begin{aligned}\sigma\tau(r) &= \sigma(r\xi^{m_\tau}) = \sigma(r)\sigma(\xi)^{m_\tau}\\ &= r\xi^{m_\sigma}\xi^{m_\tau} \qquad [\text{since } \sigma \in A(K|F(\xi))]\\ &= r\xi^{m_\sigma + m_\tau}.\end{aligned}$$

Similarly, $$\tau\sigma(r) = r\xi^{m_\sigma + m_\tau}$$

so that $\sigma\tau = \tau\sigma$ and $A(K|F(\xi))$ is abelian.

(iv) If $\sigma \in A(F(\xi)|F)$, then σ is completely determined by its value on ξ, i.e., by $\sigma(\xi)$. But again $\sigma(\xi)$ must be a root of $x^n - 1$ so that $\sigma(\xi) = \xi^{m_\sigma}$ and the proof is the same as in part (iii) above.

(v) We know that $F(\xi)$ is a normal extension of F because it is the splitting field of the separable polynomial $x^n - 1$ (see Theorem 7.5). Similarly, K is a normal extension of F because it is the splitting field of the separable polynomial $x^n - a$. By Theorem 7.6(iv) we have

$$\{e\} < A(K|F(\xi)) \triangleleft A(K|F), \tag{47}$$

and in fact

$$\frac{A(K|F)}{A(K|F(\xi))} \cong A(F(\xi)|F). \tag{48}$$

By part (iv), the group on the right in (48) is abelian. Thus (47) is a subnormal series for $A(K|F)$ with abelian quotients because by part (iii)

$$A(K|F(\xi))/\{e\} \cong A(K|F(\xi))$$

is abelian. We conclude that $A(K|F)$ is solvable. ∎

The next theorem is a fundamental result which will allow us to show that the general polynomial of degree $n \geq 5$ is not solvable by radicals.

Theorem 7.8 *Let F be a field of characteristic* 0, $p(x) \in F[x]$. *If K is a splitting field of p(x) and K is contained in a radical extension of F, then $A(K|F)$ is solvable. That is, if p(x) is solvable by radicals, then $A(K|F)$ is a solvable group.*

Proof: We emphasize again that all polynomials in $F[x]$ are separable because char $F = 0$. To say that K is contained in a radical extension of F means that there exists a sequence of fields $F = F_0$, $F_1 = F(\alpha_1)$, $F_2 = F_1(\alpha_2) = F(\alpha_1, \alpha_2), F_3 = F_2(\alpha_3) = F(\alpha_1, \alpha_2, \alpha_3), \ldots, F_r = F_{r-1}(\alpha_r) = F(\alpha_1, \ldots, \alpha_r)$, where α_i is a root of a polynomial $x^{n_i} - a_i \in F_{i-1}[x]$, $i = 1, \ldots, r$, and $K \subset F_r$.

We wish to define a new sequence of fields

$$F = K_0 \subset K_1 \subset K_2 \subset \cdots \subset K_r \tag{49}$$

where K_r is a normal extension of F,

$$F \subset K \subset F_r \subset K_r, \tag{50}$$

and each K_i is a splitting field for a polynomial of the form $x^n - a \in K_{i-1}[x]$. Once this is done (see Exercise 12), we can argue as follows. Since K_i is a splitting field for some separable polynomial of the form $x^n - a \in K_{i-1}[x]$, K_i is a normal extension of K_{i-1}, and moreover we can apply Theorem 7.7(v) to conclude that $A(K_i|K_{i-1})$ is solvable. Since K_r is a normal extension of F, it is a normal extension of each K_i [i.e., K_r being normal means (see Theorem 7.5) that it is a splitting field for a separable polynomial over F and hence over K_i]. Thus, by Theorem 7.6(iv), we obtain the subnormal series

$$\{e\} \triangleleft A(K_r|K_{r-1}) \triangleleft A(K_r|K_{r-2}) \triangleleft \cdots \triangleleft A(K_r|K_0) = A(K_r|F). \tag{51}$$

The quotients in (52) also satisfy [by Theorem 7.6(iv)]

$$\frac{A(K_r|K_{i-1})}{A(K_r|K_i)} \cong A(K_i|K_{i-1}). \tag{52}$$

As we have noted, the group on the right is solvable and hence, by Theorem

1.15(iv), Section 2.1, $A(K_r|F)$ is solvable. Since K is the splitting field of $p(x) \in F(x)$, it is normal over F so that in (50), K is a normal extension of F and K_r is a normal extension of F as well. Again by Theorem 7.6(iv),

$$A(K|F) \cong \frac{A(K_r|F)}{A(K_r|K)}. \tag{53}$$

We have observed that $A(K_r|F)$ is solvable, and (53) shows that $A(K|F)$ is isomorphic to an epimorphic image of $A(K_r|F)$, i.e., the image under the canonical map of $A(K_r|F)$ onto $A(K_r|F)/A(K_r|K)$. But then Theorem 1.15(ii), Section 2.1, tells us that $A(K|F)$ itself is solvable. ∎

In our next result we exhibit a class of polynomials over the rationals with very simple Galois groups.

Theorem 7.9 *Let $f(x) \in Q[x]$ be an irreducible polynomial of prime degree p. Assume that $f(x)$ has precisely two nonreal roots. If $K \subset \boldsymbol{C}$ is a splitting field for $f(x)$, then*

$$A(K|Q) \cong S_p. \tag{54}$$

Proof: Let $\alpha \in K$ be a root of $f(x)$ so that $[Q(\alpha):Q] = p$. Then $[Q(\alpha):Q]$ $[K:Q(\alpha)] = [K:Q]$, so $p|[K:Q]$. But

$$|A(K|Q)| = [K:Q]$$

by Corollary 1. Hence $p \mid |A(K|Q)|$. By Theorem 2.7(i), Section 2.2, $A(K|Q)$ contains a subgroup G of order p which must be cyclic because p is prime. Since $f(x) \in Q[x]$ is irreducible and char $Q = 0$, the roots of $f(x)$ are distinct (see Exercise 26, Section 3.4). By Theorem 7.3, $A(K|Q)|E < S_E$, where $E = \{\alpha_1, \ldots, \alpha_p\}$ is the set of roots of $f(x)$. Let $\sigma \in G|E$ be a generator of $G|E$, and write σ as a product of disjoint cycles:

$$\sigma = \sigma_1 \cdots \sigma_r$$

where σ_i has length m_i, $m_1 + \cdots + m_r = p$. Now p is the least power of σ equal to the identity e. Also, the least power of σ equal to e is the least common multiple of the integers $m_1, \ldots, m_r$. Thus $m_i|p$, $i = 1, \ldots, r$. It follows that $r = 1$ and $m_1 = p$. Hence $\sigma \in G|E < A(K|Q)|E$ is a p-cycle.

Consider the automorphism $\tau: \boldsymbol{C} \to \boldsymbol{C}$ given by

$$\tau(z) = \bar{z}, \qquad z \in \boldsymbol{C}.$$

Then $\tau|\boldsymbol{R} = \iota_{\boldsymbol{R}}$. The two nonreal roots of $f(x)$ must form a complex conjugate pair, i.e., they are β and $\bar{\beta}$ for some $\beta \in \boldsymbol{C}$. Since $\tau(\beta) = \bar{\beta}$ and τ holds the real roots of $f(x)$ fixed, we conclude that $\tau|K \in A(K|Q)$ and $\tau^2 = \iota_K$. It follows that $\tau|E \in A(K|Q)|E$ is a 2-cycle. But is is easy to see that if a subgroup of S_E contains a 2-cycle and a p-cycle, then it is in fact all of S_E. Thus

$$A(K|Q) \cong A(K|Q)|E = S_E \cong S_p. \quad \blacksquare$$

Corollary 2 *If $f(x) \in Q[x]$ is a polynomial of prime degree $p \geq 5$ as described in Theorem 7.9, then $f(x)$ is not solvable by radicals.*

Proof: According to Theorem 7.9, $A(K|Q) \cong S_p$. If $f(x)$ were solvable by radicals, then Theorem 7.8 states that $A(K|Q)$ is solvable. But for $p \geq 5$, S_p is not solvable (see Theorem 1.18, Section 2.1). $\blacksquare$

Example 8 Let $f(x) = x^5 - 6x + 3 \in Q[x]$. By Eisenstein's criterion with $p = 3$ (Theorem 4.6, Section 3.4), $f(x)$ is irreducible over Q and hence has five distinct roots. We see that $f'(x) = 0$ iff $x^4 = \frac{6}{5}$ iff $x = r, x = ri, x = -r, x = -ri$, where $r = (\frac{6}{5})^{1/4}$. So there are only two real points where the tangent is horizontal. It follows that $f(x)$ must have two complex conjugate roots and three distinct real roots.

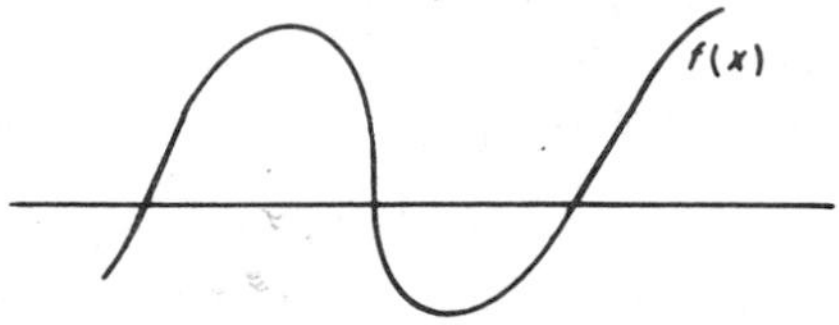

Corollary 2 applies to $f(x)$ and shows it is not solvable by radicals.

Our final concern in this section is to make precise the statement: *The general polynomial of degree $n \geq 5$ is not solvable by radicals.*

Let F be a field, and let $x_1, \ldots, x_n$ be independent indeterminates over F. Let t be an indeterminate over the field $F(x_1, \ldots, x_n)$. Then the polynomial

$$f(t) = t^n - x_1 t^{n-1} + x_2 t^{n-2} - x_3 t^{n-3} + \cdots + (-1)^n x_n$$

is called the *general polynomial of degree n over F*. Let $E_1, \ldots, E_n$ be the elementary symmetric functions of $x_1, \ldots, x_n$. We know that

$$\prod_{j=1}^{n} (t - x_j) = t^n - E_1 t^{n-1} + \cdots + (-1)^n E_n. \tag{55}$$

If we regard the right-hand side of (55) as a polynomial in $F(E_1, \ldots, E_n)[t]$, then (55) states that the roots of this polynomial are in $F(x_1, \ldots, x_n)$, i.e., they are $x_1, \ldots, x_n$. Thus $F(x_1, \ldots, x_n)$ can be thought of as the field which results from $F(E_1, \ldots, E_n)$ by adjoining $x_1, \ldots, x_n$, the roots of a polynomial of degree n. Hence (see Exercise 13)

$$[F(x_1, \ldots, x_n): F(E_1, \ldots, E_n)] \leq n! \tag{56}$$

For any $\sigma \in S_n$ we can define an automorphism $\hat{\sigma}$ of $F(x_1, \ldots, x_n)$ by

$$\hat{\sigma}(r(x_1, \ldots, x_n)) = r(x_{\sigma^{-1}(1)}, \ldots, x_{\sigma^{-1}(n)}).$$

Obviously $\hat{\sigma} = \hat{\varphi}$ iff $\sigma = \varphi$ For if $\hat{\sigma} = \hat{\varphi}$, then $x_{\sigma^{-1}(j)} = \hat{\sigma}(x_j) = \hat{\varphi}(x_j) = x_{\varphi^{-1}(j)}$ so that the independence of the indeterminates implies $\sigma^{-1}(j) = \varphi^{-1}(j)$, $j = 1, \ldots, n$, i.e., $\sigma = \varphi$. Also

$$\begin{aligned}\widehat{\sigma\varphi}(r(x_1, \ldots, x_n)) &= r(x_{(\sigma\varphi)^{-1}(1)}, \ldots, x_{(\sigma\varphi)^{-1}(n)}) \\ &= r(x_{\varphi^{-1}\sigma^{-1}(1)}, \ldots, x_{\varphi^{-1}\sigma^{-1}(n)}) \\ &= \hat{\sigma}r(x_{\varphi^{-1}(1)}, \ldots, x_{\varphi^{-1}(n)}) \\ &= \hat{\sigma}\hat{\varphi}r(x_1, \ldots, x_n).\end{aligned}$$

It is also easy to check that $\hat{\sigma}$ is additive and multiplicative on elements of $F(x_1, \ldots, x_n)$. Moreover, if

$$\hat{\sigma}(r(x_1, \ldots, x_n)) = \hat{\sigma}(s(x_1, \ldots, x_n)),$$

then obviously

$$r(x_1, \ldots, x_n) = s(x_1, \ldots, x_n).$$

In other words, the mapping

$$\sigma \rightarrow \hat{\sigma}$$

is an isomorphism between S_n and the group of automorphisms $\hat{\sigma}$ of $F(x_1, \ldots, x_n)$, $\sigma \in S_n$. Let $\hat{S}_n$ denote this group of automorphisms. From Theorem 7.1 we have

$$\begin{aligned}[F(x_1, \ldots, x_n):F(x_1, \ldots, x_n)(\hat{S}_n)] &\geq |\hat{S}_n| \\ &= n! \end{aligned} \tag{57}$$

Now obviously $E_k \in F(x_1, \ldots, x_n)(\hat{S}_n)$, $k = 1, \ldots, n$, so that

$$F(E_1, \ldots, E_n) \subset F(x_1, \ldots, x_n)(\hat{S}_n). \tag{58}$$

The polynomial (55) is over $F(E_1, \ldots, E_n)$ and all its roots, $x_1, \ldots, x_n$, are contained in $F(x_1, \ldots, x_n)$. Also, any subfield of $F(x_1, \ldots, x_n)$ containing $F(E_1, \ldots, E_n)$ and the roots of (55) must contain F and $x_1, \ldots, x_n$. Hence $F(x_1, \ldots, x_n)$ is the splitting field for (55) regarded as a polynomial over $F(E_1, \ldots, E_n)$. Note that

$$\begin{aligned}n! &\geq [F(x_1, \ldots, x_n): F(E_1, \ldots, E_n)] && \text{[by (56)]} \\ &= [F(x_1, \ldots, x_n):F(x_1, \ldots, x_n)(\hat{S}_n)][F(x_1, \ldots, x_n)(\hat{S}_n):F(E_1, \ldots, E_n)] \\ & && \text{[by (58)]} \\ &\geq n! && \text{[by (57)]}\end{aligned}$$

Thus

$$F(x_1, \ldots, x_n)(\hat{S}_n) = F(E_1, \ldots, E_n). \tag{59}$$

If $G = A(F(x_1, \ldots, x_n) \mid F(E_1, \ldots, E_n))$, then (59) implies that $\hat{S}_n < G$. Since $F(x_1, \ldots, x_n)$ is the splitting field for (55) as a polynomial over $F(E_1, \ldots, E_n)$, it follows from Theorem 7.6(i) that

$$F(x_1, \ldots, x_n)(G) = F(E_1, \ldots, E_n). \tag{60}$$

By Theorem 7.6(ii), we conclude that

$$G = \hat{S}_n. \tag{61}$$

When we say that the *general polynomial of degree* $n \geq 5$ *is not solvable by radicals,* we mean that the polynomial (55) regarded as a polynomial over $F(E_1, \ldots, E_n)$ is not solvable by radicals. The following is the great classical theorem due to N. Abel.

Theorem 7.10 *If* $n \geq 5$, *then the general polynomial of degree n is not solvable by radicals.*

Proof: According to (61) the Galois group of (55) is $\hat{S}_n$ which is isomorphic to S_n. By Theorem 1.18, Section 2.1, S_n is not solvable. Apply Theorem 7.8. ∎

Exercises 3.7

1. Prove that $A(K) < S_K$ as defined in (1). *Hint*: Any automorphism is bijective. Moreover, the composition of automorphisms is trivially an automorphism.
2. Prove that $A(K \mid F) < A(K)$ as defined in (2). *Hint*: $e \in A(K \mid F)$, and the product and inverses of two automorphisms holding F pointwise fixed also have this property.
3. If $G \subset A(K)$, prove that $K(G)$ is a subfield of K. *Hint*: Clearly 0 and 1 are in $K(G)$ since any automorphism must hold these elements fixed. If $f \in K(G)$ and $\sigma \in G$, then $\sigma(f^{-1}) \in \sigma(f)^{-1} = f^{-1}$, etc.
4. Let $\sigma_i: G \to K$, $i = 1, \ldots, n$, be distinct group characters. Show that $\sigma_1, \ldots, \sigma_n$ are l.i. over K. *Hint*: Suppose $\sigma_1, \ldots, \sigma_r$ is a basis of $\langle \sigma_1, \ldots, \sigma_n \rangle$ regarded as a subset of the vector space K^G. If $r = 1$ and $n > 1$, then $\sigma_2 = k\sigma_1$, $\sigma_2(e) = k\sigma_1(e)$, $1 = k$, and $\sigma_2 = \sigma_1$, contradicting the distinctness unless $n = 1$. Suppose now that $r < n$ so that

$$\sigma_{r+1} = \sum_{t=1}^{r} k_t \sigma_t, \qquad k_{t_0} \neq 0.$$

We have for any a and x in G that

$$\begin{aligned} \sigma_{r+1}(ax) &= \sigma_{r+1}(a)\sigma_{r+1}(x) \\ &= \sum_{t=1}^{r} k_t \sigma_{r+1}(a)\sigma_t(x). \end{aligned}$$

Also

$$\begin{aligned} \sigma_{r+1}(ax) &= \sum_{t=1}^{r} k_t \sigma_t(ax) \\ &= \sum_{t=1}^{r} k_t \sigma_t(a)\sigma_t(x), \end{aligned}$$

and subtracting these equations, we have

$$0 = \sum_{t=1}^{r} k_t(\sigma_{r+1}(a) - \sigma_t(a))\sigma_t(x).$$

Now $k_{t_0} \neq 0$ and $\sigma_{t_0} \neq \sigma_{r+1}$; so choose $a \in G$ such that $\sigma_{r+1}(a) \neq \sigma_{t_0}(a)$. Then we have exhibited a nontrivial linear combination of $\sigma_1, \ldots, \sigma_r$ which is 0, a contradiction.

5. Let K be a field, and let $t_1, \ldots, t_p$ be independent indeterminates over K. Then the rth elementary symmetric function of these indeterminates is the polynomial

$$E_r(t_1, \ldots, t_p) = \sum_{\omega \in Q_{r,p}} \prod_{j=1}^{r} t_{\omega(j)}, \qquad 1 \leq r \leq p.$$

Show that if $\sigma \in S_p$, then

$$E_r(t_{\sigma(1)}, \ldots, t_{\sigma(p)}) = E_r(t_1, \ldots, t_p).$$

6. Let $f(x) \in K[x]$ be monic, and suppose $\alpha_1, \ldots, \alpha_p$ are all the roots of $f(x)$ in some splitting field. Then $f(x) = \prod_{i=1}^{p}(x - \alpha_i)$. Show that

$$f(x) = x^p - E_1(\alpha_1, \ldots, \alpha_p)x^{p-1} + \cdots + (-1)^r E_r(\alpha_1, \ldots, \alpha_p)t^{p-r} + \cdots + (-1)^p E_p(\alpha_1, \ldots, \alpha_p).$$

7. Find a polynomial in $Q[x]$ whose roots are the numbers $\alpha_1\alpha_2$, $\alpha_1\alpha_3$, $\alpha_2\alpha_3$ where α_1, α_2, α_3 are the roots of the cubic polynomial $x^3 + 3x^2 + 2x + 1$.

8. Let $G \subset A(K)$ and let $F = K(G)$. Let $E = \{\alpha_1, \ldots, \alpha_p\} \subset K$, $|E| = p$, and assume that for every $\sigma \in G$, $\sigma(E) \subset E$. Prove that

$$E_r(\alpha_1, \ldots, \alpha_p) \in F, \qquad r = 1, \ldots, p.$$

Hint: If $\sigma \in G$, then

$$\sigma(E_r(\alpha_1, \ldots, \alpha_p)) = E_r(\sigma(\alpha_1), \ldots, \sigma(\alpha_p)) = E_r(\alpha_{\theta(1)}, \ldots, \alpha_{\theta(p)})$$

where $\theta \in S_p$. Apply Exercise 5.

9. Let K be a splitting field for a separable polynomial in $F[x]$, and assume char $F = 0$. Let $F \subset B \subset K$, B a subfield of K. Prove that B is a normal extension of F iff $A(K \mid F)B \subset B$. *Hint*: We know from Theorem 4.15, Section 3.4, that $B = F(\alpha)$, $\alpha \in K$, i.e., α is the primitive element. Consider the polynomial $p(x)$ in Theorem 7.2 in which G is taken to be $A(K \mid F)$. Then according to Theorem 7.2(ii) the polynomial $p(x)$ is monic irreducible over $K(A(K \mid F)) = F$ [by (32)], and the roots of $p(x)$ are of the form $\sigma(\alpha)$, $\sigma \in A(K \mid F)$. Hence if $\sigma(B) \subset B$, $\sigma \in A(K \mid F)$, it follows that B is the splitting field for $p(x)$ over F, and thus B is a normal extension of F by Theorem 7.5. Conversely, if B is a normal extension of F, then by Theorem 7.6(i) we have $F = B(A(B \mid F))$. Since the irreducible polynomial $p(x) \in F[x] = B(A(B \mid F))[x]$ has the root $\alpha \in B$, it follows from Theorem 7.4 that B contains all the roots of $p(x)$, i.e., $\sigma(\alpha) \in B$ for all $\sigma \in A(K \mid F)$. Hence for all $\sigma \in A(K \mid F)$, $\sigma(B) = \sigma(F(\alpha)) \subset F(\alpha) = B$.

10. Show that for the groups (46) the corresponding fields are given by

$$K(G_8) = Q,$$
$$K(G_{4,1}) = Q(i),$$

$$K(G_{4,2}) = Q(2^{1/2}),$$
$$K(G_{4,3}) = Q(i2^{1/2}),$$
$$K(G_{2,1}) = Q(i,2^{1/2}),$$
$$K(G_{2,2}) = Q(2^{1/4}),$$
$$K(G_{2,3}) = Q((1+i)2^{1/4}),$$
$$K(G_{2,4}) = Q(i2^{1/4}),$$
$$K(G_{2,5}) = Q((1-i)2^{1/4}),$$
$$K(e) = K.$$

11. Let G be a finite group whose elements belong to a field Γ and whose operation is the multiplication in Γ. Prove that G is a cyclic group. *Hint*: Obviously G is abelian. Let $n = |G|$ and write $n = p_1^{\nu_1} \cdots p_r^{\nu_r}$ where $p_1, \ldots, p_r$ are distinct primes and $\nu_1, \ldots, \nu_r$ are positive integers. The polynomial $x^{n/p_i} - 1$ cannot have more than n/p_i roots in Γ (and hence in $G \subset \Gamma$) so that there exist $g_1, \ldots, g_r$ in G such that $g_i^{n/p_i} \neq 1$. Let $h_i = g_i^{n/\mu_i}$ where $\mu_i = p_i^{\nu_i}$, $i = 1, \ldots, r$. Obviously, $h_i^{\mu_i} = g_i^{n} = 1$. Thus the *order* of h_i (i.e., the least power of h_i which equals 1) must be a divisor of μ_i, say $h_i^{\gamma_i} = 1$ where $\gamma_i = p_i^{d_i}$, $d_i \leqq \nu_i$. If $d_i < \nu_i$, then $\nu_i - d_i - 1 \geqq 0$ and we easily compute that $h_i^{\alpha_i} = 1$ where $\alpha_i = p_i^{\nu_i - 1}$. But $h_i^{\alpha_i} = g_i^{n/p_i} \neq 1$. Hence h_i has order $p_i^{\nu_i}$. It follows in general that $\xi = h_1 \cdots h_r$ has order which is the least common multiple of the orders of $h_1, \ldots, h_r$, i.e., $p_1^{\nu_1} \cdots p_r^{\nu_r} = n$. Thus $1, \xi, \ldots, \xi^{n-1}$ are distinct elements of G and thereby must comprise G.

12. In the notation of the proof of Theorem 7.8, show that the sequence of fields K_i in (49) with the properties stated in the proof exists. *Hint*: We begin by noting that K and all the fields $F = F_0, F_1, \ldots, F_r$ are subfields of F_r. If we regard $p(x)$ and all of $x^{n_i} - a_i$ as polynomials over F_r, we can construct a splitting field Ω for the product of all these polynomials which contains F_r. We want to define a sequence of fields $F = K_0 \subset K_1 \subset K_2 \subset \cdots \subset K_r$, where K_r is a normal extension of F, $F \subset K \subset F_r \subset K_r$, and each K_i is a splitting field for a polynomial of the form $x^n - a$ in $K_{i-1}[x]$. We begin by defining $K_1 \subset \Omega$ to be the splitting field of $x^{n_1} - a_1 \in K_0[x] = F[x]$. Moreover, we can assume $\alpha_1 \in K_1$ by simply beginning the construction of K_1 by adjoining the root α_1 of $x^{n_1} - a_1$ to F. Then clearly $F_1 = F(\alpha_1) \subset K_1$. To construct the field K_2, define

$$f_2(x) = \prod x^{n_2} - \sigma(a_2)$$

where the product is taken over all $\sigma \in A(K_1|F)$ which yield distinct values $\sigma(a_2)$. This makes sense because $a_2 \in F_1 \subset K_1$ [and thus $\sigma(a_2) \in K_1$ also]. As in the proof of Theorem 7.2(ii), $f_2(x) \in K_1(A(K_1|F))[x]$. Since K_1 is a normal extension of F, it follows from Theorem 7.5 that $K_1(A(K_1|F)) = F$ so that $f_2(x) \in F[x]$. Now let K_2 be a splitting field for $(x^{n_1} - a_1)f_2(x)$ over K_1 and observe three things:

(i) Since $(x^{n_1} - a_1)f_2(x) \in F[x]$, $F \subset K_1 \subset K_2$, and K_2 is a splitting field for $(x^{n_1} - a_1)f_2(x)$ regarded as a polynomial over K_1, it is also a splitting field for $(x^{n_1} - a_1)f_2(x)$ regarded as a polynomial over F, and hence K_2 is a normal extension of F by Theorem 7.5. To see this, use Theorem 4.14(ii), Section 3.4.

(ii) We can begin the construction of K_2 by adjoining α_2 to K_1 so that we can assume $F_2 = F_1(\alpha_2) \subset K_2$.

(iii) The splitting field K_2 can be obtained by a sequence of simple radical extensions of K_1. That is, we can regard each of the polynomials $x^{n_2} - \sigma(a_2)$ as being over K_1 and insert between K_1 and K_2 a sequence of splitting fields of polynomials of the form $x^{n_2} - \sigma(a_2) \in K_1[x]$: $K_1 \subset K_{11} \subset \cdots \subset K_{1p_1} \subset K_2$.

To construct K_3, let

$$f_3(x) = \prod x^{n_3} - \sigma(a_3)$$

where the product is taken over all $\sigma \in A(K_2|F)$ which yield distinct values $\sigma(a_3)$. Again, this makes sense because $a_3 \in F_2 \subset K_2$ and also, as above, $f_3(x) \in F[x]$. We construct a splitting field K_3 for $(x^{n_1} - a_1)f_2(x)f_3(x)$, regarded as a polynomial over K_2, so that $K_2 \subset K_3$. Once again, K_3 is a normal extension of F and we can assume $\alpha_3 \in K_3$ so that $F_3 \subset K_3$. Precisely as in (iii) above, K_3 can be obtained from K_2 by a sequence of simple radical extensions. Thus, so far we have produced a sequence of fields

$$F = K_0 \subset K_1 \subset K_{11} \subset \cdots \subset K_{1p_1} \subset K_2 \subset K_{21} \subset \cdots \subset K_{2p_2} \subset K_3$$

with the following properties:

(a) K_1, K_2, K_3 are normal extensions of F (this may not be true of the intermediate fields).

(b) Each field in the sequence is the splitting field of a polynomial of the form $x^n - a$ in which a lies in the immediately preceding field.

(c) $F_1 \subset K_1, \quad F_2 \subset K_2, \quad F_3 \subset K_3$.

We continue in this way through r steps, finally producing a sequence of fields beginning with F and terminating in a field K_r which contains F_r,

$$F \subset K \subset F_r \subset K_r.$$

Moreover, K_r is a normal extension of F, and again we emphasize that each field in the sequence is the splitting field of a polynomial of the form $x^n - a$, where a lies in the immediately preceding field. [We simply drop the double subscripts and label the fields $K_0 \subset K_1 \subset \cdots \subset K_r$.]

13. Let $f(x)$ be a polynomial of degree n over F. Let $\alpha_1, \ldots, \alpha_n$ be the roots of F in some splitting field for $f(x)$. Prove that

$$[F(\alpha_1, \ldots, \alpha_n): F] \leq n!$$

Hint: $[F(\alpha_1):F] \leq n$ by Theorem 4.7, Section 3.4. Then $f(x)/(x - \alpha_1)$ can be regarded as a polynomial of degree $n - 1$ over $F(\alpha_1)$, and we can use induction and Theorem 6.2, Section 3.6.

Glossary 3.7

automorphism group of K, 265
automorphism group of K over F, 265
$A(K)$, 265
$A(K|F)$, 265
character of degree 1, 264
extension by radicals, 276
fixed field of G, 265
$K(G)$, 265

4

Modules and Linear Operators

4.1 The Hermite Normal Form

The theory of matrices over a P.I.D. yields many important consequences: the similarity theory for matrices over a field, the structure theory for finitely generated abelian groups, and the structure theory for modules, to name a few. The present section is mainly devoted to a discussion of "equivalence" and "normal forms" of matrices.

Throughout, R denotes a P.I.D. and $M_{m,n}(R)$ denotes the R-module of $m \times n$ matrices whose entries are elements of R (see Example 4, Section 1.2). If $A \in M_{m,n}(R)$ and the (i, j) entry of A is a_{ij}, $1 \le i \le m$, $1 \le j \le n$, it is customary to write $A = [a_{ij}]$ or $A = (a_{ij})$. Let A and B be matrices in $M_{m,n}(R)$. We say that A is *left equivalent* to B, written $A \overset{L}{=} B$, if there exists a matrix $U \in M_m(R)$ which satisfies the following two conditions:

(i) U is a unit in the ring $M_m(R)$.
(ii) $A = UB$.

Several remarks are in order concerning this definition. First, a unit matrix U in $M_m(R)$ is one for which there exists an inverse in $M_m(R)$, that is, there exists a matrix $V \in M_m(R)$ such that

$$UV = VU = I_m . \tag{1}$$

In this case $\det(UV) = \det U \cdot \det V = 1$ so that $\det U$ is a unit in R. On the other hand, recall from elementary matrix theory that if $U \in M_m(R)$ and $\det U$ is a unit in R, then

$$V = (\det U)^{-1} \operatorname{adj} U. \tag{2}$$

The matrix V is the inverse of U in $M_m(R)$, and the matrix adj U is the *ad-*

jugate of U. The (i, j) entry of adj U is

$$(-1)^{i+j} \det U(j \mid i), \tag{3}$$

where $U(j|i)$ is the $(m-1)$-square matrix obtained from U by deleting row j and column i. Thus the unit matrices in $M_m(R)$ are precisely those matrices whose determinants are units in R; we shall refer to them as *unimodular matrices*. The multiplicative group of m-square unimodular matrices over R is denoted by $\mathrm{GL}(m, R)$ (see Example 5, Section 1.2).

It is an elementary exercise to check that in $M_{m,n}(R)$ left equivalence is an equivalence relation (see Exercise 1). Our first goal will be to determine a system of distinct representatives (S.D.R.) for the equivalence classes induced in $M_{m,n}(R)$. We begin by defining *elementary row operations* and corresponding *elementary matrices*. There are three types of *elementary row operations* that can be performed on a matrix in $M_{m,n}(R)$:

I. Interchange row i and row j.
II. To row i add (in R^n) r times row j, $i \neq j$, $r \in R$.
III. Multiply row i by a unit $u \in R$.

If any one of these operations is performed on I_m, the resulting matrix is an *elementary matrix* of the corresponding type. The three types are designated by E_{ij}, $E_{ij}(r)$, and $E_i(u)$. It is easy to see that all three types of elementary matrices are in $\mathrm{GL}(m, R)$, and in fact the following formulas for their inverses hold:

$$E_{ij}^{-1} = E_{ij}, \tag{4}$$

$$E_{ij}(r)^{-1} = E_{ij}(-r), \tag{5}$$

$$E_i(u)^{-1} = E_i(u^{-1}) \tag{6}$$

(see Exercise 2). Moreover, from elementary properties of the determinant we have

$$\det E_{ij} = -1, \tag{7}$$

$$\det E_{ij}(r) = 1, \tag{8}$$

$$\det E_{ij}(u) = u. \tag{9}$$

It is a simple problem in matrix multiplication to confirm that an elementary row operation of a given type can be accomplished by left multiplication with the corresponding type of elementary matrix (see Exercise 3).

It is easy to construct elements of $\mathrm{GL}(m, R)$.

Theorem 1.1 Let $r_1, \ldots, r_m$ *be relatively prime elements of* R. *Then there exists a matrix* $U \in \mathrm{GL}(m, R)$ *with* $[r_1, \ldots, r_m]$ *as a prescribed row or column.*

Proof: Since g.c.d. $(r_1, \ldots, r_m) = 1$, we can apply Corollary 9, Section 3.3, to obtain a matrix $V \in \mathrm{GL}(m, R)$ with $[r_1, \ldots, r_m]$ as first row and such that det V is a unit. By left multiplication by E_{ij} we obtain $E_{ij}V = U$, where $U_{(j)} = [r_1, \ldots, r_m]$ (see the notation in Example 4, Section 1.2). Also det U = (det E_{ij}) (det V) is a unit in R and hence $U \in \mathrm{GL}(m, R)$. To obtain the corresponding result about columns, simply consider the *transpose* of U, i.e., the matrix U^{T} whose (s, t) entry is the (t, s) entry of U. ∎

We recall from elementary matrix theory that the *rank* of a matrix $A \in M_{m,n}(R)$, here denoted by $\rho(A)$, is the size of the largest submatrix of A with nonzero determinant. In order to discuss submatrices of a matrix A, we require some notation, If $1 \leq p \leq m$, $1 \leq q \leq n$, $\alpha \in Q_{p,m}$, and $\beta \in Q_{q,n}$ [see formula (27), Section 3.2], then $A[\alpha|\beta]$ is the $p \times q$ submatrix of A whose (i, j) entry is $a_{\alpha(i),\beta(j)}$, $i = 1, \ldots, p$, $j = 1, \ldots, q$. If $\alpha' \in Q_{m-p,m}$ is the sequence complementary to α in $1, \ldots, m$ [e.g., $(1, 3, 5)' = (2,4)$ if $m = 5$], then $A(\alpha|\beta]$ is the $(m - p) \times q$ submatrix $A[\alpha'|\beta]$. Similarly, $A[\alpha|\beta) = A[\alpha|\beta']$ and $A(\alpha|\beta) = A[\alpha'|\beta']$.

The *Cauchy-Binet* and *Laplace Expansion* theorems for determinants are of fundamental importance here. We shall simply state these results and refer to M. Marcus and H. Minc, *Introduction to Linear Algebra* (New York; MacMillan 1965), Section 2.6, for the proofs.

Cauchy-Binet Theorem *Let $X \in M_{m,n}(R)$, $Y \in M_{n,k}(R)$, and $1 \leq p \leq \min\{m, n, k\}$. If $\alpha \in Q_{p,m}$ and $\beta \in Q_{p,k}$, then*

$$\det(XY)[\alpha|\beta] = \sum_{\omega \in Q_{p,n}} \det X[\alpha|\omega] \det Y[\omega|\beta]. \tag{10}$$

Laplace Expansion Theorem *Let $X \in M_n(R)$, $1 \leq p \leq n$, and $\alpha \in Q_{p,n}$. Then*

$$\det X = \sum_{\beta \in Q_{p,n}} (-1)^{s(\alpha)+s(\beta)} \det X[\alpha|\beta] \det X(\alpha|\beta) \tag{11}$$

where $s(\alpha) = \sum_{i=1}^{p} \alpha(i)$ for any $\alpha \in Q_{p,n}$.

It is perhaps worth remarking that both (10) and (11) are proved in the above cited reference (and most other books) for matrices over a field. The proofs over a P.I.D. require the following modifications. Let x_{ij}, $i = 1, \ldots, m$, $j = 1, \ldots, n$, y_{jl}, $j = 1, \ldots, n$, $l = 1, \ldots, k$ be a total of $mn + nk$ independent indeterminates over R, and construct the quotient field of the polynomial ring $R[x_{11}, \ldots, x_{mn}, y_{11}, \ldots, y_{nk}]$. Apply the result (10) proved for fields to $R(x_{11}, \ldots, x_{mn}, y_{11}, \ldots, y_{nk})$. Then the formulas (10) and (11) follow by specialization of the matrices to R.

It is an easy consequence of (10) that if $A \stackrel{\text{L}}{=} B$ then $\rho(A) = \rho(B)$. For if $U \in \mathrm{GL}(m, R)$, $A = UB$, and $\rho(A) = p$, then there exist $\alpha \in Q_{p,m}$, $\beta \in Q_{p,n}$ such that $\det A[\alpha \mid \beta] \neq 0$ and

$$\det A[\alpha \mid \beta] = \sum_{\omega \in Q_{p,m}} \det U[\alpha \mid \omega] \det B[\omega \mid \beta].$$

It follows that not all $\det B[\omega \mid \beta]$ can be 0. Hence $\rho(B) \geq p = \rho(A)$. The same argument applied to $B = U^{-1}A$ shows that $\rho(A) \geq \rho(B)$.

The following result is fundamental.

Theorem 1.2 (*Hermite Normal Form*) *Let* $A \in M_{m,n}(R)$, *R a* P.I.D., *and assume that* $\rho(A) = r$. *Then* $A \overset{L}{=} H$, *where* $H = [h_{ij}] \in M_{m,n}(R)$ *is a matrix having the following properties:*

(i) $H_{(i)} \neq 0$, $i = 1, \ldots, r$; $H_{(i)} = 0$, $i = r + 1, \ldots, m$.

(ii) *There is a sequence of integers* $n_1, \ldots, n_r$, $1 \leq n_1 < n_2 < \cdots < n_r \leq n$ *such that* $h_{ij} = 0$, $j = 1, \ldots, n_i - 1$, $h_{in_i} \neq 0$, $i = 1, \ldots, r$, *and* $h_{tn_i} = 0$, $t = i + 1, \ldots, m$, $i = 1, \ldots, r$.

(iii) h_{in_i} *is in a prescribed complete set of nonassociates in R*, $i = 1, \ldots, r$.

(iv) h_{tn_i}, $t = 1, \ldots, i - 1$, *is in a prescribed complete system of residues modulo* h_{in_i}, $i = 1, \ldots, r$.

Proof: Let the first nonzero column of A be column n_1. Suppose $[a_1, \ldots, a_m]$ is column n_1 of A (we write columns of a matrix horizontally to conserve space) and $d_1 = \text{g.c.d.}(a_1, \ldots, a_m)$. Then

$$1 = \text{g.c.d.}\left(\frac{a_1}{d_1}, \ldots, \frac{a_m}{d_1}\right),$$

there exist $r_1, \ldots, r_m \in R$ such that $\sum_{i=1}^{m} r_i (a_i/d_1) = 1$, and hence

$$\text{g.c.d.}(r_1, \ldots, r_m) = 1$$

(see Theorem 3.5, Section 3.3). By Theorem 1.1, construct $U \in \mathrm{GL}(m, R)$ with first row $[r_1, \ldots, r_m]$. By matrix multiplication, the $(1, n_1)$ entry of $B = UA$ is d_1, and the rest of the entries in $B^{(n_1)}$ are sums of multiples of $a_1, \ldots, a_m$ and hence divisible by d_1. By adding appropriate multiples of row 1 to rows $2, \ldots, m$ of B (type II row operations, or equivalently, left multiplications by type II elementary matrices), we can bring B to the form

$$C = \begin{array}{c} \begin{array}{cccccccccc} & & & n_1 & & & & n_2 & & \end{array} \\ \begin{bmatrix} 0 & \cdots & 0 & d_1 & * & * & * & * & \cdots & * \\ 0 & \cdots & 0 & 0 & 0 & \cdots & 0 & * & \cdots & * \\ \vdots & & \vdots & \vdots & \vdots & & \vdots & \vdots & & \vdots \\ 0 & \cdots & 0 & 0 & 0 & \cdots & 0 & * & \cdots & * \end{bmatrix} \end{array}.$$

Now scan C for the first column which has a nonzero entry below row 1, say it is column numbered n_2 of C. Let d_2 be the g.c.d. of the elements $c_1, \ldots, c_{m-1}$ in column n_2 of C below row 1, and as above let $s_1, \ldots, s_{m-1}$ be elements in R such that $\sum_{i=1}^{m-1} s_i c_i = d_2$. By Theorem 1.1, obtain $V \in \mathrm{GL}(m-1, R)$ such that row 1 of V is $[s_1, \ldots, s_{m-1}]$. Consider the matrix

$$1 \dotplus V = \begin{bmatrix} 1 & 0 & \cdots & 0 \\ 0 & & & \\ \vdots & & V & \\ 0 & & & \end{bmatrix}$$

(the notation $\dotplus$ is self-explanatory). Obviously $1 \dotplus V \in \mathrm{GL}(m, R)$ [a trivial application of (11)], and it is easy to see that $(1 \dotplus V)C = D$ has the following properties:

(i) Columns $1, \ldots, n_2 - 1$ remain unaltered.
(ii) The $(1, n_2)$ entry remains unaltered.
(iii) The $(2, n_2)$ entry is $d_2 = \sum_{i=1}^{m-1} s_i c_i$.
(iv) The remaining entries of column n_2 are replaced with sums of multiples of $c_1, \ldots, c_{m-1}$ and are thereby divisible by d_2.

Then by an obvious succession of type II row operations on rows $2, \ldots, m$ of D, we obtain an equivalent matrix of the form

$$E = \begin{bmatrix} 0 & \cdots & 0 & \overset{n_1}{d_1} & * & \cdots & * & \overset{n_2}{*} & * & \cdots & * & \overset{n_3}{*} & * & \cdots & * \\ 0 & \cdots & 0 & 0 & 0 & \cdots & 0 & d_2 & * & \cdots & * & * & * & \cdots & * \\ 0 & \cdots & 0 & 0 & 0 & \cdots & 0 & 0 & 0 & \cdots & 0 & * & * & \cdots & * \\ \vdots & & \vdots & \vdots & \vdots & & \vdots & \vdots & \vdots & & \vdots & \vdots & \vdots & & \vdots \\ 0 & \cdots & 0 & 0 & 0 & \cdots & 0 & 0 & 0 & \cdots & 0 & * & * & \cdots & * \end{bmatrix}.$$

Again scan E below row 2 for the first column having a nonzero entry below row 2, say column numbered n_3, and proceed as above. Clearly this process can be continued after $n_1, \ldots, n_{k-1}$ and $d_1, \ldots, d_{k-1}$ have been obtained iff below row $k-1$ there is a nonzero entry to the right of column n_{k-1}. Thus the process terminates with a matrix of the following form:

$$
F = \left[\begin{array}{cccccccccccccccc}
 & & & n_1 & & & & n_2 & & & & & n_p & & & \\
0 & \cdots & 0 & d_1 & * & \cdots & * & * & * & \cdots & * & * & * & * & \cdots & * \\
\cdot & & 0 & 0 & 0 & \cdots & 0 & d_2 & * & & 0 & 0 & * & * & \cdots & * \\
\cdot & & \cdot & \cdot & \cdot & & \cdot & \cdot & \cdot & & \cdot & \cdot & \cdot & \cdot & & \cdot \\
\cdot & & \cdot & \cdot & \cdot & & \cdot & \cdot & \cdot & & \cdot & \cdot & \cdot & \cdot & & \cdot \\
\cdot & & \cdot & \cdot & \cdot & & \cdot & \cdot & \cdot & & \cdot & \cdot & \cdot & \cdot & & \cdot \\
0 & \cdots & 0 & 0 & 0 & \cdots & 0 & 0 & 0 & \cdots & 0 & 0 & d_p & * & \cdots & * \\
\hdashline
\multicolumn{16}{c}{} \\
\multicolumn{16}{c}{\bigcirc} \\
\multicolumn{16}{c}{}
\end{array}\right].
$$

By left multiplication of F in succession by $E_1(u_1), \ldots, E_p(u_p)$, u_i a unit in R, we can replace d_i by any prescribed associate of d_i, $i = 1, \ldots, p$. Then by adding appropriate multiples of row 2 to row 1, row 3 to rows 2 and 1, row 4 to rows 3, 2, and 1, . . . , row p to rows $p - 1, p - 2, \ldots, 1$, an element above the (i, n_i) element in column n_i can be replaced by an element lying in a prescribed residue system modulo the (i, n_i) element, $i = 1, \ldots, p$. Call the resulting matrix H. Finally, it is clear that any $(p + 1)$-square submatrix of H must have a row of zeros and hence have determinant 0. Also, $\det H[1, \ldots, p \mid n_1, \ldots, n_p] = u_1 \cdots u_p d_1 \cdots d_p \neq 0$. Thus $\rho(H) = p$. Since $\rho(A) = \rho(H)$ (why?), we conclude that $p = r$. ▌

The matrix H in Theorem 1.2 is said to be in *Hermite normal form* or sometimes *row echelon form,* and it is called a *Hermite normal form for A:*

$$
H = \left[\begin{array}{ccccccccccccccc}
 & & & n_1 & & & & n_2 & & & & n_r & & & \\
0 & \cdots & 0 & h_{1n_1} & * & \cdots & * & * & * & \cdots & * & * & * & \cdots & * \\
0 & & 0 & 0 & 0 & \cdots & 0 & h_{2n_2} & * & \cdots & * & 0 & * & \cdots & * \\
\cdot & & \cdot & \cdot & \cdot & & \cdot & \cdot & \cdot & & \cdot & \cdot & \cdot & & \cdot \\
\cdot & & \cdot & \cdot & \cdot & & \cdot & \cdot & \cdot & & \cdot & \cdot & \cdot & & \cdot \\
\cdot & & \cdot & \cdot & \cdot & & \cdot & \cdot & \cdot & & \cdot & \cdot & \cdot & & \cdot \\
0 & \cdots & 0 & 0 & 0 & & 0 & 0 & 0 & \cdots & 0 & h_{rn_r} & * & \cdots & * \\
\hdashline
\multicolumn{15}{c}{} \\
\multicolumn{15}{c}{\bigcirc} \\
\multicolumn{15}{c}{}
\end{array}\right], \qquad (12)
$$

$\rho(A) = \rho(H) = r$. We describe H again. Rows $r + 1, \ldots, m$ are zero; the element h_{in_i} is the first nonzero element in row i, $i = 1, \ldots, r$; h_{in_i} is in a prescribed complete system of nonassociates, $i = 1, \ldots, r$; each element above h_{in_i} in column n_i is in a prescribed complete system of residues modulo h_{in_i}, $i = 1, \ldots, r$. The $*$'s indicate unspecified entries. The integers $n_1, \ldots, n_r$ are called the *indices* of H. Of course, the question of the uniqueness of H is of immediate interest. We continue to use the notation of Theorem 1.2.

Theorem 1.3 *The Hermite normal form for a matrix $A \in M_{m,n}(R)$ is unique. That is, suppose $K \overset{L}{=} A$, and*

(i) $K_{(i)} \neq 0$, $i = 1, \ldots, r'$; $K_{(i)} = 0$, $i = r' + 1, \ldots, m$.

(ii) *There is a sequence of integers $q_1, \ldots, q_{r'}$, $1 \leq q_1 < \cdots < q_{r'} \leq n$, such that $k_{ij} = 0$, $j = 1, \ldots, q_i - 1$, $k_{iq_i} \neq 0$, $i = 1, \ldots, r'$,* and $k_{tq_i} = 0$, $t = i + 1, \ldots, m$, $i = 1, \ldots, r'$.

(iii) *k_{iq_i} is in a prescribed complete system of nonassociates* (*the same such set as in Theorem* 1.2) *in R, $i = 1, \ldots, r'$.*

(iv) *k_{tq_i}, $t = 1, \ldots, i - 1$, is in a prescribed complete system of residues* (*the same such set as in Theorem 1.2*) *modulo k_{iq_i}, $i = 1, \ldots, r'$.*

Then $K = H$, where H is the Hermite normal form of Theorem 1.2.

Proof: Since $K \overset{L}{=} A \overset{L}{=} H$, we have $K \overset{L}{=} H$ and hence $r' = \rho(K) = \rho(H) = r$. Thus the number of nonzero rows in K is r, and we need no longer refer to r'. Now $K = PH$, where $P \in \mathrm{GL}(m, R)$, and since $H^{(1)} = \cdots = H^{(n_1-1)} = 0$ it follows that $K^{(j)} = PH^{(j)} = 0, j = 1, \ldots, n_1 - 1$. Also $K^{(n_1)} = PH^{(n_1)} = P[h_{1n_1}, 0, \ldots, 0] = h_{1n_1} P^{(1)}$. Now $P^{(1)} \neq 0$ because P is unimodular. Thus $K^{(n_1)} \neq 0$ and $K^{(n_1)}$ is the first nonzero column of K, i.e., $q_1 = n_1$. Moreover, $K^{(n_1)} = [k_{1n_1}, 0, \ldots, 0]$ so that $h_{1n_1}P^{(1)} = [k_{1n_1}, 0, \ldots, 0]$. It follows that

$$h_{1n_1} p_{11} = k_{1n_1} \tag{13}$$

and $P^{(1)} = [p_{11}, 0, \ldots, 0]$. Since $\det P = p_{11} \det P(1\,|\,1)$ is a unit in R, p_{11} is a unit in R. We conclude from (13) that h_{1n_1} and k_{1n_1} are associates. But these two elements are in the same complete system of nonassociates in R. Hence $h_{1n_1} = k_{1n_1}$, $p_{11} = 1$, and $P^{(1)} = [1, 0, \ldots, 0]$.

The columns of H strictly between columns n_1 and n_2 are of the form $[a, 0, \ldots, 0]$. Suppose $H^{(t)} = [a, 0, \ldots, 0]$ is one of these columns; then $K^{(t)} = PH^{(t)} = P[a, 0, \ldots, 0] = aP^{(1)} = [a, 0, \ldots, 0] = H^{(t)}$. In other words, $K^{(t)} = H^{(t)}$, $t = 1, \ldots, n_2 - 1$. Now

$$\begin{aligned} K^{(n_2)} &= PH^{(n_2)} \\ &= P[h_{1n_2}, h_{2n_2}, 0, \ldots, 0] \\ &= h_{1n_2}P^{(1)} + h_{2n_2}P^{(2)} \\ &= [h_{1n_2}, 0, \ldots, 0] + h_{2n_2}[p_{12}, p_{22}, p_{32}, \ldots, p_{m2}]. \end{aligned} \tag{14}$$

All the columns of K up to $K^{(n_2)}$ have at most the first entry nonzero, so that $K^{(n_2)}$ can have at most its first two entries nonzero. Thus $p_{32} = \cdots = p_{m2} = 0$ (since $h_{2n_2} \neq 0$), and from (14) we have

$$K^{(n_2)} = [h_{1n_2} + h_{2n_2}p_{12}, h_{2n_2}p_{22}, 0, \ldots, 0]. \tag{15}$$

Moreover P is of the form

$$P = \begin{bmatrix} 1 & p_{12} & \cdot & \cdots & \cdot \\ 0 & p_{22} & \cdot & \cdots & \cdot \\ 0 & 0 & \cdot & & \cdot \\ \cdot & \cdot & \cdot & & \cdot \\ \cdot & \cdot & \cdot & & \cdot \\ \cdot & \cdot & \cdot & & \cdot \\ 0 & 0 & \cdot & \cdots & \cdot \end{bmatrix}$$

and $\det P = p_{22} \det P(1, 2|1, 2)$ is a unit in R. Hence p_{22} is a unit in R so that $h_{2n_2}p_{22} \neq 0$. From (15), $K^{(n_2)}$ has a nonzero entry in the second position and is the first such column of K for which this is true. Hence $q_2 = n_2$ and

$$h_{2n_2}p_{22} = k_{2n_2}. \tag{16}$$

But since p_{22} is a unit, the argument we used before shows that $h_{2n_2} = k_{2n_2}$ and $p_{22} = 1$. Also, from (15),

$$h_{1n_2} + h_{2n_2}p_{12} = k_{1n_2}. \tag{17}$$

Since $k_{2n_2} = h_{2n_2}$, we conclude from (17) that h_{1n_2} and k_{1n_2} are equivalent modulo $k_{2n_2} = h_{2n_2}$. But h_{1n_2} and k_{1n_2} are in the same complete system of residues modulo $k_{2n_2} = h_{2n_2}$. Thus $h_{1n_2} = k_{1n_2}$, and since $h_{2n_2} \neq 0$, we conclude from (17) that $p_{12} = 0$. Hence P has the form

$$P = \begin{bmatrix} 1 & 0 & \cdot & \cdots & \cdot \\ 0 & 1 & \cdot & & \\ 0 & 0 & \cdot & & \\ \vdots & \vdots & & & \\ 0 & 0 & \cdot & & \cdot \end{bmatrix} \tag{18}$$

and from (15), $K^{(n_2)} = H^{(n_2)}$. We have proved so far that $q_1 = n_1$, $q_2 = n_2$, $H^{(t)} = K^{(t)}$, $t = 1, \ldots, n_2$, and P has the form (18).

A typical column of H strictly between columns n_2 and n_3 is of the form

$$H^{(t)} = [h_{1t}, h_{2t}, 0, \ldots, 0]$$

and hence

$$\begin{aligned} K^{(t)} &= PH^{(t)} = h_{1t}P^{(1)} + h_{2t}P^{(2)} \\ &= [h_{1t}, h_{2t}, 0, \ldots, 0] \\ &= H^{(t)}. \end{aligned}$$

In other words, H and K agree up to and including column $n_3 - 1$. At column n_3 we have

$$\begin{aligned} K^{(n_3)} &= PH^{(n_3)} \\ &= P[h_{1n_3}, h_{2n_3}, h_{3n_3}, 0, \ldots, 0] \\ &= h_{1n_3}P^{(1)} + h_{2n_3}P^{(2)} + h_{3n_3}P^{(3)} \\ &= [h_{1n_3}, 0, \ldots, 0] + [0, h_{2n_3}, 0, \ldots, 0] \\ &\quad + h_{3n_3}[p_{13}, p_{23}, p_{33}, p_{43}, \ldots, p_{m3}]. \end{aligned} \tag{19}$$

The columns of K up to column n_3 have at most the first two entries nonzero so that $K^{(n_3)}$ can have at most its first three entries nonzero. Since $h_{3n_3} \neq 0$, we conclude from (19) that $p_{43} = \cdots = p_{m3} = 0$. As before $\det P = p_{33} \det P(1, 2, 3 | 1, 2, 3)$ so that p_{33} is a unit. Then from (19),

$$h_{3n_3}p_{33} = k_{3n_3} \tag{20}$$

and $K^{(n_3)}$ is the first column of K with a nonzero entry below row 2. It follows that $q_3 = n_3$ and as before that $h_{3n_3} = k_{3n_3}$, $p_{33} = 1$. From (19), we have

$$h_{1n_3} + h_{3n_3}p_{13} = k_{1n_3}, \tag{21}$$

$$h_{2n_3} + h_{3n_3}p_{23} = k_{2n_3}. \tag{22}$$

Thus h_{1n_3} and k_{1n_3} are in the same residue class modulo $h_{3n_3} = k_{3n_3}$, as are h_{2n_3} and k_{2n_3}. Hence $h_{1n_3} = k_{1n_3}$, $h_{2n_3} = k_{2n_3}$ by the argument used before, and from (21) and (22) we conclude that $p_{13} = p_{23} = 0$. We now have $H^{(t)} = K^{(t)}$, $t = 1, \ldots, n_3$. Obviously this argument can be continued to conclude the following:

$$q_t = n_t, \qquad t = 1, \ldots, r, \tag{23}$$

$$K^{(t)} = H^{(t)}, \qquad t = 1, \ldots, n_r, \tag{24}$$

and

$$P = \begin{matrix} r \\ m-r \end{matrix}\left[\begin{array}{c:c} I_r & C \\ \hdashline O & Q \end{array}\right] \quad (\text{column blocks } r,\ m-r)$$

where $Q \in \mathrm{GL}(m - r, R)$. Consider the conformally partitioned block matrix multiplication for $K = PH$:

$$\begin{matrix} r \\ m-r \end{matrix}\left[\begin{array}{c:c} K_{11} & K_{12} \\ \hdashline O & O \end{array}\right] = \begin{matrix} r \\ m-r \end{matrix}\left[\begin{array}{c:c} I_r & C \\ \hdashline O & C \end{array}\right]\left[\begin{array}{c:c} H_{11} & H_{12} \\ \hdashline O & O \end{array}\right]\begin{matrix} r \\ m-r \end{matrix}. \tag{25}$$

(column blocks: n_r, $n - n_r$; r, $m - r$; n_r, $n - n_r$)

The upper right corner in the product on the right in (25) is $H_{12} = K_{12}$. Since we already know that $H_{11} = K_{11}$, we conclude that $H = K$. ∎

Corollary 1 *If R is a field and $A \in M_{m,n}(R)$, then the Hermite normal form*

H for A can be chosen so that $H^{(n_i)} = [0, \ldots, 1, 0, 0, \ldots, 0]$ *where* 1 *appears in position* i, $i = 1, \ldots, r$.

Proof: Since R is a field, every nonzero element is a unit so that every nonzero element is an associate of 1. Choose the complete system of nonassociates to be $\{1, 0\}$. Also $\{0\}$ is a complete system of residues modulo any nonzero element in a field. ▮

In view of the uniqueness of the Hermite form we shall refer to the indices $n_1, \ldots, n_r$ as the *indices of A*.

Corollary 2 *If* $A \in M_n(R)$ *is unimodular, then the Hermite normal form of A is an upper triangular matrix with units in R as main diagonal entries.*

Proof: Since $m = n = r$, it follows that the indices of A must be $1, \ldots, m$. Since det A is a unit in R, the Hermite normal form of A has units as main diagonal entries. ▮

Example 1 We compute the Hermite normal form for

$$A = \begin{bmatrix} 7 & 3 & 15 \\ 1 & 5 & 3 \end{bmatrix} \tag{26}$$

in $M_{2,3}(Z)$. The complete system of nonassociates is chosen to be the nonnegative integers, and the complete residue system modulo any $m \neq 0$ is $\{0, 1, \ldots, |m| - 1\}$. We write the corresponding unimodular matrix that accomplishes the indicated result by left multiplication:

$$A \overset{L}{=} \begin{bmatrix} 1 & 5 & 3 \\ 7 & 3 & 15 \end{bmatrix}, \qquad P_1 = \begin{bmatrix} 0 & 1 \\ 1 & 0 \end{bmatrix},$$

$$\overset{L}{=} \begin{bmatrix} 1 & 5 & 3 \\ 0 & -32 & -6 \end{bmatrix}, \qquad P_2 = \begin{bmatrix} 1 & 0 \\ -7 & 1 \end{bmatrix},$$

$$\overset{L}{=} \begin{bmatrix} 1 & 5 & 3 \\ 0 & 32 & 6 \end{bmatrix}, \qquad P_3 = \begin{bmatrix} 1 & 0 \\ 0 & -1 \end{bmatrix}.$$

Observe that $P = P_3 P_2 P_1 = \begin{bmatrix} 0 & 1 \\ -1 & 7 \end{bmatrix}$.

As a first application of the Hermite normal form we determine the structure of the left ideals in $M_n(R)$, where R is a P.I.D. Before doing this, we make the following definition in which R can be an arbitrary ring. If M is a (left) R-module (see Section 1.2), then M is said to be *generated by a subset* $S \subset M$ if no proper submodule of M contains S. Obviously, any element $m \in M$ must have the form

$$m = n_1 \cdot s_1 + \cdots + n_k \cdot s_k + r_1 s_1 + \cdots + r_k s_k \tag{27}$$

in which the n_i are integers, $s_i \in S$, $r_i \in R$, and

$$n_i \cdot s_i = \begin{cases} \overbrace{s_i + \cdots + s_i}^{n_i}, & \text{if } n_i > 0 \\ -(|n_i| \cdot s_i), & \text{if } n_i < 0 \\ 0, & \text{if } n_i = 0 \end{cases}.$$

For, the set of all elements of the form (27) is a submodule of M containing S. If R has a multiplicative identity 1, then by the usual argument [see formula (70) et seq. in Section 3.2], the terms $n_i \cdot s_i$ can be omitted in (27). If $S = \{s_1, \ldots, s_n\}$ is finite, then M is said to be a *finitely generated* R-module. If R has a multiplicative identity 1 and the representation of any element $m \in M$ as

$$m = \sum_{i=1}^{n} r_i s_i$$

is unique, then M is called a *free R-module of rank n* and $\{s_1, \ldots, s_n\}$ is called a *basis of M*, or a *set of free generators of M*. It is clear that if 0 has a unique representation in terms of the s_i, then so does any $m \in M$. In general, we write $M = \langle s_1, \ldots, s_n \rangle$ to indicate that M is generated by $\{s_1, \ldots, s_n\}$ (not necessarily a basis).

Theorem 1.4 *If R is a* P.I.D. *and M is a finitely generated R-module, then any submodule W of M is finitely generated.*

Proof: Let $\{s_1, \ldots, s_n\}$ be a finite set of generators of M, $s_i \neq 0$, $i = 1, \ldots, n$, and for $1 \leq k \leq n$ let W_k be the set of all elements in W of the form

$$w = r_k s_k + r_{k+1} s_{k+1} + \cdots + r_n s_n, \qquad r_t \in R. \tag{28}$$

Either $r_k = 0$ for every such $w \in W$ or there is at least one $w \in W$ such that $r_k \neq 0$. In the first case define $w_k = 0$. In the second case consider all w for which $r_k \neq 0$, and choose one for which the number of nonassociate prime divisors of r_k is minimal. Call this element w_k:

$$w_k = \rho_k s_k + \rho_{k+1} s_{k+1} + \cdots + \rho_n s_n. \tag{29}$$

If $w \in W_k$ is given by (28), then we assert that $\rho_k | r_k$. For if $\rho_k \nmid r_k$, then $d =$ g.c.d.$(\rho_k, r_k) \neq 0$ has fewer prime factors than does ρ_k. Let $a,b \in R$ be chosen so that $ar_k + b\rho_k = d$. Observe that

$$aw + bw_k = (ar_k + b\rho_k)s_k + (ar_{k+1} + b\rho_{k+1})s_{k+1} + \cdots + (ar_n + b\rho_n)s_n \tag{30}$$

is in W, has the form (28), and hence is in W_k. But $ar_k + b\rho_k = d$, and this contradicts the choice of ρ_k.

Now consider the set $\{w_1, \ldots, w_n\}$, and let $w \in W$. Determine $\beta_1 \in R$ such that $w - \beta_1 w_1 \in W_2$. Then determine $\beta_2 \in R$ such that $w - \beta_1 w_1 - \beta_2 w_2 \in W_3$, and so on. Finally, we obtain

$$w = \sum_{t=1}^{n} \beta_t w_t, \qquad \beta_t \in R, \ t = 1, \ldots, n.$$

We conclude that the set $\{w_1, \ldots, w_n\}$ generates W. ▮

Example 2 (a) Consider $M = Z^n$, $R = Z$, with the usual operations:

$$(a_1, \ldots, a_n) + (b_1, \ldots, b_n) = (a_1 + b_1, \ldots, a_n + b_n)$$

and

$$c(a_1, \ldots, a_n) = (ca_1, \ldots, ca_n).$$

Then Z^n is a Z-module and

$$\{e_i = (\delta_{i1}, \ldots, \delta_{in}) \mid i = 1, \ldots, n\}$$

is a basis of Z^n.

(b) If G is any abelian group, written additively, then G can be regarded as a Z-module with

$$ng = \begin{cases} \overbrace{g + \cdots + g}^{n}, & \text{if } n > 0 \\ -(|n| g), & \text{if } n < 0 \\ 0, & \text{if } n = 0 \end{cases} \quad . \tag{31}$$

To say that G is a *free abelian group of rank n* means that as a Z-module, there exist elements $g_1, \ldots, g_n$ in G such that any element $g \in G$ has a unique representation

$$g = \sum_{i=1}^{n} m_i g_i, \qquad m_i \in Z, \quad i = 1, \ldots, n.$$

(c) Let $M = Z_n$, the ring of integers modulo n, regarded as a Z-module. Then M is not free. For if $[x] \in Z_n$ then $n[x] = [0]$, so that $[0]$ could never be uniquely represented.

Theorem 1.5 (*Dedekind's Theorem*) *Let R be a* P.I.D. *and M a free* R-*module with a basis* $\{s_1, \ldots, s_n\}$.

(i) *Any basis of M must contain n elements.*

(ii) *If W is a submodule of M, then W is free of rank r,* $r \leq n$. *Moreover, after a suitable reordering of* $s_1, \ldots, s_n$ *it is possible to construct a basis* $w_1, \ldots, w_r$ *of W of the following form:*

$$\begin{aligned} w_1 &= a_{11}s_1 + \cdots + a_{1n}s_n \\ w_2 &= a_{22}s_2 + \cdots + a_{2n}s_n \\ &\vdots \\ w_r &= a_{rr}s_r + \cdots + a_{rn}s_n \end{aligned}$$

where each $a_{ii} \neq 0$ *and lies in any prescribed system of nonassociates,* $i = 1, \ldots, r$.

Proof: (i) Let $\{u_1, \ldots, u_m\}$ be another basis of M. Write

$$u_i = \sum_{j=1}^{n} a_{ij} s_j, \qquad i = 1, \ldots, m, \tag{32}$$

$a_{ij} \in R$. Also write

$$s_j = \sum_{k=1}^{m} b_{jk} u_k, \qquad j = 1, \ldots, n, \tag{33}$$

$b_{jk} \in R$. Let $A = [a_{ij}] \in M_{m,n}(R)$, $B = [b_{jk}] \in M_{n,m}(R)$. Then from (32) and (33) we have

$$s_j = \sum_{k=1}^{m} b_{jk} u_k = \sum_{k=1}^{m} b_{jk} \sum_{p=1}^{n} a_{kp} s_p$$
$$= \sum_{p=1}^{n} \left(\sum_{k=1}^{m} b_{jk} a_{kp} \right) s_p, \qquad j = 1, \ldots, n.$$

Since $\{s_1, \ldots, s_n\}$ is a basis, it follows that

$$\sum_{k=1}^{m} b_{jk} a_{kp} = \delta_{jp}, \qquad j, p = 1, \ldots, n$$

or

$$(BA)_{jp} = \delta_{jp}.$$

Thus

$$BA = I_n \tag{34}$$

and by a similar argument

$$AB = I_m. \tag{35}$$

The equations (34) and (35) imply that $m = n$ and both A and B are unimodular. For suppose $n > m$. Then since $\rho(B) \leq m$ (i.e., there are no subdeterminants of order greater than m) it follows from Theorem 1.2 that the Hermite normal form of B must have a zero last row. But then if $PB = H$ is the Hermite form of B and P is unimodular, we have from (34) that

$$P = PBA = HA$$

and HA has a zero last row (why?). This is impossible since P is unimodular. Thus $n \leq m$ and similarly from (35), $m \leq n$. This argument shows in fact that any two bases of M are connected as in (32) by a unimodular matrix (also see Exercise 5).

(ii) In Theorem 1.4 we saw that a generating set for W of the required form exists. Indeed, simply delete the zero elements w_k obtained in the proof of Theorem 1.4 and reorder the s_i. We need only show that these generating elements form a basis, i.e., are free. But suppose

$$\sum_{i=1}^{r} \alpha_i w_i = 0, \qquad \alpha_i \in R. \tag{36}$$

Then (36) becomes

$$\alpha_1(a_{11}s_1 + a_{12}s_2 + \cdots + a_{1n}s_n) + \cdots$$
$$+ \alpha_r(a_{rr}s_r + \cdots + a_{rn}s_n) = 0. \tag{37}$$

Hence $\alpha_1 a_{11} = 0$ so that $\alpha_1 = 0$. Then $\alpha_2 a_{22} = 0$ so that $\alpha_2 = 0$, and so on. ∎

Theorem 1.6 *Let $\mathfrak{A}$ be a left ideal in $M_n(R)$, where R is a* P.I.D.. *Then $\mathfrak{A}$ is a principal ideal, i.e., $\mathfrak{A}$ consists of all left multiples of some $D \in \mathfrak{A}$.*

Proof: Obviously $M_n(R)$ is a free R-module with $\{E_{pq} = [\delta_{ip}\delta_{jq}] \mid p, q =$

$1, \ldots, n\}$ as a basis. The left ideal $\mathfrak{A}$ is a submodule and hence by Theorem 1.5 is free with a basis $\{A_1, \ldots, A_r\}$, where $r \leq n^2$.

Define the $rn \times n$ matrix A by

$$A = \begin{bmatrix} A_1 \\ A_2 \\ \vdots \\ A_r \end{bmatrix} \tag{38}$$

and choose an rn-square unimodular $P \in M_{rn}(R)$ such that $PA = H$ is in Hermite normal form. Since A has n columns, $\rho(A) = \rho(H) \leq n$, so that

$$H = \begin{bmatrix} D \\ O \end{bmatrix}$$

where $D \in M_n(R)$ is in Hermite normal form and 0 is the $(nr - n) \times n$ zero matrix. Partition P into r^2 $n \times n$ blocks P_{ij} indicated notationally by $P = [P_{ij}]$. Then by block multiplication

$$D = \sum_{t=1}^{r} P_{1t}A_t$$

and hence $$D \in \mathfrak{A}.$$

Thus $(D) = M_n(R)D \subset \mathfrak{A}$. But $Q = P^{-1}$ is also unimodular and

$$A = QH = [Q_{ij}] \begin{bmatrix} D \\ O \end{bmatrix} = \begin{bmatrix} Q_{11}D \\ Q_{21}D \\ \vdots \\ Q_{r1}D \end{bmatrix}. \tag{39}$$

It follows from (38) and (39) that $A_i = Q_{i1}D$, $i = 1, \ldots, r$, and hence $\mathfrak{A} \subset (D)$. ▮

Recall (see Exercises 32 and 33 in Section 3.4) that a Euclidean domain R is a P.I.D. so that the concept of a Hermite normal form for $A \in M_{m,n}(R)$ makes sense.

Theorem 1.7 *If (R, δ) is a Euclidean domain, $A \in M_{m,n}(R)$, and H is the Hermite normal form of A, then there exists a product $P \in M_m(R)$ of elementary matrices such that $PA = H$.*

Proof: If we examine the proof of Theorem 1.2, we see that the reduction of A to H is accomplished by left multiplication by elementary matrices except at one stage: the replacement of the first element in a column of a k-rowed matrix by the g.c.d. of the elements $a_1, \ldots, a_k$ appearing in the column. In a Euclidean domain we can proceed as follows. Scan the nonzero elements among $a_1, \ldots, a_k$ for the one of least norm, say it is a_1, which lies in the first row (perform elementary type I operations to achieve this if necessary). Then from each of rows $2, \ldots, k$ subtract appropriate multiples of row 1. This results in a column of the form $a_1, r_2, \ldots, r_k$ in which $r_2, \ldots, r_k$ are all 0 or have norm strictly less than $\delta(a_1)$. Repeat the process until the elements in rows $2, \ldots, k$ are all 0. This must ultimately happen because the maximum norm of any element in the column is strictly decreased after each step. ∎

We remark that in a Euclidean domain a complete system of residues modulo $m \neq 0$ may be chosen to consist of 0 and elements of norm strictly smaller than $\delta(m)$. Thus, for example, the Hermite normal form for $A \in M_{m,n}(Z)$ may be chosen so that the first nonzero entry h_{in_i} in each row is positive and every entry above h_{in_i} is nonnegative and less than h_{in_i}. Also, any field is trivially a Euclidean domain (why?) so that from Theorem 1.7 and Corollary 1 any matrix over a field can be reduced to the Hermite normal form in Corollary 1 by elementary row operations.

Corollary 3 *If (R, δ) is a Euclidean domain, then $A \in \mathrm{GL}(n, R)$ iff A is a product of elementary matrices.*

Proof: If A is a product of elementary matrices, then of course $A \in \mathrm{GL}(n,R)$. Conversely suppose $A \in \mathrm{GL}(n,R)$. By Corollary 2, the Hermite normal form H of A is upper triangular with units along the main diagonal. We can assume these main diagonal elements are all 1, and then by elementary type II row operations we can eliminate the elements in each of the columns above the main diagonal. That is, from row i subtract a_{in} times row n, $i = n-1, n-2, \ldots, 1$; then from row i subtract a_{in-1} times row $n-1$, $i = n-2, n-3, \ldots, 1$, etc. The resulting matrix is I_n so that

$$U_1 U_2 \cdots U_p A = I_n, \qquad U_i \in \mathrm{GL}(n, R), \; i = 1, \ldots, p,$$

and

$$A = U_p^{-1} \cdots U_1^{-1}. \qquad ∎$$

The following example exhibits some of the applications of the Hermite normal form.

Example 3 (a) Express $A = \begin{bmatrix} 2 & 1 \\ 5 & 3 \end{bmatrix}$ as a product of elementary matrices in $\mathrm{GL}(2,Z)$. We have:

$$E_2(-1)E_{12}(1)E_{21}(-2)E_{12}E_{21}(-2)A = I_2$$

so that

$$\begin{aligned} A &= E_{21}(-2)^{-1}E_{12}{}^{-1}E_{21}(-2)^{-1}E_{12}(1)^{-1}E_2(-1)^{-1} \\ &= E_{21}(2)E_{12}E_{21}(2)E_{12}(-1)E_2(-1) \\ &= \begin{bmatrix} 1 & 0 \\ 2 & 1 \end{bmatrix}\begin{bmatrix} 0 & 1 \\ 1 & 0 \end{bmatrix}\begin{bmatrix} 1 & 0 \\ 2 & 1 \end{bmatrix}\begin{bmatrix} 1 & -1 \\ 0 & 1 \end{bmatrix}\begin{bmatrix} 1 & 0 \\ 0 & -1 \end{bmatrix}. \end{aligned}$$

(b) Theorem 1.6 showed that any left ideal in $M_n(R)$ is principal. More generally, let $A_1, \ldots, A_r$ be arbitrary elements in $M_{m,n}(R)$. Let $\mathfrak{A}$ denote the set of all matrices

$$\sum_{t=1}^{r} X_t A_t, \qquad X_t \in M_m(R),\ t = 1, \ldots, r.$$

We seek a $D \in \mathfrak{A}$ such that $\mathfrak{A}$ consists of all left multiples of D, say

$$D = \sum_{t=1}^{r} X_t A_t.$$

Observe that if $A_t = C_t B$, $B \in M_{m,n}(R)$, $C_t \in M_m(R)$, $t = 1, \ldots, r$, then

$$D = \left(\sum_{t=1}^{r} X_t C_t\right) B = CB, \tag{40}$$

i.e., if B is a right divisor of each A_t, then B is a right divisor of D. On the other hand, if everything in $\mathfrak{A}$ is a left multiple of D, then each A_t is. It is natural to call D a *greatest common right divisor* (g.c.r.d.) of $A_1, \ldots, A_r$, i.e., D is a right divisor of each A_t, and if $B \in M_{m,n}(R)$ is any other common right divisor of $A_1, \ldots, A_r$, then B is a right divisor of D. The procedure for computing a g.c.r.d. is suggested by the proof of Theorem 1.6 if $\rho(A) \leq m$, where

$$A = \begin{bmatrix} A_1 \\ \vdots \\ A_r \end{bmatrix} \in M_{rm,n}(R)$$

[$\rho(A) \leq m$ when, for example, $n \leq m$]. Choose an rm-square unimodular $P \in M_{rm}(R)$ such that $PA = H$ is in Hermite normal form; since $\rho(A) \leq m$, we have

$$H = \begin{bmatrix} D \\ 0 \end{bmatrix}$$

in which D is $m \times n$ and in Hermite normal form. Partition P into r^2 $m \times m$ blocks, $P = [P_{ij}]$, so that

$$D = \sum_{j=1}^{r} P_{1j} A_j \in \mathfrak{A}.$$

Then let $P^{-1} = Q = [Q_{ij}]$ be partitioned conformally with P. Since $PA = H$, we have $A = QH$ and $A_i = Q_{i1} D$, $i = 1, \ldots, r$; i.e., D is a common right divisor of each A_i. Hence D is a g.c.r.d. of $A_1, \ldots, A_r$. Note that if $\rho(A) = m$, then $\rho(D) = m$. If $B \in M_{m,n}(R)$ is a right divisor of each A_t, then from (40), $D = CB$ and hence $m = \rho(D) \leq \rho(B) = m$. Now suppose moreover that B is also a g.c.r.d. of $A_1, \ldots, A_r$; then

$$D = CB, \qquad C \in M_m(R) \tag{41}$$

and since D is a common right divisor of $A_1, \ldots, A_r$, D is a right divisor of B,

$$B = C_1 D, \qquad C_1 \in M_m(R). \tag{42}$$

Combining (41) and (42), we have

$$D = CC_1 D,$$

and since $\rho(D) = m$, it follows that $CC_1 = I_m$ (see Exercise 12), i.e., C and C_1 are unimodular. Hence $D \overset{L}{=} B$. This shows that *if*

$$\rho \begin{bmatrix} A_1 \\ \vdots \\ A_r \end{bmatrix} = m,$$

then any two g.c.r.d. *of* $A_1, \ldots, A_r$ *are left equivalent.*

(c) With the same notation as in (b) above we define a *least common left multiple* (l.c.l.m.) of $A_1, \ldots, A_r$ to be a matrix L which is a common left multiple (c.l.m.) of $A_1, \ldots, A_r$, i.e., $L = E_t A_t$, $t = 1, \ldots, r$, and if L_1 is any other c.l.m. of $A_1, \ldots, A_r$, then L_1 is a left multiple of L, i.e., $L_1 = EL$ for some $E \in M_m(R)$. Using essentially the same arguments as above we can construct a nonzero c.l.m. of A_1 and A_2 in $M_{m,n}(R)$. Again consider the $2m \times n$ matrix

$$A = \begin{bmatrix} A_1 \\ A_2 \end{bmatrix}$$

and assume that $\rho(A) \leq m$. Reduce A to Hermite normal form

$$H = \begin{bmatrix} D \\ 0 \end{bmatrix}, \qquad D \in M_{m,n}(R),$$

by premultiplication with a unimodular $2m \times 2m$ matrix $P = [P_{ij}]$, partitioned into four $m \times m$ matrices. Then

$$\begin{bmatrix} P_{11} & P_{12} \\ P_{21} & P_{22} \end{bmatrix} \begin{bmatrix} A_1 \\ A_2 \end{bmatrix} = \begin{bmatrix} D \\ 0 \end{bmatrix}, \tag{43}$$

so that

$$P_{21}A_1 + P_{22}A_2 = 0.$$

Set

$$L = P_{21}A_1 = -P_{22}A_2; \tag{44}$$

then obviously L is a c.l.m. of A_1 and A_2. We assert that *if* $\rho(A_1) = m$ *and* $\rho(A_2) = m$, *then in fact* L *is a* l.c.l.m. *of* A_1 *and* A_2. To see this, let $Q = P^{-1} = [Q_{ij}]$ be conformally partitioned with P. Then

$$\begin{aligned} PQ &= \begin{bmatrix} P_{11}Q_{11} + P_{12}Q_{21} & P_{11}Q_{12} + P_{12}Q_{22} \\ P_{21}Q_{11} + P_{22}Q_{21} & P_{21}Q_{12} + P_{22}Q_{22} \end{bmatrix} \\ &= I_m \dotplus I_m. \end{aligned}$$

[If X and Y are $m \times m$ and $n \times n$ matrices, respectively, then $X \dotplus Y$, the *direct sum* of X and Y, is the $(m + n) \times (m + n)$ matrix $\begin{bmatrix} X & 0 \\ 0 & Y \end{bmatrix}$.] Hence

$$I_m = P_{21}Q_{12} + P_{22}Q_{22}. \tag{45}$$

Suppose L_1 is a c.l.m. of A_1 and A_2 so that

$$L_1 = E_1A_1, \qquad L_1 = E_2A_2. \tag{46}$$

We want to prove that L_1 is a left multiple of L. Observe from (44) that

$$\begin{bmatrix} L \\ L_1 \end{bmatrix} = \begin{bmatrix} P_{21} & 0 \\ 0 & E_1 \end{bmatrix} \begin{bmatrix} A_1 \\ A_1 \end{bmatrix}$$

and since $\rho\left(\begin{bmatrix} A_1 \\ A_1 \end{bmatrix}\right) = \rho(A_1) = m$ (see Exercise 13), we know from part (b) (applied to L and L_1 in place of A_1 and A_2) that a g.c.r.d. of L and L_1 exists. Thus let L_2 be a g.c.r.d. of L_1 and L so that

$$L_2 = XL + YL_1. \tag{47}$$

Since both L and L_1 are c.1.m.'s of A_1 and A_2, it immediately follows from (47) that L_2 is a c.1.m. of A_1 and A_2:

$$L_2 = Z_1A_1 = Z_2A_2. \tag{48}$$

Moreover since L_2 is a g.c.r.d. of L_1 and L, we have

$$L = WL_2, \qquad L_1 = W_1L_2, \tag{49}$$

where $W, W_1 \in M_m(R)$. We compute that

$$\begin{aligned} P_{21}A_1 &= L && \text{[by (44)]} \\ &= WL_2 && \text{[by (49)]} \\ &= WZ_1A_1 && \text{[by (48)]} \end{aligned}$$

and

$$\begin{aligned} -P_{22}A_2 &= L && \text{[by (44)]} \\ &= WL_2 && \text{[by (49)]} \\ &= WZ_2A_2 && \text{[by (48)].} \end{aligned}$$

Thus since $\rho(A_1) = \rho(A_2) = m$, it follows (see Exercise 12) that

$$P_{21} = WZ_1, \qquad -P_{22} = WZ_2. \tag{50}$$

From (45) and (50) we compute

$$\begin{aligned} I_m &= P_{21}Q_{12} + P_{22}Q_{22} \\ &= WZ_1Q_{12} - WZ_2Q_{22} \\ &= W(Z_1Q_{12} - Z_2Q_{22}). \end{aligned} \tag{51}$$

In other words, W is unimodular in $M_m(R)$, and so $W^{-1} \in M_m(R)$. But then from (49),

$$L_1 = W_1L_2 = W_1W^{-1}L.$$

Thus L_1 is a left multiple of L, and hence L is a 1.c.1.m. of A_1 and A_2.

(d) Let M be a free R-module of rank n (R a P.I.D.) with a basis $\{\varepsilon_1, \ldots, \varepsilon_n\}$. Let $A \in M_{m,n}(R)$ and define

$$w_i = \sum_{j=1}^{n} a_{ij}\varepsilon_j, \qquad i = 1, \ldots, m. \tag{52}$$

Then the submodule $W = \langle w, \ldots, w_m \rangle$ (which is free by Theorem 1.5) has rank $k = \rho(A)$. To verify this, we introduce some rather convenient notation: Let the equations (52) be abbreviated

$$w = A\varepsilon. \tag{53}$$

Set $\rho(A) = k$, and choose a unimodular $Q \in M_m(R)$ such that $QA = H$ is in Hermite normal form. Then from (53)

$$Qw = QA\varepsilon = H\varepsilon$$

so that if $v = Qw$, then $w = Q^{-1}v$ and any element of W is a linear combination of the elements v. But v has only its first k components nonzero, i.e., $v_{k+1} = \cdots = v_m = 0$ because $H_{(k+1)} = \cdots = H_{(m)} = 0$. Moreover,

$$\begin{aligned} v_1 &= h_{1n_1}\varepsilon_{n_1} + \cdots + h_{1n}\varepsilon_n \\ v_2 &= \qquad h_{2n_2}\varepsilon_{n_2} + \cdots + h_{2n}\varepsilon_n \\ &\vdots \\ v_k &= \qquad\qquad h_{kn_k}\varepsilon_{nk} + \cdots + h_{kn}\varepsilon_n \end{aligned}$$

so that a relation $\sum_{i=1}^{k}\alpha_i v_i = 0$ implies (because $\varepsilon_1, \ldots, \varepsilon_n$ are free and $h_{in_i} \neq 0$, $i = 1, \ldots, k$) that in succession $\alpha_1 = 0, \alpha_2 = 0, \ldots, \alpha_k = 0$. Thus $v_1, \ldots, v_k$ are free and form a basis of W, i.e., W has rank k. We say that the matrix A is *associated with the submodule* W if $W = \langle w_1, \ldots, w_m\rangle$ and the w_i are given by (52). The basis $\{\varepsilon_1, \ldots, \varepsilon_n\}$ of M is, of course, understood in advance.

(e) Suppose both A and B are in $M_{m,n}(R)$. Let W be a submodule of M of rank m and assume that A and B are both associated with W. Then it follows that $A \stackrel{\text{L}}{=} B$. For, suppose $\{\varepsilon_1, \ldots, \varepsilon_n\}$ is a basis of M,

$$w_i = \sum_{j=1}^{n} a_{ij}\varepsilon_j, \qquad i = 1, \ldots, m, \tag{54}$$

$$u_i = \sum_{j=1}^{n} b_{ij}\varepsilon_j, \qquad i = 1, \ldots, m, \tag{55}$$

and $W = \langle w_1, \ldots, w_m\rangle = \langle u_1, \ldots, u_m\rangle$. Then by (d), the rank of W is $\rho(A)$ [and $\rho(B)$] so that $\rho(A) = \rho(B) = m$. Write (54) and (55) as

$$w = A\varepsilon \tag{56}$$

and

$$u = B\varepsilon. \tag{57}$$

Next, we assert that the sets of generators $\{w_1, \ldots, w_m\}$ and $\{u_1, \ldots, u_m\}$ of W are both free. Suppose $\sum_{i=1}^{m} \alpha_i w_i = 0$; write

$$\alpha w = \sum_{i=1}^{m} \alpha_i w_i$$

where $\alpha = [\alpha_1, \ldots, \alpha_m]$. Then $\alpha w = 0$ implies $\alpha A\varepsilon = 0$ by (56). Taking transposes, we have $\varepsilon^{\text{T}}A^{\text{T}}\alpha^{\text{T}} = 0$, and since $\varepsilon_1, \ldots, \varepsilon_n$ are free it follows that $A^{\text{T}}\alpha^{\text{T}} = 0$ (verify). Now A^{T} is $n \times m$ and $\rho(A^{\text{T}}) = \rho(A) = m$ (why?). If we reduce A^{T} to Hermite normal form $K = PA^{\text{T}}$, we have

$$\begin{aligned} PA^{\text{T}}\alpha^{\text{T}} &= 0, \\ K\alpha^{\text{T}} &= 0, \end{aligned} \tag{58}$$

and

$$K = \begin{bmatrix} D \\ 0 \end{bmatrix},$$

where $\rho(D) = m$ and D is $m \times m$. But $\det D \neq 0$ implies that over the quotient field of R, D has an inverse. Then from (58) we have

$$D\alpha^{\text{T}} = 0,$$

and hence $\alpha^{\text{T}} = 0$. Thus $w_1, \ldots, w_m$ are free and similarly $u_1, \ldots, u_m$ are free.

Each of these sets therefore form a basis of W and it follows (see Exercise 11) that

$$w = Qu$$

where $Q \in \mathrm{GL}(m,R)$. Then from (56) and (57),

$$\begin{aligned} A\varepsilon &= w = Qu \\ &= QB\varepsilon. \end{aligned} \tag{59}$$

Since $\varepsilon_1, \ldots, \varepsilon_n$ are free, (59) implies that $A = QB$, i.e., $A \stackrel{L}{=} B$.

On the other hand, suppose $A,B \in M_{m,n}(R)$ and $A \stackrel{L}{=} B$, say $A = QB$ where $Q \in \mathrm{GL}(m,R)$. If $\varepsilon = \{\varepsilon_1, \ldots, \varepsilon_n\}$ is a basis of M, $w = A\varepsilon$, and $u = B\varepsilon$, then the fact that Q is unimodular implies that $\langle u_1, \ldots, u_m\rangle = \langle w_1, \ldots, w_m\rangle = W$, so that A and B are associated with the same submodule W of M.

If we combine Theorem 1.5 with Examples 3(d) and (e) we have the following result (R continues to be a P.I.D.).

Theorem 1.8 *Let M be a free R-module with a basis $\{\varepsilon_1, \ldots, \varepsilon_n\}$. If $A \in M_{m,n}(R)$ and $\rho(A) = m$, then the submodule W generated by*

$$w_i = \sum_{j=1}^{n} a_{ij}\varepsilon_j, \qquad i = 1, \ldots, m$$

is of rank m and $\{w_1, \ldots, w_m\}$ is a basis of W. Moreover, $\{u_1, \ldots, u_m\}$ is another basis of W iff $u = B\varepsilon$, $B \in M_{m,n}(R)$, and $A \stackrel{L}{=} B$.

In view of this result we can go back and forth between matrices and submodules. That is, *every submodule of rank m of a free R-module M determines an $m \times n$ matrix A uniquely to within left equivalence, and conversely, the equivalence class* (*with respect to* $\stackrel{L}{=}$) *of $m \times n$ matrices over R determines a unique rank m submodule of M.*

Example 4 (a) Let M be a free R-module with a basis $\{\varepsilon_1, \ldots, \varepsilon_n\}$. If W is a submodule, we can obtain a matrix $A \in M_n(R)$ defined as follows: If $\{w_1, \ldots, w_m\}$ is a basis of W, then $w_i = \sum_{j=1}^{n} a_{ij}\varepsilon_j$, $i = 1, \ldots, m$; adjoin $n - m$ zero rows to $[a_{ij}]$ and call the resulting matrix A. We continue to denote the n-tuple of elements

$(w_1, \ldots, w_m, \overbrace{0, \ldots, 0}^{n-m})$ by w so that we can write

$$w = A\varepsilon. \tag{60}$$

Suppose that $D \in M_n(R)$ is a right divisor of A: $A = XD$, $X \in M_n(R)$. Let

$$u = D\varepsilon \tag{61}$$

and define a submodule N of M by $N = \langle u_1, \ldots, u_n\rangle$. Then by (60) and (61),

$$\begin{aligned} w &= A\varepsilon = XD\varepsilon \\ &= Xu \end{aligned}$$

so that $w_i \in \langle u_1, \ldots, u_n\rangle = N$, $i = 1, \ldots, n$. Hence

$$W \subset N. \tag{62}$$

Conversely, suppose $N = \langle u_1, \ldots, u_n \rangle$ is a submodule of M and $W \subset N$. Then every element of W is a linear combination of $u_1, \ldots, u_n$, i.e., $w = Xu$ for some $X \in M_n(R)$. Since $u = D\varepsilon$ for some $D \in M_n(R)$, we have

$$\begin{aligned} A\varepsilon &= w = Xu \\ &= XD\varepsilon \end{aligned}$$

and it follows (why?) that $A = XD$, i.e., that D is a right divisor of A. In other words, *if* $W = \langle A\varepsilon \rangle = \langle w_1, \ldots, w_m, 0, \ldots, 0 \rangle$ and $N = \langle D\varepsilon \rangle = \langle u_1, \ldots, u_n \rangle$, *then* $W \subset N$ *iff* D *is a right divisor of* A. We continue with this notation in the next part of the example.

(b) Suppose now that $W_1 = \langle A_1\varepsilon \rangle$, $W_2 = \langle A_2\varepsilon \rangle, \ldots, W_r = \langle A_r\varepsilon \rangle$, $A_i \in M_n(R)$. Suppose that $D \in M_n(R)$ is a g.c.r.d. of $A_1, \ldots, A_r$ and let $N = \langle D\varepsilon \rangle$. Then since D is a right divisor of each A_i, it follows from (a) above that

$$W_i \subset N, \qquad i = 1, \ldots, r.$$

Suppose that W is any submodule of M such that

$$W_i \subset W, \qquad i = 1, \ldots, r,$$

and $W = \langle \Delta\varepsilon \rangle$, $\Delta \in M_n(R)$. Then again by (a), Δ is a right divisor of $A_1, \ldots, A_r$, and hence Δ must be a right divisor of D (since D is a g.c.r.d. of $A_1, \ldots, A_r$). Thus by (a),

$$\begin{aligned} N &= \langle \mathrm{D}\varepsilon \rangle \subset \langle \Delta\varepsilon \rangle \\ &= W. \end{aligned}$$

Thus the submodule $N = \langle D\varepsilon \rangle$ has the following property: *N contains every W_i, and any submodule containing every W_i must contain N*. In other words, *N is the smallest submodule containing every W_i*. On the other hand, let

$$\sum_{i=1}^{r} W_i \tag{63}$$

be the sum (as complexes) of the submodules $W_1, \ldots, W_r$. It is obviously a submodule of M containing each W_i; moreover, any submodule W of M containing each W_i must contain $\sum_{i=1}^{r} W_i$. Hence

$$N = \sum_{i=1}^{r} W_i.$$

Of course $\sum_{i=1}^{r} W_i$ is called the *sum of the submodules* $W_1, \ldots, W_r$.

(c) The result of (b) can be used to effectively compute a basis for the sum of submodules. To illustrate this, let $R = Z$ and let $M = Z^4$ be the free R-module consisting of all 4-tuples of integers with the usual definition of addition and scalar multiplication:

$$(a_1, \ldots, a_4) + (b_1, \ldots, b_4) = (a_1 + b_1, \ldots, a_4 + b_4)$$

and

$$c(a_1, \ldots, a_4) = (ca_1, \ldots, ca_4).$$

Let $\{\varepsilon_i = (\delta_{i1}, \ldots, \delta_{i4}) \mid i = 1, \ldots, r\}$ be the obvious basis of M, and define

$$W_i = \langle A_i\varepsilon \rangle, \qquad i = 1,2,$$

where
$$A_1 = \begin{bmatrix} 2 & 1 & 4 & 2 \\ 1 & 2 & 9 & 5 \\ 2 & 1 & 4 & 2 \\ 4 & 5 & 22 & 12 \end{bmatrix}, \qquad A_2 = \begin{bmatrix} 0 & -2 & 1 & 3 \\ -1 & 3 & 4 & 0 \\ 2 & 4 & 18 & 10 \\ 1 & 7 & 22 & 10 \end{bmatrix}.$$

Set
$$A = \begin{bmatrix} A_1 \\ A_2 \end{bmatrix}$$

and by elementary row operations reduce A to Hermite normal form
$$H = \begin{bmatrix} 1 & 0 & 10 & 8 \\ 0 & 1 & 15 & 11 \\ 0 & 0 & 31 & 25 \\ 0 & 0 & 0 & 0 \\ \cdots & \cdots & \cdots & \cdots \\ & & \bigcirc & \end{bmatrix}$$

(see Exercise 14). We know from Example 3(b) that the 4×4 matrix in the first four rows of H is a g.c.r.d. of A_1 and A_2. Thus the elements
$$\begin{aligned} \varepsilon_1 + 10\varepsilon_3 + \ \ 8\varepsilon_4 &= (1,0,10,8), \\ \varepsilon_2 + 15\varepsilon_3 + 11\varepsilon_4 &= (0,1,15,11), \\ 31\varepsilon_3 + 25\varepsilon_4 &= (0,0,31,25) \end{aligned}$$
form a basis of $W_1 + W_2$.

(d) We can rephrase the result of Example 4(a) above as follows: If $W = \langle A\varepsilon \rangle$ *and* $N = \langle D\varepsilon \rangle$, *then* $W \subset N$ *iff* A *is a left multiple of* D. In the notation of Example 4(b) above suppose L is a l.c.l.m. of $A_1, \ldots, A_r$, and set $I = \langle L\varepsilon \rangle$. Then $I \subset \langle A_i\varepsilon \rangle = W_i$, $i = 1, \ldots, r$. Moreover, if $W = \langle L_1\varepsilon \rangle$ is any submodule of M contained in every W_i, then L_1 is a left multiple of each A_i and hence a left multiple of L. But then
$$W \subset I.$$
In other words, I is a submodule of M contained in every W_i, and any other submodule W of M contained in every W_i is contained in I. Thus I is the largest submodule of M contained in every W_i. On the other hand, the intersection $\bigcap_{i=1}^r W_i$ of the submodules W_i is clearly a submodule of each W_i; and if W is any submodule contained in every W_i, then W is contained in $\bigcap_{i=1}^r W_i$. Hence
$$I = \bigcap_{i=1}^r W_i.$$

(e) Example 3(c) can be used to compute a basis for the intersection of two submodules (it can be used repeatedly for more than two submodules). To illustrate, suppose M is the free module Z^2 of 2-tuples over Z with the obvious basis $\{\varepsilon_1 = (1,0), \varepsilon_2 = (0,1)\}$. Let

$$W_i = \langle A_i \varepsilon \rangle, \qquad i = 1,2,$$

where
$$A_1 = \begin{bmatrix} 2 & 0 \\ 2 & 2 \end{bmatrix}, \qquad A_2 = \begin{bmatrix} 2 & -1 \\ 3 & 2 \end{bmatrix}.$$

Both A_1 and A_2 have rank 2; so the procedure of Example 3(c) can be employed to compute the l.c.l.m. of A_1 and A_2. Thus set

$$A = \begin{bmatrix} 2 & 0 \\ 2 & 2 \\ \hdashline 2 & -1 \\ 3 & 2 \end{bmatrix}.$$

Reduce A to Hermite normal form

$$\begin{bmatrix} 1 & 0 \\ 0 & 1 \\ \hdashline 0 & 0 \\ 0 & 0 \end{bmatrix} = PA,$$

where
$$P = \begin{bmatrix} -3 & 0 & 2 & 1 \\ 1 & 0 & -1 & 0 \\ \hdashline -3 & 1 & 2 & 0 \\ -7 & 0 & 4 & 2 \end{bmatrix}.$$

We have

$$P_{21} = \begin{bmatrix} -3 & 1 \\ -7 & 0 \end{bmatrix}, \qquad P_{22} = \begin{bmatrix} 2 & 0 \\ 4 & 2 \end{bmatrix}$$

and
$$L = P_{21}A_1 = \begin{bmatrix} -3 & 1 \\ -7 & 0 \end{bmatrix}\begin{bmatrix} 2 & 0 \\ 2 & 2 \end{bmatrix}$$
$$= \begin{bmatrix} -4 & 2 \\ -14 & 0 \end{bmatrix},$$
$$-P_{22}A_2 = -\begin{bmatrix} 2 & 0 \\ 4 & 2 \end{bmatrix}\begin{bmatrix} 2 & -1 \\ 3 & 2 \end{bmatrix}$$
$$= \begin{bmatrix} -4 & 2 \\ -14 & 0 \end{bmatrix}$$

in conformity with the general computation in Example 3(c). Then L is a l.c.l.m. of A_1 and A_2. If we reduce L to its Hermite form K, the submodule $\langle K\varepsilon \rangle$ is the same as $\langle L\varepsilon \rangle$ so that $\langle K\varepsilon \rangle = W_1 \cap W_2$. We easily see that

$$K = \begin{bmatrix} 2 & 6 \\ 0 & 14 \end{bmatrix}$$

so that the elements

$$2\varepsilon_1 + 6\varepsilon_2 = (2,6),$$
$$14e_2 = (0,14)$$

form a basis of $W_1 \cap W_2$.

As a final topic in this section we show that Theorem 1.2 can be used to analyze systems of linear equations over a field. First, if M is a module over R and R is a field, then the elements $u_1, \ldots, u_n \in M$ are said to be *linearly independent* (l.i.) *over* R if they are free; otherwise $u_1, \ldots, u_n$ are said to be *linearly dependent* (l.d.) *over* R. Thus if $u_1, \ldots, u_n$ are l.i. over R, then

$$\sum_{i=1}^{n} r_i u_i = 0, \ r_i \in R, \qquad i = 1, \ldots, n \tag{64}$$

iff $r_1 = \cdots = r_n = 0$. Modules over fields are called *vector spaces* (see Section 1.2). A vector space is said to be *finite dimensional* if it has a finite generating set. The elements of a vector space over R are called *vectors*, and the elements of R are called *scalars*. In general, a finitely generated R-module M may not have a basis, e.g., let $M = Z_n$, the integers modulo n, as a Z-module. Then $n[x] = [0]$ for any $[x]$ so that a relation such as (64) can hold without all the r_i being 0. Thus a module can be finitely generated and yet have no basis. This does not happen for vector spaces.

Theorem 1.9 *Let M be a finite dimensional vector space over a field. R. Then M has a basis.*

Proof: Let $\{v_1, \ldots, v_p\}$ be a generating set for M. If these vectors are l.i., then of course they form a basis by definition. In general, we can assume (by simply reordering the v_i) that $\{v_1, \ldots, v_n\}$, $n \leq p$, is a maximal l.i. subset. If $t > n$ then $v_1, \ldots, v_n, v_t$ are l.d. so that there exist $c_1, \ldots, c_n, c$ in R, not all 0, such that

$$\sum_{j=1}^{n} c_j v_j + c v_t = 0.$$

Clearly $c \neq 0$; otherwise we would contradict the linear independence of $v_1, \ldots, v_n$. Then

$$v_t = \sum_{j=1}^{n} (-c^{-1} c_j) v_j$$
$$\in \langle v_1, \ldots, v_n \rangle.$$

In other words, $v_t \in \langle v_1, \ldots, v_n \rangle$, $t = 1, \ldots, p$, and hence $M = \langle v_1, \ldots, v_p \rangle = \langle v_1, \ldots, v_n \rangle$. ∎

In view of the proof of Theorem 1.9, it is clear that if M is a vector space over a field R, then the elements $u_1, \ldots, u_n \in M$ are linearly independent over R iff the submodule

$$W = \langle u_1, \ldots, u_n \rangle$$

of M has rank n.

The rank of a finite dimensional vector space M is usually called the *dimension of M* and is denoted by

$$\dim M.$$

Another point of departure in the general theory of modules from the theory of vector spaces is in the structure of submodules. As we saw in Example 4(e), $\{(2,6), (0,14)\}$ is a basis of a proper submodule of Z^2. (It is proper since every element in it has an even first component.) Thus a proper submodule of a free module of rank n can also be of rank n. This does not happen for vector spaces. A submodule of a vector space is called a *subspace.*

Theorem 1.10 *Let M be a finite dimensional vector space over a field R,* dim $M = n$. *Let W be a subspace. Then W is finite dimensional and*

$$\dim W \leq n. \tag{65}$$

Moreover, dim $W = n$ *iff* $W = M$.

Proof: The fact that W is finite dimensional follows from Theorem 1.5(ii); indeed, the inequality (65) is a consequence of this result. Let $\{\varepsilon_1, \ldots, \varepsilon_n\}$ be a basis of M. Then by Theorem 1.5 again we can construct a basis of W of the form

$$\begin{aligned} w_1 &= a_{11}\varepsilon_1 + \cdots + a_{1n}\varepsilon_n \\ w_2 &= \qquad a_{22}\varepsilon_2 + \cdots + a_{2n}\varepsilon_n \\ &\vdots \\ w_r &= \qquad\qquad a_{rr}\varepsilon_r + \cdots + a_{rn}\varepsilon_n, \end{aligned}$$

$\prod_{i=1}^{r} a_{ii} \neq 0$. (The ε_i may have to be reordered in the above expressions for the w_i, but there is no loss in generality in assuming this.) Clearly, if $P \in$ GL(r,R), then $\langle Pw \rangle = \langle w \rangle$, and since R is a field we can find a $P \in$ GL(r,R) such that

$$PA = [I_r \mid B]$$

where B is $r \times (n - r)$. Now if dim $W = n$, i.e., if $r = n$, then $PA = I_n$ and we have

$$W = \langle w \rangle = \langle Pw \rangle = \langle PA\varepsilon \rangle = \langle \varepsilon \rangle = M.$$

Of course, if $W = M$, then dim $W = n$. ∎

Actually the argument in the proof of Theorem 1.10 can be modified slightly to prove the following.

Corollary 4 (*Steinitz Exchange Theorem*) *If W is a subspace of M,* $\{v_1, \ldots, v_r\}$ *is a basis of W, and* $\{\varepsilon_1, \ldots, \varepsilon_n\}$ *is a basis of M, then there exists some* *choice of* $n - r$ *of the* ε_i, *say* $\varepsilon_{r+1}, \ldots, \varepsilon_n$, *such that* $\{v_1, \ldots, v_r, \varepsilon_{r+1}, \ldots, \varepsilon_n\}$ *is a basis of* M.

Proof: By Theorem 1.5, construct a basis of W of the form

$$\begin{aligned} w_1 &= a_{11}\varepsilon_1 + \cdots + a_{1n}\varepsilon_n \\ w_2 &= \quad a_{22}\varepsilon_2 + \cdots + a_{2n}\varepsilon_n \\ &\vdots \\ w_r &= \qquad a_{rr}\varepsilon_r + \cdots + a_{rn}\varepsilon_n, \end{aligned}$$

$\prod_{i=1}^{r} a_{ii} \neq 0$, reordering the ε_i if necessary. Choose $P \in \mathrm{GL}(r,R)$ such that

$$PA = [I_r \mid B]$$

where B is $r \times (n - r)$. Then if $u = Pw = PA\varepsilon = [I_r \mid B]\varepsilon$, we have

$$u_i = \varepsilon_i + v_i, \qquad i = 1, \ldots, r, \tag{66}$$

where $v_i \in \langle \varepsilon_{r+1}, \ldots, \varepsilon_n \rangle$. We claim that the elements $u_1, \ldots, u_r, \varepsilon_{r+1}, \ldots, \varepsilon_n$ are l.i.. For suppose

$$\sum_{i=1}^{r} c_i u_i + \sum_{i=r+1}^{n} d_i \varepsilon_i = 0. \tag{67}$$

Then from (66) and (67) it follows that

$$\sum_{i=1}^{r} c_i \varepsilon_i + \sum_{i=r+1}^{n} r_i \varepsilon_i = 0, \qquad r_i \in R, \quad i = r + 1, \ldots, n;$$

since $\{\varepsilon_1, \ldots, \varepsilon_n\}$ is a basis of M, we must have $c_1 = \cdots = c_r = 0$, and then (67) implies that $d_{r+1} = \cdots = d_n = 0$. Now $\{u_1, \ldots, u_r\}$ is a basis of W (see Exercise 5), so there exists a matrix $Q \in \mathrm{GL}(r,R)$ such that $v = Qu$ [see the proof of Theorem 1.5(i)]. But then $v_1, \ldots, v_r, \varepsilon_{r+1}, \ldots, \varepsilon_n$ are l.i. because

$$\begin{bmatrix} v_1 \\ \vdots \\ v_r \\ \varepsilon_{r+1} \\ \vdots \\ \varepsilon_n \end{bmatrix} = (Q + I_{n-r}) \begin{bmatrix} u_1 \\ \vdots \\ u_r \\ \varepsilon_{r+1} \\ \vdots \\ \varepsilon_n \end{bmatrix}$$

and $Q + I_{n-r} \in \mathrm{GL}(n,R)$ (see Exercise 17). Thus $\dim \langle v_1, \ldots, v_r, \varepsilon_{r+1}, \ldots, \varepsilon_n \rangle = n$. It follows from Theorem 1.10 that $M = \langle v_1, \ldots, v_r, \varepsilon_{r+1}, \ldots, \varepsilon_n \rangle$. ∎

We note here that when R is a field, matrices in $\mathrm{GL}(m,R)$ are called *non-singular* or *regular*. If $\det A = 0$, then A is called a *singular* matrix.

If R is a field and $A \in M_{m,n}(R)$, then two integers can be associated with A:

$$\dim \langle A_{(1)}, \ldots, A_{(m)} \rangle \tag{68}$$

and $$\dim\langle A^{(1)}, \ldots, A^{(n)}\rangle. \tag{69}$$

The subspace $\langle A_{(1)}, \ldots, A_{(m)}\rangle$ of R^n is called the *row space* of A, denoted by $\mathscr{R}(A)$. The subspace $\langle A^{(1)}, \ldots, A^{(n)}\rangle$ of R^m is called the *column space* of A, denoted by $\mathscr{C}(A)$. The integers (68) and (69) are called the *row* and *column ranks* of A, respectively. In general, if R is a P.I.D., then $\mathscr{R}(A) = \langle A_{(1)}, \ldots, A_{(m)}\rangle$ and $\mathscr{C}(A) = \langle A^{(1)}, \ldots, A^{(m)}\rangle$, regarded as modules over R.

Theorem 1.11 *If R is a field and $A \in M_{m,n}(R)$, then both the row and column ranks of A are equal to $\rho(A)$.*

Proof: A moment's reflection on matrix multiplication shows that if $P = [p_{ij}] \in M_m(R)$ and $B = PA$, then

$$\mathscr{R}(B) \subset \mathscr{R}(A). \tag{70}$$

In fact,

$$B_{(i)} = \sum_{k=1}^{m} p_{ik}A_{(k)}, \qquad i = 1, \ldots, m. \tag{71}$$

We also note that

$$B^{(j)} = (PA)^{(j)} = PA^{(j)}, \qquad j = 1, \ldots, n. \tag{72}$$

Although it is not relevant to the present proof, we note that if $Q = [q_{ij}] \in M_n(R)$ and $D = AQ$, then

$$\mathscr{C}(D) \subset \mathscr{C}(A) \tag{73}$$

and

$$\begin{aligned} D^{(j)} &= (AQ)^{(j)} \\ &= \sum_{k=1}^{n} q_{kj}A^{(k)}, \qquad j = 1, \ldots, n. \end{aligned} \tag{74}$$

Let $\rho(A) = r$ and by Corollary 1 choose $P \in \mathrm{GL}(m,R)$ such that

$$PA = H,$$

$H_{(i)} = 0$, $i = r+1, \ldots, m$, and $H^{(n_t)}$ is the m-tuple

$$e_t = (\delta_{t1}, \ldots, \delta_{tm}), \qquad t = 1, \ldots, r. \tag{75}$$

Obviously the r rows of H are l.i. in R^n so that

$$\dim \mathscr{R}(H) = r.$$

But from (70),

$$\begin{aligned} \mathscr{R}(H) &= \mathscr{R}(PA) \subset \mathscr{R}(A) \\ &= \mathscr{R}(I_m A) = \mathscr{R}(P^{-1}(PA)) \\ &\subset \mathscr{R}(PA) = \mathscr{R}(H). \end{aligned}$$

Thus

$$\dim \mathscr{R}(A) = r.$$

It is obvious that $\dim \mathscr{C}(H) = r$ since the vectors (75) form a basis of $\mathscr{C}(H)$. Consider columns $n_1, \ldots, n_r$ of A and from (72) observe that

$$PA^{(n_k)} = e_k, \qquad k = 1, \ldots, r.$$

Thus $PA^{(n_1)}, \ldots, PA^{(n_r)}$ are l.i. But if $c_t \in R$, $t = 1, \ldots, r$, and

$$\sum_{t=1}^{r} c_t A^{(n_t)} = 0,$$

then

$$0 = P\sum_{t=1}^{r} c_t A^{(n_t)} = \sum_{t=1}^{r} c_t PA^{(n_t)}$$
$$= \sum_{t=1}^{r} c_t e_t$$

and hence $c_1 = \cdots = c_r = 0$. In other words, columns $A^{(n_1)}, \ldots, A^{(n_r)}$ are l.i. in R^m. On the other hand, if $A^{(k)}$ is any column of A, then

$$PA^{(k)} = H^{(k)} = \sum_{t=1}^{r} h_{tk}e_t = \sum_{t=1}^{r} h_{tk}H^{(n_t)}$$
$$= \sum_{t=1}^{r} h_{tk}PA^{(n_t)} \qquad \text{[from (72)]}$$
$$= P\sum_{t=1}^{r} h_{tk}A^{(n_t)}.$$

Since P has an inverse, we conclude that

$$A^{(k)} = \sum_{t=1}^{r} h_{tk}A^{(n_t)}, \qquad k = 1, \ldots, n.$$

It follows that $\{A^{(n_1)}, \ldots, A^{(n_r)}\}$ is a basis of $\mathscr{C}(A)$, and so

$$\dim \mathscr{C}(A) = r. \quad \blacksquare$$

Observe that in the proof of Theorem 1.11 we obtained the following rather precise information.

Corollary 5 *If R is a field, $A \in M_{m,n}(R)$, $\rho(A) = r$, and $H = [h_{ij}]$ is the Hermite normal form of A with indices $n_1, \ldots, n_r$, then $A^{(n_1)}, \ldots, A^{(n_r)}$ comprise a basis of $\mathscr{C}(A)$ and in fact*

$$A^{(k)} = \sum_{t=1}^{r} h_{tk}A^{(n_t)}, \qquad k = 1, \ldots, n.$$

The theory of linear equations over a field is an easy consequence of the Hermite normal form theory. Let $A \in M_{m,n}(R)$, R a field, and let $b \in M_{m,1}(R)$ (or R^m). Then a system of linear equations is an equation of the form

$$Ax = b; \tag{76}$$

the problem is to determine all $x \in M_{n,1}(R)$ for which (76) holds. If $b = 0$, then (76) is called a *homogeneous system*. The totality of solutions to the homogeneous system is called the *null space* of A. Clearly the null space is a subspace of R^n; its dimension is called the *nullity of A*, written $\eta(A)$. The $m \times (n + 1)$ matrix obtained by appending the m-tuple b to A as an $(n + 1)$st column is called the *augmented matrix for the system* (76). It is denoted by $[A:b]$.

Theorem 1.12

(i) *The system* (76) *has a solution iff* $\rho([A:b]) = \rho(A)$.

(ii) $\eta(A) = n - \rho(A)$.

(iii) *The solutions to* (76) *are of the form* $x^0 + w$, *where* x^0 *is a particular solution to* (76) *and* w *is in the null space of* A. *Moreover, any solution to* (76) *is constructible from the Hermite normal form of* A.

Proof: Let $P \in \mathrm{GL}(m,R)$ be such that $PA = H$ is in Hermite normal form (see Corollary 1), and let $n_1, \ldots, n_r$ be the indices of A, $r = \rho(A)$. Then (76) clearly has the same set of solutions as the system

$$Hx = c, \tag{77}$$

$c = Pb$. Since $Hx = \sum_{t=1}^{n} x_t H^{(t)}$, where $x = [x_1, \ldots, x_n]$, it follows that (77) has a solution iff $c \in \mathscr{C}(H)$. But $c \in \mathscr{C}(H)$ iff $Pb \in \mathscr{C}(PA)$ iff $\rho([PA:Pb]) = \rho(PA)$ iff $\rho(P[A:b]) = \rho(PA)$ iff $\rho([A:b]) = \rho(A)$. This proves (i). If $c \in \mathscr{C}(H)$, we can easily write down a particular solution to (76). For $H^{(n_t)} = e_t = (\delta_{t1}, \ldots, \delta_{tm})$, $t = 1, \ldots, r$, so that if $c = [c_1, \ldots, c_r, 0, \ldots, 0]$ [the last $m - r$ components are 0 because $c \in \mathscr{C}(H)$], then

$$x^0 = [0, \ldots, 0, \underset{\substack{\uparrow \\ n_1}}{c_1}, 0, \ldots, 0, \underset{\substack{\uparrow \\ n_2}}{c_2}, 0, \ldots, 0, \underset{\substack{\uparrow \\ n_r}}{c_r}, 0, \ldots, 0] \tag{78}$$

is a solution of (77) and hence of (76). Moreover, (78) is constructible by simply performing the same sequence of elementary row operations on $[A:b]$ as are performed on A to reduce it to H. Then of course the last column of $P[A:b]$ is just $Pb = c$.

Observe next that if $Ax^0 = b$ and $Ax^1 = b$, then obviously $A(x^1 - x^0) = 0$ so that $x^1 = x^0 + w$ for some w in the null space of A. In other words, any solution of (76) can be obtained by adding to a particular solution x^0 of (76) a solution w of the homogeneous system $Ax = 0$. Thus the problem is to construct a basis for the null space of A. But $Ax = 0$ iff $Hx = 0$, and this latter equation becomes

$$x_{n_i} + \sum_t{}' h_{it} x_t = 0, \qquad i = 1, \ldots, r,$$

where $\sum_t{}'$ indicates that the summation is taken over all $t \neq n_1, \ldots, n_r$. Thus $Hx = 0$ iff

$$\begin{aligned} x &= (x_1, \ldots, \underset{\substack{\uparrow \\ n_1}}{\sum_t{}'(-h_{1t})x_t}, \ldots, \underset{\substack{\uparrow \\ n_r}}{\sum_t{}'(-h_{rt})x_t}, \ldots, x_n) \\ &= \sum_t{}' x_t u_t \end{aligned}$$

where
$$u_t = (0, \ldots, \overset{t}{\overset{\downarrow}{0,1}},0, \ldots, \overset{n_1}{\overset{\downarrow}{-h_{1t}}},0, \ldots, 0, \overset{n_r}{\overset{\downarrow}{-h_{rt}}},0, \ldots, 0), \quad (79)$$
i.e., the n-tuple (79) has a 1 in position t and $-h_{1t}, \ldots, -h_{rt}$ in positions $n_1, \ldots, n_r$, respectively, 0 elsewhere, and t runs over the $n - r$ integers complementary to $n_1, \ldots, n_r$ in $1, \ldots, n$. The n-tuples u_t are obviously l.i. and hence form a basis of the null space of A. ∎

Example 5 Let

$$A = \begin{bmatrix} 0 & 0 & 0 & 0 & 1 & 1 & 1 \\ 0 & 2 & 6 & 2 & 0 & 0 & 4 \\ 0 & 1 & 3 & 1 & 1 & 0 & 1 \\ 0 & 1 & 3 & 1 & 2 & 1 & 2 \end{bmatrix},$$

$$b = [-4,2,2,-2].$$

The sequence of left multiplications by elementary matrices necessary to reduce A to Hermite normal form is (in the order in which they are performed)

$$E_{13},\ E_{21}(-2),\ E_{41}(-1),\ E_2(-\tfrac{1}{2}),\ E_{32}(-1),\ E_{42}(-1),\ E_{43}(-1),\ E_{12}(-1).$$

When performed on the augmented matrix $[A : b]$, the result is

$$[H : c] = \begin{bmatrix} 0 & 1 & 3 & 1 & 0 & 0 & 2 & 1 \\ 0 & 0 & 0 & 0 & 1 & 0 & -1 & 1 \\ 0 & 0 & 0 & 0 & 0 & 1 & 2 & -5 \\ 0 & 0 & 0 & 0 & 0 & 0 & 0 & 0 \end{bmatrix}.$$

The indices of A are $n_1 = 2$, $n_2 = 5$, $n_3 = 6$. The n-tuples (79) are

$$u_1 = (1,0,0,0,0,0,0)$$
$$u_3 = (0,-3,1,0,0,0,0)$$
$$u_4 = (0,-1,0,1,0,0,0)$$
$$u_7 = (0,-2,0,0,1,-2,1).$$

The particular solution is

$$x^0 = (0,1,0,0,1,-5,0).$$

Exercises 4.1

1. (a) Show that in $M_{m,n}(R)$ left equivalence is an equivalence relation.
   (b) If *right equivalence* in $M_{m,n}(R)$ is defined by $A \overset{R}{=} B$ iff $A = BU$ for some $U \in \mathrm{GL}(n,R)$, show that right equivalence is an equivalence relation in $M_{m,n}(R)$.
   (c) If *equivalence* or *two-sided equivalence* in $M_{m,n}(R)$ is defined by $A \overset{E}{=} B$ iff

$A = VBU$ for some $V \in GL(m,R)$, $U \in GL(n,R)$, show that equivalence is an equivalence relation in $M_{m,n}(R)$.

2. (a) Show that the three types of elementary matrices E_{ij}, $E_{ij}(r)$, and $E_i(u)$ are in $GL(m,R)$, with

$$\begin{aligned} E_{ij}^{-1} &= E_{ij}, \\ E_{ij}(r)^{-1} &= E_{ij}(-r), \\ E_i(u)^{-1} &= E_i(u^{-1}), \qquad u \text{ a unit in } R. \end{aligned}$$

(b) Show that

$$\begin{aligned} \det E_{ij} &= -1; \\ \det E_{ij}(r) &= 1; \\ \det E_i(u) &= u. \end{aligned}$$

3. Show that an elementary row operation of a given type can be accomplished by left multiplication by the corresponding type of elementary matrix.

4. (a) Show that if $A \overset{R}{=} B$, then $\rho(A) = \rho(B)$.
(b) Show that if $A \overset{E}{=} B$, then $\rho(A) = \rho(B)$.

5. Let $\{s_1, \ldots, s_n\}$ be a basis of a free R-module M, and let $A \in M_n(R)$ be unimodular. Show that if

$$u_i = \sum_{j=1}^{n} a_{ij} s_j, \qquad i = 1, \ldots, n,$$

then $\{u_1, \ldots, u_n\}$ is a basis of M. *Hint*: Since A is unimodular, there exists $B \in M_n(R)$ such that $AB = BA = I_n$, $B = A^{-1} = (1/\det A)\operatorname{adj} A$. It is easy to check that

$$\sum_{k=1}^{n} b_{jk} u_k = s_j, \qquad j = 1, \ldots, n,$$

so that any element of M is of the form $\sum_{i=1}^{n} r_i u_i$. Note that if $\sum_{i=1}^{n} r_i u_i = 0$, then $\sum_{i=1}^{n} a_{ij} r_i = 0$, $j = 1, \ldots, n$, or $rA = 0$, $r = [r_1, \ldots, r_n]$. But $AB = I_n$, and hence $r = rAB = 0$.

6. Prove that $\begin{bmatrix} 1 & 1 \\ 0 & 1 \end{bmatrix}^k = \begin{bmatrix} 1 & k \\ 0 & 1 \end{bmatrix}$ for any $k \in Z$.

7. Show that if $A \in M_n(Z)$ and $\det(A) = \pm 1$, then A is a product of the following matrices:
(i) $U_1 = A(\sigma)$, where $\sigma = (1\ 2\ 3 \cdots n)$ and $A(\sigma)$ is the incidence matrix for σ (see Exercise 1, Section 1.3).
(ii) $U_2 = A(\tau)$, where $\tau = (1\ 2)$.
(iii) $U_3 = \begin{bmatrix} 1 & 1 \\ 0 & 1 \end{bmatrix} \dotplus I_{n-2}$.
(iv) $U_4 = \begin{bmatrix} -1 & 0 \\ 0 & 1 \end{bmatrix} \dotplus I_{n-2}$.

Hint: Note that $\sigma(1\ 2)\sigma^{-1} = (2\ 3)$, $\sigma(2\ 3)\sigma^{-1} = (3\ 4)$, . . . and $\sigma^{-1} = \sigma^{n-1}$ so that for any $\varphi \in S_n$, $A(\varphi)$ is a product of U_1 and U_2 (see Exercise 13, Section 1.3). Thus any permutation of the rows of a matrix is achievable by left multiplication by a product of U_1 and U_2; in particular any two rows can be brought into row positions 1 and 2 in this way. Since $\begin{bmatrix}-1 & 0\\ 0 & 1\end{bmatrix}\begin{bmatrix}1 & 1\\ 0 & 1\end{bmatrix}\begin{bmatrix}-1 & 0\\ 0 & 1\end{bmatrix}=\begin{bmatrix}1 & -1\\ 0 & 1\end{bmatrix}$, we use Exercise 6 to conclude that any matrix

$$\begin{bmatrix}1 & k\\ 0 & 1\end{bmatrix} \dotplus I_{n-2}, \qquad k \in Z$$

is a product of the matrices U_3 and U_4. Thus, in view of the preceding remark, any elementary type II operation over Z is achievable by moving the two pertinent rows to the first and second row positions, performing the required type II operation, and then moving them back. Also it is clear that the matrix $E_k(-1)$, $k = 1, \ldots, n$, is a product of U_4 and appropriate permutation matrices. Thus any elementary matrix in GL(n,Z) is a product of $U_1, \ldots, U_4$, and the result follows from Corollary 3.

8. Express $\begin{bmatrix}3 & 2\\ 7 & 5\end{bmatrix}$ as a product of elementary matrices in GL$(2,Z)$. *Hint*: See Example 3(a).

9. In $M_2(Z)$ find a g.c.r.d. of $A_1 = \begin{bmatrix}1 & 0\\ 2 & 1\end{bmatrix}$ and $A_2 = \begin{bmatrix}2 & -1\\ 2 & 2\end{bmatrix}$. *Hint*: Follow the method outlined in Example 3(b).

10. Find a l.c.l.m. of the matrices A_1 and A_2 in Exercise 9 [see formula (44) in Example 3(c)].

11. Let $\{w_1, \ldots, w_m\}$ and $\{u_1, \ldots, u_m\}$ be two bases of a free R-module W over a P.I.D. R. Show that there is a unimodular $Q \in M_m(R)$ such that $w = Qu$, i.e., $w_i = \sum_{j=1}^m q_{ij}u_j$, $i = 1, \ldots, m$. *Hint*: Let Q be defined by $w_i = \sum_{j=1}^m q_{ij}u_j$, $i = 1, \ldots, m$. Similarly let $u = Pw$. Then $w = Qu = QPw$; so $QP = I_m$ (why?). Also $u = Pw = PQu$ and $PQ = I_m$. Thus Q is unimodular.

12. Suppose $D \in M_{m,n}(R)$, R a P.I.D., $\rho(D) = m$, and $XD = YD$ for $X, Y \in M_m(R)$. Show that $X = Y$. *Hint*: D must contain a submatrix $D[1, \ldots, m|\omega]$ ($\omega \in Q_{m,n}$) of rank m. But by matrix multiplication, $XD[1, \ldots, m|\omega] = YD[1, \ldots, m|\omega]$, and $D[1, \ldots, m|\omega]$ has an inverse over the quotient field of R.

13. Let $A \in M_{m,n}(R)$, R a P.I.D., $\rho(A) = k$. Show that $\rho\left(\begin{bmatrix}A\\ A\end{bmatrix}\right) = k$. *Hint*: Let $P = \begin{bmatrix}I_m & 0\\ -I_m & I_m\end{bmatrix}$. Then $P\begin{bmatrix}A\\ A\end{bmatrix} = \begin{bmatrix}A\\ 0\end{bmatrix}$.

14. Verify that the matrix H in Example 4(c) is the Hermite normal form for the indicated matrix A. (The choices of a complete system of nonassociates and a complete system of residues modulo the first nonzero element in each row are indicated in H.)

15. Let M be a finite dimensional vector space over a field R, and let W_1 and W_2 be subspaces. Prove that

$$\dim(W_1 + W_2) = \dim W_1 + \dim W_2 - \dim W_1 \cap W_2.$$

Hint: Obtain a basis for $W_1 \cap W_2$, say $v_1, \ldots, v_r$, and augment it to bases

$v_1, \ldots, v_r, w_{r+1}, \ldots, w_p$ of W_1 and $v_1, \ldots, v_r, u_{r+1}, \ldots, u_q$ of W_2, respectively, by Corollary 4. Show that $v_1, \ldots, v_r, w_{r+1}, \ldots, w_p\, u_{r+1}, \ldots, u_q$ is a basis of $W_1 + W_2$.

16. Define the *elementary column operations* on a matrix in $M_{m,n}(R)$ as follows:

I. Interchange column i and column j.

II. Add to column i r times column j, $i \neq j$.

III. Multiply column j by a unit u in R.

(a) Show that I can be accomplished by right multiplication by E_{ij}.

(b) Show that II can be accomplished by right multiplication by $E_{ji}(r)$.

(c) Show that III can be accomplished by right multiplication by $E_i(u)$.

(d) Show that $E_{ij}{}^{\mathsf{T}} = E_{ij}$, $E_{ij}(r)^{\mathsf{T}} = E_{ji}(r)$; $E_i(u)^{\mathsf{T}} = E_i(u)$.

17. Let $u_1, \ldots, u_m$ be free in the R-module M and let $\det Q \neq 0$, $Q \in M_m(R)$. Show that if $v = Qu$, then $v_1, \ldots, v_m$ are free. *Hint*: Suppose $\sum_{i=1}^{m} \alpha_i v_i = 0$, $\alpha_i \in R$, $i = 1, \ldots, m$, i.e., $\alpha v = 0$ where $\alpha = [\alpha_1, \ldots, \alpha_m]$ and $v = [v_1, \ldots, v_m]^{\mathsf{T}}$. Then $\alpha Q u = 0$ or $\beta u = 0$, where $\beta = \alpha Q = 0$ since $u_1, \ldots, u_m$ are free. Hence $\alpha Q Q^{-1} = \alpha I_m = \alpha = 0$, i.e., $\alpha_1 = \cdots = \alpha_m = 0$. Of course, Q^{-1} is computed over the quotient field of R.

18. Solve the following system of equations over Q:

$$\begin{aligned} 2x_1 + x_2 + \ x_4 + 3x_5 &= 3 \\ x_1 + x_2 + x_3 + x_4 + x_5 &= 3 \\ 2x_1 + x_2 + x_3 + 2x_4 + x_5 &= 3 \\ 2x_1 + 2x_2 + x_3 + x_4 + 4x_5 &= 6. \end{aligned}$$

19. Express the matrix

$$\begin{bmatrix} 2 & 1 & 1 & 1 \\ 1 & 2 & 1 & 1 \\ 1 & 1 & 2 & 1 \\ 1 & 1 & 1 & 2 \end{bmatrix}$$

as a product of elementary matrices.

Glossary 4.1

4.2 The Smith Normal Form

As we noted in Exercise 1, Section 4.1, *equivalence* of two matrices in $M_{m,n}(R)$, written

$$A \overset{E}{=} B,$$

is an equivalence relation. The Smith normal form provides us with a S.D.R. in $M_{m,n}(R)$ in much the same way as the Hermite normal form produces a S.D.R. under left equivalence. The ring R continues to be a P.I.D. unless otherwise stated.

If $A \in M_{m,n}(R)$ and $1 \leq k \leq \min\{m,n\}$, we define

$$d_k(A) = \text{g.c.d.}\{\det A[\alpha|\beta] \mid \alpha \in Q_{k,m}, \beta \in Q_{k,n}\} \qquad \text{if } \rho(A) \geq k,$$

and
$$d_k(A) = 0 \qquad \text{if } \rho(A) < k.$$

The scalar $d_k(A)$ is called the kth *determinantal divisor of A*. By the Laplace expansion theorem any kth order subdeterminant of A is a sum of multiples (in R) of $(k-1)$st order subdeterminants. It follows that $d_k(A) \neq 0$ implies $d_j(A) \neq 0$, $1 \leq j \leq k-1$. Moreover, if $r = \rho(A)$, then $d_r(A) \neq 0$. Also, if $k > r$, then every k^{th} order subdeterminant of A is 0. We also conclude from the Laplace expansion theorem that if $d_k(A) \neq 0$, then

$$d_k(A) \mid d_{k+1}(A). \tag{1}$$

Assume in what follows that a fixed complete system of nonassociates containing 1 has been stipulated in R so that all g.c.d.'s are chosen from this system.

Theorem 2.1 *If $A,B \in M_{m,n}(R)$ and $A \overset{E}{=} B$, then $d_k(A) = d_k(B), k = 1, \ldots, \min\{m,n\}$.*

Proof: Suppose $A = PBQ$, $P \in \mathrm{GL}(m,R)$, $Q \in \mathrm{GL}(n,R)$. Then if $\alpha \in Q_{k,m}$, $\beta \in Q_{k,n}$ the Cauchy-Binet theorem implies that

$$\det A[\alpha \mid \beta] = \sum_{\omega \in Q_{k,m}} \sum_{\gamma \in Q_{k,n}} \det P[\alpha \mid \omega] \det B[\omega \mid \gamma] \det Q[\gamma \mid \beta]. \tag{2}$$

Thus if $d_k(B) \neq 0$, it follows from (2) that $d_k(B) \mid \det A[\alpha \mid \beta]$, for all $\alpha \in Q_{k,m}$, $\beta \in Q_{k,n}$. We conclude that $d_k(B) \mid d_k(A)$. Similarly, $d_k(A) \mid d_k(B)$; so $d_k(A)$ and $d_k(B)$ are associates and hence equal. Since $\rho(A) = \rho(B)$, $d_k(A) = 0$ iff $d_k(B) = 0$. ∎

The main result of this section can now be stated.

Theorem 2.2 (*Smith Normal Form*) *Let $A \in M_{m,n}(R)$, $\rho(A) = r$. Then*

$$A \overset{E}{=} \left[\begin{array}{cccc|c} q_1 & & & 0 & \\ & q_2 & & & 0 \\ & & \ddots & & \\ 0 & & & q_r & \\ \hline & & 0 & & 0 \end{array}\right] \tag{3}$$

where $q_i \mid q_{i+1}$, $i = 1, \ldots, r-1$, and $q_i \neq 0$, $i = 1, \ldots, r$. The matrix on the right in (3) is called the Smith normal form of A, denoted by $S(A)$. The elements q_i are uniquely determined by A to within unit multiples.

Proof: We can assume $A \neq 0$. By type I elementary row and column operations (see Exercise 16, Section 4.1) a nonzero element in A can be brought into the (1,1) position, and then it can be replaced by the g.c.d. of the entries in the first column as in Theorem 1.2, Section 4.1. After that the elements in column 1 below the first row can be made 0 by type II elementary row operations. The matrix on hand at this point has the form

$$B = \begin{bmatrix} b_{11} & b_{12} & \cdots & b_{1n} \\ 0 & b_{22} & \cdots & b_{2n} \\ \cdots & \cdots & \cdots & \cdots \\ 0 & b_{m2} & \cdots & b_{mn} \end{bmatrix}.$$

If b_{11} divides all the remaining entries in row 1, then these can be made 0 by elementary type II column operations. If not, then let $r_1, \ldots, r_n$ be elements in R such that $\sum_{j=1}^{n} b_{1j}r_j = d_1 = \text{g.c.d.}(b_{11}, \ldots, b_{1n})$ and $\text{g.c.d.}(r_1, \ldots, r_n) = 1$. Then construct a unimodular $Q \in \text{GL}(n,R)$ with first column $[r_1, \ldots, r_n]^{\text{T}}$ so that d_1 is then the (1,1) entry of BQ. If d_1 divides all the entries in the first row and column of BQ, these can all be made 0 by type II elementary row and column operations [except the (1,1) entry of course]. If not, then as before replace d_1 by the g.c.d. (call it d_2) of the elements in column 1 of BQ and make the rest of the elements in the first column equal to 0. If d_2 divides the rest of the entries now on hand in row 1, make them all 0 while leaving column 1 unaltered. Clearly this process can be continued until we obtain an equivalent matrix whose (1,1) entry divides all the other entries in row 1 and column 1. For at each stage we replace the (1,1) entry by an element with fewer prime factors; e.g., d_2 is the g.c.d. of d_1 and other elements not divisible by d_1, and since R is a U.F.D., d_2 has fewer prime factors than d_1. Thus the process must terminate with a matrix equivalent to A whose (1,1) entry divides every other entry in row 1 and column 1 (it may have "zero" prime factors, i.e., it may be a unit). Thus A is equivalent to a matrix of the form

$$C = \begin{bmatrix} c_{11} & 0 & \cdots & 0 \\ 0 & c_{22} & \cdots & c_{2n} \\ \cdots & \cdots & \cdots & \cdots \\ 0 & c_{m2} & \cdots & c_{mn} \end{bmatrix}. \tag{4}$$

Suppose that c_{ij} is an entry in $C(1\,|\,1)$ not divisible by c_{11}. Add row i to row 1 in C to put c_{ij} in the $(1,j)$ position—this leaves c_{11} alone. Then repeat the whole procedure described before to obtain a matrix equivalent to C whose (1,1) entry is a proper divisor of c_{11} and which has the same form as the matrix (4). By repeated applications of the process, we finally obtain a matrix D having the same form as (4) in which the (1,1) entry, call it q_1, divides all the other entries. For at each stage we replace the (1,1) entry by an element with fewer prime factors. It is important to observe that if $P = 1 + P_1$ and $Q = 1 + Q_1$, then

$$PDQ = \begin{bmatrix} q_1 & 0 & \cdots & 0 \\ 0 & & & \\ \vdots & & P_1 D(1 \mid 1) Q_1 & \\ 0 & & & \end{bmatrix}, \tag{5}$$

and the fact that q_1 divides all the entries in $D(1 \mid 1)$ remains true of any matrix equivalent to $D(1 \mid 1)$. Now the entire process is repeated with $D(1 \mid 1)$, and as indicated in (5) this can be done by left and right multiplications by unimodular matrices in $M_m(R)$ and $M_n(R)$, respectively, in such a way that row 1 and column 1 are unaltered. Thus A is equivalent to a matrix of the form

$$S = \left[\begin{array}{ccc:c} q_1 & & \bigcirc & \\ & \ddots & & \bigcirc \\ \bigcirc & & q_p & \\ \hdashline & \bigcirc & & \bigcirc \end{array}\right] \tag{6}$$

where $q_i \mid q_{i+1}$, $i = 1, \ldots, p - 1$. However, since rank is invariant under equivalence, it follows that $p = r$.

The divisibility properties of $q_1, \ldots, q_r$ immediately imply that

$$d_k(S) = \varepsilon_k q_1 \cdots q_k, \qquad k \leq r,$$

and

$$d_k(S) = 0, \qquad k > r,$$

where ε_k is a unit in R. But $d_k(A) = d_k(S)$ so that

$$\varepsilon_k q_1 \cdots q_k = d_k(A), \qquad k = 1, \ldots, r.$$

Hence

$$\begin{aligned} q_k &= \frac{q_1 \cdots q_{k-1} q_k}{q_1 \cdots q_{k-1}} \\ &= u_k \frac{d_k(A)}{d_{k-1}(A)}, \qquad k = 1, \ldots, r \ (d_0(A) = 1), \end{aligned}$$

where u_k is a unit in R. By obvious type III elementary row operations we can replace q_k in (6) by $u_k^{-1} q_k$, $k = 1, \ldots, r$; so we can assume that $q_k = d_k(A)/d_{k-1}(A)$. ∎

The elements $q_1, \ldots, q_r$ in $S(A)$,

$$q_k(A) = \frac{d_k(A)}{d_{k-1}(A)}, \qquad k = 1, \ldots, r = \rho(A), \tag{7}$$

are called the *invariant factors* of A. We make the convention that $S(A)$ is chosen so that (7) holds, and hence $S(A)$ is uniquely determined by A since we have stipulated the system of nonassociates in R in advance.

Since R is a P.I.D., we can write

$$q_k = \varepsilon_k p_1^{e_{k1}} p_2^{e_{k2}} \cdots p_m^{e_{km}}, \qquad k = 1, \ldots, r, \tag{8}$$

where $p_1, \ldots, p_m$ are distinct primes in the complete system of nonassociates in R, e_{kj} are nonnegative integers, $j = 1, \ldots, m$, $k = 1, \ldots, r$, and ε_k are units, $k = 1, \ldots, r$. From $q_k \mid q_{k+1}$ it follows that

$$e_{kj} \leq e_{k+1\,j}, \qquad k = 1, \ldots, r-1,\ j = 1, \ldots, m.$$

A prime power $p_j^{e_{kj}}$ in which $e_{kj} > 0$ in (8) is called an *elementary divisor* of A. The *list of elementary divisors* of A is the totality of elementary divisors, each counted the number of times it appears among the factorizations (8).

Example 1 (a) If $R = Z$ and $q_1 = 2^3 \cdot 3$, $q_2 = 2^4 \cdot 3^3 \cdot 5^2 \cdot 7$, $q_3 = 2^4 \cdot 3^3 \cdot 5^5 \cdot 7^2$, then the list of elementary divisors is

$$2^3,\ 2^4,\ 2^4,\ 3,\ 3^3,\ 3^3,\ 5^2,\ 5^5,\ 7,\ 7^2.$$

(b) If 2^3, 2^4, 2^4, 3, 3^3, 3^3, 5^2, 5^5, 7, 7^2 is the complete list of elementary divisors of $A \in M_{5,7}(Z)$ and $\rho(A) = 4$, we can construct the Smith normal form $S(A)$. First, note that since $\rho(A) = 4$, it follows that there are four invariant factors of A. Moreover, q_4 must be the product of all the highest powers of the distinct primes that appear in the list of elementary divisors of A:

$$q_4 = 2^4 \cdot 3^3 \cdot 5^5 \cdot 7^2.$$

Then q_3 must be the product of all the highest powers of the distinct primes in the list after q_4 has been constructed:

$$q_3 = 2^4 \cdot 3^3 \cdot 5^2 \cdot 7.$$

Similarly,

$$q_2 = 2^3 \cdot 3.$$

Hence q_1 must be a unit: $q_1 = 1$, if the system of nonassociates in Z is the set of nonnegative integers. Thus

$$S(A) = \begin{bmatrix} q_1 & 0 & 0 & 0 & 0 & 0 & 0 \\ 0 & q_2 & 0 & 0 & 0 & 0 & 0 \\ 0 & 0 & q_3 & 0 & 0 & 0 & 0 \\ 0 & 0 & 0 & q_4 & 0 & 0 & 0 \\ 0 & 0 & 0 & 0 & 0 & 0 & 0 \end{bmatrix}.$$

We see from Example 1(b) that if the list of elementary divisors, the rank, and the dimensions of A are given, then $S(A)$ can be reconstructed. Also note that

$$d_1(A) = q_1$$

$$d_2(A) = q_1 q_2$$
$$\cdots\cdots\cdots\cdots\cdots$$
$$d_r(A) = q_1 \cdots q_r$$

so that the determinantal divisors can be reconstructed from the list of elementary divisors. We summarize.

Corollary 1 *The rank, dimensions, and list of elementary divisors (invariant factors, determinantal divisors) comprise a complete set of invariants with respect to equivalence. That is, two rectangular matrices are equivalent over R iff they have the same rank, dimensions, and list of elementary divisors (invariant factors, determinantal divisors).*

In analogy to Theorem 1.7, Section 4.1, we have

Theorem 2.3 *If (R,δ) is a Euclidean domain and $A \in M_{m,n}(R)$, then A can be reduced to Smith normal form $S(A)$ by elementary row and column operations.*

Proof: If we examine the proof of Theorem 2.2, we see that the reduction of A to $S(A)$ is accomplished by elementary row and column operations except at one stage: the replacement of the first element in a column (or row) of a k-rowed matrix by the g.c.d. of the elements $a_1, \ldots, a_k$ appearing in the column. In a Euclidean domain we can proceed as follows: Scan the nonzero elements among $a_1, \ldots, a_k$ for the one of least norm which lies in the first row (bring it there if necessary). Then from each of rows $2, \ldots, k$, subtract appropriate multiples of row 1. This results in a column of the form $a_1, r_2, \ldots, r_k$ in which $r_2, \ldots, r_k$ are all 0 or are of norm strictly less than $\delta(a_1)$. Repeat the process until the elements in rows $2, \ldots, k$ are all 0. This must ultimately happen since the maximum norm of any element in the column is decreased after each such step. The matrix then on hand has the following form:

$$\begin{bmatrix} b_1 & b_2 & b_3 & \cdots & b_p \\ 0 & * & & & * \\ \cdots & \cdots & \cdots & \cdots & \cdots \\ 0 & * & & & * \end{bmatrix}.$$

Now replace b_1 by the element of least norm in row 1, and perform elementary column operations to ultimately reduce to 0 all the elements in row 1 except the (1,1) element. Of course, the elements in column 1 may no longer be 0 but the (1,1) entry has no larger a norm than $\delta(b_1)$. The sweep through column 1 can then be repeated, etc. Finally, an element must appear in the (1,1) entry which must divide all the other elements in row 1 and column 1, and then these can be made 0. The process can then be continued as in the proof of Theorem 2.2. ▮

Corollary 2 *If R is a field,* $A \in M_{m,n}(R)$ *and* $\rho(A) = r$, *then A can be reduced to*

$$S(A) = \left[\begin{array}{c|c} I_r & 0 \\ \hline 0 & 0 \end{array}\right]$$

by elementary row and column operations.

Proof: Any field is a Euclidean domain, and $\{0,1\}$ is a complete system of nonassociates. Apply Theorem 2.3. ▮

The following result is very helpful in determining $S(A)$ in the event A has nonzero entries only in positions (t,t), $t = 1, \ldots, k$.

Theorem 2.4 *Suppose that* $D \in M_{m,n}(R)$ *has nonzero elements* $h_1, \ldots, h_k$ *in positions* $(1,1), \ldots, (k,k)$, *respectively, and* 0 *elsewhere. Then the list of elementary divisors of D is the list of all the prime power factors of any of the* $h_1, \ldots, h_k$, *each counted the number of times it occurs.*

Proof: Let $p_1, \ldots, p_m$ be all the distinct prime divisors of any of the h_i, $i = 1, \ldots, k$. Then write

$$h_t = p_1^{e_{t1}} p_2^{e_{t2}} \cdots p_m^{e_{tm}}, \qquad t = 1, \ldots, k$$

where the e_{tj} are nonnegative integers. Let $1 \leq s \leq m$ be fixed for the following discussion. We will show that if $e_{rs} > 0$, then $p_s^{e_{rs}}$ is an elementary divisor of D. We can perform elementary type I operations on D so that we can assume that

$$e_{1s} \leq e_{2s} \leq \cdots \leq e_{ks}, \tag{9}$$

i.e., we can assume that $h_1, \ldots, h_k$ are arranged in order of increasing powers of p_s. Then

$$h_t = p_s^{e_{ts}} a_t$$

where p_s and a_t are relatively prime, $t = 1, \ldots, k$. If $1 \leq r \leq k$ and $\omega \in Q_{r,k}$, then

$$\begin{aligned} \det D[\omega|\omega] &= h_{\omega(1)} \cdots h_{\omega(r)} \\ &= p_s^{e_{\omega(1)s}} a_{\omega(1)} \cdots p_s^{e_{\omega(r)s}} a_{\omega(r)} \\ &= p_s^{m(\omega,s)}\, a_\omega \end{aligned}$$

where $$m(\omega,s) = e_{\omega(1)s} + \cdots + e_{\omega(r)s}$$

and $$a_\omega = a_{\omega(1)} \cdots a_{\omega(r)}.$$

Now p_s and a_ω are relatively prime, and obviously the power of p_s occurring in $d_r(D) = \text{g.c.d.}\{\det D[\alpha|\beta] \mid \alpha,\beta \in Q_{r,k}\}$ is

$$\min_{\omega\in Q_{r,k}} m(\omega,s).$$

But from (9),

$$\min_{\omega\in Q_{r,k}} m(\omega,s) = e_{1s} + e_{2s} + \cdots + e_{rs}.$$

Thus the power of p_s occurring in $d_r(D)$ is

$$\sum_{i=1}^{r} e_{is}.$$

From (7), the invariant factors of D are given by

$$q_r(D) = \frac{d_r(D)}{d_{r-1}(D)},$$

and hence the power of p_s occurring in $q_r(D)$ is

$$\sum_{i=1}^{r} e_{is} - \sum_{i=1}^{r-1} e_{is} = e_{rs}.$$

Thus in the factorization of $q_r(D)$, p_s appears with power e_{rs}. We conclude that the list of elementary divisors of D just consists of all $p_s^{e_{rs}}$, $e_{rs} > 0$. ▮

Example 2 If $D = \text{diag}(3^2,7^3,24)$, we find $S(D)$ over Z. We recall that if $X = [x_{ij}]$ is a square matrix with nonzero elements only on the main diagonal, then $\text{diag}(x_{11}, \ldots, x_{nn})$ is used to denote X. From Theorem 2.4 we need only factor each of the main diagonal entries to obtain the list of elementary divisors of D:

$$2^3,\ 3,\ 3^2,\ 7^3.$$

The list of invariant factors is

$$q_3 = 2^3\cdot 3^2\cdot 7^3,$$
$$q_2 = 3,$$
$$q_1 = 1,$$

Thus
$$S(D) = \text{diag}\,(1,\ 3,\ 2^3\cdot 3^2\cdot 7^3).$$

Corollary 3. *Assume that $A \in M_{m,\,n}(R)$ has the following form:*

$$A = \begin{array}{c} \\ p\{ \\ r\{ \\ \\ \end{array}\left[\begin{array}{c|c|c} \overbrace{\quad}^{q} & \overbrace{\quad}^{s} & \\ B & 0 & 0 \\ \hline 0 & C & 0 \\ \hline 0 & 0 & 0 \end{array}\right].$$

Then the list of elementary divisors of A is the combined list of elementary divisors of B and C.

Proof: Let P_1, Q_1, P_2 and Q_2 be $p \times p$, $q \times q$, $r \times r$, and $s \times s$ unimodular matrices, respectively, so chosen that

$$P_1BQ_1 = S(B) \qquad \text{and} \qquad P_2CQ_2 = S(C).$$

Let $\quad P = P_1 \dotplus P_2 \dotplus I_{m-(p+r)} \qquad$ and $\qquad Q = Q_1 \dotplus Q_2 \dotplus I_{n-(q+s)}$.

Then by the Laplace expansion theorem, P and Q are both unimodular, and by block multiplication,

$$PAQ = \begin{array}{c} p\{ \\ r\{ \\ \\ \end{array} \left[\begin{array}{c|c|c} \overbrace{S(B)}^{q} & \overbrace{0}^{s} & 0 \\ \hline 0 & S(C) & 0 \\ \hline 0 & 0 & 0 \end{array}\right].$$

By elementary type I operations, PAQ can be brought to the form

$$\begin{bmatrix} q_1(B) & & & & & & & & \\ & \ddots & & & & & & & \\ & & q_{\rho_1}(B) & & & & & & \\ & & & q_1(C) & & & & & \\ & & & & \ddots & & & & \\ & & & & & q_{\rho_2}(C) & & & \\ & & & & & & 0 & & \\ & & & & & & & \ddots & \\ & & & & & & & & 0 \end{bmatrix}, \tag{10}$$

$\rho_1 = \rho(B)$, $\rho_2 = \rho(C)$. From Theorem 2.4 the list of elementary divisors of A is just the list of all prime power factors of any of the nonzero entries appearing in the matrix (10). But this is exactly the combined list of elementary divisors of B and C. ∎

Example 3 (a) Suppose $A \in M_{m,n}(R)$ and $b \in R^m$ and it is required to solve the system of linear equations

$$Ax = b, \tag{11}$$

for $x = (x_1, \ldots, x_n) \in R^n$. This problem is more difficult than the corresponding problem for a field that we considered in Theorem 1.12, Section 4.1. However, suppose P and Q are unimodular matrices such that $PAQ = S(A)$. Let $x = Qy$ in (11) so that

$$AQy = b,$$
$$PAQy = Pb,$$

or

$$S(A)y = c, \tag{12}$$

where $c = Pb$. Then (12) becomes

$$\begin{aligned} q_1(A)y_1 &= c_1 \\ &\vdots \\ q_r(A)y_r &= c_r \\ 0 &= c_{r+1} \\ &\vdots \\ 0 &= c_m \end{aligned} \tag{13}$$

where $r = \rho(A)$. Thus we can state the following: The system of equations (11) has a solution iff $q_t(A) | c_t$, $t = 1, \ldots, r$ and $c_{r+1} = \cdots = c_m = 0$. The totality of solutions x is then obtained by letting $y_{r+1}, \ldots, y_n$ be arbitrary elements of R in

$$x = Q[c_1/q_1(A), c_2/q_2(A), \ldots, c_r/q_r(A), y_{r+1}, \ldots, y_n]^{\mathrm{T}}.$$

(b) A *linear diophantine system* is a system such as (11) in which $R = Z$. We solve the diophantine system

$$\begin{aligned} 7x_1 + 3x_2 + 15x_3 &= 100 \\ x_1 + 5x_2 + 3x_3 &= 120. \end{aligned} \tag{14}$$

The system (14) has the form

$$Ax = b,$$

where

$$A = \begin{bmatrix} 7 & 3 & 15 \\ 1 & 5 & 3 \end{bmatrix}$$

and $b = (100, 120)$. A minor calculation (see Exercise 6) shows that

$$PAQ = S(A)$$

where

$$P = \begin{bmatrix} 0 & 1 \\ -1 & 7 \end{bmatrix},$$

$$Q = \begin{bmatrix} 1 & 10 & -33 \\ 0 & 1 & -3 \\ 0 & -5 & 16 \end{bmatrix},$$

and

$$S(A) = \begin{bmatrix} 1 & 0 & 0 \\ 0 & 2 & 0 \end{bmatrix}.$$

Also

$$c = Pb = (120, 740).$$

Thus

$$\frac{c_1}{q_1(A)} = \frac{120}{1} = 120,$$

$$\frac{c_2}{q_2(A)} = \frac{740}{2} = 370.$$

Then the totality of solutions must have the form

$$x = Q[120, 370, y_3]^{\mathrm{T}}$$
$$= [3820 - 33y_3, 370 - 3y_3, -1850 + 16y_3]^{\mathrm{T}}.$$

If we set $y_3 = 116 - d$, where d runs through Z, we have a neater expression for x:

$$x_1 = -8 + 33d,$$
$$x_2 = 22 + 3d,$$
$$x_3 = 6 - 16d.$$

Our immediate goal at this point is to use the Smith normal form as the main tool in proving the fundamental theorem describing the structure of a finitely generated R-module. We will illustrate some of the aspects of the proof in the following example.

Example 4 By using the Smith normal form, we will reprove Theorem 1.9 in Section 4.1 to the effect that any finite dimensional vector space M over a field R has a basis. By definition, M possesses a finite generating (or *spanning*) set $\{s_1, \ldots, s_k\}$. Any element $\alpha \in M$ has the form

$$\alpha = r_1s_1 + \cdots + r_ks_k, \qquad r_i \in R, i = 1, \ldots, k, \tag{15}$$

but in general there is nothing necessarily unique about the representation (15) because we do not know that the $s_1, \ldots, s_k$ are l.i.. Consider the subset $W \subset R^k$ consisting of all $(r_1, \ldots, r_k)$ for which

$$\sum_{j=1}^{k} r_js_j = 0.$$

It is an easy matter (see Exercise 7) to confirm that W is a subspace of R^k. Since R^k has the obvious basis $\{\varepsilon_j = (\delta_{j1}, \ldots, \delta_{jk}) \mid j = 1, \ldots, k\}$, we can conclude by Theorem 1.4, Section 4.1 that W has a finite spanning set, say

$$\{(a_{i1}, \ldots, a_{ik}) \mid i = 1, \ldots, q\}. \tag{16}$$

Let $A \in M_{q,k}(R)$ be the matrix whose i^{th} row is $(a_{i1}, \ldots, a_{ik})$, $i = 1, \ldots, q$. Observe that the row space of A satisfies

$$\mathscr{R}(A) = W,$$

and if $P \in \mathrm{GL}(q,R)$ then $\mathscr{R}(PA) = \mathscr{R}(A) = W$. If we define

$$\gamma = Q^{-1}s,$$

where $Q \in \mathrm{GL}(q,R)$, then as we have seen a number of times before, $\{\gamma_1, \ldots, \gamma_k\}$ is also a set of generators for M. Now suppose P and Q are chosen so that $PAQ = S(A)$, and set $r = \rho(A)$. If $c = [c_1, \ldots, c_r, 0, \ldots, 0] \in \mathscr{R}(PAQ) = \mathscr{R}(S(A))$ then

$$c\gamma = cQ^{-1}s = \left(\sum_{i=1}^{r} c_i\varepsilon_i\right)Q^{-1}s$$
$$= \left[\sum_{i=1}^{r} c_i(S(A))_{(i)}\right]Q^{-1}s \qquad \text{(see Corollary 2)}$$
$$= \left[\sum_{i=1}^{r} c_i(PAQ)_{(i)}\right]Q^{-1}s = \left[\sum_{i=1}^{r} c_i(PA)_{(i)}Q\right]Q^{-1}s$$

$$= \left[\sum_{i=1}^{r} c_i(PA)_{(i)}\right]s. \tag{17}$$

But $\mathscr{R}(PA) = \mathscr{R}(A)$, and so $\sum_{i=1}^{r} c_i(PA)_{(i)} \in W$; hence (17) is 0. Thus if $c \in \mathscr{R}(PAQ) = \mathscr{R}(S(A))$, then $c\gamma = 0$. Conversely, suppose $d = [d_1, \ldots, d_k]$ and $d\gamma = 0$. Then $dQ^{-1}s = 0$, so $dQ^{-1} \in W = \mathscr{R}(A)$, i.e.,

$$dQ^{-1} = \sum_{i=1}^{q} r_i A_{(i)},$$

$$d = \sum_{i=1}^{q} r_i A_{(i)} Q = \sum_{i=1}^{q} r_i (AQ)_{(i)}$$

$$\in \mathscr{R}(AQ).$$

But $\mathscr{R}(PAQ) = \mathscr{R}(AQ)$, and hence

$$d \in \mathscr{R}(PAQ) = \mathscr{R}(S(A)).$$

Thus $\mathscr{R}(S(A))$ consists precisely of those k-tuples $[c_1, \ldots, c_r, 0, \ldots, 0]$ for which

$$\sum_{j=1}^{r} c_j \gamma_j = 0.$$

We have found a generating set $\gamma_1, \ldots, \gamma_r, \gamma_{r+1}, \ldots, \gamma_k$ of M with the property that

$$c_1\gamma_1 + \cdots + c_r\gamma_r = 0 \tag{18}$$

for arbitrary choices of $c_1, \ldots, c_r$ in R. For

$$S(A) = \left[\begin{array}{c|c} I_r & 0 \\ \hline 0 & 0 \end{array}\right],$$

and hence any $c = [c_1, \ldots, c_r, 0, \ldots, 0]$ is in $\mathscr{R}(S(A))$. Moreover,

$$d_1\gamma_1 + \cdots + d_k\gamma_k = 0$$

iff $d_{r+1} = \cdots = d_k = 0$, i.e., $d \in \mathscr{R}(S(A))$. It follows from (18) that $\gamma_1 = \cdots = \gamma_r = 0$ (choose $c_i = 1$, $c_j = 0$, $j \neq i$, $i = 1, \ldots, r$). Also, if

$$d_{r+1}\gamma_{r+1} + \cdots + d_k\gamma_k = 0,$$

then $d_{r+1} = \cdots = d_k = 0$ and hence $\gamma_{r+1}, \ldots, \gamma_k$ are l.i..

A *torsion element* α in a module M is an element for which there exists $r \in R$, $r \neq 0$, such that

$$r\alpha = 0. \tag{19}$$

In other words, α is not free. The module M is *torsion free* if $\alpha = 0$ is the only torsion element. The totality of torsion elements in M is a submodule (see Exercise 8), and it is called the *torsion submodule* of M. If $\alpha \in M$ is a torsion element, then the totality of $r \in R$ for which (19) holds forms an ideal in R (see Exercise 9) called the *order ideal* of α and denoted by

$$O(\alpha). \tag{20}$$

We can extend (20) to any element α by simply defining $O(\alpha)$ to be $\{0\}$ if α is free.

This is sensible since the only $r \in R$ for which $r\alpha = 0$ is $r = 0$ is α is free, i.e., not a torsion element. Since R is a P.I.D. and since any two generators of an ideal are associates, if follows that

$$O(\alpha) = (q)$$

and q is uniquely determined if we insist that it lie in the stipulated system of nonassociates in R. We call q the *order of* α. For example, if $R = Z$ and the system of nonassociates is $N \cup \{0\}$, then the order of a torsion element α is the least positive integer q such that $q\alpha = 0$.

More generally, if N is any subset of M, then the set of all $r \in R$ such that $rN = 0$ is an ideal in R (see Exercise 10), called the *order ideal of* N and denoted by $O(N)$. The generator of $O(N)$ is called the *order of* N.

If $W_1, \ldots, W_p$ are submodules of M such that $M = \sum_{i=1}^{p} W_i$, and moreover every element $\alpha \in M$ has a unique representation as $\alpha = w_1 + \cdots + w_p$, $w_i \in W_i$, $i = 1, \ldots, p$, then M is called the *direct sum* of $W_1, \ldots, W_p$ and we write

$$M = \sum_{i=1}^{p}{}^{\cdot} W_i = W_1 \dotplus \cdots \dotplus W_p. \tag{21}$$

The fundamental theorem concerning finitely generated modules over a P.I.D. R is the following.

Theorem 2.5 *Let M be a (nonzero) finitely generated R-module. Then there exist $s + r$ nonzero elements in M, $\alpha_1, \alpha_2, \ldots, \alpha_s, \alpha_{s+1}, \ldots, \alpha_{s+r}$ such that*

(i)
$$M = \sum_{i=1}^{s+r}{}^{\cdot} \langle \alpha_i \rangle. \tag{22}$$

(ii) *α_i is a torsion element of order μ_i, $i = 1, \ldots, s$, and $\mu_1 \mid \mu_2 \mid \cdots \mid \mu_s$.*

(iii) *$\alpha_{s+1}, \ldots, \alpha_{s+r}$ are free.*

Proof: We remark that none of the μ_i can be 1 (i.e., a unit) since $1 \cdot \alpha_i = 0$ means $\alpha_i = 0$ and none of $\alpha_1, \ldots, \alpha_s$ is 0. The argument begins in the same way as in Example 4. Let $\{s_1, \ldots, s_k\}$ be a generating set for M, and let $W \subset R^k$ be the set of all k-tuples $(c_1, \ldots, c_k)$ for which

$$cs = \sum_{j=1}^{k} c_j s_j = 0.$$

Since R^k has a basis, it follows that W has a spanning set

$$\{(a_{i1}, \ldots, a_{ik}) \mid i = 1, \ldots, m\},$$

and we define $A = [a_{ij}] \in M_{m,k}(R)$. Clearly, $c \in \mathscr{R}(A) = W$ iff $cs = 0$. Choose P and Q unimodular such that $PAQ = S(A)$, and let $q_t = q_t(A)$, $t = 1, \ldots, \rho = \rho(A)$ be the invariant factors of A:

$$S(A) = \left[\begin{array}{ccc|c} q_1 & & & \\ & \ddots & & \bigcirc \\ & & q_\rho & \\ \hline & \bigcirc & & \bigcirc \end{array}\right]. \tag{23}$$

Define $\gamma = Q^{-1}s$, also a generating set for M. We prove first that:

$$d \in \mathscr{R}(S(A)) \qquad \text{iff } d\gamma = 0. \tag{24}$$

Suppose first that $d \in \mathscr{R}(S(A))$ so that $d = \sum_{t=1}^{\rho} \delta_t S(A)_{(t)}$:

$$\begin{aligned} d\gamma = dQ^{-1}s &= \left(\sum_{t=1}^{\rho} \delta_t S(A)_{(t)}\right) Q^{-1}s \\ &= \left(\sum_{t=1}^{\rho} \delta_t (PAQ)_{(t)}\right) Q^{-1}s = \left(\sum_{t=1}^{\rho} \delta_t (PA)_{(t)} Q\right) Q^{-1}s \\ &= \left(\sum_{t=1}^{\rho} \delta_t (PA)_{(t)}\right) QQ^{-1}s = \left(\sum_{t=1}^{\rho} \delta_t (PA)_{(t)}\right) s. \end{aligned} \tag{25}$$

Now $\sum_{t=1}^{\rho} \delta_t (PA)_{(t)} \in \mathscr{R}(PA) = \mathscr{R}(A)$, and hence (25) is 0. Conversely if $d\gamma = 0$, then $dQ^{-1}s = 0$; so $dQ^{-1} \in W = \mathscr{R}(A)$, i.e.,

$$\begin{aligned} dQ^{-1} &= \sum_{i=1}^{m} r_i A_{(i)}, \\ d &= \sum_{i=1}^{m} r_i A_{(i)} Q = \sum_{i=1}^{m} r_i (AQ)_{(i)} \\ &\in \mathscr{R}(AQ) = \mathscr{R}(PAQ). \end{aligned}$$

Hence

$$d \in \mathscr{R}(S(A)). \tag{26}$$

This establishes (24). We examine the meaning of (26). From (23) we see that (26) is equivalent to the following:

$$q_i \mid d_i, \qquad i = 1, \ldots, \rho, \tag{27}$$

$$d_{\rho+1} = \cdots = d_k = 0. \tag{28}$$

Thus $d\gamma = 0$ iff both (27) and (28) hold. If we take $d = q_i e_i$, $e_i = (\delta_{i1}, \ldots, \delta_{ik})$, $i = 1, \ldots, \rho$, then d satisfies (27) and (28), and hence

$$q_i \gamma_i = 0, \qquad i = 1, \ldots, \rho.$$

Moreover, if $m\gamma_i = 0$, then $(me_i)\gamma = 0$ so that $me_i \in \mathscr{R}(S(A))$, i.e., me_i satisfies (27) and (28). Thus $q_i \mid m$. In other words, $m\gamma_i = 0$ iff $q_i \mid m$. It follows that q_i is the order of γ_i, $i = 1, \ldots, \rho$. Suppose now that $q_1 = \cdots = q_\nu = 1$, and $q_{\nu+1}, \ldots, q_\rho$ are not 1. Then since

$$q_i \gamma_i = 0$$

it follows that $\gamma_1 = \cdots = \gamma_\nu = 0$ and thus M is generated by

$$\gamma_{\nu+1}, \ldots, \gamma_\rho, \gamma_{\rho+1}, \ldots, \gamma_k,$$

and γ_t has order $q_t \neq 1$, $t = \nu + 1, \ldots, \rho$. We assert that $\gamma_{\rho+1}, \ldots, \gamma_k$ are free. For suppose

$$d_{\rho+1}\gamma_{\rho+1} + \cdots + d_k \gamma_k = 0. \tag{29}$$

Then if we set $d = [0, \ldots, 0, d_{\rho+1}, \ldots, d_k]$ it follows from $d\gamma = 0$ that $d \in \mathscr{R}(S(A))$, i.e., that (28) holds. Suppose finally that

$$d_{\nu+1}\gamma_{\nu+1} + \cdots + d_p\gamma_p + d_{p+1}\gamma_{p+1} + \cdots + d_k\gamma_k$$
$$= \delta_{\nu+1}\gamma_{\nu+1} + \cdots + \delta_p\gamma_p + \delta_{p+1}\gamma_{p+1} + \cdots + \delta_k\gamma_k. \qquad (30)$$

Then if we set $d = [0, \ldots, 0, d_{\nu+1}, \ldots, d_k]$ and $\delta = [0, \ldots, 0, \delta_{\nu+1}, \ldots, \delta_k]$, it follows that $(d - \delta)\gamma = 0$. Hence from (27) and (28) again we have

$$d_{p+1} = \delta_{p+1}, \ldots, d_k = \delta_k$$

and
$$q_t | (d_t - \delta_t), \qquad t = \nu + 1, \ldots, p. \qquad (31)$$

But since $q_t\gamma_t = 0$, $t = \nu + 1, \ldots, p$, it follows from (31) that

$$(d_t - \delta_t)\gamma_t = 0,$$
$$d_t\gamma_t = \delta_t\gamma_t, \qquad t = \nu + 1, \ldots, p.$$

In other words, any element in M can be written as an element in

$$\langle\gamma_{\nu+1}\rangle + \cdots + \langle\gamma_p\rangle + \langle\gamma_{p+1}\rangle + \cdots + \langle\gamma_k\rangle,$$

and moreover this can be done in only one way. If we set $\alpha_j = \gamma_{\nu+j}$, $j = 1, \ldots, k - \nu$, $s = p - \nu$, $r = k - p$, $\mu_j = q_{\nu+j}$, $j = 1, \ldots, s$, then (i), (ii) and (iii) immediately follow. ▮

Corollary 4 *In the notation of Theorem* 2.5, *the torsion submodule of M is precisely*

$$T = \sum_{i=1}^{s}{}^{\cdot} \langle\alpha_i\rangle. \qquad (32)$$

Moreover, $F = \sum^{\cdot\, s+r}_{i=s+1}\langle\alpha_i\rangle$ *is a free submodule of M of rank r and*

$$M = T \dotplus F. \qquad (33)$$

Proof: Clearly, $T = \sum^{\cdot\, s}_{i=1}\langle\alpha_i\rangle$ consists of torsion elements only, for obviously $\mu_s T = 0$. On the other hand, if

$$\alpha = \sum_{i=1}^{s} r_i\alpha_i + \sum_{i=s+1}^{s+r} r_i\alpha_i$$

and $c\alpha = 0$, then since the sum (22) is direct, $cr_i\alpha_i = 0$, $i = 1, \ldots, s + r$. But then $cr_i \in O(\alpha_i)$, $i = 1, \ldots, s + r$, and this in turn implies that $cr_i = 0$, $i = s + 1, \ldots, s + r$, because $\alpha_{s+1}, \ldots, \alpha_{s+r}$ are free. Hence if $c \neq 0$, then $r_i = 0$, $i = s + 1, \ldots, s + r$, i.e., $\alpha \in \sum^{\cdot\, s}_{i=1}\langle\alpha_i\rangle$. This proves (32), and (33) is obvious. ▮

Corollary 5 *The module M in Theorem* 2.5 *is free iff it is torsion free.*

Proof: If M is torsion free, then $T = 0$ and $M = F$ which is a free module by Corollary 4. Conversely, a free module M cannot contain a nonzero torsion element (see Exercise 11). ▮

In what follows we assume that the system of nonassociates in Z is $N \cup \{0\}$.

Corollary 6 *If $R = Z$, then $\langle \alpha_i \rangle$ is isomorphic to Z_{μ_i}, $i = 1, \ldots, s$, and $\langle \alpha_i \rangle$ is isomorphic to Z, $i = s + 1, \ldots, s + r$, as abelian groups.*

Proof: For $i = 1, \ldots, s$ define $\varphi_i \colon Z_{\mu_i} \to \langle \alpha_i \rangle$ by

$$\varphi_i([k]) = k\alpha_i.$$

First observe that if $[k_1] = [k_2]$, then $\mu_i | (k_1 - k_2)$ and hence $(k_1 - k_2)\alpha_i = 0$, $k_1\alpha_i = k_2\alpha_i$. Thus φ_i is well-defined. Also,

$$\begin{aligned} \varphi_i([k] + [l]) &= \varphi_i([k + l]) = (k + l)\alpha_i = k\alpha_i + l\alpha_i \\ &= \varphi([k]) + \varphi([l]). \end{aligned}$$

Note that if $\varphi_i([k]) = \varphi_i([l])$, then $(k - l)\alpha_i = 0$ and hence $\mu_i | (k - l)$, i.e., $[k] = [l]$. Since φ_i is clearly surjective, it is an isomorphism.

For $i = s + 1, \ldots, s + r$ define $\varphi_i \colon Z \to \langle \alpha_i \rangle$ by

$$\varphi_i(k) = k\alpha_i.$$

It is obvious that φ_i is a surjective homomorphism, and since α_i is free, it is clearly injective. ∎

If G is an abelian group, we can regard it as a Z-module [see Example 2(b), Section 4.1]. We then have the so-called *fundamental theorem of finitely generated abelian groups*:

Corollary 7 *If G is a finitely generated abelian group (as a Z-module), then G is the direct sum (as submodules) of s cyclic subgroups $G_1, \ldots, G_s$ of finite order and r infinite cyclic subgroups $G_{s+1}, \ldots, G_{s+r}$. Moreover, $|G_i| \,|\, |G_{i+1}|$, $i = 1, \ldots, s - 1$.*

Proof: We simply apply the decomposition of Theorem 2.5 and take

$$G_i = \langle \alpha_i \rangle, \qquad i = 1, \ldots, s + r. \quad ∎$$

Corollary 8 *If G is a finite abelian group (as a Z-module), then*

$$G = \sum_{i=1}^{s}{}^{\boldsymbol{\cdot}}\, G_i, \tag{34}$$

where $G_1, \ldots, G_s$ are finite cyclic subgroups satisfying $|G_i| \,|\, |G_{i+1}|$, $i = 1, \ldots, s - 1$. Moreover $|G| = \prod_{i=1}^{s} |G_i|$, and G_i is isomorphic to $Z_{|G_i|}$, $i = 1, \ldots, s$.

Proof: Everything has been proved except $|G| = \prod_{i=1}^{s} |G_i|$. But we know from Corollary 6 that

$$G_i = \{0, \alpha_i, 2\alpha_i, \ldots, (\mu_i - 1)\alpha_i\}, \qquad \mu_i = |G_i|, \tag{35}$$

and since the sum (34) is direct, it follows that if

$$\sum_{i=1}^{s} k_i\alpha_i = \sum_{i=1}^{s} l_i\alpha_i,$$

$0 \le k_i \le \mu_i - 1$, $0 \le l_i \le \mu_i - 1$, $i = 1, \ldots, s$, then $k_i = l_i$, $i = 1, \ldots, s$. In other words, each element of G is represented exactly once by adding together one element from each of the G_i in (35). Thus there are a total of $\mu_1 \cdots \mu_s$ elements in G. ▮

An abelian group is said to be *indecomposable* if it is not the (internal) direct sum of two nonzero subgroups (see Section 2.1). As a prelude to a further refinement of Corollary 8 we prove the following theorem.

Theorem 2.6
(i) *Z is indecomposable.*
(ii) *Z_n is indecomposable if $n = p^m$, a prime power.*
(iii) *If $n = rs$ and r and s are realtively prime, then $Z_n = H_r \dot{+} H_s$, where H_r (H_s) is isomorphic to Z_r (Z_s).*

Proof: (i) Suppose $Z = G \dot{+} H$ and there are nonzero integers $g \in G$ and $h \in H$. Then

$$gh = \begin{cases} \overbrace{h + \cdots + h}^{g}, & \text{if } g > 0 \\ -|g|h, & \text{if } g < 0 \end{cases},$$

so in any event $gh \in H$ and similarly $gh \in G$. Then $0 = 0 + 0 = gh + (-gh)$ are two distinct representations of 0 and the sum $G + H$ is not direct.

(ii) Suppose $n = p^m$, $Z_n = G \dot{+} H$, and neither subgroup is 0. Now Z_n is cyclic, and hence (see Exercise 2) G and H are cyclic and are generated by $[p^a]$ and $[p^b]$, respectively, where a and b are both less than m. If $a \le b$, say, then $[p^{b-a}][p^a] = [p^b]$, so $[p^b] \in G$ and hence $H \subset G$, a contradiction.

(iii) Consider $H_r = \{[0], [s], 2[s], \ldots, (r-1)[s]\}$ and $H_s = \{[0], [r], 2[r], \ldots, (s-1)[r]\}$. These are obviously subgroups of Z_n. Assume that H_r and H_s overlap so that $k[s] = l[r]$, where $0 \le k \le r - 1$, $0 \le l \le s - 1$. Then $[ks - lr] = [0]$, so $rs \mid (ks - lr)$. Assume $ks \ge lr$. Then since $ks < rs$ and $lr < rs$, it follows that $ks - lr < rs$ so that $rs \nmid (ks - lr)$ unless $ks = lr$. But r and s are relatively prime so that $ks = lr$ implies $r \mid k$ and hence $k = 0$ or $k \ge r$. Since $k \ge r$ is impossible, H_r and H_s do not overlap except in [0]. This implies that as Z-modules the sum $H_r + H_s$ is direct (i.e., if $x_1 + y_1 = x_2 + y_2$, then $x_1 - x_2 = y_2 - y_1$ and $x_1 - x_2 \in H_r$, $y_2 - y_1 \in H_s$ so that $x_1 - x_2 = y_2 - y_1 = 0$ and $x_1 = x_2$, $y_1 = y_2$). It is obvious that H_r is isomorphic to Z_r and H_s is isomorphic to Z_s. Now let k and l be integers for which $ks + lr = 1$. Then

$$k[s] + l[r] = [ks + lr] = [1].$$

But $k[s] \in H_r$ and $l[r] \in H_s$, and we conclude that $[1] \in H_r \dot{+} H_s$. Hence $Z_n = H_r \dot{+} H_s$. ▮

Corollary 9 *Suppose* $n = p_1^{e_1} \cdots p_k^{e_k}$, *where* $p_1, \ldots, p_k$ *are distinct primes and* $e_i > 0$, $i = 1, \ldots, k$. *Then*

$$Z_n = H_1 \dotplus \cdots \dotplus H_k, \tag{36}$$

where H_i *is a subgroup of* Z_n *isomorphic to* $Z_{p_i^{e_i}}$, $i = 1, \ldots, k$.

Proof: Write $n = rs$, $r = p_1^{e_1} \cdots p_{k-1}^{e_{k-1}}$, $s = p_k^{e_k}$. Then r and s are relatively prime, and we can apply Theorem 2.6(iii) to obtain subgroups H and H_k of Z_n such that

$$Z_n = H \dotplus H_k, \tag{37}$$

with H isomorphic to Z_r and H_k isomorphic to $Z_{p_k^{e_k}}$. By induction on k (with $k = 1$ trivial),

$$Z_r = K_1 \dotplus \cdots \dotplus K_{k-1}$$

where K_i is a subgroup of Z_r isomorphic to $Z_{p_i^{e_i}}$, $i = 1, \ldots, k-1$. Now H is isomorphic to Z_r, and hence

$$H = H_1 \dotplus \cdots \dotplus H_{k-1}, \tag{38}$$

where H_i is isomorphic to K_i and hence to $Z_{p_i^{e_i}}$, $i = 1, \ldots, k-1$. In fact, if $\nu\colon Z_r \to H$ is the isomorphism, then $\nu(K_i) = H_i$. Combining (37) and (38), we have (36). ∎

We now have our final result on the existence of decompositions of finite abelian groups.

Theorem 2.7 *Suppose G is a finite abelian group of order* $n = p_1^{e_1} p_2^{e_2} \cdots p_k^{e_k}$. *Then there exist sk nonnegative integers* ε_{ij}, $i = 1, \ldots, s$, $j = 1, \ldots, k$, *such that*

$$G = \sum_{i=1}^{s}{}^{\cdot} \sum_{j=1}^{k}{}^{\cdot} H_{ij} \tag{39}$$

where H_{ij} *is isomorphic to* $Z_{p_j^{e_{ij}}}$, $j = 1, \ldots, k$, $i = 1, \ldots, s$, $\prod_{j=1}^{k} p_j^{e_{ij}} = \mu_i$, $\mu_i \mid \mu_{i+1}$, $i = 1, \ldots, s-1$, *and* $n = \mu_1 \cdots \mu_s$.

Proof: From Corollary 8 we can write

$$G = \sum_{i=1}^{s}{}^{\cdot} G_i, \tag{40}$$

where G_i is a cyclic subgroup of order μ_i, $i = 1, \ldots, s$, $\mu_1 \mid \mu_2 \mid \cdots \mid \mu_s$, $n = \mu_1 \cdots \mu_s$, and G_i is isomorphic to Z_{μ_i}. Now $\mu_i = p_1^{e_{i1}} p_2^{e_{i2}} \cdots p_k^{e_{ik}}$ so that by Corollary 9, Z_{μ_i} is a direct sum of k subgroups of orders $p_1^{e_{i1}}, \ldots, p_k^{e_{ik}}$, respectively, and the jth one is isomorphic to $Z_{p_j^{e_{ij}}}$(if e_{ij} is 0, then Z_1 is just the 0 group). Thus

$$G_i = H_{i1} \dotplus \cdots \dotplus H_{ik} \tag{41}$$

where H_{ij} is a subgroup of G_i isomorphic to $Z_{p_j^{e_{ij}}}$. Substituting (41) into (40) produces (39). ∎

Observe that since H_{ij} is isomorphic to $Z_{p_j^{e_{ij}}}$, it follows from Theorem 2.6(ii) that H_{ij} is indecomposable.

The problem we wish to consider at this point is the following: To what extent is the decomposition in Theorem 2.5 unique? We break our arguments into several theorems. We refer to the notation of Theorem 2.5 in showing first that r and μ_s are independent of the decomposition (22). It is convenient to introduce the following idea: If $c \in R$ and N is any submodule of M, we let $k_c(N)$ denote the totality of $\alpha \in N$ for which $c\alpha = 0$. Clearly (see Exercise 12) $k_c(N)$ is a submodule of M, called the *kernel of c with respect to N.*

Theorem 2.8

(i) $T = \sum^{\cdot}{}_{i=1}^{s}\langle\alpha_i\rangle$ *is the torsion submodule of M.*

(ii) μ_s *is the order of T and hence is independent of the decomposition* (22).

(iii) $\mu_s M$ *is a free submodule of M of rank r, and hence r is independent of the decomposition* (22).

Proof: We begin the proof by observing the following two general formulas for a direct sum of submodules which will be useful now and later. If $c \in R$, then

$$c\sum_{i=1}^{p}{}^{\cdot} W_i = \sum_{i=1}^{p}{}^{\cdot} cW_i \tag{42}$$

and

$$k_c\left(\sum_{i=1}^{p}{}^{\cdot} W_i\right) = \sum_{i=1}^{p}{}^{\cdot} k_c(W_i). \tag{43}$$

Formula (42) is obvious, and (43) is contained in Exercise 15. That $T = \sum^{\cdot}{}_{i=1}^{s}\langle\alpha_i\rangle$ is the torsion submodule of M was proved in Corollary 4. Next observe that since $\mu_i \mid \mu_s$ and $\mu_i\alpha_i = 0$, it follows that $\mu_s T = \sum^{\cdot}{}_{i=1}^{s} \mu_s\langle\alpha_i\rangle = 0$. Hence $\mu_s \in O(T)$. On the other hand, if $fT = 0$, then $f\alpha_i = 0$, $i = 1, \ldots, s$, and $f \in O(\alpha_s) = (\mu_s)$. Thus $\mu_s \mid f$. In other words, μ_s is the order of T. This proves (ii).

Now write $M = T + F$ where $T = \sum^{\cdot}{}_{i=1}^{s} \langle\alpha_i\rangle$ and $F = \sum^{\cdot}{}_{i=s+1}^{s+r} \langle\alpha_i\rangle$. Then $\mu_s M = \mu_s T + \mu_s F$ by (42), and thus

$$\mu_s M = \mu_s F.$$

But

$$\mu_s F = \sum_{i=s+1}^{s+r}{}^{\cdot} \langle\mu_s\alpha_i\rangle$$

is obviously a free submodule of rank r. Hence (iii) follows. ∎

Corollary 10 *If M has a second decomposition (as in Theorem* 2.5) *of the form*

$$M = \sum_{i=1}^{s'}{}^{\cdot}\langle\beta_i\rangle + \sum_{i=s'+1}^{s'+r'}{}^{\cdot} \langle\beta_i\rangle \tag{44}$$

where the order of β_i *is* m_i, $i = 1, \ldots, s'$, $m_1 | m_2 | \cdots | m_{s'}$, *and* $\beta_{s'+1}, \ldots, \beta_{s'+r'}$ *are free, then* $m_{s'} = \mu_s$, $r' = r$, *and*

$$\sum_{i=1}^{s'} {}^{\cdot} \langle \beta_i \rangle = \sum_{i=1}^{s} {}^{\cdot} \langle \alpha_i \rangle \tag{45}$$

Proof: By Theorem 2.8, both sides of (45) are the torsion submodule T of M, and $m_{s'}$ and μ_s are both equal to the order of T and hence are equal. Finally, r and r' are both equal to the rank of $\mu_s M = m_{s'} M$. ∎

In our next result we use the following elementary ideas. If $0 \neq d \in R$, then we assert that the abelian group structure in the quotient ring $R/(d)$ can be regarded as an R-module as well, for if $a \in R$ and $b \in R$, then simply define

$$a[b + (d)] = ab + (d). \tag{46}$$

This definition is sensible if we show that $b + (d) = b_1 + (d)$ implies that $ab + (d) = ab_1 + (d)$. But $b - b_1 = xd$ clearly implies that $ab - ab_1 = a(b - b_1) = axd \in (d)$. Hence $ab + (d) = ab_1 + (d)$, and (46) makes sense.

We already know what it means for two groups to be isomorphic. The extension of the idea to the notion of *isomorphism of two R-modules* is evident: If M and N are R-modules which are isomorphic as abelian groups via $\nu: M \to N$, then they are isomorphic as R-modules if ν satisfies

$$\nu(cm) = c\nu(m) \tag{47}$$

for all $c \in R$, $m \in M$. We will use the convenient notation

$$M \cong N$$

to denote the fact that M and N are isomorphic R-modules (see Exercise 14).

The following result is a key technical tool in our analysis of uniqueness and is of interest in itself.

Theorem 2.9 *Suppose* $\alpha \in M$ *is a torsion element of order* μ *and* $c \in R$.

(i) *If* c *and* μ *are relatively prime, then* $c\langle\alpha\rangle = \langle\alpha\rangle$ *and* $k_c(\langle\alpha\rangle) = 0$.

(ii) *Assume* $c | \mu$ *so that* $\mu = cd$. *Then as R-modules* [*see* (46)] $R/(d) \cong c\langle\alpha\rangle$ *and* $k_c(\langle\alpha\rangle) \cong R/(c)$.

Proof: (i) Suppose c and μ are relatively prime so that there are a and b in R for which $ac + b\mu = 1$. Then if $w \in \langle\alpha\rangle$, we have $w = 1w = (ac + b\mu)w = acw + b\mu w$ and $\mu w = 0$. Thus $w = acw = c(aw) \in c\langle\alpha\rangle$, i.e., $\langle\alpha\rangle \subset c\langle\alpha\rangle$ and the other inclusion is trivial. Thus $\langle\alpha\rangle = c\langle\alpha\rangle$. Also if $w \in k_c(\langle\alpha\rangle)$, then $cw = 0$. But then $w = 1w = (ac + b\mu)w = a(cw) + b\mu w = 0$.

(ii) Define

$$\nu: R/(d) \to c\langle\alpha\rangle$$

by the formula

$$\nu(a + (d)) = ac\alpha. \tag{48}$$

We must verify that (48) indeed defines a function and that ν is an isomorphism of R-modules. First observe that if $a + (d) = a_1 + (d)$, then $a_1 = a + zd$, $z \in R$ so that

$$\begin{aligned} a_1 c\alpha &= (a + zd)c\alpha = ac\alpha + zdc\alpha \\ &= ac\alpha + z\mu\alpha = ac\alpha. \end{aligned}$$

Thus ν is well-defined and obviously a group homomorphism. It also follows immediately from (46) that $\nu(z(a + (d)) = \nu(za + (d)) = zac\alpha = z\nu(a + (d))$. Since ν is clearly surjective, it remains only to verify that ν is injective. Suppose that

$$\nu(a + (d)) = \nu(a_1 + (d)).$$

Then $ac\alpha = a_1 c\alpha$ and hence $\mu | (ac - a_1 c)$. But then $ac - a_1 c = z\mu = zcd$, and we can cancel the c to conclude $a - a_1 \in (d)$, i.e., $a + (d) = a_1 + (d)$. Thus

$$R/(d) \cong c\langle\alpha\rangle. \tag{49}$$

Define

$$\tau\colon\ R/(c) \to k_c(\langle\alpha\rangle)$$

by

$$\tau(a + (c)) = a(d\alpha). \tag{50}$$

First observe that τ is well-defined. For if $a_1 = a + xc$, then

$$\begin{aligned} a_1 d\alpha &= (a + xc)d\alpha = ad\alpha + xcd\alpha \\ &= ad\alpha + x\mu\alpha = ad\alpha. \end{aligned}$$

Also note that τ is surjective. For suppose $z\alpha \in k_c(\langle\alpha\rangle)$, i.e., $cz\alpha = 0$. Then $\mu | cz$ so that $cz = x\mu = xcd$ and $z = xd$. Hence $z\alpha = xd\alpha = \tau(x + (c))$. Also, τ is obviously a group homomorphism and satisfies $\tau(z(a + (c))) = \tau(za + (c)) = zad\alpha = z\tau(a + (c))$. Finally, if $\tau(a_1 + (c)) = \tau(a + (c))$, then $a_1 d\alpha = ad\alpha$, $(a_1 d - ad)\alpha = 0$, $(a_1 - a)d = x\mu = xcd$, $a_1 - a = xc$, and $a_1 + (c) = a + (c)$. Hence τ is injective, and we conclude that

$$R/(c) \cong k_c(\langle\alpha\rangle). \quad \blacksquare$$

Theorem 2.10 *Suppose that*

$$\sum_{i=1}^{s}{}^{\boldsymbol\cdot}\langle\alpha_i\rangle = \sum_{i=1}^{s'}{}^{\boldsymbol\cdot}\langle\beta_i\rangle, \tag{51}$$

the order of $\alpha_i \neq 0$ *is* μ_i, $i = 1, \ldots, s$, *the order of* $\beta_i \neq 0$ *is* m_i, $i = 1, \ldots, s'$, *and* $\mu_1 | \mu_2 | \cdots | \mu_s$, $m_1 | m_2 | \cdots | m_{s'}'$. *Then* $s = s'$ *and* $\mu_i = m_i$, $i = 1, \ldots, s$.

Proof: Let T denote the common value of the modules on either side of (51). Assume that $s \neq s'$, say $s > s'$, and argue by contradiction. Now $\mu_1 \neq 1$, i.e., μ_1 is not a unit, because $\alpha_1 \neq 0$ and the order of α_1 is μ_1. Thus there is a prime $p \in R$ such that $p | \mu_1$ and hence $p | \mu_t$, $t = 1, \ldots, s$. From formula (43) we have

$$k_p(T) \cong \sum_{i=1}^{s}{}^{\cdot} k_p(\langle \alpha_i \rangle) \tag{52}$$

and hence from Theorem 2.9(ii),

$$k_p(T) \cong \sum_{i=1}^{s}{}^{\cdot} R/(p). \tag{53}$$

The right side of (53) is a direct sum of s copies of $R/(p)$ as an R-module [see Exercise 14(h)], and this is precisely the set of all s-tuples of elements

$$(a_1 + (p), \ldots, a_s + (p)), \tag{54}$$

$a_i \in R$, $i = 1, \ldots, s$, regarded as an R-module with scalar multiplication given by [see (46)]

$$c(a_1 + (p), \ldots, a_s + (p)) = (ca_1 + (p), \ldots, ca_s + (p)), \tag{55}$$

$c \in R$. Suppose that $c - d \in (p)$. Then clearly $ca - da \in (p)$ and hence from (55),

$$c(a_1 + (p), \ldots, a_s + (p)) = d(a_1 + (p), \ldots, a_s + (p)). \tag{56}$$

In other words, any two elements which are in the same (p) coset in R yield the same answer when used as scalar multipliers on any elements in the R-module $\sum_{i=1}^{\cdot s} R/(p)$. Thus we can define a new scalar multiplication in $\sum_{i=1}^{\cdot s} R/(p)$ by

$$\{c + (p)\}(a_1 + (p), \ldots, a_s + (p)) = (ca_1 + (p), \ldots, ca_s + (p)), \tag{57}$$

and hence $\sum_{i=1}^{\cdot s} R/(p)$ can be regarded as an $R/(p)$-module. Since $k_p(T)$ is isomorphic to $\sum_{i=1}^{\cdot s} R/(p)$ as R-modules, we can also define a scalar multiplication in $k_p(T)$ using as scalars the elements in $R/(p)$, i.e., if ν: $k_p(T) \to \sum_{i=1}^{\cdot s} R/(p)$ is the isomorphism, simply define

$$(c + (p))\gamma = \nu^{-1}((c + (p))\nu(\gamma)) \tag{58}$$

for any $\gamma \in k_p(T)$ and any $c + (p) \in R/(p)$. Since ν is bijective, formula (58) is just the assertion that ν is a module isomorphism between $k_p(T)$ and $\sum_{i=1}^{\cdot s} R/(p)$ now regarded as $R/(p)$-modules. The point of all this maneuvering is that p is a prime, and hence (see Theorem 2.11, Section 3.2) $R/(p)$ is a field. Thus $k_p(T)$ as an $R/(p)$-module is isomorphic to the vector space $(R/(p))^s$, and this space is of dimension s:

$$\dim k_p(T) = s. \tag{59}$$

Next consider the equality

$$T = \sum_{i=1}^{s'}{}^{\cdot} \langle \beta_i \rangle$$

and the corresponding equality

$$k_p(T) = \sum_{i=1}^{s'}{}^{\cdot} k_p(\langle \beta_i \rangle). \tag{60}$$

Since p is a prime, there are two possibilities for each m_i: Either p and m_i are relatively prime in which case Theorem 2.9(i) asserts that

$$k_p(\langle\beta_i\rangle) = 0, \tag{61}$$

or $p|m_i$ and then Theorem 2.9(ii) tells us that

$$k_p(\langle\beta_i\rangle) \cong R/(p). \tag{62}$$

Since $m_1|m_2|\cdots|m_{s'}$, we know that if $p|m_k$, then $p|m_t$, $t \geq k$. Observe that if p divides none of the m_i, then (61) and (60) together produce the statement $k_p(T) = 0$ which of course contradicts (59). Thus there is a first m_k such that $p|m_k$ and then (60) becomes

$$k_p(T) \cong \sum_{i=k+1}^{s'}{}^{\cdot}\, R/(p). \tag{63}$$

By precisely the same argument as we used before, (63) implies that $k_p(T)$ can be regarded as a vector space over the field $R/(p)$ and

$$\dim k_p(T) = s' - k. \tag{64}$$

Since $s > s'$, (64) contradicts (59) and we conclude that $s = s'$.

We can now apply Corollary 10 to conclude that $m_s = \mu_s$, and thus if the two lists $m_1, \ldots, m_s$ and $\mu_1, \ldots, \mu_s$ do not coincide, then they disagree for the first time at some integer $k < s$, i.e., $\mu_k \neq m_k$ but $\mu_{k+1} = m_{k+1}, \ldots, \mu_s = m_s$. Without loss of generality assume that μ_k is not a multiple of m_k (they cannot both be multiples of one another; otherwise they would be equal because the system of nonassociates in R is prescribed). Hence μ_k is not a multiple of $m_{k+1}, \ldots, m_s$. Since $\mu_1|\cdots|\mu_{k-1}|\mu_k$, we have from (42)

$$\mu_k T = \sum_{i=1}^{s}{}^{\cdot}\mu_k\langle\alpha_i\rangle = \sum_{i=k+1}^{s}{}^{\cdot}\, \mu_k\langle\alpha_i\rangle. \tag{65}$$

There are at most $s - k$ nonzero summands in (65). On the other hand,

$$\mu_k T = \sum_{i=1}^{s}{}^{\cdot}\mu_k\langle\beta_i\rangle. \tag{66}$$

Since μ_k is not a multiple of any of $m_k, \ldots, m_s$, it follows that $\mu_k\beta_i \neq 0$, $i = k, \ldots, s$ [otherwise $\mu_k \in O(\beta_i) = (m_i)$], and hence there are at least $s - k + 1$ nonzero summands in (66). But it follows from the first part of the theorem that any two representations of $\mu_k T$ such as (65) and (66) must involve the same number of summands. This contradiction implies that $\mu_i = m_i$, $i = 1, \ldots, s$. ∎

If we combine Corollary 10 and Theorem 2.10, we have the following basic result.

Theorem 2.11 *Let M be a nonzero finitely generated R-module. Suppose M has two representations*

$$M = \sum_{i=1}^{s} {}^{\cdot} \langle \alpha_i \rangle \dotplus \sum_{i=s+1}^{s+r} {}^{\cdot} \langle \alpha_i \rangle$$
$$= \sum_{i=1}^{s'} {}^{\cdot} \langle \beta_i \rangle \dotplus \sum_{i=s'+1}^{s'+r'} {}^{\cdot} \langle \beta_i \rangle$$

in which

(i) α_i *is a torsion element of order* μ_i, $i = 1, \ldots, s$, *and* $\mu_1 | \mu_2 | \cdots | \mu_s$.
(ii) $\alpha_{s+1}, \ldots, \alpha_{s+r}$ *are free.*
(iii) β_i *is a torsion element of order* m_i, $i = 1, \ldots, s'$, *and* $m_1 | m_2 | \cdots | m_{s'}$.
(iv) $\beta_{s'+1}, \ldots, \beta_{s'+r'}$ *are free.*

Then

(v) $s = s'$, $r = r'$, $\mu_i = m_i$, $i = 1, \ldots, s$.
(vi) $\sum_{i=1}^{s} {}^{\cdot} \langle \alpha_i \rangle = \sum_{i=1}^{s} {}^{\cdot} \langle \beta_i \rangle$ *is the torsion submodule of* M.

We can use Theorem 2.11 to determine all nonisomorphic finite abelian groups, for Theorem 2.11 and Corollary 8 can be combined to obtain the following:

Corollary 11 *If G is a finite abelian group regarded as a Z-module, then*

$$G \cong Z_{\mu_1} \dotplus \cdots \dotplus Z_{\mu_s}$$

where $\mu_1 > 1$, $\mu_1 | \mu_2 | \cdots | \mu_s$, *and* $|G| = \mu_1 \cdots \mu_s$. *Moreover,* μ_s *is the least integer such that* $\mu_s G = 0$ *and the integers* $\mu_1, \ldots, \mu_s$ *are uniquely determined by G.*

In view of this uniqueness the integers $\mu_1, \ldots, \mu_s$ are called the *invariant factors* of G.

Example 5 We list the finite abelian groups of order 18. The divisors of 18 (other than 1) are 2, 3, 6, 9, 18. The lists of possibilities for $\mu_1, \ldots, \mu_s$ are:

$$s = 1: \qquad \mu_1 = 18$$
$$s = 2: \qquad \mu_1 = 3, \ \mu_2 = 6$$

($\mu_1 = 2$, $\mu_2 = 9$ is excluded because $2 \nmid 9$). There are no other possibilities for $s \geq 3$. Thus to within isomorphism Z_{18} and $Z_3 \dotplus Z_6$ are the only abelian groups of order 18. Of course $Z_6 \cong Z_3 \dotplus Z_2$ by Theorem 2.6, and hence Z_{18}, $Z_2 \dotplus Z_3 \dotplus Z_3$ are the only abelian groups of order 18.

We remark here that we write expressions such as $Z_3 \dotplus Z_2$ without stipulating the group in which Z_3 and Z_2 are subgroups. In general the *external direct sum M of R-modules* $W_1, \ldots, W_p$ is defined as the totality of p-tuples $(w_1, \ldots, w_p)$ with addition defined by

$$(w_1, \ldots, w_p) + (v_1, \ldots, v_p) = (w_1 + v_1, \ldots, w_p + v_p)$$

and scalar multiplication defined by

$$r(w_1, \ldots, w_p) = (rw_1, \ldots, rw_p)$$

(see Theorem 1.5, Section 2.1). It is obvious that the external direct sum M is an R-module and that $W_i' = \{(0, \ldots, 0, w_i, 0, \ldots, 0) \in M \mid w_i \in W_i\}$ is a submodule of M isomorphic to W_i. Moreover, $M = \sum^{\cdot}{}_{i=1}^{p} W_i'$ so that we abuse the notation and also write

$$M = \sum_{i=1}^{p}{}^{\cdot} W_i.$$

Thus $Z_3 \dotplus Z_2$ means the external direct sum of these groups and to say, for example, that an abelian group G of order 18 is isomorphic to $Z_2 \dotplus Z_3 \dotplus Z_3$ can be interpreted in either of two ways: Either

$$G = G_2 \dotplus G_3 \dotplus G_3',$$

where G_2, G_3, and G_3' are subgroups of G and $G_2 \cong Z_2$, $G_3 \cong Z_3$, $G_3' \cong Z_3$, or

$$G \cong Z_2 \dotplus Z_3 \dotplus Z_3,$$

where the sum on the right is an external direct sum. These are obviously equivalent statements.

Exercises 4.2

1. Let $A = \{x^0 = e, x^1, \ldots, x^{m-1}\}$ be a cyclic group, $|A| = m$. Let $\{e\} \neq B$ be a subgroup of A, and let l be the least positive integer such that $x^l \in B$. Show that $m = ql$, $|B| = q$, and $B = \{x^l, x^{2l}, \ldots, x^{(q-1)l}\}$. *Hint*: To begin with, it is obvious that $\{x^l, x^{2l}, \ldots, x^{(q-1)l}\} \subset B$ for any positive integer q. Also if $x^n \in B$, write $n = kl + s$, $0 \leq s < l$, so that $x^n = x^{kl+s} = x^{kl}x^s$. Hence $x^s = x^n x^{-kl}$ and since $x^{kl} \in B$, $x^{-kl} \in B$ and also $x^n \in B$. But then $x^s \in B$ so that the minimality of l implies $s = 0$, i.e., $n = kl$, and hence B consists of powers of x^l. Now divide m by l to obtain $m = ql + r$, $0 \leq r < l$. Again, since $x^m = e$, we have $e = x^m = x^{ql+r}$ and $x^r = x^{-ql} \in B$. But $0 \leq r < l$; so the minimality of l again implies that $r = 0$, i.e., $m = ql$. Now the elements $x^l, x^{2l}, \ldots, x^{(q-1)l}$ are distinct. For $x^{sl} = x^{tl}$, $s > t$, implies $x^{(s-t)l} = e$. But since s and t are both less than $q - 1$, $(s - t)l < ql = m$. Thus some power of x less than m is e. This contradicts the fact that $|A| = m$. Finally, we saw that B consists of powers of x^l, i.e., x^{kl}. If $k > q$, write $k = tq + \nu$, $0 \leq \nu < q$, and then $x^{kl} = x^{(t_q+\nu)l} = x^{(ql)t}x^{\nu l} = x^{mt}x^{\nu l} = x^{\nu l}$. Thus B is precisely the set $\{x^l, x^{2l}, \ldots, x^{(q-1)l}\}$.
2. Suppose G is a proper subgroup of Z_n and $n = p^m$, p a prime. Show that G is a cyclic subgroup generated by $[p^a]$, where $0 < a < m$. *Hint*: Z_n is itself a cyclic group which we write additively:

$$Z_n = \{0 \cdot [1] = [0], 2[1], \ldots, (n-1)[1]\}.$$

If G is a subgroup, then by Exercise 1 above,

$$G = \{[l], 2[l], \ldots, (q-1)[l]\}$$

and $ql = n = p^m$. But then $l = p^a$ for some $0 < a < m$.

3. Let M be a finitely generated free R-module of rank n. Let N be a submodule. Show that there exists a basis $\{\gamma_1, \ldots, \gamma_n\}$ of M and nonzero elements $q_1, \ldots, q_m$ of R, $q_1 | q_2 | \cdots | q_m$ such that $\{q_1\gamma_1, \ldots, q_m\gamma_m\}$ is a basis of N. *Hint*: Let $\{\varepsilon_1, \ldots, \varepsilon_n\}$ be a basis of M. By Theorems 1.5 and 1.8 in Section 4.1, there exists a basis $\{v_1, \ldots, v_m\}$ of N and a matrix $A \in M_{m,n}(R)$ of rank m such that $v = A\varepsilon$. Now write $A = P^{-1}S(A)Q^{-1}$, where P and Q are unimodular, and set $Q^{-1}\varepsilon = \gamma$, $Pv = u$. Then γ is a basis of M, u is a basis of N, and $u = S(A)\gamma$. But

$$S(A) = \begin{bmatrix} q_1 & 0 & \cdots & 0 & \cdots & 0 \\ \vdots & & \ddots & \vdots & & \vdots \\ 0 & \cdots & \cdots & q_m & \cdots & 0 \end{bmatrix}$$

so that $u_i = q_i\gamma_i$, $i = 1, \ldots, m$.

4. In Z^3 let N be the submodule generated by (7,3,15) and (1,5,3). Find a basis $\{\gamma_1,\gamma_2,\gamma_3\}$ of Z^3 and positive integers q_1 and q_2, $q_1 | q_2$, such that $\{q_1\gamma_1, q_2\gamma_2\}$ is a basis of N. *Hint*: Write $v_1 = (7,3,15)$, $v_2 = (1,5,3)$ so that $v = A\varepsilon$, $\varepsilon_i = (\delta_{i1},\delta_{i2},\delta_{i3})$, $i = 1, 2, 3$. Note that v is a basis of N. Now $A = P^{-1}S(A)Q^{-1}$, where

$$A = \begin{bmatrix} 7 & 3 & 15 \\ 1 & 5 & 3 \end{bmatrix}, \quad P = \begin{bmatrix} 0 & 1 \\ -1 & 7 \end{bmatrix},$$

$$Q = \begin{bmatrix} 1 & 10 & -33 \\ 0 & 1 & -3 \\ 0 & -5 & 16 \end{bmatrix}, \quad S(A) = \begin{bmatrix} 1 & 0 & 0 \\ 0 & 2 & 0 \end{bmatrix}.$$

Thus $Pv = S(A)Q^{-1}\varepsilon$, and if we set $Q^{-1}\varepsilon = \gamma$ and $u = Pv$, then γ is a basis of Z^3 and u is a basis of N. Now

$$Q^{-1} = \frac{1}{\det Q} \operatorname{adj} Q = \begin{bmatrix} 1 & 5 & 3 \\ 0 & 16 & 3 \\ 0 & 5 & 1 \end{bmatrix}.$$

Thus $\gamma_1 = (1,5,3)$, $\gamma_2 = (0,16,3)$, $\gamma_3 = (0,5,1)$. Also $u_1 = \gamma_1 = (1,5,3)$; $u_2 = 2\gamma_2 = (0,32,6)$.

5. Establish formula (43). *Hint*: Suppose $\alpha \in k_c(\sum_{i=1}^{p} W_i)$. Then $\alpha = w_1 + \cdots + w_p$ and $0 = c\alpha = cw_1 + \cdots + cw_p$. The uniqueness of the representation of an element as a sum of elements in $W_1, \ldots, W_p$ then implies that $cw_1 = \cdots = cw_p = 0$, i.e., $w_i \in k_c(W_i)$, $i = 1, \ldots, p$. Thus $k_c(\sum_{i=1}^{p} W_i) \subset \sum_{i=1}^{p} k_c(W_i)$, and the other inclusion is equally obvious.

6. Verify the computations of the matrices P, Q, and $S(A)$ in Example 3(b).

7. Show that if $s_1, \ldots, s_k$ are generators of an R-module M, then the set $W \subset R^k$ consisting of all $(r_1, \ldots, r_k) \in R^k$ for which $\sum_{j=1}^{k} r_j s_j = 0$ is a submodule of R^k. *Hint*: It is obviously closed under addition and scalar multiplication.

8. Show that if T is the set of all torsion elements in the R-module M, then T is a submodule of M (*the torsion submodule*). *Hint*: Suppose α and β are in T, $c\alpha = 0$ and $d\beta = 0$, $0 \neq c,d \in R$. Then $cd(\alpha + \beta) = 0$ and $ca\alpha = ac\alpha = 0$ for any $a \in R$.

9. Show that if α is a torsion element in M, then the totality $O(\alpha)$ of $r \in R$ for which $r\alpha = 0$ is an ideal in M (the *order ideal* of α). *Hint*: If $r_1, r_2 \in O(\alpha)$ then $(r_1 + r_2)\alpha = r_1\alpha + r_2\alpha = 0$, so $O(\alpha)$ is closed under addition. Also if $a \in R$, $r \in O(\alpha)$, then $ra\alpha = ar\alpha = 0$.
10. Show that if $N \subset M$, then the totality of $r \in R$ for which $rN = 0$ is an ideal in R. *Hint*: Same argument as in Exercise 9.
11. Show that a free module M can contain no nonzero torsion elements. *Hint*: Suppose $\{s_1, \ldots, s_k\}$ is a basis of M and $\alpha \sum_{t=1}^{k} c_t s_t$ is a torsion element. Then for some $c \neq 0$, $0 = c\alpha = \sum_{t=1}^{k} cc_t s_t$, and hence $cc_t = 0$, $t = 1, \ldots, k$. Since $c \neq 0$, it follows that $c_t = 0$, $t = 1, \ldots, k$, and hence $\alpha = 0$.
12. Let N be a submodule of M, and let $k_c(N) = \{\alpha \in N \mid c\alpha = 0\}$. Show that $k_c(N)$ is a submodule of M ($k_c(N)$ is called the *kernel of c with respect to N*). *Hint*: If $\alpha, \beta \in k_c(N)$ and $r \in R$, then $cr\alpha = rc\alpha = 0$ and $c(\alpha + \beta) = c\alpha + c\beta = 0$.
13. Verify formula (42).
14. Let M and N be R-modules, and let $\nu: M \to N$ be a homomorphism of the additive abelian group structures in M and N which satisfies $\nu(r\alpha) = r\nu(\alpha)$ for all $r \in R$, $\alpha \in M$. Then ν is a *module homomorphism*. The words *epimorphism* (surjective), *monomorphism* (injective), *isomorphism* (bijective) are also used in connection with module homomorphisms.
    (a) Show that $\iota_M: M \to M$ is a homomorphism.
    (b) If $M \subset N$ and $\iota: M \to N$ with $\iota(\alpha) = \alpha$, then ι is a homomorphism.
    (c) If $\nu: M \to N$, then im ν and ker $\nu = \{\alpha \in M \mid \nu(\alpha) = 0\}$ are submodules of N and M, respectively.
    (d) Let N be a submodule of M, and define a scalar multiplication in the quotient group M/N by $r(\alpha + N) = r\alpha + N$. Show that M/N is an R-module with this definition and that the canonical map $\nu: M \to M/N$, $\nu(\alpha) = \alpha + N$, is a module homomorphism. The R-module M/N is called the *quotient module*.
    (e) Show that if $\nu: M \to N$, then im $\nu \cong M/\ker \nu$.
    (f) Show that if $\nu: M \to N$, then ν is injective iff $\ker \nu = (0)$.
    (g) Show that if W_i, $i = 1, \ldots, p$, are submodules of M and $\sigma \in S_p$, then $\sum_{i=1}^{\cdot p} W_i \cong \sum_{i=1}^{\cdot p} W_{\sigma(i)}$.
    (h) Suppose F is a ring and M is an R-module. Assume that W_i is a submodule of M and $W_i \cong F$, $i = 1, \ldots, p$. Then $\sum_{i=1}^{\cdot p} W_i \cong \sum_{i=1}^{\cdot p} F = F^p$ as R-modules. [We abuse the notation for direct sum slightly by regarding each copy of F in $\sum_{i=1}^{\cdot p} F$ as the set of all p-tuples over F of the form $(0, \ldots, 0, f, 0, \ldots, 0)$.] The scalar multiplication in F^p by elements of R is of course $r(f_1, \ldots, f_p) = (rf_1, \ldots, rf_p)$.

    Hint: (a), (b), (c) are immediate. (d) If $\alpha - \beta \in N$, then $r(\alpha - \beta) \in N$, so $r\alpha + N = r\beta + N$. Thus the scalar multiplication is well-defined. Then $\nu(r\alpha) = r\alpha + N = r(\alpha + N) = r\nu(\alpha)$. (e) Define $\bar{\nu}: M/\ker \nu \to \operatorname{im} \nu$ by $\bar{\nu}(\alpha + \ker \nu) = \nu(\alpha)$. This is well-defined since $\alpha - \beta \in \ker \nu$ implies $\nu(\alpha) = \nu(\beta)$. Also $\bar{\nu}$ is a surjective homomorphism. If $\bar{\nu}(\alpha + \ker \nu) = 0$, then $\alpha \in \ker \nu$, so $\alpha + \ker \nu = \ker \nu$, i.e., $\bar{\nu}$ maps only the zero in $M/\ker \nu$ into 0. This makes $\bar{\nu}$ injective and hence an isomorphism. (f) This is obvious. (g) Define $\nu: \sum^{\cdot}{}_{i=1}^{p} W_i \to \sum^{\cdot}{}_{i=1}^{p} W_{\sigma(i)}$ by $\nu(w_1 + \cdots + w_p) = w_{\sigma(1)} + \cdots + w_{\sigma(p)}$. (h) Let $\varphi_i: W_i \to F$ be the isomorphism as R-modules, $i = 1, \ldots, p$. Then set $\nu(w_1 + \cdots + w_p) = (\varphi_1(w_1), \ldots, \varphi_p(w_p))$. It is trivial to verify $\nu(rw) = r\nu(w)$ and $\nu(w + u) = \nu(w) + \nu(u)$. If $\nu(w_1 + \cdots + w_p) = 0$, then $\varphi_i(w_i) = 0$, $i = 1, \ldots, p$, and hence $w_i = 0$, $i = 1, \ldots, p$, i.e., $w = w_1 + \cdots + w_p = 0$.

15. List all nonisomorphic abelian groups of order 36 (to within isomorphism). *Hint*: Z_{36}, $Z_{18} \dotplus Z_2$, $Z_{12} \dotplus Z_3$, $Z_6 \dotplus Z_6$. Also $Z_{18} \cong Z_2 \dotplus Z_9$, $Z_{12} \cong Z_3 \dotplus Z_4$ and $Z_6 \cong Z_2 \dotplus Z_3$, by Theorem 2.6. Thus $Z_{18} \dotplus Z_2 \cong Z_2 \dotplus Z_2 \dotplus Z_9$, $Z_{12} \dotplus Z_3 \cong Z_3 \dotplus Z_3 \dotplus Z_4$, $Z_6 \dotplus Z_6 \cong Z_2 \dotplus Z_2 \dotplus Z_3 \dotplus Z_3$, and Z_4, Z_9 are indecomposable by Theorem 2.6.

16. List all nonisomorphic abelian groups of order 64.

17. Suppose that W is a submodule of M (over the P.I.D. R, as usual) and $cdW = 0$, i.e., $cd \in O(W)$, where c and d are relatively prime. Prove that

$$W = k_c(W) \dotplus k_d(W).$$

Hint: There exist λ and μ in R such that $\lambda d + \mu c = 1$. Thus if $\alpha \in W$, then $\alpha = 1 \cdot \alpha = (\lambda d + \mu c)\alpha = \lambda(d\alpha) + \mu(c\alpha)$, and $c(d\alpha) = 0$, $d(c\alpha) = 0$, so that $d\alpha \in k_c(W)$ and $c\alpha \in k_d(W)$. Thus any $\alpha \in W$ is a sum of two elements, one in $k_c(W)$ and the other in $k_d(W)$. Also if $\alpha \in k_c(W) \cap k_d(W)$, then $c\alpha = 0$, $d\alpha = 0$, and hence $0 = \lambda d\alpha + \mu c\alpha = (\lambda d + \mu c)\alpha = \alpha$. Thus the sum is direct.

18. Prove that if $\alpha \in M$ has order p^k and $\beta \in M$ has order p^l, p a prime, k and l nonnegative integers, then $\alpha + \beta$ has order a divisor of p^{k+l}. *Hint*: $p^{k+l}(\alpha + \beta) = p^l p^k \alpha + p^k p^l \beta = 0$.

19. Let $T_p(W) = \{\alpha \in W \mid \alpha \text{ has order } p^k \text{ for some } k \in N \cup \{0\}\}$. Prove that if p is a prime, then $T_p(W)$ is a submodule of W. *Hint*: This follows immediately from Exercise 18. In general, a *p-module*, p a prime, is a module each of whose elements has order p^k for some $k \in N \cup \{0\}$. Thus $T_p(W)$ contains all p-submodules of W, i.e., it is the largest submodule of W which is a p-module. In general, a module is called *primary* if it is a p-module for some prime p.

20. Let $T = \sum^{\cdot}{}_{i=1}^{s} \langle \alpha_i \rangle$, where $0 \neq \alpha_i$ has order μ_i, $i = 1, \ldots, s$, and $\mu_1 | \cdots | \mu_s$. Show that T has order μ_s. *Hint*: Obviously $\mu_s T = 0$; so $\mu_s \in O(T)$. Also if $\mu T = 0$, then clearly $\mu \alpha_s = 0$, $\mu_s | \mu$. Thus $O(T) = (\mu_s)$ and μ_s is the order of T.

21. Let T be the torsion submodule of a finitely generated R-module M. Let μ be the order of T (i.e., μ is the μ_s in Exercise 20), and write $\mu = p_1^{e_1} \cdots p_k^{ek}$, where the e_i are positive integers and $p_1, \ldots, p_k$ are distinct, i.e., nonassociated primes. Prove that

$$T = T_{p_1}(T) \dotplus \cdots \dotplus T_{p_k}(T)$$

where $T_{p_i}(T)$ is the largest p_i-submodule of T. Of course, this decomposition is uniquely determined by μ since each $T_{p_i}(T)$ is the largest p_i-submodule of T. *Hint*: If $k = 1$, then T has order $p_1^{e_1}$, and by definition $T_{p_1}(T)$ just consists of all elements of T which have order a power of p_1. But since T has order $p_1^{e_1}$, every element of T has order a power of p_1. Thus by induction assume the result up to $k-1$, and write $\mu = cd$, $c = p_1^{e_1} \cdots p_{k-1}^{e_{k-1}}$, $d = p_k^{e_k}$. Then c and d are relatively prime, and since $\mu T = 0$ we use Exercise 17 to write

$$T = k_c(T) \dotplus k_d(T).$$

We first assert that $k_c(T) = \{\alpha \in T \mid c\alpha = 0\}$ has order precisely c. For suppose not. Since by definition $c\alpha = 0$ for each $\alpha \in k_c(T)$, we know that the order of $k_c(T)$ must be a divisor of $c = p_1^{e_1} \cdots p_{k-1}^{e_{k-1}}$. If it were a proper divisor, then $k_c(T)$ would have order $x = p_1^{f_1} \cdots p_{k-1}^{f_{k-1}}$, where $f_i \leq e_i$ with at least one strict inequality. Now $xp_k^{e_k}T = p_k^{e_k}xk_c(T) \dotplus xp_k^{e_k}k_d(T)$, $xk_c(T) = 0$, and $p_k^{e_k}k_d(T) = 0$. Thus $xp_k^{e_k} \in O(T)$. But then $\mu = p_1^{e_1} \cdots p_k^{e_k} \mid xp_k^{e_k} = p_1^{f_1} \cdots p_{k-1}^{f_{k-1}}p_k^{e_k}$, a contradiction. Thus by induction on k,

$$k_c(T) = T_{p_1}(k_c(T)) \dotplus \cdots \dotplus T_{p_{k-1}}(k_c(T)).$$

We assert that $T_{p_i}(k_c(T)) = T_{p_i}(T)$. The inclusion $T_{p_i}(k_c(T)) \subset T_{p_i}(T)$ is trivial because $k_c(T) \subset T$. Now suppose $\alpha \in T_{p_i}(T)$. This means that α has order p_i^m for some m. Now clearly $m \leq e_i$; otherwise $p^m \nmid \mu$ while $\mu\alpha = 0$. Hence $c\alpha = p_1^{e_1} \cdots p_{k-1}^{e_{k-1}} \alpha = 0$ and $\alpha \in T_{p_i}(k_c(T))$. Hence $k_c(T) = \sum_{i=1}^{\cdot k-1} T_{p_i}(T)$. Also we assert $k_d(T) = T_{p_k}(T)$. For the inclusion $k_d(T) \subset T_{p_k}(T)$ is trivial because every element in $k_d(T)$ has order a divisor of $p_k^{e_k}$ since by definition $p_k^{e_k}\alpha = 0$ for $\alpha \in k_d(T)$. On the other hand, if $\alpha \in T_{p_k}(T)$, then α has order p_k^m and by the argument we just made, $m \leq e_k$. Hence $p_k^{e_k}\alpha = 0$; so $\alpha \in k_d(T)$.

22. Let $W = W_1 \dotplus \cdots \dotplus W_k$, where the W_i are submodules of M containing only torsion elements. If W_i has order μ_i and g.c.d.$(\mu_i, \mu_j) = 1$, $i \neq j$, then show that W has order $\mu = \mu_1 \cdots \mu_k$. *Hint*: Clearly $\mu W = 0$. Also if $cW = 0$, then $cW_i = 0$; so $\mu_i \mid c$. Hence $\mu \mid c$.

23. Show that the submodule $T_{p_i}(T)$ in Exercise 21 has order $p_i^{e_i}$. *Hint*: It has order some power of p_i, say $p_i^{a_i}$; and since $p_1, \ldots, p_k$ are distinct, it follows by Exercise 22 that the order of T is $p_1^{a_1} \cdots p_k^{a_k}$. But T has order $p_1^{e_1} \cdots p_k^{e_k}$, and the uniqueness of factorization then implies $a_i = e_i$, $i = 1, \ldots, k$.

24. (*Cyclic Primary Decompoistion*) Let T be as in Exercise 21. Show that T is a direct sum of cyclic primary submodules and this decomposition is unique to within order. *Hint*: Each $T_{p_i}(T)$ in Exercise 21 contains only torsion elements, is of order $p_i^{e_i}$, and from Theorems 2.5 and 2.10 has a unique decomposition as a direct sum of cyclic submodules of orders $p_i^{e_{i1}}, \ldots, p_i^{e_{in_i}}$:

$$T_{p_i}(T) = \langle \alpha_{i1} \rangle \dotplus \cdots \dotplus \langle \alpha_{in_i} \rangle.$$

Combine this with Exercise 21. The orders of the various primary cyclic submodules are called the *elementary divisors* of T.

25. Let M be an abelian group, and let $\{s_1, \ldots, s_k\}$ be a generating set for M. Let $(a_{i1}, \ldots, a_{ik}) \in Z^k$, $i = 1, \ldots, m$. Then M is said to be *generated* by $\{s_1, \ldots, s_k\}$ *subject to the defining relations*

$$\sum_{j=1}^{k} a_{ij}s_j = 0, \qquad i = 1, \ldots, k,$$

if $(a_{i1}, \ldots, a_{ik})$, $i = 1, \ldots, m$, span the submodule $W \subset Z^k$ consisting of all $(c_1, \ldots, c_k)$ for which $\sum_{j=1}^{k} c_js_j = 0$. We also say that M has the *presentation*

$$\langle s_1, \ldots, s_k \mid \sum_{j=1}^{k} a_{ij}s_j = 0, \ i = 1, \ldots, m.\rangle$$

The proof of Theorem 2.5 shows how M can be decomposed into a direct sum o s torsion subgroups $\langle \alpha_1 \rangle, \ldots, \langle \alpha_s \rangle$ of orders $\mu_1, \ldots, \mu_s$ and r infinite cyclic subgroups $\langle \alpha_{s+1} \rangle, \ldots, \langle \alpha_{s+r} \rangle$. Moreover Corollary 6 showed that $\langle \alpha_i \rangle \cong Z_{\mu_i}, i = 1, \ldots, s$, and $\langle \alpha_{s+j} \rangle \cong Z, j = 1, \ldots, r$. Then Theorem 2.6 can be used to break the group up into indecomposable subgroups. Find the decomposition of each of the following groups into indecomposable cyclic groups:

(a) $\langle s_1,s_2 \mid 2s_1 + 3s_2 = 0, \ s_1 - 7s_2 = 0 \rangle$;
(b) $\langle s_1,s_2,s_3 \mid s_1 + s_2 + s_3 = 0, \ 3s_1 + s_2 + 5s_3 = 0 \rangle$;
(c) $\langle s_1,s_2,s_3 \mid -4s_1 + 2s_2 + 6s_3 = 0, \ -6s_1 + 2s_2 + 6s_3 = 0,$
$7s_1 + 4s_2 + 15s_3 = 0 \rangle$.
(d) $\langle s_1,s_2,s_3 \mid 7s_1 + 3s_2 + 15s_3 = 0, \ s_1 + 5s_2 + 3s_3 = 0 \rangle$.
(e) $\langle s_1,s_2,s_3 \mid 7s_1 + 4s_2 + s_3 = 0, \ 8s_1 + 5s_2 + 2s_3 = 0, \ 9s_1 + 6s_2 + 3s_3 = 0 \rangle$.

Hint: We work (d) and (e) following the method in the proof of Theorem 2.5.

(d) Define the *matrix of relations*

$$A = \begin{bmatrix} 7 & 3 & 15 \\ 1 & 5 & 3 \end{bmatrix}.$$

According to the proof of Theorem 2.5, we must in general reduce A to

$$S(A) = PAQ = \begin{bmatrix} q_1 & & & & & \\ & \ddots & & & & \\ & & q_\rho & & & \\ & & & 0 & & \\ & & & & \ddots & \\ & & & & & 0 \end{bmatrix},$$

where $q_1 = \cdots = q_\nu$, and $q_{\nu+1}, \ldots, q_\rho$ are not 1. We set $\gamma = Q^{-1}s$, and then $\gamma_{\nu+1}, \ldots, \gamma_\rho$ generate the finite cyclic groups isomorphic to $Z_{q_{\nu+1}}, \ldots, Z_{q_\rho}$, respectively, and $\gamma_{\rho+1}, \ldots, \gamma_k$ generate the infinite cyclic groups each isomorphic to Z. Now, in this case,

$$P = \begin{bmatrix} 0 & 1 \\ -1 & 7 \end{bmatrix},$$

$$Q = \begin{bmatrix} 1 & 10 & -33 \\ 0 & 1 & -3 \\ 0 & -5 & 16 \end{bmatrix},$$

$$S(A) = \begin{bmatrix} 1 & 0 & 0 \\ 0 & 2 & 0 \end{bmatrix}.$$

Also

$$Q^{-1} = \begin{bmatrix} 1 & 5 & 3 \\ 0 & 16 & 3 \\ 0 & 5 & 1 \end{bmatrix}.$$

Thus

$$\gamma = Q^{-1}s = (s_1 + 5s_2 + 3s_3,\ 16s_2 + 3s_3,\ 5s_2 + s_3),$$

$q_1 = 1, q_2 = 2, \rho = 2$. Hence $M = \langle \gamma_2 \rangle + \langle \gamma_3 \rangle$. Also $\langle \gamma_2 \rangle \cong Z_2$ and $\langle \gamma_3 \rangle \cong Z$. Thus $M \cong Z_2 + Z$.

(e) Here the matrix of relations is

$$A = \begin{bmatrix} 7 & 4 & 1 \\ 8 & 5 & 2 \\ 9 & 6 & 3 \end{bmatrix}.$$

We have

$$PAQ = S(A) = \text{diag}(1,3,0)$$

where
$$P = \begin{bmatrix} 1 & 0 & 0 \\ 2 & -1 & 0 \\ 1 & -2 & 1 \end{bmatrix},$$

$$Q = \begin{bmatrix} 0 & 0 & 1 \\ 0 & 1 & -2 \\ 1 & -4 & 1 \end{bmatrix}.$$

Hence
$$Q^{-1} = -\begin{bmatrix} -7 & -4 & -1 \\ -2 & -1 & 0 \\ -1 & 0 & 0 \end{bmatrix} = \begin{bmatrix} 7 & 4 & 1 \\ 2 & 1 & 0 \\ 1 & 0 & 0 \end{bmatrix}.$$

Thus $\gamma = Q^{-1}s = (7s_1 + 4s_2 + s_3,\ 2s_1 + s_2,\ s_1)$. Since $q_1 = 1$, $q_2 = 3$, $\rho = 2$, the decomposition is

$$M \cong Z_3 + Z.$$

In fact, the isomorphic copy of Z_3 in M is $\langle \gamma_2 \rangle = \langle 2s_1 + s_2 \rangle$, and the isomorphic copy of Z in M is $\langle \gamma_3 \rangle = \langle s_1 \rangle$.

Glossary 4.2

cyclic primary decomposition, 348
determinantal divisor of A, 320
$d_k(A)$, 320
diophantine system of equations, 329
direct sum, 332
elementary divisors of A, 324
elementary divisors of a torsion submodule, 348
equivalence of matrices, 320
external direct sum of R-modules, 343
group generated by $\{s_1, \ldots, s_k\}$ subject to defining relations, 348
invariant factors, 324
$q_k(A)$, 323
isomorphism of R-modules, 339
kernel of c with respect to N, 338
$k_c(N)$, 338
list of elementary divisors, 324
matrix of relations, 349
module epimorphism, 346
module homomorphism, 346
module isomorphism, 339
module monomorphism, 346
order ideal of α, 331
$O(\alpha)$, 331
order ideal of N, 332
$O(N)$, 332
order of α, 332

4.3 The Structure of Linear Transformations

Let V and W be finite dimensional vector spaces over a field R; i.e., V and W are finitely generated modules over R. Let $T: V \to W$ be a module homomorphism (see Exercise 14, Section 4.2). Then T is called a *linear transformation* (abbreviated l.t.), and the totality of such T is denoted by $\text{Hom}(V,W)$. It is obvious that if T and S are in $\text{Hom}(V,W)$, then their *sum* $T + S$, defined by

$$(T + S)v = Tv + Sv, \qquad v \in V, \tag{1}$$

is in $\text{Hom}(V,W)$. Also, if $r \in R$, then the *scalar product* $rT \in \text{Hom}(V,W)$ is the l.t. defined by

$$(rT)v = r(Tv), \qquad v \in V. \tag{2}$$

Thus $\text{Hom}(V,W)$ is a vector space. If $\{e_1, \ldots, e_n\}$ is a basis of V, then there is one and only one l.t. which satisfies

$$Te_i = w_i, \qquad i = 1, \ldots, n,$$

where w_i are prescribed vectors in W. For if $v = \sum_{i=1}^n \xi_i e_i \in V$, then

$$Tv = T\sum_{i=1}^n \xi_i e_i = \sum_{i=1}^n \xi_i Te_i$$

$$= \sum_{i=1}^n \xi_i w_i.$$

This process of defining a l.t. by stipulating its values on a basis is called *linear extension.*

Let λ be an indeterminate over R, and let $T \in \text{Hom}(V,V)$ be a fixed linear transformation. We define a scalar multiplication of vectors $v \in V$ by polynomials $f(\lambda) \in R[\lambda]$ as follows: If $f(\lambda) = a_0 + a_1\lambda + \cdots + a_k\lambda^k$, then

$$f(\lambda)v = f(T)v = (a_0 I_V + a_1 T + a_2 T^2 + \cdots + a_k T^k)v.$$

It is simple (see Exercise 2) to confirm that V is an $R[\lambda]$-module. It is, in fact, a torsion module. For by Exercise 1, $\dim \text{Hom}(V,V) = N = n^2$, where $\dim V = n$. Hence the $N + 1$ powers $I_V, T, T^2, \ldots, T^N$ must be linearly dependent in $\text{Hom}(V,V)$. Let $a_0, \ldots, a_N$ be elements of R, not all 0, for which

$$a_0 I_V + a_1 T + \cdots + a_N T^N = 0.$$

If $f(\lambda) = a_0 + a_1\lambda + \cdots + a_N\lambda^N$, then obviously

$$f(\lambda)V = 0.$$

We stipulate the monic polynomials in $R[\lambda]$ as a system of nonassociates. Then the order of V as an $R[\lambda]$-module is called the *minimal polynomial* of T; i.e., the minimal polynomial of T is the monic polynomial $m(\lambda)$ of least degree such that $m(T) = 0$.

According to Exercise 24 in Section 4.2, V can be written uniquely as a direct sum of cyclic primary $R[\lambda]$-submodules; that is, V is a direct sum of $R[\lambda]$-submodules of the form

$$\langle \alpha \rangle,$$

where $p(\lambda)^e$ is the order of α, $p(\lambda)$ is a prime in $R[\lambda]$, and e is a positive integer. To say that $\langle d \rangle$ is an $R[\lambda]$-submodule means that for any $g(\lambda) \in R[\lambda]$,

$$g(T)\alpha \in \langle \alpha \rangle.$$

Let
$$g(\lambda) = p(\lambda)^e = b_0 + b_1\lambda + \cdots + \lambda^k.$$
Then
$$b_0\alpha + b_1 T\alpha + \cdots + T^k\alpha = 0$$

so that $T^k\alpha$ is in the subspace U of V spanned by $\alpha, T\alpha, \ldots, T^{k-1}\alpha$. It follows immediately that $T^m\alpha \in U$ for any positive integer m and thus that $f(T)\alpha \in U$ for any $f(\lambda) \in R[\lambda]$. Hence $f(\lambda)\alpha \in U$ for all $f(\lambda)$, and since by definition $\langle \alpha \rangle$ consists of the totality of such $R[\lambda]$-multiples of α, it follows that $\langle \alpha \rangle$, as a subspace of V, is spanned by $\alpha, T\alpha, \ldots, T^{k-1}\alpha$. But we can also see that these vectors are 1.i. over R; otherwise there would exist $c_0, \ldots, c_{k-1}$ not all 0 such that

$$\begin{aligned} 0 &= c_0\alpha + c_1 T\alpha + \cdots + c_{k-1}T^{k-1}\alpha \\ &= (c_0 + c_1 T + \cdots + c_{k-1}T^{k-1})\alpha \\ &= h(T)\alpha \end{aligned}$$

where $h(\lambda) = c_0 + c_1\lambda + \cdots + c_{k-1}\lambda^{k-1}$. But $g(\lambda) = p(\lambda)^e$ is the order of $\langle \alpha \rangle$ which means that if $h(\lambda)\alpha = 0$, then $g(\lambda) \mid h(\lambda)$. But $\deg g(\lambda) = k$, $\deg h(\lambda) = k - 1$. Thus $\alpha, T\alpha, \ldots, T^{k-1}\alpha$ is a basis of $\langle \alpha \rangle$ as a subspace of V. We can apply Exercise 24, Section 4.2, to conclude the following result.

Theorem 3.1 (*Cyclic Invariant Subspace Decomposition*) *If* $\dim V = n$ *and* $T \in \mathrm{Hom}(V,V)$, *then there exists a unique list of prime power monic polynomials*

$$p_1(\lambda)^{e_{11}}, \ldots, p_1(\lambda)^{e_{1n_1}}, p_2(\lambda)^{e_{21}}, \ldots, p_2(\lambda)^{e_{2n_2}}, \ldots, p_r(\lambda)^{e_{r1}}, \ldots, p_r(\lambda)^{e_{rn_r}}$$

such that V *is the direct sum of subspaces* W_{ij},

$$V = \sum_{i=1}^{r}{}^{\cdot} \sum_{j=1}^{n_i}{}^{\cdot} W_{ij} \tag{2}$$

with the following properties:

(i) $T(W_{ij}) \subset W_{ij}, j = 1, \ldots, n_i, i = 1, \ldots, r.$

(ii) $p_i(\lambda)^{e_{ij}}$ *is the order of* W_{ij} *as an* $R[\lambda]$*-module and* $\dim W_{ij} = \deg p_i(\lambda)^{e_{ij}} = m_{ij}, j = 1, \ldots, n_i, i = 1, \ldots, r.$

(iii) *There exist vectors* $\alpha_{ij} \in V$ *such that*

$$\alpha_{ij},\ T\alpha_{ij},\ T^2\alpha_{ij}, \ldots, T^{m_{ij}-1}\alpha_{ij} \tag{3}$$

comprise a basis of W_{ij}.

Observe that since the sum (2) is direct, it follows that the totality of vectors (3), $j = 1, \ldots, n_i$, $i = 1, \ldots, r$, comprise a basis of V. The subspaces W_{ij} in Theorem 3.1 are called *cyclic invariant subspaces.*

Although Theorem 3.1 is essentially the main result concerning the structure of a single linear transformation T, its proof is unsatisfactory in that we have no idea of how to systematically go about constructing the subspaces W_{ij} in (2). We shall now take a somewhat different approach to overcome this difficulty.

Let $V = \langle e_1, \ldots, e_n \rangle$, $W = \langle f_1, \ldots, f_m \rangle$, where the indicated vectors constitute bases of the vector spaces V and W. If $T \in \text{Hom}(V, W)$, then there exist uniquely defined scalars $c_{ij} \in R$, $i = 1, \ldots, m$, $j = 1, \ldots, n$, such that

$$Te_j = \sum_{i=1}^{m} c_{ij} f_i, \qquad j = 1, \ldots, n.$$

The $m \times n$ matrix $[c_{ij}] = C$ is called the *matrix representation of* T *with respect to the ordered bases* $E = \{e_1, \ldots, e_n\}$ and $F = \{f_1, \ldots, f_m\}$. We use the notation

$$C = [T]_E^F.$$

There are a number of elementary results concerning matrix representations.

Theorem 3.2 *Let* U, V, *and* W *be vector spaces over* R *with ordered bases* $G = \{g_1, \ldots, g_p\}$, $E = \{e_1, \ldots, e_n\}$, *and* $F = \{f_1, \ldots, f_m\}$, *respectively.*

(a) *If* $\varphi_{E,F}$: $\text{Hom}(V, W) \to M_{m,n}(R)$ *is the function that maps each* $T \in \text{Hom}(V, W)$ *into* $[T]_E^F$, *then* $\varphi_{E,F}$ *is a vector space isomorphism.*

(b) *If* $S: U \to V$ *and* $T: V \to W$, *then*

$$[TS]_G^F = [T]_E^F [S]_G^E. \tag{4}$$

(c) $[I_U]_G^G = I_p$, *the* $p \times p$ *identity matrix* $[\delta_{ij}]$.

(d) *If* $A \in M_n(R)$, *then* A *is a unit (i.e.,* A *is nonsingular) iff there exists a basis* $H = \{h_1, \ldots, h_n\}$ *of* V *such that*

$$[I_V]_H^E = A.$$

(e) *If $T: V \to V$, then T is a unit in* $\text{Hom}(V,V)$ *iff $[T]_E^E$ is a unit in $M_n(R)$, and in this event*

$$[T^{-1}]_E^E = ([T]_E^E)^{-1}. \tag{5}$$

(f) *Let $T: V \to V$ and $A = [T]_E^E$. If $B \in M_n(R)$, then there exists a basis $H = \{h_1, \ldots, h_n\}$ of V such that $[T]_H^H = B$ iff there exists a unit matrix $P \in M_n(R)$ such that*

$$A = PBP^{-1}. \tag{6}$$

Proof: The proofs of (a), (b), (c) are left to the reader (see Exercise 3).

(d) Set $B = [I_V]_E^H$, and use (4) to compute

$$AB = [I_V]_H^E[I_V]_E^H = [I_V]_E^E = I_n,$$

and similarly $BA = I_n$. Thus if $A = [I_V]_H^E$, then A is a unit. Conversely, suppose A is a unit and define

$$h_j = \sum_{i=1}^{n} a_{ij}e_i, \qquad j = 1, \ldots, n, \tag{7}$$

i.e.,

$$h = A^{\mathrm{T}}e.$$

Let $c = [c_1, \ldots, c_n]$, $c_i \in R$, and assume that $ch = 0$. Then $cA^{\mathrm{T}}e = 0$, and since E is a basis, $cA^{\mathrm{T}} = 0$. But then $Ac^{\mathrm{T}} = 0$ and A is unit so that $c^{\mathrm{T}} = 0$. Thus the vectors $H = \{h_1, \ldots, h_n\}$ are l.i. and hence form a basis of V. The equations (7) state that

$$[I_V]_H^E = A.$$

(e) This follows immediately from (4).

(f) Suppose such a basis H exists. Then from (4) we have

$$A = [T]_E^E = [I_V]_H^E[T]_H^H[I_V]_E^H. \tag{8}$$

Now $[I_V]_H^E = P$ is a unit by (d), and by (4) again, $P^{-1} = [I_V]_E^H$. Thus (8) becomes (6). Suppose conversely that

$$A = PBP^{-1}.$$

Define vectors $h_1, \ldots, h_n$ by $h = P^{\mathrm{T}}e$. Since P is a unit so is P^{T}, and hence $h_1, \ldots, h_n$ form a basis of V. The equation $h = P^{\mathrm{T}}e$ is the statement that $[I_V]_H^E = P$. Then

$$\begin{aligned}[T]_H^H &= [I_V T I_V]_H^H = [I_V]_E^H[T]_E^E[I_V]_H^E \\ &= P^{-1}AP = B. \quad \blacksquare\end{aligned}$$

Theorem 3.2(f) is a particularly important result: It states that two matrices A and B are matrix representations for the same $T \in \text{Hom}(V,V)$ iff there exists a nonsingular P such that

$$A = PBP^{-1}. \tag{9}$$

Two matrices A and B in $M_n(R)$ are *similar* over R, written

$$A \overset{S}{=} B, \tag{10}$$

if there exists $P \in \mathrm{GL}(n,R)$ such that (9) holds. It is simple (see Exercise 4) to confirm that similarity is an equivalence relation. Obviously we can analyze the structure of a $T \in \mathrm{Hom}(V,V)$ if we can find a simple matrix representation for T. In view of (9) the problem is to reduce any matrix representation of T to something simple by similarity transformations. *The key notion in this reduction procedure is the connection between similarity over the field R and equivalence over the ring of polynomials over R.*

Let λ be an indeterminate over $M_n(R)$. We observe that $\iota: R \to M_n(R)$, $\iota(r) = rI_n$, is an injection and $\iota(R)$ is isomorphic to R. Also, λ is an indeterminate over $\iota(R)$; so by a now familiar construction there is an indeterminate λ_1 over R and a ring isomorphism $\iota_1: R[\lambda_1] \to \iota(R)[\lambda]$ such that $\iota_1(\lambda_1) = \lambda$, $\iota_1 | R = \iota$. Observe that

$$\begin{aligned} \iota_1(\lambda_1 r) = \iota_1(\lambda_1)\iota_1(r) &= \lambda\iota(r) \\ &= \lambda(rI_n). \end{aligned} \tag{11}$$

In order to avoid cumbersome notation, we shall not distinguish between λ and λ_1. Thus λ will be regarded as an indeterminate over R as well. Thus we have

$$M_n(R)[\lambda] = M_n(R[\lambda]). \tag{12}$$

Note that equivalence in $M_n(R[\lambda])$ makes perfect sense since $R[\lambda]$ is a P.I.D.; in fact $R[\lambda]$ is a Euclidean ring so that equivalence can be accomplished by elementary row and column operations (see Theorem 2.3, Section 4.2).

If $A \in M_n(R)$, then the *characteristic matrix* of A is the matrix

$$\lambda I_n - A \in M_n(R)[\lambda], \tag{13}$$

and

$$\det(\lambda I_n - A) \in R[\lambda] \tag{14}$$

is the *characteristic polynomial* of A. If F is a splitting field for the characteristic polynomial, then the roots $\alpha_1, \ldots, \alpha_n \in F$ of the characteristic polynomial are called the *characteristic roots* or *eigenvalues* (e.v.) of A. Thus

$$\det(\alpha_j I_n - A) = 0, \qquad j = 1, \ldots, n. \tag{15}$$

The splitting field F is determined to within equivalent extensions (see Theorem 4.14, Section 3.4).

The next result is fundamental.

Theorem 3.3 *Let A and B be in $M_n(R)$. Then $A \overset{S}{=} B$ iff $\lambda I_n - A \overset{E}{=} \lambda I_n - B$, where $\overset{E}{=}$ is equivalence in $M_n(R[\lambda])$.*

Proof: Assume that $\lambda I_n - A \overset{E}{=} \lambda I_n - B$, and let $P = P(\lambda)$ and $Q = Q(\lambda)$ be unimodular matrices in $M_n(R[\lambda])$ for which

$$\lambda I_n - A = P(\lambda I_n - B)Q.$$

Recall that P is unimodular iff $\det P$ is a unit in $R[\lambda]$, i.e., iff $0 \neq \det P \in R$. Let $L = Q^{-1} = M_n(R[\lambda])$, and write $L = L(\lambda)$ as a polynomial in λ with coefficients in $M_n(R)$:

$$L = \sum_{t=0}^{m} L_{m-t}\lambda^{m-t}, \qquad L_{m-t} \in M_n(R), \qquad t = 0, \ldots, m \tag{16}$$

(see Exercise 5). Then

$$I_n = Q^{-1}Q = \sum_{t=0}^{m} L_{m-t}\lambda^{m-t}Q = \sum_{t=0}^{m} L_{m-t}Q\lambda^{m-t}. \tag{17}$$

Now write $Q = Q(\lambda)$ as a polynomial in λ with coefficients in $M_n(R)$, and let

$$W = Q_r(A); \tag{18}$$

i.e., W is the right-hand value of $Q(\lambda)$ at A. Then, since (see Exercise 6)

$$\begin{aligned}(Q(\lambda)\lambda^{m-t})_r(A) &= Q_r(A)A^{m-t} \\ &= WA^{m-t},\end{aligned}$$

we can evaluate the right-hand value of (17) at A to obtain

$$\begin{aligned}I_n &= \sum_{t=0}^{m} L_{m-t}(Q(\lambda)\lambda^{m-t})_r(A) \\ &= \sum_{t=0}^{m} L_{m-t}WA^{m-t}.\end{aligned} \tag{19}$$

Now

$$\begin{aligned}P^{-1}(\lambda I_n - A) &= (\lambda I_n - B)Q(\lambda) = \lambda Q(\lambda) - BQ(\lambda) \\ &= Q(\lambda)\lambda - BQ(\lambda).\end{aligned} \tag{20}$$

It is elementary to verify (see Exercise 7) that the right-hand value of $P^{-1}(\lambda I_n - A)$ at A is 0, and thus from (20)

$$\begin{aligned}0 &= (Q(\lambda)\lambda)_r(A) - (BQ(\lambda))_r(A) = Q_r(A)A - BQ_r(A) \\ &= WA - BW,\end{aligned}$$

$$WA = BW. \tag{21}$$

Thus, from (21) we have

$$WA^{m-t} = B^{m-t}W, \qquad t = 0, \ldots, m, \tag{22}$$

and substituting (22) into (19) produces

$$I_n = \left(\sum_{t=0}^{m} L_{m-t}B^{m-t}\right)W$$

so that $W = Q_r(A)$ is nonsingular (see Exercise 8). From (21) we have

$$A = W^{-1}BW \tag{23}$$

and $A \stackrel{S}{=} B$.

Conversely, if $A \stackrel{S}{=} B$ so that $A = S^{-1}BS$, $S \in M_n(R)$, then

$$\lambda I_n - A = \lambda I_n - S^{-1}BS$$

$$= S^{-1}(\lambda I_n - B)S \tag{24}$$

and since S is nonsingular in $M_n(R)$, it is unimodular in $M_n(R[\lambda])$. But then (24) implies

$$\lambda I_n - A \overset{E}{=} \lambda I_n - B. \quad ∎$$

It should be observed that the proof of Theorem 3.3 provides a constructive procedure for obtaining a matrix W in $\mathbf{GL}(n,R)$ for which

$$A = W^{-1}BW$$

if $\lambda I_n - A \overset{E}{=} \lambda I_n - B$:

(i) Obtain $Q = Q(\lambda)$ (a product of elementary matrices) in $M_n(R[\lambda])$ for which

$$\lambda I_n - A = P(\lambda I_n - B)Q.$$

(ii) Write $Q(\lambda)$ as a polynomial in λ with coefficients in $M_n(R)$.

(iii) Evaluate $W = Q(\lambda)_r(A)$.

Example 1 We show that $A = \begin{bmatrix} 1 & 1 \\ 1 & 1 \end{bmatrix}$ and $B = \begin{bmatrix} 2 & 0 \\ 0 & 0 \end{bmatrix}$ are similar over $M_n(R)$ where R is the field of rational numbers. First,

$$\lambda I_2 - A = \begin{bmatrix} \lambda - 1 & -1 \\ -1 & \lambda - 1 \end{bmatrix}$$

and

$$\lambda I_2 - B = \begin{bmatrix} \lambda - 2 & 0 \\ 0 & \lambda \end{bmatrix}.$$

There is no difficulty in confirming that

$$E_{21}(1-\lambda)E_1(-1)E_{12}\begin{bmatrix} \lambda - 1 & -1 \\ -1 & \lambda - 1 \end{bmatrix}E_{12}(\lambda - 1) = \begin{bmatrix} 1 & 0 \\ 0 & \lambda^2 - 2\lambda \end{bmatrix}, \tag{25}$$

$$E_2(2)E_{21}(2-\lambda)E_1(-\tfrac{1}{2})E_{12}(1)E_{12}\begin{bmatrix} \lambda - 2 & 0 \\ 0 & \lambda \end{bmatrix}E_{12}E_{12}(-1)E_{12}E_{12}\left(\frac{\lambda}{2}\right)$$

$$= \begin{bmatrix} 1 & 0 \\ 0 & \lambda^2 - 2\lambda \end{bmatrix}. \tag{26}$$

From (25) and (26),

$$P_1(\lambda)(\lambda I_2 - A)Q_1(\lambda) = P_2(\lambda)(\lambda I_2 - B)Q_2(\lambda),$$

and

$$\lambda I_2 - A = P_1(\lambda)^{-1}P_2(\lambda)(\lambda I_2 - B)Q_2(\lambda)Q_1(\lambda)^{-1}.$$

Moreover from Exercise 2(a), Section 4.1,

$$Q(\lambda) = Q_2(\lambda)Q_1(\lambda)^{-1} = E_{12}E_{12}(-1)E_{12}E_{12}\left(\frac{\lambda}{2}\right)E_{12}(1-\lambda)$$

$$= \begin{bmatrix} 1 & 1 - \lambda/2 \\ -1 & \lambda/2 \end{bmatrix} = \begin{bmatrix} 1 & 1 \\ -1 & 0 \end{bmatrix} + \begin{bmatrix} 0 & -\frac{1}{2} \\ 0 & \frac{1}{2} \end{bmatrix}\lambda.$$

Then

$$\begin{aligned} W &= Q_r(\lambda)(A) \\ &= \begin{bmatrix} 1 & 1 \\ -1 & 0 \end{bmatrix} + \begin{bmatrix} 0 & -\frac{1}{2} \\ 0 & \frac{1}{2} \end{bmatrix}\begin{bmatrix} 1 & 1 \\ 1 & 1 \end{bmatrix} \\ &= \frac{1}{2}\begin{bmatrix} 1 & 1 \\ -1 & 1 \end{bmatrix}. \end{aligned}$$

We check that

$$WA = \begin{bmatrix} 1 & 1 \\ 0 & 0 \end{bmatrix} = BW$$

and since W is nonsingular, $A = W^{-1}BW$. We remark that it is not necessary to find W explicitly in order to conclude that A and B are similar over R, for $\lambda I_2 - A$ and $\lambda I_2 - B$ obviously have the same determinantal divisors and hence are equivalent over $R[\lambda]$. It follows from Theorem 3.3 that $A \overset{S}{=} B$ over R.

Let $f(\lambda) \in R[\lambda]$ be a monic polynomial,

$$f(\lambda) = \lambda^m + c_{m-1}\lambda^{m-1} + c_{m-2}\lambda^{m-2} + \cdots + c_1\lambda + c_0. \tag{27}$$

The *companion matrix* of $f(\lambda)$, denoted by $C(f(\lambda)) \in M_m(R)$, is defined by

$$C(f(\lambda)) = \begin{bmatrix} 0 & 0 & \cdots & 0 & -c_0 \\ 1 & 0 & \cdots & 0 & -c_1 \\ 0 & 1 & \cdots & 0 & -c_2 \\ \cdot & \cdot & \cdot & & \cdot \\ \cdot & \cdot & & \cdot & \cdot \\ \cdot & \cdot & & & \cdot \\ \cdot & \cdot & & 0 & -c_{m-2} \\ 0 & 0 & \cdots & 1 & -c_{m-1} \end{bmatrix} \tag{28}$$

(if $m = 1$ so that $f(\lambda) = \lambda + c_0$, then $C(f(\lambda)) = [-c_0]$). The matrix (28) is intimately connected with the idea of a torsion element in an appropriate $R[\lambda]$-module. For, suppose $T \in \text{Hom}(V,V)$ and suppose moreover that we regard V as an $R[\lambda]$-torsion module as in the discussion preceding Theorem 3.1. Let $\alpha \in V$ be a torsion element of order $f(\lambda)$. Once again, this means that

$$c_0\alpha + c_1T\alpha + c_2T^2\alpha + \cdots + c_{m-1}T^{m-1}\alpha + T^m\alpha = 0. \tag{29}$$

Let U be the subspace of V spanned by the m vectors $\alpha, T\alpha, \ldots, T^{m-1}\alpha$, i.e.,

$$U = \langle \alpha, T\alpha, \ldots, T^{m-1}\alpha \rangle \tag{30}$$

[U is a subspace of V, and (30) just denotes all linear combinations of the vectors on the right—it is not to be confused with the notation $\langle\alpha\rangle$ meaning the $R[\lambda]$-submodule—of course, they are equal here!]. Clearly $\alpha, T\alpha, \ldots, T^{m-1}\alpha$ are l.i., otherwise α would not have order $f(\lambda)$ (again, see the discussion preceding Theorem 3.1). Also, it is obvious from (29) and (30) that $T(U) \subset U$, i.e., U is an *invariant subspace under* T. Now set

$$e_1 = \alpha, \quad e_2 = T\alpha, \quad e_3 = T^2\alpha, \ldots, e_m = T^{m-1}\alpha,$$

and compute the matrix representation of $T_1 = T|U$ with respect to the ordered basis $E = \{e_1, \ldots, e_m\}$. We have

$$\begin{aligned}
T_1 e_1 &= e_2,\\
T_1 e_2 &= T^2 e_1 = e_3,\\
&\vdots\\
T_1 e_{m-1} &= T^{m-1} e_1 = e_m,\\
T_1 e_m &= T^m e_1 = T^m \alpha = -c_0\alpha - c_1 T\alpha - \cdots - c_{m-1}T^{m-1}\alpha\\
&= -c_0 e_1 - c_1 e_2 - \cdots - c_{m-1} e_m.
\end{aligned}$$

Hence from (28),

$$[T_1]_E^E = C(f(\lambda)). \tag{31}$$

We summarize these remarks.

Theorem 3.4 *Let $T \in \mathrm{Hom}(V,V)$, and regard V as an $R[\lambda]$-torsion module with $g(\lambda)v = g(T)v$, $v \in V$, $g(\lambda) \in R[\lambda]$. Let $\alpha \in V$ have order*

$$f(\lambda) = \lambda^m + c_{m-1}\lambda^{m-1} + \cdots + c_1\lambda + c_0,$$

and set $e_k = T^{k-1}\alpha$, $k = 1, \ldots, m$. Then $U \doteq \langle e_1, \ldots, e_m\rangle$ is an m-dimensional invariant subspace under T, and the matrix representation of $T|U$ is $C(f(\lambda))$.

The subspace U is called a *cyclic invariant subspace of order* $f(\lambda)$.

Our next task is to compute the determinantal divisors of $\lambda I_m - C(f(\lambda))$ as a matrix in $M_n(R[\lambda])$.

Theorem 3.5 *Let*

$$f(\lambda) = \lambda^m + c_{m-1}\lambda^{m-1} + \cdots + c_1\lambda + c_0.$$

Then the determinantal divisors of $\lambda I_m - C(f(\lambda))$ are $d_0 = \cdots = d_{m-1} = 1$ and $d_m = f(\lambda)$. Thus $\lambda I_m - C(f(\lambda))$ has only one nonunit invariant factor, namely $f(\lambda)$.

Proof: Consider

$$\lambda I_m - C(f(\lambda)) = \begin{vmatrix}
\lambda & 0 & \cdots & & 0 & c_0\\
-1 & \lambda & & \bigcirc & & c_1\\
0 & \ddots & \ddots & & & c_2\\
\vdots & & \ddots & \ddots & & \vdots\\
\vdots & \bigcirc & & & \lambda & c_{m-2}\\
0 & \cdots & & 0 & -1 & \lambda + c_{m-1}
\end{vmatrix}. \tag{32}$$

Let $X = \lambda I_m - C(f(\lambda))$, and observe that $X(1\,|\,m)$ is an upper triangular matrix with -1 down the main diagonal. Thus $\det X(1\,|\,m) = (-1)^{m-1}$, and hence $d_{m-1} = 1$. The divisibility property $d_1 \,|\, \cdots \,|\, d_{m-1}$ implies $d_1 = \cdots = d_{m-1} = 1$. We also easily see from (32) that

$$X(1\,|\,1) = \lambda I_{m-1} - C(g(\lambda)),$$

where $g(\lambda) = \lambda^{m-1} + c_{m-1}\lambda^{m-2} + \cdots + c_2\lambda + c_1$. Hence by induction on m,

$$\det X(1\,|\,1) = g(\lambda).$$

By using the Laplace expansion theorem on the first row of (32), we obtain

$$\begin{aligned}\det X &= \lambda \det X(1\,|\,1) + (-1)^{m-1}c_0 \det X(1\,|\,m)\\ &= \lambda g(\lambda) + (-1)^{m-1}(-1)^{m-1}c_0\\ &= \lambda(\lambda^{m-1} + c_{m-1}\lambda^{m-2} + \cdots + c_2\lambda + c_1) + c_0\\ &= f(\lambda). \quad \blacksquare\end{aligned}$$

The next result is of great importance in the study of linear transformations.

Theorem 3.6 (*Frobenius Normal Form*) *If $A \in M_n(R)$, then A is similar over R to the direct sum of the companion matrices of the elementary divisors of $\lambda I_n - A$. Equivalently, if $T \in \mathrm{Hom}(V,V)$ and A is any matrix representation of T, then V is a direct sum of cyclic invariant subspaces whose orders as $R[\lambda]$-torsion modules are precisely the elementary divisors of $\lambda I_n - A$.*

Proof: Let $h_1(\lambda), \ldots, h_p(\lambda)$ be the complete list of elementary divisors of $\lambda I_n - A$, each $h_t(\lambda)$ a power of a prime polynomial in $R[\lambda]$. Let $m_t = \deg h_t(\lambda)$. Since the product of all the invariant factors of a matrix is its determinant, it follows that $h_1(\lambda) \cdots h_p(\lambda)$ is the characteristic polynomial of A and hence

$$m_1 + \cdots + m_p = n.$$

By Theorem 3.5, the determinantal divisors of $\lambda I_{m_t} - C(h_t(\lambda))$ are $\overbrace{1, \ldots, 1}^{m_t - 1}$, $h_t(\lambda)$. Since $h_t(\lambda)$ is a power of a prime, it is obvious that $h_t(\lambda)$ is the single elementary divisor of $\lambda I_{m_t} - C(h_t(\lambda))$. Consider the matrix

$$\lambda I_n - \sum_{t=1}^{p}{}^{\cdot} C(h_t(\lambda)) = \sum_{t=1}^{p}{}^{\cdot} \lambda I_{m_t} - C(h_t(\lambda)). \tag{33}$$

The list of elementary divisors of the right side of (33) is precisely

$$h_1(\lambda), \ldots, h_p(\lambda)$$

by Corollary 3, Section 4.2. Thus the characteristic matrices $\lambda I_n - A$ and $\lambda I_n - \sum_{t=1}^{\cdot\,p} C(h_t(\lambda))$ have the same elementary divisors, are both n-square, and are both of rank n in $M_n(R[\lambda])$ (i.e., they both have determinant equal to the characteristic polynomial). Thus from Corollary 1, Section 4.2,

$$\lambda I_n - A \overset{\text{E}}{=} \lambda I_n - \sum_{t=1}^{p} {}^{\cdot} C(h_t(\lambda)).$$

But then Theorem 3.3 implies that $A \overset{\text{S}}{=} \sum_{t=1}^{\cdot p} C(h_t(\lambda))$ over R.

If $T \in \text{Hom}(V,V)$, let A be the matrix representation of T with respect to some basis H, $[T]_H^H = A$. Then $A \overset{\text{S}}{=} \sum_{t=1}^{\cdot p} C(h_t(\lambda))$, and hence by Theorem 3.2(f) there exists a basis E of V such that

$$[T]_E^E = \sum_{t=1}^{p} {}^{\cdot} C(h_t(\lambda)). \tag{34}$$

We see from (34) that

$$\begin{aligned} Te_1 &= e_2, \quad Te_2 = e_3, \quad \ldots, \quad Te_{m_1-1} = e_{m_1}, \\ Te_{m_1} &= -c_0e_1 - c_1e_2 - \cdots - c_{m_1-1}e_{m_1}, \end{aligned}$$

and thus $\langle e_1, \ldots, e_{m_1}\rangle$ is a cyclic invariant subspace. Similarly, $\langle e_{m_1+1}, \ldots, e_{m_1+m_2}\rangle$ is a cyclic invariant subspace, etc. If the matrix representation of an l.t. is a companion matrix of a polynomial, then the order of the space as an $R[\lambda]$-torsion module is precisely the polynomial. The proof of this elementary fact is contained in Theorem 3.7 and so will be omitted here. ∎

If $A \in M_n(R)$, then the *minimal polynomial* of A is the monic polynomial of least degree, $m(\lambda)$, for which $m(A) = 0$. We leave it to the reader to verify (see Exercise 9) that any two matrix representations of the same l.t. have the same minimal polynomial. In general, any two matrix representations of $T \in \text{Hom}(V,V)$ are similar, and we define the determinantal divisors, invariant factors, elementary divisors, characteristic polynomial, eigenvalues, etc., of T to be the corresponding terms associated with any matrix representation of T.

Theorem 3.7 *If $A \in M_n(R)$, then A is similar over R to the direct sum of the companion matrices of the nonunit invariant factors of $\lambda I_n - A$. Moreover, the minimal polynomial of A is $q_n(\lambda)$, the* nth *invariant factor of $\lambda I_n - A$.*

Proof: The blocks in a direct sum can be arranged in any desired order by a similarity (see Exercise 10). Thus we can rearrange the blocks in the Frobenius normal form so that the companion matrices of all the highest degree elementary divisors of $\lambda I_n - A$ come first; then the companion matrices of all the remaining highest degree elementary divisors come next; etc. Let A_1 denote the direct sum of this first chain of companion matrices, A_2 the direct sum of the second chain, etc., so that

$$A \overset{\text{S}}{=} A_1 \dotplus A_2 \dotplus \cdots \dotplus A_k. \tag{35}$$

The elementary divisors of $\lambda I_p - A_1$ (I_p is an appropriate size identity matrix) are (by Corollary 3, Section 4.2) all the highest degree elementary divisors of $\lambda I_n - A$. On the other hand, if $q_n(\lambda)$ is the highest degree invariant factor of $\lambda I_n - A$, then $\deg q_n(\lambda) = p$ because $q_n(\lambda)$ is the product of the highest degree elementary divisors of $\lambda I_n - A$. Also $\lambda I_p - C(q_n(\lambda))$ has the single nonunit invariant factor $q_n(\lambda)$, and hence its elementary divisors are precisely the highest degree elementary divisors of $\lambda I_n - A$. Thus $\lambda I_p - C(q_n(\lambda))$ and $\lambda I_p - A_1$ have the same elementary divisors and thereby are equivalent. Hence $A_1 \overset{S}{=} C(q_n(\lambda))$. By precisely the same argument, $A_2 \overset{S}{=} C(q_{n-1}(\lambda))$, $\ldots$, $A_k \overset{S}{=} C(q_{n-k+1}(\lambda))$, and $q_n(\lambda), \ldots, q_{n-k+1}(\lambda)$ account for all the nonunit invariant factors of $\lambda I_n - A$ (i.e., they are obtained from the list of elementary divisors of $\lambda I_n - A$). It follows that

$$A_1 \dotplus \cdots \dotplus A_k \overset{S}{=} C(q_n(\lambda)) \dotplus \cdots \dotplus C(q_{n-k+1}(\lambda)), \tag{36}$$

and (35) and (36) yield the first statement in the theorem.

To prove the second statement, let V be a vector space over R, and choose a basis $E = \{e_1, \ldots, e_n\}$ of V and a linear transformation $T \in \operatorname{Hom}(V, V)$ such that (see Exercise 11)

$$[T]_E^E = C(q_n(\lambda)) \dotplus \cdots \dotplus C(q_{n-k+1}(\lambda)). \tag{37}$$

If $p = \deg q_n(\lambda)$, then it is clear from (37) that $U = \langle e_1, \ldots, e_p \rangle$ is an invariant subspace under T and that the matrix representation of $T|U$ with respect to the basis $\{e_1, \ldots, e_p\}$ is precisely $C(q_n(\lambda))$. If

$$q_n(\lambda) = \lambda^p + c_{p-1}\lambda^{p-1} + c_{p-2}\lambda^{p-2} + \cdots + c_1\lambda + c_0,$$

then we have [from the form of $C(q_n(\lambda))$] that

$$\begin{aligned} Te_1 &= e_2 \\ &\vdots \\ Te_{p-2} &= e_{p-1} \\ Te_{p-1} &= e_p \\ Te_p &= -c_0e_1 - c_1e_2 - \cdots - c_{p-1}e_p. \end{aligned} \tag{38}$$

Now $e_2 = Te_1$, $e_3 = Te_2 = T^2e_1, \ldots, e_p = T^{p-1}e_1$ so that the last equation in (38) becomes

$$T^pe_1 = -c_0e_1 - c_1Te_1 - \cdots - c_{p-1}T^{p-1}e_1$$

or

$$q_n(T)e_1 = 0.$$

Also,

$$\begin{aligned} q_n(T)e_k &= q_n(T)T^{k-1}e_1 \\ &= T^{k-1}q_n(T)e_1 \\ &= 0, \qquad k = 1, \ldots, p. \end{aligned}$$

Thus $q_n(T)|U$ is the zero transformation. Equivalently, if we regard U as an $R[\lambda]$-module, then $q_n(\lambda) \in O(U)$. Suppose that $g(\lambda) \in O(U)$, $\deg g(\lambda) = r < p$. Then

$$g(T)e_1 = 0$$

so that writing $g(\lambda) = d_r\lambda^r + d_{r-1}\lambda^{r-1} + \cdots + d_0$, we have

$$d_rT^re_1 + d_{r-1}T^{r-1}e_1 + \cdots + d_0e_1 = 0,$$

or

$$d_re_{r+1} + d_{r-1}e_r + \cdots + d_0e_1 = 0,$$

a linear relation among $e_1, \ldots, e_{r+1}$, $r + 1 \leq p$. It follows that $q_n(\lambda)$ is a polynomial of least degree in the ideal $O(U)$ in $R(\lambda)$, and hence it is the unique monic generator of $O(U)$. Thus we have proved: *$q_n(T)|U = 0$ and if $f(T)|U = 0$, where $f(\lambda) \in R[\lambda]$, then $q_n(\lambda)|f(\lambda)$.* A repetition of this argument shows that if W is the subspace spanned by those vectors in $\{e_1, \ldots, e_n\}$ for which the matrix representation of $T|W$ is $C(q_t(\lambda))$, then $q_t(T)|W = 0$; and if $f(\lambda) \in R[\lambda], f(T)|W = 0$, then $q_t(\lambda)|f(\lambda)$. (This argument also establishes the final assertion in the proof of Theorem 3.6.) The divisibility properties of the $q_t(\lambda)$, $t = n, \ldots, n - k + 1$, immediately imply that $q_n(T)$ maps every such W into 0, i.e.,

$$q_n(T) = 0.$$

Now if $f(T) = 0$, then obviously $f(T)|U = 0$ and $q_n(\lambda)|f(\lambda)$. In other words, $q_n(\lambda)$ divides any polynomial $f(\lambda)$ for which $f(T) = 0$, and $q_n(T) = 0$. Hence $q_n(\lambda)$ is the polynomial of least degree for which $q_n(T) = 0$, i.e., $q_n(\lambda)$ is the minimal polynomial of T and hence of A. ∎

Corollary 1 *(Cayley-Hamilton Theorem) If $A \in M_n(R)$ and $f(\lambda) = \det(\lambda I_n - A)$, then $f(A) = 0$.*

Proof: The characteristic polynomial $f(\lambda)$ is the product of the nonunit invariant factors of $\lambda I_n - A$. Hence $q_n(\lambda)|f(\lambda)$, and it follows from Theorem 3.7 that $f(A) = 0$. ∎

The result of Corollary 1 is often expressed by saying: Every matrix satisfies its own characteristic equation.

The question naturally arises: To what extent is it true that the cyclic invariant subspaces described in Theorem 3.6 can be further reduced? To make this notion precise, we say that a linear transformation $T: V \to V$ is *reducible* if there exists a direct sum decomposition $V = W \dotplus U$, $W \neq 0$, $U \neq 0$, such that $T(W) \subset W$ and $T(U) \subset U$, i.e., both W and U are invariant subspaces under T. We write

$$T = T_1 \dotplus T_2, \tag{39}$$

where $T_1 = T|W$, $T_2 = T|U$, and we call T the *direct sum of T_1 and T_2*. If T is not reducible, it is called *irreducible*. If we choose a basis of V, $E = \{e_1, \ldots, e_r, e_{r+1}, \ldots, e_n\}$ such that $\{e_1, \ldots, e_r\}$ is a basis of W and $\{e_{r+1}, \ldots, e_n\}$ is a basis of U, then clearly if $[T]_E^E = A$, (see Exercise 12)

$$A = A_1 \dotplus A_2 \tag{40}$$

where $A_1 = [T_1]_{E_r}^{E_r}$, $A_2 = [T_2]_{E_{n-r}}^{E_{n-r}}$, $E_r = \{e_1, \ldots, e_r\}$, $E_{n-r} = \{e_{r+1}, \ldots, e_n\}$. Conversely (see Exercise 13), it is obvious that if $[T]_E^E = A$ and $A \stackrel{S}{=} A_1 \dotplus A_2$, then $T = T_1 \dotplus T_2$, where $T_1 = T|W$, $T_2 = T|U$ in which W and U are appropriate invariant subspaces under T. Summarizing these remarks, we have:

Theorem 3.8 *If $T \in \text{Hom}(V,V)$, $\dim V = n$, then T is reducible iff there exists a basis E of V such that $[T]_E^E = A$ is similar to a direct sum $A_1 \dotplus A_2$.*

We say $A \in M_n(R)$ is *reducible* (over R) if it is similar to a direct sum of smaller matrices, and *irreducible* otherwise.

It should not be assumed simply because V is a cyclic invariant subspace under T that T is necessarily irreducible. In view of Theorem 3.8 we need only exhibit a companion matrix which is similar to a direct sum of smaller matrices.

Example 2 Consider the 6×6 matrix over Q, $A = C(f(\lambda))$, where $f(\lambda) = (\lambda - 1)^2(\lambda^2 + 1)^2$. According to Theorem 3.5, the single nonunit invariant factor of $\lambda I_6 - A$ is $f(\lambda)$. Then the list of elementary divisors (over $Q[\lambda]$) of $\lambda I_6 - A$ is

$$h_1(\lambda) = (\lambda - 1)^2 = \lambda^2 - 2\lambda + 1,$$
$$h_2(\lambda) = (\lambda^2 + 1)^2 = \lambda^4 + 2\lambda^2 + 1.$$

Hence from Theorem 3.6, $A \stackrel{S}{=} C(h_1(\lambda)) \dotplus C(h_2(\lambda))$.

The preceding example suggests the following result.

Theorem 3.9 *A matrix $A \in M_n(R)$ is irreducible iff $\lambda I_n - A$ has precisely one elementary divisor. Equivalently, $T \in \text{Hom}(V,V)$ is irreducible iff T has precisely one elementary divisor.*

Proof: Suppose A is reducible so that $A \stackrel{S}{=} B \dotplus C$. Then by Theorem 3.3, $\lambda I_n - A \stackrel{E}{=} \lambda I_n - (B \dotplus C) = (\lambda I_p - B) \dotplus (\lambda I_q - C)$, where B is $p \times p$ and C is $q \times q$. Then by Corollary 3, Section 4.2, the elementary divisors of $\lambda I_n - A$ are those of $\lambda I_p - B$ and $\lambda I_q - C$ taken together. Hence $\lambda I_n - A$ has at least two elementary divisors. Conversely, if $\lambda I_n - A$ has at least two elementary divisors, then the Frobenius normal form of A (Theorem 3.6) is a direct sum of at least two matrices. The statement concerning $T \in \text{Hom}(V,V)$ is now an immediate consequence of Theorem 3.8. ∎

Corollary 2 *Any matrix $A \in M_n(R)$ is similar over R to a direct sum of irreducible matrices:*

$$A \stackrel{S}{=} A_1 \dotplus \cdots \dotplus A_p.$$

If

$$A \overset{S}{=} B_1 \dotplus \cdots \dotplus B_q$$

is another such representation, then $p = q$ *and there exists a* $\sigma \in S_p$ *such that* $A_i \overset{S}{=} B_{\sigma(i)}$, $i = 1, \ldots, p$.

Proof: The Frobenius normal form provides one such representation of A as a direct sum. Now suppose the Frobenius normal form of A is given by

$$A \overset{S}{=} A_1 \dotplus \cdots \dotplus A_p, \tag{41}$$

where $h_1(\lambda), \ldots, h_p(\lambda)$ are the elementary divisors of $\lambda I_n - A$ and $A_i = C(h_i(\lambda))$, $i = 1, \ldots, p$. Suppose

$$A \overset{S}{=} B_1 \dotplus \cdots \dotplus B_q, \tag{42}$$

where each B_i is irreducible and B_t is $m_t \times m_t$, $t = 1, \ldots, q$. Then by Corollary 3, Section 4.2, the elementary divisors of $\lambda I_n - A$ are the elementary divisors of all the $\lambda I_{m_t} - B_t$ taken together. By Theorem 3.9, each $\lambda I_{m_t} - B_t$ has precisely one elementary divisor, say $k_t(\lambda)$. It follows that the two lists $h_1(\lambda), \ldots, h_p(\lambda)$ and $k_1(\lambda), \ldots, k_q(\lambda)$ are identical. Thus $p = q$, and there exists $\sigma \in S_p$ such that

$$h_i(\lambda) = k_{\sigma(i)}(\lambda), \qquad i = 1, \ldots, p.$$

But $B_i \overset{S}{=} C(k_i(\lambda))$ because the characteristic matrix of B_i has $k_i(\lambda)$ as its only elementary divisor [i.e., $C(k_i(\lambda))$ is the Frobenius normal form of B_i]. Thus

$$\begin{aligned} B_{\sigma(i)} &\overset{S}{=} C(k_{\sigma(i)}(\lambda)) \\ &= C(h_i(\lambda)) \\ &\overset{S}{=} A_i, \qquad i = 1, \ldots, p. \quad \blacksquare \end{aligned}$$

It is customary to list the elementary divisors of $T \in \text{Hom}(V, V)$ [or $\lambda I_n - A$, $A \in M_n(R)$] in a standard way: $p_1(\lambda), \ldots, p_r(\lambda)$ is a list of distinct monic prime polynomials. Then

$$\begin{array}{ll} p_1(\lambda)^{e_{11}}, p_1(\lambda)^{e_{12}}, \ldots, p_1(\lambda)^{e_{1n_1}}, & 0 < e_{11} \le e_{12} \le \cdots \le e_{1n_1}, \\ p_2(\lambda)^{e_{21}}, p_2(\lambda)^{e_{22}}, \ldots, p_2(\lambda)^{e_{2n_2}}, & 0 < e_{21} \le \cdots \le e_{2n_2}, \\ \vdots & \\ p_r(\lambda)^{e_{r1}}, p_r(\lambda)^{e_{r2}}, \ldots, p_r(\lambda)^{e_{rn_r}}, & 0 < e_{r1} \le \cdots \le e_{rn_r}, \end{array} \tag{43}$$

is the complete list of elementary divisors of T. If R is a splitting field for the characteristic polynomial of T, then each $p_i(\lambda)$ is a monic binomial of the form $\lambda - \alpha_i$.

Corollary 3 *Let* $T \in \text{Hom}(V, V)$, $\dim V = n$, *and assume that* (43) *is the list of elementary divisors of* T. *If* V *is a direct sum of irreducible invariant subspaces under* T, *then the orders of these subspaces as* $R[\lambda]$*-modules are precisely the polynomials in the list* (43). *Moreover, the polynomials in* (43) *are the polynomials that appear in Theorem 3.1.*

Proof: By choosing the basis of V appropriately, we know from Theorem 3.8 that T has a matrix representation which is the direct sum of irreducible matrices. Each of the matrices in the direct sum is a matrix representation of T restricted to one of the irreducible invariant subspaces in the statement of the theorem, and in view of Theorem 3.9 the corresponding restriction of T has precisely one elementary divisor. It follows from Theorem 3.7 that this elementary divisor is the minimal polynomial of the restriction of T. Now by definition, the order of a typical invariant subspace W as an $R[\lambda]$-module is precisely the minimal polynomial of the matrix representation of $T|W$. If we show that T restricted to each of the cyclic invariant subspaces W_{ij} appearing in Theorem 3.1 is irreducible, then it will follow immediately from Corollary 2 that the orders of the subspaces described in the present corollary are precisely the polynomials in Theorem 3.1, and from the Frobenius normal form these polynomials are also those in the list (43). Now suppose W_{ij} has order $p_i(\lambda)^{e_{ij}}$. Then from Theorem 3.4, $T|W_{ij}$ has $C(p_i(\lambda)^{e_{ij}})$ as a matrix representation. If $T|W_{ij}$ were reducible, it would follow from Theorem 3.9 that the characteristic matrix of $C(p_i(\lambda)^{e_{ij}})$ would have at least two elementary divisors. But since $p_i(\lambda)$ is a prime, the characteristic matrix of $C(p_i(\lambda)^{e_{ij}})$ has only $p_i(\lambda)^{e_{ij}}$ as its single elementary divisor. ∎

Corollary 4 *If $A \in M_n(R)$, then A is similar over R to a diagonal matrix D iff the elementary divisors of $\lambda I_n - A$ are all linear.*

Proof: If the elementary divisors of $\lambda I_n - A$ are all linear, then the Frobenius normal form is a diagonal matrix. Conversely if $A \overset{S}{=} \text{diag}(\alpha_1, \ldots, \alpha_n)$, then $\lambda I_n - A \overset{E}{=} \text{diag}(\lambda - \alpha_1, \ldots, \lambda - \alpha_n)$ and the elementary divisors of $\lambda I_n - A$ are $\lambda - \alpha_1, \ldots, \lambda - \alpha_n$. ∎

Corollary 5 *If $A \in M_n(R)$, then A is similar over R to a diagonal matrix D iff the minimal polynomial of A, $m(\lambda)$, factors into distinct linear factors in $R[\lambda]$.*

Proof: We know from Theorem 3.7 that the minimal polynomial is the nth invariant factor of $\lambda I_n - A$. Thus in the notation (43),

$$m(\lambda) = p_1(\lambda)^{e_{1n_1}} p_2(\lambda)^{e_{2n_2}} \cdots p_r(\lambda)^{e_{rn_r}}.$$

Suppose $m(\lambda)$ is a product of distinct linear factors in $R[\lambda]$. Then by unique factorization, $e_{1n_1} = \cdots = e_{rn_r} = 1$ and $p_t(\lambda) = \lambda - \alpha_t$, $\alpha_i \neq \alpha_j$ for $i \neq j$. But then all the elementary divisors in the list (43) are linear, and hence by Corollary 4, A is similar to a diagonal matrix.

Conversely, if A is similar to a diagonal matrix, then all the elementary divisors of $\lambda I_n - A$ are linear from Corollary 4. The last invariant factor, i.e., $m(\lambda)$, is the product of all the distinct highest degree elementary divisors of $\lambda I_n - A$, and hence $m(\lambda)$ is a product of distinct linear factors. ∎

It should be noted that the elements $\alpha_1, \ldots, \alpha_n$ appearing along the main diagonal in Corollary 4 are precisely the characteristic roots of A.

In order to discuss another of the standard canonical forms for matrices over a field, we define the m-square *auxiliary unit matrix:*

$$U_m = \begin{bmatrix} 0 & 1 & 0 & \cdot & \cdots & 0 \\ 0 & 0 & 1 & 0 & \cdots & 0 \\ \cdot & \cdot & \cdot & \cdot & & \cdot \\ \cdot & \cdot & & \cdot & \cdot & \cdot \\ \cdot & \cdot & & & \cdot & \cdot \\ \cdot & \cdot & & & \cdot & 1 \\ 0 & 0 & \cdot & \cdots & \cdot & 0 \end{bmatrix}$$

(if $m = 1$ then U_1 is the 1×1 matrix [0]). If $g(\lambda) = (\lambda - \alpha)^m$, $\alpha \in R$, then the *hypercompanion matrix of* $g(\lambda)$ is defined as

$$H(g(\lambda)) = \alpha I_m + U_m.$$

Theorem 3.10 *If* $g(\lambda) = (\lambda - \alpha)^m$, *then*

$$H(g(\lambda)) \overset{\mathrm{S}}{=} C(g(\lambda)).$$

Proof: If $m = 1$, then $H(g(\lambda)) = [\alpha] = C(g(\lambda))$. If $m > 1$, then

$$H(g(\lambda)) = \begin{bmatrix} \alpha & 1 & 0 & \cdots & 0 \\ 0 & \cdot & \cdot & & \cdot \\ \cdot & & \cdot & \cdot & \cdot \\ \cdot & & & \cdot & \cdot \\ \cdot & & & \cdot & 0 \\ \cdot & & & \cdot & 1 \\ 0 & \cdot & \cdots & \cdot & \alpha \end{bmatrix}$$

and clearly $\det(\lambda I_m - H(g(\lambda)))(m\,|\,1) = (-1)^{m-1}$. Thus $d_0 = \cdots = d_{m-1} = 1$, where the d_i are the determinantal divisors of $\lambda I_m - H(g(\lambda))$. Also, it is obvious that $\det(\lambda I_m - H(g(\lambda))) = (\lambda - \alpha)^m = g(\lambda)$, and hence the invariant factors of $\lambda I_m - H(g(\lambda))$ are $\overbrace{1, \ldots, 1}^{m-1}, g(\lambda)$. But according to Theorem 3.5, these are precisely the invariant factors of $\lambda I_m - C(g(\lambda))$. Thus by Theorem 3.3, $H(g(\lambda)) \overset{\mathrm{S}}{=} C(g(\lambda))$. ∎

The following theorem is classical.

Theorem 3.11 (*Jordan Normal Form*) *Let* $A \in M_n(R)$, *and let* R *be a splitting field for the characteristic polynomial of* A. *Then the polynomials* $p_1(\lambda), \ldots, p_r(\lambda)$ *in the list* (43) *are of the form*

$$p_t(\lambda) = \lambda - \alpha_t, \qquad t = 1, \ldots, r, \tag{44}$$

and the matrix A is similar over R to the direct sum of the hypercompanion matrices of the polynomials (43).

Proof: The product of the elementary divisors of $\lambda I_n - A$ is the characteristic polynomial of A, and by hypothesis this factors into linear factors in $R[\lambda]$. Since $R[\lambda]$ is a U.F.D. and each of the elementary divisors is a power of a prime in $R[\lambda]$, it follows that each $p_t(\lambda)$ is linear. The Frobenius normal form of A is

$$\sum_{i=1}^{r}{}^{\cdot} \sum_{j=1}^{n_i}{}^{\cdot} C((\lambda - \alpha_i)^{e_{ij}})$$

which by Theorem 3.10 is similar to

$$\sum_{i=1}^{r}{}^{\cdot} \sum_{j=1}^{n_i}{}^{\cdot} H((\lambda - \alpha_i)^{e_{ij}}). \quad \blacksquare \tag{45}$$

The matrix (45) is called the *Jordan normal form* of A.

Example 3 (a) Suppose $R = \mathbf{R}$, the real number field. Then the irreducible monic polynomials in $\mathbf{R}[\lambda]$ are of the form $\lambda - \alpha$ and $\lambda^2 + c_1\lambda + c_0$, $c_1{}^2 - 4c_0 < 0$. Thus the possibilities for the elementary divisors of the characteristic matrix of a matrix in $M_2(\mathbf{R})$ are $\lambda - \alpha_1$, $\lambda - \alpha_2$, $\alpha_i \in \mathbf{R}$; $(\lambda - \alpha)^2$, $\alpha \in \mathbf{R}$; $\lambda^2 + c_1\lambda + c_0$, $c_1{}^2 - 4c_0 < 0$. Thus any $A \in M_2(\mathbf{R})$ is similar over $\mathbf{R}$ to one of the following three types of matrices:

$$\begin{bmatrix} \alpha_1 & 0 \\ 0 & \alpha_2 \end{bmatrix}, \quad \begin{bmatrix} \alpha & 1 \\ 0 & \alpha \end{bmatrix}, \quad \begin{bmatrix} 0 & -c_0 \\ 1 & -c_1 \end{bmatrix}.$$

(b) We compute the Jordan normal form over the complex numbers $\mathbf{C}$ of the matrix $A = C(f(\lambda))$, $f(\lambda) = (\lambda - 1)^2 (\lambda^2 + 1)$. The single nonunit determinantal divisor of $\lambda I_4 - A$ is $f(\lambda)$, and hence the only nonunit invariant factor is $f(\lambda)$. Factoring $f(\lambda)$, we have

$$f(\lambda) = (\lambda - 1)^2(\lambda + i)(\lambda - i),$$

and hence the list of elementary divisors of $\lambda I_4 - A$ is

$$\begin{gathered} (\lambda - 1)^2, \\ \lambda + i, \\ \lambda - i. \end{gathered}$$

We have

$$\begin{aligned} H((\lambda - 1)^2) &= \begin{bmatrix} 1 & 1 \\ 0 & 1 \end{bmatrix}, \\ H(\lambda + i) &= [-i], \\ H(\lambda - i) &= [i]. \end{aligned}$$

Thus the Jordan normal form of A is

$$\begin{bmatrix} 1 & 1 \\ 0 & 1 \end{bmatrix} \dotplus [-i] \dotplus [i].$$

(c) We compute the Jordan normal form over **R** of

$$A = \begin{bmatrix} 1 & 3 & -1 \\ 0 & 1 & 2 \\ 0 & 0 & 1 \end{bmatrix}.$$

The characteristic matrix is

$$\lambda I_3 - A = \begin{bmatrix} \lambda - 1 & -3 & 1 \\ 0 & \lambda - 1 & -2 \\ 0 & 0 & \lambda - 1 \end{bmatrix}.$$

Thus $d_0 = 1$, $d_1 = 1$, and two of the 2×2 subdeterminants are $(\lambda - 1)^2$ and $\lambda - 7$ (to within unit multiples). It follows that $d_2 = 1$, and clearly $d_3 = (\lambda - 1)^3$. The invariant factors are $q_1 = 1$, $q_2 = 1$, $q_3 = (\lambda - 1)^3$, and $(\lambda - 1)^3$ is the single elementary divisor. Hence

$$A \overset{S}{=} H((\lambda - 1)^3) = I_3 + U_3.$$

(d) If $A \in M_n(R)$ and $A^2 = A$, we discuss the possibilities for the Frobenius normal form of A. Since $A^2 - A = 0$, it follows that the minimal polynomial of A, $m(\lambda)$, is a divisor of $\lambda^2 - \lambda$. Thus the possibilities for $m(\lambda)$ are: (i) $m(\lambda) = \lambda$; (ii) $m(\lambda) = \lambda - 1$; (iii) $m(\lambda) = \lambda(\lambda - 1)$. In case (i), $A = 0$; in (ii), $A - I_n = 0$, i.e., $A = I_n$; in (iii), the highest degree elementary divisors are λ and $\lambda - 1$, i.e., the factors of $m(\lambda)$. Thus the Frobenius normal form of A is a direct sum of companion matrices of λ and $\lambda - 1$. Thus A is similar over R to a direct sum of 1-square matrices of the form [0] or [1]. But then

$$A \overset{S}{=} I_r \dotplus 0_{n-r},$$

where $r = \rho(A)$, I_r is the r-square identity matrix and 0_{n-r} is the $(n - r)$-square zero matrix.

(e) Let $B \in M_2(\mathbf{R})$ be the companion matrix of the irreducible binomial $f(\lambda) = \lambda^2 + c_1\lambda + c_0 \in \mathbf{R}[\lambda]$,

$$B = \begin{bmatrix} 0 & -c_0 \\ 1 & -c_1 \end{bmatrix}.$$

Define $A \in M_{2p}(\mathbf{R})$ to be the matrix

$$\begin{bmatrix} B & 1 & & & \\ & B & 1 & & \\ & & B & & \\ & & & \ddots & 1 \\ & & & & B \end{bmatrix}$$

in which there are p occurrences of B down the main diagonal of A, 1's occur in positions (2,3), (4,5), . . . , $(2p - 2, 2p - 1)$, and all other entries of A are 0. Then

$$A \overset{S}{=} C(f(\lambda)^p).$$

For it is easy to see that

$$\det (\lambda I_{2p} - A)(2p \mid 1) = c_0{}^p,$$

and since $f(\lambda)$ is irreducible, $c_0 \neq 0$. Hence $d_{2p-1} = 1$, and the determinantal divisors of $\lambda I_{2p} - A$ are $d_0 = \cdots = d_{2p-1} = 1$, $d_{2p} = \det(\lambda I_{2p} - A) = \{\det(\lambda I_2 - B)\}^p = f(\lambda)^p$. Thus $\lambda I_{2p} - A$ has the single elementary divisor $f(\lambda)^p$ as does the characteristic matrix of $C(f(\lambda)^p)$.

Our next sequence of results will help answer questions of the following kind: If we know the Frobenius normal form of $A \in M_n(R)$, then what is the Frobenius normal form of A^2 or A^3? We illustrate this kind of problem in the following example.

Example 4 (a) Let $f(\lambda) = p(\lambda)^e \in R[\lambda]$, $p(\lambda)$ a monic prime polynomial, $\deg f(\lambda) = n > 0$. Let $A = C(f(\lambda))$. Since $\det(\lambda I_n - A) = f(\lambda)$, it follows that $f(0) = (-1)^n \det A$ so that $f(0) \neq 0$ iff A is nonsingular. Now write

$$f(\lambda) = c_0 + c_1\lambda + \cdots + c_{n-1}\lambda^{n-1} + \lambda^n, \qquad c_0 \neq 0,$$

and

$$g(\lambda) = \lambda^n + \frac{c_1}{c_0}\lambda^{n-1} + \cdots + \frac{c_{n-1}}{c_0}\lambda + \frac{1}{c_0}.$$

Then

$$\begin{aligned} g(A^{-1}) &= A^{-n} + \frac{c_1}{c_0}A^{-(n-1)} + \cdots + \frac{c_{n-1}}{c_0}A^{-1} + \frac{1}{c_0}I_n \\ &= \frac{1}{c_0}A^{-n}(A^n + c_{n-1}A^{n-1} + \cdots + c_0 I_n) \\ &= \frac{1}{c_0}A^{-n}f(A) \\ &= 0 \qquad \text{(by the Cayley-Hamilton theorem).} \end{aligned}$$

It follows that if $h(\lambda)$ is the minimal polynomial of A^{-1}, then $h(\lambda) \mid g(\lambda)$. Let $g(\lambda) = h(\lambda)k(\lambda)$, $\deg h(\lambda) = p$, $\deg k(\lambda) = q$, $p + q = n$. Observe that

$$\begin{aligned} c_0\lambda^n g\left(\frac{1}{\lambda}\right) &= c_0\lambda^n\left(\frac{1}{\lambda^n} + \frac{c_1}{c_0}\frac{1}{\lambda^{n-1}} + \cdots + \frac{c_{n-1}}{c_0}\frac{1}{\lambda} + \frac{1}{c_0}\right) \\ &= c_0 + c_1\lambda + \cdots + c_{n-1}\lambda^{n-1} + \lambda^n \\ &= f(\lambda). \end{aligned}$$

Now $h(0)k(0) = g(0) = 1/c_0 \neq 0$. Thus the constant terms of both $h(\lambda)$ and $k(\lambda)$ are nonzero; call them h_0 and k_0, respectively. Then set $\varphi(\lambda) = h_0\lambda^p h(1/\lambda) \in R[\lambda]$ and $\theta(\lambda) = k_0\lambda^q k(1/\lambda) \in R[\lambda]$. From $g(\lambda) = h(\lambda)k(\lambda)$, $p + q = n$, and $c_0 = h_0 k_0$, we have

$$\begin{aligned} \varphi(\lambda)\theta(\lambda) &= h_0 k_0 \lambda^{p+q} h\left(\frac{1}{\lambda}\right) k\left(\frac{1}{\lambda}\right) = c_0\lambda^n g\left(\frac{1}{\lambda}\right) \\ &= f(\lambda). \end{aligned}$$

Thus $\varphi(\lambda) \mid f(\lambda)$. Note also that $\deg \varphi(\lambda) = p$. Now $h(\lambda)$ is the minimal polynomial of A^{-1}; so

$$\begin{aligned} 0 = h(A^{-1}) &= A^{-p} + h_{p-1}A^{-(p-1)} + \cdots + h_1 A^{-1} + h_0 I_n \\ &= A^{-p}(I_n + h_{p-1}A + \cdots + h_1 A^{p-1} + h_0 A^p). \end{aligned}$$

Hence $$I_n + h_{p-1}A + \cdots + h_1A^{p-1} + h_0A^p = 0. \tag{46}$$

But $$\frac{1}{h_0}\varphi(\lambda) = \lambda^p h\left(\frac{1}{\lambda}\right) = \lambda^p\left(h_0 + h_1\frac{1}{\lambda} + \cdots + h_{p-1}\frac{1}{\lambda^{p-1}} + \frac{1}{\lambda^p}\right)$$
$$= h_0\lambda^p + h_1\lambda^{p-1} + \cdots + h_{p-1}\lambda + 1.$$

Hence from (46),

$$\frac{1}{h_0}\varphi(A) = h_0A^p + h_1A^{p-1} + \cdots + h_{p-1}A + I_n$$
$$= 0.$$

In other words, $\varphi(A) = 0$. But since A is the companion matrix of $f(\lambda)$, the minimal polynomial of A is $f(\lambda)$. Thus $p = \deg \varphi(\lambda) = \deg h(\lambda) = n$. It follows that $g(\lambda)$ is the minimal polynomial of A^{-1}. Also since $\deg g(\lambda) = n$, $g(\lambda)$ is the characteristic polynomial of A^{-1}. Hence the invariant factors of $\lambda I_n - A^{-1}$ are $\overbrace{1, \ldots, 1}^{n-1}, g(\lambda)$. These are the invariant factors of $\lambda I_n - C(g(\lambda))$, and hence $A^{-1} \stackrel{S}{=} C(g(\lambda))$. Now, if A^{-1} were reducible, i.e., $A^{-1} \stackrel{S}{=} C \dotplus D$, then (see Exercise 14) $A \stackrel{S}{=} C^{-1} \dotplus D^{-1}$ and A would be reducible. But A is a companion matrix of a power of a prime polynomial and hence by Theorem 3.9 is irreducible. Thus $C(g(\lambda))$ is irreducible and, by Theorem 3.9 again, $\lambda I_n - C(g(\lambda))$ has precisely one elementary divisor. Hence $g(\lambda)$ is a power of a prime polynomial and $C(g(\lambda))$ is in Frobenius normal form.

We have proved: *If $f(\lambda) = p(\lambda)^e = c_0 + c_1\lambda + \cdots + c_{n-1}\lambda^{n-1} + \lambda^n$, $\deg f(\lambda) > 0$, $p(\lambda)$ a prime monic polynomial, $f(0) \neq 0$ and $A = C(f(\lambda))$, then A^{-1} exists and the Frobenius normal form of A^{-1} is*

$$C(g(\lambda)),$$

where $g(\lambda) = \lambda^n + \dfrac{c_1}{c_0}\lambda^{n-1} + \cdots + \dfrac{c_{n-1}}{c_0}\lambda + \dfrac{1}{c_0} = \dfrac{1}{c_0}\lambda^n f\left(\dfrac{1}{\lambda}\right)$.

(b) If $h_1(\lambda), \ldots, h_p(\lambda)$ are the elementary divisors of the characteristic matrix of A, $\det A \neq 0$, then the Frobenius normal form of A^{-1} is $\sum_{i=1}^{\cdot p} C(k_i(\lambda))$, where $k_i(\lambda) = (1/h_{i0})\lambda^{e_i}h_i(1/\lambda)$, $\deg h_i(\lambda) = e_i$ [h_{i0} is the constant term of $h_i(\lambda)$]. For, $A^{-1} \stackrel{S}{=} \sum_{i=1}^{\cdot p} C(k_i(\lambda))$ by part (a). Each of the $C(k_i(\lambda))$ is irreducible as we also saw in part (a). It follows from Corollary 2 that to within order the $C(k_i(\lambda))$ are similar to the blocks in the Frobenius normal form of A^{-1}. Since these are companion matrices of monic polynomials, the result follows.

(c) If A is the matrix in Example 2, find the Frobenius normal form of A^{-1} over Q. We saw in Example 2 that the Frobenius normal form for A is

$$A \stackrel{S}{=} C(h_1(\lambda)) \dotplus C(h_2(\lambda)),$$

$h_1(\lambda) = (\lambda - 1)^2 = \lambda^2 - 2\lambda + 1$, $h_2(\lambda) = \lambda^4 + 2\lambda^2 + 1$. Now

$$k_1(\lambda) = \lambda^2 - 2\lambda + 1, \qquad k_2(\lambda) = \lambda^4 + 2\lambda^2 + 1$$

so that the Frobenius normal form for A^{-1} is

$$A^{-1} \stackrel{S}{=} C(k_1(\lambda)) \dotplus C(k_2(\lambda)).$$

Observe that A^{-1} is similar to A.

Let $p(\lambda) \in R[\lambda]$. In order to compute the elementary divisors of the characteristic matrix of $p(A)$ in terms of the elementary divisors of the characteristic matrix of A, we shall require the following result.

Theorem 3.12 *Let U_n be the n-square auxiliary unit matrix, and set*

$$A = c_1U_n^{\,q} + c_2U_n^{\,q+1} + \cdots + c_{n-q}U_n^{\,n-1}$$

$$= \left[\begin{array}{ccccccc} 0 & \cdots & 0 & \overset{\text{col. } q+1}{\overset{\downarrow}{c_1}} & c_2 & \cdots & c_{n-q} \\ \vdots & & & & \ddots & \ddots & \vdots \\ \vdots & & & & & \ddots & c_2 \\ \vdots & & & & & & c_1 \\ \vdots & & & & & & 0 \\ \vdots & & & & & & \vdots \\ 0 & & \cdot & & \cdot & \cdot & 0 \end{array}\right] \begin{array}{l} \\ \\ ,\qquad c_1 \neq 0. \\ \leftarrow \text{row } n-q \\ \\ \\ \\ \end{array} \tag{47}$$

Let $n = dq + r, 0 \leq r \leq q - 1$. Then the elementary divisors of $\lambda I_n - A$ are

$$\lambda^d, \qquad q - r \quad \textit{times}$$

and

$$\lambda^{d+1}, \qquad r \textit{ times.}$$

Proof: Observe first that

$$\rho(U_n^{\,t}) = n - t, \qquad t = 1, 2, \ldots, n.$$

Now

$$A^k = c_1^{\,k}U_n^{\,kq} + (\text{linear combination of higher powers of } U_n)$$

so that

$$\rho(A^k) = n - kq, \qquad k = 1, \ldots, \left[\frac{n}{q}\right] \tag{48}$$

where $d = [n/q]$ is the largest integer in n/q, i.e., $n = dq + r, 0 \leq r \leq q - 1$. Let J be the Jordan normal form of A (possibly over an extension field of R). Since $\det(\lambda I_n - A) = \lambda^n$, all the elementary divisors of $\lambda I_n - A$ are powers of λ and hence $J \in M_n(R)$. It follows (see Exercise 15) that $A \overset{S}{=} J$ over R. Now suppose there are ν_i elementary divisors of $\lambda I_n - A$ of the form λ^i. Since $A^{d+1} = 0$, we know that there can be no hypercompanion matrices appearing in J which are larger than $(d + 1)$-square, i.e., $\nu_{d+2} = \cdots = \nu_n = 0$. The 1-square matrices in J are $H(\lambda) = [0]$. Since $\rho(H(\lambda^i)) = i - 1$, we thus have

$$\begin{aligned} \nu_2 + 2\nu_3 + \cdots + (d-1)\nu_d + d\nu_{d+1} &= \rho(J) = \rho(A) = n - q \\ \nu_3 + 2\nu_4 + \cdots + (d-2)\nu_d + (d-1)\nu_{d+1} &= \rho(J^2) = \rho(A^2) = n - 2q \\ &\ \ \vdots \\ \nu_d + 2\nu_{d+1} &= \rho(J^{d-1}) = \rho(A^{d-1}) = n - (d-1)q \\ \nu_{d+1} &= \rho(J^d) = \rho(A^d) = n - dq. \end{aligned} \tag{49}$$

The system (49) can be considered as a system of d equations for the determination of the d integers $\nu_2, \ldots, \nu_{d+1}$. Observe that the coefficient matrix has determinant 1, and hence the solutions are uniquely determined. In fact, if we set $\nu_{d+1} = r$, $\nu_d = q - r$ and $\nu_2 = \cdots = \nu_{d-1} = 0$, then the left side of the kth equation becomes

$$\begin{aligned}\nu_{k+1} + 2\nu_{k+2} + \cdots + (d-k-1)\nu_{d-1} + (d-k)\nu_d + (d-k+1)\nu_{d+1} \\ = (d-k)(q-r) + (d-k+1)r \\ = (d-k)q + r.\end{aligned}$$

But the right side of the kth equation is

$$n - kq = dq + r - kq = (d-k)q + r.$$

Thus there are r elementary divisors λ^{d+1} and $q - r$ elementary divisors λ^d. Their product is

$$\begin{aligned}(\lambda^{d+1})^r(\lambda^d)^{q-r} &= \lambda^{dr+r}\lambda^{dq-dr} = \lambda^{dq+r} \\ &= \lambda^n\end{aligned}$$

and hence we have accounted for the complete list of elementary divisors. ∎

Theorem 3.13 *If R is a field of characteristic 0, $\alpha \in R$, and $p(\lambda) \in R[\lambda]$, then*

$$p(\alpha I_n + U_n) = p(\alpha)I_n + \sum_{j=1}^{n-1} \frac{p^{(j)}(\alpha)}{j!} U_n^{\,j} \tag{50}$$

where $p^{(j)}(\alpha)$ is the jth derivative of $p(\lambda)$ evaluated at α.

Proof: Set $p(\lambda) = a_k\lambda^k + a_{k-1}\lambda^{k-1} + \cdots + a_1\lambda + a_0$ so that

$$\begin{aligned}p(\alpha I_n + U_n) &= \sum_{s=0}^{k} a_s(\alpha I_n + U_n)^s \\ &= \sum_{s=0}^{k} a_s \sum_{t=0}^{s} \tbinom{s}{t}\alpha^t U_n^{\,s-t} \qquad (U_n^{\,0} = I_n) \\ &= \sum_{j=0}^{k}\left(\sum_{s-t=j} \frac{s!}{(s-t)!\,t!} a_s\alpha^t\right) U_n^{\,j} \\ &= \sum_{j=0}^{k}\left(\sum_{s=j}^{k} \frac{s!}{j!\,(s-j)!} a_s\alpha^{s-j}\right) U_n^{\,j} \\ &= \sum_{j=0}^{k} \frac{1}{j!}\left(\sum_{s=j}^{k} \frac{s!}{(s-j)!} a_s\alpha^{s-j}\right) U_n^{\,j}.\end{aligned} \tag{51}$$

Now observe that

$$p^{(j)}(\alpha) = \sum_{s=j}^{k} \frac{s!}{(s-j)!} a_s\alpha^{s-j}$$

so that (51) becomes

$$p(\alpha I_n + U_n) = \sum_{j=0}^{k} \frac{p^{(j)}(\alpha)}{j!} U_n^{\,j}. \tag{52}$$

If $k \geq n$, then since $U_n^{\,n} = 0$, we can cut off the sum (52) at $n - 1$ to obtain (50). If $k < n$, then since $p^{(k+1)}(\lambda) = \cdots = p^{(n-1)}(\lambda) = 0$, we can trivially replace k by $n - 1$ in (52) again. Thus, under any circumstances (50) is valid. ∎

Theorem 3.14 *Let $A \in M_n(R)$, where R is a splitting field for the characteristic polynomial of A and R is of characteristic 0. Let $p(\lambda) \in R[\lambda]$. Then the elementary divisors of $\lambda I_n - A$ and $\lambda I_n - p(A)$ are related according to the following table:*

$\lambda I_n - A$	$\lambda I_n - p(A)$	
(i) $\lambda - \alpha$	$\lambda - p(\alpha)$	
(ii) $(\lambda - \alpha)^e, e > 1$	$\lambda - p(\alpha)$, *e times if* $p^{(j)}(\alpha) = 0, j = 1, \ldots, e - 1$	
(iii) $(\lambda - \alpha)^e,\ e > 1$	$(\lambda - p(\alpha))^d$, $q - r$ *times* $(\lambda - p(\alpha))^{d+1}$, r *times*	*where q is the least positive integer less than e for which $p^{(q)}(\alpha) \neq 0$, and $e = dq + r$, $0 \leq r \leq q - 1$*

Proof: We first note in general that if $a \in R$, then $h(\lambda)$ is an elementary divisor of $\lambda I_n - A$ iff $h(\lambda + a)$ is an elementary divisor of $\lambda I_n - (A - aI_n)$. For, $\lambda I_n - (A - aI_n) = (\lambda + a)I_n - A$. Set $\mu = \lambda + a$, and we know that $h(\mu)$ is an elementary divisor of $\mu I_n - A$. But $h(\mu) = h(\lambda + a)$.

We have $A \overset{S}{=} J$, the direct sum of the hypercompanion matrices of the elementary divisors of $\lambda I_n - A$. Moreover, $p(A) \overset{S}{=} p(J)$ and $p(J)$ is a direct sum of matrices of the form

$$p(H((\lambda - \alpha)^e)) = p(\alpha I_e + U_e),$$

where $(\lambda - \alpha)^e$ is an elementary divisor of $\lambda I_n - A$. But according to Theorem 3.13,

$$p(\alpha I_e + U_e) = p(\alpha)I_e + \sum_{j=1}^{e-1} \frac{p^{(j)}(\alpha)}{j!} U_e^{\,j}. \tag{53}$$

Consider the matrix

$$p(\alpha I_e + U_e) - p(\alpha)I_e = \sum_{j=1}^{e-1} \frac{p^{(j)}(\alpha)}{j!} U_e^{\,j}. \tag{54}$$

The possibilities are:

(i) $e = 1$. Then of course the matrix (53) collapses to $[p(\alpha)]$ with the corresponding elementary divisor of the characteristic matrix of $p(J)$ equal to $\lambda - p(\alpha)$.

(ii) $e > 1$ and $p^{(j)}(\alpha) = 0$, $j = 1, 2, \ldots, e - 1$. Then the matrix (53) is equal to $p(\alpha)I_e$, and hence the corresponding elementary divisors of the characteristic matrix of $p(J)$ are $\lambda - p(\alpha)$, e times.

(iii) $e > 1$ and $p^{(j)}(\alpha) \neq 0$ for some j, $1 \leq j \leq e - 1$. Then let q be the least such j, and the matrix (54) becomes

$$\sum_{j=q}^{e-1} \frac{p^{(j)}(\alpha)}{j!} U_e^{\,j}. \tag{55}$$

From Theorem 3.12 the elementary divisors of the characteristic matrix of (55) are

$$\lambda^d, \qquad q - r \text{ times}$$

and

$$\lambda^{d+1}, \qquad r \text{ times}$$

where

$$e = dq + r, \qquad 0 \leq r \leq q - 1. \tag{56}$$

Hence by the remark in the first paragraph of the proof, the elementary divisors of the characteristic matrix of (53) are

$$(\lambda - p(\alpha))^d, \qquad q - r \text{ times}$$

and

$$(\lambda - p(\alpha))^{d+1}, \qquad r \text{ times}$$

where d and r are defined in (56). ∎

If both $r > 0$ and $q - r > 0$, i.e., case (iii) in the above table, then we say that $(\lambda - \alpha)^e$ *splits* into $(\lambda - p(\alpha))^d$ and $(\lambda - p(\alpha))^{d+1}$.

Observe that it follows from Theorem 3.14 that the eigenvalues of $p(A)$ are precisely the numbers $p(\alpha)$, where α is an eigenvalue of A.

Example 5 Let m be a positive integer. We discuss the elementary divisors of the characteristic matrix of A^m over a field R of characteristic 0 containing the characteristic roots of A. We take $p(\lambda) = \lambda^m$. If $\lambda - \alpha$ is an elementary divisor of $\lambda I_n - A$, then $\lambda - \alpha^m$ is an elementary divisor of $\lambda I_n - A^m$. If $\alpha \neq 0$ and $(\lambda - \alpha)^e$, $e > 1$, is an elementary divisor of $\lambda I_n - A$, then $p'(\alpha) \neq 0$ so that by (iii) in Theorem 3.14, $q = 1$, $d = e$, $r = 0$, and $(\lambda - \alpha^m)^e$ is the corresponding elementary divisor of $\lambda I_n - A^m$. If λ^e, $e > 1$, is an elementary divisor of $\lambda I - A$ (i.e., $\alpha = 0$), then $p^{(j)}(0) = 0$, $j = 1, \ldots, m - 1$, $p^{(m)}(0) = m! \neq 0$. There are two cases: $m \leq e - 1$ and $m \geq e$. If $m \leq e - 1$, then $q = m$; if $e = md + r$, $0 \leq r \leq m - 1$, then λ^d, $m - r$ times, and λ^{d+1}, r times, are the corresponding elementary divisors of $\lambda I_n - A^m$. If $m \geq e$, then by (ii) in Theorem 3.14, λ is an elementary divisor of $\lambda I_n - A^m$, e times.

In applications of matrix theory to systems of ordinary differential equations it is important to consider a more general function of a matrix A than simply a scalar polynomial in A. Thus let $A \in M_n(R)$, where R is a subfield of the complex numbers containing the characteristic roots of A. Let $\alpha_1, \ldots, \alpha_s$ be the distinct characteristic roots of A, and let

$$m(\lambda) = (\lambda - \alpha_1)^{\mu_1}(\lambda - \alpha_2)^{\mu_2} \cdots (\lambda - \alpha_s)^{\mu_s}$$

be the minimal polynomial of A. Let $f: R \rightarrow R$ be a function which satisfies the following conditions: The function values and derivatives

$$f^{(j)}(\alpha_k), \qquad j = 0, \ldots, \mu_k - 1, \; k = 1, \ldots, s,$$

all exist where μ_k is the degree of the highest degree elementary divisor of $\lambda I_n - A$ of the form $(\lambda - \alpha_k)^e$. If α_k is real, then $f^{(j)}$ is the jth derivative of f as a function of a real variable; if α_k is complex, then $f^{(j)}$ is the jth derivative as a function of a complex variable. Under these circumstances we say that *f is defined at A*. Let $S \in M_n(R)$ satisfy

$$S^{-1}AS = J = \sum_{i=1}^{r}{}^{\cdot} H_i,$$

where J is the Jordan normal form of A and the H_i are the hypercompanion matrices of the elementary divisors of $\lambda I_n - A$. If $H_i = H((\lambda - \alpha_k)^e)$, $1 \le e \le \mu_k$, then define the value $f(H_i)$ by

$$f(H_i) = \sum_{j=0}^{e-1} \frac{f^{(j)}(\alpha_k)}{j!} U_e^{\,j}. \tag{57}$$

Set

$$f(A) = S \sum_{i=1}^{r}{}^{\cdot} f(H_i) S^{-1}. \tag{58}$$

The matrix $f(A)$ in (58) is called the *value of f at A*. We show in our next result that $f(A)$ is independent of the matrix S that brings A to Jordan normal form.

Theorem 3.15 *Let f be a function defined at A. Let $p(\lambda)$ be a polynomial in $R[\lambda]$ for which*

$$p^{(j)}(\alpha_k) = f^{(j)}(\alpha_k), \qquad j = 0, \ldots, \mu_k - 1, \quad k = 1, \ldots, s, \tag{59}$$

where μ_k is the degree of the highest degree elementary divisor of $\lambda I_n - A$ involving α_k. Then

$$p(A) = f(A). \tag{60}$$

Moreover, the value $f(A)$ is independent of the matrix S in (58) that brings A to Jordan normal form.

Proof: It follows immediately from (57), (59), and Theorem 3.13 that

$$p(H_i) = f(H_i)$$

and thus

$$\begin{aligned} f(A) &= S \sum_{i=1}^{r}{}^{\cdot} f(H_i) S^{-1} = S \sum_{i=1}^{r}{}^{\cdot} p(H_i) S^{-1} \\ &= Sp\left(\sum_{i=1}^{r}{}^{\cdot} H_i\right) S^{-1} = Sp(J)S^{-1} = p(SJS^{-1}) \\ &= p(A). \end{aligned}$$

Since $p(A)$ obviously does not depend on S, it follows that if a polynomial $p(\lambda)$ exists for which the conditions (59) obtain, then $f(A)$ is independent of S. The proof of the existence of such a polynomial is found in Exercises 16 to 18, and it is shown there that in fact $p(\lambda)$ can be chosen so that $\deg p(\lambda) < \deg m(\lambda)$. ∎

Observe that Theorem 3.15 tells us that if f is defined at A, then f is defined at any matrix similar to A and

$$f(Q^{-1}AQ) = p(Q^{-1}AQ) = Q^{-1}p(A)Q$$
$$= Q^{-1}f(A)Q. \tag{61}$$

Also note that the value $f(A)$ depends only on the values $f^{(j)}(\alpha_k)$ so that if $g^{(j)}(\alpha_k) = f^{(j)}(\alpha_k)$, $j = 0, \ldots, \mu_k - 1$, $k = 1, \ldots, s$, then $g(A) = f(A)$.

Example 6 (a) Let i and j be positive integers, $1 \leq i \leq s$, $1 \leq j \leq \mu_i$, and let $f_{ij}(\lambda)$ be a polynomial which satisfies

$$f_{ij}^{(t)}(\alpha_k) = \delta_{ik}\delta_{t,j-1}, \qquad k = 1, \ldots, s, \quad t = 1, \ldots, \mu_k - 1.$$

As we noted at the end of Theorem 3.15, $f_{ij}(\lambda)$ can be chosen so that $\deg f_{ij}(\lambda) < \deg m(\lambda)$ (see Exercise 16). In other words, $f_{ij}^{(j-1)}(\alpha_i) = 1$, but the other values of $f_{ij}^{(t)}(\alpha_k)$ are all 0. We define the matrices

$$Z_{ij} = f_{ij}(A). \tag{62}$$

The matrices Z_{ij} are called the *component matrices* of A.

(b) If f is any function defined at A, consider the polynomial

$$g(\lambda) = \sum_{i=1}^{s}\sum_{j=1}^{\mu_i} f^{(j-1)}(\alpha_i)f_{ij}(\lambda).$$

Then for $0 \leq t \leq \mu_k - 1$, we have

$$g^{(t)}(\alpha_k) = \sum_{i=1}^{s}\sum_{j=1}^{\mu_i} f^{(j-1)}(\alpha_i)f_{ij}^{(t)}(\alpha_k) = \sum_{i=1}^{s}\sum_{j=1}^{\mu_i} f^{(j-1)}(\alpha_i)\delta_{ik}\delta_{t,j-1}$$
$$= f^{(t)}(\alpha_k).$$

In other words, the functions g and f and their derivatives agree on the eigenvalues of A. It follows that

$$f(A) = g(A) = \sum_{i=1}^{s}\sum_{j=1}^{\mu_i} f^{(j-1)}(\alpha_i)f_{ij}(A)$$
$$= \sum_{i=1}^{s}\sum_{j=1}^{\mu_i} f^{(j-1)}(\alpha_i)Z_{ij}. \tag{63}$$

(c) As we noted at the end of the proof of Theorem 3.15 (see Exercise 18), we can find a polynomial $p(\lambda)$ such that $p(A) = f(A)$ and $\deg p(\lambda) < \deg m(\lambda)$. Now suppose

$$\sum_{i=1}^{s}\sum_{j=1}^{\mu_i} c_{ij}Z_{ij} = 0.$$

Then

$$\sum_{i=1}^{s}\sum_{j=1}^{\mu_i} c_{ij}f_{ij}(A) = 0,$$

so that

$$g(\lambda) = \sum_{i=1}^{s}\sum_{j=1}^{\mu_i} c_{ij}f_{ij}(\lambda)$$

satisfies
$$g(A) = 0.$$
Since $\deg f_{ij}(\lambda) < \deg m(\lambda)$, we conclude that $g(\lambda) = 0$. But then for each k, $1 \leq k \leq s$, and $t = 0, \ldots, \mu_k - 1$ we have

$$0 = \sum_{i=1}^{s}\sum_{j=1}^{\mu_i} c_{ij} f_{ij}^{(t)}(\alpha_k) = \sum_{i=1}^{s}\sum_{j=1}^{\mu_i} c_{ij}\delta_{ik}\delta_{t,j-1}$$
$$= c_{k,t+1}, \qquad t = 0, \ldots, \mu_k - 1,$$

i.e., $c_{ij} = 0$, $i = 1, \ldots, s$, $j = 1, \ldots, \mu_i$. Thus the component matrices Z_{ij} are l.i. over R.

(d) Let $f(\lambda) = e^{t\lambda}$ and $A = \begin{bmatrix} 3 & -1 & 1 \\ 2 & 0 & 1 \\ 1 & -1 & 2 \end{bmatrix}$. We compute $f(A)$. The first problem, if we wish to use (63), is to compute the elementary divisors of $\lambda I_3 - A$. Now

$$\lambda I_3 - A = \begin{bmatrix} \lambda - 3 & 1 & -1 \\ -2 & \lambda & -1 \\ -1 & 1 & \lambda - 2 \end{bmatrix}.$$

Then $d_0 = 1$, $d_1 = 1$ and among the 2×2 subdeterminants are the relatively prime polynomials $\lambda - 1$ and $\lambda - 2$. Thus $d_2 = 1$. We also compute that $d_3 = (\lambda - 1)(\lambda - 2)^2$. Thus, $q_1 = q_2 = 1$, $q_3 = (\lambda - 1)(\lambda - 2)^2$, and the elementary divisors are $\lambda - 1$ and $(\lambda - 2)^2$. We have $s = 2$, $\alpha_1 = 1$, $\alpha_2 = 2$, $\mu_1 = 1$, $\mu_2 = 2$, and if $f(\lambda)$ is any function defined at A, then

$$f(A) = f(\alpha_1)Z_{11} + f(\alpha_2)Z_{21} + f'(\alpha_2)Z_{22}. \tag{64}$$

First, set $f(\lambda) = 1$ so that $f(A) = I_3$. Then

$$I_3 = Z_{11} + Z_{21}. \tag{65}$$

Next, set $f(\lambda) = \lambda - 2$ so that $f(A) = A - 2I_3$ and

$$A - 2I_3 = -Z_{11} + Z_{22}. \tag{66}$$

Finally, set $f(\lambda) = (\lambda - 2)^2$ so that $f(\alpha_1) = 1, f(\alpha_2) = 0, f'(\alpha_2) = 0$, and

$$(A - 2I_3)^2 = Z_{11}. \tag{67}$$

Now
$$A - 2I_3 = \begin{bmatrix} 1 & -1 & 1 \\ 2 & -2 & 1 \\ 1 & -1 & 0 \end{bmatrix}, \tag{68}$$

$$(A - 2I_3)^2 = \begin{bmatrix} 0 & 0 & 0 \\ -1 & 1 & 0 \\ -1 & 1 & 0 \end{bmatrix}. \tag{69}$$

Thus Z_{11} is the matrix (69) and from (66)

$$Z_{22} = A - 2I_3 + Z_{11} = \begin{bmatrix} 1 & -1 & 1 \\ 1 & -1 & 1 \\ 0 & 0 & 0 \end{bmatrix}. \tag{70}$$

Also from (65),

$$Z_{21} = I_3 - Z_{11} = \begin{bmatrix} 1 & 0 & 0 \\ 1 & 0 & 0 \\ 1 & -1 & 1 \end{bmatrix}. \tag{71}$$

Putting (69) to (71) in (64) and using $f(\lambda) = e^{t\lambda}$, we have

$$f(A) = e^{tA} = e^t \begin{bmatrix} 0 & 0 & 0 \\ -1 & 1 & 0 \\ -1 & 1 & 0 \end{bmatrix} + e^{2t} \begin{bmatrix} 1 & 0 & 0 \\ 1 & 0 & 0 \\ 1 & -1 & 1 \end{bmatrix} + te^{2t} \begin{bmatrix} 1 & -1 & 1 \\ 1 & -1 & 1 \\ 0 & 0 & 0 \end{bmatrix}$$
$$= \begin{bmatrix} (t+1)e^{2t} & -te^{2t} & te^{2t} \\ -e^t + (1+t)e^{2t} & e^t - te^{2t} & te^{2t} \\ -e^t + e^{2t} & e^t - e^{2t} & e^{2t} \end{bmatrix}. \tag{72}$$

(e) If we combine (b) and (c) above, we can immediately conclude the following: *If $f(\lambda)$ and $g(\lambda)$ are defined at A, then $f(A) = g(A)$ iff $f^{(j-1)}(\alpha_i) = g^{(j-1)}(\alpha_i)$, $j = 1, \ldots, \mu_i$, $i = 1, \ldots, s$.* For, the component matrices Z_{ij} are independent of both $f(\lambda)$ and $g(\lambda)$ and if the preceding conditions hold, then $f(A) = g(A)$ follows from (63). Conversely, if

$$\sum_{i=1}^{s} \sum_{j=1}^{\mu_i} f^{(j-1)}(\alpha_i) Z_{ij} = f(A) = g(A) = \sum_{i=1}^{s} \sum_{j=1}^{\mu_i} g^{(j-1)}(\alpha_i) Z_{ij}$$

then the linear independence of the Z_{ij} implies that

$$f^{(j-1)}(\alpha_i) = g^{(j-1)}(\alpha_i), \qquad j = 1, \ldots, \mu_i, \quad i = 1, \ldots, s.$$

If $X(t) = [x_{ij}(t)]$ is a matrix whose entries are differentiable functions of a parameter t, then by the *derivative* of $X(t)$, $\dot{X}(t)$, we mean the matrix

$$\dot{X}(t) = [\dot{x}_{ij}(t)].$$

The following interesting result is the fundamental theorem concerning systems of first-order linear differential equations with constant coefficients.

Theorem 3.16 *If $A \in M_n(C)$ and $X(t) = e^{tA}$, where t is a complex parameter, then*

$$\dot{X}(t) = Ae^{tA} = AX(t). \tag{73}$$

Proof: Clearly, since $f(\lambda) = e^{t\lambda}$ has derivatives of all orders, $f(A) = e^{tA}$ is well-defined. In fact, from the representation (63) and the fact that $f^{(j-1)}(\lambda) = t^{j-1}e^{t\lambda}$, we have

$$X(t) = e^{tA} = f(A)$$
$$= \sum_{i=1}^{s} \sum_{j=1}^{\mu_i} t^{j-1} e^{t\alpha_i} Z_{ij}$$
$$= \sum_{i=1}^{s} e^{t\alpha_i} \sum_{j=1}^{\mu_i} t^{j-1} Z_{ij}. \tag{74}$$

Since the Z_{ij} are constant matrices (independent of t), we can differentiate both sides of (74) to obtain

$$\begin{aligned}\dot{X}(t) &= \sum_{i=1}^{s}\left[\alpha_i e^{t\alpha_i}\sum_{j=1}^{\mu_i} t^{j-1}Z_{ij} + e^{t\alpha_i}\sum_{j=2}^{\mu_i}(j-1)t^{j-2}Z_{ij}\right]\\ &= \sum_{i=1}^{s}[e^{t\alpha_i}\alpha_i Z_{i1} + e^{t\alpha_i}(t\alpha_i + 1)Z_{i2} + e^{t\alpha_i}(t^2\alpha_i + 2t)Z_{i3}\\ &\qquad + \cdots + e^{t\alpha_i}(t^{\mu_i-1}\alpha_i + (\mu_i - 1)t^{\mu_i-2})Z_{i\mu_i}]\\ &= \sum_{i=1}^{s}\sum_{j=1}^{\mu_i} e^{t\alpha_i}(t^{j-1}\alpha_i + (j-1)t^{j-2})Z_{ij}. \qquad (75)\end{aligned}$$

Now consider the function $g(\lambda) = \lambda e^{t\lambda}$. An elementary induction on r shows that

$$g^{(r)}(\lambda) = e^{t\lambda}(t^r\lambda + rt^{r-1}), \qquad r = 1,2, \ldots . \qquad (76)$$

It is also immediate (see Exercise 19) that

$$g(A) = Ae^{tA}. \qquad (77)$$

But from (76),

$$g^{(j-1)}(\alpha_i) = e^{t\alpha_i}(t^{j-1}\alpha_i + (j-1)t^{j-2})$$

and thus from (63) we have

$$g(A) = \sum_{i=1}^{s}\sum_{j=1}^{\mu_i} e^{t\alpha_i}(t^{j-1}\alpha_i + (j-1)t^{j-2})Z_{ij}. \qquad (78)$$

If we compare (75) and (78) and use (77), we obtain (73). ∎

A homogeneous first-order system of linear ordinary differential equations with constant coefficients has the following form:

$$\dot{x}(t) = Ax(t), \qquad (79)$$

where $x(t) = (x_1(t), \ldots, x_n(t))$ and $\dot{x}(t) = (\dot{x}_1(t), \ldots, \dot{x}_n(t))$. The *initial value problem* in connection with (79) is to determine a solution $x(t)$ for which $x(t_0) = x_0$, a given n-tuple. Now consider the matrix $X(t) = e^{tA}$ and form the n-tuple of functions

$$x(t) = X(t - t_0)x_0.$$

Since $e^0 = I_n$ (see Exercise 20), we have

$$x(t_0) = X(0)x_0 = x_0.$$

From Theorem 3.16 we have

$$\begin{aligned}\dot{x}(t) &= \dot{X}(t - t_0)x_0 = AX(t - t_0)x_0\\ &= Ax(t).\end{aligned}$$

Thus the solution to (79) which takes on the value x_0 when $t = t_0$ is

$$x(t) = e^{(t-t_0)A}x_0. \qquad (80)$$

Example 7 We solve the system

$$\begin{aligned}\dot{x}_1 &= 3x - x_2 + x_3\\ \dot{x}_2 &= 2x_1 + x_3\\ \dot{x}_3 &= x_1 - x_2 + 2x_3\end{aligned}$$

with the condition $x(0) = x_0 = (1,-1,0)$. We rewrite (81) in the form

$$\dot{x} = Ax$$

where

$$A = \begin{bmatrix} 3 & -1 & 1 \\ 2 & 0 & 1 \\ 1 & -1 & 2 \end{bmatrix}.$$

The matrix e^{tA} was computed in (72). Evaluating (80) in this case, we have

$$\begin{aligned} x(t) &= e^{tA}x_0 \\ &= ((t+1)e^{2t} + te^{2t},\ -e^t + (1+t)e^{2t} - e^t + te^{2t},\ -e^t + e^{2t} - e^t + e^{2t}) \\ &= (e^{2t}(2t+1),\ -2e^t + (2t+1)e^{2t},\ -2e^t + 2e^{2t}) \\ &= e^{2t}(2t+1,\ 2t+1,\ 2) + e^t(0,-2,-2). \end{aligned}$$

Exercises 4.3

1. Show that dim Hom(V,W) = dim V dim W, and exhibit a basis of Hom(V,W). *Hint*: Let $\{e_1, \ldots, e_n\}$ be a basis of V, $\{f_1, \ldots, f_m\}$ a basis of W and define $T_{ij} \in$ Hom(V,W) by $T_{ij}e_i = f_j$, $T_{ij}e_k = 0$, $k \neq i$, and linear extension. Then the T_{ij} form a basis of Hom(V,W).
2. Let $T \in$ Hom(V,V). For each $f(\lambda) \in R[\lambda]$ define $f(\lambda)v = f(T)v$. Show that V is an $R[\lambda]$-module.
3. Prove Theorem 3.2(a) to (c).
4. Let R be a P.I.D. If A and B are in $M_n(R)$, then A is *similar* to B, $A \overset{S}{=} B$, if there exists $P \in M_n(R)$ such that P is unimodular and $A = PBP^{-1}$. Prove that $A \overset{S}{=} B$ is an equivalence relation in $M_n(R)$. The equivalence classes are called *similarity classes*.
5. Let $a_{ij}(\lambda) \in R[\lambda]$, $i,j = 1, \ldots, n$. By adjoining 0 coefficients if necessary, write

$$a_{ij}(\lambda) = \sum_{k=0}^{m} a_{ij}{}^k \lambda^k$$

where m does not depend on i and j. If $L(\lambda) = [a_{ij}(\lambda)] \in M_n(R[\lambda])$, show that $L(\lambda)$ can be regarded as an element of $M_n(R)[\lambda]$ by writing

$$\begin{aligned} L(\lambda) &= \sum_{k=0}^{m} [a_{ij}{}^k]\lambda^k \\ &= \sum_{k=0}^{m} A_k\lambda^k, \qquad A_k = [a_{ij}{}^k],\ k = 0, \ldots, m. \end{aligned}$$

Also prove that if $M(\lambda) = [b_{ij}(\lambda)] = \sum_{k=0}^{m}[b_{ij}{}^k]\lambda^k = \sum_{k=0}^{m} B_k\lambda^k$, then

$$L(\lambda)M(\lambda) = \left[\sum_{\nu=1}^{n} a_{i\nu}(\lambda)b_{\nu j}(\lambda)\right] = \sum_{s=0}^{2m}\left(\sum_{\nu=0}^{s} A_{s-\nu}B_\nu\right)\lambda^s.$$

Hint:
$$\sum_{\nu=1}^{n} a_{i\nu}(\lambda)b_{\nu j}(\lambda) = \sum_{\nu=1}^{n}\sum_{k=0}^{m} a_{i\nu}{}^{k}\lambda^{k} \sum_{t=0}^{m} b_{\nu j}{}^{t}\lambda^{t}$$
$$= \sum_{k=0}^{m}\sum_{t=0}^{m}\sum_{\nu=1}^{n} a_{i\nu}{}^{k}b_{\nu j}{}^{t}\lambda^{k+t} = \sum_{k=0}^{m}\sum_{t=0}^{m} (A_k B_t)_{ij}\lambda^{k+t}$$
$$= \sum_{s=0}^{2m}\sum_{\nu=0}^{s} (A_{s-\nu}B_{\nu})_{ij}\lambda^{s}.$$

6. Let $f(\lambda) \in M_n(R)[\lambda]$, $A \in M_n(R)$. Show that
$$(f(\lambda)\lambda^k)_r(A) = f_r(A)A^k.$$
Hint: Write $f(\lambda) = A_0 + A_1\lambda + \cdots + A_m\lambda^m$. Then
$$\begin{aligned}(f(\lambda)\lambda^k)_r(A) &= (A_0\lambda^k + A_1\lambda^{k+1} + \cdots + A_m\lambda^{k+m})_r(A)\\ &= A_0A^k + A_1A^{k+1} + \cdots + A_mA^{k+m}\\ &= (A_0 + A_1A + \cdots + A_mA^m)A^k\\ &= f_r(A)A^k.\end{aligned}$$

7. Let $f(\lambda) \in M_n(R)[\lambda]$. Show that
$$(f(\lambda)(\lambda I_n - A))_r(A) = 0.$$
Hint: Write $f(\lambda) = A_0 + A_1\lambda + \cdots + A_{m-1}\lambda^{m-1} + A_m\lambda^m$ so that $f(\lambda)(\lambda I_m - A)$ $= A_0\lambda + A_1\lambda^2 + \cdots + A_{m-1}\lambda^m + A_m\lambda^{m+1} - A_0A - A_1A\lambda - \cdots - A_{m-1}A\lambda^{m-1}$ $- A_mA\lambda^m$. Then $(f(\lambda)(\lambda I_m - A))_r(A) = A_0A + A_1A^2 + \cdots + A_{m-1}A^m$ $+ A_mA^{m+1} - A_0A - A_1A^2 - \cdots - A_{m-1}A^m - A_mA^{m+1} = 0$.

8. Show that if A and B are in $M_n(R)$, R a P.I.D., and $AB = I_n$, then $BA = I_n$. *Hint*: $\det AB = \det A \det B = 1$; so $\det A$ is a unit. Thus $(\det A)^{-1} \operatorname{adj} A = A^{-1}$ $\in M_n(R)$, and from $AB = I_n$ we have $B = A^{-1}$ so that $BA = I_n$.

9. Show that any two similar matrices have the same minimal polynomial. Show that if $T \in \mathrm{Hom}(V, V)$, then the minimal polynomial of T [i.e., the monic polynomial $m(\lambda)$ of least degree for which $m(T) = 0$] is equal to the minimal polynomial of any matrix representation of T. *Hint*: This first part follows easily from the fact that $p(SAS^{-1}) = Sp(A)S^{-1}$ for any polynomial $p(\lambda) \in R[\lambda]$. The second statement is obvious from properties of matrix representations.

10. Let A be an n-square matrix partitioned as follows:

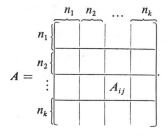

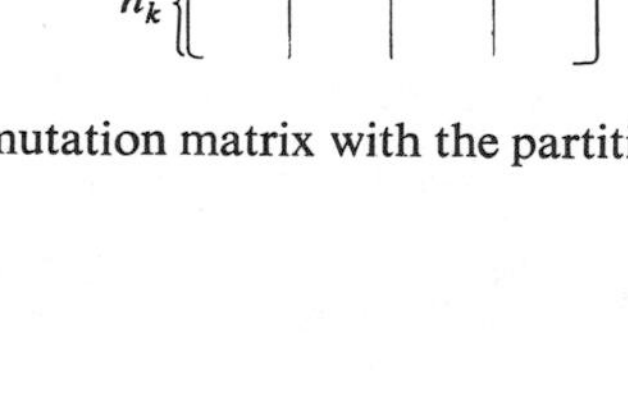

Let P be an $n \times n$ permutation matrix with the partitioning

$P =$

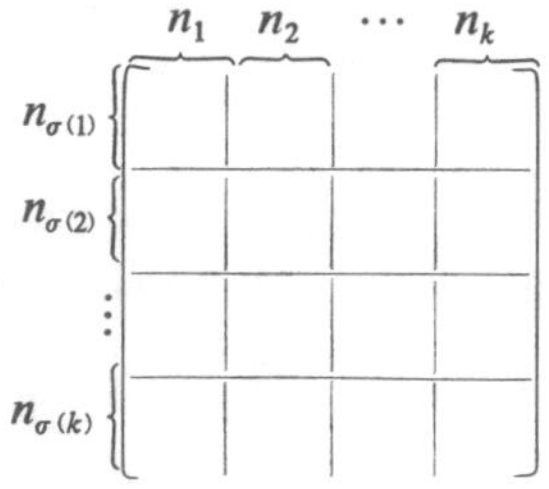

in which $\sigma \in S_k$ and in block row s every block is 0 except for the block in block column $t = \sigma(s)$, and there the identity matrix $I_{n_t} = I_{n_{\sigma(s)}}$ is found. Similarly, for $\mu \in S_k$ define an n-square permutation matrix Q by the partitioning

$Q =$

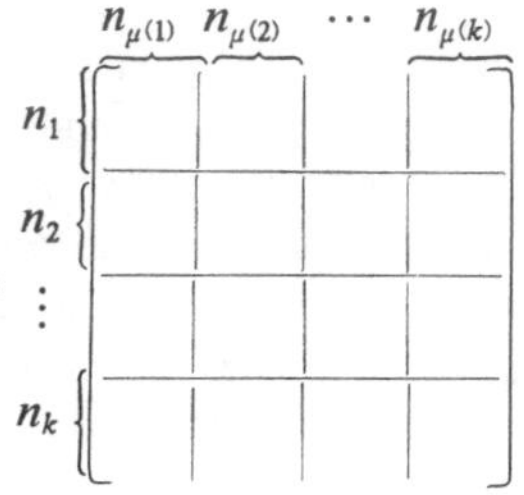

wherein block column t has 0 blocks in it except in block row $s = \mu(t)$ and there the identity matrix $I_{n_s} = I_{n_{\mu(t)}}$ is found. Show that the (s,t) block in the partitioned matrix PAQ is $A_{\sigma(s),\mu(t)}$, $s,t = 1, \ldots, k$.

11. Show that if $A \in M_{m,n}(R)$, dim $V = n$, and dim $U = m$, then there exists $T \in \text{Hom}(V,U)$ such that $[T]_E^F = A$ where E and F are bases of V and U, respectively. *Hint*: Define T by $Te_j = \sum_{i=1}^{m} a_{ij} f_i$, $j = 1, \ldots, n$ and linear extension.

12. Let $V = W \dotplus U$, where W and U are invariant subspaces under $T \in \text{Hom}(V,V)$. Let $E = \{e_1, \ldots, e_r, e_{r+1}, \ldots, e_n\}$ be a basis of V such that $E_r = \{e_1, \ldots, e_r\}$ and $E_{n-r} = \{e_{r+1}, \ldots, e_n\}$ are bases of W and U, respectively. Show that $[T]_E^E = A_1 \dotplus A_2$, where $A_1 = [T_1]_{E_r}^{E_r}$, $A_2 = [T_2]_{E_{n-r}}^{E_{n-r}}$, and $T_1 = T \mid W$, $T_2 = T \mid U$. *Hint*: This is an immediate consequence of the definition of a matrix representation.

13. State and prove the converse of the result in Exercise 12.

14. Show that if $A \overset{S}{=} C \dotplus D$ and $A \in GL(n,R)$, then $A^{-1} \overset{S}{=} C^{-1} \dotplus D^{-1}$.

15. Let A and B be matrices in $M_n(R)$. Let F be an extension field of R, and suppose that there exists a nonsingular matrix $S \in M_n(F)$ such that $S^{-1}AS = B$. Then show that there exists a nonsingular matrix $P \in M_n(R)$ such that $P^{-1}AP = B$. *Hint*: The characteristic matrices $\lambda I_n - A$ and $\lambda I_n - B$ are equivalent as matrices over $F[\lambda]$ and hence have the same determinantal divisors. However these determinantal divisors are g.c.d.'s of polynomials in $R[\lambda]$ and hence lie in $R[\lambda]$. Thus as matrices over $R[\lambda]$, $\lambda I_n - A$ and $\lambda I_n = B$ have the same determinantal divisors. Thus they are equivalent over $R[\lambda]$ so that by Theorem 3.3, A and B are similar over R.

16. Let $\alpha_1, \ldots, \alpha_s$ be distinct elements in a field R of characteristic 0, and let $\gamma_1, \ldots, \gamma_s$ be nonnegative integers. Let

$$b_{k0}, \ldots, b_{k,\gamma_k}, \qquad k = 1, \ldots, s$$

be prescribed elements of R, and let $\alpha \in R$. Define the integer $m = (s-1) + \sum_{k=1}^{s} \gamma_k$. Then prove that there exists precisely one polynomial

$$p(\lambda) = \sum_{j=0}^{m} a_j(\lambda - \alpha)^j$$

of degree at most m for which

$$p^{(j)}(\alpha_k) = b_{kj}, \qquad j = 0, \ldots, \gamma_k, \quad k = 1, \ldots, s.$$

Hint: There are a total of $\gamma_k + 1$ linear conditions imposed on $a_0, \ldots, a_m$ by the equalities $p^{(j)}(\alpha_k) = b_{kj}$, $j = 0, \ldots, \gamma_k$. Thus altogether there are $\sum_{k=1}^{s}(\gamma_k + 1) = s + \sum_{k=1}^{s} \gamma_k = m + 1$ linear conditions on the coefficients $a_0, \ldots, a_m$. Now imagine every $b_{kj} = 0$ so that the resulting system of $m + 1$ linear equations in $a_0, \ldots, a_m$ is homogeneous. Any solution defines a polynomial $p(\lambda)$ of degree at most m for which α_k is a root of multiplicity at least $\gamma_k + 1$, $k = 1, \ldots, s$. Thus $p(\lambda)$ has a total of at least $\sum_{k=1}^{s}(\gamma_k + 1) = \sum_{k=1}^{s}\gamma_k + s = m + 1$ roots, counting multiplicities. But $p(\lambda)$ has degree at most m and hence must be 0. In other words, the only solution $a_0, \ldots, a_m$ to the system of homogeneous linear equations must be $a_0 = \cdots = a_m = 0$. But this happens iff the coefficient matrix is nonsingular. Thus the system has a unique solution for any b_{kj}.

17. Let $p(\lambda) \in R[\lambda]$, and let $m(\lambda)$ be the minimal polynomial of $A \in M_n(R)$. Prove that if $r = \deg m(\lambda)$, then there exists $d(\lambda) \in R[\lambda]$ with $\deg d(\lambda) \leq r - 1$ such that $p(A) = d(A)$. *Hint*: Write

$$p(\lambda) = m(\lambda)q(\lambda) + d(\lambda), \qquad \deg d(\lambda) \leq r - 1;$$

then $p(A) = d(A)$.

18. Let $m(\lambda) = (\lambda - \alpha_1)^{\mu_1} \cdots (\lambda - \alpha_s)^{\mu_s}$ be the minimal polynomial of $A \in M_n(R)$ ($\alpha_1, \ldots, \alpha_s$ are the distinct eigenvalues of A). Let $f(\lambda)$ be a function defined at A. Show that there exists a polynomial $p(\lambda) \in R[\lambda]$ such that $p(A) = f(A)$ and $\deg p(\lambda) < \deg m(\lambda)$. *Hint*: We need only exhibit a polynomial $p(\lambda)$ of degree less than $\deg m(\lambda)$ for which

$$p^{(j)}(\alpha_k) = f^{(j)}(\alpha_k), \qquad j = 0, \ldots, \mu_k - 1, \; k = 1, \ldots, s.$$

Apply Exercises 16 and 17.

19. Let $f_1, \ldots, f_m$ be functions defined at A, $G(z_1, \ldots, z_m)$ a polynomial in $z_1, \ldots z_m$, and let

$$h(\lambda) = G(f_1(\lambda), \ldots, f_m(\lambda)).$$

Show that $h(A) = G(f_1(A), \ldots, f_m(A))$. *Hint*: Choose polynomials $p_1(\lambda), \ldots, p_m(\lambda)$ such that $p_i(A) = f_i(A)$, $i = 1, \ldots, m$, and let $r(\lambda) = G(p_1(\lambda), \ldots, p_m(\lambda))$, a polynomial. By composite differentiation, $r^{(j)}(\lambda)$ is a sum of products involving partial derivatives of G and derivatives of the form $p_i^{(k)}(\lambda)$, $0 \leq k \leq j$, $1 \leq i \leq m$. Now $p_i^{(j)}(\alpha_k) = f^{(j)}(\alpha_k)$, where $\alpha_1, \ldots, \alpha_s$ are the distinct eigenvalues of A. The jth derivative of $h(\lambda) = G(f_1(\lambda), \ldots, f_m(\lambda))$ is expressed in terms of the partials of G and the derivatives of the f_i in exactly the same way as $r^{(j)}(\lambda)$. Thus $r^{(j)}(\alpha_k) = h^{(j)}(\alpha_k)$. Thus h is defined at A and by the remark following formula (61), $h(A) = r(A)$. But $r(A) = G(p_1(A), \ldots, p_m(A))$ because these are all scalar polynomials. Since $p_j(A) = f_j(A)$, we are done.

20. Show that $e^0 = I_n$. *Hint*: The eigenvalues of 0 are all 0, and every elementary divisor is linear. Thus $s = 1$ (there is only one eigenvalue) and $\mu_1 = 1$. The polynomial $f_{11}(\lambda)$ in Example 6(a) used to define the component matrix Z_{11} must satisfy $f_{11}(0) = 1$; so we can take $f_{11}(\lambda) = 1$. Then $Z_{11} = I_n$. If $f(\lambda) = e^\lambda$, then $e^0 = e^0 Z_{11} = I_n$ [see example 6(b)].

21. Assume f and h are functions defined at A, g is defined at $f(A)$, and g has derivatives of order $\mu_i - 1$ at $f(\alpha_i)$, $i = 1, \ldots, s$. Let $h = g \circ f$ be the composite function. Then show that $h(A) = g(f(A))$. *Hint*: Let $p(\lambda)$ and $q(\lambda)$ be polynomials such that $p(A) = f(A)$ and $q(f(A)) = g(f(A))$. Let $r(\lambda) = q(p(\lambda))$. Then $r^{(j)}(\lambda)$ is a sum of products of factors of the form $q^{(i)}(p(\lambda))$ and $p^{(l)}(\lambda)$, $i,l \le j$. Also $p^{(l)}(\alpha_k) = f^{(l)}(\alpha_k)$ and $q^{(i)}(p(\alpha_k)) = g^{(i)}(f(\alpha_k))$ for any eigenvalue α_k of A. If we differentiate $h(\lambda) = g(f(\lambda))$ j times, we get a sum of products of terms such as $g^{(i)}(f(\lambda))$ and $f^{(l)}(\lambda)$, $i,l \le j$, and this expression is the same as the expression for $r^{(j)}(\lambda)$ in terms of $q^{(i)}(p(\lambda))$ and $p^{(l)}(\lambda)$. Hence $r^{(j)}(\alpha_k) = h^{(j)}(\alpha_k)$ so that $h(A) = r(A) = q(p(A)) = q(f(A)) = g(f(A))$.

22. Let V be an n-square *unitary matrix*, i.e., $V^*V = I_n$, where $V^* = [\bar{v}_{ji}]$. Show that there exists a skew-hermitian matrix X such that $e^X = V$. *Hint*: Let $QVQ^{-1} = \operatorname{diag}(e^{i\theta_1}, \ldots, e^{i\theta_n})$, Q unitary. Let $D = \operatorname{diag}(i\theta_1, \ldots, i\theta_n)$. Clearly, e^λ is defined at any A so that from (61) we have $e^{Q^{-1}DQ} = Q^{-1}e^DQ$. Now $\lambda I_n - D$ has linear elementary divisors so if we obtain a polynomial $p(\lambda)$ so that $p(i\theta_k) = e^{i\theta_k}$, then $p(D) = e^D$. But of course $p(D) = \operatorname{diag}(p(i\theta_1), \ldots, p(i\theta_n)) = \operatorname{diag}(e^{i\theta_1}, \ldots, e^{i\theta_n}) = QVQ^{-1}$. Note that $X = Q^{-1}DQ$ is skew-hermitian.

23. Let $f(\lambda)$ and $g(\lambda)$ be functions defined at A, $h(\lambda) = f(\lambda)g(\lambda)$, $m(\lambda) = f(\lambda) + g(\lambda)$. Show that $h(A) = f(A)g(A)$, $m(A) = f(A) + g(A)$. *Hint*: Let $G(z_1,z_2) = z_1 + z_2$, $G(z_1,z_2) = z_1z_2$ in Exercise 19.

24. Show that $\det(e^A) = e^{tr(A)}$. *Hint*: The eigenvalues of e^A are $e^{\alpha_1}, \ldots, e^{\alpha_n}$. Then $\det(e^A) = e^{\alpha_1} \cdots e^{\alpha_n} = e^{\alpha_1 + \cdots + \alpha_n} = e^{tr(A)}$.

25. Let f be defined at A. Show that $f(A^\top) = f(A)^\top$. If, in addition, $f^{(j)}(\alpha_k) = \overline{f^{(j)}(\bar{\alpha}_k)}$, $j = 0, \ldots, \mu_t - 1$, $k = 1, \ldots, s$, then $f(A^*) = f(A)^*$. *Hint*: Consider the formula (63) for $f(A)$ in terms of the component matrices: $f(A) = \sum_{i=1}^{s} \sum_{j=1}^{\mu_i} f^{(j-1)}(\alpha_i) f_{ij}(A)$. Now $A^\top$ is similar to A (why?), and hence from Example 6(a) the polynomials $f_{ij}(\lambda)$ for $A^\top$ are the same as those for A. Since $p(A)^\top = p(A^\top)$ for any polynomial $p(\lambda)$,

$$
\begin{aligned}
f(A^\top) &= \sum_{i=1}^{s} \sum_{j=1}^{\mu_i} f^{(j-1)}(\alpha_i) f_{ij}(A^\top) \\
&= \sum_{i=1}^{s} \sum_{j=1}^{\mu_i} f^{(j-1)}(\alpha_i)(f_{ij}(A))^\top = \left[\sum_{i=1}^{s} \sum_{j=1}^{\mu_i} f^{(j-1)}(\alpha_i) f_{ij}(A)\right]^\top \\
&= (f(A))^\top.
\end{aligned}
$$

Also, $f(A^*) = f(\bar{A}^{\mathrm{T}}) = (f(\bar{A}))^{\mathrm{T}}$. Thus we need only show that $f(\bar{A}) = \overline{f(A)}$. If J is the Jordan normal form of A, then a moment's reflection on the form of $f(J)$ shows that the condition $\overline{f^{(j)}(\alpha_k)} = f^{(j)}(\bar{\alpha}_k)$ implies that $f(\bar{J}) = \overline{f(J)}$, and hence that $\overline{f(A)} = \overline{f(S^{-1}JS)} = \overline{S^{-1}f(J)S} = \bar{S}^{-1}\overline{f(J)}\bar{S} = f(\bar{S}^{-1}\bar{J}\bar{S}) = f(\overline{S^{-1}JS}) = f(\bar{A})$.

26. Let $A_k = [a_{ij}{}^k]$, $k = 1, 2, \ldots$ be a sequence of matrices in $M_n(\mathbf{C})$. Then we write $\lim_{k\to\infty} A_k = A$ and say A_k *converges to* $A \in M_n(\mathbf{C})$ if each sequence $a_{ij}{}^k$, $k = 1, 2, \ldots$, converges to a_{ij}, i.e., $\lim_{k\to\infty} a_{ij}{}^k = a_{ij}$. Show that if $\lim_{k\to\infty} A_k = A$ and $\lim_{k\to\infty} B_k = B$, then $\lim_{k\to\infty}(A_k + B_k) = A + B$ and $\lim_{k\to\infty} A_k B_k = AB$. *Hint*: Use properties of limits.

27. Let f_k, $k = 1, 2, \ldots$, be a sequence of functions each defined at $A \in M_n(\mathbf{C})$. If $\lim_{k\to\infty} f_k(A) = B$, then $f_k(A)$ is said to *converge to B*. Prove that $f_k(A)$ converges to some matrix iff each of the limits

$$\lim_{k\to\infty} f_k{}^{(j)}(\alpha_t), \qquad j = 0, \ldots, \mu_t - 1, t = 1, \ldots, s$$

exists. *Hint*: $f_k(A)$ is by definition similar (via a fixed matrix S independent of k) to a direct sum of matrices of the type

$$\sum_{j=0}^{e-1} \frac{f_k{}^{(j)}(\alpha_t)U_e{}^j}{j!},$$

where $1 \leq e \leq \mu_t$ and U_e is an auxiliary unit matrix. Thus, if each $\lim_{k\to\infty} f_k{}^{(j)}(\alpha_t)$ exists, then clearly $\lim_{k\to\infty} f_k(A)$ exists. Conversely, each $f_k{}^{(j)}(\alpha_t)$ can be expressed as a fixed linear combination of the elements of $f_k(A)$, and hence if $\lim_{k\to\infty} f_k(A)$ exists, then so does each $\lim_{k\to\infty} f_k{}^{(j)}(\alpha_t)$.

28. Let $f_k(\lambda) = \sum_{i=0}^{k} c_i(\lambda - a)^i$ be the partial sum in a power series whose circle of convergence contains all of the eigenvalues of $A \in M_n(\mathbf{C})$ in its interior. Suppose $f(\lambda) = \lim_{k\to\infty} f_k(\lambda)$. Then prove that f is defined at A and $\lim_{k\to\infty} f_k(A) = f(A)$. *Hint*: Obviously each $f_k(\lambda)$ is defined at A because $f_k(\lambda)$ is a polynomial. Also, a convergent power series can be differentiated term by term in the interior of its circle of convergence and hence $f^{(j)}(\lambda) = \lim_{k\to\infty} f_k{}^{(j)}(\lambda)$ for any λ in the interior of that circle. Hence f is defined at A. Now from formula (63),

$$f_k(A) = \sum_{i=1}^{s}\sum_{j=1}^{\mu_i} f_k{}^{(j-1)}(\alpha_i)Z_{ij},$$

and the component matrices Z_{ij} do not depend on $f_k(\lambda)$. Also $\lim_{k\to\infty} f_k{}^{(j)}(\alpha_i) = f^{(j)}(\alpha_i)$ because each α_i is in the interior of the circle of convergence. It follows that

$$\lim_{k\to\infty} f_k(A) = \sum_{i=1}^{s}\sum_{j=1}^{\mu_i} \lim_{k\to\infty} f_k{}^{(j-1)}(\alpha_i)Z_{ij} = \sum_{i=1}^{s}\sum_{j=1}^{\mu_i} f^{(j-1)}(\alpha_i)Z_{ij}$$
$$= f(A).$$

29. Prove: $e^A = I_n + A + \dfrac{A^2}{2!} + \dfrac{A^3}{3!} + \cdots$. *Hint*: Let $f_k(\lambda) = 1 + \lambda + \dfrac{\lambda^2}{2!} + \cdots + \dfrac{\lambda^k}{k!}$ and apply Exercise 28.

30. Prove: If $P = \begin{bmatrix} 0 & 1 \\ -1 & 0 \end{bmatrix}$, then $e^{i\varphi P} = \begin{bmatrix} \cosh\varphi & i\sinh\varphi \\ -i\sinh\varphi & \cosh\varphi \end{bmatrix}$. *Hint*: Choose a unitary V such that $V^*PV = \begin{bmatrix} i & 0 \\ 0 & -i \end{bmatrix}$, e.g. $V = \begin{bmatrix} -i/\sqrt{2} & 1/\sqrt{2} \\ 1/\sqrt{2} & -i/\sqrt{2} \end{bmatrix}$.

Then $e^{i\varphi P} = e^{i\varphi V\left[\begin{smallmatrix} i & 0 \\ 0 & -i\end{smallmatrix}\right]V^*} = Ve^{i\varphi\left[\begin{smallmatrix} i & 0 \\ 0 & -i\end{smallmatrix}\right]}V^* = V\operatorname{diag}(e^{-\varphi},e^{\varphi})V^*$

$$= \frac{1}{2}\begin{bmatrix} -i & 1 \\ 1 & -i \end{bmatrix}\begin{bmatrix} e^{-\varphi} & 0 \\ 0 & e^{\varphi} \end{bmatrix}\begin{bmatrix} i & 1 \\ 1 & i \end{bmatrix}$$

$$= \frac{1}{2}\begin{bmatrix} -i & 1 \\ 1 & -i \end{bmatrix}\begin{bmatrix} ie^{-\varphi} & e^{-\varphi} \\ e^{\varphi} & ie^{\varphi} \end{bmatrix} = \begin{bmatrix} \dfrac{e^{\varphi}+e^{-\varphi}}{2} & i\dfrac{e^{\varphi}-e^{-\varphi}}{2} \\ i\dfrac{e^{-\varphi}-e^{\varphi}}{2} & \dfrac{e^{\varphi}+e^{-\varphi}}{2} \end{bmatrix}$$

$$= \begin{bmatrix} \cosh\varphi & i\sinh\varphi \\ -i\sinh\varphi & \cosh\varphi \end{bmatrix}.$$

31. Assume $\operatorname{Log}\lambda$ is defined at A where we use the principal value of the logarithm. Show that $e^{\operatorname{Log} A} = A$. *Hint*: Use Exercise 21.

32. Let R be a field, $A,B \in M_n(R)$, and let $f(\lambda)$, $g(\lambda)$ be the characteristic polynomials of A, B. Prove: If g.c.d.$(f(\lambda), g(\lambda)) = 1$, then

$$S((\lambda I_n - A)(\lambda I_n - B)) = S(\lambda I_n - A)S(\lambda I_n - B).$$

Hint: It suffices to prove that the product of the kth determinantal divisors $d_{A,k}(\lambda)$ of $\lambda I_n - A$ and $d_{B,k}(\lambda)$ of $\lambda I_n - B$ equals the kth determinantal divisor $d_{AB,k}(\lambda)$ of the product $(\lambda I_n - A)(\lambda I_n - B)$. Let

$$\mathscr{S}_A = \begin{bmatrix} q_1(\lambda) & & & \\ & q_2(\lambda) & & \\ & & \ddots & \\ & & & q_n(\lambda) \end{bmatrix}, \qquad \mathscr{S}_B = \begin{bmatrix} r_1(\lambda) & & & \\ & r_2(\lambda) & & \\ & & \ddots & \\ & & & r_n(\lambda) \end{bmatrix}$$

denote, respectively, the Smith normal forms of $\lambda I_n - A$ and $\lambda I_n - B$. There exist unimodular S, T, V, W in $M_n(R[\lambda])$ so that $\lambda I_n - A = S\mathscr{S}_A T$ and $\lambda I_n - B = V\mathscr{S}_B W$. Thus

$$(\lambda I_n - A)(\lambda I_n - B) = S\mathscr{S}_A TV\mathscr{S}_B W \overset{\text{E}}{=} \mathscr{S}_A U \mathscr{S}_B,$$

where $U = TV$ is unimodular. Using the Cauchy-Binet theorem,

$$\begin{aligned} d_{AB,k}(\lambda) &= \underset{\alpha,\beta\in Q_{k,n}}{\text{g.c.d.}} \Big\{ \sum_{\gamma,\rho\in Q_{k,n}} \det \mathscr{S}_A[\alpha|\gamma] \det U[\gamma|\rho] \det \mathscr{S}_B[\rho|\beta]\Big\} \\ &= \underset{\alpha,\beta\in Q_{k,n}}{\text{g.c.d.}} \{\det \mathscr{S}_A[\alpha|\alpha] \det U[\alpha|\beta] \det \mathscr{S}_B[\beta|\beta]\} \\ &= \underset{\alpha,\beta\in Q_{k,n}}{\text{g.c.d.}} \left\{\left(\prod_{i=1}^{k} q_{\alpha(i)}(\lambda)\right) \det U[\alpha|\beta] \left(\prod_{i=1}^{k} r_{\beta(i)}(\lambda)\right)\right\}. \end{aligned} \tag{82}$$

The polynomials

$$d_{A,k}(\lambda) = q_1(\lambda) \cdots q_k(\lambda)$$

and

$$d_{B,k}(\lambda) = r_1(\lambda) \cdots r_k(\lambda)$$

divide, respectively, the first and third factors in each of the terms in (82). Hence

$$d_{A,k}(\lambda)d_{B,k}(\lambda) \mid d_{AB,k}(\lambda).$$

Suppose that there exists some (irreducible) polynomial $h(\lambda)$ in $R[\lambda]$ so that $h(\lambda)d_{A,k}(\lambda)d_{B,k}(\lambda)$ divides each of the terms in (82). Then, in particular:

(a) It divides each element of

$$\left\{q_1(\lambda) \cdots q_k(\lambda) \det U[1 \cdots k \mid \beta]\left(\prod_{i=1}^{k} r_{\beta(i)}(\lambda)\right) \mid \beta \in Q_{k,n}\right\}.$$

(b) It divides each element of

$$\left\{\left(\prod_{i=1}^{k} q_{\alpha(i)}(\lambda) \det U[\alpha \mid 1 \cdots k] \, r_1(\lambda) \cdots r_k(\lambda) \mid \alpha \in Q_{k,n}\right\}.$$

Using the Laplace expansion theorem on the first k rows (and then on the first k columns) of the unimodular matrix U, we see that each of the sets

$$\{\det U[1, \ldots, k \mid \beta] \mid \beta \in Q_{k,n}\}$$

and

$$\{\det U[\alpha \mid 1, \ldots, k] \mid \alpha \in Q_{k,n}\}$$

consist of relatively prime elements in $R[\lambda]$. So from (a) and the fact that the determinantal divisors form a divisibility chain, the irreducible polynomial $h(\lambda)$ divides some product $\prod_{i=1}^{k} r_{\beta(i)}(\lambda)$, and from (b), $h(\lambda)$ also divides some product $\prod_{i=1}^{k} q_{\alpha(i)}(\lambda)$. By hypothesis, these are relatively prime, and so we have a contradiction. It follows that $d_{A,k}(\lambda)d_{B,k}(\lambda) = d_{AB,k}(\lambda)$.

33. Suppose that A and B have the same irreducible characteristic polynomial. Prove that A and B are similar. *Hint*: If a matrix has an irreducible characteristic polynomial, then that polynomial is the only nontrivial invariant factor of the characteristic matrix. Thus $\lambda I_n - A$ and $\lambda I_n - B$ have the same invariant factors, i.e., they are similar.

34. Let A have the form

$$A = \begin{bmatrix} A_1 & A_3 \\ 0 & A_2 \end{bmatrix} \in M_{m+n}(R),$$

where $A_1 \in M_m(R)$, $A_2 \in M_n(R)$. Suppose further that A_1 and A_2 have no eigenvalues in common. Prove that $A \overset{S}{=} A_1 \dotplus A_2$. *Hint*: Take unimodular $P(\lambda)$ and $Q(\lambda)$ in $M_m(R[\lambda])$ such that

$$P(\lambda)\,(\lambda I_m - A_1)Q(\lambda) = S(\lambda I_m - A_1),$$

the Smith normal form. Similarly, take unimodular $D(\lambda)$ and $F(\lambda)$ in $M_n(R[\lambda])$ such that

$$D(\lambda)(\lambda I_n - A_2)F(\lambda) = S(\lambda I_n - A_2).$$

By block multiplication,

$$\begin{bmatrix} P(\lambda) & 0 \\ 0 & D(\lambda) \end{bmatrix}\begin{bmatrix} \lambda I_m - A_1 & -A_3 \\ 0 & \lambda I_n - A_2 \end{bmatrix}\begin{bmatrix} Q(\lambda) & 0 \\ 0 & F(\lambda) \end{bmatrix}$$

$$= \begin{bmatrix} S(\lambda I_m - A_1) & B(\lambda) \\ 0 & S(\lambda I_n - A_2) \end{bmatrix}$$

where $B(\lambda) \in M_{m,n}(R[\lambda])$. Let

$$N = \begin{bmatrix} S(\lambda I_m - A_1) & B(\lambda) \\ 0 & S(\lambda I_n - A_2) \end{bmatrix}.$$

We show that by a sequence of elementary row and column operations on N all of the entries of $B(\lambda)$ can be made zero. Take a nonzero element of $B(\lambda)$, say $b_{ij}(\lambda)$. Since A_1 and A_2 have no eigenvalues in common,

$$\text{g.c.d.}(q_k(\lambda I_m - A_1), q_t(\lambda I_n - A_2)) = 1, \qquad k = 1, \ldots, m,\ t = 1, \ldots, n.$$

Thus there exist $p_1(\lambda), p_2(\lambda) \in R[\lambda]$ such that

$$p_1(\lambda)q_i(\lambda I_m - A_1) + p_2(\lambda)q_j(\lambda I_n - A_2) = 1.$$

Hence $\quad -b_{ij}(\lambda)p_1(\lambda)q_i(\lambda I_m - A_1) - b_{ij}(\lambda)p_2(\lambda)q_j(\lambda I_n - A_2) = -b_{ij}(\lambda)$ and it follows that the $(i, m+j)$ entry of

$$E_{i,m+j}(-b_{ij}(\lambda)p_2(\lambda))NE_{i,m+j}(-b_{ij}(\lambda)p_1(\lambda))$$

is zero. We have eliminated the $b_{ij}(\lambda)$ entry without disturbing any of the other elements of N so we can repeat the process, ultimately eliminating all of the nonzero entries of $B(\lambda)$. But then $(\lambda I_{m+n} - A) \stackrel{E}{=} (\lambda I_m - A_1) \dotplus (\lambda I_n - A_2)$, and so $A \stackrel{S}{=} A_1 \dotplus A_2$.

35. Let

$$A = \begin{bmatrix} A_1 & A_3 \\ 0 & A_2 \end{bmatrix} \in M_{m+n}(R),$$

where A_1, A_2 are m- and n-square, respectively. Assume α is an eigenvalue of A_1 but not of A_2. Show that elementary divisors of $\lambda I_{m+n} - A$ and $\lambda I_m - A_1$ of the form

$$(\lambda - \alpha)^e$$

are the same. *Hint*: Choose nonsingular P, Q in $M_m(R)$, $M_n(R)$, respectively, such that $PA_1P^{-1} = J_1$, $QA_2Q^{-1} = J_2$, where J_i, $i = 1, 2$, are Jordan normal forms and furthermore the Jordan blocks of A_1 involving the eigenvalue α are in the first k rows (i.e., α is a root of multiplicity k of the characteristic polynomial of A_1). By block multiplication

$$\begin{bmatrix} P & 0 \\ 0 & Q \end{bmatrix}\begin{bmatrix} A_1 & A_3 \\ 0 & A_2 \end{bmatrix}\begin{bmatrix} P^{-1} & 0 \\ 0 & Q^{-1} \end{bmatrix} = \begin{bmatrix} J_1 & B \\ 0 & J_2 \end{bmatrix},$$

where $B \in M_{m,n}(R)$. Let

$$W = \begin{bmatrix} J_1 & B \\ 0 & J_2 \end{bmatrix},$$

and re-partition the matrix W so that

$$W = \begin{bmatrix} B_2 & B_4 \\ 0 & B_3 \end{bmatrix}$$

where B_2 is $k \times k$ (i.e., with the k occurrences of α down the main diagonal) and B_3 is upper triangular. Apply Exercise 34 to W, i.e.,

$$W \stackrel{S}{=} B_2 \dotplus B_3.$$

But B_3 is upper triangular; so the eigenvalues are displayed on the main diagonal, and none of them are α. Now since the complete list of elementary divisors of a direct sum is the combined lists of the summands, the elementary divisors of the characteristic matrix of A of the form $(\lambda - \alpha)^e$ are exactly those of the characteristic matrix of A_1.

36. Let A and B be rectangular matrices. Define $A \otimes B = [a_{ij}B]$, that is, $A \otimes B$ is a block matrix whose (i,j) block is $a_{ij}B$. Prove that $(C \otimes D)(A \otimes B) = CA \otimes DB$ (assume that the products CA and DB are defined). The matrix $A \otimes B$ is called the *Kronecker product* of A and B. *Hint*: By block multiplication, the (i,j) block in the product

$$\begin{bmatrix} c_{11}D & c_{12}D & \cdots & \\ c_{21}D & \cdot & & \\ \vdots & & \cdot & \\ & & & c_{mn}D \end{bmatrix} \begin{bmatrix} a_{11}B & a_{12}B & \cdots & \\ a_{21}B & \cdot & & \\ \vdots & & \cdot & \\ & & & a_{ns}B \end{bmatrix}$$

is

$$\left(\sum_{k=1}^{n} c_{ik}a_{kj}\right) DB,$$

which is precisely the (i,j) block in the matrix

$$CA \otimes DB.$$

37. Let R be a P.I.D., and let $A \in M_m(R)$, $B \in M_n(R)$. Let $p \in R$ be a prime, and suppose $p^{e_1}, \ldots, p^{e_r}$ are all the elementary divisors of A involving p while $p^{f_1}, \ldots, p^{f_s}$ are all the elementary divisors of B involving p. Prove that the elementary divisors of the Kronecker product $A \otimes B$ involving p are $p^{e_i+f_j}$, $i = 1, \ldots, r$, $j = 1, \ldots, s$. *Hint*: First note that $A \otimes B \stackrel{E}{=} S_A \otimes S_B$ where S_A, S_B are in Smith normal form over R. For suppose $PAQ = S_A$ and $DBG = S_B$. Then by Exercise 36, $(P \otimes D)(A \otimes B)(Q \otimes G) = S_A \otimes S_B$. But $S_A \otimes S_B$ is diagonal; so its elementary divisors are just the prime power factors of the main diagonal elements. Hence the elementary divisors of $A \otimes B$ involving p are $p^{e_i+f_j}$, $i = 1, \ldots, r$, $j = 1, \ldots, s$. (Here we must allow the exponents e_i, f_j to be zero, if necessary.)

38. Suppose $A_1 \in M_m(R)$, $A_2 \in M_n(R)$. Define a linear map $L: M_{m,n}(R) \to M_{m,n}(R)$ by $L(X) = A_1X - XA_2$, $X \in M_{m,n}(R)$. Prove that a necessary and sufficient condition that L be nonsingular is that A_1 and A_2 have no eigenvalues in common. Also show that

$$\det L = \prod_{i=1,j=1}^{m,n} (\alpha_i - \beta_j),$$

where the α_i are the eigenvalues of A_1 and the β_j are the eigenvalues of A_2. *Hint*:

The matrix representation of L with respect to the basis $\{E_{ij}\}$, ordered lexicographically in the pairs (i,j), is

$$A_1 \otimes I_n - I_m \otimes A_2^{\mathsf{T}}.$$

The eigenvalues of $A_1 \otimes I_n$ are the eigenvalues of A_1 listed n times, and similarly the eigenvalues of $I_m \otimes A_2^{\mathsf{T}}$ are the eigenvalues of A_2^{T} (i.e., the eigenvalues of A_2) listed m times. Now $A_1 \otimes I_n$ and $I_m \otimes A_2^{\mathsf{T}}$ commute, and hence (why?) the eigenvalues of L are $\{\alpha_i - \beta_j \mid 1 \le i \le m, 1 \le j \le n\}$. Thus

$$\det L = \prod_{i=1,j=1}^{m,n} (\alpha_i - \beta_j).$$

Hence L is nonsingular iff $\det L \neq 0$, iff A_1 and A_2 have no eigenvalues in common.

39. Let $A \in M_n(\mathbf{C})$, and consider the differential equation

$$\dot{x}(t) = Ax(t) + f(t), \qquad x(t_0) = c,$$

i.e.,

$$\dot{x}_i(t) = \sum_{j=1}^{n} a_{ij}x_j(t) + f_i(t), \; x_i(t_0) = c_i.$$

Show that

$$x(t) = e^{(t-t_0)A}c + \int_{s=t_0}^{s=t} e^{(t-s)A} f(s)\, ds.$$

Hint: We need only show that if

$$x(t) = e^{(t-t_0)A}c + \int_{s=t_0}^{s=t} e^{(t-s)A} f(s)\, ds,$$

then

$$\dot{x}(t) = Ax(t) + f(t) \qquad \text{and } x(t_0) = c.$$

But

$$\begin{aligned}\dot{x}(t) &= Ae^{(t-t_0)A}c + \frac{d}{dt}\left[\int_{s=t_0}^{s=t} e^{(t-s)A} f(s)\, ds\right]\\ &= Ae^{(t-t_0)A}c + \frac{d}{dt}\left[e^{tA}\int_{s=t_0}^{s=t} e^{-sA} f(s)\, ds\right]\\ &= Ae^{(t-t_0)A}c + Ae^{tA}\int_{s=t_0}^{s=t} e^{-sA} f(s)\, ds + e^{tA}(e^{-tA}f(t))\\ &= Ax(t) + f(t)\end{aligned}$$

and $x(t_0) = e^0 c = c$.

40. Prove that if $A,B \in M_n(\mathbf{C})$ and $e^{\lambda(A+B)} = e^{\lambda A}e^{\lambda B}$ for all $\lambda \in \mathbf{C}$, then $AB = BA$. *Hint*: If $e^{\lambda(A+B)} = e^{\lambda A}e^{\lambda B}$ for all $\lambda \in \mathbf{C}$, then

$$\frac{d}{d\lambda}e^{\lambda(A+B)} = (A+B)e^{\lambda(A+B)} = Ae^{\lambda A}e^{\lambda B} + e^{\lambda A}Be^{\lambda B}$$

and

$$\frac{d^2}{d\lambda^2}e^{\lambda(A+B)} = (A+B)^2e^{\lambda(A+B)} = A^2e^{\lambda A}e^{\lambda B} + Ae^{\lambda A}Be^{\lambda B} + Ae^{\lambda A}Be^{\lambda B} + e^{\lambda A}B^2e^{\lambda B}.$$

Evaluating at $\lambda = 0$, we have

$$(A+B)^2 = A^2 + 2AB + B^2$$

or $$A^2 + AB + BA + B^2 = A^2 + AB + AB + B^2.$$

Hence $$AB = BA.$$

41. Let A be an n-square upper triangular matrix, and let a be a characteristic root of multiplicity r. Assume that

$$A[i, i+1 \,|\, i, i+1] = \begin{bmatrix} a & x \\ 0 & a \end{bmatrix}, \qquad x \neq 0.$$

Prove that $\lambda I_n - A$ has an elementary divisor of the form $(\lambda - a)^e$, $e \geq 2$. *Hint*: $\lambda I_n - A$ is upper triangular, and so

$$\det(\lambda I_n - A) = (\lambda - a)^r p(\lambda),$$

where g.c.d. $(\lambda - a, p(\lambda)) = 1$ and $r \geq 2$. Observe that $(\lambda I_n - A)(i+1 \,|\, i)$ is upper triangular, and

$$d_{n-1}(\lambda) \,|\, \det(\lambda I_n - A)(i+1 \,|\, i) = -x(\lambda - a)^{r-2} p(\lambda).$$

Hence $$(\lambda - a)^2 \,|\, q_n(\lambda) = \frac{d_n(\lambda)}{d_{n-1}(\lambda)}.$$

42. Let $f(\lambda)$ and $g(\lambda)$ be monic, g.c.d. $(f(\lambda), g(\lambda)) = 1$ in $R[\lambda]$. Prove that

$$C(f(\lambda)g(\lambda)) \overset{S}{=} C(f(\lambda)) \dotplus C(g(\lambda)).$$

Hint: The list of elementary divisors of the characteristic matrix of $C(f(\lambda)\, g(\lambda))$ is obtained by factoring its single nontrivial invariant factor, $f(\lambda)\, g(\lambda)$. The list of elementary divisors of the characteristic matrix of $C(f(\lambda)) \dotplus C(g(\lambda))$ is obtained by putting together the lists of elementary divisors of the characteristic matrices of $C(f(\lambda))$ and $C(g(\lambda))$, i.e., by listing the factors of $f(\lambda)$ and those of $g(\lambda)$. Since g.c.d.$(f(\lambda), g(\lambda)) = 1$, these lists are the same.,

43. Discuss the elementary divisors over $Q[\lambda]$ of the characteristic matrix of a permutation matrix $A(\sigma)$, $\sigma \in S_n$. *Hint*: Let $\sigma \in S_n$ be a permutation with cycle structure $[1^{\lambda_1}, 2^{\lambda_2}, \ldots, n^{\lambda_n}]$, i.e., σ is a product of λ_1 1-cycles, λ_2 2-cycles, . . . , λ_n n-cycles, all of these cycles being disjoint (some of the λ_i are 0). If $\tau \in S_n$ is any other permutation with the same cycle structure as σ, then there exists $\theta \in S_n$ such that $\tau = \theta\sigma\theta^{-1}$. Thus

$$\begin{aligned} A(\tau) &= A(\theta\sigma\theta^{-1}) = A(\theta)A(\sigma)A(\theta^{-1}) \\ &= A(\theta)A(\sigma)A(\theta)^{-1}. \end{aligned}$$

This means that permutation matrices corresponding to permutations with the same cycle structure are similar. Hence their characteristic matrices have the same elementary divisors. Consider the following permutation τ with the same cycle structure as σ:

$$\tau = \underbrace{(1)(2)\cdots(\lambda_1)}_{\lambda_1}\ \underbrace{(\lambda_1+1, \lambda_1+2)(\lambda_1+3, \lambda_1+4)\cdots}_{\lambda_2}\ \cdots.$$

Then

$$A(\tau) = \left[\begin{array}{c|c|c} \lambda_1\left\{\begin{matrix}1 & & & \\ & 1 & & \\ & & 1 & \\ & & & \ddots \\ & & & & 1\end{matrix}\right. & & \\ \hline & 2\lambda_2\left\{\begin{matrix} \begin{bmatrix}0 & 1\\ 1 & 0\end{bmatrix} & & & \\ & \cdot & & \\ & & \cdot & \\ & & & \begin{bmatrix}0 & 1\\ 1 & 0\end{bmatrix}\end{matrix}\right. & \\ \hline & & \ddots \end{array}\right],$$

i.e., $A(\tau)$ is a direct sum of λ_1 1×1 blocks, λ_2 2×2 blocks, . . . , λ_n $n \times n$ blocks, where the typical $i \times i$ block is

$$A_i = \begin{bmatrix} 0 & & & & 1 \\ 1 & 0 & & \bigcirc & \\ & 1 & \ddots & & \\ & \bigcirc & \ddots & \ddots & \\ & & & 1 & 0 \end{bmatrix}.$$

The complete list of elementary divisors of $\lambda I - A(\tau)$ is the combined lists of elementary divisors of the characteristic matrices of the direct summands. However, $\lambda I_i - A_i = \lambda I_i - C(\lambda^i - 1)$ has as its only invariant factor $\lambda^i - 1$. Thus the elementary divisors of $\lambda I_i - A_i$ are the prime power factors of $\lambda^i - 1$ over Q. But

$$\lambda^i - 1 = \prod_{d \mid i} \Phi_d(\lambda)$$

where Φ_d is the dth *cyclotomic polynomial*; i.e., the $\Phi_d(\lambda)$ are the irreducible factors of $\lambda^i - 1$ over $Q[\lambda]$. Therefore the elementary divisors of $\lambda I_i - A_i$ are the cyclotomic polynomials $\Phi_d(\lambda)$, where $d \mid i$.

44. If $A(\sigma)$ and $A(\varphi)$ are similar permutation matrices over Q, prove that there exists $\theta \in S_n$ such that $A(\theta^{-1}\sigma\theta) = A(\varphi)$.

45. Let X be an n-square matrix of functions of t which satisfies

$$\dot{X}(t) = AX + XB, \qquad X(0) = C.$$

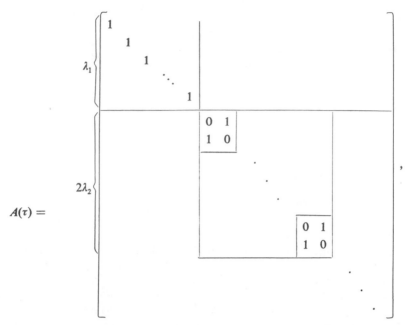

Express the solution in terms of A, B, C. *Hint*: Try $e^{tA}Ce^{tB}$. Remember B commutes with any function evaluated at B.

46. (Hard!) Let A and B be hermitian matrices. Prove

$$\operatorname{tr}(e^{A+B}) \leq \operatorname{tr}(e^{A}e^{B}).$$

(Note that e^A, e^B are positive definite hermitian matrices.)

47. If $A \in M_n(R)$ and Z_{ij} are the component matrices of A, show that

$$\sum_{i=1}^{s} Z_{i1} = I_n.$$

Hint: Let $Z_{ij}(A)$ be the i,j component matrix of A, and $Z_{ij}(J_A)$ be the i,j component matrix of the Jordan form of A. Observe that

$$\begin{aligned} Z_{ij}(A) &= f_{ij}(A) = f_{ij}(SJ_AS^{-1}) \\ &= Sf_{ij}(J_A)S^{-1} = S(Z_{ij}(J_A))S^{-1}, \end{aligned}$$

where $Z_{ij}(J_A)$ is a direct sum of zero blocks [corresponding to the elementary divisors $(\lambda - \alpha_k)^e$, $k \neq i$, since $f_{ij}^{(t)}(\alpha_k) = 0$ when $0 \leq t \leq e - 1$ and $k \neq i$] together with nonzero blocks corresponding to the elementary divisors $(\lambda - \alpha_i)^e$. In each of these nonzero blocks there is exactly one nonzero stripe, i.e., $1/(j-1)!$ appears as an entry in the positions $(k, j + k - 1)$, $k = 1, \ldots,$ $n - j + 1$. This follows directly from the definition of a function defined at a matrix, and from the definition of f_{ij}. Hence $Z_{i1}(J_A)$ has identity subblocks corresponding to elementary divisors involving $(\lambda - \alpha_i)^e$, with zero blocks elsewhere. So, clearly,

$$\sum_{i=1}^{s} f_{i1}(J_A) = I_n$$

since the blocks are disjoint. But then

$$\sum_{i=1}^{s} Z_{i1}(A) = S\left(\sum_{i=1}^{n} Z_{i1}(J_A)\right)S^{-1} = SI_nS^{-1} = I_n.$$

48. Show that

$$Z_{ij}Z_{lt} = 0 \qquad \text{for } i \neq l.$$

Hint: In the notation of Exercise 47 we first consider $Z_{ij}(J_A)Z_{lt}(J_A)$, $i \neq l$. $Z_{ij}(J_A)$ and $Z_{lt}(J_A)$ are block diagonal matrices; so $Z_{ij}(J_A)Z_{lt}(J_A)$ is also block diagonal where the kth diagonal block of $Z_{ij}(J_A)Z_{lt}(J_A)$ is the product of the kth diagonal blocks of $Z_{ij}(J_A)$ and $Z_{lt}(J_A)$. Thus the kth diagonal block of $Z_{ij}(J_A)Z_{lt}(J_A)$ can be nonzero only if both the kth diagonal block of $Z_{ij}(J_A)$ and that of $Z_{lt}(J_A)$ are nonzero. This is impossible if $i \neq l$ since the only nonzero blocks of $Z_{ij}(J_A)$ are those corresponding to elementary divisors of the form $(\lambda - \alpha_i)^e$, and the only nonzero blocks of $Z_{lt}(J_A)$ are those corresponding to elementary divisors of the form $(\lambda - \alpha_l)^e$. Therefore $Z_{ij}(J_A)Z_{lt}(J_A) = 0$. Now, for some nonsingular S,

$$\begin{aligned} Z_{ij}(A)Z_{lt}(A) &= SZ_{ij}(J_A)S^{-1}SZ_{lt}(J_A)S^{-1} \\ &= SZ_{ij}(J_A)Z_{lt}(J_A)S^{-1} \\ &= 0. \end{aligned}$$

49. Show that $Z_{i1}{}^2 = Z_{i1}$. *Hint*: First, observe that

$$(Z_{i1}(A))^2 = (SZ_{i1}(J_A)S^{-1})^2 = S(Z_{i1}(J_A))^2 S^{-1}.$$

But $Z_{i1}(J_A)$ is the direct sum of zero blocks and identity blocks, and so $(Z_{i1}(J_A))^2 = Z_{i1}(J_A)$.

50. Show that $Z_{k1}Z_{kr} = Z_{kr}$. *Hint*: By block multiplication,

$$Z_{k1}(J_A)Z_{kr}(J_A) = Z_{kr}(J_A)$$

since nonzero blocks in $Z_{k1}(J_A)$ are identity matrices. But then

$$\begin{aligned} Z_{k1}(A)Z_{kr}(A) &= S(Z_{k1}(J_A))S^{-1}S(Z_{kr}(J_A))S^{-1} \\ &= S(Z_{k1}(J_A)Z_{kr}(J_A))S^{-1} \\ &= S(Z_{kr}(J_A))S^{-1} = Z_{kr}(A). \end{aligned}$$

51. Show $Z_{ij} = \dfrac{1}{(j-1)!}(A - \alpha_i I_n)^{j-1} Z_{i1}$. *Hint*: Again, in the notation of Exercise 47, we consider $Z_{ij}(J_A)$. Observe that $J_A - \alpha_i I_n$ is a block diagonal matrix with U_e as the block corresponding to the elementary divisors $(\lambda - \alpha_i)^e$. Thus in $\dfrac{1}{(j-1)!}(J_A - \alpha_i I_n)^{j-1}$ the block corresponding to $(\lambda - \alpha_i)^e$ is

$j - 1$ stripe ↘

$$\frac{1}{(j-1)!}U_e^{j-1} = \begin{bmatrix} 0 & \cdots & 0 & \frac{1}{(j-1)!} & & & \\ & \ddots & 0 & 0 & \frac{1}{(j-1)!} & & \\ & & & \ddots & \ddots & \ddots & \\ & & & & & \ddots & 0 \\ & \bigcirc & & & & & 0 \end{bmatrix}.$$

Now $Z_{i1}(J_A)$ has I_e as the block corresponding to $(\lambda - \alpha_i)^e$, and 0 blocks elsewhere. Thus, by block multiplication, $\dfrac{1}{(j-1)!}(J_A - \alpha_i I_n)^{j-1} Z_{i1}(J_A)$ has $\dfrac{1}{(j-1)!}U_e^{j-1}$ as the block corresponding to $(\lambda - \alpha_i)^e$, and 0 blocks elsewhere. Therefore $\dfrac{1}{(j-1)!}(J_A - \alpha_i I_n)^{j-1} Z_{i1}(J_A) = Z_{ij}(J_A)$. Finally,

$$\begin{aligned} Z_{ij}(A) &= SZ_{ij}(J_A)S^{-1} \\ &= S\left[\frac{1}{(j-1)!}(J_A - \alpha_i I_n)^{j-1} Z_{i1}(J_A)\right]S^{-1} \\ &= \frac{1}{(j-1)!}(SJ_AS^{-1} - \alpha_i I_n)^{j-1}(SZ_{i1}(J_A)S^{-1}) \\ &= \frac{1}{(j-1)!}(A - \alpha_i I_n)^{j-1} Z_{i1}(A). \end{aligned}$$

Glossary 4.3

auxiliary unit matrix, 367
U_m, 367
Cayley-Hamilton theorem, 363
characteristic matrix of A, 355
characteristic polynomial of A, 355
characteristic roots of A, 355
component matrices, 377
Z_{ij}, 377
companion matrix, 358
$C(f(\lambda))$, 358
cyclic invariant subspace, 353
cyclic invariant subspace decomposition, 352
cyclotomic polynomial, 393
$\Phi_d(\lambda)$, 393
derivative of a matrix, 379
$\dot{X}(t)$, 379
direct sum of transformations, 363
$T_1 \dotplus T_2$, 363
eigenvalues, 355
e.v., 355
Frobenius normal form, 360
f is defined at A, 376
hypercompanion matrix, 367
$H(g(\lambda))$, 367
invariant subspace under T, 359
irreducible matrix, 364
irreducible transformation, 363
Jordan normal form, 368
Kronecker product, 390
$A \otimes B$, 390
linear extension, 351
linear transformation, 351
l.t., 351
matrix convergence, 386
matrix representation of T, 353
$C = [T]_E^F$, 353
minimal polynomial, 361
reducible matrix, 364
reducible transformation, 363
scalar product, 351
rT, 351
sequence of matrices, 386
similar matrices, 354
$A \overset{S}{=} B$, 355
similarity classes, 381
splitting of an elementary divisor, 375
unitary matrix, 385
V^*, 385
value of f at A, 376

4.4 Introduction to Multilinear Algebra

Let $V_1, \ldots, V_m$ and U be vector spaces over the same field R. A function $\varphi: \times_{i=1}^m V_i \to U$ is said to be *multilinear* if it is linear separately in each vector variable. This means that

$$\varphi(\ldots, cv_i + dw_i, \ldots) = c\varphi(\ldots, v_i, \ldots) + d\varphi(\ldots, w_i, \ldots) \quad (1)$$

for all vectors $v_i, w_i \in V_i$, $i = 1, \ldots, m$, and all scalars $c,d \in R$. The dots indicate the same fixed but unspecified vectors on both sides of (1). Multilinear algebra is concerned with the study of such multilinear functions.

For example, if $V_1 = M_{1,m}(R)$, $V_2 = M_{1,n}(R)$, and $U = M_{m,n}(R)$, define

$$\varphi: \quad V_1 \times V_2 \to U$$

by
$$\varphi(v_1,v_2) = [\xi_i\eta_j] = v_1{}^{\mathsf{T}}v_2,$$
where $v_1 = (\xi_1, \ldots, \xi_m)$ and $v_2 = (\eta_1, \ldots, \eta_n)$. Observe that φ is linear separately in v_1 and v_2; also im φ consists of all rank 1 matrices in $M_{m,n}(R)$ together with the 0 matrix, so that
$$\langle \text{im } \varphi \rangle = U.$$
However, im φ itself is not a vector space unless $m = 1$ or $n = 1$ (why?).

Sometimes problems of symmetry are important. For example, if
$$\varphi: \overbrace{R^m \times \cdots \times R^m}^{m} \to R$$
is defined by
$$\varphi(v_1, \ldots, v_m) = \det(v_1, \ldots, v_m),$$
then for any $\sigma \in S_m$,
$$\begin{aligned}\varphi(v_{\sigma(1)}, \ldots, v_{\sigma(m)}) &= \det(v_{\sigma(1)}, \ldots, v_{\sigma(m)})\\ &= \varepsilon(\sigma)\det(v_1, \ldots, v_m)\\ &= \varepsilon(\sigma)\varphi(v_1, \ldots, v_m).\end{aligned}$$
This situation is described by saying that φ is "symmetric with respect to the group S_m and the character ε."

For the purposes of this introduction we shall only discuss multilinear functions φ, defined on the cartesian product of a finite dimensional vector space V with itself m times, with values in some vector space U, i.e.,
$$\varphi: \mathop{\times}_1^m V \to U.$$
The theory, however, is easily extended to multilinear functions defined on cartesian products of arbitrary finite dimensional vector spaces. The underlying field R is always assumed to have characteristic 0.

The initial problem is to show how the study of multilinear functions can be referred to the simpler study of linear functions. The concept required for this is that of the *tensor product.*

Definition 4.1 (*Tensor Product*) Let V be a vector space over R. The pair (P,ν) is called a *tensor product* of V with itself m times if the following two conditions are satisfied.

(i) The map $\nu: \times_1^m V \to P$ is a multilinear function with values in the vector space P over R such that
$$\langle \text{im } \nu \rangle = P. \tag{2}$$

(ii) (*Universal Factorization Property*) If U is any vector space over R and $\varphi: \times_1^m V \to U$ is any multilinear function, then there exists a linear function $h: P \to U$ such that

$$\varphi = h\nu,$$

i.e.,
$$\varphi(v_1, \ldots, v_m) = h\nu(v_1, \ldots, v_m)$$

for all vectors $v_1, \ldots, v_m \in V$. Thus the following diagram is required to be commutative:

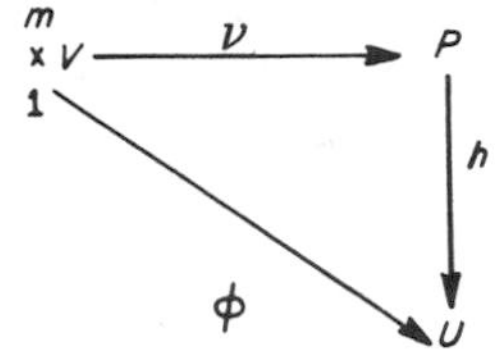

Our first and most important problem is to show that a tensor product of V with itself m times always exists. To do this we let $M_m(V:U)$ denote the totality of multilinear functions $\varphi: \times_1^m V \to U$. It is trivial to verify that $M_m(V:U)$ is a vector space over R with addition and scalar multiplication defined by

$$(\varphi + \theta)(v_1, \ldots, v_m) = \varphi(v_1, \ldots, v_m) + \theta(v_1, \ldots, v_m),$$
$$(c\varphi)(v_1, \ldots, v_m) = c\varphi(v_1, \ldots, v_m);$$

here $\varphi,\theta \in M_m(V:U)$, $v_i \in V$ for $i = 1, \ldots, m$, and $c \in R$.

Theorem 4.1 *The vector space $M_m(V:U)$ is finite dimensional with*

$$\dim M_m(V:U) = (\dim V)^m \dim U.$$

Proof: We remind the reader that $\Gamma_m(n)$ denotes the totality of n^m sequences of length m of integers chosen from $\{1, 2, \ldots, n\} = [1,n]$:

$$\Gamma_m(n) = \{\omega: [1,m] \to [1,n]\}$$

[see Section 3.2, formula (23)].

Suppose $f_1, \ldots, f_m$ are elements of $V^* = \{f: V \to R \mid f \text{ is linear}\}$, the *dual space* of V; thus each $f_j: V \to R$ is a linear transformation or *linear functional* on V to R. Then their *product*

$$\varphi = f_1 \cdots f_m$$

is the element of $M_m(V:R)$ defined by

$$\varphi(v_1, \ldots, v_m) = f_1(v_1) \cdots f_m(v_m), \qquad v_i \in V,\ i = 1, \ldots, m. \tag{3}$$

Now let $E = \{e_1, \ldots, e_n\}$ be a basis of V and $W = \{w_1, \ldots, w_p\}$ a basis of U. Let $\{f_1, \ldots, f_n\}$ be the *basis of V^* dual to E*, so that $f_i(e_j) = \delta_{ij}$, $i,j =$

$1, \ldots, n$ (see Exercise 1). For each $\alpha \in \Gamma_m(n)$ and $1 \leq j \leq p$, define $\varphi_{\alpha,j}: \times_1^m V \to U$ by

$$\varphi_{\alpha,j}(v_1, \ldots, v_m) = \left(\prod_{k=1}^{m} f_{\alpha(k)}(v_k)\right) w_j. \tag{4}$$

It is easy to see that $\varphi_{\alpha,j} \in M_m(V:U)$. We claim that $\{\varphi_{\alpha,j} \mid \alpha \in \Gamma_m(n), j = 1, \ldots, p\}$ is a basis of $M_m(V:U)$. First, the $\varphi_{\alpha,j}$ are linearly independent. For if $c_{\alpha,j} \in R$, $\alpha \in \Gamma_m(n)$, $j = 1, \ldots, p$, are scalars such that

$$\sum_{\alpha \in \Gamma_m(n)} \sum_{j=1}^{p} c_{\alpha j} \varphi_{\alpha,j} = 0,$$

then for $\beta \in \Gamma_m(n)$ we have

$$\begin{aligned} 0 &= \sum_{\alpha,j} c_{\alpha j} \varphi_{\alpha,j}(e_{\beta(1)}, \ldots, e_{\beta(m)}) \\ &= \sum_{\alpha,j} c_{\alpha j} \left[\prod_{k=1}^{m} f_{\alpha(k)}(e_{\beta(k)})\right] w_j \\ &= \sum_{\alpha,j} c_{\alpha j} \delta_{\alpha,\beta} w_j \\ &= \sum_{j=1}^{p} c_{\beta j} w_j. \end{aligned} \tag{5}$$

The linear independence of the w_j now implies that $c_{\beta j} = 0$, $j = 1, \ldots, p$. Finally, if $\varphi \in M_m(V:U)$, then for each $\gamma \in \Gamma_m(n)$ there exist scalars $c_{\gamma j} \in R$, $j = 1, \ldots, p$ such that

$$\varphi(e_\gamma) = \varphi(e_{\gamma(1)}, \ldots, e_{\gamma(m)}) = \sum_{j=1}^{p} c_{\gamma j} w_j.$$

Let

$$\theta = \sum_{\alpha \in \Gamma_m(n)} \sum_{j=1}^{p} c_{\alpha j} \varphi_{\alpha,j} \tag{6}$$

and note, as in the calculation (5), that

$$\theta(e_\gamma) = \sum_{j=1}^{p} c_{\gamma j} w_j = \varphi(e_\gamma).$$

Since this holds for arbitrary $\gamma \in \Gamma_m(n)$ and φ and θ are multilinear, it follows that $\varphi = \theta$. Thus $\{\varphi_{\alpha,j} \mid \alpha \in \Gamma_m(n), j = 1, \ldots, p\}$ is a basis of $M_m(V:U)$. It is said to be *dual to the bases E and W*. The facts that $|\Gamma_m(n)| = n^m$ and $\dim U = p$ yield the result. ∎

Note that in the event $U = R$, Theorem 4.1 states that

$$\dim M_m(V:R) = (\dim V)^m.$$

Let

$$P = M_m(V:R)^*$$

be the dual space of $M_m(V:R)$, and let $\nu \in M_m(V:P)$ be defined by

$$\nu(v_1, \ldots, v_m)(\varphi) = \varphi(v_1, \ldots, v_m), \qquad v_1, \ldots, v_m \in V, \tag{7}$$

for any $\varphi \in M_m(V:R)$. The equation (7) makes sense: im $\nu \subset P$ means that $\nu(v_1, \ldots, v_m)$ must be a linear functional on $M_m(V:R)$, i.e., the value $\nu(v_1, \ldots, v_m)(\varphi)$ must be specified for each $\varphi \in M_m(V:R)$.

Theorem 4.2 *The pair (P,ν) is a tensor product of V with itself m times.*

Proof: We must verify that (P,ν) satisfies conditions (i) and (ii) in Definition 4.1. Let $E = \{e_1, \ldots, e_n\}$ be a basis of V, and let $\{f_1, \ldots, f_n\}$ be a basis of V^* dual to E. We know from the proof of Theorem 4.1 that

$$\{\varphi_\alpha = \prod_{t=1}^{m} f_{\alpha(t)} \mid \alpha \in \Gamma_m(n)\}$$

is a basis of $M_m(V:R)$ and

$$\begin{aligned}\varphi_\alpha(e_\beta) &= \varphi_\alpha(e_{\beta(1)}, \ldots, e_{\beta(m)})\\ &= \delta_{\alpha\beta}, \qquad \alpha,\beta \in \Gamma_m(n).\end{aligned} \tag{8}$$

Then the definition (7) of $\nu \in M_m(V:P)$ yields

$$\nu(e_\beta)(\varphi_\alpha) = \nu(e_{\beta(1)}, \ldots, e_{\beta(m)})(\varphi_\alpha) = \delta_{\alpha\beta}, \qquad \alpha,\beta \in \Gamma_m(n). \tag{9}$$

Thus the linear functionals $\nu(e_\beta) \in P = M_m(V:R)^*$ form a basis of P dual to the basis $\{\varphi_\alpha | \alpha \in \Gamma_m(n)\}$ of $M_m(V:R)$ (see Exercise 1); in particular,

$$\langle \text{im } \nu \rangle = P.$$

To confirm that the universal factorization property holds, we must show that given $\varphi \in M_m(V:U)$, the commutative diagram

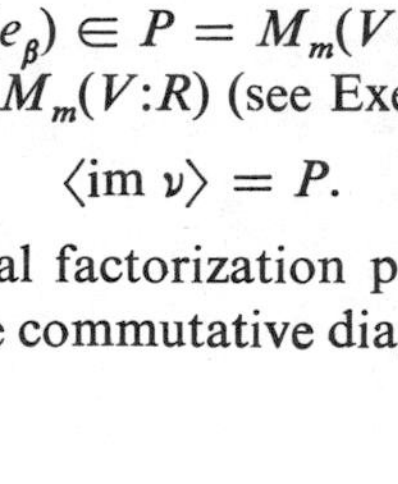

can be completed with a linear h. For this purpose simply define $h: P \to U$ on the basis $\{\nu(e_\alpha) | \alpha \in \Gamma_m(n)\}$ of P by

$$\begin{aligned}h(\nu(e_\alpha)) &= \varphi(e_\alpha)\\ &= \varphi(e_{\alpha(1)}, \ldots, e_{\alpha(m)}), \qquad \alpha \in \Gamma_m(n),\end{aligned}$$

and extend h linearly to all of P. Then if $v_t = \sum_{j=1}^{n} \xi_{tj} e_j \in V$, $t = 1, \ldots, m$, we have

$$\begin{aligned}
h(\nu(v_1, \ldots, v_m)) &= h\Big(\sum_{\gamma\in\Gamma_m(n)} \xi_{1\gamma(1)} \cdots \xi_{m\gamma(m)} \nu(e_\gamma)\Big) \\
&= \sum_{\gamma\in\Gamma_m(n)} \left(\prod_{t=1}^{m} \xi_{t\gamma(t)}\right) h(\nu(e_\gamma)) \\
&= \sum_{\gamma\in\Gamma_m(n)} \left(\prod_{t=1}^{m} \xi_{t\gamma(t)}\right) \varphi(e_{\gamma(1)}, \ldots, e_{\gamma(m)}) \\
&= \varphi\left(\sum_{j=1}^{n} \xi_{1j}e_j, \ldots, \sum_{j=1}^{n} \xi_{mj}e_j\right) \\
&= \varphi(v_1, \ldots, v_m).
\end{aligned}$$

Thus
$$h\nu = \varphi,$$
and the proof is complete. ∎

It should be noted that if (P,ν) and (Q,μ) are both tensor products of V with itself m times, then there exists precisely one linear bijection $h\colon P \to Q$ such that

$$h\nu = \mu. \tag{10}$$

Indeed, since (P,ν) and (Q,μ) have the universal factorization property, we obtain the diagram

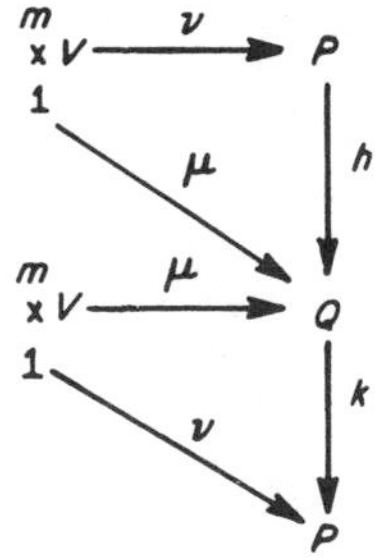

Then $h\nu = \mu$ and $k\mu = \nu$ imply $kh\nu = \nu$. Since $\langle \text{im } \nu \rangle = P$, it follows that $kh = I_P$. Similarly $hk = I_Q$, so that h is a bijection. Finally, if $h_1\colon P \to Q$ and $h_1\nu = \mu = h\nu$, then $\langle \text{im } \nu \rangle = P$ implies that $h_1 = h$.

The same kind of argument shows that the linear map h in the diagram

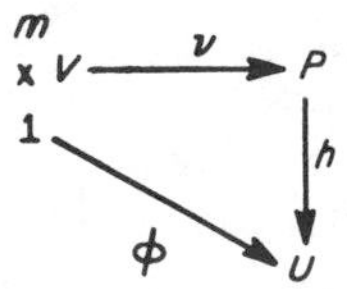

of Definition 4.1(ii) is uniquely determined by φ: We have $h\nu = \varphi$ and the values of ν span P.

In the definition of a tensor product the assumption is made that $\nu \in M_m(V:P)$ satisfies $\langle \text{im } \nu \rangle = P$. It seems to be a very difficult question to characterize those $\varphi \in M_m(V:U)$ for which

$$\text{im } \varphi = U.$$

In other words, when is the range of a multilinear function a vector space? This problem is in sharp contrast to the situation for linear maps; the range of a linear map is always a vector space.

As an example of a tensor product we observe that the space $M_n(R)$ of $n \times n$ matrices over R is a tensor product of $M_{1,n}(R)$ with itself. Define

$$\nu: M_{1,n}(R) \times M_{1,n}(R) \to M_n(R)$$

by

$$\begin{aligned}\nu(\xi,\eta) &= [\xi_i \eta_j] \\ &= \xi^{\mathrm{T}}\eta,\end{aligned}$$

where $\xi = [\xi_1, \ldots, \xi_n]$ and $\eta = [\eta_1, \ldots, \eta_n]$. Notice that if $e_i = [\delta_{i1}, \ldots, \delta_{in}]$, $i = 1, \ldots, n$, then

$$\{\nu(e_i, e_j) \mid i,j = 1, \ldots, n\}$$

is a basis for $M_n(R)$. Clearly ν is bilinear and $\langle \text{im } \nu \rangle = M_n(R)$. Also, if $\varphi: M_{1,n}(R) \times M_{1,n}(R) \to U$ is any bilinear function, define $h: M_n(R) \to U$ by

$$h\nu(e_i,e_j) = \varphi(e_i,e_j) \tag{11}$$

and linear extension. We have for all $\xi,\eta \in M_{1,n}(R)$ that

$$\begin{aligned}h\nu(\xi,\eta) &= h\left(\sum_{i,j=1}^{n} \xi_i\eta_j\nu(e_i,e_j)\right) \\ &= \sum_{i,j=1}^{n} \xi_i\eta_j h\nu(e_i,e_j) \\ &= \sum_{i,j=1}^{n} \xi_i\eta_j\varphi(e_i,e_j) \\ &= \varphi(\xi,\eta);\end{aligned}$$

thus $h\nu = \varphi$.

In view of the uniqueness properties of a tensor product (P,ν) of V with itself m times (see the remarks following the proof of Theorem 4.2), it is customary to write

$$P = \overbrace{V \otimes \cdots \otimes V}^{m} = \bigotimes_{1}^{m} V$$

and $\quad \nu(v_1, \ldots, v_m) = v_1 \otimes \cdots \otimes v_m, \qquad v_i \in V, i = 1, \ldots, m.$

An element in $\otimes_1^m V$ of the form $v_1 \otimes \cdots \otimes v_m$ is called *decomposable*. Observe, for example, that not every element in $M_n(R) = M_{1,n}(R) \otimes M_{1,n}(R)$

is decomposable, i.e., not every $n \times n$ matrix is of the form $[\xi_i \eta_j]$ for $\xi = [\xi_1, \ldots, \xi_n]$, $\eta = [\eta_1, \ldots, \eta_n] \in M_{1,n}(R)$ (the matrices of this form are precisely the rank 1 matrices together with the zero matrix).

It follows from Theorem 4.1, Theorem 4.2, and Exercise 1 that if $\dim V = n$, then $\dim \bigotimes_1^m V = n^m$. Moreover, if $\{e_1, \ldots, e_n\}$ is a basis of V and $\{f_1, \ldots, f_n\}$ is a dual basis of V^*, then the equation (8) states that

$$\left\{\varphi_\alpha = \prod_{t=1}^m f_{\alpha(t)} \mid \alpha \in \Gamma_m(n)\right\}$$

and

$$\{e_{\alpha(1)} \otimes \cdots \otimes e_{\alpha(m)} \mid \alpha \in \Gamma_m(n)\} \tag{12}$$

are dual bases of $M_m(V{:}R)$ and $\bigotimes_1^m V$, respectively. We abbreviate the notation for the elements in (12) to

$$e_\alpha{}^\otimes, \qquad \alpha \in \Gamma_m(n).$$

In general, if $v_1, \ldots, v_m$ belong to V we will sometimes write $v^\otimes$ for $v_1 \otimes \cdots \otimes v_m$.

If V is a unitary space, there is a natural way of defining an inner product in $\bigotimes_1^m V$ which makes the tensor product into a unitary space. Let $\{e_1, \ldots, e_n\}$ be an orthonormal basis of V; then the decomposable tensors $e_\alpha{}^\otimes$, $\alpha \in \Gamma_m(n)$ comprise a basis of $\bigotimes_1^m V$. For arbitrary $z = \sum_\alpha c_\alpha e_\alpha{}^\otimes$ and $w = \sum_\alpha d_\alpha e_\alpha{}^\otimes$ in $\bigotimes_1^m V$, define

$$(z,w) = \sum_{\alpha \in \Gamma_m(n)} c_\alpha \bar{d}_\alpha. \tag{13}$$

It is obvious that the definition (13) produces an inner product in $\bigotimes_1^m V$, called the *induced inner product*. The only question is whether this inner product is independent of the o.n. basis of V used to define it. This is resolved in the following theorem.

Theorem 4.3 *Let $u_i, v_i \in V$, $i = 1, \ldots, m$. Then if $(\cdot\,,\cdot)$ denotes the inner product in both V and $\bigotimes_1^m V$, we have*

$$(u^\otimes, v^\otimes) = \prod_{i=1}^m (u_i, v_i). \tag{14}$$

In particular, the right side of (14) is independent of the o.n. basis $\{e_1, \ldots, e_n\}$ of V used to define the inner product in (13). Thus, since the decomposable elements span $\bigotimes_1^m V$, the definition (13) is independent of $\{e_1, \ldots, e_n\}$.

Proof: Formula (13) implies that $(e_\alpha{}^\otimes, e_\beta{}^\otimes) = \delta_{\alpha\beta}$ for $\alpha,\beta \in \Gamma_m(n)$. Setting $u_i = \sum_{j=1}^n \xi_{ij} e_j$ and $v_i = \sum_{j=1}^n \eta_{ij} e_j$, $i = 1, \ldots, m$, we compute that

$$u^\otimes = \sum_{\omega \in \Gamma_m(n)} \xi_\omega e_\omega{}^\otimes \qquad \left(\xi_\omega = \prod_{i=1}^m \xi_{i\omega(i)}\right),$$

$$v^\otimes = \sum_{\omega \in \Gamma_m(n)} \eta_\omega e_\omega{}^\otimes \qquad \left(\eta_\omega = \prod_{i=1}^m \eta_{i\omega(i)}\right),$$

and hence
$$(u^{\otimes},v^{\otimes}) = \sum_{\omega \in \Gamma_m(n)} \xi_\omega \bar{\eta}_\omega = \sum_{\omega \in \Gamma_m(n)} \prod_{i=1}^{m} (\xi_{i\omega(i)} \bar{\eta}_{i\omega(i)})$$
$$= \prod_{i=1}^{m} \sum_{j=1}^{n} \xi_{ij}\bar{\eta}_{ij} = \prod_{i=1}^{m} (u_i,v_i). \quad \blacksquare$$

As an example of this construction consider $M_n(\mathbf{C})$ as the tensor product
$$M_n(\mathbf{C}) = M_{1,n}(\mathbf{C}) \otimes M_{1,n}(\mathbf{C})$$
with $\xi \otimes \eta = [\xi_i \eta_j]$, $\quad \xi = [\xi_1, \ldots, \xi_n], \eta = [\eta_1, \ldots, \eta_n] \in M_{1,n}(\mathbf{C})$. Using the standard inner product $(u,v) = \sum_{t=1}^{n} u_t \bar{v}_t$ in $M_{1,n}(\mathbf{C})$, we see by Theorem 4.3 that
$$(x \otimes y, \xi \otimes \eta) = (x,\xi)(y,\eta) = \sum_{t=1}^{n} x_t \bar{\xi}_t \sum_{t=1}^{n} y_t \bar{\eta}_t. \tag{15}$$
On the other hand,
$$\begin{aligned}\operatorname{tr}((\xi \otimes \eta)^*(x \otimes y)) &= \operatorname{tr}([\xi_i \eta_j]^*[x_i y_j]) \\ &= \operatorname{tr}([\overline{\xi_j \eta_i}][x_i y_j]) \\ &= \sum_{k=1}^{n} \sum_{l=1}^{n} \bar{\eta}_k \bar{\xi}_l x_l y_k \\ &= \sum_{l=1}^{n} x_l \bar{\xi}_l \sum_{k=1}^{n} y_k \bar{\eta}_k \\ &= (x \otimes y, \xi \otimes \eta) \qquad \text{[from (15)]}.\end{aligned}$$
Since any element in $M_n(\mathbf{C})$ is a sum of decomposable elements, it follows that the induced inner product is the usual inner product in $M_n(\mathbf{C})$ given by
$$(A, B) = \operatorname{tr}(B^*A), \qquad A,B \in M_n(\mathbf{C}).$$

Much of multilinear algebra is devoted to the study of multilinear functions which have certain symmetries. For example, the multilinear function $\varphi \in M_n(M_{1,n}(R): R)$ defined by
$$\varphi(v_1, \ldots, v_n) = \det[\xi_{ij}], \qquad v_i = [\xi_{i1}, \ldots, \xi_{in}] \in M_{1,n}(R),$$
$$i = 1, \ldots, n$$
satisfies
$$\varphi(v_{\sigma(1)}, \ldots, v_{\sigma(n)}) = \varepsilon(\sigma)\varphi(v_1, \ldots, v_n)$$
for all $\sigma \in S_n$. More generally, let $H < S_m$ and let $\chi: H \to R$ be a homomorphism of H into the multiplicative group of nonzero elements of R. Then $\varphi \in M_m(V: U)$ is said to be *symmetric with respect to H and χ* if
$$\varphi(v_{\sigma(1)}, \ldots, v_{\sigma(m)}) = \chi(\sigma)\varphi(v_1, \ldots, v_m)$$
holds for all $\sigma \in H$ and $v_i \in V$, $i = 1, \ldots, m$. As another example,
$$\varphi \in M_2(M_{1,n}(R): M_n(R))$$

defined by

$$\begin{aligned}\varphi(\xi,\eta) &= [\xi_i\eta_j] - [\xi_i\eta_j]^{\mathrm{T}} \\ &= \xi^{\mathrm{T}}\eta - \eta^{\mathrm{T}}\xi, \qquad \xi = [\xi_1, \ldots, \xi_n], \eta = [\eta_1, \ldots, \eta_n] \in M_{1,n}(R)\end{aligned}$$

is symmetric with respect to S_2 and ε.

We now introduce the important concept of a *symmetry class of tensors* in terms of certain universal properties.

Definition 4.2 (*Symmetry Class of Tensors*) Let V and P be vector spaces over R, $H < S_m$, and $\chi: H \to R$ a homomorphism of H into the multiplicative group of nonzero elements of R. Assume that $\nu \in M_m(V : P)$ is symmetric with respect to H and χ and satisfies the following two conditions:

(i) $\langle \text{im } \nu \rangle = P$.

(ii) For any vector space U over R and any $\varphi \in M_m(V : U)$, symmetric with respect to H and χ, there exists a linear function $h : P \to U$ which makes the diagram

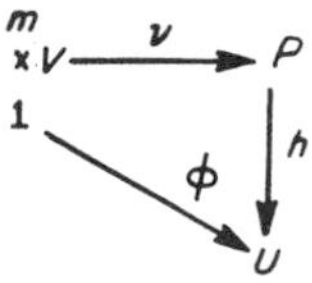

commutative, i.e.,

$$\varphi = h\nu. \tag{16}$$

Then the pair (P,ν) is called a *symmetry class of tensors associated with H and χ*.

It will be shown in Theorem 4.4 that a symmetry class of tensors associated with H and χ always exists. We need some preliminary definitions. Let $\sigma \in S_m$, and define

$$\varphi \in M_m(V: \bigotimes_1^m V)$$

by $\varphi(v_1, \ldots, v_m) = v_{\sigma^{-1}(1)} \otimes \cdots \otimes v_{\sigma^{-1}(m)}, \qquad v_i \in V, i = 1, \ldots, m.$

By the universal factorization property there exists a unique linear function $P(\sigma): \otimes_1^m V \to \otimes_1^m V$ which makes the diagram

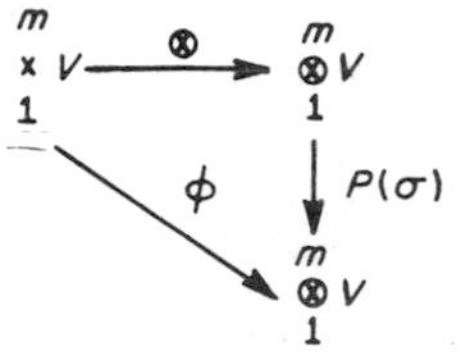

commute. Thus

$$P(\sigma)v_1 \otimes \cdots \otimes v_m = v_{\sigma^{-1}(1)} \otimes \cdots \otimes v_{\sigma^{-1}(m)},$$
$$v_i \in V, i = 1, \ldots, m.$$

The linear transformation $P(\sigma)$ is called a *permutation operator* on $\otimes_1^m V$. Any linear combination of permutation operators

$$S = \sum_{\sigma \in S_m} c_\sigma P(\sigma) : \bigotimes_1^m V \to \bigotimes_1^m V$$

is called a *symmetrizer* or a *symmetry operator* on $\otimes_1^m V$ (here $c_\sigma \in R$ for all $\sigma \in S_m$). In particular, the *symmetry operator defined by* $H < S_m$ *and* $\chi : H \to R$ is the linear map

$$S = \frac{1}{|H|} \sum_{\sigma \in H} \chi(\sigma)P(\sigma) : \bigotimes_1^m V \to \bigotimes_1^m V. \tag{17}$$

We have the following important theorem.

Theorem 4.4 *Let S be the symmetry operator on* $\otimes_1^m V$ *defined by* $H < S_m$ *and* $\chi : H \to R$. *Set* $P = \operatorname{im} S \subset \otimes_1^m V$ *and define* $\nu \in M_m(V : P)$ *by*

$$\nu(v_1, \ldots, v_m) = Sv^{\otimes}, \qquad v_i \in V, i = 1, \ldots, m.$$

Then the pair (P, ν) *is a symmetry class of tensors associated with H and* χ.

Proof: Since the decomposable elements $v^{\otimes}$ span $\otimes_1^m V$, it is clear that $\langle \operatorname{im} \nu \rangle = \operatorname{im} S = P$ and condition (i) of Definition 4.2 is satisfied. Next note that for any $\theta \in H$,

$$P(\theta)S = SP(\theta) = \chi(\theta^{-1})S. \tag{18}$$

To verify (18), we compute that

$$\begin{aligned} P(\theta)S &= \frac{1}{|H|} \sum_{\sigma \in H} \chi(\sigma)P(\theta)P(\sigma) = \frac{1}{|H|} \sum_{\sigma \in H} \chi(\sigma)P(\theta\sigma) \\ &= \frac{1}{|H|} \sum_{\tau \in H} \chi(\theta^{-1}\tau)P(\tau) = \frac{1}{|H|} \sum_{\tau \in H} \chi(\theta^{-1})\chi(\tau)P(\tau) \\ &= \chi(\theta^{-1})S, \end{aligned}$$

and a similar computation yields $SP(\theta) = \chi(\theta^{-1})S$.

Thus
$$\nu(v_{\theta(1)}, \ldots, v_{\theta(m)}) = Sv_{\theta(1)} \otimes \cdots \otimes v_{\theta(m)} = SP(\theta^{-1})v^{\otimes}$$
$$= \chi(\theta)\, Sv^{\otimes} = \chi(\theta)\nu(v_1, \ldots, v_m),$$

and so ν is symmetric with respect to H and χ.

It remains to confirm that condition (ii) of Definition 4.2 is satisfied. Let $\varphi \in M_m(V : U)$ be symmetric with respect to H and χ; then the identity

$$\varphi(v_1, \ldots, v_m) = \frac{1}{|H|} \sum_{\sigma \in H} \chi(\sigma)\varphi(v_{\sigma^{-1}(1)}, \ldots, v_{\sigma^{-1}(m)}) \tag{19}$$

is easily established. Now let $k\colon \otimes_1^m V \to U$ factor φ, i.e.,

$$k \otimes = \varphi, \tag{20}$$

and let $h = k \mid P$ be the restriction of k to P. Then

$$\begin{aligned} h\nu(v_1, \ldots, v_m) &= k\nu(v_1, \ldots, v_m) \\ &= kSv^{\otimes} \\ &= k\left\{\frac{1}{|H|} \sum_{\sigma \in H} \chi(\sigma)v_{\sigma^{-1}(1)} \otimes \cdots \otimes v_{\sigma^{-1}(m)}\right\} \\ &= \frac{1}{|H|} \sum_{\sigma \in H} \chi(\sigma)\varphi(v_{\sigma^{-1}(1)}, \ldots, v_{\sigma^{-1}(m)}) \qquad \text{[by (20)]} \\ &= \varphi(v_1, \ldots, v_m) \qquad \text{[by (19)]}. \end{aligned}$$

This completes the proof. ∎

It should be observed that formula (18) immediately implies $S^2 = S$. Indeed, since $\chi(\theta)P(\theta)S = S$ for all $\theta \in H$, we have

$$\begin{aligned} S &= \frac{1}{|H|} \sum_{\theta \in H} S = \frac{1}{|H|} \sum_{\theta \in H} \chi(\theta)P(\theta)S \\ &= S^2. \end{aligned}$$

We usually denote an element $\nu(v_1, \ldots, v_m)$ in a symmetry class of tensors by

$$v_1 * \cdots * v_m \tag{21}$$

(or v^*) and call it a *decomposable element* or the *star product* of $v_1, \ldots, v_m \in V$. Note that the symmetry of $\nu \in M_m(V : P)$ with respect to H and χ can be expressed as follows in our new notation:

$$v_{\theta(1)} * \cdots * v_{\theta(m)} = \chi(\theta)v_1 * \cdots * v_m \tag{22}$$

for all $\theta \in H$ and $v_i \in V$, $i = 1, \ldots, m$. Also, if $\alpha \in \Gamma_m(n)$, we sometimes write

$$v_\alpha^* = v_{\alpha(1)} * \cdots * v_{\alpha(m)}. \tag{23}$$

The three most important symmetry classes of tensors are

$\bigotimes_1^m V$, associated with $H = \{e\}$ (and $\chi \equiv 1$),

$\bigwedge^m V$, associated with $H = S_m$ and $\chi = \varepsilon$,

and $V^{(m)}$, associated with $H = S_m$ and $\chi \equiv 1$.

The space $\bigwedge^m V$ is called the mth *Grassmann space* or the mth *exterior space* or the mth *skew-symmetric space over* V; a decomposable element in $\bigwedge^m V$ is customarily denoted by

$$v_1 \wedge \cdots \wedge v_m \tag{24}$$

(or $v^\wedge$) and called the *exterior product* of $v_1, \ldots, v_m \in V$. The space $V^{(m)}$ is called the mth *completely symmetric space over* V; a decomposable element in $V^{(m)}$ is customarily denoted by

$$v_1 \bullet \cdots \bullet v_m \tag{25}$$

(or $v^\bullet$) and called the *symmetric product* of $v_1, \ldots, v_m \in V$. As in (23) we sometimes abbreviate $v_{\alpha(1)} \wedge \cdots \wedge v_{\alpha(m)}$ to $v_\alpha^\wedge$ and $v_{\alpha(1)} \bullet \cdots \bullet v_{\alpha(m)}$ to $v_\alpha^\bullet$ where $\alpha \in \Gamma_m(n)$.

The following result provides an analytic expression for the inner product in an arbitrary symmetry class of tensors.

Theorem 4.5 *Let $x_1, \ldots, x_m$ and $y_1, \ldots, y_m$ be vectors in a unitary space V. Denote both the inner product in V and the induced inner product in $\otimes_1^m V$ by $(\cdot,\cdot)$. Let $P \subset \otimes_1^m V$ be the symmetry class of tensors associated with $H < S_m$ and χ: $H \to R$. Then*

$$(x_1 * \cdots * x_m, y_1 * \cdots * y_m) = \frac{1}{|H|} \sum_{\sigma \in H} \chi(\sigma) \prod_{i=1}^m (x_i, y_{\sigma(i)}). \tag{26}$$

Proof: Let

$$S = \frac{1}{|H|} \sum_{\sigma \in H} \chi(\sigma) P(\sigma)$$

be the symmetry operator on $\otimes_1^m V$ defined by H and χ. As we saw in the proof of Theorem 4.4,

$$S^2 = S.$$

Observe also that for any $\sigma \in S_m$,

$$\begin{aligned}(P(\sigma)x^\otimes, y^\otimes) &= \prod_{i=1}^m (x_{\sigma^{-1}(i)}, y_i) = \prod_{i=1}^m (x_i, y_{\sigma(i)}) \\ &= (x^\otimes, P(\sigma^{-1})y^\otimes).\end{aligned}$$

[Note: since this holds for any decomposable elements $x^\otimes$, $y^\otimes \in \bigotimes_1^m V$, and since the decomposable elements span $\bigotimes_1^m V$, we conclude that $(P(\sigma)u, v) = (u, P(\sigma^{-1})v)$ for all $u, v \in \bigotimes_1^m V$. Thus

$$P(\sigma)^* = P(\sigma^{-1}), \tag{27}$$

where $P(\sigma)^*$ is the conjugate dual of $P(\sigma)$ with respect to the induced inner product in $\bigotimes_1^m V$.] Hence

$$\begin{aligned}(x^*, y^*) &= (Sx^\otimes, Sy^\otimes) = (S^*Sx^\otimes, y^\otimes)\\ &= (S^2x^\otimes, y^\otimes) = (Sx^\otimes, y^\otimes)\\ &= \frac{1}{|H|}\sum_{\sigma\in H}\chi(\sigma)(P(\sigma)x^\otimes, y^\otimes)\\ &= \frac{1}{|H|}\sum_{\sigma\in H}\chi(\sigma)(x^\otimes, P(\sigma^{-1})y^\otimes)\\ &= \frac{1}{|H|}\sum_{\sigma\in H}\chi(\sigma)\prod_{i=1}^m (x_i, y_{\sigma(i)}). \quad \blacksquare\end{aligned}$$

The formulas

$$(x_1 \wedge \cdots \wedge x_m, y_1 \wedge \cdots \wedge y_m) = \frac{1}{m!}\det([(x_i, y_j)]) \tag{28}$$

and

$$(x_1 \cdot \cdots \cdot x_m, y_1 \cdot \cdots \cdot y_m) = \frac{1}{m!}\operatorname{per}([(x_i, y_j)]) \tag{29}$$

are particular instances of (26). The *permanent* of an $m \times m$ matrix $A = [a_{ij}]$ is defined by

$$\operatorname{per} A = \sum_{\sigma\in S_m}\prod_{i=1}^m a_{i\sigma(i)}.$$

As an interesting application of the result (29), we can resolve the positive semidefinite case of the famous *van der Waerden conjecture for doubly stochastic matrices.* A *doubly stochastic* matrix is a matrix with nonnegative entries each of whose row and column sums is 1. In 1926 B. L. van der Waerden conjectured that if the matrix A is n-square doubly stochastic, then

$$\operatorname{per} A \geq \operatorname{per} J_n = n!/n^n, \tag{30}$$

where the n-square matrix J_n has every entry equal to $1/n$. Despite efforts by many workers over the last half century, the only general case of (30) presently known is for A positive semidefinite hermitian. An excellent survey of the problem is found in the article by H. J. Ryser, "Permanents and Systems of Distinct Representatives," *The University of North Carolina Monograph Series in Probability and Statistics* (1969), pp. 55–70, The University of North Carolina Press, Chapel Hill.

Theorem 4.6 (*van der Waerden Conjecture*) *Let A be an n-square positive semidefinite hermitian doubly stochastic matrix. Then*

$$\text{per } A \geq n!/n^n.$$

Proof: We first show that the positive semidefinite determination of $A^{1/2}$ has every row and column sum equal to 1. For as we know from Theorem 3.15, $A^{1/2}$ is a polynomial in A:

$$A^{1/2} = c_0 I_n + \cdots + c_p A^p.$$

Therefore

$$A^{1/2} J_n = J_n A^{1/2} = c J_n$$

where $c = c_0 + \cdots + c_p$; the second equality follows since A is doubly stochastic and hence $J_n A = J_n$. Now

$$\begin{aligned} J_n &= J_n A J_n \\ &= (J_n A^{1/2})(A^{1/2} J_n) \\ &= c^2 J_n \end{aligned}$$

so that $c^2 = 1$. On the other hand,

$$\begin{aligned} J_n A^{1/2} J_n &= c J_n^{\,2} \\ &= c J_n \end{aligned}$$

and hence $c \geq 0$. Thus $c = 1$.

Denote by $(\cdot,\cdot)$ the standard inner product in $V = \mathbf{C}^n$. From (29) with $m=n$ and the Cauchy-Schwarz inequality we have, setting $e_i = (\delta_{i1}, \ldots, \delta_{in})$, $i = 1, \ldots, n$, that

$$\begin{aligned} \left(\frac{1}{n!}\right)^2 |\text{per}([(A^{1/2}e_i, J_n e_j)])|^2 &= |(A^{1/2}e_1 \cdot \cdots \cdot A^{1/2}e_n, J_n e_1 \cdot \cdots \cdot J_n e_n)|^2 \\ &\leq \|A^{1/2}e_1 \cdot \cdots \cdot A^{1/2}e_n\|^2 \|J_n e_1 \cdot \cdots \cdot J_n e_n\|^2 \\ &= \frac{1}{n!} \text{per}([(A^{1/2}e_i, A^{1/2}e_j)]) \frac{1}{n!} \text{per}([(J_n e_i, J_n e_j)]). \end{aligned} \tag{31}$$

Now $(J_n e_i, J_n e_j) = 1/n$ for $i, j = 1, \ldots, n$, since $J_n e_i = (1/n)\, e$ where $e = (1, \ldots, 1)$. Also

$$\begin{aligned} (A^{1/2}e_i, A^{1/2}e_j) &= (Ae_i, e_j) \\ &= a_{ji}, \qquad i, j = 1, \ldots, n, \end{aligned}$$

and

$$\begin{aligned} (A^{1/2}e_i, J_n e_j) &= (J_n A^{1/2} e_i, e_j) \\ &= (J_n e_i, e_j) \qquad (\text{since } J_n A^{1/2} = J_n) \\ &= \frac{1}{n}, \qquad i, j = 1, \ldots, n. \end{aligned}$$

Thus (31) simplifies to

$$(\text{per } J_n)^2 \leq \text{per}(A^{\mathsf{T}}) \text{ per } J_n,$$

and $\text{per}(A^{\mathsf{T}}) = \text{per } A$ yields

$$\text{per } A \geq \text{per } J_n = n!/n^n. \quad \blacksquare$$

Exercises 4.4

1. Let V be a vector space over a field R, $\dim V = n$, and let V^* be the dual space of V, i.e., $V^* = \mathrm{Hom}(V,R)$. Show that $\dim V = \dim V^*$. *Hint*: Let $\{e_1, \ldots, e_n\}$ be a basis of V. For $i = 1, \ldots, n$ define $f_i \in V^*$ by $f_i(e_j) = \delta_{ij}, j = 1, \ldots, n$, and linear extension, i.e., $f_i(\sum_{j=1}^n \xi_j e_j) = \sum_{j=1}^n \xi_j f_i(e_j) = \xi_i$. Verify that $\{f_1, \ldots, f_n\}$ is a basis of V^*. The basis $\{f_1, \ldots, f_n\}$ is said to be *dual to* $\{e_1, \ldots, e_n\}$.
2. Let V be a vector space over a field R, $\dim V = n$, and let $\{f_1, \ldots, f_n\}$ be a basis of V^*. Show that there exists a basis $\{e_1, \ldots, e_n\}$ of V such that $\{f_1, \ldots, f_n\}$ is dual to $\{e_1, \ldots, e_n\}$. *Hint*: Let $\{v_1, \ldots, v_n\}$ be any basis of V. We seek scalars $\xi_{ij} \in R$ such that $\{e_i = \sum_{j=1}^n \xi_{ij} v_j \mid i = 1, \ldots, n\}$ is a basis of V dual to $\{f_1, \ldots, f_n\}$. Let $\{g_1, \ldots, g_n\}$ be the basis of V^* dual to $\{v_1, \ldots, v_n\}$ and write $f_i = \sum_{k=1}^n \eta_{ik} g_k$, $i = 1, \ldots, n$. We want

$$\begin{aligned}\delta_{ij} = f_i(e_j) = f_i\left(\sum_{t=1}^n \xi_{jt} v_t\right) &= \sum_{t=1}^n \xi_{jt} f_i(v_t)\\ = \sum_{t=1}^n \xi_{jt} \sum_{k=1}^n \eta_{ik} g_k(v_t) &= \sum_{t=1}^n \sum_{k=1}^n \xi_{jt}\eta_{ik}\delta_{kt}\\ = \sum_{t=1}^n \xi_{jt}\eta_{it}.\end{aligned}$$

The matrix $B = [\eta_{it}]$ is nonsingular (why?). If we set $A = [\xi_{jt}]$, then the above calculation requires that A satisfy

$$\delta_{ij} = (AB^{\mathsf{T}})_{ji},$$

i.e.,

$$AB^{\mathsf{T}} = I_n.$$

Thus if we choose $A = B^{\mathsf{T}^{-1}}$, the entries of A provide the scalars $\xi_{ij} \in R$ needed to define $\{e_1, \ldots, e_n\}$. The matrices A and B are said to be *contragredient*.
3. Let $\chi\colon G \to \mathbf{C}$ be a a homomorphism of the finite group G into the group of nonzero complex numbers. Prove that for all $\sigma \in G$, $|\chi(\sigma)| = 1$ and $\chi(\sigma^{-1}) = \overline{\chi(\sigma)}$.
4. Show that the linear map h in the diagram in Definition 4.2(ii) is uniquely determined by φ.
5. Prove that if (P, ν) and (Q, μ) are two symmetry classes of tensors associated with H and χ, then there exists precisely one linear bijection $h\colon P \to Q$ such that $h\nu = \mu$.
6. Suppose $\dim V \geq 4$ and let e_1, e_2, e_3, e_4 be linearly independent elements in V. Show that $e_1 \otimes e_2 + e_3 \otimes e_4$ is not a decomposable element of $V \otimes V$.
7. Let $v_1, \ldots, v_m \in V$. Show that $v_1 \otimes \cdots \otimes v_m = 0$ iff some $v_i = 0$.
8. Let $v_i, u_i \in V$, $i = 1, \ldots, m$. Show that if $v_1 \otimes \cdots \otimes v_m \neq 0$, then $v_1 \otimes \cdots \otimes v_m = u_1 \otimes \cdots \otimes u_m$ iff $v_i = c_i u_i$, $i = 1, \ldots, m$, and $\prod_{i=1}^m c_i = 1$. *Hint*: Assume $0 \neq v_1 \otimes \cdots \otimes v_m = u_1 \otimes \cdots \otimes u_m$. Then by Exercise 7, each v_i and u_i is nonzero. For a fixed k, let $f_k \in V^*$ be arbitrary and choose $f_i \in V^*$, $i \neq k$, such that $f_i(v_i) = 1$. Then there exists a linear $h\colon \otimes_1^m V \to R$ such that

$$h(v_1 \otimes \cdots \otimes v_m) = \prod_{i=1}^{m} f_i(v_i) = f_k(v_k).$$

Also, $h(u_1 \otimes \cdots \otimes u_m) = f_k(u_k) \prod_{i \neq k} f_i(u_i)$. Setting $c_k = \prod_{i \neq k} f_i(u_i)$, we have $f_k(v_k) = h(v_1 \otimes \cdots \otimes v_m) = h(u_1 \otimes \cdots \otimes u_m) = c_k f_k(u_k) = f_k(c_k u_k)$. Thus $f_k(v_k - c_k u_k) = 0$, and since the choice of f_k was arbitrary, it follows that $v_k - c_k u_k = 0$, i.e., $v_k = c_k u_k$. Finally, observe that $0 \neq u_1 \otimes \cdots \otimes u_m = v_1 \otimes \cdots \otimes v_m = (\prod_{i=1}^m c_i) u_1 \otimes \cdots \otimes u_m$ implies $\prod_{i=1}^m c_i = 1$. The converse statement is obvious.

9. Let $z \in V \otimes V$, and let k be the least integer for which z has a representation as a sum of k decomposable elements, $z = \sum_{t=1}^k u_t \otimes v_t$. Prove that both $\{u_1, \ldots, u_k\}$ and $\{v_1, \ldots, v_k\}$ are linearly independent sets. *Hint:* Suppose $u_1, \ldots, u_k$ were linearly dependent, say $u_k = \sum_{t=1}^{k-1} c_t u_t$. Then

$$z = \sum_{t=1}^{k} u_t \otimes v_t = \sum_{t=1}^{k-1} u_t \otimes v_t + \left(\sum_{t=1}^{k-1} c_t u_t\right) \otimes v_k$$

$$= \sum_{t=1}^{k-1} (u_t \otimes v_t + u_t \otimes c_t v_k) = \sum_{t=1}^{k-1} u_t \otimes (v_t + c_t v_k).$$

But this exhibits z as a sum of $k - 1$ decomposable elements, contradicting the minimality of k. A similar argument shows that $v_1, \ldots, v_k$ are linearly independent.

Glossary 4.4

contragredient, 411
decomposable element, 402, 407
doubly stochastic matrix, 409
dual basis, 398
dual space of V, 398
exterior product, 408
J_n, 409
linear functional, 398
mth completely symmetric space over V, 408
mth exterior space over V, 408
$M_m(V:U)$, 398
permanent, 409
permutation operator, 406
φ is symmetric with respect to H and χ, 404
product of linear functionals, 398
star product, 407
symmetric product, 408
symmetry class of tensors, 405
symmetrizer, 406
symmetry operator, 406
tensor product, 397
universal factorization property, 397
van der Waerden conjecture, 409
$\bigotimes_1^m V$, 402
$v_1 \otimes \cdots \otimes v_m$, 402
$v^\otimes$, 403
$e_\alpha^\otimes$, 403
$v_1 * \cdots * v_m$, 407
v^*, 407
v_α^*, 408
$\bigwedge^m V$, 408
$v_1 \wedge \cdots \wedge v_m$, 408
$v^\wedge$, 408
$v_\alpha^\wedge$, 408
$V^{(m)}$, 408
$v_1 \bullet \cdots \bullet v_m$, 408
$v^\bullet$, 408
$v_\alpha^\bullet$, 408

5

Representations of Groups

5.1 Representations

Let S and G be groupoids with the operations in both denoted by juxtaposition (see Section 1.2). If a function

$$L: S \to G$$

satisfies

$$L(s_1 s_2) = L(s_1)L(s_2) \qquad \text{for all } s_1, s_2 \in S, \tag{1}$$

then im L is called a *representation of S in G* and L is called a *representation function,* or by abuse of language simply a *representation.* If L is injective, the representation (function) is said to be *faithful.* Let V be a vector space over R. The *general linear group* GL(V) *over* V is the multiplicative group of all units in Hom(V, V). In case dim $V = n$, we write $\text{GL}(V) = GL_n(V)$. If $L: S \to \text{Hom}(V, V)$ is a representation function, then V is called a *representation module for S* or simply an *S-module,* and V is a *proper representation module for S* if im $L \subset \text{GL}(V)$. A representation function $L: S \to M_n(R)$ is called a *matrix representation* (*function*) *for S,* and L is a *proper matrix representation* if im $L \subset \text{GL}(n, R)$.

Example 1 (a) Let $S = S_n$ be the symmetric group of degree n, and define

$$L: S \to \text{GL}(n, R)$$

by

$$L(\sigma) = A(\sigma) = [\delta_{i\sigma(j)}], \qquad \sigma \in S,$$

so that $L(\sigma)$ is the permutation matrix associated with σ. Then L is a faithful proper matrix representation function for S (see Section 1.3, Exercise 1).

(b) Let V be a vector space of dimension n over a field R, and suppose E is a basis of V. Let $S = \text{Hom}(V,V)$, and define

$$L: S \to M_n(R)$$

by

$$L(T) = [T]_E^E, \qquad T \in S.$$

Then L is a faithful matrix representation function for S (see Section 4.3, Theorem 3.2).

Recall that if R is a field and S is a groupoid, then $R[S]$ is the totality of functions $f \in R^S$ which are nonzero for at most a finite number of elements of S. In Section 1.2, Example 9, we saw that operations of addition and multiplication can be defined in $R[S]$ by the formulas

$$(f + g)(s) = f(s) + g(s), \qquad s \in S \tag{2}$$

and

$$(f \cdot g)(s) = \sum_{rt=s} f(r)g(t), \qquad s \in S, \tag{3}$$

where $f,g \in R[S]$. (If no pair $r,t \in S$ exists for which $rt = s$, then $(f \cdot g)(s) = 0$ by definition.) It was proved in the above cited reference that $R[S]$ is a ring with respect to these operations, called the *groupoid ring of S over R*. If S is a semigroup (so that the operation in S is associative), then $R[S]$ is an associative ring, called the *semigroup ring of S over R*. We can also define a scalar multiplication in $R[S]$ by the obvious formula

$$(rf)(s) = rf(s), \qquad f \in R[S],\ r \in R; \tag{4}$$

it is then trivial to verify that $R[S]$ is a linear algebra over R (see Example 3, Section 1.2), called the *groupoid algebra of S over R*. If S is a group, then $R[S]$ is of course called the *group algebra of S over R*. For any $s \in S$, the *indicator function $f_s \in R[S]$ of s* is defined by

$$f_s(t) = \delta_{st}, \qquad t \in S. \tag{5}$$

Theorem 1.1 *If the groupoid S possesses an identity, then so does the groupoid algebra $R[S]$. If $\nu: S \to R[S]$ is defined by*

$$\nu(s) = f_s, \qquad s \in S, \tag{6}$$

then ν is a faithful representation of S in $R[S]$ and

$$R[S] = \langle \text{im } \nu \rangle. \tag{7}$$

Proof: For s, t, and x in S we have

$$\nu(st)(x) = f_{st}(x) = \delta_{st,x} \tag{8}$$

while

$$\begin{aligned}(\nu(s)\nu(t))(x) &= (f_s \cdot f_t)(x)\\ &= \sum_{uv=x} f_s(u)f_t(v)\\ &= \sum_{uv=x} \delta_{su}\delta_{tv}.\end{aligned} \tag{9}$$

This last sum is 0 unless there exist $u,v \in S$ such that $uv = x$, $u = s$, and

$t = v$, i.e., the value of the sum (9) is $\delta_{st,x}$. Thus (8) and (9) coincide for each $x \in S$, and hence

$$\nu(st) = \nu(s)\nu(t), \qquad s,t \in S,$$

showing that ν is a representation. If $\nu(s) = \nu(t)$, then $f_s = f_t$ and

$$\begin{aligned} 1 = \delta_{ss} &= f_s(s) \\ &= f_t(s) = \delta_{ts} \end{aligned}$$

implies $t = s$. Hence ν is faithful. It is easy to see that for any $f \in R[S]$ we have

$$f = \sum_{s \in S} f(s) f_s, \tag{10}$$

whence (7) follows. Observe that if $a_s \in R$, $s \in S$ are such that $\sum_{s\in S} a_s f_s = 0$, then $0 = \sum_{s\in S} a_s f_s(x) = \sum_{s\in S} a_s \delta_{sx} = a_x$ for each $x \in S$. Thus the indicator functions f_s are l.i. in $R[S]$ and consequently comprise a basis for $R[S]$ over R. If e is the identity of S, then

$$\begin{aligned} f_s f_e &= \nu(s)\nu(e) = \nu(se) \\ &= \nu(s) = f_s, \end{aligned}$$

and similarly,

$$f_e f_s = f_s.$$

Thus from (10) f_e is the identity in $R[S]$. ∎

In view of the fact that $\nu\colon S \to R[S]$ is an injection, we can apply the argument that appears in the proof of the ring extension theorem to obtain a linear algebra $\mathfrak{A}$ containing S and an isomorphism $\bar{\nu}\colon \mathfrak{A} \to R[S]$ such that $\bar{\nu}\,|\,S = \nu$. This construction is almost identical to the one in the ring extension theorem (see Section 3.2, Theorem 2.3). In outline, we let $\mathfrak{A} = S \cup (R[S] - \nu(S))$ and define $\bar{\nu}\colon \mathfrak{A} \to R[S]$ to be ν on S and the identity on $R[S] - \nu(S)$. To ensure that $\bar{\nu}$ is well-defined, it should be observed that $S \cap (R[S] - \nu(S)) = \varnothing$. In fact, we claim that $S \cap R[S] = \varnothing$. For suppose $s \in S$ and $s = f$ for some $f \in R[S]$. Then

$$s = \{(x, f(x)) \mid x \in S\}$$

and hence in particular

$$(s, f(s)) \in s,$$

or

$$s \in \{s\} \in \{\{s\}, \{s, f(s)\}\} = (s, f(s)) \in s. \tag{11}$$

But (11) is disallowed by the regularity axiom (see Section 1.1). Thus $S \cap R[S] = \varnothing$ and $\bar{\nu}\colon \mathfrak{A} \to R[S]$ is well-defined. It is also easy to see that $\bar{\nu}$ is a bijection. Now define addition, multiplication, and scalar multiplication in $\mathfrak{A}$ by

$$u + v = \bar{\nu}^{-1}(\bar{\nu}(u) + \bar{\nu}(v)), \qquad u,v \in \mathfrak{A},$$

$$uv = \bar{\nu}^{-1}(\bar{\nu}(u)\bar{\nu}(v)), \qquad u,v \in \mathfrak{A},$$

and

$$ru = \bar{\nu}^{-1}(r\bar{\nu}(u)), \qquad u \in \mathfrak{A},\ r \in R,$$

respectively. We conclude that *the groupoid S is a subset of a linear algebra $\mathfrak{A}$ which is isomorphic to the groupoid ring $R[S]$, and $R[S]$ has as a basis the indicator functions $f_s = \nu(s)$, $s \in S$*. In view of this isomorphism we shall regard $\mathfrak{A}$ itself as the groupoid algebra and identify each $s \in S$ with the corresponding indicator function f_s. Thus any element of $\mathfrak{A}$ can be written according to (10) as a unique linear combination

$$\sum_{s \in S} a_s s \tag{12}$$

of groupoid elements (it is understood in this sum that at most finitely many $a_s \in R$ are nonzero). In other words, the groupoid algebra $\mathfrak{A}$ is a linear algebra containing the groupoid S and having S as a basis. Moreover, it is immediate that the multiplication in $\mathfrak{A}$ is completely determined by the multiplication in S, and thus $\mathfrak{A}$ is uniquely determined.

In general, if $\mathfrak{B}$ is any linear algebra over a field R, V is a vector space over R, and

$$\mathscr{L}\colon \mathfrak{B} \to \operatorname{Hom}(V,V)$$

is a linear transformation which satisfies

$$\mathscr{L}(\beta_1\beta_2) = \mathscr{L}(\beta_1)\mathscr{L}(\beta_2) \qquad \text{for all } \beta_1,\beta_2 \in \mathfrak{B}, \tag{13}$$

then $\mathscr{L}$ is called a *representation function* and im $\mathscr{L}$ is a *representation of* $\mathfrak{B}$. If $\mathscr{L}$ is injective, the representation (function) is said to be *faithful*. The vector space V is called a *representation module for* $\mathfrak{B}$ or simply a $\mathfrak{B}$*-module*.

If V is a representation module for the groupoid S, i.e., if $L\colon S \to \operatorname{Hom}(V,V)$ is a representation function, it is frequently convenient to write sv instead of $L(s)v$, $s \in S$, $v \in V$. Then

$$L(s_1s_2)v = L(s_1)(L(s_2)v)$$

becomes

$$(s_1s_2)v = s_1(s_2v), \tag{14}$$

$$L(s)(v_1 + v_2) = L(s)v_1 + L(s)v_2$$

becomes

$$s(v_1 + v_2) = sv_1 + sv_2, \tag{15}$$

and

$$L(s)(rv) = rL(s)v$$

becomes

$$s(rv) = r(sv), \tag{16}$$

for all s, s_1, $s_2 \in S$, v, v_1, $v_2 \in V$, and $r \in R$.

Similarly, if V is a representation module for the linear algebra $\mathfrak{B}$, we write βv for $\mathscr{L}(\beta)v$, $\beta \in \mathfrak{B}$, $V \in V$. We then have the formulas

$$(\beta_1\beta_2)v = \beta_1(\beta_2v) \qquad [\text{i.e., } \mathscr{L}(\beta_1\beta_2)v = \mathscr{L}(\beta_1)(\mathscr{L}(\beta_2)v)],$$

$$(\beta_1 + \beta_2)v = \beta_1v + \beta_2v \qquad [\text{i.e., } \mathscr{L}(\beta_1 + \beta_2)v = \mathscr{L}(\beta_1)v + \mathscr{L}(\beta_2)v],$$

$$(r\beta)v = r(\beta v) \qquad [\text{i.e.,} \quad \mathscr{L}(r\beta)v = r\mathscr{L}(\beta)v], \tag{17}$$

$$\beta(v_1 + v_2) = \beta v_1 + \beta v_2 \qquad [\text{i.e.,} \quad \mathscr{L}(\beta)(v_1 + v_2) = \mathscr{L}(\beta)v_1 + \mathscr{L}(\beta)v_2],$$

and
$$\beta(rv) = r(\beta v) \qquad [\text{i.e.,} \quad \mathscr{L}(\beta)(rv) = r\mathscr{L}(\beta)v],$$

for all $\beta,\beta_1,\beta_2 \in \mathfrak{B}$, v, v_1, $v_2 \in V$, and $r \in R$.

Theorem 1.2 *Let V be a representation module for the groupoid S and let $E = \{e_1, \ldots, e_n\}$ be a basis of V. For each $s \in S$, define*

$$\Delta(s) = [\Delta_{ij}(s)] \in M_n(R)$$

by
$$se_j = \sum_{i=1}^{n} \Delta_{ij}(s)e_i, \qquad j = 1, \ldots, n. \tag{18}$$

Then $\Delta: S \to M_n(R)$ is a matrix representation of S. If V is a proper representation module for S, then Δ is a proper matrix representation of S.

Conversely, suppose $\Delta: S \to M_n(R)$ is a matrix representation of S. If $s \in S$, define se_j by (18) for $j = 1, \ldots, n$, and define sv for $v \in V$ by linear extension. Then V is a representation module for S. If Δ is a proper matrix representation of S, then V is a proper representation module for S.

Proof: Let $s_1,s_2 \in S$ and $1 \leq j \leq n$. Using (18), compute that

$$\begin{aligned}
\sum_{i=1}^{n} \Delta_{ij}(s_1s_2)e_i &= (s_1s_2)e_j = s_1(s_2e_j) \\
&= s_1\left(\sum_{i=1}^{n} \Delta_{ij}(s_2)e_i\right) = \sum_{i=1}^{n} \Delta_{ij}(s_2)s_1e_i \\
&= \sum_{i=1}^{n} \Delta_{ij}(s_2) \sum_{t=1}^{n} \Delta_{ti}(s_1)e_t = \sum_{t=1}^{n}\left(\sum_{i=1}^{n} \Delta_{ti}(s_1)\Delta_{ij}(s_2)\right)e_t \\
&= \sum_{t=1}^{n} (\Delta(s_1)\Delta(s_2))_{tj}e_t.
\end{aligned}$$

It follows by the linear independence of $e_1, \ldots, e_n$ that

$$\Delta_{ij}(s_1s_2) = (\Delta(s_1)\Delta(s_2))_{ij}, \qquad i = 1, \ldots, n.$$

Since the choice of j was arbitrary, we conclude that

$$\Delta(s_1s_2) = \Delta(s_1)\Delta(s_2).$$

Thus $\Delta: S \to M_n(R)$ is a matrix representation of S.

Observe that if $L: S \to \text{Hom}\,(V,V)$ is the given representation function, then

$$\Delta(s) = [L(s)]_E^E, \qquad s \in S. \tag{19}$$

Thus if V is a proper representation module for S, then $\Delta(s) \in \text{GL}(n,R)$ for all $s \in S$ [see Section 4.3, Theorem 3.2(e)], i.e., Δ is a proper matrix representation of S.

To prove the converse, we need only compute that

$$(s_1s_2)e_j = s_1(s_2e_j), \qquad j = 1, \ldots, n,$$

and this follows easily from (18). Indeed,

$$\begin{aligned} s_1(s_2e_j) &= s_1 \sum_{i=1}^{n} \Delta_{ij}(s_2)e_i \\ &= \sum_{i=1}^{n} \Delta_{ij}(s_2)s_1e_i \qquad \text{(by the definition of } sv\text{)} \\ &= \sum_{i=1}^{n} \Delta_{ij}(s_2) \sum_{t=1}^{n} \Delta_{ti}(s_1)e_t \\ &= \sum_{t=1}^{n} \left(\sum_{i=1}^{n} \Delta_{ti}(s_1)\Delta_{ij}(s_2) \right) e_t \\ &= \sum_{t=1}^{n} (\Delta(s_1)\Delta(s_2))_{tj}e_t \\ &= \sum_{t=1}^{n} (\Delta(s_1s_2))_{tj}e_t \qquad \text{(since } \Delta \text{ is a representation)} \\ &= (s_1s_2)e_j. \end{aligned}$$

Finally, if $\Delta\colon S \to \mathrm{GL}(n,R)$ is a proper matrix representation of S and we write $L(s)v = sv$, $s \in S$, $v \in V$, then

$$[L(s)]_E^E = \Delta(s) \qquad \text{for all } s \in S$$

and hence $L(s) \in \mathrm{GL}_n(V)$ for all $s \in S$, i.e., V is a proper representation module for S. ∎

We note that if V is a proper representation module for a groupoid S and S has an identity e, then

$$ev = v, \qquad v \in V. \tag{20}$$

For let $L\colon S \to \mathrm{GL}(V)$ be the representation function. Then

$$\begin{aligned} L(e) &= L(e^2) = L(ee) \\ &= L(e)L(e), \end{aligned}$$

and $L(e)$ has an inverse so that

$$L(e) = I_V.$$

The following fundamental result shows that a representation module for a groupoid S can be regarded as a representation module for the groupoid algebra $\mathfrak{A}$ of S over R.

Theorem 1.3 *Let S be a groupoid, and V an S-module. Let $\mathfrak{A}$ be the groupoid algebra of S over R. Then V is made into an $\mathfrak{A}$-module by setting*

$$fv = \sum_{s \in S} a_s(sv) \tag{21}$$

for each $f = \sum_{s \in S} a_s s \in \mathfrak{A}$.

Proof: Each $f \in \mathfrak{A}$ has a unique representation of the form $f = \sum_{s \in S} a_s s$ (only a finite number of the $a_s \in R$ are different from 0), and so the vector $fv \in V$ is well-defined by (21). We must verify that

$$(fg)v = f(gv), \tag{22}$$

$$(f + g)v = fv + gv, \tag{23}$$

$$(rf)v = r(fv), \tag{24}$$

$$f(v + w) = fv + fw, \tag{25}$$

and

$$f(rv) = r(fv) \tag{26}$$

for all $f,g \in \mathfrak{A}$, $v,w \in V$, and $r \in R$. If $g = \sum_{s \in S} b_s s$, we have

$$\begin{aligned} f(gv) &= \sum_{x \in S} a_x x \left(\sum_{y \in S} b_y (yv)\right) = \sum_{x \in S} \sum_{y \in S} a_x b_y x(yv) \\ &= \sum_{x,y \in S} a_x b_y (xy)v \qquad (V \text{ is an } S\text{-module}) \\ &= \sum_{s \in S} \left(\sum_{xy=s} a_x b_y\right)(sv) \\ &= (fg)v \qquad \left[\text{since } fg = \sum_{s \in S} \left(\sum_{xy=s} a_x b_y\right)\right]. \end{aligned}$$

This verifies (22). The much simpler verifications of (23) to (26) are found in Exercise 1. ∎

If $\mathfrak{B}$ is any linear associative algebra over R, then $\mathfrak{B}$ itself may be regarded as a $\mathfrak{B}$-module by simply defining

$$fv, \qquad f,v \in \mathfrak{B}$$

to be the product in $\mathfrak{B}$ of f and v. Explicitly, we have the representation

$$\rho\colon \mathfrak{B} \to \text{Hom}\,(\mathfrak{B}, \mathfrak{B})$$

given by

$$\rho(f)v = fv, \qquad f,v \in \mathfrak{B};$$

ρ is called the (*left*) *regular representation of* $\mathfrak{B}$. (The notation ρ is standard and should not be confused with the notation for rank.)

In particular, if S is a semigroup and $\mathfrak{A}$ is the semigroup algebra of S over R, then $\mathfrak{A}$ may be regarded as an S-module by defining

$$sv, \qquad s \in S,\ v \in \mathfrak{A}$$

to be the product of s and v in $\mathfrak{A}$. Explicitly, the representation

$$\rho\colon S \to \text{Hom}\,(\mathfrak{A}, \mathfrak{A})$$

is given by

$$\rho(s)v = sv, \qquad s \in S,\ v \in \mathfrak{A}.$$

Observe that if S is a group and $s \in S$, then $\rho(s)x = sx$ runs over S exactly once as x does. Since S is a basis of $\mathfrak{A}$ over R, it follows that $\rho(s)$ is nonsingular, i.e., $\rho(s) \in \text{GL}(\mathfrak{A})$. In other words, *the regular representation* $\rho\colon S \to \text{GL}(\mathfrak{A})$ *is proper when* S *is a group.*

Observe also that if $\mathfrak{B}$ is an associative linear algebra over R and $f,g \in \mathfrak{B}$, then $fv = gv$ for all $v \in \mathfrak{B}$ iff $(f - g)v = 0$ for all $v \in \mathfrak{B}$. Thus the regular representation $\rho: \mathfrak{B} \to \text{Hom}(\mathfrak{B},\mathfrak{B})$ is faithful iff whenever $hv = 0$ for all $v \in \mathfrak{B}$ we have $h = 0$.

Example 2 (a) Let S be the cyclic group of order 3: $S = \{e,g,g^2\}$. The group algebra $\mathfrak{A}$ of S over R consists of all elements of the form

$$f = ae + bg + cg^2, \qquad a,b,c \in R,$$

and $S = \{e,g,g^2\}$ is a basis of $\mathfrak{A}$. The regular representation is defined by left multiplication by each of the elements of S; and since S is a group, the regular representation is proper. We can compute the matrices $\Delta(e)$, $\Delta(g)$, $\Delta(g^2)$ of Theorem 1.2 to obtain an explicit proper matrix representation of S. Clearly $\Delta(e) = I_3$. To compute $\Delta(g)$, we use the fact (see the proof of Theorem 1.2) that

$$\Delta(g) = [\rho(g)]_E^E,$$

where $E = \{e,g,g^2\}$. Since $\rho(g)e = g$, $\rho(g)g = g^2$, and $\rho(g)\,g^2 = g^3 = e$, we have

$$\Delta(g) = \begin{bmatrix} 0 & 0 & 1 \\ 1 & 0 & 0 \\ 0 & 1 & 0 \end{bmatrix}.$$

Similarly,

$$\Delta(g^2) = \begin{bmatrix} 0 & 1 & 0 \\ 0 & 0 & 1 \\ 1 & 0 & 0 \end{bmatrix}.$$

(b) If $S = \{g_1, \ldots, g_n\}$ is a finite group and $\mathfrak{A}$ is the group algebra of S over R, then for each $g_k \in S$ there exists $\sigma_k \in S_n$ such that

$$g_k g_t = g_{\sigma k(t)}, \qquad t = 1, \ldots, n.$$

Thus the matrix representation $\Delta: S \to \text{GL}(n,R)$ associated with the regular representation $\rho: S \to \text{GL}(\mathfrak{A})$ of S has the form

$$\Delta(g_k) = [\delta_{i\sigma k(j)}], \qquad k = 1, \ldots, n$$

(see Theorem 1.2).

If $f = \sum_{k=1}^n c_k g_k \in \mathfrak{A}$, then $\rho(f)v = (\sum_{k=1}^n c_k g_k)v = \sum_{k=1}^n c_k g_k v = \sum_{k=1}^n c_k\, \rho(g_k)v = (\sum_{k=1}^n c_k \rho(g_k))v$ for any $v \in \mathfrak{A}$. Hence letting $E = \{g_1, \ldots, g_n\}$ denote S as a basis of $\mathfrak{A}$, we have

$$[\rho(f)]_E^E = \sum_{k=1}^n c_k [\rho(g_k)]_E^E.$$

Since $\rho(f_1 f_2) = \rho(f_1)\rho(f_2)$ and $\rho(r_1 f_1 + r_2 f_2) = r_1\rho(f_1) + r_2\rho(f_2)$, if we define $\Delta(f) = [\rho(f)]_E^E$, we can conclude that $\Delta: \mathfrak{A} \to M_n(R)$ is a *matrix representation* of $\mathfrak{A}$ in the sense that

$$\Delta(f_1 f_2) = \Delta(f_1)\Delta(f_2)$$

and

$$\Delta(r_1 f_1 + r_2 f_2) = r_1\Delta(f_1) + r_2\Delta(f_2)$$

for all $f_1,f_2 \in \mathfrak{A}$ and $r_1,r_2 \in R$. Observe that for any $f \in \mathfrak{A}$, $\Delta(f)$ is a matrix all of whose row and column sums are equal (see Exercise 2).

(c) Let $S = S_3$ be the symmetric group of degree 3. Then $V = R^3$ can be regarded as a proper representation module for S_3 by defining

$$\sigma x = A(\sigma)x, \qquad \sigma \in S_3, \quad x \in R^3,$$

where $A(\sigma)$ is the permutation matrix $A(\sigma) = [\delta_{i\sigma(j)}]$; thus

$$\sigma x = (x_{\sigma^{-1}(1)}, x_{\sigma^{-1}(2)}, x_{\sigma^{-1}(3)}).$$

Indeed, for $\sigma, \theta \in S_3$ we have

$$\begin{aligned}(\sigma\theta)x &= A(\sigma\theta)x = (A(\sigma)A(\theta))x \\ &= A(\sigma)(A(\theta)x) = A(\sigma)(\theta x) \\ &= \sigma(\theta x).\end{aligned}$$

[See Exercise 1, Section 1.3 for the proof that $A(\sigma\theta) = A(\sigma)A(\theta)$.]

Now let $z = (z_1, z_2, z_3)$ be a fixed vector in $V = R^3$ such that $z_1 + z_2 + z_3 = 0$, and define V_0 to be the subspace

$$V_0 = \langle\{\sigma z \mid \sigma \in S_3\}\rangle$$

of V. Clearly V_0 is a proper subspace of V. Moreover, if $\theta \in S_3$ and $v \in V_0$, then $\theta v \in V_0$; thus

$$\theta V_0 \subset V_0 \qquad \text{for all} \quad \theta \in S_3,$$

or more simply,

$$S_3 V_0 \subset V_0.$$

It is also obvious that if $\mathfrak{A}$ is the group algebra of S_3 over R and V is regarded as an $\mathfrak{A}$-module (see Theorem 1.3), then

$$fV_0 \subset V_0 \qquad \text{for all } f \in \mathfrak{A},$$

or

$$\mathfrak{A} V_0 \subset V_0.$$

In other words, V_0 is a proper subspace of V which is held invariant by every element of the group algebra $\mathfrak{A}$.

The notion of an invariant subspace introduced in Example 2(c) is crucial. We proceed to develop it in a general setting. Let S be a groupoid and V a representation module for S. If $V_0 \subset V$ and

$$SV_0 = \{sv \mid s \in S, \ v \in V_0\} \subset V_0,$$

then V_0 is called an *invariant subset for the representation of* S. If V is regarded as a representation module for the groupoid algebra $\mathfrak{A}$ of S over R (see Theorem 1.3) and $V_0 \subset V$ is such that

$$\mathfrak{A} V_0 = \{fv \mid f \in \mathfrak{A}, \ v \in V_0\} \subset V_0,$$

then V_0 is an *invariant subset for the representation of* $\mathfrak{A}$. The case of greatest interest is where V_0 is a (proper) subspace of V.

If V is a vector space over R and W_1, W_2 are invariant subspaces of V under $T \in \operatorname{Hom}(V,V)$ [so $T(W_i) \subset W_i$, $i = 1, 2$] such that $V = W_1 + W_2$, then we say that the pair (W_1, W_2) *reduces* T.

Recall that if V is a module and W is a submodule of V, then V/W is a module and the canonical module homomorphism or quotient map $q\colon V \to V/W$ is defined by

$$q(v) = v + W, \qquad v \in V \tag{27}$$

[see Section 4.2, Exercise 14(d)].

Let V be a finite dimensional vector space over R and W a subspace of V; then $\dim (V/W) = \dim V - \dim W$. In fact, if $\dim W = r$, let $\{v_1, \ldots, v_r, v_{r+1}, \ldots, v_n\}$ be a basis of V chosen so that $\{v_1, \ldots v_r\}$ is a basis of W. Then it is elementary to confirm that $\{q(v_{r+1}), \ldots, q(v_n)\}$ is a basis of V/W (see Exercise 3). Let $T \in \operatorname{Hom}(V,V)$, and suppose W_1 is an invariant subspace of V under T. Then the function $T_1\colon V/W_1 \to V/W_1$ given by

$$T_1 q(v) = q(Tv), \qquad v \in V \tag{28}$$

is well-defined and linear (see Exercise 4). Moreover, if the pair (W_1, W_2) of subspaces reduces T, then

$$T_1 = QPQ^{-1}, \tag{29}$$

where $Q = q\,|\,W_2$ and $P = T|\,W_2$ (see Exercise 5). Finally, it is not difficult to see that W_1 is not strictly contained in any proper invariant subspace of V under T iff W_2 does not strictly contain any proper invariant subspace of V under T (see Exercise 6).

If V is a representation module for the groupoid S and V_0 is an invariant subspace of V for the representation of S, then V_0 is said to be *minimal* (*maximal*) if the only invariant subspace strictly contained in V_0 (strictly containing V_0) is $\{0\}$ (V). The representation and the representation module are said to be *irreducible* if V is itself a minimal invariant subspace.

Theorem 1.4 *Let S be a groupoid, $\mathfrak{A}$ the groupoid algebra of S over R, V an S-module, and $v_0 \in V$. Then*

(i) $\langle Sv_0\rangle = \mathfrak{A}v_0$.

(ii) *$\mathfrak{A}v_0$ is an invariant subspace of V for the representation of $\mathfrak{A}$.*

(iii) *If $\mathfrak{A}v = V$ for all $v \in V$, $v \neq 0$, then the representation of S is irreducible.*

(iv) *If the representation of S is irreducible, then $\mathfrak{A}v = V$ for all $v \in V$, $v \neq 0$, unless $SV = \{0\}$ and* $\dim V = 1$. (This latter representation is called the *trivial representation of S*.)

Proof: (i) The elements of $\langle Sv_0\rangle$ are of the form $\sum_{s\in S} a_s(sv_0) = (\sum_{s\in S} a_s s)v_0$, a typical element of $\mathfrak{A}v_0$.

(ii) If $f \in \mathfrak{A}$ and $gv_0 \in \mathfrak{A}v_0$, then $f(gv_0) = (fg)v_0 \in \mathfrak{A}v_0$.

(iii) Suppose W is a proper invariant subspace of V for the representation of S, and let $0 \neq v \in W$. Then $V = \mathfrak{A}v \subset W$, contradicting the fact that W is proper. Thus the representation of S is irreducible.

(iv) If $0 \neq v \in V$, then by (ii), $\mathfrak{A}v$ is an invariant subspace for the representation of $\mathfrak{A}$ and thus automatically for the representation of S. Since the representation of S is irreducible, we have $\mathfrak{A}v = V$ or $\mathfrak{A}v = \{0\}$. Suppose $\mathfrak{A}v = \{0\}$ for some $0 \neq v \in V$; then of course $Sv = \{0\}$. If dim $V > 1$, then $W = \langle v \rangle$ is a proper subspace of V and $SW = \{0\} \subset W$, contradiciting the irreducibility of the representation of S. Thus dim $V = 1$ and $SV = Sv = \{0\}$. ∎

Let S be a groupoid and V an S-module. If there exists a proper invariant subspace of V for the representation of S, then the representation and the representation module are said to be *reducible*. If for any proper invariant subspace $W_1 \subset V$ there exists an invariant subspace $W_2 \subset V$ such that $V = W_1 + W_2$, then the representation and the representation module are said to be *completely* or *fully reducible*. It should be noted that if V is irreducible, then V is fully reducible (the defining condition is fulfilled vacuously).

Theorem 1.5 *Let S be a groupoid and V an S-module of dimension n over R.*

(i) *The representation of S is reducible iff there exists a basis $E = \{e_1, \ldots, e_n\}$ of V such that if $\Delta: S \to M_n(R)$ is the associated matrix representation of S defined in* (18), *then*

$$\Delta(s) = \begin{bmatrix} \Delta_{11}(s) & 0 \\ \Delta_{21}(s) & \Delta_{22}(s) \end{bmatrix}, \qquad s \in S, \tag{30}$$

where $\Delta_{11}(s) \in M_r(R)$ and $\Delta_{22}(s) \in M_{n-r}(R)$ for some fixed integer $1 \leq r < n$; in this event we have

$$\Delta_{11}(s_1 s_2) = \Delta_{11}(s_1)\Delta_{11}(s_2) \tag{31}$$

and

$$\Delta_{22}(s_1 s_2) = \Delta_{22}(s_1)\Delta_{22}(s_2) \tag{32}$$

for all $s_1, s_2 \in S$.

(ii) *The representation of S is fully reducible iff whenever $E = \{e_1, \ldots, e_n\}$ is a basis of V for which (30) holds with the fixed integer $1 \leq r < n$, there exists a basis $E' = \{e'_1, \ldots, e'_n\}$ of V such that $\langle e'_{r+1}, \ldots, e'_n \rangle = \langle e_{r+1}, \ldots, e_n \rangle$ and the associated matrix representation $\Delta': S \to M_n(R)$ of S has the form*

$$\Delta'(s) = \begin{bmatrix} \Delta'_{11}(s) & 0 \\ 0 & \Delta'_{22}(s) \end{bmatrix}, \qquad s \in S, \tag{33}$$

where $\Delta'_{11}(s) \in M_r(R)$ and $\Delta'_{22}(s) \in M_{n-r}(R)$; notice that in this case the matrices $\Delta'_{22}(s)$ are "simultaneously similar" to the matrices $\Delta_{22}(s)$ in $M_{n-r}(R)$.

Proof: (i) We first note that if $\Delta(s)$ has the form (30) for all $s \in S$, then it is a trivial exercise in block matrix multiplication to confirm (31) and (32). Suppose now that W is a proper invariant subspace of V for the representation of S. Choose a basis $E = \{e_1, \ldots, e_r, e_{r+1}, \ldots, e_n\}$ of V such that $W =$

$\langle e_{r+1}, \ldots, e_n \rangle$. Letting $L: S \to \text{Hom}(V, V)$ denote the representation of S, it is obvious that $\Delta(s) = [L(s)]_E^E$ has the form (30) for each $s \in S$. Conversely, if $E = \{e_1, \ldots, e_n\}$ is a basis of V such that $\Delta(s) = [L(s)]_E^E$ has the form (30) for all $s \in S$, where $\Delta_{11}(s) \in M_r(R)$ $(1 \leq r < n)$, then $W = \langle e_{r+1}, \ldots, e_n \rangle$ is a proper invariant subspace of V for the representation of S.

(ii) Let $L: S \to \text{Hom}(V,V)$ denote the representation of S. Assume first that L is fully reducible, and let $E = \{e_1, \ldots, e_n\}$ be a basis of V for which (30) holds with $\Delta_{11}(s) \in M_r(T)$ for all $s \in S$ $(1 \leq r < n)$. Let $W_1 = \langle e_{r+1}, \ldots, e_n \rangle$; then W_1 is a proper invariant subspace of V for the representation of S, so by the full reducibility of L there exists an invariant subspace W_2 of V such that $V = W_1 \dotplus W_2$. Let $\{e_1', \ldots, e_r'\}$ be a basis of W_2. Then $E' = \{e_1', \ldots, e_r', e_{r+1}, \ldots, e_n\}$ is a basis of V of the required type and such that the associated matrix representation $\Delta': S \to M_n(R)$ of S has the form

$$\Delta'(s) = \begin{bmatrix} \Delta'_{11}(s) & 0 \\ 0 & \Delta_{22}(s) \end{bmatrix}, \qquad s \in S.$$

Conversely, let us assume the condition and show that the representation L of S is fully reducible. Suppose W_1 is a proper invariant subspace of V for the representation of S. Let $\{e_{r+1}, \ldots, e_n\}$ $(1 \leq r < n)$ be a basis of W_1, and augment it to a basis $E = \{e_1, \ldots, e_n\}$ of V. Then the matrix representation Δ of S associated with L and E has the form (30); so by hypothesis there exists a basis $E' = \{e_1', \ldots, e_n'\}$ of V such that $\langle e'_{r+1}, \ldots, e_n' \rangle = \langle e_{r+1}, \ldots, e_n \rangle$ and the associated matrix representation Δ' of S has the form

$$\Delta'(s) = \begin{bmatrix} \Delta'_{11}(s) & 0 \\ 0 & \Delta'_{22}(s) \end{bmatrix}, \qquad s \in S,$$

where $\Delta'_{11}(s) \in M_r(R)$ and $\Delta'_{22}(s) \in M_{n-r}(R)$. Now let $W_2 = \langle e_1', \ldots, e_r' \rangle$. Then clearly W_2 is an invariant subspace of V for the representation of S, and since $W_1 = \langle e_{r+1}, \ldots, e_n \rangle = \langle e'_{r+1}, \ldots, e_n' \rangle$, we have $V = W_1 \dotplus W_2$. Thus $L: S \to \text{Hom}(V,V)$ is fully reducible. ∎

A more general concept of reducibility is available for an arbitrary set of matrices $M \subset M_n(R)$. The set M is said to be *reducible* if there exist a fixed nonsingular matrix $B \in \text{GL}(n,R)$ and fixed integers p and q, $1 \leq p < n$, such that

$$B^{-1}AB = \begin{bmatrix} A_{11} & 0 \\ A_{21} & A_{22} \end{bmatrix}, \qquad A \in M, \tag{34}$$

where $A_{11} \in M_p(R)$ and $A_{22} \in M_q(R)$. In other words, the matrices in M are required to be "simultaneously similar" to matrices of the form (34).

Let S be a groupoid and V an S-module of dimension n over R. Let $L: S \to \text{Hom}(V,V)$ denote the representation of S. If $\Delta: S \to M_n(R)$ and

$\kappa: S \to M_n(R)$ are the associated matrix representations with respect to bases E and F of V, respectively, then for all $s \in S$ we have

$$\begin{aligned} \Delta(s) &= [L(s)]_E^E = [I_V]_F^E[L(s)]_F^F[I_V]_E^F \\ &= B^{-1}\kappa(s)B \end{aligned}$$

where $B = [I_V]_E^F \in \mathrm{GL}(n,R)$. Thus Theorem 1.5(i) can be reformulated as follows: *A representation of S is reducible iff the matrices in any associated matrix representation of S form a reducible set.*

Observe that Theorem 1.5(ii) says that *a representation of S is fully reducible iff whenever $\Delta: S \to M_n(R)$ is an associated matrix representation of S of the form*

$$\Delta(s) = \begin{bmatrix} \Delta_{11}(s) & 0 \\ \Delta_{21}(s) & \Delta_{22}(s) \end{bmatrix}, \qquad s \in S$$

where $\Delta_{11}(s) \in M_r(R)$ and $\Delta_{22}(s) \in M_{n-r}(R)$ for some fixed integer $1 \le r < n$, there exists a fixed nonsingular matrix

$$P = \begin{bmatrix} P_{11} & 0 \\ P_{21} & P_{22} \end{bmatrix} \in \mathrm{GL}(n,R)$$

such that $P_{11} \in M_r(R)$ and

$$P^{-1}\Delta(s)P = \begin{bmatrix} \Delta'_{11}(s) & 0 \\ 0 & \Delta'_{22}(s) \end{bmatrix}, \qquad s \in S,$$

where $\Delta'_{11}(s) \in M_r(R)$ and $\Delta'_{22}(s) \in M_{n-r}(R)$; in this case we have $P_{22}{}^{-1}\Delta_{22}(s)P_{22} = \Delta'_{22}(s)$ for all $s \in S$.

In general, a set of matrices $M \subset M_n(R)$ is said to be *fully reducible* if whenever (34) holds there exists a fixed nonsingular matrix $C \in \mathrm{GL}(n,R)$ such that

$$C^{-1}AC = \begin{bmatrix} D_{11} & 0 \\ 0 & D_{22} \end{bmatrix}, \qquad A \in M, \tag{35}$$

where $D_{11} \in M_p(R)$ and $D_{22} \in M_q(R)$. If follows from the above reformulation of Theorem 1.5(ii) that *if a representation of S is fully reducible, then the matrices in any associated matrix representation of S form a fully reducible set.* As a final remark we note that any irreducible (i.e., not reducible) set of matrices is fully reducible, since the defining condition is fulfilled vacuously.

Example 3 (a) Let S be the multiplicative group of nonzero complex numbers. Let $E = \{e_1,e_2\}$ be a basis of the two-dimensional vector space $V = \mathbf{R}^2$, and for $x + iy \in S$ define $L(x + iy) \in \mathrm{GL}(V)$ by

$$\begin{aligned} L(x+iy)e_1 &= xe_1 + ye_2, \\ L(x+iy)e_2 &= -ye_1 + xe_2 \end{aligned}$$

and linear extension [notice that det $L(x+iy) = x^2+y^2 \neq 0$, i.e., $L(x+iy)$ is indeed nonsingular]. Then set

$$\kappa(x+iy) = [L(x+iy)]_E^E = \begin{bmatrix} x & -y \\ y & x \end{bmatrix}.$$

Observe that if $s_1 = x_1 + iy_1$, $s_2 = x_2 + iy_2 \in S$, then $s_1 s_2 = (x_1x_2 - y_1y_2) + i(x_1y_2 + x_2y_1)$ so that

$$\kappa(s_1s_2) = \begin{bmatrix} x_1x_2 - y_1y_2 & -(x_1y_2 + x_2y_1) \\ x_1y_2 + x_2y_1 & x_1x_2 - y_1y_2 \end{bmatrix}.$$

On the other hand,

$$\begin{aligned}\kappa(s_1)\kappa(s_2) &= \begin{bmatrix} x_1 & -y_1 \\ y_1 & x_1 \end{bmatrix}\begin{bmatrix} x_2 & -y_2 \\ y_2 & x_2 \end{bmatrix} \\ &= \begin{bmatrix} x_1x_2 - y_1y_2 & -(x_1y_2 + x_2y_1) \\ x_1y_2 + x_2y_1 & x_1x_2 - y_1y_2 \end{bmatrix} \\ &= \kappa(s_1s_2).\end{aligned}$$

Thus $L\colon S \to \mathrm{GL}(V)$ is a representation function for S. It is obvious that the representation is faithful. Now set $Q = \begin{bmatrix} 0 & -1 \\ 1 & 0 \end{bmatrix}$; then for every $s = x + iy \in S$ and $P \in \mathrm{GL}(2,\mathbf{R})$ or $P \in \mathrm{GL}(2,\mathbf{C})$, we have

$$P^{-1}\kappa(s)P = P^{-1}(xI_2 + yQ)P = xI_2 + yP^{-1}QP.$$

It follows that V is reducible as a module over **R** (**C**) if and only if the matrix Q is similar over **R** (**C**) to a matrix with (1,2) entry equal to 0. Since the characteristic roots of Q are $\pm i$, we know that Q is not similar over **R** to a matrix with (1,2) entry equal to 0. Thus V is irreducible regarded as a module over **R**. However, over **C** the elementary divisors of $\lambda I_2 - Q$ are the linear polynomials $\lambda + i$ and $\lambda - i$, so there exists a matrix $P \in \mathrm{GL}(2,\mathbf{C})$ such that

$$P^{-1}QP = \begin{bmatrix} -i & 0 \\ 0 & i \end{bmatrix}.$$

In fact, if we let

$$P = \begin{bmatrix} 1 & i \\ i & 1 \end{bmatrix} \in \mathrm{GL}(2,\mathbf{C}),$$

then

$$P^{-1} = \frac{1}{2}\begin{bmatrix} 1 & -i \\ -i & 1 \end{bmatrix}$$

and

$$P^{-1}QP = \frac{1}{2}\begin{bmatrix} 1 & -i \\ -i & 1 \end{bmatrix}\begin{bmatrix} 0 & -1 \\ 1 & 0 \end{bmatrix}\begin{bmatrix} 1 & i \\ i & 1 \end{bmatrix} = \begin{bmatrix} -i & 0 \\ 0 & i \end{bmatrix}.$$

Hence V is reducible regarded as a module over **C**. Thus the question of reducibility sometimes depends on the underlying field.

(b) Suppose V is a two-dimensional vector space over R, and let $S = \mathrm{GL}(V)$. Then V can be regarded trivially as an S-module by defining the representation function $L: S \to \mathrm{GL}_2(V)$ to be the identity. Once a basis E of V is chosen, the associated matrix representation $\Delta: S \to \mathrm{GL}(2,R)$ is simply given by

$$\Delta(s) = [s]_E^E, \qquad s \in S = \mathrm{GL}_2(V).$$

If the representation were reducible, there would exist a matrix $P \in \mathrm{GL}(2,R)$ such that

$$P^{-1}AP = \begin{bmatrix} b_{11} & 0 \\ b_{21} & b_{22} \end{bmatrix} = B$$

for all matrices $A \in \mathrm{GL}(2,R)$. Then $AP = PB$ so that $AP^{(2)} = PB^{(2)} = b_{22}P^{(2)}$. Thus for any nonsingular A, $AP^{(2)} \in \langle P^{(2)} \rangle$. On the other hand, a nonsingular A can be chosen such that $AP^{(2)} = P^{(1)}$ and $AP^{(1)} = P^{(2)}$; e.g., simply let $A = P\begin{bmatrix} 0 & 1 \\ 1 & 0 \end{bmatrix}P^{-1}$. But for this A we have $AP^{(2)} \notin \langle P^{(2)} \rangle$, since the columns $P^{(1)}$ and $P^{(2)}$ of the nonsingular matrix P are linearly independent. We conclude that the representation is irreducible.

The reader should recall the following concepts from elementary linear algebra. If V is a vector space over $\mathbf{C}$ and $\beta: V \times V \to \mathbf{C}$ is a function which satisfies (i) $\beta(x,y) = \overline{\beta(y,x)}$, $x,y \in V$; (ii) $\beta(ax_1 + bx_2, y) = a\beta(x_1, y) + b\beta(x_2,y)$, $x_1,x_2,y \in V$, $a,b \in \mathbf{C}$; (iii) $\beta(x,x) \geq 0$, $x \in V$, with equality iff $x = 0$, then β is called an *inner product on* V and V is called a *unitary space*. If V is a vector space over $\mathbf{R}$ and $\beta: V \times V \to \mathbf{R}$ satisfies $\beta(x,y) = \beta(y,x)$, $x,y \in V$, as well as (ii) and (iii) above, then β is an *inner product on* V and V is a *Euclidean space*. If V is a unitary space, $T \in \mathrm{Hom}(V,V)$, and $\beta(Tx,Ty) = \beta(x,y)$ for all $x,y \in V$, then T is called a *unitary transformation*. If V is a Euclidean space and $T \in \mathrm{Hom}(V,V)$ satisfies $\beta(Tx,Ty) = \beta(x,y)$ for all $x,y \in V$, then T is an *orthogonal transformation*. An *orthonormal* (o.n.) *basis* (*with respect to* β) of a unitary (or Euclidean) space V is a basis $E = \{e_1, \ldots, e_n\}$ of V such that $\beta(e_i,e_j) = \delta_{ij}$, $i,j = 1, \ldots, n$. The *conjugate dual* (*with respect to* β) of $T \in \mathrm{Hom}(V,V)$ is the unique transformation $T^* \in \mathrm{Hom}(V,V)$ satisfying

$$\beta(Tx,y) = \beta(x,T^*y), \qquad x,y \in V.$$

It is a standard elementary result that T is unitary (or orthogonal) iff $T^*T = TT^* = I_V$. Moreover, if $T \in \mathrm{Hom}(V,V)$ and E is an o.n. basis of V, then T is unitary (orthogonal) iff $U = [T]_E^E$ satisfies $UU^* = U^*U = I_n$ ($UU^\top = U^\top U = I_n$); such a matrix U is called *unitary* (*orthogonal*). Here $U^* = [\bar{u}_{ji}]$ is the *conjugate transpose of* U.

Let S be a groupoid, and V a unitary (Euclidean) space. If $L: S \to \mathrm{Hom}(V,V)$ is a representation function such that im L consists of unitary (orthogonal) transformations, then V is called a *unitary* (*Euclidean*) *representation module for* S. If $L: S \to M_n(\mathbf{C})$ [$L: S \to M_n(\mathbf{R})$] is a matrix representation such that im L consists of unitary (orthogonal) matrices, then L is called a *unitary* (*orthogonal*) *matrix representation of* S.

Theorem 1.6 *Any set* $M \subset M_n(\mathbf{C})$ *of unitary matrices is fully reducible. Moreover, if* S *is a groupoid and* V *is a unitary representation module for* S, *then* V *is fully reducible.*

Proof: If the set $M \subset M_n(\mathbf{C})$ of unitary matrices is irreducible, we are done. So assume M is reducible. Let E be an o.n. basis of the unitary space V, and let Ω be the set of all unitary transformations $T \in \text{Hom}(V,V)$ such that $[T]_E^E = U$ for some $U \in M$. Since M is reducible, there exists a proper subspace W of V which is invariant under all $T \in \Omega$. Choose an o.n. basis $\{w_{r+1}, \ldots, w_n\}$ of W $(1 \le r < n)$, and augment it to an o.n. basis $F = \{w_1, \ldots, w_r, w_{r+1}, \ldots, w_n\}$ of V. Then since $T(W) \subset W$ for each $T \in \Omega$, we have

$$[T]_F^F = \begin{bmatrix} U_{11} & 0 \\ U_{21} & U_{22} \end{bmatrix}, \qquad T \in \Omega.$$

Now
$$[T]_F^F = [I_V T I_V]_F^F = [I_V]_F^E [T]_E^E [I_V]_F^E,$$

and it is easy to see (Exercise 7) that $P = [I_V]_F^E$ is unitary with inverse $P^* = [I_V]_E^F$. Therefore each $[T]_F^F$ is unitary, so that

$$\begin{bmatrix} U_{11} & 0 \\ U_{21} & U_{22} \end{bmatrix} \begin{bmatrix} U_{11}^* & U_{21}^* \\ 0 & U_{22}^* \end{bmatrix} = \begin{bmatrix} I_r & 0 \\ 0 & I_{n-r} \end{bmatrix}.$$

Hence $U_{11}U_{21}^* = 0$; but U_{11} is nonsingular since $U_{11}U_{11}^* = I_r$, and it follows that $U_{21} = 0$. Thus

$$P^{-1}AP$$

is a direct sum (i.e., $U_{21} = 0$) for each $A \in M$.

Let $L: S \to \text{GL}_n(V)$ be the representation of S in which each $L(s)$ is unitary. Let W be an invariant subspace common to all the $L(s)$, $s \in S$, and assume dim $W = n - r$. Select an o.n. basis of V, $F = \{w_1, \ldots, w_r, w_{r+1}, \ldots, w_n\}$, precisely as in the proof of the first assertion in the statement of the theorem. Then exactly as before, the matrices $[L(s)]_F^F$ are unitary and have the form

$$\begin{bmatrix} U_{11} & 0 \\ U_{21} & U_{22} \end{bmatrix}$$

in which U_{22} is $(n - r)$-square. Hence, as above, $U_{21} = 0$. Thus $\langle w_1, \ldots, w_r \rangle$ is an invariant subspace of every $L(s)$, $s \in S$, and the proof is complete. ∎

We remark that Theorem 1.6 holds, of course, if $\mathbf{C}$ is replaced by $\mathbf{R}$ and "unitary" is replaced by "orthogonal."

Example 4 (a) Let $S = \{g_1, g_2, \ldots, g_n\}$ be a finite group and let $\rho: S \to \text{GL}(\mathfrak{A})$

be the regular representation of S, where $\mathfrak{A}$ is the group algebra of S over either **C** or **R**. Then ρ is a fully reducible representation. To see this, denote the basis $\{g_1, \ldots, g_n\}$ of $\mathfrak{A}$ by E, and define an inner product β on $\mathfrak{A}$ by

$$\beta\left(\sum_{k=1}^{n} a_k g_k, \sum_{k=1}^{n} b_k g_k\right) = \sum_{k=1}^{n} a_k \bar{b}_k.$$

Then E is clearly an orthonormal basis of $\mathfrak{A}$ with respect to β. For each $r = 1, \ldots, n$, define $\sigma_r \in S_n$ by $g_r g_k = g_{\sigma_r(k)}$, $k = 1, \ldots, n$. We compute that

$$\begin{aligned}\beta\left(\rho(g_r)\sum_{k=1}^{n} a_k g_k, \rho(g_r)\sum_{k=1}^{n} b_k g_k\right) &= \beta\left(\sum_{k=1}^{n} a_k g_r g_k, \sum_{k=1}^{n} b_k g_r g_k\right)\\ &= \beta\left(\sum_{k=1}^{n} a_k g_{\sigma_r(k)}, \sum_{k=1}^{n} b_k g_{\sigma_r(k)}\right)\\ &= \sum_{k=1}^{n} a_k \bar{b}_k\\ &= \beta\left(\sum_{k=1}^{n} a_k g_k, \sum_{k=1}^{n} b_k g_k\right), \qquad r = 1, \ldots, n.\end{aligned}$$

Thus each $\rho(g_r)$ is unitary (orthogonal if the underlying field is **R**). It follows from Theorem 1.6 that ρ is fully reducible.

(b) Let V be a finite dimensional unitary space with inner product β, and let $T \in \operatorname{Hom}(V,V)$. Recall that T is said to be *hermitian* if $T^* = T$. Also, T is said to be *positive semidefinite* if $\beta(Tx,x) \geq 0$ for all $x \in V$, and T is *positive definite* if in addition equality holds iff $x = 0$. It is true that T is positive semidefinite iff the eigenvalues of T are nonnegative, and T is positive definite iff the eigenvalues of T are positive. A standard result in elementary linear algebra states that if T is hermitian, then V has an orthonormal basis consisting of eigenvectors of T. Moreover, if E is any o.n. basis of V and $[T]_E^E = A$, then T is hermitian iff the matrix A satisfies $A^* = A$ (i.e., A is *hermitian*). If T is hermitian and $X \in \operatorname{GL}(V)$ satisfies $X^*TX = T$, then X is called an *automorph of* T. Observe that if X and Y are automorphs of T, we have

$$(XY)^*T(XY) = Y^*(X^*TX)Y = Y^*TY = T$$

and XY is an automorph of T.

(c) Let S be a finite group, and let V be a proper S-module over **C**. Let β be an inner product on V (we are *not* assuming that V is necessarily a unitary S-module, i.e., that the transformations in the representation are unitary with respect to β). Denote by $L\colon S \to \operatorname{GL}(V)$ the representation function for S. Let $T \in \operatorname{Hom}(V,V)$ be positive definite hermitian and define

$$H = \sum_{s \in S} L(s)^*TL(s) \in \operatorname{Hom}(V,V). \tag{36}$$

Notice that H is positive definite hermitian, being a sum of positive definite hermitian transformations. Then for $t \in S$ we compute that

$$\begin{aligned}L(t)^*HL(t) &= \sum_{s \in S} L(t)^*L(s)^*TL(s)L(t) = \sum_{s \in S} (L(s)L(t))^*TL(s)L(t)\\ &= \sum_{s \in S} L(st)^*TL(st) = H,\end{aligned}$$

where the final equality follows since st runs over S as s does. Thus $L(t)$ is an automorph of H for each $t \in S$. Now define a new inner product on V, denoted simply $(\cdot\,,\cdot)$, by the formula

$$(x,y) = \beta(Hx,y), \qquad x,y \in V. \tag{37}$$

We have

$$\begin{aligned}(L(s)x,L(s)y) &= \beta(HL(s)x,L(s)y)\\ &= \beta(L(s)^*HL(s)x,y)\\ &= \beta(Hx,y)\\ &= (x,y), \qquad s \in S.\end{aligned}$$

In other words, each $L(s) \in \operatorname{im} L$ is unitary with respect to the inner product given in (37). Thus V together with the inner product given in (37) is a unitary S-module, and hence V is fully reducible by Theorem 1.6. Since any vector space V over **C** can be equipped with an inner product β (see Exercise 8), we conclude that *if S is a finite group and V is a proper S-module over* **C**, *then V is fully reducible*. The same argument shows that *this result also holds if* **C** *is replaced by* **R**.

(d) Let $\Delta\colon S \to \mathrm{GL}(n,\mathbf{C})$ be a matrix representation of the finite group S. Then there exists a fixed nonsingular matrix $P \in \mathrm{GL}(n,\mathbf{C})$ such that $P^{-1}\Delta(s)P$ is unitary for every $s \in S$. To see this, let V be an n-dimensional vector space over **C** with basis E, and define the representation $L\colon S \to \mathrm{GL}(V)$ of S by $[L(s)]_E^E = \Delta(s)$, $s \in S$. In (c) above we saw that there exists an inner product $(\cdot,\cdot)$ on V which makes V into a unitary S-module, i.e., each of the transformations $L(s) \in \operatorname{im} L$ is unitary with respect to $(\cdot,\cdot)$. Now let F be an o.n. basis of V with respect to $(\cdot,\cdot)$; then $[L(s)]_F^F = \kappa(s)$ is a unitary matrix for each $s \in S$. Finally,

$$\begin{aligned}\kappa(s) &= [L(s)]_F^F\\ &= P^{-1}\Delta(s)P, \qquad s \in S,\end{aligned}$$

where $P = [I_V]_F^E \in \mathrm{GL}(n,\mathbf{C})$. An analogous result holds if **C** is replaced by **R** and "unitary" is replaced by "orthogonal."

Actually, formula (36) suggests how to construct a matrix $P \in \mathrm{GL}(n,\mathbf{C})$ such that $P^{-1}\Delta(s)P$ is a unitary matrix for every $s \in S$. Let

$$H = \sum_{s \in S} \Delta(s)^*\,\Delta(s) \in \mathrm{GL}(n,\mathbf{C});$$

for each $s \in S$, $\Delta(s)^*\Delta(s)$ is positive definite hermitian since $\Delta(s)$ is nonsingular, and hence H is positive definite hermitian. Now let $P = H^{-1/2}$ be the inverse of the positive definite hermitian square root of H. Then for every $s \in S$,

$$\begin{aligned}(P^{-1}\Delta(s)P)^*(P^{-1}\Delta(s)P) &= P^*\Delta(s)^*P^{-1*}P^{-1}\Delta(s)P\\ &= H^{-1/2}\Delta(s)^*H\Delta(s)H^{-1/2}\\ &= H^{-1/2}HH^{-1/2} \qquad (\text{since } \Delta(s)^*H\Delta(s) = H)\\ &= I_n,\end{aligned}$$

and so $P^{-1}\Delta(s)P$ is a unitary matrix. This construction is illustrated in Exercise 9.

Let S be a groupoid and V an S-module over R. Let W be an invariant subspace of V for the representation of S, and denote by $q\colon V \to V/W$ the canonical quotient map. Then the quotient space V/W can be made into an S-module by defining

$$sq(v) = q(sv), \qquad s \in S,\ v \in V. \tag{38}$$

To see that (38) makes sense, observe that $q(v_1) = q(v_2)$ implies $v_1 - v_2 \in W$ and hence $s(v_1 - v_2) \in W$ for $s \in S$; but then $q(sv_1) = q(sv_2)$. Also observe that for $s,t \in S$ and $v \in V$,

$$\begin{aligned} s(tq(v)) &= sq(tv) = q(s(tv)) \\ &= q((st)v) = stq(v); \end{aligned}$$

so (38) indeed defines a representation of S. It is called the *quotient representation of S (with respect to W)*.

Suppose V has dimension n over R. Let $\{e_{r+1}, \ldots, e_n\}$ $(1 \leq r \leq n)$ be a basis of W and augment it to a basis $E = \{e_1, \ldots, e_n\}$ of V. Then the associated matrix representation $\Delta: S \to M_n(R)$ of S has the form

$$\Delta(s) = \begin{bmatrix} \Delta_{11}(s) & 0 \\ \Delta_{21}(s) & \Delta_{22}(s) \end{bmatrix}, \qquad s \in S,$$

where $\Delta_{11}(s) \in M_r(R)$ and $\Delta_{22}(s) \in M_{n-r}(R)$. As we saw in Theorem 1.5, the matrices $\Delta_{11}(s)$ and $\Delta_{22}(s)$ constitute respective matrix representations $\Delta_{11}: S \to M_r(R)$ and $\Delta_{22}: S \to M_{n-r}(R)$ of S [see formulas (31) and (32)]. Now it is clear that *the invariant subspace W of the S-module V itself becomes an S-module if we simply restrict the original representation transformations to W, and* $\Delta_{22}: S \to M_{n-r}(R)$ *is precisely the associated matrix representation of S relative to the basis* $\{e_{r+1}, \ldots, e_n\}$ *of W*. Moreover, for $s \in S$ and $1 \leq k \leq r$ we have

$$\begin{aligned} sq(e_k) &= q(se_k) = q\left(\sum_{i=1}^{n} \Delta_{ik}(s)e_i\right) \\ &= \sum_{i=1}^{n} \Delta_{ik}(s)q(e_i) \\ &= \sum_{i=1}^{r} \Delta_{ik}(s)q(e_i) \qquad [\text{since } q(e_{r+1}) = \cdots = q(e_n) = 0]. \end{aligned}$$

It follows that $\Delta_{11}: S \to M_r(R)$ *is the matrix representation associated with the quotient representation of S relative to the basis* $\{q(e_1), \ldots, q(e_r)\}$ *of* V/W.

We have the following interesting result.

Theorem 1.7 *Let S be a groupoid and V an S-module.*

(i) *If W is a minimal invariant subspace of V for the representation of S, then W is an irreducible S-module.*

(ii) *If* $W \neq V$ *is an invariant subspace of V for the representation of S, then V/W is an irreducible S-module (under the quotient representation) iff W is a maximal invariant subspace.*

Proof: (i) This is obvious.

(ii) Assume V/W is an irreducible S-module (under the quotient representation). If W is not a maximal invariant subspace of V for the representation of S, there exists an invariant subspace U of V such that $W \subset U \subset V$ and the inclusions are strict. Let $q: V \to V/W$ denote the quotient map, and set $W_1 = q(U)$. Clearly W_1 is a proper subspace of V/W (see Exercise 11). Moreover, for $s \in S$ and $u \in U$ we have

$$sq(u) = q(su) \in W_1$$

($su \in U$ because U is an invariant subspace of V). Thus W_1 is a proper invariant subspace of the S-module V/W, contradicting the irreducibility of V/W. We conclude that W is a maximal invariant subspace of V.

Conversely, assume W is a maximal invariant subspace of V for the representation of S. Suppose W_1 is an invariant subspace of the S-module V/W. Then $W_1 = q(U)$, where $U = q^{-1}(W_1)$ is a subspace of V containing W (see Exercise 10). If $s \in S$ and $u \in U$, then $q(u) \in W_1$ and the invariance of W_1 for the quotient representation of S implies that $sq(u) \in W_1$. Since $sq(u) = q(su)$ we have $q(su) \in W_1$, i.e., $su \in U$. Hence U is an invariant subspace of V for the representation of S, and since $W \subset U$ the maximality of W implies that $U = W$ or $U = V$. If $U = W$, then $W_1 = q(U) = \{0\}$. If $U = V$, then $W_1 = q(U) = V/W$. This shows that V/W has only $\{0\}$ and V/W as invariant subspaces for the quotient representation of S. Thus V/W is an irreducible S-module. ∎

Theorem 1.7 tells us the following. If S is a groupoid and V an S-module, then both minimal and maximal invariant subspaces of V give rise to irreducible representations of S: A minimal invariant subspace W is made into an irreducible S-module by simply restricting the original representation transformations to W; and if W is a maximal invariant subspace, then the quotient representation of S in V/W is irreducible.

We next introduce the important notion of equivalence. Let S be a groupoid, and let V and U be S-modules. If there exists a bijective $T \in \mathrm{Hom}(U,V)$ such that

$$sTu = Tsu, \qquad s \in S, u \in U, \tag{39}$$

then the two S-modules V and U (or the corresponding representations) are said to be *equivalent*. If $L: S \to \mathrm{Hom}(V,V)$ and $M: S \to \mathrm{Hom}(U,U)$ are the respective representations, then (39) states that

$$L(s)Tu = TM(s)u, \qquad s \in S, u \in U,$$

or

$$L(s)T = TM(s), \qquad s \in S. \tag{40}$$

Even if $T \in \mathrm{Hom}(U,V)$ is not necessarily bijective in (40), we say that *L and M are linked by T (or that U and V are linked by T)*. The linking is of course trivial if $T = 0$. When L and M are equivalent representations, we write $L \sim M$. It is easy to check that in the class of all representations of a groupoid S, $\sim$ is an equivalence relation (see Exercise 12).

There is a similar notion of linking sets of matrices. Let $\Omega \subset M_n(R)$, $\Gamma \subset M_m(R)$, and suppose $M \in M_{m,n}(R)$. If for each $A \in \Omega$ there exists a $B \in \Gamma$ such that

$$MA = BM \tag{41}$$

and also for each $B \in \Gamma$ there exists an $A \in \Omega$ such that (41) holds, then M is said to *link* Ω and Γ.

Finally, two matrix representations $\Delta: S \to M_n(R)$ and $\kappa: S \to M_n(R)$ of a groupoid S are said to be *equivalent* if there exists a nonsingular matrix $P \in GL(n,R)$ such that

$$\Delta(s)P = P\kappa(s), \qquad s \in S.$$

Once again, we write $\Delta \sim \kappa$, and $\sim$ is an equivalence relation in the class of all matrix representations of S.

The next result shows that any nontrivial irreducible representation of a semigroup over a field is equivalent to a quotient representation of the regular representation with respect to a maximal invariant subspace of the semigroup algebra over the field. (The *trivial* irreducible representation is the 0 representation in a space of dimension 1.)

Theorem 1.8 *Let S be a semigroup, $\mathfrak{A}$ the semigroup algebra of S over a field R, and $L: S \to \mathrm{Hom}(V,V)$ a nontrivial irreducible representation of S over R. Then there exists a maximal invariant subspace A of $\mathfrak{A}$ for the regular representation $\rho: S \to Hom(\mathfrak{A}, \mathfrak{A})$ such that the corresponding irreducible quotient representation $\rho_1: S \to \mathrm{Hom}(\mathfrak{A}/A, \mathfrak{A}/A)$ is equivalent to L.*

Proof: Let $v_0 \in V$, $v_0 \neq 0$. Then $\mathfrak{A}v_0 = V$ by Theorem 1.4(iv). Let $A = \{f \in \mathfrak{A} \mid L(f)v_0 = 0\}$. It is obvious that $A \neq \mathfrak{A}$ is a subspace of $\mathfrak{A}$, and in fact A is invariant under ρ; for if $f \in A$ and $g \in \mathfrak{A}$, then

$$\begin{aligned} L(gf)v_0 &= L(g)L(f)v_0 \\ &= L(g)0 \qquad [\text{since } L(f)v_0 = 0] \\ &= 0, \end{aligned}$$

so that $\rho(g)f = gf \in A$. We obtain the quotient representation

$$\rho_1: S \to \mathrm{Hom}(\mathfrak{A}/A, \mathfrak{A}/A)$$

given by

$$\begin{aligned} \rho_1(s)q(f) &= q(\rho(s)(f)) \\ &= q(sf), \qquad s \in S, f \in \mathfrak{A}, \end{aligned}$$

where $q: \mathfrak{A} \to \mathfrak{A}/A$ is the quotient map. Define the function

$$T: V \to \mathfrak{A}/A$$

by

$$Tv = q(f), \qquad v \in V, \tag{42}$$

where $f \in \mathfrak{A}$ is such that $L(f)v_0 = v$ (such an $f \in \mathfrak{A}$ exists because $\mathfrak{A}v_0 = V$). We must confirm that T is well-defined: If $L(f)v_0 = v = L(g)v_0$, then

$L(f-g)v_0 = 0$ so that $f - g \in A$, and hence $q(f) = q(g)$. Observe also that T is linear. For if $Tu = q(f)$, $Tv = q(g)$ and $r,k \in R$, then (using L extended to $\mathfrak{A}$) we have

$$L(rf + kg)v_0 = rL(f)v_0 + kL(g)v_0 = ru + kv$$

[since $Tu = q(f)$ means $L(f)v_0 = u$ and $Tv = q(g)$ means $L(g)v_0 = v$], and hence

$$\begin{aligned} T(ru + kv) &= q(rf + kg) = rq(f) + kq(g) \\ &= rTu + kTv. \end{aligned}$$

Next note that T is injective. For if $Tu = q(f) = q(g) = Tv$, then $f - g \in A$, $L(f-g)v_0 = 0$, $L(f)v_0 = L(g)v_0$, and since $L(f)v_0 = u$ and $L(g)v_0 = v$, we obtain $u = v$. Clearly if $f \in \mathfrak{A}$ and $L(f)v_0 = v$, then $Tv = q(f)$ by definition; so T is surjective as well. Finally, observe that T links L and ρ_1. Indeed, suppose $s \in S$, $v \in V$, and $f \in \mathfrak{A}$ is chosen so that $L(f)v_0 = v$. Then

$$\begin{aligned} TL(s)v &= TL(s)L(f)v_0 \\ &= T(L(sf)v_0) \\ &= q(sf) && \text{[by the definition (42)]} \\ &= \rho_1(s)q(f) && \text{[by the definition of } \rho_1\text{]} \\ &= \rho_1(s)Tv && \text{[by the definition (42)].} \end{aligned}$$

Thus $L \sim \rho_1$. It is easy to check (see Exercise 13) that since L is irreducible and ρ_1 is equivalent to L, ρ_1 is irreducible. Hence by Theorem 1.7(ii), A is a maximal invariant subspace of $\mathfrak{A}$ for the regular representation $\rho: S \to \mathrm{Hom}(\mathfrak{A}, \mathfrak{A})$. ▮

Theorem 1.8 has the following immediate consequence.

Corollary 1 *Let S be a semigroup, $\mathfrak{A}$ the semigroup algebra of S over a field R, and $L: S \to \mathrm{Hom}(V,V)$ a nontrivial irreducible representation of S over R. Assume that the regular representation $\rho: S \to \mathrm{Hom}(\mathfrak{A}, \mathfrak{A})$ of S is fully redccible. Then there exists a minimal invariant subspace B of $\mathfrak{A}$ for ρ such that the restriction $\rho|B$ of ρ to B [i.e., $(\rho|B)(s) = \rho(s)|B$, $s \in S$] is equivalent to L.*

Proof: By Theorem 1.8, there exists a maximal invariant subspace A of $\mathfrak{A}$ for the regular representation ρ such that $L \sim \rho_1$, where $\rho_1: S \to \mathrm{Hom}(\mathfrak{A}/A, \mathfrak{A}/A)$ is the corresponding quotient representation of S. Since ρ is fully reducible by hypothesis, there exists an invariant subspace B of $\mathfrak{A}$ for ρ such that $\mathfrak{A} = A \dotplus B$. Then B is a minimal invariant subspace of $\mathfrak{A}$ (see Exercise 14). Let Q be the restriction of the quotient map $q: \mathfrak{A} \to \mathfrak{A}/A$ to B, i.e., $Q = q|B$. Then Q is nonsingular, and the statement (29) specializes to

$$\rho_1(s) = Q(\rho(s)|B)Q^{-1}, \qquad s \in S. \tag{43}$$

But (43) says precisely that $\rho|B \sim \rho_1$, and since $\rho_1 \sim L$, we obtain $L \sim \rho|B$. ▮

Example 5 We determine all the irreducible representations over **C** of a cyclic group $S = \{e = g^0, g, \ldots, g^{n-1}\}$ of order n. Let $\mathfrak{A}$ be the group algebra of S over **C**. By Example 4(a) we know that the regular representation $\rho: S \to \mathrm{GL}(\mathfrak{A})$ of S is fully reducible. Suppose $L: S \to \mathrm{Hom}(V, V)$ is a nontrivial irreducible representation of S over **C**. It follows from Corollary 1 that $L \sim \rho | B$, where B is a minimal invariant subspace of $\mathfrak{A}$ for ρ. Now let $b \in B$ be an eigenvector of $\rho(g)$. Since $\rho(g^k) = \rho(g)^k$, $k \in Z$, b is an eigenvector of each $\rho(g^k)$. Hence $\langle b \rangle \subset B$ is a one-dimensional invariant subspace of $\mathfrak{A}$ for ρ, and the minimality of B implies that $\langle b \rangle = B$. But $L \sim \rho | B$ entails in particular that dim $V =$ dim B; so dim $V = 1$. Write $V = \langle v \rangle$, where $0 \neq v \in V$, and let $\lambda \in \mathbf{C}$ be such that $L(g)v = \lambda v$. Then $L(g^k)v = L(g)^k v = \lambda^k v$, $k \in Z$, and since $g^n = e$, we have $\lambda^n = 1$.

In Theorem 1.8 we saw that for a semigroup S, any nontrivial irreducible representation L of S over a field R is equivalent to a quotient representation of the regular representation ρ of S with respect to a maximal invariant subspace A of the semigroup algebra $\mathfrak{A}$ of S over R. Moreover, Corollary 1 told us that if the regular representation ρ is fully reducible, then L is equivalent to ρ acting on a minimal invariant subspace B of $\mathfrak{A}$. It is important to observe that *an invariant subspace A of $\mathfrak{A}$ for the regular representation ρ is simply a left ideal in $\mathfrak{A}$*; i.e., A satisfies

$$\mathfrak{A}A \subset A.$$

We define *minimal (maximal) left ideals* in $\mathfrak{A}$ to be minimal (maximal) invariant subspaces of $\mathfrak{A}$ for ρ. Thus A is a maximal left ideal in $\mathfrak{A}$ iff A is a left ideal not strictly contained in any other left ideal in $\mathfrak{A}$ other than $\mathfrak{A}$ itself. Similarly, a minimal left ideal in $\mathfrak{A}$ is a left ideal not strictly containing any left ideal in $\mathfrak{A}$ other than $\{0\}$. *The study of representations over R of a semigroup S is equivalent to the study of the ideal structure in the linear associative semigroup algebra $\mathfrak{A}$ of S over R.*

In the next section we shall study representations of groups (mostly finite groups) mainly through the use of matrices. There are several reasons for proceeding in this way. First, the matrix theoretic approach is easily understood. Second, the originators of group representation theory (Burnside, Frobenius, Schur, Maschke) developed the theory as a part of matrix theory. Third, the results are very specific and suitable for applications. Finally, the main results are quickly accessible.

Exercises 5.1

1. Verify (23) to (26). *Hint*: For example, $f(rv) = (\sum_{x \in S} a_x x)(rv) = \sum_{x \in S} a_x x(rv) = \sum_{x \in S} a_x r(xv) = r \sum_{x \in S} a_x(xv) = r(fv)$.
2. Referring to Example 2(b), prove that for each $f \in \mathfrak{A}$ the row and column sums of the matrix $\Delta(f)$ are all equal. *Hint*: $\Delta(f) = \sum_{k=1}^{n} c_k \Delta(g_k)$. The $\Delta(g_k)$. are

permutation matrices. Thus $c_1 + \cdots + c_n$ is the common value of the row and column sums.

3. Prove the assertion following (27) that $\{q(v_{r+1}), \ldots, q(v_n)\}$ is a basis of V/W. *Hint*: If $0 = \sum_{t=r+1}^{n} c_t q(v_t) = q(\sum_{t=r+1}^{n} c_t v_t)$, then $\sum_{t=r+1}^{n} c_t v_t \in W = \langle v_1, \ldots, v_r \rangle$ and hence $c_{r+1} = \cdots = c_n = 0$. Also, since $q(v_i) = 0$ for $i = 1, \ldots, r$, it follows that $V/W = q(V)$ is spanned by the indicated set.

4. Prove that the function $T_1: V/W_1 \to V/W_1$ defined in (28) is well-defined and linear. *Hint*: If $q(v_1) = q(v_2)$, then $v_1 - v_2 \in W_1$ so that $T(v_1 - v_2) \in W_1$. Hence $Tv_1 - Tv_2 \in W_1$ and $q(Tv_1) = q(Tv_2)$. Thus T_1 is well-defined. Also for $r,s \in R$ we have $T_1(q(rv_1 + sv_2)) = q(T(rv_1 + sv_2)) = q(rTv_1 + sTv_2) = rq(Tv_1) + sq(Tv_2) = rT_1q(v_1) + sT_1q(v_2)$, showing that T_1 is linear.

5. Prove that (29) is valid. *Hint*: First observe that $Q: W_2 \to V/W_1$ is bijective. For if $Qw_2 = 0$, then $q(w_2) = 0$ and $w_2 \in W_1 \cap W_2 = \{0\}$. Thus Q is injective. Also if $v = w_1 + w_2 \in V$, then $q(v) = q(w_1) + q(w_2) = q(w_2) = Qw_2$. Thus $V/W_1 = \operatorname{im} q = \operatorname{im} Q$ and Q is surjective. Finally, if $v = w_1 + w_2 \in V$, we have $QPQ^{-1}q(v) = QPQ^{-1}q(w_2) = QPQ^{-1}Qw_2 = QPw_2 = QTw_2 = q(Tw_2) = q(Tw_1) + q(Tw_2) = q(Tv) = T_1q(v)$.

6. Prove the assertion following (29) concerning W_1 and W_2. *Hint*: Assume W_1 is not strictly contained in any proper invariant subspace of V under T. Suppose W_3 is a proper invariant subspace of V under T strictly contained in W_2. Then $W_1 \dotplus W_3$ is an invariant subspace of V under T. Since the inclusions $\{0\} \subset W_3 \subset W_2$ are strict, so are the inclusions $W_1 \subset W_1 \dotplus W_3 \subset W_1 \dotplus W_2 = V$. This contradicts our initial assumption.

Conversely, assume W_2 does not strictly contain any proper invariant subspace of V under T. Suppose W_3 is a proper invariant subspace of V under T which strictly contains W_1. Then $W_3 \cap W_2$ is an invariant subspace of V under T, and since the inclusions $W_1 \subset W_3 \subset V$ are strict, so are the inclusions $\{0\} \subset W_3 \cap W_2 \subset W_2$. Again, this provides a contradiction.

7. Let V be a unitary space. Prove that if E and F are o.n. bases of V, then $[I_V]_E^F$ is a unitary matrix. *Hint*: Let $A = [I_V]_E^F$. Then $e_j = I_V e_j = \sum_{i=1}^{n} a_{ij} f_i$, $j = 1, \ldots, n$, so that

$$\begin{aligned}\delta_{pq} = \beta(e_p, e_q) &= \beta\left(\sum_{i=1}^{n} a_{ip} f_i, \sum_{i=1}^{n} a_{iq} f_i\right)\\ &= \sum_{s,t=1}^{n} a_{sp}\bar{a}_{tq}\beta(f_s, f_t) = \sum_{s,t=1}^{n} a_{sp}\bar{a}_{tq}\delta_{st}\\ &= \sum_{t=1}^{r} (A^*)_{qt}a_{tp} = (A^*A)_{qp}.\end{aligned}$$

Hence $A^*A = I_n$ and A is unitary.

8. Let V be a vector space over either **R** or **C**. Prove that it is always possible to define an inner product β on V. *Hint*: Let $E = \{e_1, \ldots, e_n\}$ be a basis of V and define $\beta(\sum_{i=1}^{n} \xi_i e_i, \sum_{i=1}^{n} \eta_i e_i) = \sum_{i=1}^{n} \xi_i \bar{\eta}_i$, $\sum_{i=1}^{n} \xi_i e_i, \sum_{i=1}^{n} \eta_i e_i \in V$.

9. Let $S \subset M_2(\mathbf{C})$ be the group of matrices

$$S = \left\{\begin{bmatrix}1 & 0\\ 0 & 1\end{bmatrix}, \begin{bmatrix}0 & 1\\ 1 & 0\end{bmatrix}, \begin{bmatrix}-1 & -1\\ 0 & 1\end{bmatrix}, \begin{bmatrix}0 & 1\\ -1 & -1\end{bmatrix}, \begin{bmatrix}-1 & -1\\ 1 & 0\end{bmatrix}, \begin{bmatrix}1 & 0\\ -1 & -1\end{bmatrix}\right\}.$$

(a) Show that the matrices in S constitute a faithful irreducible matrix representation of S_3.
(b) Find a matrix $P \in \mathrm{GL}(2,\mathbf{C})$ such that $P^{-1}\Delta(s)P$ is a unitary matrix for each matrix $\Delta(s)$ in S.

Hint: (a) To see that the matrices in S constitute a faithful representation of S_3, construct a 6×6 multiplication table for each group and compare the two. Suppose S were reducible over $\mathbf{C}$ and $Q \in \mathrm{GL}(2,\mathbf{C})$ were such that $Q^{-1}\Delta(s)Q$ is lower triangular for each matrix $\Delta(s)$ in S. Then it would follow that $Q^{(2)}$ is an eigenvector of each $\Delta(s)$. The eigenvectors of $\begin{bmatrix} 0 & 1 \\ 1 & 0 \end{bmatrix}$ are (1,1) and (1,–1) (and scalar multiples of these). However,

$$\begin{bmatrix} -1 & -1 \\ 0 & 1 \end{bmatrix}(1,1) \quad = (-2,1)$$

and

$$\begin{bmatrix} -1 & -1 \\ 0 & 1 \end{bmatrix}(1,-1) = (0,-1),$$

so that $\begin{bmatrix} 0 & 1 \\ 1 & 0 \end{bmatrix}$ and $\begin{bmatrix} -1 & -1 \\ 0 & 1 \end{bmatrix}$ do not have a common eigenvector. Thus S must be irreducible over $\mathbf{C}$.

(b) First form the sum $H = \sum_{s \in S} \Delta(s)^* \Delta(s)$. Then find a 2×2 unitary matrix U such that $U^{-1}HU = \mathrm{diag}(\alpha_1,\alpha_2)$. Define $P = U\,\mathrm{diag}(\alpha_1{}^{-1/2},\alpha_2{}^{-1/2})U^{-1}$.

10. Let W be a subspace of the vector space V, and let $q\colon V \to V/W$ be the quotient map. Show that any subspace W_1 of V/W has the form $W_1 = q(U)$, where U is a subspace of V containing W. *Hint*: Since $q\colon V \to V/W$ is linear, it follows that $U = q^{-1}(W_1) = \{x \in V \mid q(x) \in W_1\}$ is a subspace of V, and obviously $q(U) = W_1$. Note that $W = q^{-1}(\{0\}) \subset q^{-1}(W_1) = U$.

11. Let W and U be subspaces of the vector space V. Suppose $W \subset U \subset V$ and the inclusions are strict. Let $q\colon V \to V/W$ be the quotient map. Show that $q(U)$ is a proper subspace of V/W. *Hint*: Since U is a subspace of V and $q\colon V \to V/W$ is a linear map, $q(U)$ is a subspace of V/W. If $q(U) = \{0\}$, then $q(u) = 0$ for all $u \in U$ and hence $U \subset W$, which is impossible. Next, choose $v_0 \in V - U$. Then if $q(v_0) \in q(U)$, it would follow that $q(v_0) = q(u)$ for some $u \in U$, and hence $v_0 = u + w$ for some $w \in W$. Since $w \in W \subset U$, we would have $v_0 \in U$, a contradiction. Thus $q(v_0) \notin q(U)$ and $q(U)$ is a proper subspace of V/W.

12. Show that if S is a groupoid, then $\sim$ is an equivalence relation in the class of all representations of S. *Hint*: Let $L\colon S \to \mathrm{Hom}(V,V)$, $M\colon S \to \mathrm{Hom}(U,U)$, and $K\colon S \to \mathrm{Hom}(W,W)$ be representations of S. Clearly $I_V L(s) = L(s) I_V$ for $s \in S$; so $L \sim L$. If $T \in \mathrm{Hom}(U,V)$ is a bijection such that $L(s)T = TM(s)$, $s \in S$, then $T^{-1} \in \mathrm{Hom}(V,U)$ and $M(s)T^{-1} = T^{-1}L(s)$, $s \in S$. Hence $L \sim M$ implies $M \sim L$. If $T \in \mathrm{Hom}(U,V)$ is a bijection linking L and M, and if $D \in \mathrm{Hom}(W,U)$ is a bijection linking M and K, then $TD \in \mathrm{Hom}(W,V)$ is a bijection and $L(s)TD = TM(s)D = TDK(s)$, $s \in S$. Thus $L \sim M$ and $M \sim K$ imply $L \sim K$.

13. Let S be a groupoid, and let V and U be equivalent S-modules; say the respective representations $L\colon S \to \mathrm{Hom}(V,V)$ and $M\colon S \to \mathrm{Hom}(U,U)$ are linked by the bijection $T \in \text{Hom}(U,V)$. Show that W is a minimal (maximal) invariant subspace of U iff $T(W)$ is a minimal (maximal) invariant subspace of V.

14. Let S be a groupoid and $L: S \to \text{Hom}(V,V)$ a representation of S. Let W_1 and W_2 be subspaces of V such that the pair (W_1,W_2) reduces each $L(s) \in \text{im } L$. Prove that W_1 is a maximal invariant subspace of V for the representation of S iff W_2 is a minimal invariant subspace of V for the representation of S. *Hint*: This is essentially Exercise 6.

15. Let S be a group, and V an S-module of dimension n over R. Let e be the identity of S. Denote by $L: S \to \text{Hom}(V,V)$ the representation function for S. Show that
(a) $eV = L(e)V$ is a proper S-module.
(b) $V = eV \dotplus (I_V - L(e))V$.
(c) $L(s) \mid (I_V - L(e))V = 0$ for each $s \in S$.
(d) There exists a basis E of V such that if $\Delta(s) = [L(s)]_E^E$, $s \in S$, then

$$\Delta(s) = \begin{bmatrix} \kappa(s) & 0 \\ 0 & 0 \end{bmatrix},$$

where $\kappa: S \to \text{GL}(r,R)$ is a proper matrix representation of S and $r = \dim eV$.

Hint: (a) If $s \in S$, then $s(ev) = (se)v = (es)v = e(sv)$ for all $v \in V$, and hence $s(eV) \subset eV$. Also $e(ev) = e^2v = ev$ for all $v \in V$, so that $L(e)|eV = I_{eV}$. In general, *if $M: S \to \text{Hom}(W,W)$ is a representation of the group S and $M(e) = I_W$, then M is proper*. Indeed, for each $s \in S$ we have $I_W = M(e) = M(ss^{-1}) = M(s)M(s^{-1})$, and so $M(s)$ has an inverse. Thus each $L(s)|eV$ has an inverse, and it follows that eV is a proper S-module for which the representation function is simply $L|eV$.

(b) If $w \in eV \cap (I_V - L(e))V$, then $w = ev = u - eu$ for some $u,v \in V$. We have $ew = e(ev) = ev = w$ and $ew = e(u - eu) = eu - e^2u = eu - eu = 0$. Hence $w = 0$. Also, each $v \in V$ can be written as $v = ev + (v - ev) \in eV + (I_V - L(e))V$.

(c) $L(s)(I_V - L(e))v = L(s)v - L(s)L(e)v = L(s)v - L(se)v = 0$, $s \in S$, $v \in V$.

(d) Let $E = \{e_1, \ldots, e_n\}$ be a basis of V so chosen that $\{e_1, \ldots, e_r\}$ is a basis of eV and $\{e_{r+1}, \ldots, e_n\}$ is a basis of $(I_V - L(e))V$. Obviously each $[L(s)]_E^E$ has the indicated form. In particular, we have proved that *if $L: S \to \text{Hom}(V,V)$ is a representation of a group S, then all the transformations $L(s)$, $s \in S$, have the same rank r.*

16. Let M be a multiplicative group of (not necessarily nonsingular) matrices in $M_n(R)$, where R is a field. Prove the following statements.
(a) There exists an integer r such that the rank of every matrix $A \in M$ is r.
(b) There exists a fixed nonsingular matrix $P \in \text{GL}(n,R)$ such that for each $A \in M$, there is a $B \in M_r(R)$ with $P^{-1}AP = B \dotplus 0_{n-r}$, where 0_{n-r} is the $(n - r)$-square zero matrix.
(c) If H is the identity in M, then $P^{-1}HP = I_r \dotplus 0_{n-r}$.

Hint: Let V be an n-dimensional vector space over R, F a basis of V, and S the group of transformations $T \in \text{Hom}(V,V)$ such that $[T]_F^F = A$ for some $A \in M$.

Define $L: S \to \text{Hom}(V,V)$ by $L(T) = T$, $T \in S$; then V becomes an S-module. According to Exercise 15, there exists a basis E of V such that if $\Delta(T) = [L(T)]_E^E = [T]_E^E$, $T \in S$, then $\Delta(T) = \kappa(T) \dotplus 0_{n-r}$, where $\kappa: S \to \text{GL}(r,R)$ is a proper matrix representation of S. But for each $T \in S$, $\Delta(T) = [T]_E^E = [I_V]_F^E[T]_F^F[I_V]_E^F$ and hence $P^{-1}AP = B \dotplus 0_{n-r}$, where $P = [I_V]_E^F$, $A = [T]_F^F \in M$, and $B = \kappa(T)$. Since κ is proper, each $\kappa(T) = B$ has rank r so that each matrix A has rank r. This proves (a) and (b). If H is the identity in M, the fact that κ is proper also implies $\kappa(H) = I_r$. This proves (c).

17. Let R be a field of characteristic not 2. Let M be the set of all matrices of the form $c\begin{bmatrix}1 & 1\\1 & 1\end{bmatrix}$, $0 \neq c \in R$.

(a) Show that M is a group with respect to matrix multiplication.

(b) Find a fixed matrix $P \in \text{GL}(2,R)$ such that $P^{-1}AP = \begin{bmatrix}b & 0\\0 & 0\end{bmatrix}$ for every $A \in M$ (b depends on A).

Hint: (a) M is obviously closed with respect to matrix multiplication. The identity in M is $\frac{1}{2}\begin{bmatrix}1 & 1\\1 & 1\end{bmatrix}$, and the inverse of $c\begin{bmatrix}1 & 1\\1 & 1\end{bmatrix} \in M$ is $\frac{1}{4c}\begin{bmatrix}1 & 1\\1 & 1\end{bmatrix}$.

(b) Let $P = \begin{bmatrix}1 & -1\\1 & 1\end{bmatrix} \in \text{GL}(2,R)$; then $P^{-1} = \frac{1}{2}\begin{bmatrix}1 & 1\\-1 & 1\end{bmatrix}$ and

$$P^{-1}c\begin{bmatrix}1 & 1\\1 & 1\end{bmatrix}P = \begin{bmatrix}2c & 0\\0 & 0\end{bmatrix},\ 0 \neq c \in R.$$

Glossary 5.1

automorph, 429
completely reducible representation (module), 423
conjugate dual, 427
conjugate transpose, 427
equivalent representations, 432
Euclidean space, 427
Euclidean representation module, 427
fully reducible representation (module), 423
fully reducible set of matrices, 425
general linear group, 413
GL(V), 413
group algebra, 414
groupoid algebra, 414
groupoid ring, 414
hermitian matrix, 429,
hermitian transformation, 429
indicator function, 414
f_s, 414
inner product, 427
invariant subset, 421
irreducible representation (module), 422
linked transformations, 432
matrix representation, 413
maximal invariant subspace, 422
maximal left ideal, 435
minimal invariant subspace, 422
minimal left ideal, 435
orthogonal matrix, 427
orthogonal matrix representation, 428
orthogonal transformation, 427
orthonormal basis, 427
o.n., 427
positive definite hermitian transformation, 429
positive semidefinite hermitian transformation, 429
proper matrix representation, 413

5.2 Matrix Representations

Let S be a group and V an S-module. It is shown in Exercise 15, Section 5.1, that if e is the identity in S, then eV is a proper representation module for S. Moreover, there exists a basis $E = \{e_1, \ldots, e_n\}$ of V such that if $L: S \to \text{Hom}(V,V)$ is the representation function and $\Delta(s) = [L(s)]_E^E$, $s \in S$, then

$$\Delta(s) = \kappa(s) \dotplus 0_{n-r},$$

where $r = \dim eV$ and $\kappa: S \to \text{GL}(r,R)$ is a proper matrix representation with $\kappa(s) = [L(s)\,|\,eV]_{E_r}^{E_r}$, $E_r = \{e_1, \ldots, e_r\}$. Thus it suffices to consider proper representations when dealing with group representations. *Henceforth in this section we shall assume that all representations of groups are proper.*

The *degree of a representation* $L: S \to \text{Hom}(V,V)$ (or the *degree of the S-module* V) is the dimension of V, and if $\Delta: S \to \text{GL}(n,R)$ is a matrix representation, then n is the *degree of* Δ.

Following is a basic result in the representation theory for finite groups.

Theorem 2.1 (*Maschke's Theorem*) *Let S be a finite group, and V an S-module of dimension n over R. Let $h = |S|$, and assume that $h = h \cdot 1 \neq 0$ in R. Then V is fully reducible.*

Proof: Let $L: S \to \text{GL}_n(V)$ denote the representation function. Suppose E is a basis of V such that the associated matrix representation $\Delta: S \to \text{GL}(n,R)$ has the form

$$\Delta(s) = \begin{bmatrix} \Delta_{11}(s) & 0 \\ \Delta_{21}(s) & \Delta_{22}(s) \end{bmatrix}, \qquad s \in S, \tag{1}$$

where $\Delta_{11}(s) \in \text{GL}(p,R)$ and $\Delta_{22}(s) \in \text{GL}(q,R)$ for some fixed integers p and q, $1 \leq p < n$, $p + q = n$. The theorem will be proved once we exhibit a fixed matrix $P = \begin{bmatrix} P_{11} & 0 \\ P_{21} & P_{22} \end{bmatrix} \in \text{GL}(n,R)$ such that $P_{11} \in M_p(R)$ and

$$P^{-1}\Delta(s)P = \kappa(s) \dotplus \mu(s), \qquad s \in S,$$

where $\kappa(s) \in M_p(R)$ and $\mu(s) \in M_q(R)$ (recall the discussion following Theorem 1.5).

Let

$$C = \sum_{s \in S} \Delta_{21}(s)\Delta_{11}(s)^{-1} \in M_{q,p}(R)$$

and set
$$P = \begin{bmatrix} I_p & 0 \\ \frac{1}{h}C & I_q \end{bmatrix} \in \mathrm{GL}(n,R).$$

By block multiplication, we confirm directly from (1) and the equality $\Delta(st) = \Delta(s)\,\Delta(t)$, that for all $s,t \in S$,

$$\Delta_{11}(st) = \Delta_{11}(s)\,\Delta_{11}(t),$$
$$\Delta_{22}(st) = \Delta_{22}(s)\,\Delta_{22}(t),$$

and
$$\Delta_{21}(st) = \Delta_{21}(s)\,\Delta_{11}(t) + \Delta_{22}(s)\,\Delta_{21}(t). \tag{2}$$

From (2), we have

$$\Delta_{21}(s)\,\Delta_{11}(t) = \Delta_{21}(st) - \Delta_{22}(s)\,\Delta_{21}(t),$$

and multiplying both sides on the right by $\Delta_{11}(t)^{-1}$ yields

$$\begin{aligned}\Delta_{21}(s) &= \Delta_{21}(st)\,\Delta_{11}(t)^{-1} - \Delta_{22}(s)\,\Delta_{21}(t)\,\Delta_{11}(t)^{-1} \\ &= \Delta_{21}(st)\,\Delta_{11}(st)^{-1}\,\Delta_{11}(s) - \Delta_{22}(s)\,\Delta_{21}(t)\,\Delta_{11}(t)^{-1}.\end{aligned} \tag{3}$$

Sum both sides of (3) over all $t \in S$ to obtain

$$h\,\Delta_{21}(s) = \Big(\sum_{t \in S} \Delta_{21}(st)\,\Delta_{11}(st)^{-1}\Big)\,\Delta_{11}(s) - \Delta_{22}(s)\sum_{t \in S} \Delta_{21}(t)\,\Delta_{11}(t)^{-1}. \tag{4}$$

Now as t runs over S, st runs over S, and recalling the definition of C we see that the equation (4) becomes

$$h\,\Delta_{21}(s) = C\Delta_{11}(s) - \Delta_{22}(s)C. \tag{5}$$

Finally, since

$$P^{-1} = \begin{bmatrix} I_p & 0 \\ -\frac{1}{h}C & I_q \end{bmatrix},$$

we compute by block multiplication that

$$\begin{aligned} P^{-1}\Delta(s)P &= \begin{bmatrix} I_p & 0 \\ -\frac{1}{h}C & I_q \end{bmatrix}\begin{bmatrix} \Delta_{11}(s) & 0 \\ \Delta_{21}(s) & \Delta_{22}(s) \end{bmatrix}\begin{bmatrix} I_p & 0 \\ \frac{1}{h}C & I_q \end{bmatrix} \\ &= \begin{bmatrix} \Delta_{11}(s) & 0 \\ -\frac{1}{h}C\,\Delta_{11}(s) + \Delta_{21}(s) & \Delta_{22}(s) \end{bmatrix}\begin{bmatrix} I_p & 0 \\ \frac{1}{h}C & I_q \end{bmatrix} \\ &= \begin{bmatrix} \Delta_{11}(s) & 0 \\ -\frac{1}{h}C\Delta_{11}(s) + \Delta_{21}(s) + \frac{1}{h}\Delta_{22}(s)C & \Delta_{22}(s) \end{bmatrix} \end{aligned}$$

$$= \begin{bmatrix} \Delta_{11}(s) & 0 \\ 0 & \Delta_{22}(s) \end{bmatrix}, \qquad s \in S \qquad \text{[by (5)]}.$$

Thus, in fact, $P^{-1}\Delta(s)P = \Delta_{11}(s) \dotplus \Delta_{22}(s)$, $s \in S$. This completes the proof. ∎

The next result due to I. Schur is used throughout representation theory.

Theorem 2.2 (*Schur's Lemma*)

(a) *Let $\Omega \subset M_n(R)$ and $\Gamma \subset M_m(R)$ be irreducible (i.e., not reducible) sets of matrices. If $M \in M_{m,n}(R)$ links Ω and Γ, then either $M = 0$ or $m = n$ and M is nonsingular.*

(b) *Let S be a groupoid, and let L: $S \to \mathrm{Hom}(V,V)$ and K: $S \to \mathrm{Hom}(U,U)$ be irreducible representations of S. If $T \in \mathrm{Hom}(U,V)$ links L and K, then either $T = 0$ or $\dim U = \dim V$, T is a bijection, and $L \sim K$.*

Proof: (a) To say that $M \in M_{m,n}(R)$ links Ω and Γ means that for each $A \in \Omega$ there exists a $B \in \Gamma$ such that

$$MA = BM, \tag{6}$$

and for each $B \in \Gamma$ there exists an $A \in \Omega$ such that (6) holds. Assume that $M \neq 0$. Let $\mathscr{C}(M) \subset R^m$ be the column space of M, and observe that if A and B satisfy (6), then

$$BM^{(t)} = MA^{(t)} = \sum_{i=1}^{n} M^{(i)} a_{it} \in \mathscr{C}(M), \qquad t = 1, \ldots, n,$$

so that

$$B\mathscr{C}(M) \subset \mathscr{C}(M). \tag{7}$$

Since for any $B \in \Gamma$ there exists an $A \in \Omega$ such that (6) holds, we conclude that (7) holds for all $B \in \Gamma$, i.e.,

$$\Gamma\mathscr{C}(M) \subset \mathscr{C}(M). \tag{8}$$

Now suppose $\dim \mathscr{C}(M) = s < m$. Clearly (8) implies that there exists a nonsingular $P \in \mathrm{GL}(m,R)$ such that

$$P^{-1}BP = \begin{bmatrix} B_{11} & 0 \\ B_{21} & B_{22} \end{bmatrix}, \qquad B \in \Gamma, \tag{9}$$

where $B_{22} \in M_s(R)$ and $B_{11} \in M_{m-s}(R)$. But (9) contradicts the irreducibility of Γ. Thus

$$\dim \mathscr{C}(M) = m. \tag{10}$$

Since M has n columns, we conclude from (10) that $m \leq n$.

By taking transposes in (6), we have the following situation. For each $A^{\mathsf{T}} \in \Omega^{\mathsf{T}}$ there is a $B^{\mathsf{T}} \in \Gamma^{\mathsf{T}}$ such that

$$M^{\top}B^{\top} = A^{\top}M^{\top} \tag{11}$$

[and, of course, for each $B^{\top} \in \Gamma^{\top}$ there is an $A^{\top} \in \Omega^{\top}$ such that (11) holds]; the notations $\Omega^{\top}$ and $\Gamma^{\top}$ have the obvious interpretation. Repeating the above argument using equation (11), we conclude that

$$\dim \mathscr{C}(M^{\top}) = n$$

or equivalently,

$$\dim \mathscr{R}(M) = n, \tag{12}$$

where $\mathscr{R}(M)$ is the row space of M. Since M has m rows, it follows from (12) that $n \leq m$. Thus $m = n$ and from (10) or (12), $\rho(M) = n$ and M is nonsingular.

(b) To say that $T \in \text{Hom}(U, V)$ links L and K means that

$$L(s)T = TK(s), \qquad s \in S. \tag{13}$$

Observe that if $u \in \ker T$, then for all $s \in S$ we have

$$TK(s)u = L(s)Tu = L(s)0 = 0,$$

i.e., $K(s)u \in \ker T$ for all $s \in S$. Thus $\ker T$ is an invariant subspace of U for the representation K of S. Since K is an irreducible representation, it follows that

$$\ker T = \{0\} \qquad \text{or} \qquad \ker T = U. \tag{14}$$

On the other hand, if $v \in \text{im } T$, then for all $s \in S$ we have

$$\begin{aligned} L(s)v &= L(s)Tu \qquad (\text{for some } u \in U) \\ &= TK(s)u \in \text{im } T. \end{aligned}$$

Thus $\text{im } T$ is an invariant subspace of V for the representation L of S, and again, since L is an irreducible representation, it follows that

$$\text{im } T = \{0\} \qquad \text{or} \qquad \text{im } T = V. \tag{15}$$

Now assume that $T \neq 0$. Then $\ker T \neq U$ and $\text{im } T \neq \{0\}$; so from (14) and (15) we obtain $\ker T = \{0\}$ and $\text{im } T = V$. Therefore $T \in \text{Hom}(U, V)$ is a bijection, and hence $\dim U = \dim V$ and $L \sim K$. ∎

Corollary 1 (a) *Let $\Omega \subset M_n(R)$ be an irreducible set of matrices, and assume that $M \in M_n(R)$ commutes with each $A \in \Omega$:*

$$MA = AM, \qquad A \in \Omega. \tag{16}$$

Then exactly one of the following alternatives holds:

(i) $M = 0$.
(ii) $M \neq 0$, *R contains an eigenvalue α of M,* and $M = \alpha I_n$.
(iii) $M \neq 0$, *R contains no eigenvalue of M, and if α is any eigenvalue of M, then Ω is reducible over the extension field $R(\alpha)$.*

(b) *Let* $L: S \to \mathrm{Hom}(V,V)$ *be an irreducible representation of a groupoid* S*, where* V *is a finite dimensional vector space over* R*, and assume that* $T \in \mathrm{Hom}(V,V)$ *commutes with each* $L(s) \in \mathrm{im}\, L$:

$$TL(s) = L(s)T, \qquad s \in S. \tag{17}$$

Then exactly one of the following alternatives holds:

(i) $T = 0$.

(ii) $T \neq 0$, *R contains an eigenvalue α of T, and* $T = \alpha I_V$.

(iii) $T \neq 0$, *R contains no eigenvalue of T, and if α is any eigenvalue of T, then L is reducible when V is regarded as a vector space over the extension field* $R(\alpha)$. (See Exercise 2.)

Proof: (a) Suppose $M \neq 0$. If α is any eigenvalue of M, the matrix $M - \alpha I_n$ is singular as a matrix in $M_n(R(\alpha))$ and (16) implies that

$$(M - \alpha I_n)A = A(M - \alpha I_n), \qquad A \in \Omega. \tag{18}$$

If $\alpha \in R$, then by Theorem 2.2(a) (with $m = n$, $\Omega = \Gamma$) we conclude that $M - \alpha I_n = 0$, so that (ii) holds. On the other hand, assume $\alpha \notin R$. If Ω were irreducible over $R(\alpha)$ it would again follow from Theorem 2.2(a) and (18) that $M = \alpha I_n$. But this is impossible since $\alpha \notin R$ and $M \in M_n(R)$. Thus Ω is reducible over the extension field $R(\alpha)$.

(b) This follows immediately from (a) by taking matrix representations in (17) (see Exercise 1). ▮

Corollary 2 *Let S be an abelian group, and $L: S \to \mathrm{GL}_n(V)$ an irreducible representation of S of degree n. Assume that the underlying field R is one over which the characteristic polynomial of each $L(s) \in \mathrm{im}\, L$ splits. Then $n = 1$. In other words, if R contains the eigenvalues of each $L(s) \in \mathrm{im}\, L$, then each $L(s)$ is simply "multiplication by a scalar $\lambda(s)$."*

Proof: Let $s_0 \in S$. Then $L(s)L(s_0) = L(ss_0) = L(s_0s) = L(s_0)L(s)$ for all $s \in S$. Let $\lambda(s_0) \in R$ be an eigenvalue of $L(s_0)$; it follows from Corollary 1 that $L(s_0) = \lambda(s_0)I_V$. Since this is true for each $s_0 \in S$, the irreducibility of L implies that $n = 1$. ▮

Let V be an n-dimensional vector space over a field R. If S is a groupoid, a representation $L: S \to \mathrm{GL}_n(V)$ is said to be *absolutely irreducible over* R under the following circumstances. Let F be any extension field of R, regard V as a vector space V_1 over F, and regard V_1 as an S-module with the representation function $L: S \to \mathrm{GL}_n(V_1)$ (see Exercise 2). Then V_1 is an irreducible S-module, i.e., there are no proper invariant subspaces of V_1 for the representation $L: S \to \mathrm{GL}_n(V_1)$ of S. In terms of matrices, this means that if $\Delta: S \to \mathrm{GL}(n,R)$ is any associated matrix representation of S and F is any extension field of R, there exists no matrix $B \in \mathrm{GL}(n,F)$ such that

$$B^{-1}\Delta(s)B = \begin{bmatrix} \Delta_{11}(s) & 0 \\ \Delta_{21}(s) & \Delta_{22}(s) \end{bmatrix}, \qquad s \in S,$$

where $\Delta_{11}(s)$ and $\Delta_{22}(s)$ are square matrices of fixed size. More generally, a set of matrices $\Omega \subset M_n(R)$ is said to be *absolutely irreducible over R* if there is no extension field F of R for which there exists a matrix $B \in \mathrm{GL}(n,F)$ such that

$$B^{-1}AB = \begin{bmatrix} A_{11} & 0 \\ A_{21} & A_{22} \end{bmatrix}, \qquad A \in \Omega,$$

where A_{11} and A_{22} are square matrices of fixed size.

Recall that if S is a finite abelian group, then S is the direct sum of cyclic subgroups of prime power order (see Theorem 2.7, Section 4.2). We change the notation slightly to conform to our present treatment. Namely, we write the operation in S multiplicatively instead of additively. Thus S is the *direct product* of prime power order cyclic subgroups $H_1, \ldots, H_k$:

$$S = H_1 \cdots H_k \tag{19}$$

(the H_i here correspond to the H_{ij} in the statement of Theorem 2.7, Section 4.2). Let $h_1, \ldots, h_k$ be generators of $H_1, \ldots, H_k$, respectively, i.e., $H_i = [h_i]$, $i = 1, \ldots, k$. Then if e is the identity in S and $|H_i| = p_i^{e_i}$, where p_i is a prime and e_i is a positive integer, we have $h_i^{p_i^{e_i}} = e$, $i = 1, \ldots, k$. Let $n = p_1^{e_1} \cdots p_k^{e_k} = |S|$ and let $L\colon S \to \mathrm{GL}(V)$ be a representation of S. If $s = h_1^{m_1} \cdots h_k^{m_k} \in S$, then $s^n = (h_1^n)^{m_1} \cdots (h_k^n)^{m_k}$. Since $p_i^{e_i} \mid n$, $i = 1, \ldots, k$, we conclude that $s^n = e$. Thus $L(s)^n = L(s^n) = L(e) = I_V$. In other words, the eigenvalues of $L(s)$ are roots of the polynomial $\lambda^n - 1$.

Corollary 3 *Let S be a finite abelian group of order $n = p_1^{e_1} \cdots p_k^{e_k}$, where p_i is a prime and e_i is a positive integer, $i = 1, \ldots, k$. Let R be a splitting field for the polynomial $\lambda^n - 1$ and assume that $n \cdot 1 \neq 0$ in R. Then any irreducible representation $L\colon S \to \mathrm{GL}(V)$, where V is a finite dimensional vector space over R, is in fact of degree* 1. *Moreover, the number of such pairwise inequivalent irreducible representations of S is precisely n.*

Proof: Let $L\colon S \to \mathrm{GL}(V)$ be an irreducible representation of S, where V is a finite dimensional vector space over R. Since R is a splitting field for the polynomial $\lambda^n - 1$, it follows from the remarks immediately preceding the statement of Corollary 3 that the characteristic polynomial of each $L(s) \in \operatorname{im} L$ splits in R. Hence by Corollary 2, L is of degree 1; for each $s \in S$, there exists a scalar $\lambda(s) \in R$ such that

$$L(s)v = \lambda(s)v, \qquad v \in V.$$

Every element $s \in S$ is achieved precisely once as

$$s = h_1^{m_1} h_2^{m_2} \cdots h_k^{m_k} \tag{20}$$

for $1 \leq m_i \leq p_i^{e_i}$, $i = 1, \ldots, k$; this is just a consequence of the fact that S is a direct product of the $H_i = [h_i]$. The values $\lambda(s)$ for all $s \in S$ are thus completely determined by the values $\lambda(h_1), \ldots, \lambda(h_k)$. Since $h_i^{p_i^{e_i}} = e$ we we must have $\lambda(h_i)^{p_i^{e_i}} = 1$, $i = 1, \ldots, k$, so each $\lambda(h_i)$ is a $p_i^{e_i}$-th root of 1, and any $p_i^{e_i}$-th root of 1 is an nth root of 1. Now $n\lambda^{n-1} \neq 0$ and $\lambda^n - 1$ have no common factors, so the roots of $\lambda^n - 1$ are distinct. Thus formally there are $n = p_1^{e_1} \cdots p_k^{e_k}$ ways in which the function $\lambda: S \to R$ can be defined. But of course if two assignments of values to the $\lambda(h_i)$ are different, then we are obviously dealing with distinct representations (they do not agree on all the h_i, $i = 1, \ldots, k$). The proof is completed by observing that representations of degree 1 are inequivalent if and only if they are distinct. ∎

Example 1 We list all the irreducible representations over **C** of all abelian groups of order 18. By Example 5, Section 4.2, the nonisomorphic abelian groups of order 18 are Z_{18} and $Z_2 + Z_3 + Z_3$. Multiplicatively we have

$$S = C_{18} \tag{21}$$

or

$$S = C_2 \cdot C_3 \cdot C_3 \tag{22}$$

where C_p is a cyclic group of order p. In the case (21) we can define

$$\lambda(h) = \xi$$

where ξ is any eighteenth root of 1 and h is a generator of C_{18}. In the case (22) we can define

$$\lambda(h_j) = \xi_j, \qquad j = 1, 2, 3$$

where $\xi_1 = \pm 1$, ξ_2 and ξ_3 are any two cube roots of 1, and h_1, h_2, h_3 are generators of C_2, C_3, and C_3 respectively.

Much of what is known about representations of a group depends on the following important result.

Theorem 2.3 (*Burnside's Theorem*) *Let S be a group, V an n-dimensional vector space over R, and $L: S \to \mathrm{GL}_n(V)$ an absolutely irreducible representation of S. Then*

$$\dim \langle \mathrm{im}\, L \rangle = n^2. \tag{23}$$

In other words, $\langle \mathrm{im}\, L \rangle = \mathrm{Hom}(V,V)$.

Proof: Let $\Delta: S \to \mathrm{GL}(n,R)$ be an associated matrix representation of S. The theorem will be proved once we show that $\langle \mathrm{im}\, \Delta \rangle = M_n(R)$, i.e.,

$$\dim \langle \mathrm{im}\, \Delta \rangle = n^2. \tag{24}$$

To simplify notation, we will denote a typical matrix in $\mathrm{im}\, \Delta$ by A. Assume that $\dim \langle \mathrm{im}\, \Delta \rangle = r < n^2$, and let $\{A_1, \ldots, A_r\}$ be a basis of $\langle \mathrm{im}\, \Delta \rangle$. Let us think of each matrix $A \in \mathrm{im}\, \Delta$ as an n^2-tuple $(a_{11}, a_{12}, \ldots, a_{1n}, a_{21}, \ldots,$

$a_{2n}, \ldots, a_{n1}, \ldots, a_{nn}) \in R^{n^2}$. Consider the matrix $M \in M_{r,n^2}(R)$ whose ith row is A_i written as an n^2-tuple. Since $\{A_1, \ldots, A_r\}$ is a basis of $\langle \text{im}\,\Delta \rangle$, M has rank $r < n^2$, and hence the null space W of M is a nonzero subspace of R^{n^2}. In fact, the dimension of W is precisely $s = n^2 - r \geq 1$. By identifying column n^2-tuples with matrices in $M_n(R)$ in the obvious way, regard W as a subspace of $M_n(R)$, and let $\{K_1, \ldots, K_s\} \subset M_n(R)$ be a basis of W. Then $\text{tr}(K_jA_i) = \text{tr}(A_iK_j) = 0$, $i = 1, \ldots, r$, $j = 1, \ldots, s$, and since the A_i span $\langle \text{im}\,\Delta \rangle$ and the K_j span W, it follows that

$$\text{tr}(KA) = 0, \qquad A \in \text{im}\,\Delta,\ K \in W. \tag{25}$$

Moreover, if $A \in \text{im}\,\Delta$ and $K \in W$, then for $B \in \text{im}\,\Delta$ we have

$$\text{tr}((KA)B) = \text{tr}(K(AB)) = 0$$

because $AB \in \text{im}\,\Delta$. Thus $KA \in W$ whenever $A \in \text{im}\,\Delta$ and $K \in W$. For a fixed matrix $A \in \text{im}\,\Delta$ write

$$K_pA = \sum_{q=1}^{s} u_{pq}K_q, \qquad 1 \leq p \leq s \tag{26}$$

(the coefficients u_{pq} depend on the matrix $A \in \text{im}\,\Delta$). Set

$$K_p = [k_{ij}^p] \in M_n(R), \qquad 1 \leq p \leq s,$$

and compare the (i,j) entries on either side of (26) to obtain

$$\sum_{t=1}^{n} k_{it}^p a_{tj} = \sum_{q=1}^{s} u_{pq}k_{ij}^q, \qquad 1 \leq p \leq s,\ 1 \leq i,j \leq n. \tag{27}$$

Let $U \in M_s(R)$ be the matrix $[u_{pq}]$, and for each $i = 1, \ldots, n$ let $P_i \in M_{s,n}(R)$ be the matrix whose (p,t) entry is k_{it}^p, $p = 1, \ldots, s$, $t = 1, \ldots, n$. We can then rewrite (27) as

$$(P_iA)_{pj} = (UP_i)_{pj}, \qquad p = 1, \ldots, s, \quad j = 1, \ldots, n,$$

or

$$P_iA = UP_i, \qquad i = 1, \ldots, n. \tag{28}$$

In other words, each of the matrices $P_i \in M_{s,n}(R)$ links $\text{im}\,\Delta$ and the set Γ of matrices $U \in M_s(R)$ obtained by varying A over $\text{im}\,\Delta$. Next, observe that not every P_i is the 0 matrix, for otherwise

$$k_{it}^p = 0, \qquad i = 1, \ldots, n, \quad t = 1, \ldots, n, \quad p = 1, \ldots, s$$

and hence $K_1 = \cdots = K_s = 0$, contradicting the fact that $\{K_1, \ldots, K_s\}$ is a basis of $W \neq 0$. Assume $P_{i_0} \neq 0$; from (28) we have

$$P_{i_0}A = UP_{i_0}, \qquad A \in \text{im}\,\Delta. \tag{29}$$

There are now two cases to be considered: Either the set of matrices Γ is irreducible over R, or Γ is reducible over R.

Case 1. Γ is irreducible over R.

Since $\text{im}\,\Delta$ is absolutely irreducible (and hence irreducible) over R, we conclude by Theorem 2.2(a) that $s = n$ and P_{i_0} is nonsingular. Set $Q = [q_{pt}] = P_{i_0}$; then (29) becomes

$$QA = UQ, \qquad A \in \text{im } \Delta$$

and from (28) we obtain

$$P_iA = QAQ^{-1}P_i, \qquad A \in \text{im } \Delta, \quad i = 1, \ldots, n,$$

or

$$(Q^{-1}P_i)A = A(Q^{-1}P_i), \qquad A \in \text{im } \Delta, \quad i = 1, \ldots, n. \tag{30}$$

Again, since im Δ is absolutely irreducible over R, Corollary 1 implies that for each $i = 1, \ldots, n$, $Q^{-1}P_i = \alpha_i I_n$ for some $\alpha_i \in R$, i.e.,

$$P_i = \alpha_i Q, \qquad i = 1, \ldots, n. \tag{31}$$

Recall that k_{it}^p is the (p,t) entry of P_i, so (31) yields

$$k_{it}^p = \alpha_i q_{pt}, \qquad i = 1, \ldots, n, p = 1, \ldots, s = n, \quad t = 1, \ldots, n.$$

Thus for $p = 1, \ldots, s = n$ and $A \in \text{im } \Delta$, the equation

$$tr(K_pA) = 0$$

becomes

$$0 = \sum_{i,t=1}^{n} k_{it}^p a_{ti} = \sum_{i,t=1}^{n} \alpha_i q_{pt} a_{ti} = \sum_{i=1}^{n} (QA)_{pi}\alpha_i.$$

In other words, for $\alpha = (\alpha_1, \ldots, \alpha_n) \in R^n$ we have

$$QA\alpha = 0, \qquad A \in \text{im } \Delta,$$

and since Q and every $A \in \text{im } \Delta$ are nonsingular, it follows that $\alpha = 0$, i.e., $\alpha_1 = \cdots = \alpha_n = 0$. But then (31) yields $P_1 = \cdots = P_n = 0$, and we have already seen that this is untenable because it implies $K_1 = \cdots = K_s = 0$. Thus Γ cannot be irreducible over R.

Case 2. Γ is reducible over R.

There exist a fixed nonsingular matrix $N \in \text{GL}(s,R)$ and a fixed integer d, $1 \leq d \leq s$, such that

$$NUN^{-1} = \begin{bmatrix} X & 0 \\ Y & Z \end{bmatrix}, \qquad U \in \Gamma, \tag{32}$$

where $X \in M_d(R)$. Also, we can assume that either the set Ω of matrices X that appear in (32) in irreducible over R, or the matrices X are all the zero matrix. For each $U \in \Gamma$, let $D = \begin{bmatrix} X & 0 \\ Y & Z \end{bmatrix}$. Write $N = [n_{p\nu}]$, and define matrices

$$K_p' = \sum_{\nu=1}^{s} n_{p\nu} K_\nu, \qquad p = 1, \ldots, s. \tag{33}$$

Because of the nonsingularity of N, $\{K_p' \mid p = 1, \ldots, s\}$ is a basis of $W \neq$

0 and hence $K_p' \neq 0$, $p = 1, \ldots, s$. We compute for $A \in \operatorname{im} \Delta$ and $p = 1, \ldots, s$ that

$$\begin{aligned}
K_p'A &= \sum_{\nu=1}^{s} n_{p\nu} K_\nu A \\
&= \sum_{\nu=1}^{s} n_{p\nu} \sum_{q=1}^{s} u_{\nu q} K_q \qquad \text{[from (26)]} \\
&= \sum_{q=1}^{s} \left(\sum_{\nu=1}^{s} n_{p\nu} u_{\nu q} \right) K_q = \sum_{q=1}^{s} (NU)_{pq} K_q \\
&= \sum_{q=1}^{s} (DN)_{pq} K_q \qquad \text{[from (32)]} \\
&= \sum_{q=1}^{s} \left(\sum_{\mu=1}^{s} d_{p\mu} n_{\mu q} \right) K_q = \sum_{\mu=1}^{s} d_{p\mu} \left(\sum_{q=1}^{s} n_{\mu q} K_q \right) \\
&= \sum_{\mu=1}^{s} d_{p\mu} K_\mu' \qquad \text{[from (33)].}
\end{aligned}$$

Now $d_{p\mu} = 0$ when $p = 1, \ldots, d$, $\mu = d + 1, \ldots, s$, and $d_{p\mu} = x_{p\mu}$ when $p = 1, \ldots, d$, $\mu = 1, \ldots, d$, since

$$D = \begin{matrix} & \overbrace{}^{d} & \\ d\{ & \begin{bmatrix} X & 0 \\ Y & Z \end{bmatrix} \end{matrix}.$$

Hence for $A \in \operatorname{im} \Delta$,

$$K_p'A = \sum_{\mu=1}^{d} x_{p\mu} K_\mu', \qquad p = 1, \ldots, d. \tag{34}$$

This is the same situation as in (26), with s replaced by d and the matrices $X \in \Omega$ playing the role of the matrices U. We have assumed that either every $X \in \Omega$ is in fact 0, or Ω is irreducible. If the first alternative were to hold, then (34) would imply that $K_p'A = 0$, $p = 1, \ldots, d$, and since every $A \in \operatorname{im} \Delta$ is nonsingular that $K_p' = 0$, $p = 1, \ldots, d$. This is of course a contradiction (we have already observed that no K_p' is 0 since $\{K_p' \mid p = 1, \ldots, s\}$ is a basis of W). If, on the other hand, the set of matrices Ω is irreducible over R, we can repeat the argument given in Case 1 to show that $K_1' = \cdots = K_d' = 0$, again a contradiction.

Thus the assumption that $\dim\langle \operatorname{im} \Delta \rangle = r < n^2$ is untenable. We conclude that $\dim\langle \operatorname{im} \Delta \rangle = n^2 = \dim\langle \operatorname{im} L \rangle$, completing the proof. ∎

Example 2 The hypothesis in Burnside's theorem (Theorem 2.3) that the representation $L: S \to GL_n(V)$ is absolutely irreducible is essential. In Example 3, Section 5.1, we saw that if S is the multiplicative group of nonzero complex numbers, then the matrix representation $\kappa: S \to GL(2,\mathbf{R})$ given by

$$\kappa(x + iy) = \begin{bmatrix} x & -y \\ y & x \end{bmatrix}, \qquad x + iy \in S \tag{35}$$

is irreducible over **R** but not over **C**. Observe that any real linear combination of matrices of the form (35) is also of this form. But certainly not every matrix in $M_2(\mathbf{R})$ has the form (35). Thus $\langle \text{im } K\rangle \neq M_2(\mathbf{R})$.

Theorem 2.4 (*Burnside–Frobenius–Schur*) *Let S be a group, V_k an n_k-dimensional vector space over R for $k = 1, \ldots, m$, and suppose $V = \dot{\sum}_{k=1}^{m} V_k$. Let $L_k\colon S \to \mathrm{GL}_{nk}(V_k)$, $k = 1, \ldots, m$ be pairwise inequivalent absolutely irreducible representations of S. Set $n = n_1 + \cdots + n_m$, and let $L\colon S \to \mathrm{GL}_n(V)$ be the representation of S defined by*

$$L(s) = \dot{\sum_{k=1}^{m}} L_k(s), \qquad s \in S.$$

Then
$$\dim \langle \text{im } L\rangle = n_1^{\,2} + \cdots + n_m^{\,2}. \tag{36}$$

Proof: For $k = 1, \ldots, m$, let $\Delta_k\colon S \to \mathrm{GL}(n_k,R)$ be a matrix representation of S associated with $L_k\colon S \to \mathrm{GL}_{n_k}(V_k)$. We obtain a matrix representation $\Delta\colon S \to \mathrm{GL}(n,R)$ associated with $L\colon S \to \mathrm{GL}_n(V)$:

$$\Delta(x) = \dot{\sum_{k=1}^{m}} \Delta_k(x), \qquad x \in S. \tag{37}$$

The theorem will be proved once we show that

$$\dim \langle \text{im } \Delta\rangle = n_1^{\,2} + \cdots + n_m^{\,2}. \tag{38}$$

Assume (38) does not hold. Then exactly as in the proof of Theorem 2.3, there exists a nonzero matrix $K \in M_n(R)$ of the form

$$K = \dot{\sum_{k=1}^{m}} K^k, \qquad K^k \in M_{n_k}(R), \qquad k = 1, \ldots, m, \tag{39}$$

such that
$$\begin{aligned} \operatorname{tr} K\Delta(x) &= \sum_{k=1}^{m} \operatorname{tr}(K^k \Delta_k(x)) \\ &= 0, \qquad x \in S. \end{aligned} \tag{40}$$

Observe that if (40) holds for a matrix K of the form (39), then (as in the proof of Theorem 2.3) it holds for any matrix of the form

$$\dot{\sum_{k=1}^{m}} K^k \Delta_k(y), \qquad K^k \in M_{n_k}(R), \qquad k = 1, \ldots, m, \qquad y \in S. \tag{41}$$

The totality W of solutions K of the form (39) to the equations (40) is a vector space of dimension $s > 0$. Let

$$\left\{K_p = \dot{\sum_{k=1}^{m}} K_p^k \mid p = 1, \ldots, s\right\} \tag{42}$$

be a basis of W. Then by (41), for any $x \in S$ we have

$$\dot{\sum_{k=1}^{m}} K_p^k \Delta_k(x) \in W, \qquad p = 1, \ldots, s. \tag{43}$$

Hence for any $x \in S$,

$$\sum_{k=1}^{m}{}^{\cdot}\, K_p^k \Delta_k(x) = \sum_{q=1}^{s} u_{pq} K_q = \sum_{q=1}^{s} u_{pq}\left(\sum_{k=1}^{m}{}^{\cdot}\, K_q^k\right)$$
$$= \sum_{k=1}^{m}{}^{\cdot} \sum_{q=1}^{s} u_{pq} K_q^k, \qquad p = 1, \ldots, s, \tag{44}$$

so that for each $k = 1, \ldots, m$ we have

$$K_p^k \Delta_k(x) = \sum_{q=1}^{s} u_{pq} K_q^k, \qquad p = 1, \ldots, s, \tag{45}$$

where $U = [u_{pq}] \in M_s(R)$ depends on $x \in S$. Let Γ be the set of s-square matrices U obtained by varying x over S. Suppose Γ is reducible over R. Then there exist a fixed nonsingular matrix $N \in \mathrm{GL}(s,R)$ and a fixed integer d, $1 \le d < s$, such that

$$NUN^{-1} = \begin{bmatrix} X & 0 \\ Y & Z \end{bmatrix}, \qquad U \in \Gamma, \tag{46}$$

where $X \in M_d(R)$. Moreover, we can assume that either the set Ω of matrices X that appear in (46) is irreducible over R, or the matrices X are all the 0 matrix. Write $N = [n_{p\nu}]$ and for each $k = 1, \ldots, m$ define matrices

$$K_p^{k\prime} = \sum_{\nu=1}^{s} n_{p\nu} K_\nu^k, \qquad p = 1, \ldots, s. \tag{47}$$

An argument analogous to the one preceding (34) shows that for $x \in S$,

$$K_p^{k\prime} \Delta_k(x) = \sum_{\mu=1}^{d} x_{p\mu} K_\mu^{k\prime}, \qquad k = 1, \ldots, m,\ p = 1, \ldots, d. \tag{48}$$

If all the matrices $X \in \Omega$ are 0, then (48) and the nonsingularity of every $\Delta_k(x) \in \operatorname{im} \Delta_k$ imply

$$K_p^{k\prime} = 0, \qquad k = 1, \ldots, m,\ p = 1, \ldots, d. \tag{49}$$

Set

$$K_p{}' = \sum_{k=1}^{m}{}^{\cdot}\, K_p^{k\prime}, \qquad p = 1, \ldots, s,$$

and rewrite (47) as

$$K_p{}' = \sum_{\nu=1}^{s} n_{p\nu} K_\nu, \qquad p = 1, \ldots, s.$$

Since $\{K_1, \ldots, K_s\}$ is a basis of W and N is nonsingular, it follows that $\{K_1{}', \ldots, K_s{}'\}$ is a basis of W. But (49) implies $K_1{}' = \cdots = K_d{}' = 0$, providing a contradiction. Thus the matrices in Ω are not all 0, so Ω is irreducible over R. We can now return to the equations (45) [instead of using (48)] and simply assume to begin with that Γ is irreducible over R.

Proceeding as in the proof of Case 1 in Theorem 2.3, we conclude from (45) that for each fixed $k = 1, \ldots, m$, either

$$K_p^k = 0, \qquad p = 1, \ldots, s, \tag{50}$$

or $s = n_k$ and for some p, K_p^k is a nonsingular matrix linking im Δ_k and Γ. Now (50) fails for at least one value of k since $W \neq 0$. On the other hand, the representations $L_1, \ldots, L_m$ are pairwise inequivalent, so (50) can fail for at most one value of k. Say $k = k_0$ is that value of k for which not all the matrices $K_p^{k_0}$, $p = 1, \ldots, s$, are 0. Then for $p = 1, \ldots, s$, (40) becomes

$$\operatorname{tr}\,(K_p^{k_0}\Delta_{k_0}(s)) = 0, \qquad s \in S,$$

and we can apply Theorem 2.3 itself to conclude that $K_p^{k_0} = 0$, $p = 1, \ldots, s$, contradicting the choice of k_0. This completes the proof. ▮

Let S be a groupoid, V an S-module over R, and denote by $L: S \to \operatorname{Hom}(V,V)$ the representation of S. The *character of* L, denoted by χ_L, is the trace function of L. That is, $\chi_L: S \to R$ is defined by

$$\chi_L(s) = \operatorname{tr} L(s), \qquad s \in S.$$

We sometimes say that the representation L *affords* the character χ_L. If $\Delta: S \to M_n(R)$ is a matrix representation of the groupoid S, then the *character* χ_Δ *of* Δ is defined by

$$\chi_\Delta(s) = \operatorname{tr} \Delta(s), \qquad s \in S.$$

Again, we say that the matrix representation Δ *affords* the character χ_Δ. We also refer to characters as *characters of* S.

The following facts concerning characters are easily proved (see Exercises 3 to 5). If L is a representation of the groupoid S, $s \in S$, and $\lambda_i(s)$, $i = 1, \ldots, n$ are the eigenvalues of $L(s)$, then

$$\chi_L(s) = \sum_{i=1}^{n} \lambda_i(s). \tag{51}$$

If L and M are equivalent representations of S, then

$$\chi_L = \chi_M. \tag{52}$$

If S is a group, L is a proper representation of S over $R = \mathbf{C}$, and $s \in S$ is an element of *finite period*, i.e., $s^k = e$ (the identity in S) for some positive integer k, then

$$\chi_L(s^{-1}) = \overline{\chi_L(s)}. \tag{53}$$

Let $L: S \to \operatorname{Hom}(V,V)$ be a representation over a field R of the groupoid S, W an invariant subspace of V for L, and $q: V \to V/W$ the quotient map. We saw in Section 5.1 that two representations of S can be obtained in terms of L: Define

$$L_1: S \to \operatorname{Hom}(W,W)$$

by

$$L_1(s) = L(s)\,|\,W, \qquad s \in S,$$

and define

$$L_2: S \to \operatorname{Hom}(V/W, V/W)$$

by

$$L_2(s)q(v) = q(L(s)v), \qquad s \in S,\ v \in V.$$

If $\{e_1, \ldots, e_n\}$ is a basis of V such that $\{e_{r+1}, \ldots, e_n\}$ is a basis of W, the associated matrix representation $\Delta: S \to M_n(R)$ of S has the form

$$\Delta(s) = \left[\begin{array}{c|c} A(s) & 0 \\ \hline * & B(s) \end{array}\right], \qquad s \in S, \tag{54}$$

where $A(s) \in M_r(R)$ and $B(s) \in M_{n-r}(R)$. It was observed immediately prior to Theorem 1.7 that $B: S \to M_{n-r}(R)$ is a matrix representation of S associated with L_1, and $A: S \to M_r(R)$ is a matrix representation of S associated with L_2. Now if L_1 and L_2 are themselves reducible, there exist nonsingular matrices P and Q of appropriate sizes such that

$$P^{-1}A(s)P = \left[\begin{array}{c|c} C(s) & 0 \\ \hline * & D(s) \end{array}\right], \qquad s \in S$$

and

$$Q^{-1}B(s)Q = \left[\begin{array}{c|c} E(s) & 0 \\ \hline * & F(s) \end{array}\right], \qquad s \in S.$$

Then since $(P \dotplus Q)^{-1} = P^{-1} \dotplus Q^{-1}$, from (54) we have

$$(P \dotplus Q)^{-1}\Delta(s)(P \dotplus Q) = \left[\begin{array}{c|c|c|c} C(s) & 0 & 0 & 0 \\ \hline * & D(s) & 0 & 0 \\ \hline * & * & E(s) & 0 \\ \hline * & * & * & F(s) \end{array}\right], \qquad s \in S. \tag{55}$$

Proceeding in this fashion, we finally obtain a matrix $U \in GL(n,R)$ such that

$$U^{-1}\Delta(s)U = \left[\begin{array}{c|c|c|c|c} \Delta_{11}(s) & 0 & 0 & \cdots & 0 \\ \hline * & \Delta_{22}(s) & 0 & \cdots & 0 \\ \hline * & * & \Delta_{33}(s) & \cdots & 0 \\ \hline \vdots & \vdots & \vdots & \ddots & \vdots \\ \hline * & * & * & \cdots & \Delta_{pp}(s) \end{array}\right], \qquad s \in S, \tag{56}$$

and for each $i = 1, \ldots, p$ the matrices $\Delta_{ii}(s)$ constitute an irreducible matrix representation of S. In terms of L, we conclude that there exists a basis E of V such that for each $s \in S$, $[L(s)]_E^E$ is the partitioned matrix on the right in (56). Next define a chain of invariant subspaces of V for the representation L as follows: V_p is the space spanned by the basis vectors in E that correspond to the Δ_{pp} block; V_{p-1} is the space spanned by the basis vectors in E that correspond to the Δ_{pp} and $\Delta_{p-1,p-1}$ blocks; V_{p-2} is the space spanned by the basis vectors in E that correspond to the Δ_{pp}, the $\Delta_{p-1,p-1}$, and $\Delta_{p-2,p-2}$ blocks, and so on. Then clearly

$$V_p \subset V_{p-1} \subset V_{p-2} \subset \cdots \subset V_1 = V. \tag{57}$$

The invariant subspace V_p is an irreducible S-module because the matrix representation Δ_{pp} of S is irreducible. Moreover, we know that $\Delta_{p-1,\,p-1}$ is a matrix representation of S associated with the quotient representation of S in V_{p-1}/V_p. Since $\Delta_{p-1,\,p-1}$ is irreducible, it follows that this quotient representation is irreducible and hence (see Theorem 1.7, Section 5.1) that V_p is a maximal invariant subspace of V_{p-1}. In the same way, for each $i = 1, \ldots, p-1$ the quotient representation of S in V_i/V_{i+1} is irreducible with associated matrix representation Δ_{ii}, and V_{i+1} is a maximal invariant subspace of V_i. These representations of S in V_i/V_{i+1}, $i = 1, \ldots, p$ (set $V_{p+1} = \{0\}$ so that $V_p/\{0\} = V_p$) are called irreducible *constituents* or *components of* L, and a chain of subspaces such as (57) is called a *reduction of* V (or L). Similarly, the irreducible matrix representations $\Delta_{11}, \Delta_{22}, \ldots, \Delta_{pp}$ of S are called irreducible *constituents* or *components* of Δ, and the matrix (56) is called a *reduction of* Δ *into irreducible components.* The integer p is called the *length* of the reduction.

If S is a group, the Burnside–Frobenius–Schur theorem (Theorem 2.4) allows us to prove the essential uniqueness of these components in a surprisingly easy way.

Theorem 2.5 *Let S be a group, and V_k an n_k-dimensional vector space over R for $k = 1, \ldots, m$. Let $L_k\colon S \to \mathrm{GL}_{nk}(V_k)$, $k = 1, \ldots, m$, be pairwise inequivalent absolutely irreducible representations of S. Then the characters $\chi_k = \chi_{L_k}\colon S \to R$ are linearly independent elements of the vector space R^S over R.*

Proof: Suppose $c_k \in R$, $k = 1, \ldots, m$ are scalars such that

$$\sum_{k=1}^{m} c_k \chi_k = 0. \tag{58}$$

Choose a basis for each vector space V_k, let $\Delta_k\colon S \to \mathrm{GL}(n_k, R)$ denote the matrix representation of S associated with L_k relative to the chosen basis of V_k, and set $\Delta_k(s) = [a_{ij}^k(s)]$, $s \in S$. Then for each $s \in S$ we have

$$0 = \sum_{k=1}^{m} c_k \chi_k(s) = \sum_{k=1}^{m} c_k tr \Delta_k(s)$$
$$= \sum_{k=1}^{m} c_k \sum_{i=1}^{n_k} a_{ii}^k(s). \tag{59}$$

Now it follows from (the proof of) Theorem 2.4 that the $n_1^2 + \cdots + n_m^2$ "entry functions" $a_{ij}^k: S \to R, i, j = 1, \ldots, n_k, k = 1, \ldots, m$ are linearly independent in R^S. Otherwise there would exist a nonzero matrix K of the form

$$K = \sum_{k=1}^{m} {}^{\cdot} K^k, \qquad K^k \in M_{n_k}(R), k = 1, \ldots, m$$

such that

$$\text{tr } K\Delta(s) = \sum_{k=1}^{m} \text{tr}(K^k \Delta_k(s)) = \sum_{k=1}^{m} \sum_{i,j=1}^{n_k} K_{ji}^k a_{ij}^k(s)$$
$$= 0, \qquad s \in S,$$

and this in turn would imply that

$$\dim \langle \text{im } \Delta \rangle < n_1^2 + \cdots + n_m^2$$

[where $\Delta(s) = \sum_{k=1}^{\cdot m} \Delta_k(s)$, $s \in S$], in contradiction to (38). Thus, since (59) says that

$$c_1(a_{11}^1 + a_{22}^1 + \cdots + a_{n_1 n_1}^1) + \cdots + c_m(a_{11}^m + \cdots + a_{n_m n_m}^m) = 0$$

we must have $c_1 = \cdots = c_m = 0$. ∎

Corollary 4 *Let S be a group, and V and W two proper finite dimensional S-modules over R with the respective representations* $L: S \to \text{GL}(V)$ *and* $M: S \to \text{GL}(W)$. *Suppose that* $\chi_L = \chi_M$. *Then* $\dim V = \dim W$, *any reduction of L into absolutely irreducible components has the same length as any reduction of M into absolutely irreducible components, and the components of any two such reductions can be paired off into equivalent pairs.*

Proof: Let bases of V and W be given such that the matrix representations Δ and κ of S associated with L and M, respectively, have the form

$$\Delta(s) = \begin{bmatrix} \Delta_{11}(s) & 0 & 0 & \cdots & 0 \\ * & \Delta_{22}(s) & 0 & \cdots & 0 \\ * & * & \Delta_{33}(s) & \cdots & 0 \\ \vdots & \vdots & \vdots & \ddots & \vdots \\ * & * & * & \cdots & \Delta_{pp}(s) \end{bmatrix}, \qquad s \in S \tag{60}$$

and

$$\kappa(s) = \begin{bmatrix} \kappa_{11}(s) & 0 & 0 & \cdots & 0 \\ * & \kappa_{22}(s) & 0 & \cdots & 0 \\ * & * & \kappa_{33}(s) & \cdots & 0 \\ \vdots & \vdots & \vdots & \ddots & \vdots \\ * & * & * & \cdots & k_{qq}(s) \end{bmatrix}, \qquad s \in S, \tag{61}$$

where the component representations

$$\Delta_{11}, \ldots, \Delta_{pp}, \kappa_{11}, \ldots, \kappa_{qq} \tag{62}$$

are all absolutely irreducible. Let $\tau_1, \ldots, \tau_r$ be the complete list of pairwise inequivalent matrix representations occurring among the representations (62), and let χ_i be the character of τ_i, $i = 1, \ldots, r$. Suppose each τ_i is equivalent to δ_i of the representations $\Delta_{11}, \ldots, \Delta_{pp}$ and k_i of the representations $\kappa_{11}, \ldots, \kappa_{qq}$. Then

$$\chi_L(s) = \operatorname{tr} \Delta(s) = \sum_{i=1}^{p} \operatorname{tr} \Delta_{ii}(s) = \sum_{i=1}^{r} \delta_i \operatorname{tr} \tau_i(s)$$

$$= \sum_{i=1}^{r} \delta_i \chi_i(s), \qquad s \in S,$$

and similarly

$$\chi_M(s) = \sum_{i=1}^{r} k_i \chi_i(s), \qquad s \in S.$$

Since $\chi_L = \chi_M$, we have

$$\sum_{i=1}^{r} (\delta_i - k_i)\chi_i = 0. \tag{63}$$

The characters $\chi_1, \ldots, \chi_r$ are linearly independent by Theorem 2.5, so that

$$\delta_i = k_i, \qquad i = 1, \ldots, r.$$

This means that the number of representations in $\{\Delta_{11}, \ldots, \Delta_{pp}\}$ equivalent to τ_i is the same as the number of representations in $\{\kappa_{11}, \ldots, \kappa_{qq}\}$ equivalent to τ_i, $i = 1, \ldots, r$. It follows that $\dim V = \dim W$, $p = q$, and except for order the components of the representations Δ and κ are pairwise equivalent. ∎

Corollary 5 *Let S be a group, and V a proper finite dimensional S-module with representation function $L: S \to \mathrm{GL}(V)$. Then any two reductions of L into absolutely irreducible components have the same length, and apart from order the components are pairwise equivalent.*

Proof: Apply Corollary 4 with $L = M$. ∎

Corollary 6 *Let S be a group, and V and W two proper finite dimensional S-modules over R. Suppose the respective representations $L: S \to \mathrm{GL}(V)$ and $M: S \to \mathrm{GL}(W)$ are absolutely irreducible. Then $L \sim M$ iff $\chi_L = \chi_M$.*

Proof: If $L \sim M$, we know from general considerations that $\chi_L = \chi_M$ [see (52)]. If $\chi_L = \chi_M$, then by Corollary 4 (since the representations L and M are already reduced and the reductions have length 1) we have $L \sim M$. ▮

Corollary 7 *Let S be a finite group with $|S| = h$, R a field in which $h \cdot 1 \neq 0$, and V a proper finite dimensional S-module over R. Denote by $L: S \to \mathrm{GL}(V)$ the representation function for S. Then there exists a direct sum decomposition (perhaps over an extension field of R) of the S-module V into invariant subspaces $V_1, \ldots, V_p$ of V for the representation L,*

$$V = \sum_{k=1}^{p}{}^{\cdot}\, V_k,$$

such that for each $k = 1, \ldots, p$ the representation $L_k = L|V_k$ of S is absolutely irreducible. Moreover, if

$$V = \sum_{k=1}^{q}{}^{\cdot}\, U_k$$

is another such decomposition of V, then $p = q$ and for some $\sigma \in S_p$ we have $\dim V_k = \dim U_{\sigma(k)}$ *and $L|V_k \sim L|U_{\sigma(k)}$, $k = 1, \ldots, p$.*

Proof: Let a basis of V be chosen so that the matrix representation Δ of S associated with L has the form (60) in which $\Delta_{11}, \ldots, \Delta_{pp}$ are absolutely irreducible components (it may be necessary to go to an extension field of R). For each $s \in S$, let $A(s)$ denote the principal submatrix of $\Delta(s)$ immediately below and to the right of $\Delta_{11}(s)$; this provides a matrix representation A of S. By the proof of Maschke's theorem (Theorem 2.1), we know that

$$\Delta \sim \Delta_{11} \dotplus A.$$

Applying the same reasoning to A (or using induction), we obtain

$$\Delta \sim \Delta_{11} \dotplus \Delta_{22} \dotplus \cdots \dotplus \Delta_{pp}. \tag{64}$$

In other words, there exists a basis E of V such that

$$[L(s)]_E^E = \Delta_{11}(s) \dotplus \cdots \dotplus \Delta_{pp}(s), \qquad s \in S. \tag{65}$$

Now define V_k to be the subspace of V spanned by the vectors in E corresponding to Δ_{kk}, $k = 1, \ldots, p$. Obviously (65) implies that each V_k is an invariant subspace of V for the representation L. Moreover, since Δ_{kk} is a matrix representation of S associated with $L_k = L|V_k$, it follows that L_k is absolutely irreducible, $k = 1, \ldots, p$. The final statement of the result follows from Corollary 5. ▮

Example 3 Let S be a group with identity e, and let r be a positive integer. The group S is said to have *exponent* r if $s^r = e$ for all $s \in S$. We shall show in this example that if R is a field in which $r \cdot 1 \neq 0$ and $S < \mathrm{GL}(n,R)$ has exponent r, then S is a finite group. In other words, *if in a group of nonsingular n-square matrices the* rth *power of every matrix is* I_n, *then the group is finite*. This result is due to Burnside.

Case 1 S is absolutely irreducible as a set of matrices. If $A \in S$, then since $A^r = I_n$, the eigenvalues of A are roots of the polynomial $\lambda^r - 1$. This polynomial has r roots, and it follows that tr A takes on at most a finite number of distinct values as A ranges over S. In fact, there are r possible choices for each λ_i in $\mathrm{tr}\, A = \lambda_1 + \cdots + \lambda_n$, so that tr A takes on at most r^n distinct values as A ranges over S. Now the group S provides an absolutely irreducible matrix representation of itself of degree n, and so $\dim\langle A \mid A \in S\rangle = n^2$ by Theorem 2.3. Set $N = n^2$, and let $\{A_1, \ldots, A_N\}$ be a basis of $M_n(R)$ chosen from S. If $A \in S$, then $A_kA \in S$, $k = 1, \ldots, N$ so that

$$\mathrm{tr}(A_kA) = t_k, \qquad k = 1, \ldots, N \tag{66}$$

where t_k is one of the at most r^n distinct values that tr A takes on. Let $\tau = (t_1, \ldots, t_N)$, and observe that there are at most $(r^n)^N$ possibilities for the N-tuple τ. The equations (66) can be regarded as a system of linear equations for the determination of A with a coefficient matrix M whose rows are the A_k strung out as $n^2 = N$ tuples, $k = 1, \ldots, N$. Since the A_k are linearly independent, this coefficient matrix M is nonsingular. Then (66) reads

$$MA = \tau$$

or

$$A = M^{-1}\tau$$

where A has its rows strung out to form an N-tuple. Since τ is one of at most $(r^n)^N = (r^n)^{n^2} = r^{n^3}$ N-tuples, it follows that A can be one of at most r^{n^3} matrices. Thus $|S| \leq r^{n^3}$.

Case 2 S is an arbitrary subgroup of $\mathrm{GL}(n,R)$ of exponent r. The argument here is by induction on n, with nothing to prove for $n = 1$. If S is absolutely irreducible, we can apply Case 1. Assume then that S is reducible over some extension field of R (observe that we need only go to a splitting field for $\lambda^r - 1$). Then there exists a nonsingular matrix V such that

$$V^{-1}AV = \left[\begin{array}{c|c} A_{11} & 0 \\ \hline A_{21} & A_{22} \end{array}\right], \quad A \in S, \tag{67}$$

where A_{11} is $p \times p$ and A_{22} is $q \times q$. Note that $\mathfrak{A}_i = \{A_{ii} \mid A \in S\}$, $i = 1, 2$, are obviously groups of exponent r and hence are finite by induction. Define S' to be the set of matrices on the right in (67); S' is clearly a group. For $i = 1, 2$, define $\varphi_i \colon S' \to \mathfrak{A}_i$ by

$$\varphi_i\left(\left[\begin{array}{c|c} A_{11} & 0 \\ \hline A_{21} & A_{22} \end{array}\right]\right) = A_{ii}.$$

It is obvious that φ_i is a surjective homomorphism of S' onto $\mathfrak{A}_i$, $i = 1,2$. Moreover, it is also obvious that

$$\ker \varphi_1 = \left\{ \left[\begin{array}{c|c} I_p & 0 \\ \hline A_{21} & A_{22} \end{array} \right] \right\}$$

and
$$\ker \varphi_2 = \left\{ \left[\begin{array}{c|c} A_{11} & 0 \\ \hline A_{21} & I_q \end{array} \right] \right\}.$$

By the first isomorphism theorem for groups, $S'/\ker \varphi_i \cong \mathfrak{A}_i$, $i = 1, 2$. By the second isomorphism theorem for groups,

$$\ker \varphi_1/(\ker \varphi_1 \cap \ker \varphi_2) \cong (\ker \varphi_1 \ker \varphi_2)/\ker \varphi_2.$$

But $(\ker \varphi_1 \ker \varphi_2)/\ker \varphi_2 \subset S'/\ker \varphi_2 \cong \mathfrak{A}_2$, and we have observed that $\mathfrak{A}_2$ is finite. Thus

$$\ker \varphi_1/(\ker \varphi_1 \cap \ker \varphi_2) \tag{68}$$

is a finite group. Suppose $X \in \ker \varphi_1 \cap \ker \varphi_2$ so that

$$X = \left[\begin{array}{c|c} I_p & 0 \\ \hline A_{21} & I_q \end{array} \right].$$

Now

$$X^r = \left[\begin{array}{c|c} I_p & 0 \\ \hline rA_{21} & I_q \end{array} \right] = I_n$$

so $rA_{21} = 0$, i.e., $A_{21} = 0$ and $X = I_n$. Thus $|\ker \varphi_1 \cap \ker \varphi_2| = 1$. But since the group (68) is finite, it follows that $\ker \varphi_1$ is finite (why?). Finally, since $S'/\ker \varphi_1 \cong \mathfrak{A}_1$ is finite and $\ker \varphi_1$ is finite, we conclude that S' is finite, and therefore S is finite.

Example 4 Let $A \in \mathrm{GL}(n,R)$ satisfy $A^k = I_n$ for some positive integer k, i.e., suppose A has finite period k, and assume $k \cdot 1 \neq 0$. We assert that A is similar over an extension field of R to a diagonal matrix. Let r be the least integer for which $A^r = I_n$; then $S = \{I_n, A, A^2, \ldots, A^{r-1}\}$ is a finite cyclic group. Now obviously $r \mid k$, so $r \cdot 1 \neq 0$. The group S provides a matrix representation $\Delta: S \to S$ of itself, i.e., $\Delta(s) = s$, $s \in S$. By the proof of Maschke's theorem (see also the proof of Corollary 7), $\Delta \sim \Delta_{11} \dotplus \cdots \dotplus \Delta_{pp}$, where the Δ_{ii} are absolutely irreducible over some extension field of R. Now the Δ_{ii} are absolutely irreducible representations of the finite abelian (in fact, cyclic) group S, so that by Corollary 3 each Δ_{ii} is of degree 1. We have proved that over some extension field (actually the splitting field of $\lambda^r - 1$ will do), there is a nonsingular matrix P such that $P^{-1}sP$ is a diagonal matrix for each $s \in S$.

It should be observed that this fact does not really require the use of representation theory to establish. Indeed, $\lambda^r - 1$ has distinct linear factors in its splitting field because $r\lambda^{r-1} \neq 0$ and $\lambda^r - 1$ are relatively prime. Hence A is similar to a diagonal matrix because the minimal polynomial of A must divide $\lambda^r - 1$.

Also note that if

$$A = \begin{bmatrix} 0 & 1 & 0 & 0 \\ 0 & 0 & 1 & 0 \\ 0 & 0 & 0 & 1 \\ 1 & 0 & 0 & 0 \end{bmatrix} \in \mathrm{GL}(4,\mathbf{R}),$$

then the minimal polynomial of A is $\lambda^4 - 1$, and to reduce A to a diagonal matrix, we must go to the extension field **C**.

In the remainder of the present section we shall investigate some of the relationships that obtain among characters of a finite group. If h is the order of a finite group under consideration and R is the underlying field of a representation of the group, it will always be assumed that $h \cdot 1 \neq 0$ in R. Such representations are called *ordinary* representations. Representations for which $h \cdot 1 = 0$ in R are called *modular*. We study only ordinary representations here.

An *irreducible character of a group S* is a character of an irreducible representation of S. An *absolutely irreducible character*, or *simple character of S*, is a character of an absolutely irreducible representation of S. The *degree of a character of S* is just the degree of the corresponding representation. We remark that if χ and ψ are two characters of S and $\chi \neq \psi$, then the corresponding representations are inequivalent. The *principal character of S* is the character of degree 1 which is identically equal to 1 on S. Notice that any character of S of degree 1 is absolutely irreducible.

Let S be a finite group with $|S| = h$, and let $\mathfrak{A}$ be the group algebra of S over a field R. For

$$f = \sum_{s \in S} f(s)s \in \mathfrak{A}$$

and

$$g = \sum_{s \in S} g(s)s \in \mathfrak{A},$$

we define the *scalar product* (f,g) *of f and g* by

$$(f,g) = \frac{1}{h} \sum_{s \in S} f(s)g(s^{-1}) \in R. \tag{69}$$

Note that $(f,g) = (g,f)$, $(f_1 + f_2, g) = (f_1,g) + (f_2,g)$, $(f, g_1 + g_2) = (f,g_1) + (f,g_2)$, and $(af,bg) = ab(f,g)$ for all $f,f_1,f_2 \in \mathfrak{A}$, $g,g_1,g_2 \in \mathfrak{A}$, and $a,b \in R$.

Theorem 2.6 Let $\Delta = [\Delta_{ij}]$ *and* $\kappa = [\kappa_{ij}]$ *be proper irreducible matrix representations over R of the finite group S of degrees m and n, respectively. If* Δ *and* κ *are inequivalent, then*

$$(\Delta_{ip},\kappa_{qj}) = 0, \qquad i,p = 1, \ldots, m, \; j,q = 1, \ldots, n. \tag{70}$$

If Δ is absolutely irreducible, then

$$(\Delta_{ip},\Delta_{qj}) = \frac{1}{m}\,\delta_{ij}\delta_{pq}, \qquad i,p,j,q = 1, \ldots, m. \tag{71}$$

Proof: Let $U = [u_{pq}] \in M_{m,n}(R)$ be any matrix, and define $V \in M_{m,n}(R)$ by

$$V = \sum_{s \in S} \Delta(s) U \kappa(s^{-1}). \tag{72}$$

Then for any $t \in S$,

$$\Delta(t) V \kappa(t^{-1}) = V$$

so that

$$\Delta(t) V = V \kappa(t), \tag{73}$$

and hence V links Δ and κ. If Δ and κ are inequivalent, it follows by Schur's lemma (Theorem 2.2) that $V = 0$. Then from (72) we have

$$\begin{aligned} 0 = v_{ij} &= \sum_{s \in S} \sum_{p=1}^{m} \sum_{q=1}^{n} \Delta_{ip}(s) u_{pq} \kappa_{qj}(s^{-1}) \\ &= \sum_{p,q} u_{pq} h(\Delta_{ip},\kappa_{qj}), \qquad i = 1, \ldots, m, \ j = 1, \ldots, n, \end{aligned}$$

where $h = |S|$. Since the choice of $U \in M_{m,n}(R)$ was arbitrary, we conclude that

$$(\Delta_{ip},\kappa_{qj}) = 0, \qquad i,p = 1, \ldots, m, \ \ j,q = 1, \ldots, n.$$

This establishes (70).

Now set $\Delta = \kappa$, so that (73) becomes $\Delta(t)V = V\Delta(t)$, $t \in S$. If Δ is absolutely irreducible, then Corollary 1(a) implies $V = \alpha I_m$ for some $\alpha \in R$. From (72) (with $\Delta = \kappa$) we have

$$\begin{aligned} m\alpha = \operatorname{tr} V &= \sum_{s \in S} \operatorname{tr}(\Delta(s) U \Delta(s^{-1})) \\ &= h(\operatorname{tr} U) \end{aligned}$$

and hence

$$\alpha = \frac{h}{m}(\operatorname{tr} U).$$

Thus for $i,j = 1, \ldots, m$,

$$\begin{aligned} \frac{h}{m}(\operatorname{tr} U)\delta_{ij} &= \alpha\delta_{ij} \\ &= v_{ij} \qquad (\text{since } V = \alpha I_m) \\ &= \sum_{p,q=1}^{m} u_{pq} h(\Delta_{ip},\Delta_{qj}), \end{aligned}$$

or

$$\frac{h}{m} \sum_{p=1}^{m} u_{pp}\delta_{ij} = \sum_{p,q=1}^{m} u_{pq} h(\Delta_{ip},\Delta_{qj}). \tag{74}$$

Since the choice of $U \in M_m(R)$ was arbitrary, it follows from (74) that

$$(\Delta_{ip},\Delta_{qj}) = \frac{1}{m}\delta_{ij}\delta_{pq}, \qquad i,p,j,q = 1, \ldots, m.$$

This establishes (71). ∎

Corollary 8 *Let $\chi: S \to R$ and $\psi: S \to R$ be distinct irreducible characters of the finite group S. Then*

$$(\chi,\psi) = 0. \tag{75}$$

If χ is an absolutely irreducible character of S, then

$$(\chi,\chi) = 1. \tag{76}$$

Proof: Let the proper irreducible matrix representations $\Delta = [\Delta_{ij}]$ and $\kappa = [\kappa_{ij}]$ of S over R, of respective degrees m and n, afford the characters χ and ψ, respectively. Since $\chi \neq \psi$, Δ and κ are inequivalent. Now $\chi = \sum_{i=1}^{m} \Delta_{ii}$ and $\psi = \sum_{j=1}^{n} \kappa_{jj}$, so that

$$\begin{aligned}(\chi,\psi) &= \left(\sum_{i=1}^{m} \Delta_{ii}, \sum_{j=1}^{n} \kappa_{jj}\right) = \sum_{i=1}^{m}\sum_{j=1}^{n} (\Delta_{ii},\kappa_{jj}) \\ &= 0 \qquad \text{[by (70)]}.\end{aligned}$$

To establish (76), observe that if χ is an absolutely irreducible character, i.e., if Δ is absolutely irreducible, then formula (71) yields $(\Delta_{ii},\Delta_{ii}) = 1/m$, $i = 1, \ldots, m$, and hence

$$\begin{aligned}(\chi,\chi) &= \sum_{i=1}^{m} (\Delta_{ii},\Delta_{ii}) \\ &= 1. \ ∎\end{aligned}$$

The formulas (75) and (76) are called the *character orthogonality relationships of the first kind.*

Corollary 9 *Let $\chi: S \to R$ be an absolutely irreducible character of degree m of the finite group S. Then*

$$\sum_{s\in S} \chi(st)\chi(s^{-1}) = \frac{h}{m}\chi(t), \qquad t \in S \tag{77}$$

and

$$\sum_{s\in S} \chi(ts^{-1})\chi(s) = \frac{h}{m}\chi(t), \qquad t \in S. \tag{78}$$

Proof: Let Δ be a proper absolutely irreducible matrix representation of S over R of degree m which affords the character χ. Then (71) says that

$$\frac{1}{h}\sum_{s\in S} \Delta_{ip}(s)\,\Delta_{qj}(s^{-1}) = \frac{1}{m}\delta_{ij}\delta_{pq}, \qquad i,p,j,q = 1, \ldots, m,$$

where $h = |S|$. Fix $t \in S$. For any $k = 1, \ldots, m$, multiply both sides of this last equation by $\Delta_{pk}(t)$, and sum on p to obtain

$$\frac{1}{h}\sum_{s\in S}\left(\sum_{p=1}^{m}\Delta_{ip}(s)\,\Delta_{pk}(t)\right)\Delta_{qj}(s^{-1}) = \frac{1}{m}\,\delta_{ij}\sum_{p=1}^{m}\Delta_{pk}(t)\delta_{pq},$$

or
$$\frac{1}{h}\sum_{s\in S}(\Delta(s)\Delta(t))_{ik}\Delta_{qj}(s^{-1}) = \frac{1}{m}\,\delta_{ij}\,\Delta_{qk}(t), \qquad i,p,j,q = 1, \ldots, m.$$

Set $k = i$ and $q = j$ in the preceding equation to obtain

$$\frac{1}{h}\sum_{s\in S}\Delta_{ii}(st)\Delta_{jj}(s^{-1}) = \frac{1}{m}\,\delta_{ij}\Delta_{ji}(t) = \frac{1}{m}\,\delta_{ij}\Delta_{ii}(t), \qquad i,j = 1, \ldots, m.$$

Finally, summing over all i and j yields

$$\frac{1}{h}\sum_{s\in S}\chi(st)\chi(s^{-1}) = \frac{1}{m}\,\chi(t)$$

[since $\sum_{i=1}^{m}(\sum_{j=1}^{m}\delta_{ij})\Delta_{ii}(t) = \sum_{i=1}^{m}\Delta_{ii}(t) = \chi(t)$], and (77) is proved.

To prove (78), replace s by s^{-1} in (77) to obtain

$$\sum_{s\in S}\chi(s^{-1}t)\chi(s) = \frac{h}{m}\,\chi(t), \qquad t \in S$$

and then use the fact that $\chi(s^{-1}t) = \chi(ts^{-1})$ for all $s,t \in S$ (see Exercise 7). ∎

A proof similar to that of Corollary 9 establishes the following fact: *If $\chi\colon S \to R$ and $\psi\colon S \to R$ are distinct irreducible characters of the finite group S, then*

$$\sum_{s\in S}\chi(st)\psi(s^{-1}) = 0, \qquad t \in S$$

and
$$\sum_{s\in S}\chi(ts^{-1})\psi(s) = 0, \qquad t \in S.$$

Corollary 10 *Let S be a finite group with $|S| = h$, and let $\chi\colon S \to R$ be the principal character of S. Then*

$$\sum_{s\in S}\chi(s) = h. \tag{79}$$

If $\psi\colon S \to R$ is an irreducible character of S and $\psi \neq \chi$, Then

$$\sum_{s\in S}\psi(s) = 0. \tag{80}$$

Proof: The equality (79) is obvious because $\chi \equiv 1$. If ψ is an irreducible character of S different from the principal character χ, then (80) follows from (75). ∎

Corollary 11 (a) *Let $\chi_1, \ldots, \chi_k \in R^S$ be distinct absolutely irreducible characters of the finite group S, and let $\chi = \sum_{i=1}^{k}a_i\chi_i$ and $\psi = \sum_{i=1}^{k}b_i\chi_i$, $a_i,b_i \in R$, $i = 1, \ldots, k$. Then*

$$(\chi,\psi) = \sum_{i=1}^{k}a_ib_i. \tag{81}$$

(b) *Under the assumption that* $h \cdot 1 \neq 0$ *in* R, *where* $h = |S|$, *a character* $\chi: S \to R$ *of the finite group* S *is absolutely irreducible iff*

$$(\chi,\chi) = 1. \tag{82}$$

Proof: (a) Assertion (a) follows immediately from (75), (76), and the properties of the scalar product.

(b) Let Δ be a proper matrix representation of S over R which affords the character χ. It follows from Maschke's theorem (Theorem 2.1) that by extending the field R if necessary, we can assume

$$\Delta \sim \Delta_{11} \dotplus \cdots \dotplus \Delta_{pp} \tag{83}$$

where each Δ_{kk} is an absolutely irreducible matrix representation of S (this is essentially Corollary 7; see also Exercise 8). Grouping the equivalent components in (83) together, we can change the notation slightly to obtain

$$\Delta \sim m_1\Delta_1 \dotplus \cdots \dotplus m_r\Delta_r, \tag{84}$$

where $\Delta_1, \ldots, \Delta_r$ are pairwise inequivalent absolutely irreducible matrix representations of S and the m_i are positive integers ($m_i\Delta_i$ means $\overbrace{\Delta_i \dotplus \cdots \dotplus \Delta_i}^{m_i}$). Then if χ_i is the character of Δ_i, $i = 1, \ldots, r$, we have

$$\chi = \sum_{i=1}^{r} m_i\chi_i. \tag{85}$$

The absolutely irreducible characters χ_i are distinct by Corollary 6; so Corollary 8 implies that

$$(\chi,\chi) = \sum_{i=1}^{r} m_i^2. \tag{86}$$

Now if the character χ is absolutely irreducible, then $r = 1$ and $m_1 = 1$ [from (84)] so that $(\chi,\chi) = 1$. Conversely, if $(\chi,\chi) = 1$, then (86) says

$$\sum_{i=1}^{r} m_i^2 = 1;$$

thus some m_i is 1 and the rest are 0. Hence by (85) we have $\chi = \chi_i$ for some i, so χ is absolutely irreducible. ▮

Example 5 Consider an equilateral triangle T in the Euclidean plane $\mathbf{R}^2$. The three vertices of T are labeled a, b, c, and the three medians of T, labeled 1, 2, 3, intersect at the origin $\mathcal{O}$ of $\mathbf{R}^2$. A *symmetry of* T is a rigid motion of the triangle onto itself; in particular, it is a linear operator on $\mathbf{R}^2$. The set of all symmetries of T is customarily denoted by C_{3v}. The elements of C_{3v} are:

e = identity symmetry,
C_3 = clockwise rotation of 120° about $\mathcal{O}$,
C_3^2 = clockwise rotation of 240° about $\mathcal{O}$,

σ_i = reflection through axis i, $\quad i = 1, 2, 3$.

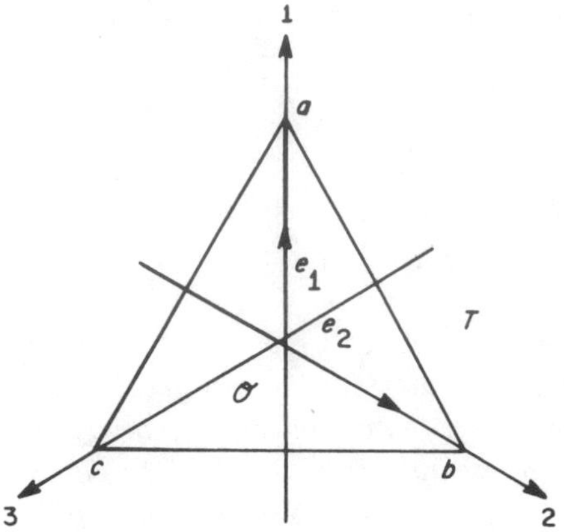

If S_3 is the symmetric group on the three symbols a, b, c, then operating on T with C_{3v} is the same as operating on $\{a,b,c\}$ with S. Using cycle notation, we have the following 1–1 correspondence between C_{3v} and S_3:

$$e \leftrightarrow (a)(b)(c), \quad C_3 \leftrightarrow (abc), \quad C_3{}^2 \leftrightarrow (acb),$$
$$\sigma_1 \leftrightarrow (bc), \quad \sigma_2 \leftrightarrow (ac), \quad \sigma_3 \leftrightarrow (ab).$$

In fact, C_{3v} is a subgroup of $\mathrm{GL}_2(\mathbf{R}^2)$ which is isomorphic to S_3.

Consider the representation $L\colon C_{3v} \to \mathrm{GL}_2(\mathbf{R}^2)$ of C_{3v} given simply by $L(s) = s$, $s \in C_{3v}$. We claim that L is absolutely irreducible. Choose a basis $E = \{e_1, e_2\}$ of $\mathbf{R}^2$ as indicated in the above figure, and let $\Delta\colon C_{3v} \to \mathrm{GL}(2,\mathbf{R})$ be the matrix representation of C_{3v} associated with L relative to the basis E. We compute that

$$\Delta(e) = I_2, \quad \Delta(C_3) = \begin{bmatrix} 0 & -1 \\ 1 & -1 \end{bmatrix}, \quad \Delta(C_3{}^2) = \begin{bmatrix} -1 & 1 \\ -1 & 0 \end{bmatrix},$$

$$\Delta(\sigma_1) = \begin{bmatrix} 1 & -1 \\ 0 & -1 \end{bmatrix}, \quad \Delta(\sigma_2) = \begin{bmatrix} -1 & 0 \\ -1 & 1 \end{bmatrix}, \quad \Delta(\sigma_3) = \begin{bmatrix} 0 & 1 \\ 1 & 0 \end{bmatrix}.$$

Let $\chi\colon C_{3v} \to \mathbf{R}$ be the character of L (and of Δ). Then $\chi(e) = 2$, $\chi(C_3) = -1$, $\chi(C_3{}^2) = -1$, $\chi(\sigma_1) = 0$, $\chi(\sigma_2) = 0$, $\chi(\sigma_3) = 0$. We have

$$\begin{aligned}(\chi,\chi) &= \frac{1}{6} \sum_{s \in C_{3v}} \chi(s)\chi(s^{-1}) \\ &= \frac{1}{6} \sum_{s \in C_{3v}} \chi(s)\overline{\chi(s)} \qquad \text{[by (53)]} \\ &= \frac{1}{6} \sum_{s \in C_{3v}} |\chi(s)|^2\end{aligned}$$

$$= \frac{1}{6}\{2^2 + (-1)^2 + (-1)^2\}$$
$$= 1.$$

It follows from Corollary 11(b) that the representation L is absolutely irreducible.

The following major result explicitly describes the central role played by the regular representation of a finite group.

Theorem 2.7 *Let S be a finite group with $|S| = h$. Let $\mathfrak{A}$ be the group algebra of S over a field R in which $h \cdot 1 \neq 0$.*

(a) *The regular representation $\rho: S \to \mathrm{GL}(\mathfrak{A})$ is fully reducible.*

(b) *If $L: S \to \mathrm{GL}(V)$ is an irreducible representation of S over R (of finite degree), then L is equivalent over R to some irreducible component of ρ.*

(c) *There are only a finite number of pairwise inequivalent absolutely irreducible (proper) representations of S. More precisely: If L is an absolutely irreducible (proper) representation of S over some extension field D of R, then L is equivalent over a finite algebraic extension K of D to an absolutely irreducible component of ρ, and moreover this component can be chosen to be over a finite algebraic extension F of R.*

(d) *Let $L_1, \ldots, L_r$ be a complete list of pairwise inequivalent absolutely irreducible (proper) representations of S over F as given by (c) (i.e., $L_1, \ldots, L_r$ are the distinct absolutely irreducible components of ρ over F, a finite algebraic extension field of R). Suppose that the degree of L_i is n_i, $i = 1, \ldots, r$. Then each L_i is equivalent over F to precisely n_i absolutely irreducible components of ρ, and*

$$\sum_{i=1}^{r} n_i^2 = h. \tag{87}$$

Thus each L_i occurs n_i times among all the absolutely irreducible components of ρ.

(e) *If L is any (proper) representation of S over R (of finite degree), then*

$$L \sim \sum_{i=1}^{r} {}^{\cdot}\, c_i L_i$$

where $c_i = (\chi_L, \chi_{L_i})$, $i = 1, \ldots, r$.

(f) *Let $\Delta^1, \ldots, \Delta^r$ be matrix representations of S over F associated with $L_1, \ldots, L_r$, respectively. Then*

$$\dim\langle \{\Delta^t_{ij} \mid i,j = 1, \ldots, n_t,\ t = 1, \ldots, r\}\rangle = h; \tag{88}$$

so the $n_1^2 + \cdots + n_r^2$ functions Δ^t_{ij}, $i,j = 1, \ldots, n_t$, $t = 1, \ldots, r$ comprise a basis of $\mathfrak{A}$ regarded as a vector space over F.

Proof: (a) This is an application of Maschke's theorem (Theorem 2.1).

(b) This follows from (a) and Corollary 1 of Section 5.1.

(c) Over some appropriate finite algebraic extension field $F = R(\theta_1, \ldots, \theta_k)$ of R (the θ_i are algebraic over R), the regular representation ρ has a reduction into absolutely irreducible components (see Exercise 8). By Corollary 5, these components are uniquely determined to within order and equivalence. Let L be an absolutely irreducible (proper) representation of S over some extension field D of R. Regard both ρ and L as representations over the finite algebraic extension field $K = D(\theta_1, \ldots, \theta_k)$ of D. Then by (b) we know that L is equivalent over K to one of the absolutely irreducible components of ρ. These components of ρ are uniquely determined to within order and equivalence, as we have just noted, and are over F.

(d) Suppose that L_i occurs d_i times among the absolutely irreducible components of ρ over F, $i = 1, \ldots, r$. Clearly,

$$\chi_\rho = \sum_{i=1}^{r} d_i \chi_i$$

where $\chi_i = \chi_{L_i}$, $i = 1, \ldots, r$. Then by Corollary 6 and Corollary 8,

$$\begin{aligned}(\chi_\rho, \chi_k) &= \sum_{i=1}^{r} d_i(\chi_i, \chi_k) \\ &= d_k, \qquad \mathrm{k} = 1, \ldots, r. \end{aligned} \tag{89}$$

On the other hand,

$$(\chi_\rho, \chi_k) = \frac{1}{h} \sum_{s \in S} \chi_\rho(s)\chi_k(s^{-1}), \qquad k = 1, \ldots, r \tag{90}$$

by the definition of the scalar product. We assert that

$$\chi_\rho(s) = \begin{cases} h, & s = e \\ 0, & s \neq e. \end{cases}$$

Indeed, the group algebra $\mathfrak{A}$ has dimension h over R, and since $\rho(e) = I_{\mathfrak{A}}$, it follows that $\chi_\rho(e) = h$. Let $E = \{s_1, \ldots, s_h\}$ be a basis for $\mathfrak{A}$ consisting of the elements of S in some fixed order. Then $\rho(s_k)s_j = s_j$ iff $s_k s_j = s_j$, i.e., iff $s_k = e$. Hence $[\rho(s_k)]_E^E$ has zeros along the main diagonal if $s_k \neq e$. It follows that $\chi_\rho(s) = 0$ if $s \neq e$, and the assertion is established. Returning to (89) and (90), we now have

$$\begin{aligned} d_k = (\chi_\rho, \chi_k) &= \frac{1}{h} \chi_\rho(e)\chi_k(e) = \frac{h}{h} \chi_k(e) \\ &= \chi_k(e), \qquad k = 1, \ldots, r. \end{aligned}$$

The degree of each L_k is n_k, so $\chi_k(e)$ is the trace of the identity transformation on a space of dimension n_k. Hence $d_k = n_k$, $k = 1, \ldots, r$.

Next, since

$$\chi_\rho = \sum_{i=1}^{r} n_i\chi_i,$$

we compute that

$$\begin{aligned}(\chi_\rho,\chi_\rho) &= \sum_{i,j=1}^{r} n_i n_j(\chi_i,\chi_j)\\ &= \sum_{i=1}^{r} n_i^2(\chi_i,\chi_i) \qquad \text{[by (75)]}\\ &= \sum_{i=1}^{r} n_i^2 \qquad \text{[by (76)].}\end{aligned}$$

On the other hand, since

$$\chi_\rho(s) = \begin{cases} h, & s = e \\ 0, & s \neq e \end{cases},$$

we have

$$\begin{aligned}(\chi_\rho,\chi_\rho) &= \frac{1}{h}\sum_{s\in S}\chi_\rho(s)\chi_\rho(s^{-1}) = \frac{1}{h}\chi_\rho(e)\chi_\rho(e) = \frac{h^2}{h}\\ &= h.\end{aligned}$$

This establishes (87).

(e) Let L be a (proper) representation of S over R. We know that L is equivalent, perhaps over an extension field of R, to a direct sum of absolutely irreducible components (see Corollary 7). According to (c), each of these components is equivalent to precisely one of the absolutely irreducible components $L_1, \ldots, L_r$ of ρ. Let c_i be the number of components of L equivalent to L_i, $i = 1, \ldots, r$. Then

$$L \sim \sum_{i=1}^{r}{}^{\cdot}\, c_i L_i.$$

Since

$$\chi_L = \sum_{i=1}^{r} c_i\chi_{L_i},$$

it follows from Corollary 8 that

$$\begin{aligned}(\chi_L,\chi_{L_k}) &= \sum_{i=1}^{r} c_i(\chi_{L_i},\chi_{L_k})\\ &= c_k, \qquad k = 1, \ldots, r.\end{aligned}$$

(f) We first remark that the dimension of the group algebra $\mathfrak{A}$ of S is h whether $\mathfrak{A}$ is considered over R or over F, since S is always a basis of $\mathfrak{A}$ (see also Exercise 2). Next, observe that the "entry functions" $\Delta^t_{ij}: S \to F$, $i,j = 1, \ldots, n_t$, $t = 1, \ldots, r$ are linearly independent elements of $\mathfrak{A}$ over F; this follows from the proof of Theorem 2.4 [see the discussion following (59)]. Thus

$$\dim\langle\{\Delta^t_{ij} \mid i,j = 1, \ldots, n_t,\ t = 1, \ldots, r\}\rangle = n_1^{\,2} + \cdots + n_r^{\,2}$$
$$= h \qquad \text{[by (87)]}.$$

Since $\mathfrak{A}$ has dimension h over F, these entry functions form a basis of $\mathfrak{A}$ over F. ∎

Let S be a group. Recall that if two elements $x,y \in S$ are related by $zxz^{-1} = y$ for some $z \in S$, then x and y are said to be *conjugate elements*; conjugacy is an equivalence relation which partitions S into equivalence classes called *conjugacy classes* (see Exercise 9). If $X \subset S$ $(X \neq \varnothing)$, then the *normalizer of X* is the set

$$N_X = \{s \in S \mid sX = Xs\}.$$

If $x \in S$ and $X = \{x\}$, we write $N_X = N_x$.

In the next theorem we summarize some elementary properties of these items. The proof is left as Exercise 10.

Theorem 2.8 *Let S be a group.*

(a) *If C_1 is the conjugacy class in S to which the identity e of S belongs, then $C_1 = \{e\}$.*

(b) *If C is a conjugacy class in S, then $C^{-1} = \{s \in S \mid s^{-1} \in C\}$ is a conjugacy class in S.*

(c) *If $\varnothing \neq X \subset S$, then $N_X < S$.*

(d) *If S is a finite group and $x \in S$, then the number of distinct conjugates of x in S, i.e., the number of elements in the conjugacy class to which x belongs, is the index of N_x in S.*

(e) *If S is a finite group with $|S| = h$ and C is a conjugacy class in S, then $|C| \mid h$.*

(f) *If χ is a character of a (proper) representation of S, then χ is constant on each conjugacy class in S.*

The following result is of fundamental importance because it explains precisely the role played by the characters in the group algebra of a finite group.

Theorem 2.9 *Let S be a finite group with $|S| = h$, and let $C_1, \ldots, C_p$ be the conjugacy classes in S. Let $\mathfrak{A}$ be the group algebra of S over a field R in which $h \cdot 1 \neq 0$. Set $\mathscr{C} = \{f \in \mathfrak{A} \mid f(zsz^{-1}) = f(s) \text{ for all } s,z \in S\}$.*

(a) *$\mathscr{C}$ is a subspace of $\mathfrak{A}$, and $\dim \mathscr{C} = p$.*

(b) *Let $\chi^1, \ldots, \chi^r$ be the characters of the representations $L_1, \ldots, L_r$ of S given by Theorem 2.7(d). Then $r = p$ and $\chi^1, \ldots, \chi^r$ form a basis of $\mathscr{C}$(over F). Thus the number of pairwise inequivalent absolutely irreducible (proper) representations of S is the number of conjugacy classes in S.*

(c) *Let* $f \in \mathfrak{A}$. *Then* $f \in \mathscr{C}$ *iff* $fg = gf$ *for all* $g \in \mathfrak{A}$. *Thus* $\mathscr{C}$ *is the center of* $\mathfrak{A}$.

Proof: A few preliminary remarks are in order. The elements of the group algebra $\mathfrak{A}$ are functions from S to R. The set $\mathscr{C}$ is the totality of elements $\sum_{s \in S} a_s s \in \mathfrak{A}$ satisfying $a_{zsz^{-1}} = a_s$ for all $s,z \in S$. The characters $\chi^1, \ldots, \chi^r$ are functions from S to F, where F is the finite algebraic extension field of R over which the complete list of pairwise inequivalent absolutely irreducible representations of S are defined. Of course, the elements of $\mathfrak{A}$ can also be thought of as functions from S to F, and the dimension of $\mathfrak{A}$ regarded as a vector space over F is still h.

(a) That $\mathscr{C}$ is a subspace of $\mathfrak{A}$ is obvious. For each $t = 1, \ldots, p$, let $\gamma_t = \sum_{s \in C_t} 1 \cdot s$, i.e., $\gamma_t \in \mathscr{C}$ takes on the value 1 at each $s \in C_t$ and 0 otherwise. Since the conjugacy classes C_t are disjoint, it is clear that the γ_t are l.i. in $\mathfrak{A}$ (over either R or F). Moreover, if $f \in \mathscr{C}$, then f is constant on each C_t, say $f = \sum_{t=1}^{p} a_t(\sum_{s \in C_t} 1 \cdot s)$. But this is simply another way of writing

$$f = \sum_{t=1}^{p} a_t \gamma_t.$$

Thus $\{\gamma_1, \ldots, \gamma_p\}$ is a basis of $\mathscr{C}$ (over either R or F), so dim $\mathscr{C} = p$.

(b) It follows from properties of the trace function that $\chi^t \in \mathscr{C}$, $t = 1, \ldots, r$. We know from Theorem 2.5 that the characters $\chi^1, \ldots, \chi^r$ are l.i. in $\mathfrak{A}$ over F. It remains to show that $\mathscr{C}$ is spanned (as a vector space over F) by $\chi^1, \ldots, \chi^r$. By Theorem 2.7(f), any element of $\mathfrak{A}$ is a unique linear combination of the entry functions Δ^t_{ij}, $i,j = 1, \ldots, n_t$, $t = 1, \ldots, r$, where Δ^t is a matrix representation of S over F associated with L_t, $t = 1, \ldots, r$. Now let $f \in \mathscr{C}$ and choose scalars $c^t_{ij} \in F$ such that $f = \sum_{t=1}^{r} \sum_{i,j=1}^{n_t} c^t_{ij} \Delta^t_{ij}$. We compute that for all $s \in S$,

$$\begin{aligned}
f(s) &= \frac{1}{h} \sum_{z \in S} f(zsz^{-1}) \qquad [f \in \mathscr{C} \text{ means that } f(zsz^{-1}) = f(s)] \\
&= \frac{1}{h} \sum_{z \in S} \sum_{t=1}^{r} \sum_{i,j=1}^{n_t} c^t_{ij} \Delta^t_{ij}(zsz^{-1}) \\
&= \frac{1}{h} \sum_{z \in S} \sum_{t=1}^{r} \sum_{i,j=1}^{n_t} c^t_{ij} \sum_{\alpha,\beta=1}^{n_t} \Delta^t_{i\alpha}(z) \Delta^t_{\alpha\beta}(s) \Delta^t_{\beta j}(z^{-1}) \\
&= \sum_{t=1}^{r} \sum_{i,j,\alpha,\beta=1}^{n_t} c^t_{ij} \Delta^t_{\alpha\beta}(s) \left(\frac{1}{h} \sum_{z \in S} \Delta^t_{i\alpha}(z) \Delta^t_{\beta j}(z^{-1}) \right) \\
&= \sum_{t=1}^{r} \sum_{i,j,\alpha,\beta=1}^{nt} c^t_{ij} \Delta^t_{\alpha\beta}(s) (\Delta^t_{i\alpha}, \Delta^t_{\beta j}) \\
&= \sum_{t=1}^{r} \sum_{i,j,\alpha,\beta=1}^{n_t} c^t_{ij} \Delta^t_{\alpha\beta}(s) \frac{1}{n_t} \delta_{ij} \delta_{\alpha\beta} \qquad \text{(by Theorem 2.6)} \\
&= \sum_{t=1}^{r} \sum_{t,\alpha=1}^{n_t} c^t_{ii} \Delta^t_{\alpha\alpha}(s) \frac{1}{n_t}
\end{aligned}$$

$$= \sum_{t=1}^{r} \frac{1}{n_t} \sum_{i=1}^{n_t} c_{ii}^t \sum_{\alpha=1}^{n_t} \Delta_{\alpha\alpha}^t(s)$$

$$= \sum_{t=1}^{r} \left(\frac{1}{n_t} \sum_{i=1}^{n_t} c_{ii}^t \right) \chi^t(s).$$

Thus f is a linear combination over F of $\chi^1, \ldots, \chi^r$. This shows that $\{\chi^1, \ldots, \chi^r\}$ is a basis of $\mathscr{C}$ over F. In particular, (a) implies that $r = p$.

(c) Assume $fg = gf$ for all $g \in \mathfrak{A}$. Writing $f = \sum_{s \in S} a_s s$, we have for all $z \in S$ that

$$\sum_{s \in S} a_s s = f = z^{-1} f z = \sum_{s \in S} a_s z^{-1} s z$$
$$= \sum_{t \in S} a_{ztz^{-1}} t = \sum_{s \in S} a_{zsz^{-1}} s.$$

Thus $a_{zsz^{-1}} = a_s$ for all $s,z \in S$, i.e., $f \in \mathscr{C}$. Conversely, if $f \in \mathscr{C}$, then the same argument in reverse shows that f commutes with every $z \in S$ and hence with every $g \in \mathfrak{A}$. ∎

The characters $\chi^1, \ldots, \chi^p$ $(p = r)$ in Theorem 2.9 are sometimes referred to as a *complete list of absolutely irreducible characters for the finite group S*. The second set of orthogonality relations for the characters $\chi^1, \ldots, \chi^p$ [see (75) and (76)] can now be derived.

Theorem 2.10 *Let S be a finite group with $|S| = h$. Let $\mathfrak{A}$ be the group algebra of S over a field R in which $h \cdot 1 \neq 0$. In the notation of Theorem 2.9, let $|C_t| = h_t$, $t = 1, \ldots, p$, and let χ_t^k be the (constant) value of χ^k on C_t, $t,k = 1, \ldots, p$. Let $\bar{\chi}_t^k$ be the value of χ^k on C_t^{-1} [which by Theorem 2.8(b) is also a conjugacy class in S], $t,k = 1, \ldots, p$. Then*

$$\sum_{t=1}^{p} \frac{h_t}{h} \chi_t^k \bar{\chi}_t^l = \delta_{kl}, \qquad k,l = 1, \ldots, p \tag{91}$$

and

$$\sum_{k=1}^{p} \frac{h_t}{h} \chi_t^k \bar{\chi}_s^k = \delta_{ts}, \qquad t,s = 1, \ldots, p. \tag{92}$$

Proof: Suppose we have established (91). If A is the $p \times p$ matrix whose (k,t) entry is $(h_t/h)\chi_t^k$ and B is the $p \times p$ matrix whose (t,l) entry is $\bar{\chi}_t^l$, then (91) says that $AB = I_p$. But this implies $BA = I_p$, i.e.,

$$\delta_{ts} = (BA)_{st} = \sum_{k=1}^{p} B_{sk} A_{kt} = \sum_{k=1}^{p} \bar{\chi}_s^k \frac{h_t}{h} \chi_t^k$$
$$= \sum_{k=1}^{p} \frac{h_t}{h} \chi_s^k \bar{\chi}_s^k, \qquad t,s = 1, \ldots, p,$$

which is precisely (92). Thus it suffices to prove (91). By the character orthogonality relationships of the first kind (see Corollary 8), we have

$$(\chi^k, \chi^l) = \delta_{kl}, \qquad k,l = 1, \ldots, p,$$

or $$\frac{1}{h}\sum_{z\in S}\chi^k(z)\chi^l(z^{-1}) = \delta_{kl}, \qquad k,l = 1, \ldots, p.$$

Now S is the disjoint union of the conjugacy classes $C_1, \ldots, C_p$, each χ^k is constant on each C_t with value χ_t^k, and the value of each χ^l on each C_t^{-1} is $\bar{\chi}_t^l$. It follows that

$$\begin{aligned}\delta_{kl} &= \frac{1}{h}\sum_{t=1}^{p}\sum_{z\in C_t}\chi^k(z)\chi^l(z^{-1}) \\ &= \frac{1}{h}\sum_{t=1}^{p} h_t\chi_t^k\bar{\chi}_t^l, \qquad k,l = 1, \ldots, p,\end{aligned}$$

proving (91). ▮

The formulas (91) and (92) are called the *character orthogonality relationships of the second kind.* They allow us to prove a result of considerable value in computing the degrees of the absolutely irreducible representations of a finite group S.

Theorem 2.11 *Let S be a finite group with $|S| = h$, and let $C_1, \ldots, C_p$ be the conjugacy classes in S. Let $\mathfrak{A}$ be the group algebra of S over a field R of characteristic* 0. *Let $L_1, \ldots, L_p$ be a complete list of pairwise inequivalent absolutely irreducible* (*proper*) *representations of S over a finite algebraic extension field F of R, and suppose the degree of L_k is n_k, $k = 1, \ldots, p$. Then*

$$n_k \mid h, \qquad k = 1, \ldots, p.$$

Proof: Fix k, $1 \le k \le p$. As in the proof of Theorem 2.9(a), for each $t = 1, \ldots, p$, let $\gamma_t = \sum_{s\in C_t} 1\cdot s$; we have seen that $\{\gamma_1, \ldots, \gamma_p\}$ is a basis of the center $\mathscr{C}$ of the group algebra $\mathfrak{A}$. Regarding L_k as extended to $\mathfrak{A}$, it is obvious that $L_k(\gamma_t)$ commutes with $L_k(s)$, $t = 1, \ldots, p$, $s \in S$. Hence by Corollary 1(b),

$$L_k(\gamma_t) = \eta_t^k I, \qquad t = 1, \ldots, p,$$

where I is the identity transformation on some appropriate vector space of dimension n_k over F and $\eta_t^k \in F$, $t = 1, \ldots, p$.

In the notation of Theorem 2.10, we have

$$\begin{aligned}\eta_t^k n_k &= \operatorname{tr}(\eta_t^k I) \\ &= \operatorname{tr} L_k(\gamma_t) \\ &= \operatorname{tr}\Big(\sum_{s\in C_t} L_k(s)\Big) \\ &= \sum_{s\in C_t}\chi^k(s) \\ &= h_t\chi_t^k\end{aligned}$$

and hence $$\eta_t^k = \frac{h_t}{n_k}\chi_t^k, \qquad t = 1, \ldots, p. \tag{93}$$

We may assume that C_1 is the conjugacy class in S to which the identity e of S belongs; then by Theorem 2.8(a), $C_1 = \{e\}$. Since $\chi_1^k = \chi^k(e) = n_k$ and $h_1 = 1$, (93) yields

$$\eta_1^k = 1. \tag{94}$$

It is clear that $\gamma_s\gamma_t \in \mathscr{C}$ for any $s,t = 1, \ldots, p$, so there exist scalars $c_{tj}^s \in R$ such that

$$\gamma_s\gamma_t = \sum_{j=1}^{p} c_{tj}^s \gamma_j, \qquad s,t = 1, \ldots, p. \tag{95}$$

Evaluate L_k at both sides of (95) to obtain

$$\eta_s^k\eta_t^k I = \sum_{j=1}^{p} c_{tj}^s \eta_j^k I$$

and hence

$$\eta_s^k\eta_t^k = \sum_{j=1}^{p} c_{tj}^s \eta_j^k, \qquad s,t = 1, \ldots, p. \tag{96}$$

Define matrices $\Gamma^s = [c_{tj}^s]$, $s = 1, \ldots, p$, and let $\eta^k = (\eta_1^k, \ldots, \eta_p^k)$. Then (96) implies

$$\Gamma^s\eta^k = \eta_s^k\eta^k, \qquad s = 1, \ldots, p. \tag{97}$$

Note that η^k is not the zero p-tuple by (94). Also note that the entries of each Γ^s, i.e., the scalars c_{tj}^s, are nonnegative integral multiples of $1 \in R$. Indeed, if $z \in C_\nu$, then $(\gamma_s\gamma_t)(z) = \sum_{j=1}^{p} c_{tj}^s\gamma_j(z) = c_{t\nu}^s$; but $(\gamma_s\gamma_t)(z)$ is a convolution of values of γ_s and γ_t, and these are all 0 or 1. Next, apply Γ^t to both sides of (97) to obtain

$$\begin{aligned} \Gamma^t\Gamma^s\eta^k &= \eta_s^k\Gamma^t\eta^k \\ &= (\eta_s^k\eta_t^k)\eta^k, \qquad s,t = 1, \ldots, p. \end{aligned} \tag{98}$$

Choose $\varphi \in S_p$ so that $C_{\varphi(t)} = C_t^{-1}$, $t = 1, \ldots, p$. Set $s = \varphi(t)$ in (98), multiply both sides by h/h_t, and sum on $t = 1, \ldots, p$ to obtain

$$\left(\sum_{t=1}^{p} \frac{h}{h_t} \Gamma^t\Gamma^{\varphi(t)}\right)\eta^k = \left(\sum_{t=1}^{p} \frac{h}{h_t} \eta_{\varphi(t)}^k\eta_t^k\right)\eta^k. \tag{99}$$

Now

$$\begin{aligned} \sum_{t=1}^{p} \frac{h}{h_t} \eta_{\varphi(t)}^k\eta_t^k &= \sum_{t=1}^{p} \frac{h}{h_t} \frac{h_{\varphi(t)}}{n_k} \frac{h_t}{n_k} \chi_{\varphi(t)}^k\chi_t^k && \text{[by (93)]} \\ &= \sum_{t=1}^{p} \frac{h}{h_t} \frac{h_t^2}{n_k^2} \bar{\chi}_t^k\chi_t^k && \text{[since } h_{\varphi(t)} = h_t \text{ and } \chi_{\varphi(t)}^k = \bar{\chi}_t^k\text{]} \\ &= \frac{h}{n_k^2} \sum_{t=1}^{p} h_t\bar{\chi}_t^k\chi_t^k \\ &= \frac{h}{n_k^2} \cdot h && \text{[by (91)]} \\ &= \left(\frac{h}{n_k}\right)^2. \end{aligned}$$

Thus from (99), we have

$$\left(\sum_{t=1}^{p} \frac{h}{h_t} \Gamma^t \Gamma^{\varphi(t)}\right) \eta^k = \left(\frac{h}{n_k}\right)^2 \eta^k. \tag{100}$$

Since $h_t | h$ for $t = 1, \ldots, p$ by Theorem 2.8(e), the matrix on the left in (100) has entries which are nonnegative integral multiples of $1 \in R$. Since $\eta^k \neq 0$, $\left(\frac{h}{n_k}\right)^2 \left[= \left(\frac{h \cdot 1}{n_k \cdot 1}\right)^2\right]$ is an eigenvalue of this matrix. Then $(h/n_k)^2$ is a root of a monic polynomial with integral coefficients and hence is itself an integer (see Exercise 11). It follows that h/n_k is an integer (see Exercise 12). ∎

Example 6 Let S be a finite group with $|S| = h = m^2$, where m is a prime. We will use Theorem 2.11 to show that S is abelian (of course, this result also follows from the Sylow theorems).

Let $n_1, \ldots, n_p$ be the complete list of degrees of the irreducible characters of S over **C** (note that we do not need to preface "irreducible" with "absolutely" since **C** is an algebraically closed field); these n_k are not necessarily all distinct, of course. By Theorem 2.11, $n_k | m^2$ for $k = 1, \ldots, p$. Since m is a prime, it follows that $n_k | m$ and hence $n_k = m$ or $n_k = 1$, $k = 1, \ldots, p$. Suppose $n_k = m$ for some k. Then by Theorem 2.7(d), we have

$$n_1^2 + \cdots + n_{k-1}^2 + m^2 + n_{k+1}^2 + \cdots + n_p^2 = m^2$$

or

$$n_1^2 + \cdots + n_{k-1}^2 + n_{k+1}^2 + \cdots + n_p^2 = 0,$$

and this last equality is impossible because the principal character of S has degree 1. Thus $n_k = 1$, $k = 1, \ldots, p$, i.e., the irreducible components of the regular representation ρ of S over **C** are all of degree 1. It follows that im ρ is abelian. Finally, the regular representation ρ of the finite group S is faithful, and we conclude that S is abelian.

Exercises 5.2

1. Complete the proof of Corollary 1(b).
2. Let V be a vector space of dimension n over a field R. Let K be an extension field of R. Show how V can be regarded as (more precisely, extended to) a vector space over K of dimension n. Also show that if $T \in \mathrm{Hom}(V, V)$, then T can be regarded as a linear map on V over the extended field K. *Hint*: Let $E = \{e_1, \ldots, e_n\}$ be a basis of V over R, and define $\nu: V \to R^n$ by

$$\nu\left(\sum_{i=1}^{n} r_i e_i\right) = (r_1, \ldots, r_n), \qquad r_i \in R,\ i \in 1, \ldots, n.$$

Obviously ν is a vector space isomorphism. Observe that $R^n \subset K^n$, and addition and scalar multiplication in K^n confined to scalars in R and vectors in R^n are the same as in R^n. Let $\Delta = (K^n - R^n) \times \{V\}$, and set $V_1 = V \cup \Delta$ (since the second member of the ordered pairs in Δ is always V, it follows from the regularity axiom that $V \cap \Delta = \emptyset$). Now define $\theta: V_1 \to K^n$ as follows: $\theta | V = \nu$, while $\theta | \Delta$ is given by

$$\theta((k_1, \ldots, k_n), V) = (k_1, \ldots, k_n), \quad (k_1, \ldots, k_n) \in K^n - R^n.$$

It is easy to see that θ is a bijection, and we define addition and scalar multiplication in V_1 by

$$x + y = \theta^{-1}(\theta(x) + \theta(y)), \qquad x,y \in V_1,$$
$$kx = \theta^{-1}(k\theta(x)), \quad x \in V_1, \qquad k \in K,$$

where the addition and scalar multiplication on the right in the above equations occur in K^n. This makes V_1 into a vector space over K. If $T \in \mathrm{Hom}(V,V)$ and $[T]_E^E = A$, define $T_1 \in \mathrm{Hom}(V_1,V_1)$ by

$$T_1 w = \theta^{-1}(A\theta(w)), \qquad w \in V_1.$$

3. Prove (51). *Hint*: The trace of any linear transformation is the sum of its eigenvalues.

4. Prove (52). *Hint*: Similar matrices have the same eigenvalues, and hence the same trace.

5. Prove (53). *Hint*: Since $s^k = e$, it follows that $L(s)^k = L(e) = I_V$. Thus the minimal polynomial of $L(s)$ is a divisor of $\lambda^k - 1$, and every eigenvalue of $L(s)$ is a root of $\lambda^k - 1$. In particular, every eigenvalue of $L(s)$ is a complex number of modulus 1 whose inverse is simply its conjugate. Using the fact that the eigenvalues $\lambda_i(s^{-1})$ of $L(s^{-1}) = L(s)^{-1}$ are the inverses of the eigenvalues $\lambda_i(s)$ of $L(s)$, we compute that

$$\chi_L(s^{-1}) = \sum_{i=1}^{n} \lambda_i(s^{-1}) = \sum_{i=1}^{n} \lambda_i(s)^{-1} = \sum_{i=1}^{n} \overline{\lambda_i(s)}$$
$$= \overline{\sum_{i=1}^{n} \lambda_i(s)} = \overline{\chi_L(s)}.$$

6. Let S be a group (written multiplicatively) and $H < S$. Then the number of left cosets sH is called the *index of H in S*. Prove that if H_1 and H_2 are subgroups of finite index in S, then $H_1 \cap H_2$ is of finite index in S. *Hint*: We first assert that if $s,t \in S$ are in different (left) $(H_1 \cap H_2)$-cosets, then they are in different H_1-cosets or they are in different H_2-cosets. For otherwise $s^{-1}t \in H_1$, $s^{-1}t \in H_2$, and hence $s^{-1}t \in H_1 \cap H_2$. Let the index of H_1 in S be p, and write $S = C_1 \cup \cdots \cup C_p$, a disjoint union of p left H_1-cosets. Similarly, let the index of H_2 in S be q, and write $S = D_1 \cup \cdots \cup D_q$, a disjoint union of q left H_2-cosets. Then for any $t = 1, \ldots, p$ we have

$$C_t = S \cap C_t = (D_1 \cap C_t) \cup \cdots \cup (D_q \cap C_t).$$

Now assume that $H_1 \cap H_2$ is not of finite index in S, so that the number of left $(H_1 \cap H_2)$-cosets in S is certainly greater than pq. Then there exists a sequence of $N > pq$ elements of S, say $s_1, \ldots, s_N$, which lie in different $(H_1 \cap H_2)$-cosets. Each s_i lies in some set $D_{m_i} \cap C_{n_i}$:

$$s_i \in D_{m_i} \cap C_{n_i}, \qquad i = 1, \ldots, N.$$

Suppose that $(m_i,n_i) = (m_j,n_j)$ for $i \neq j$. Then

$$s_i \in D_{m_i} \cap C_{n_i}$$

and

$$s_j \in D_{m_i} \cap C_{n_i}.$$

By the first statement in the present argument, s_i and s_j must lie in the same $(H_1 \cap H_2)$-coset, and this contradicts the choice of the sequence $s_1, \ldots, s_N$. Thus the pairs $(m_1,n_1), \ldots, (m_N,n_N)$ are distinct. But $m_i \in \{1, \ldots, p\}$, $n_i \in \{1, \ldots, q\}$ and hence there are only pq distinct pairs (m_i,n_i) possible, contradicting $N > pq$.

7. Let $\chi: S \to R$ be a character of the groupoid S. Show that $\chi(st) = \chi(ts)$ for all $s,t \in S$. *Hint*: Let the matrix representation Δ of S over R afford χ. Then for all $s,t \in S$ we have $\chi(st) = \text{tr } \Delta(st) = \text{tr } \Delta(s)\Delta(t) = \text{tr } \Delta(t)\Delta(s) = \text{tr } \Delta(ts) = \chi(ts)$.

8. Let S be a finite group with $|S| = h$, and let R be a field in which $h \cdot 1 \neq 0$. Let $\Delta: S \to \text{GL}(n,R)$ be a proper matrix representation of S over R. Prove that there exists a finite algebraic (and hence simple algebraic) extension field F of R such that $\Delta \sim \Delta_1 \dotplus \cdots \dotplus \Delta_p$ over F and the representations $\Delta_1, \ldots, \Delta_p$ of S over F are absolutely irreducible. *Hint*: Let $\mathscr{C}$ be the totality of matrices $X \in M_n(R)$ which satisfy $XA = AX$ for all $A \in \text{im } \Delta$. If Δ is not itself absolutely irreducible, by Theorem 2.1 there exists some extension field D of R and some $B \in \text{GL}(n,D)$ such that for any $A \in \text{im } \Delta$ we have $B^{-1}AB = \begin{bmatrix} A_{11} & 0 \\ 0 & A_{22} \end{bmatrix}$, where A_{11} is $p \times p$ and A_{22} is $q \times q$. Then for any $\alpha,\beta \in D$, $M_{\alpha,\beta} = B(\alpha I_p \dotplus \beta I_q)B^{-1}$ commutes with every $A \in \text{im } \Delta$. Think of $AX - XA = 0$, $A \in \text{im } \Delta$, as a system of linear homogeneous equations in X (S is finite, so im Δ is finite). This system of equations has a null space of dimension at least 2 over D and hence over R. It follows that there exists $M \in M_n(R)$ such that $M \neq \gamma I_n$ for any $\gamma \in R$ and $MA = AM$ for all $A \in \text{im } \Delta$. let μ be an eigenvalue of M, set $W = M - \mu I_n$, and note that $W \neq 0$ and W commutes with every $A \in \text{im } \Delta$. Also $R(\mu)$ is a finite algebraic extension of R. Since W is singular, obtain matrices $P,Q \in \text{GL}(n,R(\mu))$ such that $PWQ = I_k \dotplus 0$, where $0 < k = \rho(W) < n$. Since $WA = AW$, it follows by block multiplication from the equation $PAP^{-1}PWQ = PWQQ^{-1}AQ$ that if $Q^{-1}AQ = \begin{bmatrix} K_{11} & K_{12} \\ K_{21} & K_{22} \end{bmatrix} = K$ then $K_{12} = 0$, for any $A \in \text{im } \Delta$. Thus over $R(\mu)$, Δ is equivalent to the representation K. Then by Maschke's theorem, $K \sim K_{11} \dotplus K_{22}$ over $R(\mu)$, and hence $\Delta \sim K_{11} \dotplus K_{22}$ over $R(\mu)$. Repeat the procedure with K_{11} and K_{22}, etc., each time extending the field in question to a finite algebraic extension, until absolutely irreducible components are obtained.

9. Prove that conjugacy in a group is an equivalence relation.

10. Prove Theorem 2.8. *Hint*: (a) Trivial. (b) Clearly $zxz^{-1} = y$ iff $zx^{-1}z^{-1} = y^{-1}$. (c) $sX = Xs$ and $tX = Xt$ imply $Xs^{-1} = s^{-1}X$ and $(st)X = s(tX) = s(Xt) = (sX)t = (Xs)t = X(st)$; so N_X is a group. (d) Let $u_1N_x, \ldots, u_kN_x$ be all the left cosets of N_x in S. The k conjugates $u_ixu_i^{-1}$ are distinct. Indeed, $u_ixu_i^{-1} = u_jxu_j^{-1}$ implies $u_j^{-1}u_i \in N_x$, i.e., $u_i \in u_jN_x$. Thus the number of distinct conjugates of x is at least k. On the other hand, if $z^{-1}xz$ is any conjugate of x, then z^{-1} must lie in some left coset, say $z^{-1} = u_ty^{-1}$, $y^{-1} \in N_x$. Hence $z^{-1}xz = u_ty^{-1}xyu_t^{-1} = u_txu_t^{-1}$. Thus there are at most k conjugates of x. (e) Let C be the conjugacy class of x, i.e., C consists of all the distinct conjugates of x. According to part (d), $|C|$ is the index of N_x in S and hence must divide h. (f) $\chi(zxz^{-1}) = \text{tr } \Delta(zxz^{-1}) = \text{tr } \Delta(z)\Delta(x)\Delta(z)^{-1} = \text{tr } \Delta(x) = \chi(x)$.

11. Let R be a field of characteristic 0, and suppose p and q are relatively prime integers. Show that if $\dfrac{p \cdot 1}{q \cdot 1}$ is a root of the monic polynomial $\lambda^n + a_{n-1}\lambda^{n-1} + \cdots + a_1\lambda + a_0 \in R[\lambda]$ in which the $a_i \in R$ are integral multiples of $1 \in R$, then p/q is an integer. *Hint*: We have $\dfrac{p^n \cdot 1}{q^n \cdot 1} + a_{n-1}\dfrac{p^{n-1}}{q_{n-1}} + \cdots + a_1\dfrac{p}{q} + a_0 = 0$ so that $p^n \cdot 1 + q(a_{n-1}p^{n-1} + a_{n-2}p^{n-2}q + \cdots + a_1q^{n-2} + a_0q^{n-1}) = 0$. Then $p^n \cdot 1 - (q \cdot m) \cdot 1 = 0$, where m is an integer, and since R has characteristic 0 this implies $p^n - qm = 0$. It follows that $q \mid p^n$; since q and p have no common factors, we obtain $q = \pm 1$.

12. Show that in a field of characteristic 0, $\left(\dfrac{p}{q}\right)^2 = \left(\dfrac{p \cdot 1}{q \cdot 1}\right)^2$ is an integer iff p/q is an integer. *Hint*: $p^2/q^2 = m$ means $p^2 \cdot 1 = q^2 m \cdot 1$ or $p^2 - q^2 m = 0$. We can assume p and q are relatively prime; so $q^2 \mid p^2$ implies that $q \mid p$ and hence $q = \pm 1$.

13. The *prime subfield* π *of a field* R is the smallest subfield of R containing 0 and 1. Prove that if S is a finite group with $|S| = h$ such that $h \cdot 1 \neq 0$ in R, then the absolutely irreducible components of a proper matrix representation Δ of S over R can be chosen to be over a finite algebraic (and hence simple algebraic) extension of π. *Hint*: First reduce Δ into irreducible components over R; so we may as well start by assuming Δ itself is irreducible over R. By Theorem 2.7(b), Δ is equivalent over R to a component ρ_1 of the regular representation ρ. The entries in the matrices in im ρ are 0 and 1 when S itself is used as a basis of the group algebra $\mathfrak{A}$ of S over R. Hence ρ_1 is over π, since the proof of Maschke's theorem shows that the components of ρ are obtained using an equivalence over the field in which the entries lie. Thus $\Delta \sim \rho_1$ over R, and ρ_1 is over π. Now apply Exercise 10 to conclude that the absolutely irreducible components of ρ_1 can be chosen to be over a finite algebraic extension of π.

Glossary 5.2

absolutely irreducible character, 460
absolutely irreducible representation, 444
absolutely irreducible set of matrices, 445
Burnside's theorem, 446
Burnside–Frobenius–Schur theorem, 450
character of a representation, 452
character orthogonality relationships of the first kind, 462
character orthogonality relationships of the second kind, 472
complete list of absolutely irreducible characters for a finite group, 471
component of a representation, 454
conjugacy classes, 469
conjugate elements, 469
constituent of representation, 454
degree of a character, 460
degree of a representation, 440
element of finite period, 452
exponent of a group, 458
irreducible character, 460
length of a reduction, 454
Maschke's theorem, 440
modular representation, 460
normalizer, 469

Index